1996

BIODIVERSITY OF THE SOUTHEASTERN UNITED STATES
Aquatic Communities

BIODIVERSITY OF THE SOUTHEASTERN UNITED STATES

Aquatic Communities

Edited by

Courtney T. Hackney
S. Marshall Adams
William H. Martin

JOHN WILEY & SONS, INC.

New York • Chichester • Brisbane • Toronto • Singapore

In recognition of the importance of preserving what has been written, it is a policy of John Wiley & Sons, Inc., to have books of enduring value published in the United States printed on acid-free paper, and we exert our best efforts to that end.

Library of Congress Cataloging in Publication Data:

Biodiversity of the southeastern United States : aquatic communities /
   Courtney T. Hackney. Marshall Adams, & William Martin, editors.
      p.   cm.

   Includes bibliographical references.
   ISBN 0-471-62884-0

   1. Aquatic ecology—Southern States.   2. Biotic communities—
Southern States.   3. Biological diversity—Southern States.
I. Hackney, Courtney Thomas. II. Adams, Marshall. III. Martin,
William A.

QH104.5.S59B56   1992
574.5′263′0975—dc20   91-24296
                                                            CIP

Printed in the United States of America

10 9 8 7 6 5 4 3 2

CONTENTS

PREFACE

This volume represents years of work by authors, editors, reviewers, and many other individuals of the Southeastern Section of the Ecological Society of America. "In the beginning" the prospect of summarizing the state of knowledge of Southeastern communities seemed a straightforward task. The five intervening years are testament to our naivete. As you read this volume (and the companion terrestrial volume) you should rapidly become aware of the artificiality of segmenting the southeastern landscape into traditional ecological subdisciplines. Our early approach was to divide the Southeast into major drainage basins, but very quickly the difficulty of such a logical approach manifested itself. Not only did it involve the repetition of the attributes of one community found in many watersheds, but it involved finding individuals and/or groups of individuals willing to cover the entire gamut of communities from high-gradient streams in oak–hickory forests to low-salinity marshes and cypress swamps. In addition, some drainage basins are so poorly studied that any discussion of them involved borrowing heavily from adjacent watersheds. Ultimately, no matter how artificial, we were forced to delineate aquatic habitats. As editors, we take full responsibility for forcing this approach on our authors and thank each for providing what we wanted even if it was not what they originally wanted to do.

The organization of each chapter consists of an introduction to the scope and nature of the community, a section on unique attributes relative to the physical environment that structure and define each community, and a description of the plant and animal communities associated with each system. Each chapter also contains a subsection on resource use and management and concludes with future research and management problems. Most chapters follow this format, except Chapters 1, 2, and 15, which are introductory or historical in nature and necessarily deviate from the basic chapter plan. The objective of each chapter was to present an integrated overview of each aquatic system in the Southeast organized around functional processes. Many chapters use specific sites as examples. In a few cases this was due to space limitations, but in the vast majority of chapters it was necessitated by the lack of information for many aquatic systems in the Southeast.

The original plan for the Aquatic volume included the Mississippi River and tributaries. It quickly became evident that information on the Mississippi River alone could easily fill a volume. Thus, except for some comparative information the Mississippi has been dropped from this volume.

It is disappointing to realize the actual paucity of data on acquatic communities in the Southeast, especially at a time when the South's population is growing so rapidly. It is our hope that this publication will stimulate and expand efforts to understand the nature and intricate balance between humans and the landscape we call the Southeast.

Many individuals contributed to the completion of this volume and are recognized by chapter authors. During the final stages of this publication, Ms. Rosemarie Ganucheau took on the arduous task of reviewing all galley page proofs and found numerous errors and omissions that we the editors and authors overlooked. For this we are forever in her debt.

THE EDITORS

CONTRIBUTORS

S. MARSHALL ADAMS, Ph.D. University of North Carolina, 1974, is currently a research staff member in the Environmental Sciences Division of Oak Ridge National Laboratory and serves on the adjunct faculty in the Ecology Department at the University of Tennessee. His research interests include predator–prey and trophic dynamics relationships in aquatic systems and the effects of environmental stress on fish populations and aquatic food webs. He is an active member of the American Fisheries Society and has recently published a book through that society entitled *Biological Indicators of Stress in Fish*.

CHARLES M. COOPER, Ph.D. University of Mississippi, 1976, is Research Leader for the Water Quality and Ecology Research Unit at the U.S. Department of Agriculture Sedimentation Laboratory in Oxford, Mississippi. Dr. Cooper is also an Adjunct Professor of Biology at the University of Mississippi, where he lectures on special topics and serves on graduate student advisory committees. His professional interests center around environmental and water quality off-site damages from agriculture and recovery of damaged ecosystems. He has authored over 80 research publications and is a member of the International Society of Limnology, Ecological Society of America, American Society of Limnology and Oceanography, and the North American Benthological Society.

THOMAS L. CRISMAN is a professor in the Department of Environmental Engineering Sciences at the University of Florida, where he has been since 1977. He holds a Ph.D. in zoology from Indiana University and has postdoctoral experience at the Limnological Research Center of the University of Minnesota. Dr. Crisman's research has concentrated on subtropical limnology, lake and wetland management, paleolimnology, and the response of aquatic biota to acidic deposition. He is the author of over 60 articles and has been a visiting professor at the University of Joensuu in Finland, the University of Copenhagen, and Makerere University in Uganda. He has served as scientific advisor to the national academies of sciences of Australia and Poland.

MICHAEL R. DARDEAU, M.S. University of South Alabama, 1982, is currently a Marine Scientist at the Alabama Marine Sciences Consortium located at the Dauphin Island Sea Lab. Research interests include taxonomy, community structure, and ecological relationships of marine invertebrates, particularly food web interactions in both hard and soft bottom communities.

ix

JAMES D. FELLEY,   Ph.D. University of Oklahoma, 1980, is currently with the Office of Information Resource Management at the Smithsonian Institution, working with the Research Support group. His research interests center on fishes, including community ecology, population genetics, ecological–morphological relationships of stream fishes, and relationships between morphology and genetics in hybrid fishes.

J. F. FITZPATRICK, JR.,   Ph.D. University of Virginia, 1964, is Professor of Biology at the University of South Alabama. His area of interest is the systematics and evolution of freshwater crustaceans, especially crawfishes. He is the author of numerous articles on crawfish taxonomy and of the book, *How to Know the Freshwater Crustacea*. He has served as an officer in many regional, national, and international organizations and has been elected a Fellow of the American Association for the Advancement of Science.

GREG C. GARMAN,   Ph.D. University of Maine, 1984, is currently Assistant Professor of Biology at Virginia Commonwealth University in Richmond. His research interests include ecological structure and function of large coastal rivers, trophic structure of lotic fish communities, and the ecological role of anadromous fishes in tidal and nontidal freshwater ecosystems.

ROBERT F. GAUGUSH,   Ph.D. Kent State University, 1983, is currently a hydrologist with the Limnological Studies Team of the Environmental Laboratory at the U.S. Army Waterways Experiment Station in Vicksburg, Mississippi. His research interests include reservoir limnology, empirical trophic state modeling, and water quality management.

ELLEN GILINSKY,   Ph.D. University of North Carolina, 1981, is currently Senior Environmental Scientist at Resource International, Ltd. in Ashland, Virginia, where she directs wetlands investigations and Chesapeake Bay projects. She also serves as an Adjunct Faculty member in the Department of Biology at Virginia Commonwealth University.

COURTNEY T. HACKNEY,   Ph.D. Mississippi State University, 1977, is currently a Professor of Biological Sciences at the University of North Carolina at Wilmington. His research interests include speciation of estuarine animals, energy flux in tidal wetlands, and the impacts of sea level rise on coastal ecosystems. Courtney is past president of the Society of Wetland Scientists and is currently chairman of the Southeastern Section of the Ecological Society of America.

HORTON H. HOBBS III,   Ph.D. Indiana University, 1973, is a Professor of Biology at Wittenberg University. His current research interests include stream benthic community dynamics, ecology and systematics of decapod crustaceans, cave ecosystem dynamics, study of the caves and cave fauna of Costa Rica, and conservation and management of cave ecosystems. Horton is a former member of the Board of Governors and Fellow of the National Speleological Society, a former Council Member of The Crustacean Society, a member of the Cave Research Foundation, is listed in Who's Who in American Men of Science, is currently Director of the Ohio Cave Survey, is a member of the Ohio Non-Game Wildlife Technical Advisory Committee, serves on the Board of Trustees of the Island Cave Research Center, and is a member of the Planning Committee of the Karst Waters Institute.

WAYNE C. ISPHORDING, Ph.D. Rutgers University, 1966, is Professor of Geology at the University of South Alabama. His primary interests are in the areas of heavy metal contamination of marine, estuarine, and lacustrine sediments and the importance of ion site partitioning in the evaluation of environmental contamination. He is the author of numerous articles describing the sedimentology and chemistry of Gulf Coast estuaries and on the application of computer graphics to geological problems. He is a member of a number of scientific societies and a Fellow of the Geological Society of America. He has twice been awarded his university's "Outstanding Faculty Member" award.

ROBERT H. KENNEDY, Ph.D. Kent State University, 1978, is currently Leader, Limnological Studies Team, of the Environmental Laboratory at the U.S. Army Waterways Experiment Station in Vicksburg, Mississippi. His research interests include nutrient cycling, reservoir limnology, and water quality management for reservoirs.

BRUCE L. KIMMEL, Ph.D. University of California at Davis, 1977, is a Research Staff Member and Program Manager in the Environmental Sciences Division of the Oak Ridge National Laboratory (ORNL). His research interests include lake and reservoir productivity, water quality, contaminant transport and fate, and water resources management. He currently serves on the editorial board of *Limnology and Oceanography*, a professional journal published by the American Society of Limnology and Oceanography, and as a member of the Advisory Panel for the National Science Foundation's Ecosystem Studies Program.

DAVID R. LENAT, M.S.P.H. University of North Carolina at Chapel Hill, is an environmental biologist with the North Carolina Division of Environmental Management, where he uses stream macroinvertebrates to evaluate water quality. His research interests include effects of point and nonpoint source pollutants, development of sampling methods, and taxonomy of southeastern Chironomidae.

ROBERT J. LIVINGSTON, Ph.D. Florida State University, is currently Professor and Director of the Center for Aquatic Research. His research interests include ecology of aquatic systems, human effects on such systems, the application of research to management decisions, long-term changes in river–estuarine systems. "Skip" is currently finishing a 20-year multidisciplinary research program concerning seven river–estuarine systems in the SE United States. The data are being analyzed, modeled, and published by a team of investigators.

REX L. LOWE, Ph.D. Iowa State University, 1970, is currently a Professor of Biological Science at Bowling Green State University in Ohio. Dr. Lowe has been a Visiting Professor at the University of Michigan Biological Station for the past 17 summers, where he teaches a course on freshwater algae and conducts research into benthic algal ecology in the upper Great Lakes region.

WILLIAM H. MARTIN, Ph.D. University of Tennessee, is Professor of Biology and Director of the Division of Natural Areas at Eastern Kentucky University. He has conducted research in upland grasslands and deciduous forests of the Southeast, with particular emphasis on the dynamics of old-growth forests in the southern Appalachian highlands. He has served as the Chairman of the Editorial Board for the aquatic and terrestrial volumes of Biodiversity of the Southeastern United States.

RONALD G. MENZEL, Ph.D. University of Wisconsin, 1950, is currently a Research Soil Scientist with the U.S. Department of Agriculture Water Quality and Watershed Research Laboratory in Durant, Oklahoma. His research interests include water quality and biological productivity of small impoundments, particularly as related to sediment properties and turbidity. Dr. Menzel has served as chairman of the Division of Environmental Quality of the American Society of Agronomy, and was Editor of the *Journal of Environmental Quality* from 1977 to 1983.

RICHARD F. MODLIN, Ph.D. University of Connecticut, 1976, is an Associate Professor of Biological Sciences at the University of Alabama at Huntsville. He studies the taxonomy, evolution, and ecology of peracarid crustaceans and life history strategies of the inhabitants of temporary ponds. Richard is the liaison officer to the Alabama Marine Environmental Sciences Consortium for UAH.

RICHARD H. MOORE, Ph.D. University of Texas at Austin, 1973, is currently Professor of Biology and Assistant Vice Chancellor for Grants and Sponsored Research at Coastal Carolina College as well as an Associate of the Belle W. Baruch Institute for Marine Biology and Coastal Research, University of South Carolina. His research interests include the distribution and ecology of marine and freshwater fishes, physiology and energetics of fish populations, and zoogeography of the southeastern United States.

PATRICK J. MULHOLLAND, Ph.D. University of North Carolina at Chapel Hill, 1979, is currently a Research Staff Member in the Environmental Sciences Division, Oak Ridge National Laboratory. His research interests include primary production, nutrient cycling and organic matter flux in streams, rivers, and wetlands, and land–water interactions in watershed biogeochemical cycles. Patrick currently serves as an associate editor for the journals *Ecology* and *Ecological Monographs*.

LARRY A. NIELSEN, Ph.D. Cornell University, 1976, is Professor and Head of the Department of Fisheries and Wildlife Sciences at Virginia Polytechnic Institute and State University. His research involves large rivers and the sociology of fisheries and wildlife management. Dr. Nielsen is President of the American Fisheries Society and currently co-chairs the World Fisheries Congress.

WILLIAM W. SCHROEDER, Ph.D. Texas A & M University, 1971, is a Professor of Marine Science with the University of Alabama and a Senior Marine Scientist at the Alabama Marine Environmental Sciences Consortium located on Dauphin Island. His research interests focus on interdisciplinary oceanography of estuaries and coastal ocean environments.

LEONARD A. SMOCK, Ph.D. University of North Carolina, 1979, is currently a Professor and Chairman of the Department of Biology at Virginia Commonwealth University. His research focuses on invertebrate community ecology, trophic ecology, and organic matter dynamics of blackwater streams of the southeastern United States. His work has taken him to blackwater systems in Virginia, North and South Carolina, and Georgia.

DAVID M. SOBALLE, Ph.D. Iowa State University, 1981, is currently a senior environmental scientist at the South Florida Water Management District in West Palm Beach. Formerly a Faculty Research Associate with the University of Tennessee and consulting limnologist to the Reservoir Research Program at Oak Ridge National Laboratory, Dr. Soballe has conducted limnological research on numerous reservoirs in the midwest and southeastern United States. His research focuses currently on eutrophication processes and algal bloom formation in Lake Okeechobee, Florida.

JUDY P. STOUT, Ph.D. University of Alabama, 1978, is a Senior Marine Scientist and Director of Academic Affairs at the Alabama Marine Environmental Sciences Consortium located at the Dauphin Island Sea Lab and an Associate Professor at the University of South Alabama in Mobile. Her research emphasis is the plant ecology of coastal and shallow-water habitats, especially salt marshes, estuarine grassbeds, and beach/dune systems. Study efforts include primary productivity and growth related to environmental factors, temporal and spatial variability, plant–animal interactions, and restoration of habitats.

J. BRUCE WALLACE, Ph.D. Virginia Polytechnic Institute and State University, 1967, is currently Professor of Entomology and a member of Graduate Faculty of Ecology at the University of Georgia, Athens. His research interests include stream ecosystem processes, biology and secondary production of invertebrates, and the influence of disturbance on streams. He is president-elect of the North American Benthological Society.

JACKSON R. WEBSTER, Ph.D. University of Georgia, 1975, is Professor of Biology at Virginia Polytechnic Institute and State University, Blacksburg. His research interest is the function of stream ecosystems, including decomposition, seston transport, nutrient dynamics, and response to watershed disturbance. He is also involved in modeling stream ecosystems.

BIODIVERSITY OF THE SOUTHEASTERN UNITED STATES
Aquatic Communities

PART I
Introduction

1 Ecological Processes in Southeastern United States Aquatic Ecosystems

S. MARSHALL ADAMS

Environmental Sciences Division, Oak Ridge National Laboratory, Oak Ridge, TN 37831

COURTNEY T. HACKNEY

Department of Biological Sciences, University of North Carolina, Wilmington, NC 28403

Aquatic ecosystems of the Southeastern United States (SE-US) represent a diverse and complex array of natural and man-made systems whose ecology is poorly understood. Although some systematic multidisciplinary studies have been conducted on a limited number of ecosystems, most of these have focused on descriptive components such as species occurrence. Studies that delimit functional and controlling components are less common, primarily because of the spatial and temporal complexity of most SE-US aquatic ecosystems and their array of diverse habitats.

Principles underlying our current understanding of the ecological behavior of many SE-US aquatic systems (e.g., streams, rivers, reservoirs) derive from the river continuum concept of Vannote et al. (1980) and Minshall et al. (1983). This concept, applied to river habitat gradients from headwaters to estuaries, is a process-oriented model based on organic matter input and consumer responses along gradients of fluvial geomorphic characteristics and physical–chemical conditions. Aquatic systems display continuous changes in physical and chemical characteristics from headwaters to the mouth, which influence the structure and function of biotic communities along this continuum. In addition, upstream to downstream linkages in nutrient dynamics are the basis of the nutrient spiraling concept of Webster and Patten (1979) and Newbold et al. (1982). The serial discontinuity concept (Ward and Stanford 1983) has been used to explain the spatial patterns of species composition

Oak Ridge National Laboratory is managed by Martin Marietta Energy Systems, Inc. under Contract DE-AC05-84OR2100 with the U.S. Department of Energy. Publication No. 3289, Environmental Sciences Division, Oak Ridge National Laboratory.

and abundance in river–reservoir systems. Despite these generalized concepts, so few systems have been studied in detail that our knowledge of process-oriented functions in SE-US aquatic systems precludes all but the most general application of these theories. Consequently, the primary objective in this volume is to summarize current understanding of SE aquatic ecosystems. We hope this synthesis will serve as the basis for future research and ultimately lead to improved understanding and better management of these valuable ecosystems.

The basic ecological processes and controlling mechanisms are similar for most aquatic systems, with differences occurring mainly in the specific mechanism that exerts dominant control and the rates at which these processes occur. Irregular and extreme variability in some environmental factors and anthropogenic effects can confound typical continuity and successional patterns in the biota in many SE-US systems. Environmental extremes, for example, may restrict organisms to those with broad physiological tolerance ranges and wide behavioral adaptations. Despite differences in the biota along an upstream to downstream gradient and in different drainages, fundamental process-based interrelationships between individuals, communities, and ecosystems remain the same.

Most aquatic systems (with the possible exception of some small ponds and springs) are hydrologically linked in a longitudinal gradient from headwaters to the upper end of estuaries. Therefore, the processes occurring upstream in various aquatic systems influence, to varying degrees, the physical, chemical, and biological processes downstream. A full understanding of structural and functional aspects of any aquatic system must necessarily include a basic awareness of the influence of other systems along this continuum. Ecological interactions that occur between aquatic systems along this longitudinal gradient (i.e., headwater streams and eventually estuaries) are interconnected to a greater extent than most terrestrial systems because of the more isolated nature of terrestrial systems. Similarities and differences among biotic components of different drainage systems depend not only on site-specific conditions and process-regulating mechanisms, but also on the degree of interconnectiveness with and the relative influence of upstream systems. Superimposed on this influence of linkages between systems are site-specific conditions along each drainage, which includes basic geological and physicochemical characteristics of the watershed.

If basic ecological processes in most SE aquatic systems differ only in their magnitude or rates, then identification and quantification of process-regulating parameters should describe the range of variability in structural and functional components possible in SE aquatic communities. The complexity and wide diversity in SE-US aquatic systems almost guarantee exceptions to any generalization, especially relative to regulating processes and their relation to community structure and function. We realize that generalizations can result in oversimplifications. However, this generalized approach should provide a basic understanding of how SE-US aquatic systems behave as integrated ecosystems.

The uniqueness of SE-US aquatic systems can be attributed primarily to the geologic history and climatic conditions that have had major influences on the development of biotic communities. The Southeast was unglaciated during the last glacial period and, as a result, present-day landform diversity and soil characteristics are the product of a long history of geotechnic and pedological processes. For example, compared with glaciated areas, soils of the SE-US are largely residual, being relatively deep and highly structured vertically. Soil structure can affect hydrologic and chemical properties of streams and rivers by routing runoff along preferential flow paths (macropores). The lower horizon of SE-US soils are also rich in iron and aluminum oxides, which can strongly influence stream water chemistry via efficient retention of many negatively charged solutes (e.g., SO_4^{2-}, PO_4^{3-}, DOC). Another unique geological feature of the SE-US is the occurrence of large portions of three very different physiographic provinces, the Mountain, Piedmont, and Coastal Plain. The types of aquatic systems and their associated communities that have developed in these provinces have been influenced to a large degree by the differences in their geology, including soil and rock types and gradient and relief characteristics. Many of the large SE drainages span all three provinces, with headwaters in the mountains and increasing influences of Piedmont and Coastal Plain characteristics in the middle and lower portions of the drainages. Estuaries are all bar-built or drowned river valleys with extensive infilling from both terrestrial and marine sources. As a consequence, virtually no hard substrates are found in SE estuaries.

Climate, the other major unique feature of the SE-US, has a major controlling influence on the structure and dynamics of biotic communities. The warmer climate, compared with most other regions, affects both the life history and distribution patterns of organisms and results in increased biological activity, which has a distinct temporal component. The lack of glaciation was much more important to aquatic community diversity than it was to the adjacent terrestrial fauna and flora. Most terrestrial communities "migrated" south then north with the onset and ending of glaciation in the North American continent. Aquatic forms, especially mollusks and fishes, had limited movement and either adapted to the changing climate or became extinct. The high diversity of aquatic organisms in the Southeast contrasts greatly with similar habitats further north (see Chapter 2).

PROCESS-REGULATING PARAMETERS

Many of the process-regulating parameters in SE aquatic systems are interrelated and influence each other to varying degrees. Six factors that regulate or influence the nature of biotic communities in SE-US aquatic systems are discussed in the following sections: (1) degree of land–water interaction, (2) relative importance of allochthonous and autochthonous energy sources, (3) flushing rate or water residence time, (4) physical structure and system complexity, (5) temporal pulsing of physicochemical factors (e.g., temperature, sediment loads), and (6) anthropogenic effects. It should be recognized, however, that these mechanisms are not independent, but interact to determine community structure in SE-US aquatic ecosystems. The relative importance of environmental factors in regulating processes of the major SE aquatic ecosystems is summarized in Table 1.

Land–Water Interaction

The degree of land–water interaction and the level of interaction between the terrestrial environment and the aquatic environment have major influences on both ecological processes and the structure of the food web. The aquatic–terrestrial interface and the ratio of drainage basin area to aquatic system surface area strongly influence the quantity and quality of water, nutrients, and other materials transported from the surrounding watershed. In the continuum from headwater streams to estuaries, small upland streams have the greatest interconnection with the terrestrial environment, whereas most caves and large natural lakes with small drainage basins would, in theory, have the least degree of land–water interaction (Fig. 1A). Intermediate between these two extremes would be rivers, reservoirs, estuaries, and lagoons that vary in their level of interaction with the terrestrial environment, depending on the perimeter to surface area ratio. For example, the more dendritic the shoreline of a system, the higher the value of this ratio.

The extent and slope of adjacent floodplain systems influence the degree of terrestrial–water interaction in some rivers and estuaries. For example, many rivers of the lower Gulf and Atlantic coastal plains have extensive floodplains that greatly influence the ecological functioning of these rivers. Most Atlantic and Gulf estuaries and lagoon systems are dominated by extensive emergent vegetation (marshes) that not only influence physicochemical factors but also affect energy flow and trophic relationships. Allochthonous organic matter typically dominates in aquatic systems that have high levels of land–water connectivity. Autochthonous organic matter production is relatively more important in aquatic systems that have less interaction with the terrestrial environment (Fig. 1A).

Allochthonous vs. Autochthonous Organic Matter

Aquatic systems are linked to their watersheds by the input of terrestrially generated energy and material. In the continuum of headwater streams to the estuary, strongly heterotrophic headwater streams grade into midorder streams and downstream reservoirs where the importance of autochthonous energy increases (Fig. 1A). Because of the extensive and relatively flat coastal plain, the energy base of larger river systems in the Atlantic and Gulf coastal plains may again return to allochthonous material through large periodic inputs from bordering floodplains and swamps (Fig. 1A). In estuaries and lagoons, with their high-standing crops of emergent vegetation and connections to major rivers, food webs may also rely heavily on allochthonous

TABLE 1 General Importance of the Major Process Regulating Factors in Each of the Major Types of Aquatic Ecosystems in the Southeastern United States

	Process Regulating Factors				
Ecosystem	Water–Land Interaction	Allochthonous Input	Flushing Rate	Physical Complexity	Pulsing Physicochemical Factors
Streams	H	H	M	H	M
Rivers	H–M	H	H	H–M	H
Reservoirs	M–L	H–M	H–M	M	M
Coastal plain rivers	H	H	M–L	M	M–L
Lakes	M–L	M–L	L	M	L
Small impoundments	M	M–L	L	M	M–L
Cave-spring	L	H	H–L	M	M–L
Estuaries	L	M	M	M–L	M
Lagoons	M	H	M	M–L	H–M

Note. H = high importance of a factor for regulating or controlling the magnitude and intensity of ecological processes, M = medium importance, and L = low importance.

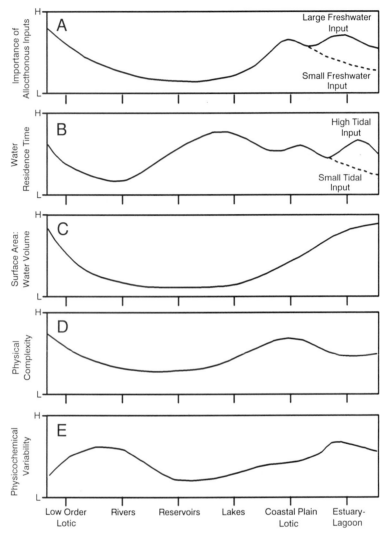

FIGURE 1. General importance of several environmental factors in regulating eco-logical processes in Southeastern aquatic ecosystems along a continuum from upland low-order streams to coastal estuaries and lagoons.

material (Fig. 1*A*). The magnitude of this dependence on allochthonous ma-terial is determined by the linkage between the estuary and the adjacent marine system, the size of the open water habitat relative to its perimeter, and the significance of river inputs to the estuary. The energy base of sub-terranean (cave) food webs is based almost exclusively on exogenous inputs of organic matter, much of which enters these systems during flood events. In contrast, the most autochthonous-based aquatic systems in the SE-US are small natural lakes with small watersheds and high residence times in which phytoplankton form the organic matter base of the food web.

The pulsing or temporal dynamics of energy input into aquatic systems influences the structure of their foodchains. In headwater streams, a majority of the allochthonous organic matter enters the stream as direct leaf-litter input. In addition to in-place utilization of this energy source by heterotrophs, high standing crops of macroinvertebrate shredders convert the course particulate organic matter into fine particulate organic matter which is then transported downstream and utilized by downstream heterotrophs. Autochthonous production is limited in these streams primarily because of the dense shading by riparian vegetation in combination with some nutrient limitation and top-down foodchain regulation by grazers.

In more lentic habitats downstream, the relative importance of autochthonous organic matter increases as nutrient loads increase, residence time increases (allowing for development of planktonic communities), and available sunlight increases. Autochthonous production dominates over allochthonous inputs in small impoundments and natural lakes because of the contribution by phytoplankton and macropytes. In rivers of the Gulf and Atlantic coastal plains, annual flooding of the adjacent floodplains provides the mechanism for mobilization and transport of terrestrial detritus, thereby increasing the importance of allochthonous organic matter in these systems. Transport of detritus from adjacent marsh areas into estuaries and lagoons during periodic tidal flooding and contribution by tidal creeks provides a substantial amount of allochthonous material to the estuarine food webs, although organic imports via rivers can also be important in some estuaries. Phytoplankton production in open water areas may also play important roles in the energetics of some estuaries.

Water Residence Time

The flushing rate or water residence time of a particular system may be the dominant process-regulating factor involved in shaping the structure and function of aquatic communities (Soballe and Kimmel 1987). Residence time influences communities directly through the effects of water renewal on plankton populations and other components of the biota and indirectly by its relationship with other important limnological variables such as nutrient and light availability, turbidity, and mixing regime.

Contrary to popular belief, the velocity of flow in many upland streams is actually lower than in downstream areas (Ledger 1981). Therefore, water residence times in upland streams may actually be higher than in some downstream lotic systems (Fig. 1B). In larger streams and rivers, shorter residence times and, in some cases, high turbidities generally lower overall system productivity and preclude the development of phytoplankton and zooplankton communities (Fig. 1B). In rapidly flushed systems, planktonic organisms are quickly washed downstream, and in these situations, attached plants (periphyton and macrophytes) that can maintain their position have increased significance to total system productivity. Longer residence times, combined with higher hydraulic and nutrient loading in more lentic systems, allow planktonic populations to grow rapidly and persist. The structure of the consumer community is dependent to varying degrees on the nature and extent of the plant communities that develop. In lower-order streams and rivers, plankton are virtually nonexistent, with consumers in these systems depending on either detrital materials (deposit feeders and detritivores) or periphyton production on immobile substrates (grazers). In larger streams and rivers, the food web generally depends on allochthonous material transported from upstream areas, littoral periphyton production, or nearshore emergent vegetation and macro-phyte production. In addition to high flushing rates in many river systems, increased turbidities resulting from heavy loads of suspended organic and inorganic matter often restrict light availability for plankton production.

Phytoplankton and macrophytes serve as the primary energy base of the foodchain in systems with long residence times such as reservoirs, lakes, and small impoundments (e.g., farm ponds, USDA impoundments). Although many coastal plain rivers have relatively slow flushing rates because of low gradients and extensive floodplains (Fig. 1B), the importance of plankton and macrophyte production is minimized because of the high dissolved organic carbon in the water which restricts light penetration and thus primary pro-duction. Residence times of estuaries, however, are intermediate between those of reservoirs and rivers, being controlled to a large extent by tidal cycling (Fig. 1B). Estuaries along the Gulf coast have minimal tidal flushing as does the Pamlico–Albemarle system in North Carolina, whereas most South Atlantic coast estuaries are flushed by 1- to 3-m semidiurnal tides (except south Florida). Plankton, benthic algae, and detrital production are important energy pathways to consumers in these systems (Haines 1979).

Physical Structure and Complexity

The physical structure and complexity of an aquatic system can affect the nature and types of biotic communities that develop within that system. The physical structure of an ecosystem pertains primarily to basin morphology, slope, geology, and character of adjacent soils, whereas physical complexity involves the quantity and quality of physical structure available to organisms.

Physical characteristics related to basin morphology can affect biological processes in complex ways, the most important of which relate to light and nutrient availability. The level of water–land interaction influences the quality and amount of nutrients and other material transported into a system. Irreg-ularly shaped and highly dendritic aquatic systems, for example, are more closely linked to the watershed, and therefore the relative importance of allochthonous inputs is increased. In cave ecosystems, the number of sinkholes is important in influencing the contribution of allochthonous material avail-able to the foodchain.

Various indices of basin morphology, such as the bottom area:water vol-ume or the surface area:water volume, reflect the relative importance of light and nutrients to process-regulating mechanisms. A low ratio of surface area or bottom area to the water volume would indicate, for example, that benthic–pelagic coupling is relatively low and that the importance of sediments in nutrient regeneration of the system is minimal (Fig. 1C). A high depth:surface area ratio would indicate, for example, that light could be limiting in lower parts of the water column. The morphoedaphic index (MEI), defined as the total dissolved solids concentration divided by mean depth, has been used as a relative indicator or predictor of fisheries yield in a variety of freshwater environments (Ryder et al. 1974, Jenkins 1982). Presumably, the MEI is broadly indicative of both nutrient availability in a system (as reflected by the total dissolved solids concentration) and the capacity of the system to process energy and materials efficiently, as reflected by the mean depth (Oglesby 1977). The MEI reflects a complex aggregation of covarying internal factors, processes, and feedback mechanisms, which may integrate to influence total system productivity. Recently, however, Downing et al. (1990) reported that fishery production in lakes is highly correlated with primary productivity levels, not with MEI.

In general, lotic, estuarine, and lagoonal ecosystems have high surface area:water volume ratios (Fig. 1C), which infers that benthic–pelagic coupling plays an important role in nutrient dynamics and total system metabolism. Even in the turbid Georgia estuaries, their shallow depth and high tidal amplitude results in strong sediment–water column interactions (Haines and Montague 1979). The relative importance of benthos to food web structure and dynamics is greater in these systems than in most lentic ecosystems (e.g., large reservoirs). In addition, a high value for the surface area:water volume ratio suggests that light is not a major limiting factor. Exceptions to this generalization are highly turbid rivers and reservoirs and dark coastal plain rivers where surface light attenuation is high.

The physical complexity within a particular aquatic system influences the distribution, type, and productivity of various organisms. The most important aspect of physical complexity is the availability of a variety of stable substrates that provide habitat and attachment surfaces (i.e., cover, protection, energetic benefits) for animals. In the higher-gradient systems, rocks and woody debris provide a majority of the physical complexity (Fig. 1D). Rocks help define and create the physical complexity, such as flow and sediment type, but fine-grain sediments also may comprise an important part of the complexity in these systems. Many organisms in upland streams are intimately associated with rocks, being attached directly to the surface (periphyton), grazing on the rock surfaces (macroinvertebrate grazers, snails), living among rocks and debris (shredders, detritivores), or hiding behind or between rocks and debris for protection and to minimize energetic costs (fish). Woody debris dams also provide attachment surfaces for organisms, retain organic and inorganic material, and alter the ratio of course particulate organic matter/fine particulate organic matter supplied to downstream areas. The availability of stable bottom substrates in rivers and lower-gradient streams is low (Fig. 1D) primarily because of the instability of silt and sand bottoms and the alternating burial and exposure processes of these sediments. High turbidities and unstable substrates severely limit primary production (periphyton and macrophytes) and secondary production (macroinvertebrates) in these systems.

In contrast to the larger streams and rivers of the Piedmont area, woody debris in coastal plain rivers and streams has a major influence on ecological processes and community structure (Fig. 1D) (see Chapter 7). Not only do these structures provide a stable habitat for attachment of organisms, they are also sites of high biological activity. Debris dams have a higher detritus retention capacity than sediments and also have an important influence on stream energetics, material spiraling, and hydrodynamics within streams. Debris dams and snag habitats provide a localized concentration of food organisms that, in comparison with dispersed food resources found in plankton environs, may be more readily utilized by consumers such as fish. In addition, snag habitats can be important downstream sources of macroinvertebrate drift for populations of fish downstream (Benke et al. 1986).

Stable substrates in lentic systems such as reservoirs, lakes, and small impoundments consist primarily of rock surfaces that fringe the shoreline, fallen or submerged timber and tree tops, or surfaces of submerged macrophytes. These structures provide attachment substrates for periphyton and macroinvertebrates and refuges for various species of fish and invertebrates. The relative contribution of these habitats to community structure in many lentic systems is much less than in streams and rivers, however, because of the increased relative importance of plankton communities.

The physical complexity of estuaries and lagoons is relatively low because of the lack of stable substrates (Fig. 1D). Tidal action and winds in these shallow systems continually remobilize and deposit sediments, limiting the development of attachment surfaces for organisms. Some stable substrates exist around and in salt marshes where oyster bars and stems of marsh grass provide attachment surfaces for periphyton and mollusks. Oyster reefs, in fact, are considered to be metabolic "hot spots" with high species diversity (Bahr and Lanier 1981). In shallow higher-salinity areas of some estuaries, submerged seagrass communities provide not only attachment surfaces for periphyton and invertebrates, but also serve as important feeding and refuge areas for various species of fish and shellfish. Salt marshes and seagrass beds also provide substantial amounts of organic material to support the detrital foodchains of estuaries (see Chapters 13 and 14).

Pulsing of Physicochemical Factors

One of the major factors complicating our understanding of processes in SE-US aquatic systems is the irregular and extreme variations in the chemical and physical environment. The extent and nature of these variations can have large influences in both the short- and long-term scales on the structure and function of biotic communities. Irregularly pulsed nutrient and sediment loading, for example, influence the magnitude and rate of process-regulating factors and affect system trophic dynamics.

Variability of physicochemical factors influences not only the types of organisms that live in a particular system but also the magnitude of process rates. Ecosystems that experience regular (seasonal) and irregular precipitation-driven pulses, for example, tend to contain species with broad physiological tolerance ranges and wide behavioral flexibility. For example, eurythermal biota dominate in systems, such as streams and rivers, that display large annual and diel variations in temperature. Pulsed events, such as floods, destabilize these systems and provide a recurrent series of disturbances that maintain high productivity and biotic diversity. Floods, however, are a major factor in the ecology and the trophic relationships of cave ecosystems. Floods operate as agents to transport allochthonous material into cave environments, serve as mechanisms of distribution and colonization of organisms, stimulate reproductive activity, and trigger molting cycles in certain organisms (see Chapter 3).

Because of their high connectivity with the land, lotic environments and especially lower-order streams and rivers may have high spatial variability and experience wide variability in temporal events (Fig. 1E) (see Chapters 4, 5, and 6). Small upland streams, however, are probably less variable in some physicochemical factors such as temperature because of the relative constant groundwater influence (Fig. 1E). High water velocities and suspended loads of organic material dislodge and scour organisms, sometimes resulting in a high degree of system instability. Lakes, reservoirs, and other lentic systems that have less interaction with the terrestrial environment and longer water residence times are better buffered against these pulsed events and develop more stable communities (Fig. 1E).

The extensive floodplain communities associated with rivers of the Gulf and Atlantic coastal plains minimize the effects of episodic events on organisms by stabilizing pulsed movements of water and by modifying the hydraulic regime (Fig. 1E) (see Chapters 7 and 8). Dams and other structures may also alter the influence of episodic events downstream (see Chapters 11 and 15). Small impoundments, such as farm ponds (Chapter 10) and beaver dams (Chapter 15), probably serve a similar function. Inflows from precipitation events that are not intercepted by ponds (Chapter 10), energy-dispersive floodplains (wetlands), and littoral interface regimes result in irregularly pulsed nutrient and sediment loading. In contrast, the trophic organization and distribution of estuarine organisms are highly dependent on seasonal and interannual variability of factors such as temperature, salinity, tidal amplitude and frequency, wave action, and the availability of larvae from the nearshore environment (Weinstein 1988). The structure of plant and animal communities in lagoonal systems is based on a mixture of species drawn from the surrounding fresh and brackish water habitats, which exhibit an ability to tolerate wide salinity variations. Variability in physicochemical factors is less extreme in estuaries than in lagoonal systems because of the regularity of tidal flushing in the former.

The biotic vs. abiotic control of community structure has been addressed by Peckarsky (1983), who suggested that lotic systems can be placed on a gradient from harsh to benign environmental conditions. Physicochemical factors appear to influence composition of invertebrates in harsh environments, whereas biotic factors such as competition and predation may be major controlling factors for community structure in benign environments. There is no strong evidence that predators can significantly influence the structure of lotic invertebrate communities (Allan 1983). Because of the greater physicochemical variability of streams and rivers, abiotic factors tend to have greater influence on their community structure than on communities of lentic systems. Indeed, one of the major mechanisms that has been proposed for influencing community structure in lentic systems is top-down (predator) regulation of food chains (Carpenter et al. 1985).

Anthropogenic Effects

Nearly all major stream systems of the SE-US have been impounded or modified in some way by dam construction, channelization, or other engineering activities (see Chapter 15), with many of these impoundments being arranged in a series of regulated pools. Most major rivers of the SE-US have therefore been converted from lotic environments to stair-step chains of lentic or semilentic systems (see Chapter 11). The construction and operation of dams on these rivers strongly influence the structure and function of ecological communities, not only within the reservoir itself, but also in downstream environments. Water releases from dams create conditions downstream that differ markedly from unmodified waters.

Although reservoir operations often serve to dampen long-term (seasonal) fluctuations in streamflow, they may also greatly enhance short-term (daily) changes in flow. The combination of hypolimnetic discharges, temporally variable hydropower generation, and serial arrangement of these impoundments creates complex hydrodynamic conditions that effect physical, chemical, and biological processes downstream. The operational regimes of impoundments can affect a wide range of process-regulating factors such as temperature and water-level fluctuations, water quality, water flow, nutrient and light availability, siltation, trapping and discharge of organic matter, and anoxia. These factors can have impacts on downstream community structure by producing shifts in functional feeding groups and composition of invertebrates (Chapter 11). The abundance and species composition of higher consumers may also reflect shifts in the macroinvertebrate community.

Water level fluctuations prevent development of littoral zone communities downstream of impoundments because of periodic drying and flooding of the littoral zone. Consequently, foodchains of downstream communities may be based primarily on pelagic production despite the presence of significant littoral areas. This situation is in contrast to natural lakes and small impoundments where littoral producers are central to the food web. Particulate and dissolved nutrients that accumulate in bottom waters of reservoirs increase nutrient availability and enhance primary production downstream of hypolimnetic releases.

There can be several direct effects of reservoir operations on fish community dynamics and structure. Rapidly rising waters that inundate terrestrial areas can temporarily increase the resources available to fish, such as invertebrate food, whereas large drawdowns can concentrate prey and temporally increase predator feeding success. Drawdown, however, can also result in loss of spawning habitat, thereby having a negative affect on reproduction and egg survivorship of nonpelagic spawners. Cool-water fisheries in tailwater areas can be enhanced by hypolimnetic discharges, but this can have a negative influence on warm-water fisheries. Dam construction also produces structural and thermal barriers to fish movement in impounded rivers, and in some cases, artificial stocking and removal of natural physical barriers (e.g., Tombigbee project, see Chapter 15) have introduced new species into some SE-US drainage basins. Therefore, significant changes in the distribution and structure of fish communities have often followed human disturbance in many of these systems.

SYNTHESIS

The uniqueness of SE-US aquatic systems can be attributed primarily to their geologic history and climatic conditions which have had a major influence on the types of biotic communities that have developed within these drainages. The structure and types of communities in each type of system depend on a variety of physicochemical and biotic factors that vary both spatially and temporally. Variability in the nature and extent of physicochemical factors dictate, to a large degree, differences in communities and process rates among systems. The environmental conditions in some streams, rivers, and even estuaries, tend toward large, rapid, and erratic fluctuations in factors such as water flow, suspended material, temperature, dissolved solids, and nutrients. Irregularly pulsed events can destabilize systems, maintaining them in various stages of biotic succession relative to diversity and system productivity. System instability typically results in biotic communities in which diversity is low but organisms are well adapted with broad physiological tolerance limits. In this situation, specialization is typically low but growth and productivity may be high. In more stable systems, such as reservoirs and lakes that experience less variability in physicochemical factors, stability may be higher but overall system productivity tends to be lower.

Generalizations about patterns in fish community structure between systems are difficult not only because of the interconnected nature of most aquatic systems but also because of the highly interconnected nature of grazer and detrital pathways and the pivotal role of benthic detritivores in aquatic food webs (Goldman and Kimmel 1978, Adams et al. 1983). Interactions among organic matter sources, trophic pathways, varying physicochemical factors, and basin morphology may underlie the conservative nature of fish production systems described by Kerr (1982). However, there appears to be a general pattern of increasing fish abundance and diversity from lower-order to higher-order rivers. Unaltered coastal plain rivers and streams have a relatively more abundant and diverse fish community compared to that of low- and mid-order systems. This trend is probably mainly the result of the greater variety and availability of habitats and food resources in extensive flood plain habitats and the important spawning and nursery areas associated with these systems. Lower-order streams tend to have more species unique to each drainage (see Chapter 2 and Hocutt and Wiley 1986).

Adequate understanding of how SE-US aquatic systems operate ecologically is dependent on our ability to predict their behavior based on biological unity (Wetzel 1990). The irregular and extreme variations in physical and chemical factors in many aquatic systems complicate our search both for unity and order in ecological behavior and, therefore, the development of generalized management strategies for these ecosystems. To obtain reasonable predictability of biotic and water quality responses of aquatic communities to irregular variations in physicochemical factors, much information is required on individual ecosystem properties. Because SE-US aquatic systems are highly linked to each other and to their watersheds, proper resource use and management may be more effectively achieved by using an integrated watershed approach for understanding ecological processes in these systems and how these processes shape community structure. This type of approach is especially important because the geological history (Chapter 2) and human modifications (Chapter 15) may have produced very different systems. Kerr (1982) concluded that "internal" (i.e., reductionist) methods of ecosystem analysis are unlikely to succeed unless preceded by relevant observation and theory at the "external" (holistic) level. To this end, the following chapters summarize our current knowledge of the composition, structure, and function of the major aquatic ecosystem types in the SE-US. Each chapter also provides specific information on some representative aquatic communities in each type of ecosystem and gives recommendations for future research and management in each ecosystem.

REFERENCES

Adams, S. M., B. L. Kimmel, and G. R. Ploskey. 1983. Sources of organic matter for reservoir fish production: a trophic-dynamic analysis. *Can. J. Fisheries Aquatic Sci.* 40:1480–1495.

Allan, J. D. 1983. Predator–prey relationships in streams. In J. R. Barnes and G. W. Minshall (eds.), *Stream Ecology.* New York: Plenum, pp. 191–229.

Bahr, L. M., and W. P. Lanier. 1981. The ecology of intertidal oyster reefs of the South Atlantic coast: a community profile. FWS/OBS-81-15, U.S. Fish and Wildlife Service.

Benke, A. C., R. J. Hunter, and F. K. Parrish. 1986. Invertebrate drift dynamics in a subtropical blackwater river. *J. N. Am. Benthol. Soc.* 5:173–190.

Carpenter, S. R., J. F. Kitchell, and J. R. Hodgson. 1985. Cascading trophic interactions and lake ecosystem productivity. *Bioscience* 35:635–639.

Downing, J. A., C. Plante, and S. Lalonde. 1990. Fish production correlated with primary productivity, not the morphoedaphic index. *Can. J. Fisheries Aquatic Sci.* 47:1929–1936.

Goldman, C. R., and B. L. Kimmel. 1978. Biological processes associated with suspended sediment and detritus in lakes and reservoirs. In J. Cairns, E. F. Benfield, and J. R. Webster (eds.), *Current Perspectives on River-Reservoir Ecosystems.* North American Benthological Soc. Pub. 1. Blacksburg, VA: Virginia Polytechnic Institute and State University, pp. 19–44.

Haines, E. B. 1979. Interactions between Georgia salt marshes and coastal waters. In R. J. Livingston (ed.), *Ecological Processes in Coastal and Marine Systems.* New York: Plenum, pp. 35–46.

Haines, E. B., and C. L. Montague. 1979. Food sources of estuarine invertebrates analyzed using $^{13}C/^{12}C$ ratios. *Ecology* 60:48–56.

Hocutt, C. H., and E. O. Wiley (eds.). 1986. *The Zoography of North American Freshwater Fishes.* New York: Wiley.

Jenkins, R. M. 1982. The morphoedaphic index and reservoir fish production. *Trans. Am. Fisheries Soc.* 111:133–140.

Kerr, S. R. 1982. The role of external analysis in fishery science. *Trans. Am. Fisheries Soc.* 111:165–170.

Ledger, D. C. 1981. The velocity of the River Tweed and its tributaries. *Freshwater Biol.* 11:1–10.

Minshall, G. W., R. C. Petersen, K. W. Cummins, T. L. Bott, J. R. Sedell, C. E. Cushing, and R. L. Vannote. 1983. Interbiome comparison of stream ecosystem dynamics. *Ecol. Monogr.* 53:1–25.

Newbold, J. D., R. V. O'Neill, J. W. Elwood, and W. Van Vinkle. 1982. Nutrient spiralling in streams: implications for nutrient limitation and invertebrate activity. *Am. Natur.* 120:628–652.

Oglesby, R. T. 1977. Relationships of fish yield to lake phytoplankton standing crop, production, and morphoedaphic factors. *J. Fisheries Res. Board Can.* 34:2271–2279.

Peckarsky, B. L. 1983. Biotic interaction or abiotic limitations? A model of lotic community structure. In T. D. Fontaine and S. M. Bartell (eds.), *Dynamics of Lotic Ecosystems.* Ann Arbor, MI: Ann Arbor Science, pp. 303–324.

Ryder, R. A., S. R. Kerr, K. H. Loftus, and H. A. Regier. 1974. The morphoedaphic index, a fish yield estimator-review and evaluation. *J. Fisheries Res. Board of Can.* 38:663–688.

Soballe, D. M., and B. L. Kimmel. 1987. A large-scale comparison of factors influencing phytoplankton abundance in rivers, lakes, and impoundments. *Ecology* 68:1943–1954.

Vannote, R. L., G. W. Minshall, K. W. Cummings, J. R. Sedell, and C. E. Cushing. 1980. The river continuum concept. *Can. J. Fisheries Aquatic Sci.* 37:130–137.

Ward, J. V., and J. A. Stanford. 1983. The serial discontinuity concept of lotic ecosystems. In T. D. Fontaine and S. M. Bartell (eds.), *Dynamics of Lotic Ecosystems.* Ann Arbor, MI: Ann Arbor Science, pp. 29–42.

Webster, J. R., and B. C. Patten. 1979. Effects of watershed perturbation on stream potassium and calcium dynamics. *Ecol. Monogr.* 49:51–72.

Weinstein, M. P. (ed.). 1988. Larval fish and shellfish transport through inlets. *American Fisheries Society Symposium No. 3.* Bethesda, MD: American Fisheries Society.

Wetzel, R. G. 1990. Reservoir ecosystems: conclusions and speculations. In K. W. Thornton, B. L. Kimmel, and F. E. Payne (eds.), *Reservoir Limnology: Ecological Perspectives.* New York: Wiley-Interscience.

2 Geologic and Evolutionary History of Drainage Systems in the Southeastern United States

WAYNE C. ISPHORDING

Department of Geology–Geography, University of South Alabama, Mobile, AL 36688

J. F. FITZPATRICK, JR.

Department of Biological Sciences, University of South Alabama, Mobile, AL 36688

Water is the basis for all life on this planet. While many environmental factors influence the abundance and distribution of both plants and animals, the most obvious is availability of water. The availability of water not only exerts major controls over the world's biota, but is equally important in the weathering of rocks and development of the landscape. Weathering is a phenomenon caused by both chemical and mechanical processes. Mechanical weathering includes those activities that simply break rocks down into smaller particles (frost wedging, expansion due to pressure release, thermal expansion, etc.). These processes are less important on a worldwide scale, however, than those associated with chemical weathering. Chemical weathering is by far the dominant process but is abetted by mechanical weathering because chemical processes are more effective on smaller particles (because of their greater surface area).

Chemical weathering involves complex processes that alter the internal structure of minerals by removing and/or adding elements. Because of this, the original rock decomposes into substances that tend toward equilibria with the surface environment. Essentially every chemical process from simple hydrolysis, which converts silicate minerals into hydrated oxides (causing the orange, yellow, red, and brown colors of soils), to more complex chemical reactions responsible for breakdown of primary minerals and formation of new phases is abetted by water. Indeed, even in desert and polar environments, water can be shown to be the most important agent controlling development of the landscape. Under such conditions, it is the action of *running water* rather than chemical reaction that causes rock degradation.

A discussion of the evolutionary history of drainage systems of the southeastern United States must necessarily include an assessment of the abundance, mode of occurrence, and interaction of water with both the surface and near-surface environment. Similarly, because the attitude and type of rock strongly influence the rate at which weathering takes place, a discussion of the drainage characteristics of the southeastern United States must first be prefaced with a discussion of the soils and general geology of the area.

SOILS

No attempt will be made in this chapter to describe in detail the complex distribution of soil types that exist in the southeastern United States. A broad overview is necessary because soil types are a reflection of rainfall, vegetation, slope, water infiltration, rock type, and runoff characteristics. Each of these factors also can be shown to influence development of a region's drainage systems.

The soil classification system now used throughout the world was first proposed in 1951 to permit detailed soil surveys to be compiled for any given region. Many revisions and modifications have been made through the years and, periodically, the International Society of Soil Science meets to evaluate and approve changes and additions to the classification. The version presently used, the 7th Approximation issued in 1968, first broadly groups a soil into one of 10 Orders (Table 1). These in turn are further divided into suborders, great groups, subgroups, families, and series. Only the lowest category in the taxonomic system, the series, has survived from older, pre-1951 classification systems and over 9000 different series are now recognized in the United States (Soil Survey Staff 1975).

Three major soil groups are dominant in the southeastern United States (Fig. 1). Soils belonging to the Ultisol group (Table 1) cover the great bulk of the Atlantic and Gulf coastal plains and the Blue Ridge and Piedmont provinces. Older classifications referred to these simply as "Pedalfers." Inceptisols are extensively developed in the Valley and Ridge province and in the Appalachian plateaus. Inceptisols are also well developed in the Mississippi River valley, the Pearl River valley system of Mississippi, and the Mobile–Alabama river systems in Alabama. Alfisols, the third major group, are well represented in the Texas Coastal Plain and marginal to the Inceptisols of the Mississippi River valley. Other major soil groups are also present (e.g., Histosols in the Florida everglades and Louisiana coastal region, Vertisols in the Texas Coastal Plain, Spodosols in Florida, and Entisols locally in Georgia, Florida, and Alabama), but are much less prevalent. Of the 10 major soil groups all are represented in the southeastern United States except Aridosols and Oxisols. In fact, small areas of each of the latter are also present but cannot be shown on a map of the scale of Fig. 1.

GEOLOGICAL OVERVIEW

The United States is divided into 34 physiographic provinces, grouped into 11 major divisions, based largely upon underlying geologic structures (see

TABLE 1   Description of Major Soil Orders Used in the Present Soil Taxonomic Classification

Group Name	Description
Entisols	Recent soil; little or no change of parent material. Lacking well-developed subhorizons. Common on gently sloping to steep, actively eroding slopes in semiarid to humid climates.
Vertisols	Clayey soils developed on gentle slopes either in calcareous clay or in residuum weathered from soft, calcareous sedimentary rocks. Characterized by wide, deep cracks at some times of year. Found in semiarid to humid climates.
Inceptisols	Some change or alteration of parent material. Young soils lacking horizons of clay (argillic) accumulation. Found on gentle to steep slopes in all climatic regimes except arid.
Aridosols	Arid climates, most often on gentle slopes. Light-colored soils of desert shrub and desert grassland areas. Possess zones of soft to indurated calcium carbonate (caliche). The dominant soil type in the United States west of the 101st meridian.
Mollisols	Soils having a thick, dark-colored surface horizon, high in bases. Widely developed on grasslands in semiarid to humid climates. Present on steep slopes, toe slopes, uplands, terraces, and flood plains; in residuum, from limestone, soft shales, sandstone, or aeolian mantles over residuum or in other transported material.
Spodosols	Soils with a spodic horizon (i.e., one in which active amorphous materials composed of organic matter and aluminum, with or without iron, have accumulated). "Active," signifying high exchange capacity, large surface area and high water retention. Extensive development in Atlantic and Gulf Coastal Plains.
Alfisols	Light-colored soils of humid temperate regions. Found bordering the alluvial plains of the Mississippi and Ohio Rivers and also in much of Texas and Oklahoma. Similar to Mollisols except for color of surface layer, preference for more moist climates; low in organic matter.
Ultisols	Light-colored soils of humid warm temperate forest lands. Extensively developed in the southeastern United States. Characterized by sandy or loamy surface horizons and subsurface horizons are loamy or clayey in texture; acid soils with low base saturation. Frequently deeply weathered with well-developed A and Bt horizons.
Oxisols	Highly weathered red soils of the tropics. Formerly termed "laterites" and "terra rosas"; featureless profiles with little horizon differentiation; low fertility.
Histosols	Soils characterized by over 20% organic matter in upper portion of profile (peats). In flat regions that may be continually or periodically covered with water. Common in Florida Everglades and along Louisiana Gulf Coast.

Source. Data from Buol (1973).

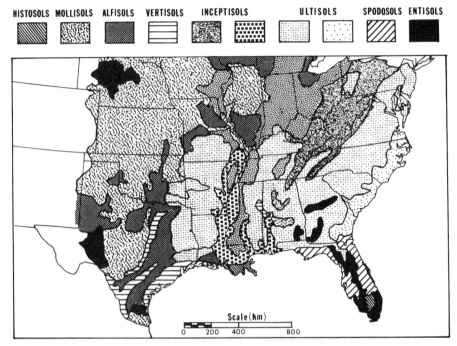

FIGURE 1. Map showing distribution of major soil orders in the southeastern United States. (Modified after Buol 1973.)

Fenneman 1937, Hunt 1967). Each province is characterized by differences in climate, soils, vegetation, lithology, plate and microplate tectonic history, and the structural attitude of the bedrock. Province boundaries are, for the most part, sharply defined, indicating that the major differences are lithologic and structural in nature (Fig. 2). Discussions in the following pages concentrate on three of these divisions (Appalachian Highlands, Atlantic and Gulf plains, and Interior Highlands) and on the physiographic provinces included within them (Table 2).

Appalachian Highlands

The Appalachian mountain system extends along the eastern side of the United States from New York to Alabama (Fig. 3) and is paralleled for most of its length by the adjacent Piedmont and Atlantic Coastal Plain provinces. Inland, the relatively stable interior of the continent is represented by the Appalachian Plateau province, which is composed of nearly horizontal strata of Paleozoic age. These rocks have been heavily dissected and, as a result, the area is characterized by rugged terrain and moderately high relief.

Topographically, the Appalachian Highland region (Valley and Ridge, Blue Ridge, and Piedmont provinces) is characterized by folded and thrust-faulted rocks whose weathering has formed a number of flat-topped, parallel,

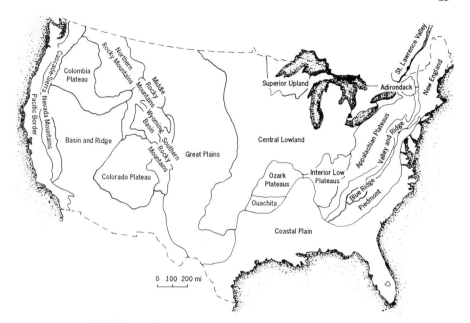

FIGURE 2. Physiographic provinces of the United States.

and subparallel ridges and valleys. The ridge and valley systems are a reflection of a series of underlying anticlines, synclines, and thrust faults that developed, for the most part, during the continental collision and orogenic (mountain building) activity that occurred during the Paleozoic era. Many rocks in the Blue Ridge have undergone moderate to high grade metamorphism and massive igneous intrusions of stock and batholithic size are common. In the Piedmont province, which lies to the east of the Blue Ridge, the rock structures also reflect the strong folding, intrusion, and metamorphism that took place during the Paleozoic. These rocks are more heavily faulted than those in the Blue Ridge province and have been considerably reduced in height by erosion. Most of the strata in the Appalachian Highlands are of Paleozoic age but older Precambrian rocks have been identified in the Blue Ridge province and rocks of Mesozoic (Triassic) age are well represented in several large block fault basins in the Piedmont province. The rocks in the Appalachian Plateau region range in age from Mississippian to Permian and are gently downwarped into an elongate syncline (or synclinorium), which parallels the more tightly folded structures of the Ridge and Valley province to the east. The topographic boundary between the two provinces is a prominent escarpment known at the north (in Maryland, West Virginia, Virginia, and Pennsylvania) as the Allegheny Front and to the south (in Tennessee and Kentucky) as the Cumberland Front (Thornbury 1954).

Climatically, the Appalachian Highland region can be described as humid, with cold winters in the north and mild winters in the south. Weathering of rocks during the latter part of the Pleistocene epoch has developed soils

TABLE 2   Major Divisions and Physiographic Provinces of these United States

Major Division	Provinces	Characteristics
Appalachian Highlands	Piedmont	Rolling upland; altitudes 500–2000 ft (150–600 m) in the south; surface slopes northeast and altitudes in northern region are below 500 ft (150 m).
	Blue Ridge	Easternmost ridge of the Appalachian Highlands; altitudes to over 6000 ft (1000 m).
	Valley and Ridge	Parallel valleys and mountainous ridges; altitudes mostly between 1000 and 3000 ft (300–900 m).
	Appalachian plateaus	Plateau surfaces largely between 2000 and 3000 feet (600–900 m); surfaces slope to west and are deeply incised by winding stream valleys; relief locally high with steep hillsides.
Atlantic and Gulf Plain	Coastal Plain	Broad plain rising inland; shores mostly sandy beaches backed by estuaries and marshes; mud flats at mouth of Mississippi River; some limestone bluffs on the west coast of Florida; inland ridges parallel the coast; altitudes less than 500 ft (150 m); surface slopes to the northeast; on Atlantic coast and northern valleys form tidal inlets near coast.
	Continental Shelf	Submarine plain sloping seaward to depth of approximately 600 ft (200 m); submerged portion of Coastal Plain.
Interior Highlands	Ozark	Rolling upland; mostly above 1000 ft (300 m).
	Ouachita	Similar to Valley and Ridge; altitudes 500–2000 ft (150–600 m).
Interior Plains	Interior Low plateaus	Plateau surfaces less than 1000 ft (300 m) in elevation; rolling uplands with moderate relief.

Source. Modified after Hunt (1967).

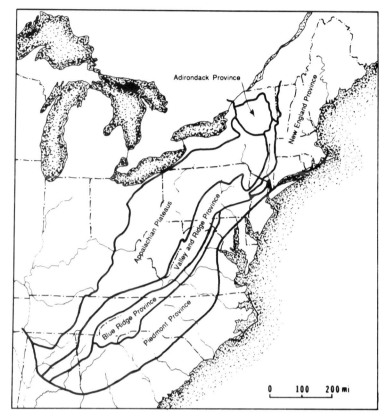

FIGURE 3. Physiographic provinces of the Appalachian Highlands.

(Fig. 1) belonging to the order Alfisols (suborder Udalfs) throughout much of the northern part of the Appalachian Plateau region, especially on glacial drift sediments; to the east, in the Blue Ridge and Valley and Ridge provinces, Inceptisols (suborder Ochrepts) are dominant, particularly those in the great group Dystrochrepts. Piedmont province rocks, in contrast, possess generally thick soils that are classified as Ultisols (suborder Udults). These are the most extensive soil type in the southern United States and most soils in the Piedmont region are assigned to the suborder Hapludult.

Interior Low Plateaus

The Interior Low Plateau area (formerly called the Cumberland Plateau) lies south of the Ohio River and extends to the Coastal Plain overlap in northern Alabama, Mississippi, Tennessee, and Kentucky (Fig. 2). The region is bounded in the east by the provinces of the Appalachian Highlands. Structurally, the area is part of a broad anticline (the Cincinnati Arch) that forms the eastern flank of a broad shallow structural basin under southern Illinois and the Mississippi River alluvial section (Fig. 4). The Interior Low Plateau province

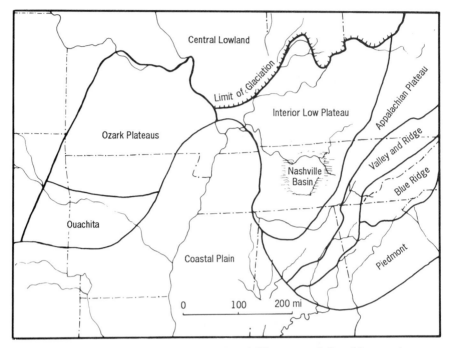

FIGURE 4. Interior Lowlands province, Ozark Plateau, and Ouachita province.

lies south of the limit of late Pleistocene glaciation and is divided into four subunits. These include the bluegrass area of Kentucky (the Lexington Plain), a large structural dome that forms a topographic basin (the Nashville Basin), a plateau region surrounding the Nashville Basin (the Highland Rim), and a highly dissected plateau region that extends from western Kentucky through central Tennessee and ends in northern Alabama. Lithologically, the entire area is underlain by Paleozoic-age sedimentary rocks, with limestones predominating over much of the area. Hence, much of the region is characterized by the abundant development of karst features, such as caves, sinkholes, and underground drainage. The soils are typically members of the order Ultisols (suborder Udults) or Alfisols (suborder Udalfs). Residual weathering of the limestones has also led to local development of Oxisols (terra rosa soils) in some areas of northern Alabama, Tennessee, and Kentucky.

Ozark Plateaus and Ouachita Province

The Ozark plateaus are similar to the Appalachian plateaus both structurally and topographically but differ in the high percentage of limestone, rather than coal-bearing sandstones and shales. The Ouachita Mountains are similar to the Appalachian Mountains in topographic appearance, although they are not nearly as high. Their level-topped, subparallel east–west ridges rise to heights of only 250 ft (75 m) near Little Rock, Arkansas, but gradually increase in height to a maximum of 2600 ft (790 m) near the Arkansas–Oklahoma border. Both mountain systems are made up of strongly deformed Paleozoic-age rocks, and are closely related in terms of their historical development and tectonic history. Both regions, similarly, are characterized by extensive thicknesses of Pennsylvanian-age, coal-bearing strata. As with the folded Appalachians, low-angle faulting is common in the region and the major thrust sheets have been thrust from south to north, reflecting continental collision that took place in the late Paleozoic. To the north of the thrust zone lies a series of folded structures that gradually decrease in relief. Beneath the Arkansas River valley, the strata thin and form a structural basin that continues northward until it is replaced by a broad upwarp known as the Ozark plateaus.

Soils of the region largely reflect the climate, but also the composition of the underlying bedrock. Ultisols (suborder Udults) dominate as the major soil order throughout much of the region. In the wet lowland area, Inceptisols (suborder Aquepts) predominate, especially on the recent alluvial plain of the Mississippi River. Clay-rich, Oxisols (terra rosa) are locally present in the Ozark plateaus where the underlying bedrock is largely limestone; Hunt (1967) noted that the reddish color of soils found south of the Arkansas River is related to both the warmer, moist climate and the fact that the red prairie soils and red chestnut soils (Mollisols, suborder Ustolls of modern classification) of the region developed from older formations that were themselves reddish in color.

Coastal Plain Province

The Atlantic and Gulf Coastal Plain provinces of the United States stretch in a broad arcuate belt from Cape Cod to Tampico, Mexico, and total nearly 10% of the land surface of the continental United States (Fig. 5). This region is some 3000 miles (4800 km) long and widens from a few miles in width in its northernmost extremity to an average of 250 miles (400 km) in the Texas Gulf Coastal Plain. The peninsula of Florida is included in this province and occurs as a partially emerged, 400-mile (650-km) -long limestone platform covered with a thin veneer of Cenozoic-age sands, clays, and gravels.

Throughout the Coastal Plain province Cretaceous, Tertiary, and Quaternary sediments dip gently toward the Atlantic Ocean and Gulf of Mexico. The plain is locally interrupted by low, inland-facing cuestas and entrenched streams, and many streams (particularly in the Atlantic Coastal Plain) are estuarine and consist of drowned river valleys in which tide waters reach completely across the plain to the edge of the Piedmont province. The Coastal Plain extends beneath the Atlantic Ocean and Gulf of Mexico to the outer margin of the continental shelf. The shelf edge is located at a distance ranging from less than 50 miles (80 km) to over 200 miles (320 km) offshore. Seaward from this point, the ocean bottom plunges rapidly, forming the steep continental slope that descends to depths of 2 miles (3.2 km) or more.

Throughout most of the Coastal Plain province, the strata dip seaward at less than one degree. This seaward inclination is the result of a slight dip

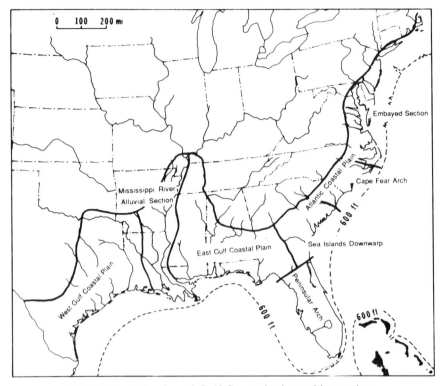

FIGURE 5. Atlantic and Gulf Coast physiographic province.

related to the original sedimentation surface and of subsequent tilting as the crust subsided. A number of faults are present and it is likely that faulting accompanied both the tilting and subsidence of the units (Eardley 1951).

Several large structural features (and a number of smaller features) occur within the Coastal Plain province that have acted to modify regional trends in the deposition and thickness of sediments. Of particular significance are the Rio Grande Embayment, the Mississippi Embayment, the Cape Fear Arch, and the Florida Peninsula.

The Rio Grande Embayment is a large basin that extends from Corpus Christi, Texas, northwestward for 200 miles (320 km) and was formed by regional downwarp that occurred in the Eocene period. Most of the thick sediments in this embayment are deltaic in origin and were carried from the interior of the continent by rivers ancestral to the present Rio Grande, Pecos, and Nueces systems (Storm 1945).

The Mississippi Embayment is the largest feature observed in the province and was formed by downwarp of the continental margin that began in the Cretaceous period and continued throughout the Tertiary. The embayment, from the delta mouth to Cairo, Illinois, is 575 miles long and serves as a natural feature to separate the Gulf Coastal Plain into a West Gulf Coastal

Plain and an East Gulf Coastal Plain. Sediments deposited in the embayment during the interval from the late Cretaceous through the Pleistocene total over 40 000 ft (12 100 m) in thickness. Much of the area is covered by recent sediments of the Mississippi River Alluvial Plain (Fig. 5), which is some 500 miles (800 km) long and 50–100 miles (80–160 km) wide. Hunt (1967) divides the plain into five distinct basins and notes that these are probably the result of irregular deposition by the Mississippi River and partly the result of down-faulting or downfolding of the Coastal Plain formations and basement rocks that underlie the alluvial fill.

The most prominent structural feature within the Atlantic Coastal Plain is the Cape Fear Arch. This structure extends from the vicinity of the Neuse River in North Carolina to the Santee River in South Carolina and represents some 2500 ft (760 m) of structural uplift. Near these rivers, Miocene-age strata extend far inland and unconformably overlap the Cretaceous forma-tions. The Cape Fear section has a distinctive coastline that is marked by three large, smooth, scalloped shorelines located between Capes Hatteras, Lookout, Fear, and Romain. Erosion is prevalent along the scalloped coasts and both the smoothness of the shorelines and the fact that they are being eroded indicates that uplift is continuing in this area (Hunt 1967).

The peninsula that forms the bulk of Florida occurs as a seaward pro-grading, partially emerged limestone platform (Fig. 5) whose earliest record dates from the Cretaceous when the western half of the peninsula was sub-merged and the eastern half was land (Eardley 1951). Since then, deposition has resulted in sediment thicknesses varying from nearly 4400 ft (1340 m) in southeastern Georgia to over 11 500 ft (3500 m) near the southern end of the peninsula. The chief subsurface structural feature is the Peninsular Arch in the north-central part of the state, which was a topographic high during Cretaceous time. This anticlinal structure is underlain by more than 10 000 ft (3050 m) of Cretaceous-age limestones and shale, confirming that during the Cretaceous the present site of Florida was a marine basin (Puri and Vernon 1964). Since then, Florida has largely existed as a shallow shelf area and at various times (including the present) was elevated above sea level.

The Coastal Plain province varies widely in climate. In the northeast, temperate climates prevail creating wet, cold winters; throughout most of the Gulf coast and the Florida peninsula subtropical climatic conditions are pre-sent with mild, wet winters and hot, rainy summers. The southwestern portion of the Coastal Plain is semiarid and characterized by mild winters and hot summers.

A variety of types of soil are found in the Coastal Plain province and these reflect the local climate and underlying rock. The dominant soils of the Coastal Plain are those belonging to the order Ultisols (suborder Udults). These soils were formerly termed "red and yellow podzols" and are found abundantly developed from east Texas north to New Jersey (Fig. 1). Along the Atlantic coast from New Jersey to southern Georgia, and extending inland for a few tens of miles, a second type of Ultisol (suborder Aquults) is found. These soils are commonly developed on poorly drained, Quaternary formations and were designated "groundwater podzols" in older reports. Inceptisols (suborder Aquepts) are also common and are found on the alluvial deposits of the Mississippi, Pearl, Escatawpa, and Mobile river systems. Members of the order Spodsols (suborder Aquods), Entisols (suborders Psamments and Aquents), and Histosols are prominent in peninsular Florida; the latter is also extensively developed in the marshes of coastal Louisiana. Texas is characterized by a variety of soils, with Mollisols (suborder Ustolls) and Alfisols (suborder Ustalfs) dominant and lesser development of Vertisols (suborders Uderts and Usterts) and Ultisols (suborder Aquults).

The age of many Coastal Plain soils can often be estimated with a fair degree of accuracy. Hunt (1967) stated that archaeological evidence indicated that groundwater podzols (Ultisols, suborder Aquults) found along the Atlantic Coast were at least 2000 years old. Soils found on most Coastal Plain surficial deposits and older formations display deep weathering that predates the last ice age (Wisconsin). Evidence for this lies in the fact that some weathered *saprolite* units can be traced northward until they are overlapped by terminal moraines and outwash of Wisconsin age. Similarly, along the coast, weathered red and yellow podzols are found that are overlain by elevated shell beds of known Pleistocene age (Hunt 1967).

BIOLOGICAL OVERVIEW

The southeastern United States includes some of the most evolutionary significant areas on the continent. This region exhibits the greatest known diversity of several important groups. Crawfish diversity in the Southeast is unparalleled (Hobbs 1989); salamanders exist in many forms—more families, more genera, and more species than anywhere else (Blair et al. 1968). Frey (1986) reexamined Cladocera, especially Chydoridae, and concluded that they, too, have had significant evolution associated with the southeastern drainage patterns.

Fishes exist in equal variety, with the family Cyprinidae being the most diverse. It is represented by about 20 genera and 139 species; more than half (85) of the species are shiners, genus *Notropis* (Lee et al. 1980 et seq.). The salamanders, including the three species of sirens, number seven families (three of which are endemic), 19 genera, and 51 species, 31 of which are assigned to the Plethodontidae (Blair et al. 1968). Among invertebrates, crawfishes (family Cambaridae) are represented by 11 genera and more than 300 species (Hobbs 1989). Pelecypod unionid mollusks of three families, 42 genera, and 186 species can be found; the most diverse genus is *Pleuroblema* with 42 species (Burch, 1973). The spheriacean clams, preferring cooler waters generally, add two families, four genera, and 18 species (Burch 1972). The tremendous diversity of aquatic insects precludes any attempt at referencing them here. Many other invertebrate groups occur with equal, or perhaps greater, variety, but they are poorly known taxonomically. The Southeast is a veritable laboratory for systematists, and all groups are continuously being revised as more data become available. For example, in crawfishes alone, Hobbs' first checklist (1974) reported nine genera and 200 species; compared with 11 genera and more than 300 species known 15 years later (Hobbs 1989).

DEVELOPMENT OF DRAINAGE SYSTEMS IN HUMID TEMPERATE CLIMATES

Historically, geologists have attributed the development of surficial drainage systems largely to the effects of weathering and erosive processes occurring on the earth's surface. Hursh (1944) noted, however, that over large areas of the earth's surface, the infiltration capacity of the soil almost always exceeds rainfall intensity, with the result that most runoff travels beneath the soil surface rather than over the surface. Hence, the great majority of water falling on the ground moves below the surface and influences development of the topography in nonobservable ways. In recent years it has become apparent that development of landforms and drainage systems is at least partially controlled by groundwater processes. Thus, while groundwater has long been identified as a causative factor in the weathering of rocks and as an agent responsible for slope failure and formation of karst topography, only in the last decade has its importance in the shaping of landforms and development of drainage systems been acknowledged. To explain the origin of the drainage patterns that characterize the various physiographic provinces of the southeastern United States, one must consider not only surface erosive phenomena but also the influence of groundwater in these areas. The geotechnical and chemical response of soils to groundwater and subsequent mass wasting that results in topographic development has been well documented (see Horton 1945, Poland and Davis 1969, Garner 1974, Rogers and Pyles 1979) but is beyond the scope of this chapter. The remaining discussion will be restricted largely to a description of the drainage systems that exist in the various physiographic regions that make up the southeastern United States and factors influencing or controlling their development.

Groundwater as an Agent in Landform Development

Rain falling on the ground is initially absorbed by the soil. If the rainfall occurs rapidly, or persists for an extended period, the infiltration capacity may be exceeded and surface runoff takes place. Some of the water will evaporate from the vegetation and from the ground itself. Thus, precipitation can be removed by infiltration, runoff, and evapotranspiration.

Seasonal variation of rainfall and the loss of water by evapotranspiration in temperate climates has a strong influence on soil drying, the development of macropores in the soil, and recharge of groundwater in underlying strata. Because of the characteristic humid climate in the eastern United States, the water present in the soil zone may control the volume and rate at which groundwater recharge takes place and, thereby, influence further soil formation and landform development. Higgins et al. (1988) have noted that in the Piedmont region of the southeastern United States the clay-rich soils underlying relatively flat drainage divides permit loss of as much as 60% of total annual precipitation by evapotranspiration. Of the remaining precipitation, approximately 50% is lost as runoff because of the low infiltration capacity of the soil. What remains is thus available to recharge the groundwater in the regolith. Therefore, the soil itself limits the volume of water that can infiltrate and reach the regolith, and thus acts to control the rates of subjacent rock weathering. The widespread occurrence of flattened upland landscape in the Piedmont and in the Interior Low plateaus is therefore a result of continuing physical and chemical weathering in part controlled by soil groundwater processes. Higgins et al. (1988) observed that the leaching of silica from these soils, with concomitant loss of volume as the soils are transformed to the typical clay-rich, red-yellow podzols (Ultisols, suborder Udults), is responsible for loss of mass and volume that has lowered land surfaces and upland divides throughout Quaternary time. Thus, these soil-forming processes have been important in the formation of the broad, flat divides known to the early physiographers as "peneplains." These surfaces subsequently have exerted their own influence upon the regional drainage systems.

In addition to the soil-forming processes, a number of other phenomena related to the movement of groundwater are also active below the surface, which contribute to, or act to modify, surface topography and drainage. These include seepage erosion (Zaslavsky and Kassiff 1965, Higgins 1984), tunnel erosion (Sherard and Decker 1977), and the more general process of piping (Parker and Jenne 1967, Jones 1981, Bryan and Yair 1982).

Higgins et al. (1988) describe seepage erosion as resulting from diffuse or concentrated outflow of groundwater from porous, granular soils. The water entrains and carries away particles at sites where it emerges on hillsides. Removal of fine-grained soil particles by this process can seriously undermine the stability of slopes and may rival, or even exceed, the effects of running water in the sculpting of the land surface and the development of gully systems on hillsides.

Tunnel erosion, by which underground water moves through macropores in the soil in sufficient quantity and velocity to shear sediment from the walls of the conduit, is probably less important because most soil waters move at velocities too low to produce this phenomenon. In recent years, however, it has become clear that the shear stress required to erode particles from sediment pores can be considerably reduced if the soil consists of dispersive clays. Under such circumstances, water will be drawn into pores and cracks in response to capillary and osmotic potential gradients with the result that the soil expands and fractures, and particles are swept along in a colloidal suspension or, if there is an open connection to the atmosphere and sufficient pressure gradient, as suspended or bedload sediment (see Sherard and Decker 1977, Higgins et al. 1988).

Piping (Jones 1981) is a particularly effective subsurface erosive mechanism and can be active in both soils and the underlying consolidated bedrock. The phenomenon is most common where weathering has attacked soils and bedrock rich in montmorillonite and illitic clays (Parker and Jenne 1967). Rapid underground erosion may result in the development of interconnected vertical or horizontal pipes from one to several feet in diameter that may be several hundred feet in length. Water moving through the pipes causes mechanical abrasion and removal of sediment particles, followed by development of pipes whose openings to the surface can frequently be observed along the banks of steep walled streams. This phenomenon may be initiated by burrowing of animals, by rainwater following root systems underground, by water moving along cracks in the soil formed during periods of dehydration, or by water moving along the contact between sediments having different degrees of permeability (Butzer 1976).

Where piping becomes highly developed, the formation of underground drainage channels can result in formation of a number of karst-like features, such as caves, sinkholes, and natural bridges (see Twidale 1984). Higgins et al. (1988) observe that such "pseudokarst" topography will mimic the true karst terranes (developed on limestones and dolomites) by the presence of ephemeral subsurface streams and sinkhole-scarred landscapes. They further note, however, that such terrain features lack the durability of those found in true karst regions. As such, this variety of pseudokarst may develop or change radically as the result of a single thunderstorm event.

The type of pseudokarst generally attributed to the effects of piping has also been shown in recent years to be attributable to the solution of noncarbonate sediments. Isphording (1984) described the widespread development of sinkholes on clayey sands of the southern Alabama and Mississippi Coastal Plain and attributed them to a combination of piping and the incongruent dissolution of kaolinitic clay. Acid leaching of the clay component of the soils by groundwater was shown to result in a 35% volume loss of sediments and cause the formation of true sinkhole features (see also Twidale 1987, Isphording and Flowers 1988).

Surface Water and Landform Development

Worldwide, running water is the most effective agent in the development of landforms. When the soil is unable to absorb a continuing rainfall, overland flow and associated sheetwash take place. The water then flows over the surface until a temporary or permanent stream channel is encountered. It then moves along an evolving path, ultimately to the sea. The combination of sheet erosion and channelized erosion thus acts to shape and modify the surface topography.

The effectiveness of surface erosion is controlled by a number of factors. These include steepness and length of the slope, composition of the underlying rock, duration of rainfall events, soil thickness and rate of infiltration of rainfall into the soil, extent of vegetation cover, and structural attitude of the underlying rocks. These factors (along with subsurface influences) act to control the type of drainage that will develop within a given region.

Examination of topographic maps and aerial photographs have indicated that only a relatively few geometric forms of drainage will develop in response to the above factors. Most of these drainage patterns largely reflect the attitude of the underlying bedrock and/or its composition (see Garner 1974). Table 3 provides a geometric description of each pattern, and their general map appearance is shown in Fig. 6. Essentially all can be found in the various physiographic provinces of the southeastern United States. Because of the prevalence of certain structural or other controlling features, one (or more) of the patterns may be dominant in a given region.

Coastal Plain–Piedmont Province

The drainage systems of the Coastal Plain and Piedmont provinces are combined in this discussion because, although some rivers do arise entirely in the

TABLE 3   Interpretation of Drainage Patterns Based on Lithology and Underlying Structure of Bedrock

Pattern Type	Substrate Composition or Structure
Dendritic	Homogenous crystalline rocks or flat-lying sedimentary rock
Rectangular	Jointed rock with intersecting bedrock fractures; may be crystalline rock or, commonly, carbonate rocks (limestones or dolomites)
Trellis	Folded sedimentary rock
Annular	Crystalline or sedimentary rock having the form of a structural dome, basin, or anticline
Centripetal	Drainage into a closed depression; may be present in karst terranes, structural basins, volcanic areas, or areas of extensive development of dunes
Radial	Normally found developed on flanks of volcanic structures or other exposed stock-like intrusions
Parallel	Associated with large scale sand dune systems
Pinnate	Very fine-grained sedimentary rock, e.g., loess
Distributary	Diversion channels developed in deltas, alluvial channels, and riverine plains
Braided	Interlacing stream channels caused by excess sediment being carried by stream

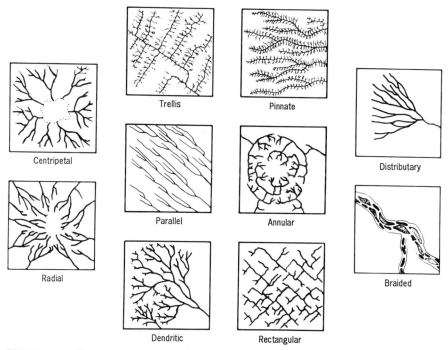

FIGURE 6. Drainage patterns developed on different terranes or geological structures.

Piedmont province itself, most either originate from eastward flowing streams in foothills of the Blue Ridge province or flow, for their greater length, across the Coastal Plain province on their journey to the sea (see Fig. 7). Examples include the Coosa River in Alabama, the Savanna River in Georgia, the Cape Fear and Roanoke rivers in North Carolina, and the Potomac, Susquehanna, and Delaware Rivers farther to the north in Virginia, Maryland, and Pennsylvania.

Piedmont Province. Rocks of the Piedmont extend nearly 1000 miles (1600 km) from southern New York to Alabama and reach a maximum width of about 125 miles (200 km) (Hunt 1967). They are chiefly metamorphic in origin but large areas exist where igneous rocks have intruded the sequence and where metasedimentary rocks of early Paleozoic age have become infolded. Several block-fault basins are also present in the Piedmont, containing extensive thicknesses of Triassic-age sediments. The boundary between the Piedmont and Coastal Plain provinces is located at the fall line where transgressing seas during the Cretaceous period deposited sediments that now overlap, and cover older Piedmont rocks. To the west, the boundary between the Piedmont and Blue Ridge provinces is marked by a major thrust fault (termed the Martic in Pennsylvania and Maryland, and the Brevard in North Carolina). To the south, the boundary is similarly abrupt and in Alabama

FIGURE 7. Map showing major river systems and drainage basins in the southeastern United States. 1, Potomac; 2, New–Kanawha; 3, James; 4, Chowan–Roanoke; 5, Tar–Neuse; 6, Cape Fear; 7, Pee Dee; 8, Santee; 9, Edisto–Coosawhatchee; 10, Savannah; 11, Ogeechee; 12, Altamaha; 13, Satilla; 14, St. Mary's; 15, St. John's; 16, Peninsular Florida; 17, Suwanee; 18, Aucilla–Ochlocknee; 19, Apalachicola; 20, Eastern Gulf; 21, Mobile; 22, Upper Tennessee; 23, Lower Tennessee; 24, Duck; 25, Cumberland; 26, Ohio; 27, Mississippi; 28, Mississippi Sound; 29, Pearl–Lake Pontchartrain; 30, Trans-Mississippi (a, White; b, Arkansas; c, Ouachita; d, Red; e, Louisiana Gulf; f, Calcasieu; g, Sabine).

and Georgia is delimited by the Cartersville fault, along which metamorphic rocks have been thrust westward onto folded Paleozoic rocks.

In spite of the great variety of rocks in the Piedmont region, there is little structural control of drainage, although, locally, rivers do follow faults, joint systems, and soft beds. Cleaves (1989) has described the evolution of drainage in the Piedmont and noted that during the Holocene the region was an upland surface of low relief undergoing dissection. Major rivers, such as the Potomac and Susquehanna, are incised into a landscape on which the landforms show "a delicate adjustment to lithology and structure. In places jointing may control stream orientation and in others thinly interdigitated layers of quartzite and schist determine stream orientation" (Cleaves 1989, p. 162). The Fall Zone is interpreted as a feature that has evolved from late Miocene time in response to Holocene warping and faulting, as well as differential erosion. Topographically, the Fall Zone can be traced from the James River in Virginia north to the Delaware River. As such, it constitutes a zone of transition from Piedmont uplands, underlain by crystalline rocks, to the Coastal Plain province, underlain by unconsolidated sediments (Cleaves 1989).

Atlantic and Gulf Plain. The Coastal Plain province of the eastern United States is underlain by a thick sequence of very gently dipping, late Mesozoic to recent sedimentary rocks. Because of the structural attitude of most coastal plain rocks and sediments, many of the rivers tend to develop a classic dendritic drainage pattern on the gently sloping surface. The Mississippi Embayment has an average slope of approximately 0.5 ft per mile that raises its surface to slightly over 200 ft (60 m) more than 400 miles (640 km) inland. The remainder of the Coastal Plain, from Texas to New Jersey, is similar in gradient.

The importance of the rivers that discharge into the Atlantic and Gulf of Mexico cannot be overstated. The total discharge to the sea of rivers draining the entire contiguous United States amounts to approximately 2×10^6 cfs (56 600 m³/s). Some 21% of this flows into the Atlantic Ocean, and about 29% flows into the Pacific. The remaining 50% empties into the Gulf of Mexico. Thus, each second, on a yearly average, about 1×10^6 ft³ (28 320 m³) of water enters the Gulf of Mexico through numerous bays and estuaries that are found along the Gulf Coast from south Texas to south Florida (see Isphording et al. 1989). The quantity of runoff in this region is unevenly distributed and ranges from high volumes for rivers in Louisiana, Mississippi, Alabama, and western Florida to low volumes from drier regions in south Texas. The Mississippi River, which drains a watershed of nearly 1.2×10^6 square miles (3 100 000 km²), discharges an annual average of 744 000 cfs (21 070 m³/s) (74% of the Gulf Coast total). The area to the west of the Mississippi River basin represents nearly 20% of the Gulf Coastal Plain drainage area but, being climatically dry, contributes only 6% (60 500 cfs or 1,713 m³/s) of the total discharge to the Gulf. East of the Mississippi basin, and continuing to the eastern margin of the Apalachicola River basin in Florida and Georgia, lies the wettest region in the Gulf. This area contains 7% (106 700 squares miles or 276 350 km²) of the total drainage area and contributes an annual average of 178 000 cfs (5040 m³/s) (18% of the total Gulf inflow). The remaining drainage area to the Gulf, situated along the western portion of the Florida peninsula, contains approximately 3% (41 200 square miles or 106 700 km²) of the total area and discharges annually an average of 25 500 cfs (722 m³/s).

The present shoreline is the result of complex erosional and depositional processes but can be generally described as a "submergent" coast, using the classification of Johnson (1919) and Shepard (1952). Realistically, the coastal plain should be separated into three subregions: the Eastern Coastal Plain, the Gulf Coastal Plain, and the Florida Platform. While certain generalizations can be made for the region as a whole, important differences exist to differentiate the three areas. As an example, it is well established that since the end of the last glacial episode, sediment influx into oceans has generally declined along many of the world's coastlines (Garner 1974). As such, less sediment is currently being brought into the oceans by rivers than is being removed to the deep oceanic region. While this statement is operable for the eastern coast of the United States (e.g., the Pee Dee, Santee, Tar, and Appomatox rivers) it is certainly not so for the Gulf of Mexico region. Here a number of major rivers continue to bring significantly greater quantities of sediment to the shore zone than can be removed to the ocean deeps. Consequently, the barrier island systems in this region are continuing to develop and expand. Examples of such rivers would include the Mississippi, Mobile, and Pearl in the northern Gulf of Mexico, and the Atchafalaya, Rio Grande, Nueces, Guadalupe, Trinity, Sabine, and Colorado Rivers in the western Gulf.

Florida, in contrast, enjoys certain unique features not shared by the other two areas. In addition to a terrigenous coast along its western panhandle, it possesses an organic "reef-buttressed" coast in the northeastern portion of the peninsula, a mangrove coast in the southwestern part of the state, and a classic "zero energy" shoreline east of Apalachicola Bay. The latter is a reflection of the many rivers in the northwestern portion of the state (Waccasassa, Econfina, Suwannee, Wacissa, Fenholloway, etc.) that carry little suspended sediment or bedload. Several other rivers draining south through Alabama, Georgia, and Florida also share this characteristic (Withlacoochee, Alapaha, Ochlockonee, Styx, Blackwater, etc.).

Excluding the Mississippi River, which is the largest river in North America and whose sources lie well outside the Coastal Plain province in the continental interior and beyond, most of the other major rivers in the Coastal Plain province lie entirely within the province itself. Exceptions would include the Mobile and Rio Grande river systems.

The Mobile River and its major tributaries (the Alabama, Coosa, and Tombigbee rivers) drain an area of 44 600 square miles (115 500 km²). The river ranks as the nation's sixth largest system in watershed area, draining two-thirds of the state of Alabama and portions of neighboring Mississippi, Tennessee, and Georgia as well. The system is fourth largest in terms of the average annual inflow of the river (some 79 300 cfs (2250 m³/s) into Mobile Bay), exceeded only by the Mississippi, Columbia, and Yukon. Major tributaries of the Mobile River system originate within the Interior Low Plateau province in Tennessee, and within both the Valley and Ridge and Piedmont provinces of northern Alabama and western Georgia.

The Rio Grande River, similarly, originates far beyond the boundary of the Coastal Plain province (see Fig. 2) and is fed by streams draining the eastern slopes of the Sangre de Cristo Mountains in southern Colorado. As such, its watershed area of 133 000 square miles (335 500 square miles including Mexico drainage) is exceeded in size only by the Mississippi River system. Of interest is the fact that at Elephant Butte Reservoir in southern New Mexico, the Rio Grande lies less than 75 miles (120 km) from the drainage of the Gila River that drains into the Pacific Ocean.

Appalachian Highlands

Fenneman (1937) described the physiographic areas of the eastern United States and subdivided the Appalachian Highlands into the four physiographic regions known as the Piedmont, the Blue Ridge, the Ridge and Valley, and the Appalachian Plateaus provinces. The two easternmost provinces (the Piedmont and Blue Ridge) were referred to as the "Older Appalachians," whereas the two western provinces (the Appalachian Plateaus and Ridge and Valley) were termed the "Newer Appalachians." The origins of both the topographic and drainage features in these areas have been the subject of extended debate for many years. Clark (1989) states that earlier workers in the Appalachian Highlands were puzzled by three aspects of drainage of rivers in this region: (1) the reversal of the original northwestward-flowing drainage that apparently existed during deposition of Paleozoic-age strata, (2) the great length, relative straightness, and parallelism of present southeastward-flowing rivers, and (3) drainage transverse to structural-lithologic trends where alternative courses were assumed to have been available. Answers were proposed for some questions in the classic papers by Davis (1888, 1889). In these papers Davis first propounded his belief that accordant summits existed that testified to the existence of large-scale "peneplains" in the region during the Tertiary and that the cardinal principles governing landscape development involved "structure, process, and stage." In recent years alternatives to the Davis theories have been offered. Hack (1960, 1980, 1982) has done much to demonstrate that the present landscape can be explained by a balance of forces involving uplift, erosion, and rock resistance. The interested reader is also directed to additional literature dealing with the origin of landforms and drainage systems in the Appalachian Highlands in papers by Flemal (1971), Judson (1975), Sevon et al. (1983), Clark (1989), Cleaves (1989), Poag and Sevon (1989), and Morisawa (1989).

Ridge and Valley Province

The Ridge and Valley province possesses a number of striking geomorphic features, many of which act to control drainage systems that have developed within the area (see Thornbury 1954). These include (1) a marked northeast–southwest-trending parallelism of the ridges and valleys, (2) the existence of the previously mentioned major streams with valleys transverse to the regional structure (the Delaware, Potomac, and Susquehanna), (3) many lesser stream valleys controlled by the geologic structure and possessing a marked trellis drainage pattern (Cowpasture and Greenbrier, in Virginia and West Virginia;

Holston and Clinch, in Tennessee and Kentucky), and (4) hundreds of wind gaps that attest to drainage changes and, in many instances, exhibit a striking alignment. The latter suggests either former transverse stream courses that have been subsequently captured by stream piracy or dismembered, or alignment along bedrock fracture traces or geologic lineaments.

The origin of the transverse drainage observed in the Appalachian Highlands is still the subject of spirited debate. Bethune (1948) concluded that geomorphic evidence allowed only two possible explanations for the origin of major transverse streams in the Ridge and Valley province: (1) stream piracy or (2) superposition. Stream piracy was rejected as a possible explanation because the water and wind gaps were not located in positions that would correspond to sites of weakness resulting from either faulting or the presence of highly jointed rock. Rather, Bethune concluded that the streams predated uplift that occurred in the area and simply continued to erode their channels as the area was uplifted and, therefore, constituted "superimposed" streams. Liebling and Scherp (1984) disagreed and argued that piracy is not only effective in explaining some transverse drainage in the Ridge and Valley province, but is also vigorous in certain areas. Outerbridge (1987), similarly presented evidence of recent piracy in the Appalachian Plateaus province and identified streams that will likely be captured in the future. Epstein (1966), however, concluded that structural weaknesses were a more reasonable explanation for the feature known as the Delaware Water Gap. Hoskins (1987) described complex fracture systems, faults, folds, and overturned bedding in a number of water gaps in the Pennsylvanian area and, similarly, believed they were responsible for localizing structural weaknesses that could lead to selective erosion and explain the transverse nature of these gaps.

Blue Ridge Province

The Blue Ridge province includes all high mountains underlain by crystalline rocks from northwest Georgia to Pennsylvania. The rocks of this province extend from Georgia to Pennsylvania (Fig. 3) and are largely of early Paleozoic age. These strata consist largely of metasedimentary and metaigneous rocks formed on a basement complex of Precambrian granites and gneisses. A detailed description of the rocks may be found in Hatcher et al. (1988). The province ranges from 5 to over 50 miles (8–80 km) in width and varies, in places, from a single ridge to a complex made up of several closely spaced ridges. Hack (1982) divided the province into two parts based on topographic origin: one, located north of the Roanoke River, that consists of a narrow range of high mountains (the Northern Blue Ridge) and a second, south of the Roanoke River, that he termed the Southern Blue Ridge province. The Northern Blue Ridge Mountains are the frontal mountain range of the Valley and Ridge province and form the northwest limb of the Blue Ridge anticlinorium. The rocks of this unit appear to be part of an erosional system in an uplifted area that includes the mountain ranges of the Valley and Ridge province that rise to approximately the same elevations. The Southern Blue

Ridge province extends from the Roanoke River southwestward into north Georgia, ending at Pine Log Mountain north of Cartersville, Georgia. This sequence includes several ranges of high mountains that exceed 5000 ft (1500 m) in altitude (e.g., the highest mountain in the eastern United States, Mount Mitchell, 6684 ft or 2040 m). Rock control of topography is apparent in the western portion of this zone where resistant quartzites often form the high ridges. Similar rock control of topography exists in the central core, where several peaks in the Great Smoky Mountains exceed 6000 ft (1800 m) and are underlain by massive conglomerates and sandstones. In the central region many intermontane basins exist that form deep valleys ("trench valleys") that cut across drainage divides at high angles. Hack (1982) concluded that this suggested an origin due in part to control by nonresistant rocks and in part to tectonic influences. The tectonic influences may have involved faulting in late geologic time or, more likely, differential erosion along brittle fracture zones associated with older faults.

Essentially all streams in the Blue Ridge province drain either eastward into the Piedmont (and beyond, across the Coastal Plain to the Atlantic Ocean) or westward into the Appalachian plateaus, and ultimately into the Gulf of Mexico. Transverse water gap development occurs only northeast of Roanoke. The Roanoke River itself was considered by Wright (1934) to be a relatively new, headward-eroding drainage basin. South of Roanoke Valley, the Blue Ridge province is not bisected by rivers. Transverse drainage (e.g., New River) originates within it and flows northwest into the Appalachian Plateau. Clark (1989, p. 217) notes that "south of New River drainage the main transverse drainages in Virginia and Tennessee are tributary to the Tennessee River system. Major cross-strike tributaries include the French Broad, Pigeon, Little Tennessee, and Hiwassee rivers that exit the Southern Blue Ridge in deep gorges. Headwater tributaries of some rivers have lost drainage by piracy along the Blue Ridge Escarpment."

Appalachian Plateaus and Interior Low Plateaus

The Appalachian Plateaus province includes two main areas: the Cumberland Plateau on the south and the Allegheny Plateau on the north. Eardley (1951, p. 71) describes the region as one of "mature or submature dissection" and one standing for the most part "higher than its neighbors." The plateaus form a broad, gentle synclinal basin and cover an area roughly equal to that of the combined Piedmont, Blue Ridge, and Valley and Ridge provinces. Both the southern and northern plateaus are composed of nearly flat-lying strata, but the northern plateau displays considerably greater dissection. Altitudes range from approximately 1000 ft (300 m) along the western edge to slightly greater than 3000 ft (900 m) at a southeast-facing escarpment that overlooks the Valley and Ridge province (the Allegheny Front). Hunt (1967, p. 173) notes that the Allegheny Front and its extensions "form one of the most persistent and most striking topographic breaks in the country. From Alabama to northern Pennsylvania, a distance of 700 miles (1125 km), the escarpment is 500–

1000 ft (150–300 m) high. Its extension along the Hudson River valley, at the eastern front of the Catskill Mountains, is more than 3000 ft (900 m) high." The Allegheny Plateau is continuous with the Cumberland plateau and Eardley (1951) states that any boundary would be arbitrary. The area formerly known as the Cumberland Plateau was ultimately given its own "province status" (Hunt 1967) and has been termed the Interior Low Plateaus province and is now included in the broad physiographic division known as the Interior Plains.

Major drainage systems within this area reflect both the structural attitude and lithology of the underlying bedrock. Dendritic stream patterns typify the major stream systems that have developed on the flat-lying or broadly arched Paleozoic sediments, but structural control is also evident where joint systems have exerted control, locally, on the paths of some rivers (e.g., the Kentucky River). This is especially true in areas underlain by limestone. West of the Lexington Plain and Nashville Basin the dissected plain slopes slightly to the west and rivers on this surface (the Cumberland and Green rivers) meander strongly and are deeply entrenched. A significant portion of the region is underlain by limestone. Karst features are common and some small streams simply disappear into sinkholes developed on the underlying cavernous rock (e.g., Little Sinking Creek, near Mammouth Cave, Kentucky).

Ozark Plateau and Ouachita Province

The Ouachita Mountains occupy a narrow belt approximately 50–60 miles (80–100 km) wide and extend in an east–west direction for about 200 miles (320 km) in northern Arkansas and eastern Oklahoma. When the Ozark Plateaus are included, the combined areas total nearly 100 000 square miles (260 000 km^2) (see Fig. 4). The Ouachita Mountains are similar in topographic expression and structure to the Appalachian Mountains, but are not nearly as high. The Ozark plateaus are similar in structure and topography to the Appalachian plateaus, but are composed largely of early and middle Paleozoic limestones rather than the Pennsylvanian-age coal-bearing rocks characteristic of the Appalachians.

The plateaus area is drained by few large streams (e.g., the Osage and St. Francis rivers), but numerous smaller streams have extensively dissected the region. Near Heber Springs, Arkansas, for example, the Boston Mountains Plateau region has an average elevation of nearly 2000 ft (6500 m) and is incised by V-shaped valleys of the Little Red River system, some of which are more than 1000 ft (3280 m) deep. Because dolomite and limestone are the dominant bedrock throughout the plateau, karst features, such as caves, sinkholes, and terra rosa soils, are especially common. The abundant carbonate formations have strongly influenced the local topography and drainage systems, not just in the Ozark Plateaus, but in the neighboring Ouachita Mountains as well. This is especially true where ancient patch reef carbonates display a greater resistance to erosion than the shaley sedimentary rocks that were deposited simultaneously in the back reef environment (Garner 1974).

Drainage in the Ouachita Mountains is similar to that in the Ozark plateaus in that few major streams are present. The largest streams traversing the region are the Arkansas and Ouachita rivers, whose valleys are largely controlled by the east–west folds of the Ouachita Mountain system. Many smaller streams are present, with watercourses either parallel to the principal ridges or draining from these ridges and entering larger streams at nearly right angles. Further, because the region lies roughly midway between the humid climatic zone that typifies the southern Appalachian and the more arid climates that characterize the southwestern United States, evidence is present to show that alternating changes in climate from humid to more arid have existed during relatively recent times. This, combined with lithologic and structural influences, has locally imprinted a relict humid drainage network on portions of this region that were formerly partially alluviated under more arid conditions (see Garner 1974).

ZOOGEOGRAPHIC IMPLICATIONS OF SOUTHEASTERN DRAINAGES

It is difficult to construct a zoogeographic discussion that parallels the geologic divisions. From a biological perspective, the mobility of organisms, especially animals, creates complexities of distribution not offered by purely physical factors. Additionally, organisms differ, even within closely related assemblages, in their ability to migrate and to respond to the exigencies generated by environmental change. Thus, this discussion is based, in some cases on broader, and in others, more restricted criteria than the geological discussion.

The Atlantic Coastal Plain

From a biological view, there are sharp distinctions between the lower reaches of aquatic systems associated with the Atlantic Coast and those of the Gulf of Mexico. Part of this difference is related to their geologic history, but part is due to the nature of their biological past. Ichthyologists recognize two Atlantic zones in the Southeast: one extends south to the Savannah River, and the other extends down to the tip of Florida. The latter zone continues about the coast to Lake Pontchartrain; the area west of it is known as the Mississippi River basin. The pattern of fish distribution suggests a greater affinity between the central Appalachians and the central Atlantic slope than between the central and southern parts of the plain. One way to examine the zoogeography of a region is to detail the situation in the several drainage systems and then to look for relationships between them. Another is to seek broad patterns from the start and to indicate details principally by reference to the literature. To offer exposure to the two techniques, the Atlantic Coastal Plain will be examined with a modified version of the former, and modifications of the latter will be used to discuss the other zones.

The James/Roanoke Basin shows signs of multiple invasions from the old Teays system, beginning in the Pliocene. The greatest number of endemic species on the Atlantic slope occurs in the Roanoke. There is some sharing of fishes between the Tar and New systems, although they each have a small fauna, mostly of Teays origin. There is one species of shiner known only from these basins, and an endemic race of another occurs in them. The zoogeographically uneventful Cape Fear likewise shows associations with the Roanoke.

Both the Peedee and Santee have a large ichthyofauna, the latter having more endemics and the greatest total number of species (90) on the Atlantic slope. Certainly the more significant of the two is the Santee, which seems to have had more invasions from the Savannah and from the Highlands section of the Tennessee. The purely coastal plain Waccamaw is considered separate from the Peedee and is noted for its endemics, several of which are lacustrine. Finally, the Edisto, another purely coastal plain drainage, seems to be populated from the laterally adjacent streams. Only four of the 35 freshwater fishes have a montane-upland affinity (Hocutt et al. 1986).

Evidence derived from study of unionid molluscs seems to support the position of the ichthyologists. Johnson (1970) believed that the entire Unionidae element is of Interior Basin origin, but he recognized two distinct elements. His "Southern Atlantic slope region" extends from the James River south to the Altamaha. Such support is not unexpected, however. The principal mechanism of dispersal available to unionid clams is the attachment of their glochidial larvae to fishes. More significant is the evidence provided by crawfishes and cladocerans. Fitzpatrick (1967) demonstrated that a species of Orconectes from the Chowan had definite affinities with the New River system and postulated a stream capture associated with the Roanoke headwaters, possibly after the Illinoian. Prominent elements in the benthos of the slope are members of the genus Cambarus, especially in the north; these came from the southern Appalachians (Hobbs 1969). Most of the coastal plain species, however, belong to Procambarus. Hobbs (1984) envisioned them to be laterally migrant rather than wholly dependent on stream capture. Frey (1986), working with chydorid cladocerans, envisioned a fluent north–south migration of species, in which movements were associated with Pleistocene glaciation. Evaluating environmental stresses experienced by species, he noted that the two assemblages that he proposed—northern and southern—face diametrically opposed seasons of maximum pressure. In the north, winter is the period of extreme stress, with most or all adults dying and the species overwintering as propagules. In contrast, in the south the generally cool water preferring Cladocera find the summer heat a time of intense pressure, and the autumn season of gametogenesis of the north is replaced by a spring season in the south. Further, in those species occurring in both zones, a seasonal alternation of southern with northern is common.

Less stable is the consensus among ichthyologists regarding the Savannah to Lake Pontchartrain segment of the Coastal Plain. Swift et al. (1986) examined the ichthyofauna of this region. Unfortunately, their study is apparently clouded by a precommitment to vicariance biogeography and phenetic methodology, techniques not universally accepted by systematists. Those having difficulty following the plethora of literature generated by the debate may find the relatively balanced review by Dupis (1984) helpful. *Swift et al.* envisioned a rather uniform fauna from the Savannah to the St. Marys, more or less allied, or at least in species occurring below the Fall Line, to the more northern drainages. A relative paucity of species in the Satilla (44) and St. Marys (45) was attributed to the acid–water drainage from the Okefenokee Swamp. Similarly small numbers are found in the Edisto/Combahee that do not have swamp outflow. On the other hand, those drainages with 1.3–2.0 times as many species cross the Fall Line, a marked faunal division, irrespective of the origins of their water. The Ochlocknee and Suwanee drain into the Gulf of Mexico, but they are more closely allied to the Atlantic than to the eastern Gulf streams.

Considerable data on crawfishes are available for this region from Hobbs (1981). He envisioned a center of origin somewhere in the general vicinity of what is now northeastern Georgia in late Cretaceous or early Cenozoic times. This center was not only for the Georgia fauna but for all of eastern North American crawfishes. The Coastal Plain was populated from this center as it emerged. He believed that the initial stock was lotic, but that lentic specializations of some groups were acquired early. Both he and Frey (1986) placed more emphasis on major climatic changes of the Cenozoic than did Swift et al. (1986), who relied heavily on sea level changes and the fluctuating salinity of what is now Coastal Plain freshwater to explain changes in the fauna. In his analysis of unionid bivalves, Johnson (1970) observed that the molluscan fauna is old (probably older than any other group here discussed) and envisioned, once again, a source in the southern Appalachians for these animals. He used marine inundations and obliteration of freshwater organisms only to explain some irregularities in smaller river systems and their lack of endemism.

Eastern Gulf Coastal Plain

Most of this area represents the incursion of the old Mississippi Embayment. The Apalachicola seems to be a dividing basin. This is reflected in the fishes (Swift et al. 1986), mollusks (Johnson 1970), and crawfishes (Hobbs 1974, 1981, 1989). Swift et al. found that endemics (8) are, with only one exception, more closely related to lateral drainages than to the Tennessee. A similar lateral relationship, albeit above the Fall Line, was recognized in mollusks (Johnson 1970). Hobbs (1981) implied a similar situation in crawfishes. From the Perdido to Mobile Bay the fishes are the only significant fauna to show some restriction. Several lowland species are restricted to the area and other widespread species of such environmental needs likewise occur. Fishes imply that many upland connections bypassed streams of the lower coastal plain because the transverse relationships between river systems are much closer than downstream ones.

The Mobile Bay drainage system is much more elaborate and, accordingly, much more complex. The principal effluent, the Mobile River, results from the confluence of two major systems, the Alabama–Coosa and the Tombigbee–Black Warrior. This drainage is the third richest in North America in its fish fauna. Some species, including some very primitive ones, are related to the Mississippi River proper; others are tied to the Tennessee. An excellent detailed analysis of the fishes was prepared recently (Mettee et al. 1989) for the western segment (i.e., Tombigbee) and offered slightly different zoogeographic interpretations than those of Swift et al. (1986). A comprehensive analysis of mollusks is lacking, but crawfishes show strong affinities to both the southern Appalachians and the trans-Mississippi fauna. West of this, the principal river is the Pearl. Swift et al. (1986) indicated two distinct components, upper and lower, in the fishes. Chien (1969) found a similar division, although not as pronounced, in the Cladocera. Crawfishes, on the other hand, are less clearly divided, but most have a trans-Mississippi relationship. The most striking feature perhaps is the indication that the Pearl–lower Mississippi can be used as a sharp line of demarcation between closely related taxa (Swift et al. 1986, Fitzpatrick 1977a, Fitzpatrick and Hobbs 1968).

The association of many faunal elements with the Tennessee system makes the history of that river important. A consensus exists that there are two segments, divided approximately by the Walden Gorge, which are not of identical descent. Precisely how and when the present configuration was established becomes prominent in the evolutionary history of the fauna of the old Mississippi Embayment. Some differences of opinion exist both among geologists and zoogeographers; allusions to these appear in the following paragraphs.

The Southern Appalachians and Adjacent Areas

This region of the Southeast is one of the most significant in establishing the faunal and floral history of this part of the continent. A series of symposia devoted to the examination of the events associated with the evolutionary phenomena originating here (Holt et al. 1969, 1970, 1971) revealed much about the region, aquatic and terrestrial. Ross (1971) articulated the standard interpretation of geologic events. In this view, the upper segment of the Tennessee (= the Appalachian River) had its outlet into the Gulf via the Coosa–Alabama basin. Although there are differences of opinion with respect to datings, most ichthyologists follow this thesis. Recently, however, Starnes and Etnier (1986) suggested that this idea was less tenable, on the basis of analysis of Mobile drainage fishes, than previously supposed. Fitzpatrick (1986a), using crawfish distribution, also favored a more westerly course for the river up to at least the Pliocene. Milici (1968) determined that the Highland Rim, Cumberland Escarpment, and the Cumberland Plateau existed at least since the late Cretaceous; he suggested that the old Tennessee drained southward off the Nashville Dome.

On the other hand, the ideas expressed in Ross' (1971) summary are more accepted for the headwaters areas of most of the systems. Again, details differ, but most workers accept relationships between the Tennessee and the western edge of the Highland Rim. In these scenarios, small stream captures in the headwaters areas led to faunal exchanges between what is now the upper Tennessee and the Buffalo/Duck/Cumberland/Green/Barren drainages. These assumptions make hypotheses about the relationships of lower Tennessee faunal elements difficult. Fitzpatrick (1987) did note that the major crawfish genera, *Orconectes* and *Cambarus*, both of which apparently originated nearly contemporaneously on the southeastern edge of the Cumberland Plateau, had complimentary distributions. Earlier, Fitzpatrick (1977b) suggested that they also arose at about the same time as *Hobbseus*, a genus (with two small exceptions in the upper Pearl and another in the upper Yazoo) confined to the upper and middle Tombigbee.

Similar captures can be associated with drainages to the north and east. Some references to the Atlantic slope are given in earlier paragraphs. Yet very interesting to the zoogeographer is the fact that, although the headwaters of the Chattahoochee (Apalachicolan) and Hiwasee (Tennessee) drainages are adjacent, no evidence of fish exchange exists (Starnes and Etnier 1986). Although not specifically noted, Hobbs (1981, pp. 22–23) indicated a similar situation in crawfishes. In contrast, the Coosa system, also adjacent to both, shares several species with the Tennessee, although not specifically with the Hiwassee. Johnson (1970) argued for a Savannah River source for the Apalachicola unionids. In contrast, he definitely associated the Coosa and Cumberland with the Tennessee in a study of *Plagiola* (Johnson 1978).

Some shared species indicate stream piracy between the Tennessee and New rivers. The New, called the "upper Kanawha" by Hocutt et al. (1986), is sharply divided from the lower by the Kanawah Falls and is unique in being the only river to traverse the Appalachian chain. A high degree of endemism is characteristic, but generally the fauna is scant (Hocutt et al. 1986). The Lower Kanawha is more speciose but is clearly related to the Ohio drainage. This is true for mussels and crawfishes. Mollusks reflect the fish diversity differences between the two segments. Neves (1983) reported that the New drainage had only one-third the number of mussel species as the region below the Falls. A Teays–Ohio association similarly distinguishes the Little Kanawha River fauna (Hocutt et al. 1986, Hobbs 1989, Fitzpatrick 1967).

Peninsular Florida

A singular geology and geologic history profoundly affects the Florida Peninsula. Essentially a cavernicolous limestone extrusion causes caves, spring-fed streams, and hard waters to offer kinds and redundancy of habitats unknown in such profusion elsewhere. Many troglobites can be found (see Chapter 3). All crawfishes, except one, are related to the more primitive members of the family and represent descendants from at least six different surface ancestors (Hobbs and Franz 1986). (Elsewhere, crawfish troglobites and troglophiles belong to relatively advanced genera, although they often exhibit character-istics that are more closely allied to what are believed to be the plesiomorphic condition.) There is one monotypic genus (close to the primitive stock) that has peculiar habits, and the 10 species can be divided into "high-energy" and "low-energy" species, whose life-style affects the dispersal abilities of each (Franz and Lee 1982).

The redeye chub (*Notropis harperi*) and yellow bullhead (*Ictalurus natilis*) are among the fishes that frequent underground waters. The former is re-stricted to springs and runoffs of the region; the latter occurs throughout the eastern half of North America. No fishes are endemic to central or southern Florida, although several can be said to be endemic to the Florida area; likewise, many have well-defined subspecies on the peninsula. In the south particularly, a tropical–Caribbean element enters the fauna. One striking feature is the regular invasion of freshwater by 16 marine/estuarine fishes. This phenomenon is much more a tropical one than a temperate one.

The Great River Systems

As long as the Mississippi and Ohio rivers are included in a zone, these rivers, two of the world's largest, must receive attention. Besides their major tri-butaries, each is provided with numerous smaller feeders and a distinct flood plain more extensive than most. Robison (1986) placed the Mississippi Basin in perspective as a zoogeographic unit and made it clear that the system cannot be examined independently of the vast areas outside of the study zone that form its watershed. For example, the upper Missouri, now contributing heavily to the Mississippi, was associated in preglacial times with Hudson Bay. Portions of the upper Ohio were a part of a Laurentian system. The discussion has already alluded to the Teays, a major pre-Pleistocene south-ward-draining stream. Glaciers wreaked havoc with these patterns, and a long series of papers have used the coincident dissections and rearrangements of drainage segments to explain the present distribution of organisms. Popula-tions of the Ozark, the Ohio Valley, West Virginia, the Mississippi Embay-ment, and segments of the Atlantic slope are all in some way related to these events. Burr and Page (1986) addressed the zoogeography of the two rivers and presented another overview of the history of these systems. Most of the specifics of their article are outside the concerns of this discussion, but are essential background information for interested readers.

Branson (1985) used fishes and gastropods to analyze aquatic distributions of the Interior Low Plateau. He clearly demonstrated the paucity of our understandings of this region, and he illustrated that the history of this region is just beginning to be understood. Even so, a great need for data still exists, and this need is made more urgent as the encroachments of man make great modifications of this complex faunal/floral area. There are many relicts of fishes and gastropods (Branson 1985) and decapod crustaceans (Fitzpatrick 1867, 1987). Isom (1969) over 30 years ago was able already to demonstrate significant changes in the mussel fauna. References to the lists of Lee et al.

(1980 et seq.), Hobbs (1989), Burch (1972, 1973), and Stansbery (1970) will identify the fauna, but the evaluation is incomplete.

A two-way exchange of aquatics between the headwaters of the Cumberland and the Kentucky rivers has been documented (Starnes and Etnier 1986). Equivalent data can be used to indicate that most of the adjacent drainages had such exchanges. A lack of high waterfalls—effective barriers to upstream migrations—characterizes most of this Great River system. The fish fauna apparently was developed in a drainage pattern connecting the present lower segments of the Tennessee and Ohio with the Mississippi independently of, and much to the south of, the Teays confluence with the present Mississippi (Burr and Page 1986).

The Trans-Mississippi

A large block of territory is included in this category. Physically, it varies from the Ozark–Ouachita highlands to the extensive coastal swamps of Louisiana. These diverse areas are lumped together largely because of ignorance of the precise faunistic composition. The fishes are the best known element, a typical situation in the Southeast. Cross et al. (1986) addressed the *ichthyofauna* of the lower Mississippi mainstream (to the Yazoo) and the southern Missouri/Arkansas/northern Louisiana drainages. A small amount of coverage of these volumes is included in Conner and Suttkus (1986). Again, preglacial and Pleistocene connections of drainages were markedly different than present ones. For example, central Kansas, now attached to the Arkansas drainage, had its outlet into the Gulf directed almost due south during the Nebraskan. Cross et al. (1986) used glacial events to explain the presence of Interior Highland species in nearly all Ozark streams, despite the stability of Highland drainage patterns. They identified archaic elements in the fish fauna; just outside this study area they indicated several species that they believed were residual Arctic components. The portion of the Arkansas River under consideration has the highest percentage of recently introduced species (13%), and there is a direct correlation between stream size and numbers of introduced species. The White River has the greatest diversity (150 native species) and the highest endemism (5 species). A high degree of species sharing exists between the White and St. Francis, Ouachita, and lower Red rivers, in that order, and an identical ranking with respect to diversity can be made. A striking feature of the lower Red River diversity is the presence of at least 13 species usually associated with the Coastal Plain.

Clear connections between the Ozark–Ouachita highlands and the Cumberland Plateau exist in fishes (Conner and Suttkus 1986, Starnes and Etnier 1986, Mayden 1985), decapod crustaceans (Fitzpatrick 1987), and salamanders (Mayden 1985), among others. Interestingly, Mayden (1985) produced a different route for the pre-Pliocene Tennessee than commonly accepted plus a suggestion that, prior to glaciation and its accompanying drift deposition, a continuum of high-gradient, silt-free, cool, clear streams existed from eastern Kansas to the Appalachians. Thus, by his thinking, much of the fauna outside the Piedmont and lower country in the southeast is descended from an expansive Highlands fauna that populated these streams.

The edge of the Tertiary deposits in Arkansas and Louisiana seems to have been a source of origin for several groups of decapod crustaceans (Fitzpatrick 1983, 1986, 1987, Hobbs and Robison 1985). A special kind of habitat of roadside ditches occurs south of this line west of the Mississippi and along the coast to Mobile Bay. It is populated by a taxonomically distinct group of fauna that seem to have been able to expand their ranges well outside the area, in one case as far as the Broad River basin in South Carolina (Fitzpatrick 1986). Fishes are less demonstrative of any particular relationship such as this, probably because they were older and much more mobile as Quaternary events developed. The pelecypods were examined recently by Gordon et al. (1979), who identified four associations. Although their report is restricted to Arkansas, it can be expanded to apply throughout the region. There is a distinct Appalachian element, possibly continuing as relicts of a Cretaceous linkage. A second group, in the southern Interior Basin and Gulf systems, enters from the south. Gordon et al. postulated that these species used the Interior Highlands as refugia during incursions of the sea via the embayments of the Cretaceous and Tertiary. They based this on the near congruence of their distributional limits with the shorelines of those periods. The third element represents taxa migrating from the northern Interior basin. These were seen as possible holdovers from use of the Highlands as a refugium during the glacial periods. A fourth group consists of ubiquitous Interior Basin species; these occurrences were attributed to changing drainage patterns and connections associated with the Pleistocene.

Although details of the trans-Mississippi area are sometimes lacking or confusing, there are some clear conclusions. The faunal elements are diverse and apparently are associated with emigration and/or immigration in all directions. Endemism is more directed toward specific populations isolated from older stocks than with the origin of major faunal groups. A possible exception to this is the decapod crustaceans, because of their comparatively recent emergence and diversification. Quite important to the study of the evolutionarily dynamic Southeast will be data gathered from lesser known, smaller invertebrates and integration of the strictly taxonomic information with detailed autoecologic and synecologic studies. Only by this interaction will we be able to understand why this region has become so much of a laboratory for evolutionary development.

SUMMARY

The development of the drainage systems that now characterize the various physiographic regions of the eastern United States had a complex history, with some events dating back nearly one-half billion years. Few streams are present whose position and age predate the glacial episodes of the Quaternary. The types of streams that characterize an area are strongly influenced by the region's depositional history, subsequent structural events, and ongoing neotectonic adjustments. Thus, although mountains uplifted during the Paleozoic have long since vanished, the fold, fault, and joint structures developed during these times of orogeny can exert significant controls on the courses of modern day streams, even when covered by thick soils and alluvium.

A number of other variables is equally important and, in some provinces, may exert even greater control than the structural attitude of the strata on development of drainage systems. Climate, for example, is arguably the most important agent controlling drainage, and imprints its effects on every stream system. In recent years this factor has been accorded even greater importance, especially with respect to the geomorphological response of rocks to periodic changes in climate. Garner (1974) has coined the term "environmental dynamism" to characterize the geomorphic implications of changes in climate and how such climatic changes can control the types and manner by which erosion will take place in an area. Also important in explaining regional and local drainage systems are variables such as rock type, the subsurface hydrology, the physical dynamics of running water, the types and degree of vegetative cover, and eustatic rise and fall of sea level.

The southeastern United States, a major area for evolutionary and faunistic studies, can be divided into several major regions based on the aquatic fauna: the Atlantic Coastal Plain, the Eastern Gulf Coastal Plain, the Southern Appalachians and adjacent areas, peninsular Florida, the Great River (Ohio–Mississippi) systems, and the Trans-Mississippi area. Each of these is characterized by a particular fauna, the best known elements of which are the fishes, the bivalve mollusks, and the crawfishes. The constraints of a single chapter preclude detailed descriptions of the faunal units, but they are easily accessible in the literature. Although zoogeographic divisions do not exactly parallel the geologic ones, they are closely related, the differences being due principally to the comparative antiquity of several events and the greater plasticity of organisms in their ability to respond to changes in the physical milieu. An integrated evaluation of taxonomic, zoogeographic, environmental, geologic, and historical information is critical to an understanding of the biotic communities of this area.

REFERENCES

Bethune, P. 1948. Geomorphic studies in the Appalachians of Pennsylvania. *Am. J. Sci.* 246:1–22.

Blair, W. F., A. P. Blair, P. Brodkorb, F. R. Cagle, and G. A. Moore. 1968. *Vertebrates of the United States*, 2d ed. New York: McGraw-Hill.

Branson, B. A. 1985. Aquatic distributional patterns in the Interior Low Plateau. *Brimleyana* 11:169–189.

Bryan, R. B., and A. Yair (eds.). 1982. *Badland Geomorphology and Piping.* Norwich, UK: GeoBooks.

Buol, S. W. 1973. Soils of the southern States and Puerto Rico. Southern Cooperative Series Bull. 174, U.S. Dept. Agriculture, Soil Conservation Service.

Burch, J. B. 1972. Freshwater sphaericean clams (Mollusca: Pelecypoda) of North America. Identification Manual 3, *Biota of Freshwater Ecosystems*. Washington, DC: U.S. Environmental Protection Agency.

Burch, J. B. 1973. Freshwater uniionacean clams (Mollusca: Pelecypoda) of North America. Identification Manual 11, *Biota of Freshwater Ecosystems*. Washington, DC: U.S. Environmental Protection Agency.

Burr, B. M., and L. M. Page. 1986. Zoogeography of fishes of the lower Ohio–upper Mississippi basin. In C. H. Hocutt and E. O. Wiley (eds.), *The Zoogeography of North American Freshwater Fishes*. New York: Wiley, pp. 287–324.

Butzer, K. W. 1976. *Geomorphology from the Earth*. New York: Harper & Row.

Chien, S. M. 1969. Summer Cladocera of the Pearl River system, Mississippi. Thesis, Mississippi State University, State College.

Clark, G. M. 1989. Central and Southern Appalachian water and wind gap origins: review and new data. *Geomorphology* 2:209–232.

Cleaves, E. T. 1989. Appalachian Piedmont landscapes from the Permian to the Holocene. *Geomorphology* 2:159–179.

Conner, J. V., and R. D. Suttkus. 1986. Zoogeography of freshwater fishes of the western Gulf slope of North America. In C. H. Hocutt and E. O. Wiley (eds.), *The Zoogeography of North American Freshwater Fishes*. New York: Wiley, pp. 413–456.

Cross, F. B., R. L. Mayden, and J. D. Stewart. 1986. Fishes in the western Mississippi Basin (Missouri, Arkansas and Red rivers). In C. H. Hocutt and E. O. Wiley (eds.), *The Zoogeography of North American Freshwater Fishes*. New York: Wiley, pp. 363–412.

Davis, W. M. 1888. Geographic methods in geologic investigations. *Natl. Geogr. Mag.* 1:11–26.

Davis, W. M. 1889. The rivers and valleys of Pennsylvania. *Natl. Geogr. Mag.* 1:183–253.

Dupis, C. 1984. Willi Hennig's impact on taxonomic thought. *Annu. Rev. Ecol. System.* 15:1–24.

Eardley, A. J. 1951. *Structural Geology of North America*. New York: Harper.

Edmondson, W. T. (ed.). 1959. *Ward and Whipple's Freshwater Biology*, 2d ed. New York: Wiley.

Epstein, J. B. 1966. Structural control of wind gaps and water gaps and of stream capture in the Stroudsburg area, Pennsylvania and New Jersey. *U.S. Geol. Sur. Prof. Paper* 550B, B80–B86.

Fenneman, N. M. 1937. *Physiography of the Eastern United States*. New York: McGraw-Hill.

Fitzpatrick, J. F., Jr. 1967. The Propinquus group of the crawfish genus *Orconectes* (Decapoda: Astacidae). *Ohio J. Sci.* 67:129–172.

Fitzpatrick, J. F., Jr. 1977a. Distribution of the subgenus *Pennides* of the genus *Procambarus* in Mississippi (Decapoda, Cambaridae). *A.S.B. Bull.* 24:51.

Fitzpatrick, J. F., Jr. 1977b. A new crawfish of the genus *Hobbseus* from northern Mississippi, with notes on the origin of the genus. *Proc. Biol. Soc. Washington* 90:367–374.

Fitzpatrick, J. F., Jr. 1983. A revision of the dwarf crawfishes (Cambaridae, Cambarellinae). *J. Crust. Biol.* 3:266–277.

Fitzpatrick, J. F., Jr. 1986. The pre-Pliocene Tennessee River and its bearing on crawfish distribution (Decapoda: Cambaridae). *Brimleyana* 12:123–146.

Fitzpatrick, J. F., Jr. 1987. The subgenera of the crawfish genus *Orconectes* (Decapoda: Cambaridae). *Proc. Biol. Soc. Washington* 101:44–74.

Fitzpatrick, J. F., Jr., and H. H. Hobbs III. 1968. The Mississippi River as a barrier to crawfish dispersal. *Am. Zool.* 8:807.

Flemal, R. C. 1971. The attack on the Davisian system of geomorphology: a symposium. *J. Geol. Educ.* 19:3–13.

Franz, R., and D. Lee. 1982. Distribution and evolutioin of Florida's troglobitic crayfishes. *Bull. Florida St. Mus., Biol. Sci.* 28:53–78.

Frey, D. G. 1986. The non-cosmopolitanism of chydorid Cladocera: implications for biogeography and evolution. *Crust. Issues* 4:237–256.

Garner, H. F. 1974. *The Origin of Landscapes.* London: Oxford UP.

Gordon, M. E., L. R. Kraemer, and A. V. Brown. 1979. Unionacean of Arkansas: historical review, checklist, and observations on distributional patterns. *Bull. Am. Malacol. Union* 79:31–37.

Hack, J. T. 1960. Interpretation of erosional topography in humid temperate regions: *Am. J. Sci.* 258A:80–97.

Hack, J. T. 1980. Rock control and tectonism: their importance in shaping the Appalachian Highlands. *U.S. Geol. Sur. Prof. Paper* 1126B:B1–B17.

Hack, J. T. 1982. Physiographic divisions and differential uplift in the Piedmont and Blue Ridge. *U.S. Geol. Sur. Prof. Paper* 1265.

Hatcher, R. D., Jr., G. W. Viele, and W. A. Thomas. 1988. *The Appalachian–Ouachita Orogen in the United States.* Geol. Soc. Am. *The Geology of North America,* Vol. F-2.

Higgins, C. G. 1984. Piping and sapping: development of landforms by groundwater outflow. In R. G. LaFleur (ed.), *Groundwater as a Geomorphic Agent.* Boston: Allen & Unwin, pp. 18–58.

Higgins, C. G., D. Coats, V. Baker, W. Dietrich, T. Dunne, E. Keller, R. Norris, G. Parker, M. Pavich, T. Pewe, J. Robb, J. Rogers, and C. Sloan. 1988 Landform Development. In W. Back, J. Rosenshein, and P. Seaber (eds.), *Hydrogeology: The Geology of North America,* Vol. O-2. Denver: Geological Society of America, pp. 383–400.

Hobbs, H. H., Jr. 1969. On the distribution and phylogeny of the crayfish genus *Cambarus.* In P. C. Holt, R. L. Hoffman, and C. W. Hart, Jr. (eds.), *The Distributional History of the Southern Appalachians.* Res. Div. Monogr. 1, Virginia Polytech. Inst. St. Univ., Blacksburg, pp. 11–42.

Hobbs, H. H., Jr. 1974. A checklist of the North and American crayfishes (Decapoda: Astacidae and Cambaridae). *Smithsonian Contrib. Zool.* 166.

Hobbs, H. H., Jr. 1981. The crayfishes of Georgia. *Smithsonian Contrib. Zool.* 318.

Hobbs, H. H., Jr. 1984. On the distribution of the crayfish genus *Procambarus* (Decapoda: Cambaridae). *J. Crust. Biol.* 4:12–24.

Hobbs, H. H., Jr. 1989. An illustrated checklist of the American crayfishes (Decapoda: Astacidae, Cambaridae, and Parastacidae). *Smithsonian Contrib. Zool.* 480.

Hobbs, H. H., Jr., and R. Franz. 1986. New troglobitic crayfish with comments on its relationship to epigean and other hypogean crayfishes of Florida. *J. Crust. Biol.* 6:509–519.

Hobbs, H. H., Jr., and H. W. Robison. 1985. A new burrowing crayfish (Decapoda: Cambaridae) from southwestern Arkansas. *Proc. Biol. Soc. Washington* 98:1035–1041.

Hocutt, C. H., R. F. Denoncourt, and J. R. Stauffer, Jr. 1978. Fishes of the Greebrier River, West Virginia, with drainage history of the central Appalachians. *Brimlyeana* 1:47–80.

Hocutt, C. H., R. E. Jenkins, and J. R. Stauffer, Jr. 1986. Zoogeography of the fishes of the central Appalachians and central Atlantic Coastal Plain. In C. H. Hocutt and E. O. Wiley (eds.), *The Zoogeography of North American Freshwater Fishes*. New York: Wiley, pp. 161–211.

Holt, P. C., and R. A. Patterson (eds.). 1970. The distributional history of the biota of the southern Appalachians, II: flora. *Res. Div. Monogr. 2, Virginia Polytech. Inst. St. Univ., Blacksburg.*

Holt, P. C., R. L. Hoffman, and C. W. Hart, Jr. (eds.). 1969. The distributional history of the biota of the southern Appalachians, I: invertebrates. *Res. Div. Monogr. 1, Virginia Polytech. Inst. St. Univ., Blacksburg.*

Holt, P. C., R. A. Patterson, and J. P. Hubbard (eds.). 1971. The distributional history of the biota of the southern Appalachians, III: vertebrates. *Res. Div. Monogr. 3, Virginia Polytech. Inst. St. Univ., Blacksburg.*

Horton, R. E. 1945. Erosional development of streams and their drainage basins. *Geol. Soc. Am. Bull.* 56:275–370.

Hoskins, D. M. 1987. The Susquehanna River water gaps near Harrisburg, Pennsylvania. *Geol. Soc. Am., Cent. Field Guide*, 5:47–50.

Hunt, C. B. 1967. *Physiography of the United States.* San Francisco: Freeman.

Hursh, C. R. 1944. Report of the sub-committee on subsurface flow. *EOS Am. Geophys. Union Trans.* 25:743–746.

Isom, B. G. 1969. The mussel resources of the Tennessee River. *Malacologia* 7:397–425.

Isphording, W. C. 1984. Sand craters in Gulf Coastal Plain sediments: an extension of the Carolina Bays phenomenon? *Geol. Soc. Am., Abstracts with Programs, Southeastern-Northcentral Section Meeting, Lexington, KY*, p. 148 (Abstract).

Isphording, W. C., and G. C. Flowers. 1988. Karst development in Coastal Plain sands: a "new" problem in foundation engineering. *Bull. Assoc. Eng. Geol.* 25:95–104.

Isphording, W. C., F. D. Imsand, and G. C. Flowers. 1989, Physical characteristics and aging of Gulf Coast estuaries. *Trans. Gulf Coast Assn. Geol. Soc.* 39:387–401.

Johnson, D. W. 1919. *Shore Processes and Shoreline Development* New York: Wiley.

Johnson, R. I. 1970. The systematics and zoogeography of the Unionidae (Mollusca: Bivalvia) of the southern Atlantic Slope region. *Bull. Mus. Comp. Zool., Harvard U.* 140:263–450.

Johnson, R. I. 1978. Systematics and zoogeography of *Plagiola* (= *Dysnomia* = *Epioblasma*), an almost extinct genus of freshwater mussels (Bivalvia: Unionidae) from middle North America. *Bull. Mus. Comp. Zool., Harvard U.* 148:239–320.

Jones, F. A. 1981. The nature of soil piping: a review. *British Geomorphological Research Group Monograph No. 3.* Norwich, UK: GeoBooks.

Judson, S. 1975. Evolution of Appalachian topography. In W. N. Melhorn and R. C. Flemal (eds.). *Theories of Landform Development*. Boston: Allen & Unwin, pp. 29–44.

Lee, D. S., C. R. Gilbert, C. H. Hocutt, R. E. Jenkins, D. E. McAlliser, and J. R. Stauffer, Jr. 1980 et seq. *Atlas of North American Freshwater Fishes*. Raleigh: North Carolina St. Mus. Nat. Hist.

Liebling, R. S., and Scherp, H. S. 1984. Stream piracy at the base of the Allegheny Front, central Pennsylvania. *Northeastern Geol.* 6:131–134.

Mayden, R. L. 1985. Biogeography of the Ouachita Highlands. *Southwest. Nat.* 30:195–211.

Mettee, M. F., P. E. O'Neill, J. M. Pierson, and R. D. Suttkus. 1989. Fishes of the western Mobile River basin in Alabama and Mississippi. *Geol. Surv. Alabama Atlas* 24.

Milici, R. C. 1968. Mesozoic and Cenozoic physiographic development of the lower Tennessee River: in terms of the dynamic equilibrium concept. *J. Geol.* 76:472–479.

Morisawa, M. 1989. Rivers and valleys of Pennsylvania, revisited. *Geomorphology* 2:1–22.

Neves, R. 1983. Zoogeography of the mussel fauna in the New–Kanawah River. *Virginia J. Sci.* 34:131.

Outerbridge, W. F. 1987. The Logan Plateau, a young physiographic region in West Virginia, Kentucky, Virginia, and Tennessee. *U.S. Geol. Sur. Bull.* 1620.

Parker, G. G., and E. Jenne. 1967. Structural failure of western highways caused by piping. *U.S. Highway Research Board, Highway Research Record* No. 203:57–76.

Pennak, R. W. 1978. *Freshwater Invertebrates of the United States*, 2d ed. New York: Wiley.

Poag, C. W., and W. D. Sevon. 1989. A record of Appalachian denudation in post-rift Mesozoic and Cenozoic sedimentary deposits of the U.S. middle Atlantic continental margin. *Appalachian Geomorphol.* 2:119–157.

Poland, J. F., and G. H. Davis. 1969. Land subsidence due to withdrawal of fluids. *Geol. Soc. Am. Rev. Eng. Geol.* 2:187–269.

Puri, H. S., and R. O. Vernon. 1964. *Summary of the Geology of Florida and a Guidebook to the Classic Exposures*. Tallahassee: Florida Geol. Sur. Spec. Publ. No. 5.

Robison, H. W. 1986. Zoogeographic implications of the Mississippi River basin. In C. H. Hocutt and E. O. Wiley (eds.), *The Zoogeography of North American Freshwater Fishes*. New York: Wiley, pp. 267–285.

Rogers, J. D., and M. R. Pyles. 1979. Evidence of cataclysmic erosional events in the Grand Canyon of the Colorado River, Arizona. In *Proceedings of the Second Conference on Research in the National Parks, San Francisco*. Washington, DC: National Park Service, Physical Sciences 5:392–454.

Ross, H. D. 1971. The drainage history of the Tennessee River. In P. C. Holt, R. A. Patterson, and J. P. Hubbard (eds.), The distributional history of the biota of the southern Appalachians, III: vertebrates. *Res. Div. Monogr. 3, Virginia Polytech. Inst. St. Univ., Blacksburg*, pp. 11–42.

Sevon, W. D., N. Potter, Jr., and G. H. Crowl. 1983. Appalachian peneplains: an historical review. *Earth Soi. Hist.* 2:156–164.

Shepard, F. P. 1952. Composite origin of submarine canyons. *J. Geol.* 60:84–96.

Sherard, J. L., and R. S. Decker. 1977. Dispersive clays related to piping, and erosion in geotechnical projects. *Am. Soc. Testing Mater. Special Tech. Paper* 623.

Soil Survey Staff. 1975. *Agricultural Handbook 436*. Washington DC: U.S. Department of Agriculture, Soil Conservation Service.

Stansbery, D. H. 1970. Eastern freshwater molluscs, I: the Mississippi and St. Lawrence River systems. *Malacologia* 10:9–22.

Starnes, W. C., and D. A. Etnier. 1986. Drainage evolution and fish biogeography of the Tennessee and Cumberland rivers drainage realm. In C. H. Hocutt and E. O. Wiley (eds.), *The Zoogeography of North American Freshwater Fishes*. New York: Wiley, pp. 325–361.

Storm, L. W. 1945. Resume of facts and opinions on sedimentation in the Gulf Coast region of Texas and Louisiana. *Bull. Am. Assoc. Petrol. Geol.* 29:1304–1335.

Swift, C. C., C. R. Gilbert, S. A. Bortone, G. H. Burgess, and R. W. Yeager. 1986. Zoogeography of the freshwater fishes of the southeastern United States: Savannah River to Lake Pontchartrain. In C. H. Hocutt and E. O. Wiley (eds.), *The Zoogeography of North American Freshwater Fishes*. New York: Wiley, pp. 213–265.

Thornbury, W. C. 1954. *Principles of Geomorphology*. New York: Wiley.

Twidale, C. R. 1984. Role of subterranean water in landform development in tropical and subtropical regions. In R. G. LaFleur (ed.), *Groundwater as a Geomorphic Agent*. Boston: Allen & Unwin, pp. 91–134.

Twidale, C. R. 1987. Sinkholes (dolines) in lateritised sediments, western Sturt Plateau, Northern Territory, Australia. *Geomorphology* 1:33–52.

Wright, F. J. 1934. The newer Appalachians of the South, I: between the Potomac and New Rivers. Denison Univ. Bull. *J. Sci. Lab.* 29:1–105.

Zaslavsky, D., and G. Kassiff. 1965. Theoretical formulation of piping mechanism in cohesive soils. *Geotechnique* 15:305–310.

PART II
Lotic Systems

3 Caves and Springs

H. H. HOBBS III

Department of Biology, Wittenberg University, Springfield, OH 45501

North Central Florida has more spectacular, indeed more incredible, springs than any other region of which I have ever heard. It is impossible here to describe these springs in detail, but it can truthfully be said that, like the stars, one differeth from another in glory.—Barbour 1944, p. 116

Caves and springs and their biotic communities are widely distributed over the southeastern United States. They are characterized as being extremely stable in character, yet caves and springs are quite diverse. Spring communities range from deep artesian systems to minor seeps, and differ considerably in complexity, sunlight being one of the most important controlling variables. Canopy-covered seeps are relatively unproductive, being allochthonous–detritus-based systems, whereas open-seep communities are diverse and dominated by mosses, grasses, and pitcher plants. Shaded bubbling and Vauclusian springs support mosses and a modest variety of invertebrates, while open springs are very productive and support complex biotic assemblages.

Approximately 15% of the continental United States, exclusive of Alaska, has gypsum, limestone, or other soluble rocks at or near the surface (Davies and LeGrand 1972). Within the southeastern United States (see Fig. 1) approximately 20 000 caves and countless thousands of springs and seeps are known to exist. The unique topography of limestone areas, characterized by sinkholes, subterranean drainage, and thin argillaceus soil often interrupted by rock outcrops, is known as karst topography. Major karst areas in the United States are located in the plateaus and lowlands of western and central Texas and eastern New Mexico, the Ozark Highlands of Arkansas and Missouri; Florida and the coastal plain of Georgia; the Interior Lowlands and plateaus of Kentucky, Indiana, and Tennessee; and the Appalachian Mountains from Alabama to Pennsylvania (see Davies and LeGrand 1972). The caves and springs discussed in this chapter are found within several major physiographic provinces (see Table 1).

Of the 311 longest caves (>3 km) in the United States, 213 are located in the Southeast (data from Gulden 1985), with the Mammoth Cave System of Kentucky being the longest in the world (>485 km of surveyed passages, see Palmer 1981). Gulden (1984) also reported that 25 of the 49 deepest caves

59

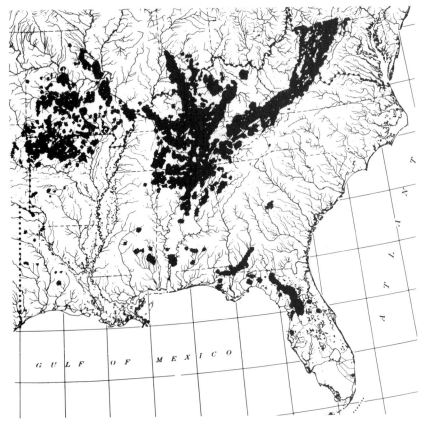

FIGURE 1. Distribution of caves and karst in the southeastern United States and some adjoining states.

in the United States occur in the Southeast. Ellison's Cave in Walker County Georgia, contains the two deepest, free-fall, vertical drops in United States caves: Incredible Dome-pit (134 m) and Fantastic Pit (178 m), which, at 331 m, is the deepest cave in the southeast (Anonymous 1974). Traditionally, political boundaries have been used in reporting karst features (Tables 2, 3).

PHYSICAL ENVIRONMENT

Caves and springs are only two of the major features characteristic of karst landscapes. The Western Slovenian term *karst* is a comprehensive term applied to areas underlain by evaporites (gypsum, anhydrite, rock salt), carbonate rocks (limestone, dolomite), and quartzite (only under extreme tropical humidity). The formation of karst landscapes and karst hydrography is dependent upon these rocks having a high degree of solubility. They must be soluble and leave little residue such that interstices are widened and remain

TABLE 1   Physiographic Provinces and Karst in the Southeastern United States

Physiographic Province	Karst	Principal Areas
Southeastern Coastal Plain	Tertiary and Pleistocene limestones	Alabama, Georgia, Florida, South Carolina
Ridge and Valley[a]	Paleozoic limestones	Alabama, Georgia, Tennessee, Virginia, West Virginia
Appalachian Plateaus		
Allegheny Mountain section	Mississippian limestones	Northeastern West Virginia
Cumberland Plateau section	Mississippian limestones	Alabama, Kentucky, Tennessee
Interior Low Province (including Nashville Basin, Highland Rim, Lexington Plain)	Ordovician, Devonian, and Mississippian limestones	Alabama, Kentucky, Tennessee
Ozark plateaus	Ordovician to Mississippian limestones	Northern Arkansas, (south-central Missouri)
Central Coastal Plain	Limestones and quartzite Sand–clay	Mississippi Louisiana, eastern Texas

[a]The Appalachian Valley contains the most extensive karst area in the eastern United States.

open, and that underground drainages evolve, replacing, at least in part, surface ones. Although the main physical characteristic of karst areas is underground drainage in soluble rock, other types of landscape also demonstrate subterranean drainage (e.g., pumice, lava, valley fill, glaciers, loess areas) (Halliday 1954, 1960, Moore 1954, Parker et al. 1964, White et al. 1970, Otvos 1976, Ogden 1980, Holler 1981), yet fail to develop other attributes common in karst (see Jennings 1971, 1985, Sweeting 1973, Mylroie 1978, Bogli 1980 for detailed discussions of karst features, e.g., sinkhole, ponor, blind valley, karst window, swallow hole).

Caves

Although caves are perceived as overhangs, shelters, and "dark holes in the ground," a widely accepted working definition is the following: a natural opening in the ground extending beyond the zone of light and, from the human viewpoint, large enough to permit entry. Caves occur in a variety of rock types, their formation (speleogenesis) is governed by numerous geological processes, and they range in size from a single small tube or room to interconnecting passages many kilometers in length and on various levels.

TABLE 2   Numerical Listing of Caves by State

State	Number of Caves	Major References
Alabama	~2000	Veitch 1967, Armstrong and Williams 1971, Varnedoe 1973, 1980
Arkansas	1000	Warden 1971, Ogden 1978, 1981, Ogden et al. 1981, Barlow and Ogden 1982
Florida	380 "dry caves"	Warren 1962, Lane 1986
Georgia	318	McCallie 1908
Kentucky	3770	George 1985
Louisiana	5	Warden 1971, Sevenair and Williamson 1983
Mississippi	43–50	Knight and Irby 1972, Knight et al. 1974, Cliburn and Middleton 1983
North Carolina	1000	Holler 1975, Maness and Holler 1979, Holler, personal communication
South Carolina	<10	MacNale 1956, Farlow 1987, Holler, personal communication
Tennessee	~3000	Barr 1961, Matthews 1971, Armstrong and Williams 1971, Wilson 1980
Texas	2000 (few in east Texas)	Smith 1971, Warden 1971, Fieseler 1978
Virginia	2502	Douglas 1964, Holsinger 1975b, Anonymous 1986
West Virginia	~2000	Anonymous 1957, Davies 1965, Medville and Medville 1971, Hemple 1975, Jones 1981, Dasher 1983

They tend to demonstrate curvilinear or rectilinear plans closely related to joint patterns in the bedrock and vary considerably in their vertical and horizontal complexity.

Caves have been classified according to several criteria, such as rock type (e.g., lava, limestone, marble), stream type, and origin. Dry caves lack streams; all caves with streams are called active caves and may be influent, effluent, or through caves. Influent caves have streams flowing into them, an insurgence, swallow hole, blind valley (e.g., Black Hole, Pocahontas Co., WV; Obe Lee Cave, Overton Co., TN; Mill Cave, Cumberland Co., TN; see Fig. 2). Effluent caves have streams flowing out of an entrance (e.g., Blanchard Springs Cavern, Stone Co., AR; see Fig. 3; Echo River and River Styx, Mammoth Cave, Edmonson Co., KY). Through caves are those having streams that enter and exit (e.g., Sinks of Gandy Creek, Randolph Co., WV; Natural Tunnel, Scott Co., VA; see Fig. 4). Tectonic caves are those developed along fractures or faults (e.g., New River Cave, Giles Co., VA; Ellison's Cave, in part, Walker Co., GA; numerous caves in North Carolina). See Bogli (1980) for other classifications of caves.

TABLE 3 Selected Spring Studies of the Southeast

State	Source
Alabama	Meinzer 1927b, Armstrong and Williams 1971
Arkansas	Fuller 1905, Purdue and Miser 1916, Bryan 1924, Meinzer 1927b, Beckman and Hinchey 1944, Waring 1965, Vineyard and Feder 1974, Vineyard 1981
Florida	Odum 1957a,b, Meinzer 1927b, Ferguson et al. 1947, Wetterhall 1965, Stringfield 1966, Maegerlein 1970, Rosenau et al. 1977, Deloach 1978, Exley and Deloach 1981, Lane 1986
Georgia	McCallie 1980, Stephenson and Veatch 1915, Watson 1924, Meinzer 1927b, Callahan 1964, Waring 1965, Stringfield 1966, Beck and Arden 1984
Kentucky	Matson 1909, VanCouvering 1962, George 1972a, Saunders 1973, Hess et al. 1974, Spangler and Thrailkill 1981, Thrailkill 1985
Louisiana	—
Mississippi	—
North Carolina	Watson 1924, Waring 1965
South Carolina	Cooke 1936
Tennessee	Whitlatch 1943, Sun et al. 1963, Armstrong and Williams 1971
Texas	Brune 1975
Virginia	Watson 1924, Collins et al. 1930, Reeves 1932, Waring 1965, Saunders et al. 1981
West Virginia	Price et al. 1936, Waring 1965, Jones 1973, 1981, Ogden 1982

FIGURE 2. Entrance to Mill Cave, Cumberland County, Tennessee.

FIGURE 3. Entrance to Blanchard Springs Cavern, Stone County, Arkansas.

FIGURE 4. Natural Tunnel, Scott County, Virginia.

Springs

A spring is a natural discharge of water as leakage or overflow from an aquifer through a natural opening in the soil/rock onto the land surface or into a body of water. The opening can be very small such that water trickles or creates only a wet seep, or it can be so large that the spring flow is the source of a sizeable river (e.g., Wakulla Springs, Wakulla Co., FL; see Olsen 1958 and Stone 1989). The natural flow of springs is controlled by hydrologic and geologic factors, such as the amount and frequency of rainfall, the porosity and permeability of the aquifer, the hydrostatic head pressure within the aquifer, and the hydraulic gradient.

Numerous descriptive terms for springs have evolved and no single basis of classification has been adopted. Commonly, definitions are based on magnitude of discharge, type of aquifer, chemical characteristics of spring water, temperature of the water, and hydraulic characteristics. Probably the most widely used spring classification in the United States is that proposed by Meinzer (1927b, pp. 2, 3) in which springs are ordered from one to eight on the basis of volume of discharge (Table 4). Rosenau et al. (1977, p. 5) indicated a total of 78 first-magnitude springs in the United States, discharging from limestone, dolomite, basalt, or sandstone aquifers (Florida, 27; Idaho, 14; Oregon, 15; Missouri, 8; California, 4; Hawaii, 3; Montana, 3; Texas, 2; Arkansas, 1).

Based on temperature, springs may be categorized as thermal or nonthermal (Stearns et al. 1937). Thermal springs are further identified as hot springs (temperatures higher than 98 °F) or warm springs (temperatures higher than the local mean annual temperature of atmosphere but lower than 98 °F). The global distribution of the majority of such springs is closely associated with areas of volcanoes of present or geologically recent activity. In the southeastern United States they occur in areas where rocks, regardless of their character and age, have been faulted and folded in geologically recent time (Fig. 5). Many are described as artesian, with waters rising from deep strata along faults and fissures. The heat of springs associated with volcanic rock

TABLE 4 Spring Classification Based on Mean Annual Flow

Magnitude	Mean Flow (Discharge)
1	>100 cfs
2	10–100 cfs
3	1–10 cfs
4	100 gal/min–1 cfs
5	10–100 gal/min
6	1–10 gal/min
7	1 pint–1 gal/min
8	1 pint/min

Source. (Meinzer 1927b).

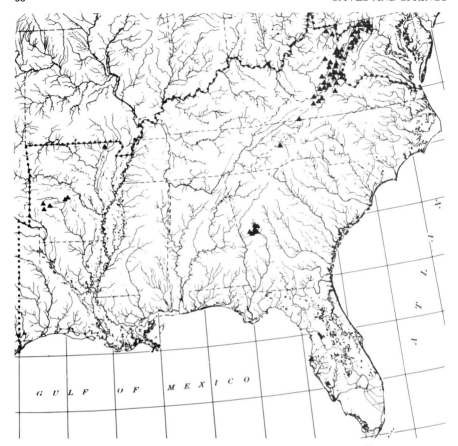

FIGURE 5. Distribution of thermal springs in the southeastern United States. (After Stearns et al. 1937.)

has been postulated to be of volcanic origin or resulting from subterranean chemical reactions (e.g., oxidation of iron pyrite) and disintegration of radioactive substances. In areas far removed from volcanoes, elevated water temperatures probably result from the great depths from which they rise (Waring 1965). The geothermal gradient for groundwater is approximately 1 °F/19.8 m of depth (1 °C/30.5 m) (Collins 1925, p. 98).

Springs are sometimes categorized according to the type of aquifer from which their water is derived (hydraulic characteristics): water table (free flow) or artesian. The free-flow springs have gravity drainage with small but variable flow and discharge. Rain that percolates through permeable sediments ultimately may reach a relatively impermeable bed (e.g., clay or shale). Water then moves down gradient along the top of the impermeable unit to a place of outcrop where it emerges (continuously or intermittently) as a spring or seep (free flow). In karst regions, such springs often emerge from caves (Fig. 3). Artesian or Vauclusian springs (Sweeting 1973, p. 210) arise from large underground reservoirs, and waters issue from the cover rock under hydro-static pressure usually in steeply inclined channels (e.g., Radium Springs, Albany, GA; Alexander Springs, Lake Co., FL). Two types of Vauclusian springs are recognized: those in which water issues from a tube cut into the bedrock (e.g., karst springs, Silver Glen Springs, Marion Co., FL) or those where water wells up through a rock tube but issues at the surface through alluvium or drift (e.g., Cold Spring, Meriwether Co., GA, and Turnhole Spring, Edmonson Co., KY; see Fig. 6).

Springs are also classified by the type of opening through which the water surfaces. In seepage springs water percolates from numerous small openings in the permeable material distributed over much of the southeast. Tubular springs are characterized by an opening of a more or less rounded channel, such as a solution passage ("tube") in limestone. In fracture springs the openings occur at joints or other fractures, such as fissures and faults. There are also submarine springs along coastal areas (e.g., Ray Hole Spring, 90 km SSE of Tallahassee, FL; see Rosenau et al. 1977).

A distinction is made between resurgences and exsurgences. A resurgence is the reemergence of a stream that originates on the surface but enters a limestone aquifer and resumes its surface course at the rise (e.g., Roger's Mill Spring, Monroe Co., WV; many springs in the Central Kentucky karst). An exsurgence is the rise of a subterranean stream containing water that is derived from meteoric water that percolates into the limestone mass (e.g., Ichetucknee Springs, Suwannee Co., FL). An estevelle is an opening that at one time of the year functions as a swallow hole and at another as a spring (e.g., Echo River, Mammoth Cave, KY).

FIGURE 6. Turnhole Spring and Green River, Edmonson County, Kentucky, during spring flood.

For further information concerning the classification (salt, magnesia, sulfur), characteristics (chemical, physical, and biological), and associated terminology of springs (particularly karst springs), see Odum (1956), Sloan (1956), Whitford (1956), Jennings (1971), Sweeting (1973), Rosenau et al. (1977), Bogli (1980), and Hess et al. (1989). One particular location, Salt Spring in Marion County, Florida, supports a brackish water/marine community (e.g., *Gammarus mucronatus*, *Orchestia platensis*, and *Grandidierella* sp.); see Odum (1953) for a discussion of marine species invasion into Florida freshwaters.

Hydrology

The hydrologic cycle is multidimensional and complex, involving precipitation, evapotranspiration, sublimation, and other physical, chemical, and biological interactions. Drainage systems evolve as waters sculpture surface and subsurface gutters and conduits that function to evacuate surplus water most efficiently. Geological parameters (e.g., rock type, porosity, permeability), vegetation (e.g., ground cover, root density, species makeup), and local climate determine the nature and density of this network.

Some meteoric water collects in depressions (e.g., lakes) or surface-water drains and flows as rivers (ultimately) to the sea. Some of it evaporates from surface-water sources, and another portion is transpired from plants. Additional meteoric water infiltrates the ground, recharging the groundwater supply. This groundwater continually moves from areas of replenishment to places of discharge. In some systems, discharge is beneath the sea, indicating a lower sea level at some previous time, for example, the blue holes of the Bahamas (Palmer 1985, 1989), the submerged caves of Bermuda (Hobbs 1985), and Crescent Beach Spring, southeast of St. Augustine, Florida (Brooks 1961). A rock formation that holds and transmits water through its interstices and channels is known as an aquifer; refer to Smith et al. (1976) for additional structures and terminology used in groundwater hydrology.

Springs in limestone regions have a greater fluctuation of discharge than springs in any other type of rock (Meinzer 1927a), although this is not always the case (see White and Reich 1970). The immense fluctuations of volume relate directly to the fact that most springs in carbonate rocks issue from tubular conduits of caverns. Rather than having characteristics of groundwater flowing through a permeable material with finite permeability, these conduits have more of the characteristics of surface water in open channels (Meinzer 1923, p. 134).

There are differences of opinion as to the character of the water table in karst regions, dating back to Grund (1903) and Martel (1910). Some maintain there is no water table in the sense of a continuous surface separating saturated and nonsaturated zones. Rather, they picture the groundwater as being discontinuous and confined to joint and bedding plane openings. In fact, limestone drainages occur both by conduit and diffuse flow, which interact to form a single integrated system. The opposing models of Grund and Martel are merely extreme cases of a spectrum of possible drainage systems (see Moneymaker 1941, Rhoades and Sinacori 1941, Smith et al. 1976, Sweeting 1973).

Limestone Solution

One of the most important factors in the formation of karst landforms is the solution of limestone (calcite, alagonite). Water with a pH of greater than 7.0 has little effect on limestone, yet acidic water readily dissolves the rock. The most available acid is carbonic, derived from the absorption of carbon dioxide from the atmosphere and biogenic CO_2 (e.g., respiration, decaying organic matter) which is transported to the rock by meteoric water. The carbon dioxide–water–calcium carbonate system (the so-called CO_2 complex) is central to an understanding of karst.

Infiltrating surface water and groundwater flow along joints and bedding planes are of critical importance to the ontogeny of caves. Atkinson and Smith (1976, p. 155) cite four factors that affect the rate at which limestone is removed: composition and resistance of the rock itself, amount of runoff from the basin, temperature of the water, and prevailing levels of CO_2. They present a good discussion of the interrelations of these variables and the effects of each. Basically, cavern origin and development are the result of corrosion, but corrasion and incasion (Fig. 7) also play varying roles (see Bogli 1980). A good overview and comparison of karst development in various areas of the Southeast are presented by Davies and LeGrand (1972).

Waters in limestone aquifers generally have Ca^{2+} and HCO_3^- as the major dissolved species, and are generally close to equilibrium with calcite. The Ca^{2+} concentration depends on the P_{CO_2} of the water, which is usually controlled by soil atmosphere in the recharge area. The primary exceptions to these generalizations are waters in deep aquifers, which have received solutes by mixing with waters from other sources, or which have received solutes by dissolution of halite or gypsum (Drever 1982).

A very brief discussion is presented herein of the dissolution processes of the solution of limestone (the most widespread karst rock). The discussion and reaction cited below by no means represent a complete summary of the chemistry of calcium carbonate solution; nor are they adequate presentations of the dissolution processes of all types of rocks yielding karst topography. The reader is referred to Sweeting (1973), Picknett et al. (1976), Bogli (1980), Drever (1982), Jennings (1985), Trudgill (1985), White 1988, and to Hess and White (1989) for more in-depth treatment. Limestone (calcite, $CaCO_3$) is only slightly soluble in pure water (12.7 mg/L $CaCO_3$ in CO_2-free water; Hutchinson 1957). Since much greater concentrations are common in natural waters, the solvent most important in dissolving this rock is carbonic acid (Adams and Swinnerton 1937, Smith and Mead 1962), although organic acids (Muxart et al. 1968) and sulfuric acids (Pohl and White 1965, Morehouse 1968, Jennings 1971) also play a role in the dissolution of carbonate rocks.

FIGURE 7. Large passage in Lost Creek Cave (=Dodson Cave), White County, Tennessee, resulting from corrosion, corrasion, and incasion.

The process of calcium carbonate dissolution is represented by the following simplified reaction:

$$CaCO_3 + H_2O + CO_2 \text{ (dissolved)} \leftrightarrow Ca^{2+} + HCO_3^-$$

The highest rates of dissolution occur in the upper part of the phreatic zone, rather than at random depths within it. This is where downward-percolating water is constantly mixed with slowly moving groundwater (see Herrick and LeGrand 1964). The amount of $CaCO_3$ dissolved by percolating water depends critically on whether the water is in communication with a gas phase while the mineral is dissolving. Measured P_{CO_2} values of groundwaters in limestone aquifers are nearly always above atmospheric level (Back and Hanshaw 1970, Langmuir 1971). In vadose water the degree of saturation is dominated by carbon dioxide transfer between solution and air. Phreatic water, on the other hand, has had adequate time to attain limestone saturation and temperature equilibrium. Cavern development in phreatic zones would be difficult to explain had not Bogli (1971) discovered "mixing corrosion." He proposed that if two saturated solutions of calcium carbonate are mixed, it is possible for the resulting water to be aggressive (to dissolve calcium carbonate).

Origin and Development of Caves

Slightly acidic groundwater moves through joints and fractures (often in separations between bedding planes) in the bedrock. These fractures are solutionally enlarged by groundwater initially moving very slowly through the rock. In a given time, those fissures conducting water will undergo the most solution (Ford 1965). Once the openings are sufficiently large such that turbulent flow results (5 mm diameter; White and Longyear 1962), the solution is greatly enhanced (the rate of solution is proportional to the velocity of the solvent; Kaye 1957). The velocity within the system of joints and bedding planes can result from steep gradients within the system or from hydrostatic pressures.

Numerous theories have been proposed to explain the development of solution caves. It is generally recognized that some caves arose through solution and abrasion by groundwater or surface streams flowing in the vadose zone, and that others were formed as a result of solution by phreatic groundwater flowing under artesian pressure (Howard 1963). Numerous caves demonstrate readily that both processes acted at different times during their development.

Until the 20th century, caves were generally thought to be formed by underground streams. In particular, it was accepted that caves were developed above the water table by diverted surface waters, that circulation and solution below the water were insignificant, and that corrasion was insignificant in cavern enlargement. The Austrian geologist, Grund (1903), and the American geomorphologist, Davis (1930), first presented arguments against these assumptions. For a historical review and current views on cavern development, the reader is referred to Davis 1930 (two-cycle theory), Swinnerton 1932 (water-table theory), Gardner 1935 (one-cycle theory), Malott 1937 (invasion theory), and Bretz 1942, 1953 (three-stage theory).

Davies (1950) studied caves that were formed in areas of folded rock. The development of horizontal passages across inclined strata and parallel to the strike (e.g., Appalachian caves) were said to be related to river terraces and other surface erosional features. The horizontal development of passages in folded rock were said to be a result of maximum solution in a small vertical zone directly beneath the water table, this occurring during periods when the water table was relatively stable (White 1960, Deike 1960, Wolfe 1964).

It becomes clear that each cave area, and certainly each cave, presents an individuality for which speleogenetic theories do not always supply answers. Bogli (1980) provided a good summary schematic, speleogenetic organization for cave development.

No discussion of limestone caves would be complete without mentioning speleothems ("secondary mineral deposits formed in caves"; Moore 1952). A secondary mineral is formed by a chemical reaction from a primary mineral (e.g., calcite) in bedrock, detritus, or vein material. The number of minerals commonly occurring in caves is small, but because of a variety of depositional mechanisms (dripping water, evaporation, flowing water, pools of standing water, microorganism activities, etc.), there exists a large number of characteristic shapes of which the following are examples: stalactites, stalagmites, helictites, rimstone, flowstone, and moonmilk (see Fig. 8). Distinct from speleothem is speleogen, a surface feature of the bedrock usually formed by solution (Lange 1953). Examples of speleogens are scallops, rills, flutes, solution pockets, anastomoses, ceiling domes, and kettles. For further discussions of speleothems, speleogens, cave minerals, etc., see White (1976), Hill (1976), and Hill and Forti (1986).

Cave Environment

Generally, the physical environment of caves varies far less than that of the surrounding surface. Although many investigators have emphasized the relative constancy of the cave environment, close examination reveals that physical conditions do vary both temporally and spatially within a single cave and between caves. Relative to epigean (surface) conditions, however, there is considerably less variation in hypogean (below surface) environments (see Vandel 1965).

Caves frequently are divisible into a series of zones, each of which exhibits distinct sets of biological, chemical, and physical characteristics: the *threshold* zone, which can be subdivided into the entrance and twilight regions, and the *dark* zone in which variable-temperature and constant-temperature regions are recognized. The threshold zone (sometimes called the "light zone"; Jefferson 1976) extends from the surface opening of the cave to the farthest place to which daylight can penetrate (see Fig. 2). It is an area where physical factors are relatively variable. The entrance environment (an ecotone) is controlled by prevailing local climatic and meterological conditions and exhibits characteristics of both epigean systems and the cave interior (see Culver and Poulson 1970). Light intensity diminishes rapidly into the twilight zone (Fig. 9), varying with time of day, season, configuration of the entrance, and external conditions. Other parameters, such as temperature and humidity, are also variable. Even though these fluctuations occur, the threshold zone is considerably less variable than the exterior environment. The biota of these regions is the most diverse of any area of the cave (see Poulson and White 1969).

The environmental conditions of the variable-temperature dark zone are much more constant than those of the preceding zones, darkness being the most important feature ecologically. Temperatures fluctuate (see Cropley 1965), species diversity decreases, and lower biomass reflects both the scarcity of organisms and their small size.

The constant-temperature dark zone (Fig. 10) is characterized by relatively stable temperatures (air temperatures rarely vary more than 1 °C throughout the year at any one place), which approximate the mean annual surface temperatures for the area, being largely dependent on latitude and altitude. Water temperatures are subject to change as a result of ice and snow melt and can be altered as flood waters flow through underground aquifers; fluc-

FIGURE 8. View of stalactites and flowstone from bottom of 28-m drop in Hooper's Well, Madison County, Alabama.

FIGURE 9. View of the "twilight zone" in Laurel Creek Cave, Monroe County, West Virginia.

FIGURE 10. Dark zone of stream passage in Jewel Cave, Greenbrier County, West Virginia.

tuations of 8 °C or more may occur. The air is not stagnant; most caves ventilate continuously. However, in areas of caves isolated from moving water, elevated levels of CO_2 are occasionally found associated with clay beds, at bases of shafts, in passages where there is little air movement, and where there are deposits of organic debris.

It should be noted that the horizontal zonation described herein also can be applied to the vertical scale, but the zones are not so well defined. Specific zonation, particularly within the threshold area, occurs as one passes down from the pit entrance (Fig. 11) into the cave below. Even submerged caves (e.g., Gopher Sink, Leon County, FL; Fig. 12) demonstrate light and, thus, biotic zonation. Senger (1980) discusses the relationships between cave morphology and cave climate.

Waters in limestone/dolomite caves show similar patterns of relative stability, yet their characteristics are more variable and change more rapidly than the cave atmosphere. The hypogean waters of pools can be quite distinct

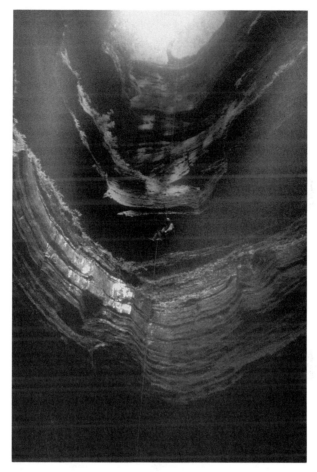

FIGURE 11. View from the bottom of 55-m-deep Neversink Pit, Jackson County, Alabama.

from those of underground streams, and in those in which water is supplied directly by the sinking of surface streams, both the chemical and physical properties will be affected to some extent by surface conditions. Such streams generally contain higher concentrations of organic matter (allochthonous, coarse particulate organic matter (CPOM), fine particulate organic matter (FPOM), and dissolved organic matter (DOM)), most of which is transported in from the surface, both as living organisms and as detritus, which can provide a source of nutrients for cave dwellers. Most cave waters tend to be alkaline (pH ranging from 7 to 8+) and have a high alkaline hardness. The biochemical oxygen demand is generally low, and, even in static pools, oxygen is at or near saturation values. Specific conductance (μmhos/cm) varies with temperature change, but generally ranges from 50 to 600 μmhos/cm at 25 °C. High values of phosphorous (50 μg PO_4-P/L) and nitrate (100 μg NO_3-N/L)

FIGURE 12. Gopher Sink (submerged cave), Leon County, Florida.

usually indicate contamination by sewage or excessive use of fertilizers. Expected values of solutes and additional physicochemical data in unpolluted waters in limestone areas may be found in Bray (1969), Jacobson and Langmuir (1970), Langmuir (1971), Shuster and White (1971), Barr and Kuehne (1971), Jones (1973), and Vineyard and Feder (1974).

PLANT COMMUNITIES

Because of the severe light restrictions characteristic of cave environments, plants occupying caves are generally representative of local epigean species that demonstrate some degree of shade tolerance, many of which have developed remarkable characteristics of their own (Tomaselli 1951, p. 67). Transects of plant species taken horizontally or vertically from the entrance into the cave interior indicate that the more advanced species (tracheophytes) occupy the entrance areas but quickly disappear as the light diminishes. Virtually no higher plants are able to survive in the permanently dark zone of caves. Bryophytes are found farther into the threshold zone, but the thallophytes are the most resistant to decreasing light and thus penetrate farther into the cave than other green plants (Tomaselli 1947, Dalby 1966). Refer to Kofoid (1900), Scott (1909), Barr (1967b, pp. 178–184), and Leitheruser et al. (1985) for additional information concerning algae in caves.

Light reduction has been shown to initiate structural changes in tissues of green plants (Dalby 1966). The effects of pH, humidity, substrate, and other physical factors that create spatial heterogeneity even at the microclimate level have been discussed by Mason-Williams and Benson-Evans (1958). Seeds of angiosperms are often washed into caves, and etiolated seedlings are commonly encountered, and appear long, thin, and pale. These life forms contribute a source of energy to the intricate food webs within cave ecosystems.

Both heterotrophic and autotrophic bacteria are found throughout caves, although not usually together in the same microbial community. Numerous forms of microbes are carried into the hypogean environment via air circulation, by animals (including man), and by seepage and streams.

Cave environments have numerous forms of chemosynthetic microbes, which utilize a variety of energy sources. Autotrophic forms are represented by iron bacteria, which derive their energy from the simple oxidation of iron compounds. These bacteria, like the majority of bacteria currently known from caves, are not confined to grottoes, but are also found in many surface soils. A possible exception is the iron bacterium *Perabacterium spelei*, which may prove to be a true cave species (Caumartin 1959). Brock et al. (1973) noted that obligate psychrophilic bacteria have not evolved in or colonized the constantly cool waters of caves. Sulfur bacteria, nitrifying bacteria, and gram-negative microbes, such as *Azotobacter* sp. and *Clostridium* sp., are significant components of cave microbial communities (see Cubbon 1969, Mason-Williams and Benson-Evans 1958, Gounot 1967). Heterotrophic bacteria degrade complex organic materials and liberate simpler substances that have potential food value for other organisms (see Fliermans and Schmidt 1977). Organic debris may be imported by flowing water (see Paul et al. 1983) or by visiting animals, and waste materials deposited by cavernicoles serve as nutrient reservoirs (see Lavoie 1980a). Some pathogenic microorganisms are known to inhabit caves. Holsinger (1966), Wells (1973), Wagner et al. (1976), and Brucker (1979) reported sewage pollution in various cave systems. The occurrence of coliform bacteria is becoming more common in cave systems. This is apparent when one compares the results of Gardiner's (1971) and Hoey's (1976) investigations of groundwaters in the environs of Bloomington, Indiana, which are contaminated with fecal coliforms and fecal streptococci, the densities of which have shown marked increases in the five-year time span that separated their studies. The reader is referred to Prager (1972) for an overview of groundwater pollution in U.S. karst regions, and Minear and Patterson (1973) for a discussion of groundwater contamination resulting from septic tank system failure. In addition to treating pollution, Wilson (1977) briefly demonstrated the effects of cultural eutrophication and certain caver activities on cave ecosystems. A summary of the literature concerning microscopic organisms in cave environments may be found in Caumartin (1963), Barr (1967a), Gittleson and Hoover (1969, 1970), Dickson (1975), and Dickson and Kirk (1976). Table 5 lists representative studies of species found in southeastern caves.

Intermediate in character between bacteria and fungi are the actinomycete microflora (mold-like filamentous bacteria) inhabiting caves (Lovett 1949). These are ubiquitous soil organisms about which little is known except their potential role related to antibiotics. Caumartin (1963) and Picknett (1967) suggested that the peculiar and distinctive "earthy" odor of caves is produced in part by cave actinomycetes.

The occurrence and dynamics of fungi in caves have been studied by Tomaselli (1953), Hazelton and Glennie (1962), Caumartin (1963), Mason-Williams (1965), Cubbon (1969), and Hunter and Thomas (1975). The majority of fungi found in caves are of epigean origin; however, Tomaselli (1956) described a number of forms that may be highly specialized cavernicoles. Upon entering caves, fungi must find a suitable substrate (organic material, living or dead) to survive. The debris and animal and plant life that are brought into caves undoubtedly also have a host of microfungal flora associated with them. These fungi will continue to grow as long as a substrate exists. They may completely utilize that substrate or become established members of the cave ecosystem. The ultimate outcome is dependent on the types and counts of nutrients, as well as on their tolerance to the physical and chemical conditions within the cave. Regardless, in time fungi make available (direct or indirect) food sources (such as nutrients) for some organisms already present in the cave community. Dickson (1975) indicated that bacterial and fungal populations not only may serve as basic food sources in caves (see also Dickson and Kirk 1976), but also may influence the distribution of specialized invertebrate cavernicoles (organisms occurring in caves). Both autotrophic and heterotrophic (see Kirk 1973) microorganisms are virtually all over caves (e.g., mud, water, dung, living and dead organisms, and even speleothems).

Because light is a major limiting factor, the plant communities of caves are important only in entrance areas (excluding plant debris washed into caves) and are dominated by lower forms, such as mosses, liverworts, and some algae, that possess limited photosynthetic ability. Few studies have been made of plant communities in springs and spring runs (see Odum 1956, 1957a,b, Whitford 1956, Minckley 1963); here, also, light is the primary limiting factor.

ANIMAL COMMUNITIES

Cavernicoles are represented by a wide range of systematic groups, and are customarily classified according to the proportion of their life spent in caves and the degree of their specialized adaptations. Some species are found only

TABLE 5   Studies of Selected Cave, Spring, and Seep Biota

Taxon	Reference
NORTH AMERICA	
Cave microbes	Caumartin 1963
Pitcher plant communities	Folkerts 1982
Cave protozoa	Gittleson and Hoover 1970
Troglobites	Nicholas 1960
Spring, cave planaria	Kenk 1972
Cave planaria	Carpenter 1982, Kenk 1984
Cave collembola	Christiansen 1966, 1982; Christiansen and Bellinger 1980
Crustacea of springs, spring-runs, caves of the SE	Holsinger and Culver 1970
Cave isopods	Steeves 1963, 1966, Fleming 1973
Seep, spring, well, cave amphipods	Holsinger 1967, 1972, 1977, 1978
Seep, cave crayfishes	Hobbs 1974
Troglobitic crayfishes	Hobbs et al. 1977
Cave crayfishes and branchiobdellids	Holt 1973
Cave vertebrates	Dearolf 1956
Amblyopsid fishes	Poulson 1963, Woods and Inger 1958, Cooper 1978
Cave salamanders	Brandon 1966, 1971, Cooper 1968, Peck 1974, Cooper 1978
ALABAMA	
Shelta Cave biology	Cooper 1975
Cave crayfishes and branciobdellids	Holt 1973
Phreatobitic isopod	Modlin 1986
Cave isopods	Steeves 1963
Cave amphipods	Holsinger 1978
Cave and spring fishes	Howell and Caldwell 1965, Williams 1968, Armstrong and Williams 1971, Cooper and Keuhne 1974, Cooper 1985
Cave salamanders	Cooper and Cooper 1968, Lee 1969b, Peck and Richardson 1976
Cave frogs	Lee 1969b
Spring community	U.S. Fish and Wildlife Service 1984
ARKANSAS	
Invertebrates	McDaniel and Smith 1976
Cave invertebrates	Youngsteadt and Youngsteadt 1978
Invertebrates and vertebrates	McDaniel et al. 1979
Cave, seep, and spring fauna	Robison and Smith 1982
Cave and spring planaria	Kenk 1973
Cave snail	Hubricht 1979
Cave collembola	Smith 1977
Cave, cistern isopods	Steeves 1966, Smith 1977, Schram 1980, Lewis 1981

TABLE 5 (*Continued*)

Taxon	Reference
Cave amphipod	Smith 1977, Holsinger 1978
Troglobitic crayfish	Smith 1984, Hobbs 1987
Vertebrates	McDaniel and Gardner 1977
Cave fish	Willis and Brown 1985, Willis 1986
Cave salamander	Grove 1974, Smith 1977
FLORIDA	
Seepage community	Hobbs 1938, 1945
Spring community	Odum 1957a, b
Spring, spring-run algae	Whitford 1956
Cave, spring troglobitic fauna	Warren 1961, Franz 1982
Spring insects	Sloan 1956
Cave isopods	Bowman and Sket 1985
Spring-run shrimp	Beck 1979
Troglobitic crayfish and branchiobdellids	Holt 1973
Troglobitic crayfish	Relyea and Sutton 1973b, Relyea et al. 1976, Franz, and Lee 1982, Franz Hobbs 1983
Cave, spring, spring-run fishes	Hubbs and Allen 1943, Marshall 1947, Relyea and Sutton 1972, 1973a
Troglobitic salamander	Pylka and Warren 1958, Lee 1969a
GEORGIA	
Invertebrate cave fauna	Holsinger and Peck 1971
Seep, spring crayfishes	Hobbs 1981
Cave crayfishes and branciobdellids	Holt 1973
Cave amphipods	Holsinger 1978
KENTUCKY	
Spring-run ecology	Minckley 1963
Cave biota	Barr 1967a, 1985, Barr and Kuehne 1971
Cave protozoa	Gittleson and Hoover 1963
Cave planaria	Carpenter 1970
Cave fish cestode	Whittaker and Hill 1968
Cave isopods	Steeves 1963
Doe Run Isopods	Walker 1961
Spring-run amphipods	Minckley and Cole 1963
Cave shrimp	Liskowski 1983, Leitheuser et al. 1985
Spring-run crayfish	Prins 1968
Cave crayfish and branciobdellids	Holt 1973
Cave sculpin	Bryant et al. 1970
LOUISIANA	
Seep isopods	Hubricht and Mackin 1949
MISSISSIPPI	
Cave vertebrates	Cliburn and Middleton 1983

TABLE 5 (*Continued*)

Taxon	Reference
NORTH CAROLINA	
Cave planaria	Kenk 1979, 1984
Cave isopod	Lewis and Bowman 1977
TENNESSEE	
Cave biota	Cope and Packard 1881, Barr 1961
Groundwater bacteria	Brown and Broughton 1981
Cave cyclopoids	Yeatman 1964
Cave isopods	Steeves 1966, Bosnak and Morgan 1981a
Cave amphipods	Holsinger 1978
Cave crayfish and branchiobdellids	Holt 1973
Cave crayfish	Matthews et al. 1975
Spring and cave fishes	Armstrong and Williams 1971, Williams and Etnier 1982
Cave salamander	Peck and Richardson 1976
VIRGINIA	
Cave fauna	Holsinger 1963, 1982
Cave microbes	Holsinger 1966
Cave microorganisms	Dickson 1975, Dickson and Kirk 1976
Groundwater planaria	Kenk 1984
Cave collembola	Wray 1952
Spring, cave snails	Hubricht 1976
Cave, spring, groundwater isopods	Holsinger and Bowman 1973, Estes and Holsinger 1978, 1982, Lewis 1980, Culver 1981a, b, Lewis and Holsinger 1985
Cave amphipods	Culver 1981a, b, Holsinger 1978, Dickson and Holsinger 1981
Cave crayfishes and branchiobdellids	Holt 1973
Cave salamander	Culver 1973b
WEST VIRGINIA	
Cave fauna	Rutherford and Handley 1976
Cave invertebrates	Holsinger et al. 1976
Cave snails	Hubricht 1976
Cave crustaceans	Culver et al. 1973
Cave isopods	Steeves 1963, Culver and Ehlinger 1980, Culver 1970a, 1971a, 1981a
Cave amphipods	Culver 1970a, 1971a, 1981a, 1987, Culver and Poulson 1971, Holsinger 1978
Cave crayfishes and branchiobdellids	Holt 1973
Cave crayfish	Van Luik 1981
Cave sculpin	Williams and Howell 1979
Troglobitic salamander	Besharse and Holsinger 1977

in underground habitats, while others also frequent other environments. The geographical distribution of aquatic cavernicoles is generally much wider than that of terrestrial forms, principally because their dispersal potential is much greater. The unique characteristics of the cave environment include a relative constancy of chemical and physical conditions and limited food. In general, troglobitic (obligate, see below) species exhibit K-selected population characteristics (MacArthur and Wilson 1967), including small population size, late maturity, low reproductive rates, large size at hatching, and increased longevity. The analogy made between caves and islands is reviewed by Culver (1970b, 1971b) and Crawford (1981).

A classification system for cave-inhabiting animals (= cavernicoles, stygobionts) has been proposed numerous times (see review in Hamilton-Smith 1971). The most commonly used system places animals into one of four ecological/evolutionary categories (Barr 1963, 1968):

1. *Troglobite* (see Holsinger 1988). Obligatory cave species that are morphologically specialized (troglomorphy) for, and restricted to, the cave habitat; they are unable to exist in epigean habitats.

2. *Troglophiles*. Facultative cave species that frequently inhabit caves and are capable of completing their entire life histories there, but may occupy ecologically similar habitats outside of the cave environment.

3. *Trogloxenes*. Species often occurring in caves, but incapable of completing their entire life history in the cave environment, generally having to exit for feeding and or mating purposes.

4. *Accidentals*. Species that accidently wander, wash, or fall into caves and can exist there only temporarily; they may serve as food sources for regular cavernicoles, yet they are of no importance in distributional or evolutionary analyses of cave fauna.

Two other terms are also employed for certain cave (and surface) animals:

1. *Edaphobites*. Species that are obligatory deep-soil dwelling forms that may occasionally occur in caves.

2. *Phreatobites*. Species that are obligatory to groundwater habitats, and are often found in slowly moving interstitial groundwaters. They are not necessarily found in caves but are frequently found in seeps, springs, and wells.

The distribution and ecology of many cavernicoles are inadequately known, and their assignment to one of the ecological–evolutionary categories is often tenuous. Some groups are characteristically placed in specific categories, such as phreatobites: some copepods, isopods, amphipods; edaphobites: earthworms; trogloxenes: bats, bears, raccoons, moths, mosquitoes; troglophiles: some salamanders, beetles, crustaceans; troglobites: blind cave fishes, some flatworms, isopods, amphipods, decapods, pseudoscorpions, spiders, millipedes, and a large number of insects. For a more detailed treatment of cavernicoles, see Vandel (1965), Barr (1967a), Jefferson (1976), and Botosaneanu (1986).

Of specific interest is the troglobitic group of cavernicoles. These organisms are highly specialized and show varying degrees of adaptation for existing in the cave environment (see Holsinger (1988) for a discussion of the evolution of troglobites). For example, there are nearly 60 species of troglobitic decapods known from the Americas (Hobbs et al. 1977; see also Hobbs 1989), and many of the adaptations common to all troglobites are recognized in this group. The most obvious characteristic that is common to virtually all troglobites is a strong reduction in or total absence of pigmentation (Fig. 13). Also conspicuous are the reduced eyes and, in the case of many troglobitic decapods, a lack of both pigment and faceted corneae. Attenuated appendages are characteristic of many forms, and numerous troglobites tend to be smaller, or at least superficially more delicately constructed, than their epigean relatives. Although few data are available, a lower basic metabolic rate ("metabolic economy," Poulson 1963, 1964) is suggested for many troglobites. See Schlagel and Breder (1947), Burbank et al. (1948), Troiani (1954), and Culver and Poulson (1971) for comparative results of oxygen consumption in cave and surface biota. Cooper and Cooper (1976) and Cooper and Cooper (1978) presented data that indicate that certain troglobitic crayfishes have considerably longer life spans than have been attributed to any other cave species (see Culver 1982, p. 52). Ginet (1960, 1969) discussed longevity in the amphipod *Niphargus*. These examples suggest a correlation with lowered metabolism. Production of fewer eggs in troglobitic decapods when compared

FIGURE 13. *Procambarus (O.) l. lucifugus* from the type locality, Sweet Gum Cave (= Gum Cave, Bat Cave), Citrus County, Florida. (Photo by Barry Mansell.)

with related epigean species (see Bechler 1981) is usually associated with lesser available energy in the cave environment. Also, Hobbs (1973) suggests that individual females within a given population do not necessarily reproduce annually but resorb oocytes and reproduce only on a staggered basis. Thus, in addition to obvious morphological adaptations (troglomorphy), many troglobites demonstrate low reproductive rates, extended life expectancies, and extreme resource efficiency.

Animal species richness in caves and dispersal potential, among other factors, are greatly influenced by the geological formations of the area in which caves are developed (e.g., continuity and separation of limestone units). "In the Appalachian Valley, where limestone is exposed in many narrow, anticlinal strike belts, species density per unit area is high, and dispersal of troglobites through subterranean channels is severely restricted by geologic structure. In the Mississippi Plateaus, where thick caverniferous limestone is widely and continuously exposed, there are fewer species per unit area; and subterranean dispersal has taken place over considerable distances" (Barr 1967a, p. 488).

Cave Community Energetics

Generally, cave communities are regarded as relatively simple systems having few species and low productivity (Barr 1968, Poulson and White 1969, Culver 1976). Notable exceptions to this are the approximately 200 species of animals (many not troglobites) and plants in the Mammoth Cave system in Kentucky (Barr 1967b), the rich aquatic fauna of the cave communities of the Edwards Aquifer (22 troglobites in the artesian well in San Marcos, Texas; Holsinger and Longley 1980), and the diverse aquatic community of Shelta Cave, Madison County, Alabama (Cooper 1975).

Because cave environments essentially lack an autotrophic component, all cave communities must depend on exogenous organic material to be transferred from the surface. Every cave or cave system will demonstrate variances in energy input–output dynamics over time and space. Hawes (1939) indicated the importance of the flood factor in the ecology of caves, particularly with reference to food input. In addition, he, and others, pointed out that floods may operate as agents of distribution and colonization, stimulate reproductive activity, and trigger molting cycles in certain organisms (see Jegla 1966, 1969, Jegla and Poulson 1970).

Poulson (1978, p. 94) has proposed that "Energy availability depends on rigor, variability, and predictability of energy concentration, renewal, and quality." Obviously, all organic materials do not have equal caloric values (see references in Paul et al. 1983); therefore, energy availability of certain foods is greater than that of others. For example, raccoon feces are "high payoff" foods and leaf litter is a "low payoff" food. Undoubtedly energy availability can greatly affect the numbers and biomass that can be supported, foraging behaviors, overall energetics, life histories, and, ultimately, community organization. Additional information concerning effect of energy availability can be obtained in Poulson (1979), Franz and Lee (1982), and Culver (1985).

Most caves receive potential food in the form of DOM, organic litter and detritus, bacteria, protozoans, and other organisms that are washed, blown, or carried in. Temporal variations in quality and quantity are evident, with greatest inputs occurring in late winter and spring. Microflora are responsible for decomposition and transformation of this allochthonous material, yet they themselves are sources of energy when detritus is consumed (the "peanut butter" on the detrital "cracker"!). These food sources support both the terrestrial and aquatic communities (see Kostalos and Seymour 1976; Culver 1985).

Guano is another major food source of many cave organisms (cricket and bat guano in particular) (Park and Barr 1961 and Reichle et al. 1965). Guano piles also function as distinct and complex ecosystems. Spatial variation in guano piles leads to sharp microzonation; thus, there is spatial as well as temporal variability within the ecosystem. The increase in food input (food pulse) that is initiated at each year when bats return to a cave greatly affects energy availability and also the community of guanobites. For further information concerning bat guano ecosystems see Harris (1970), Mitchell (1970), Richards (1971), Peck (1971), Horst (1972), Poulson (1972), Fletcher (1976), Martin (1977), Franklin (1978), and Hill (1981).

In addition to cricket and bat guano, dung of larger vertebrates (pack rats and raccoons) is another source of energy to the cave system. The heterotrophic decomposition of this material is successional and somewhat predictable, as is discussed by Lavoie (1981a). For further information on dung ecosystems see Lavoie (1980a, 1981b).

Representative Animal Communities

There are several groups of cavernicoles that typically occur in aquatic cave environments of the southeastern United States and these are represented by a large number of species, many of which are troglobites (see Botosaneanu 1986).

Planaria and Collembola   Cave planarians are not restricted to quiet water habitats, but seem to be found in a variety of ecological situations: springs, rimstone pools, epi- and hyponeuston, seeps within caves, hyporheic zone (interstitial habitats), and silt-mud substrates. For a summary of the biology of turbellarians inhabiting caves the reader is referred to Hyman (1939, 1951), Kenk (1970, 1972, 1973), and Carpenter (1982). Collembola are traditionally considered to be terrestrial or semiaquatic apterygous insects, but often they are found on the surface of pools in caves (e.g., *Arrhopalites* spp.) and thus are mentioned as part of the aquatic fauna. Good summaries of the distribution and ecology of cave collembola may be found in Christiansen (1966, 1982), Christiansen and Culver (1968), and Christiansen and Bellinger (1980).

Mollusca (Gastropoda)   Numerous species of snails are found in caves of the southeast, but detailed taxonomic and ecological studies on this important group are sparse. Although few troglobitic species are known at present, undoubtedly many more will be revealed with future research. Barr (1967a) mentioned *Antroselates spiralis* from the Mammoth Cave region in Kentucky where it is rare and found only under large stones in shallow riffles. Hubricht (1976) described and discussed troglobitic species of the genus *Fontigens* from Appalachian cave streams and/or springs and also described (1979) *Amnicola cora* from a cave stream in northeastern Arkansas.

Isopoda and Amphipoda   Isopods (Fig. 14) are among the richest and most studied of North American freshwater crustacea. Many of the studies have been taxonomic and/or distributional in nature (Hubricht and Mackin 1949; see references in Lewis 1980), and only in recent years has the ecology of cave/spring species been examined (see Smith 1977, Schram 1980, Culver and Ehlinger 1980, 1982, Culver 1981b, Estes and Holsinger 1982). Amphipods are common inhabitants of springs, small spring-fed streams, and caves of the southeast (Fig. 15). These crustaceans are one of the largest and most prominent groups of freshwater invertebrates inhabiting interstices of stream gravel, mud-bottomed pools, drip pools, and rimstone pools (Bousfield 1958, Holsinger 1967, 1972, 1975a, 1978). Holsinger and Culver (1970) noted differences in the morphology of *Gammarus minus* Say from populations in the central Appalachians and concluded that it is a variable species capable of

FIGURE 14. *Caecidotea stygia* from Mammoth Cave, Edmonson County, Kentucky.

FIGURE 15. *Crangonyx floridanus* from China Cave, Jackson County, Florida. (Photo by Barry Mansell.)

occupying a variety of habitats, ranging from surface springs to small or large cave systems. Dickson (1977) also noted morphological variation among populations of the troglobitic amphipod *Crangonyx antennatus* Packard and indicated that the type and amount of food are the most important environmental parameters affecting form. Culver (1971b) showed that the drop in spring abundance of populations of *Stygonectes* (= *Stygobromus) spinatus* and *Gammarus minus* in West Virginia caves was due to washout (density-independent for *S. spinatus* and primarily density-dependent for *G. minus*). Other studies concerning spring-fed streams and subterranean amphipod ecology are Minckley and Cole (1963), Culver (1970a, 1971b, 1973a, 1976, 1981a), Culver and Poulson (1971), Holsinger and Dickson (1978), Dickson and Holsinger (1981), and Bechler and Fernandez (1981).

Decapoda Only six of the 32 recognized troglobitic decapods of North America are found outside of the area treated in this volume. Since the work of Hobbs et al. (1977), three additional species of troglobitic decapods, *Procambarus (Ortmannicus) franzi* Hobbs and Lee (1976), *P. (O.) leitheuseri* Franz and Hobbs (1983), and *P. (O.) delicatus* Hobbs and Franz (1986) have been described from the southeastern United States (a total of three troglobitic shrimps and 23 crayfishes; see Fig. 16). Hobbs et al. reviewed the biology of American troglobitic crayfishes and shrimps (see for indepth treatment of species from the Southeast).

FIGURE 16. Undescribed species of troglobitic crayfish from northern Florida. (Photo by Barry Mansell.)

The high species richness demonstrated by Florida's troglobitic crayfish fauna is unique. An explanation was suggested by Relyea and Sutton (1973b) and Releyea et al. (1976) and was more fully investigated by Franz and Lee (1982). Franz and Lee (1982, p. 76) proposed multiple invasions of Florida's aquifers by epigean crayfishes and summarized that "once underground, certain crayfishes, possibly resulting from sympatry, engaged in occupying differing parts of the inhabitable cave environment. Some species (e.g., members of the *P. lucifugus* complex) inhabit the energy rich areas (i.e., in large sinks and under bat roosts), while others are not so dependent on these accumulations. The low energy species (e.g., members of the *P. pallidus* complex and *Troglocambarus maclanei*) have greater dispersal abilities, and today occupy the greatest geographic ranges." Franz (1982) discusses rare and endangered invertebrates from Florida, and several crayfishes and a shrimp are included.

Detailed, long-term population studies have been conducted only in Shelta Cave, Madison County, Alabama (Cooper and Cooper 1971, Cooper 1975) and in West Virginia caves (Van Luik 1981). Other lengthy studies of troglobitic crayfishes have been conducted in areas located outside of the geographical boundaries delineated in this book (see Jegla 1964, Hobbs 1973, Weingartner 1977).

Included with this discussion of cave crayfishes is a brief summary of branchiobdellid annelids found as epizoites on isopods and crayfishes in caves of the Southeast. Holt (1973) points out that the majority of the records of these worms are of representatives of common epigean forms epizootic on cray-fishes. The troglobitic worms appear to be associated with troglobitic isopods (e.g., *Cambarincola marthae*) and possibly a few troglobitic crayfishes from Florida, Kentucky, and Tennessee (e.g., *Cambarincola leoni*).

The troglobitic shrimp *Palaemonias ganteri* Hay is known only from base level cave passages in the Mammoth Cave System in Kentucky. Between 1967 and 1979 this shrimp was not observed, despite searches in historic localities. Efforts since 1979 have led to the alarming conclusion that, although the species is still present, it "is absent from prime habitat where it used to be common, and rare in other habitats" (Liskowski 1983, p. 91); see also Lei-theuser et al. (1985). Efforts by the National Speleological Society to deter-mine the status of the endangered shrimp *P. alabamae* in Shelta Cave were initiated in Spring 1986; none have been observed since November 1973 (see Hobbs and Bagley 1989).

Fishes Numerous species of fishes occur in caves throughout the world, however, only six troglobites are known from the United States and only four are found in the southeastern United States, all assigned to the Family Am-blyopsidae (see Fig. 17): *Amblyopsis spelaea, A. rosae, Typhlichtyhys sub-terraneus* (Fig. 18), and *Speoplatyrhinus poulsoni. Chologaster agassizi* is also an amblyopsid (troglophile) and is found mostly in sinking streams, springs, and occasionally wells (see Cooper 1985b). Poulson (1963), the only long-term population study of troglobitic fishes, summarizes the basic ecology/biology of these species, except for *S. poulsoni*. Available data for this blind fish are presented by Cooper and Kuehne (1974).

Of the five species adapted to cave life, *S. poulsoni* shows the strongest degree of regressive evolution (increased pigment loss and maximum eye degeneration), followed in decreasing order by *A. rosae, A. spelaea, T. sub-terraneus*, and *C. agassizi*. Poulson (1963) discussed the ecology of each and noted (1967, 1968) that in areas of Mammoth Cave where species occurrence overlaps, *A. spelaea* is restricted to the deep, food-poor rivers at or below base level, whereas *T. subterraneus* is found in the shallower, rocky, food-rich streams that drain vertical shafts above base level. Near entrances where food is abundant, *C. agassazi* may replace both species such as in the Echo and Styx rivers and Cedar Sink. Bechler (1981) demonstrated that the diversity of agonistic behavior in amblyopsid fishes decreases with increasing cave adaptation and that this is paralleled by a concommitant improvement in the efficiency of energy utilization.

A. rosae is now federally listed as threatened and its continued existence may depend on protection of critical habitats (such as Cave Spring Cave, Benton County, Arkansas, recently purchased by The Nature Conservancy and to be transferred to the Arkansas Natural Heritage Commission for management). Populations of this species are comparatively large in caves inhabited by the federally listed endangered bat, *Myotis grisescens* (guano enrichment of streams). Cave Springs Cave currently houses a maternal colony of this bat and also the largest known population of *A. rosae* (approximately

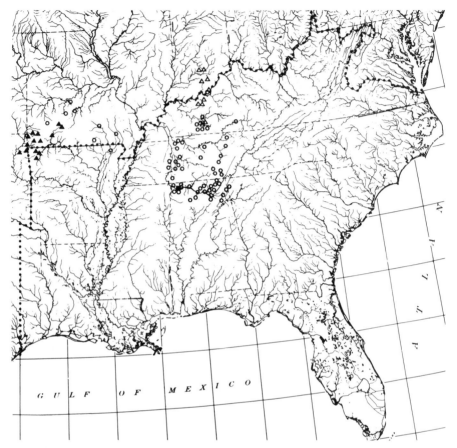

FIGURE 17. Distribution of the troglobitic amblyopsid fishes: △, *Amblyopsis spelaea*; ▲, *A. rosae*; ○, *Typhlichthys subterraneus*; ■, *Speoplatyrhinus poulsoni*. (In part, from Cooper 1985 and Willis 1986.)

300 individuals or about 60% of the estimated total number of members of this species) (1984 *Endangered Species Technical Bulletin*, 9(10):5; see also Willis and Brown 1985).

Additional fish species are found in caves and springs of the southeast that are either moderately modified for spelean existence or are epigean forms that occasionally enter subterranean streams. References to these fishes are presented in Table 5.

Amphibia   There are 11 described species and subspecies of troglobitic salamanders known in the world and all except one, *Proteus anguinus* from karst regions in Yugoslavia, occur in the United States. The following are reported from the southeastern states: *Typhlotriton spelaeus* (Arkansas), *Gyrinophilus palleucus* (3 subspecies in Alabama, Georgia, and Tennessee), *Gyrinophilus*

FIGURE 18. *Typhlichthys subterraneus* from Mammoth Cave, Edmonson County, Kentucky.

subterraneus (Greenbrier County, West Virginia), and *Haidetriton wallacei* (Georgia, Florida; Fig. 19). Summaries of the life histories, ecology, and distribution for these species are found in Brandon (1966, 1971), Cooper (1968), Cooper and Cooper (1968), Lee (1969a), Peck (1973), and Besharse and Holsinger (1977). A useful key to the salamander larvae and larviform adults of the United States is presented by Altig and Ireland (1984).

Trogloxenic, troglophilic, as well as surface salamanders are often encountered in caves (see Dearolf 1956, Cliburn and Middleton 1983). The feeding habits of a number of salamanders occupying southeastern caves are reported by Culver (1973b) (see his references), Peck (1974), and Peck and Richardson (1976). Dearolf (1956, p. 206) cites six species of frogs/toads inhabiting caves in the southeast, and Lee (1969b) noted the ability of *Rana catesbeiana* to feed and survive in subterranean environments.

REPRESENTATIVE AQUATIC COMMUNITIES

In a simplistic manner, both caves and springs can be reduced to "constant temperature systems" (see Sloan 1956, Teal 1957, Odum 1957a,b, Barr 1967a,b, Tilly 1968). Other potentially controlling variables are minimized or dampened by these relatively simple environments (see above). Presented below are brief descriptions of specific cave and spring systems that are representative of these types of environments in the southeastern United States.

FIGURE 19. *Haidetriton wallacei* from China Cave, Jackson County, Florida. (Photo by Barry Mansell.)

Mammoth Cave

The Mammoth Cave system in Mammoth Cave National Park is the most extensive in the world. There are 300 miles (483 km) of surveyed passages. The cave is developed in the Girkin, Ste. Genevieve, and St. Louis limestones. The lowest level (in the St. Louis) drains into the Green River (local base level) via springs, most prominent of which are Turnhole Spring, Pike Spring, and the Echo and Styx River outlets. This immense subterranean environment supports one of the largest assemblages of aquatic and terrestrial organisms anywhere. The high species richness is not merely an artifact of extensive collecting, but is due to niche availability resulting from the high spatial heterogeneity. In addition, long time periods have been available for species to evolve within the system (e.g., *Palaemonias ganteri*) and to be invaded by subterranean species from other cave regions (e.g., blind fishes).

Energy enters the cave via three primary sources: (1) guano and dead bodies of the trogloxenic cave cricket (*Hadenoecus subterraneus*), which feeds outside the cave; (2) plant debris and meso- and nannoplankton brought in by streams; and (3) wood and artifacts of cave commercialization. Aquatic habitats range from upper level drip and rimstone pools inhabited, for example, by the triclad *Sphalloplana percoeca*, which fluctuate and periodically become dry, to perennially wet base level sections such as the Echo and Roaring rivers, River Styx, Mystic River, and Eyeless Fish Trail. These lower levels sometimes support large populations of the crayfish *O. pellucidus*, and the fishes *Typhlichthys subterraneus* and *Amblyopsis spelaea*. Crystal Lake (surface area 120 m², maximum depth 3.5 m) is above flood level and is fed by ceiling drip and wall flow. It supports low biomass—various plankton, *O. pellucidus*, *Caecidotea stygia* (isopod), and the amphipod *Stygobromus vitreus*. Streams in the intermediate cave levels are short, flow with moderate to slow velocities over bedrock and gravel substrates, and support small to moderately large populations of crayfish, isopods, and amphipods. The presence of many species in a variety of habitats within this immense system provides stimuli for study and statistical analyses of data gathered from communities under different physical and chemical conditions (yet lacking constraints of a variety of epigean variables). The reader is directed to Barr (1967a) and Barr and Kuehne (1971) for discussions of the aquatic fauna of Mammoth Cave and to White and White (1989) for an indepth treatment of karst hydrology in the Mammoth Cave area.

Shelta Cave

Shelta Cave, in northwest Huntsville, Madison County, Alabama, consists of three large, interconnected rooms. The floor is covered by an extensive lake, which is partially dry both spatially and temporally; lake levels vary as much as 7.5 m. The cave has a particularly rich assemblage of freshwater troglobites (11 species; Cooper 1975, p. 53). The triclad *Sphalloplana percoeca* and two species of *Stygobromus* (amphipod) are found in Shelta Cave. It is one of a few known caves housing three syntopic crayfish species, one of which remains undescribed (*Orconectes a. australis*, *Cambarus (Aviticambarus) jonesi*, and *Orconectes* sp). In addition, the crayfishes are host to the branchiobdellid *Cambarincola sheltensis* and the ostracod *Sagittocythere barri*. The endangered atyid shrimp *Palaemonias alabamae* Smalley inhabits the cave and two troglobitic predators are also part of the community: the rare neotenic salamander *Gyrinophilus palleucus* and the relatively abundant blind fish *T. subterraneus*. Cooper (1975) conducted an intensive ecological and behavioral study of the decapod assemblage from 1968 to 1969, with periodic sampling through late 1973. Hobbs and Bagley (1989) proposed a management plan for the cave and reviewed the unique biological community and the potential threats to its future existence.

Silver Springs

Silver Springs, like many other large karst springs in central Florida, is a giant steady-state laboratory. Odum (1957b), in a now classical study, investigated community metabolism and productivity of the spring and the spring run. He found that chemical, physical, and biological parameters as a whole are constant in quantity and species. Flow averages 600 million gallons per day. Annual variation is greatest with light and thus production in the springtime is two to three times as great as that of winter. Gross annual primary production averaged 6390 g/m², while daily production varied from approximately 10 g/m² (winter) to greater than 30 g/m² (summer).

The trophic structure of the spring is based on the macrophyte *Sagittaria lorata* and associated Aufwuchs (periphyton). Although *Sagittaria* accounts for about 70% of the total production biomass, 70% of the total production is from the Aufwuchs. Bacteria, like the Aufwuchs, are also characterized by small biomass (0.3–0.6 g/m^2), yet have a large metabolic demand (0.079 g oxygen/m^2 h^{-1}). Numerous organisms occupy various trophic levels and species lists and a trophic classification for this system is presented by Odum (1957b). Some of the prominent components of the community are represented by primary producers (*Sagittaria, Lobelia, Ceratophyllum, Vallisneria, Potamogeton, Ludwigia*), herbivores (*Viviparus, Oxytrema, Palaemonetes, Pseudemys, Mugil*), and carnivores (*Lepomis, Micropterus, Lepisosteus, Alligator*). The disappearance of the mullet (and catfishes) from the spring during the last 15 years may be attributed to the construction of the Rodman Dam on the Oklawaha River (see Knight 1983). Using the upstream–downstream method (see Odum 1956), Odum sampled a number of Florida springs and found that production values ranged from 1 to 59 g oxygen/m^2 day^{-1} (Odum 1956, 1957a). These production values are typically lower than those of many other aquatic ecosystems. See Hubbs and Allen (1943), Sloan (1956), Whitford (1956), and Rosenau et al. (1977) for additional studies concerning organism distribution, production, structure, and types (e.g., oligohaline, mesohaline) of springs in Florida.

Spring-fed Streams

Doe Run Groundwaters commonly emerge and flow down gradient as a surface stream. That section of stream below the emergence (rise, boil) is called a *run*. Doe Run represents a typical system of this type. This stream, located in eastern Meade County, Kentucky, issues from a collapsed cave in the St. Louis limestone and flows north-northeast (Fig. 20) for 15.6 km (9.7 miles) where it empties into the Ohio River. From February 1959 until May 1962, Minckley (1963) conducted a baseline study on this unpolluted, yet modified, spring run. In addition, Minckley and Cole (1963) and Prins (1968) carried out detailed studies of amphipods and crayfishes, respectively; both macroinvertebrates are significant components of the stream benthic community. This spring-fed stream has received considerable study (see references in Minckley 1963) and provides significant data concerning the gradient of community composition, distribution, trophic relationships, etc., for spring runs.

Florida Runs The runs of some of the large springs in Florida (e.g., Ichetucknee, Wacissa, Silver, Wakulla) are extremely rich biotically, yet harbor few endemics. Around the immediate area of the springs there is little vegetation, but immediately beyond the periphery of the "boil," *Ceratophyllum, Myriophyllum, Naias, Chara, Sagittaria, Isnardia*, algae, and mosses can be very abundant. Often the surface of the vegetation near the spring becomes covered with marl as CO_2 is released from the bicarbonate-rich waters. The

FIGURE 20. Heavy discharge from head of Doe Run, Meade County, Kentucky.

snails *Goniobasis, Pomacea, Oxytrema*, and *Physa* can exhibit high densities on the vegetation or on the calcareous bottom sands. Numerous fishes, turtles, and a variety of invertebrates are components of a complex trophic system, and some species are migratory (e.g., various birds, mullets). The alkaline water is generally clear except for those runs heavy with swamp input, in which the water can be tea-colored. Thousands of small springs contribute mineral-rich waters to streams over the Southeast. Immediately below the outlet the stream is often choked with water cress, which provides a habitat for large populations of snails, amphipods, and isopods.

Seepage Communities

Seepage areas are characterized by poorly drained soils, where groundwater surfaces and slowly moves down slope. Although those areas are quite variable with respect to slope, soil makeup, volume of flow, community composition, and so on, four basic types are recognized: open seeps, canopied seeps, montane seeps, and pitcher plant seeps.

Open seepage areas are characterized by spongy, sandy soil covered by grasses and sedges; in many areas, open seeps are indeed pitcher plant seeps (see below). Numerous species of primary burrowing crayfishes inhabit these habitats (e.g., *Cambarus (Depressicambarus) reflexus*; see Hobbs 1981, p. 27). Canopied seeps are shaded areas having considerable leaf-litter input. Various ferns, mosses, and shrubs make up the understory, while, for example, *Cornus florida, Leriodendron tulipifera, Acer* sp., and *Quercus* sp., are conspicuous elements of the canopy. *C. (D.) harti* is another burrowing crayfish inhabiting this type of seepage area (see Hobbs 1981, p. 104). Montane canopied seeps are quite variable but usually are found on mountain slopes and are spring heads with rock, sand, or shallow mucky peat substrates. Little primary production occurs and much allochthonous material contributes to the organic budget of these shaded systems. Such areas are frequented by a variety of amphibians (e.g., *Desmognathus, Pseudotriton, Gyrinophilus*) and crayfishes also burrow into the organically rich substrates (e.g., *C. (Jugicambarus) asperimanus, C. (J.) carolinus*, and *C. (J.) dubius*). In Ouachita Mountain seeps (Montgomery and Garland counties, Arkansas) hydrophytic, acid seep, and disjunct plants (northern eastern and southern taxa) are characteristic components.

Pitcher plant seeps are generally found in poorly drained soils (coarse, sandy clay often overlain by "muck" or sandy loam, sometimes underlain by hardpan). Some pines and various grasses as well as the abundant semiaquatic pitcher plants, sundews, clubmosses, and hatpins make up the plant community. In the Florida panhandle three subspecies of the crayfish *Procambarus (Hagenides) rogersi* burrow into the soil and in these complex burrows are often found *Crangonyx* sp., *Caecidotea hobbsi, Bidessus rogersi*, and *Cyclops* sp. (see Hobbs 1945 and Folkerts 1982 for further discussion of this most interesting community).

RESOURCE USE AND MANAGEMENT EFFECTS

Groundwater constitutes approximately 4% of the water in the hydrologic cycle, and approximately 30% of the stream flow of the United States is supplied by groundwater emerging as natural springs or various seepage areas. Nearly half the United States population uses groundwater from springs or wells as the primary source of drinking water. Its quality, therefore, is an important issue. Clearly, groundwater contamination is not a new or a unique occurrence, yet increased population density, industrialization, and heightened agricultural activities have greatly exacerbated the problem in many areas. Pye and Patrick (1983, p. 717) report that estimates of contamination of this nation's usable groundwater range from 1 to 2%. Potable groundwater occurs in most areas of the Southeast and can often be used as a raw resource without pretreatment. These aquifers are frequently underlain by deeper saline aquifers. Pye and Patrick (1983, p. 714) outline five types of naturally occurring groundwater that often demonstrate total dissolved solids (TDS) readings exceeding 10 000 ppm (connate water, magmatic and geothermal water, intruded seawater, water affected by evapotranspiration, and water affected by salt leaching). These saline waters are generally unusable due to high salinities or to the presence of naturally occurring toxic substances (e.g., radioactivity from uranium in Texas, arsenic in some thermal springs).

The types of contamination currently found in groundwater range from simple inorganic ions (e.g., nitrate, chloride, and heavy metals) to complex synthetic organic chemicals as well as pathogens (viruses and bacteria). Waste disposal is one of the major source centers of groundwater contamination and causes a variety of problems (see Langmuir 1972). Domestic waste includes individual sewage disposal systems, collection and treatment of municipal wastewater, and land disposal of liquid and solid wastes. Disposal of wastes from animal feed lots may contribute to contamination of groundwater. Industrial activities result in a wide variety of pollution sources, such as brine associated with petroleum industry, deep well disposal of liquid wastes, industrial and wastewater impoundments, and disposal of mine wastes. Low- and high-level radioactive wastes resulting from private and governmental sources are potential contaminants of groundwater.

Sources unrelated to disposal activities include such potential nonpoint sources of groundwater pollution as minute amounts of cadmium generated from wear of vulcanized tires on highways, agricultural activities (e.g., application of fertilizers and pesticides, vegetative land cover change, irrigation), atmospheric contaminants (e.g., lead from leaded gasoline combustion along highways; the effects that acid rain has on cave ecosystems and limestone dissolution virtually has not been addressed), and the use of highway deicing salts. Stieglitz and Schuster (1985) note some of the problems and fears rural landowners face concerning well contamination.

Potential point-source contamination results from mining operations (particularly "dewatering" and abandonment), improper well construction, accidental spills and leaks (e.g., petroleum products; see Garton 1983), and infiltration from polluted surface water. In addition, poorly developed groundwater use plans can lead to contamination (e.g., coastal communities overpump freshwater aquifers causing saltwater intrusion into the aquifers). All of these affect the quality and quantity of potable water, but, more to the point, drastically alter the habitats for subterranean communities. Waters emerging from caves/springs are also used as a power source for hydroelectric power plants (e.g., Blue Springs, Jackson County, Florida, and Mammoth Spring, Fulton County, Arkansas, now inactive).

Discussed below are selected examples of inadequate protection of karst areas in the Southeast. The vast majority of examples assess groundwater pollution, but some also examine effects of overextraction of subsurface water, surface land-use changes, and various construction, development (suburban expansion), and management problems (see Dilamarter and Csallany 1986).

One of the most common, and often most serious, karst-related problems is pollution of groundwater (see LeGrand 1973). Serious pollution problems result whenever contaminants enter groundwater at rates exceeding acceptable attenuation. Estes (1984) reported considerable surface water and groundwater contamination in Lee and Powell counties, Kentucky. The TDS from Quicksand Cave and Flood Cave are 12 100 and 16 900 ppm, respectively. These high levels suggest that connate brine is leaking into the streams from uncapped and abandoned oil and gas wells and that these contaminants are being distributed to local potable water supplies via undelineated cave systems. Estes also suggests that locally owned oil companies are washing out oil storage tanks and dumping the sludge into nearby sinkholes. Quinlin (1983) reported that the combination of sewage, creamery waste, and heavy metals (up to 24 ppm chromium and 13 ppm nickel) resulted in the demise of blind fish, blind crayfish, and other cave fauna in Hidden River Cave beneath Horse Cave (town), Hart County, Kentucky. He noted that these contaminated waters flow 7 km underground to 46 springs at 16 locations along an 8-km stretch of the Green River. Pushkar (1985) reported on water quality assessment studies being conducted in the Sloan's Valley Cave System in Pulaski County, Kentucky. Concern over leachate from a recently opened (1982) municipal landfill (former coal strip mine) into the cave system precipitated these studies (this is a habitat for the troglobitic fish *Typhlichthys subterraneus*, the troglobitic crayfish *Orconectes australis packardi*, the copepod *Attheyella pilosa*, the troglobitic isopod *Caecidotea stygia*, the amphipod *Crangonyx packardi*, and the troglophilic crayfish *Cambarus tenebrosus*; see Cooper and Beiter 1972). Results indicate that the Dixie Bend Landfill runoff is not contaminating the cave system with heavy metals (except in the railroad tunnel entrance), but the concentration of Fe and Mn, as well as Pb and Cd, exceeds recommended levels in isolated sites. Stecko (1985) presented an update concerning the quality of industrial waste being placed at the landfill: washer sludge, steel shot, and asbestos from one company, and fiberglass from another. A further account of Kentucky cave pollution was presented by Schindel (1984). He stated that the Lost River Cave system, Warren County, received benzene methylene chloride during late 1982. In addition, a surface pond (shown to drain into State Trooper Cave, part of the Lost River Cave drainage) contained methylene chloride, 1,1-dichloroethane, 1,1,1-trichlorethane, benzene, toluene, xylene, arsenic, and barium.

Spigner and Graves (1983) reported that in 1974 a shallow unconfined karst aquifer (Bangor Limestone) was contaminated by chromium, resulting in the pollution and abandonment of a municipal well at Irondale, Jefferson County, Alabama. Critical water shortage was avoided by drilling into the deeper, confined karst Fort Payne–Tuscumbia aquifer, which provided maximum protection against contamination and subsidence (see below).

Large aquatic troglobites (cambarid crayfishes and amblyopsid fishes) are limited to rather stable environments, have low reproductive capabilities, and may require many years (decades? see Cooper and Cooper 1978) to attain sexual maturity. Their longevity may be their demise. For example, minute concentrations of heavy metals acquired each year may ultimately reach lethal levels. It is quite possible that many of the highly specialized aquatic cavernicoles could become extinct, not from a single, massive dose of pollutants, but slowly and insidiously from low levels of toxic metals (see Dickson et al. 1979, Dickson and Franz 1980, Bosnak and Morgan 1981a,b).

Septic tanks and subsurface waste disposal fields are being employed on an unprecedented scale, this being particularly obvious on the periphery of municipalities where new subdivisions have outdistanced extensions of sewage collection systems. Estimates of the number of septic tanks performing inadequately range up to 50% (conservatively) of those currently in use. These malfunctioning systems constitute a severe health hazard as well as a major source of contamination to the surface water and groundwater environments (Minear and Patterson 1973). Vineyard (1981) reviewed the sinkhole collapse and subsequent release of several million gallons of partially treated and raw sewage into groundwaters at West Plains, Missouri. Dye tracing indicated that the sewage drained into Mammoth Spring to the south, the largest spring in Arkansas. An enrichment role can be played by this kind of contamination as well. Holsinger (1966) suggested that the increase in organic material and bacterial populations in Banners Corner Cave, Virginia, accounted for the unusually high troglobitic invertebrate numbers observed, particularly in pools.

Robbins and Krig (1973) list five major categories of agricultural groundwater pollutants: animal wastes, fertilizers, pesticides, plant residues, and saline irrigation–waste waters. A review of the factors surrounding each category is presented by them, and additional pertinent information can be obtained from Mallmann and Mack (1961) and Gardner (1986, p. 19). Keith and Poulson (1981) discuss exposure to pesticides as a cause for "broken back syndrome" in a population of *Amblyopsis spelaea* from the Donaldson-Twin Cave system in southern Indiana. Such problems may prove to be more widespread than is currently apparent.

Road salt contamination of groundwater springs has been noted in several instances in West Virginia (see Wilmoth 1972, Ogden and Flint 1976, Werner 1983a,b). Werner (1983a) reported chloride ion levels as high as 55.6 ppm in springs at Edray in Pocahontas County. He noted that the source of the chloride is derived from salt that is applied regularly to deice U.S. Hwy. 219. The very high, short-period levels that occur during snowmelt result from runoff from the road. The chloride ions pass through the conduit flow system of the karst aquifer and exit rapidly at the springs. However, some contaminated water infiltrates from the land surface into the diffuse flow system of the aquifer and is stored there for long periods. It is discharged slowly into the conduit flow system of the cave passage, thus maintaining chloride levels higher than background at all times. Wilmoth (1972) and Ogden and Flint (1976) in Monroe County demonstrated a long-term storage of chloride in a diffuse system: Six years following the removal of the source the levels still remain high. Liskowski and Poulson (1981) also point out that brines from oil field development can be introduced into caves.

Another source of salt contamination of groundwater occurs when freshwater aquifers are invaded by saltwater bodies, usually occurring in coastal aquifers. Seawater intrusion historically results primarily from overpumping of the freshwater aquifer, causing replacement by saltwater (see Kohout et al. 1975, Popenoe et al. 1984). The reader is referred to Black et al. (1953) and to Kashef (1971, 1973) for more complete descriptions of saltwater invasion.

In addition to saltwater intrusion, groundwater withdrawal can lead to spring flow cessation (e.g., Peek 1951, Brune 1975, Rosenau et al. 1977). Waters pumped to the surface for industrial, domestic, municipal, and agricultural uses often result in groundwater levels being systematically drawn down (e.g., several localities in Georgia and South Carolina; Stringfield 1966). Peek (1951) attributed the cessation of flow in February 1950 of Kissengen Spring, Polk County, Florida, to increased pumping of water from the Floridan Aquifer. This spring had a maximum recorded flow of 43.6 cfs (October 1933) and was the first known major Florida spring to cease flowing because of groundwater withdrawal from wells (Rosenau et al. 1977, p. 307).

Brune (1975) discussed causes of spring flow decline in Texas and postulated that the first historical effect upon groundwater levels and spring flow resulted from deforestation by the early white settlers. This practice, coupled with heavy grazing by introduced livestock, left the soil so impervious and compacted that soils could take in only a small fraction of the recharge formerly transported to the groundwater. By the mid 1800s deep wells were drilled, resulting in the release of tremendous artesian pressures. Most wells ran continuously until the piezometric heads were exhausted and the wells stopped flowing. This was followed by pumping, initially for municipal and industrial use and more recently for irrigation. Additionally, paving urban areas has reduced the volume of recharge to some aquifers and many springs have been inundated by man-made reservoirs. Examples of springs in eastern Texas that no longer flow are Chalybeate and Thrasher Springs in Cass County, Coushatta Spring in Harrison County, and Nacogdoches Spring in Nacogdoches County (Brune 1975).

During the late 1930s and early 1940s open pine flatwoods slightly north of Blountstown, Calhoun County, Florida, supported dense stands of pitcher plants. Water from the saturated, sandy, clay soil slowly trickled down gentle slopes, and grasses and *Sarracenia psittacina* and *S. drummondi* dominated the community (Hobbs 1938). The primary burrowing crayfish, *Procambarus (H.) r. rogersi*, was abundant in this seep community and cohabiting the burrows were *Crangonyx sp.*, *Caecidotea* sp., and copepods. When the site was visited in December 1985, there were few pitcher plants to be seen. The water table, which formerly was near the surface, had been lowered by as much as 1 m due to ditching. The pitcher plant community had been greatly altered; a general decline of pitcher plant communities in Florida was also noted by Folkerts (1982).

Changes in land use that have resulted in a variety of impacts on karst regions are well documented (e.g., Beck 1980, Dougherty 1981, 1983, Huppert et al. 1983, Crawford 1981, Thrailkill et al. 1983). Sinking Valley, an 88-km^2 enclosed drainage area situated in Pulaski County, Kentucky, was typical in its land use and hydrological regimen of karst regions on the Cumberland Plateau. Dougherty (1981, 1983; see also Johnson 1981) studied this area and noted that it had undergone a multiple land-use change: Early settlers removed forest cover, which resulted in an economy based on clean tilled crops (e.g., corn, tobacco). A recent shift in its use to grazing retrograded toward a return to its former nature (abandoned farms and increased vegetative cover). Dougherty demonstrated that spring peak runoff is significant, while yearly totals remain relatively constant in areas of differing land use. Historically, the heavy agricultural period shows a greater spring flood than that of the pre-settlement forest and the modern transition period exhibits some-what intermediate flood levels. Causes of floods were traced to increased runoff from denuded soil areas where poor agricultural techniques were prac-ticed. Intensity of floods was aggravated by increased sediment loads in the runoff which blocked the already insufficient subterranean drainage conduits. Recent land-use regression to an increased vegetative cover has resulted in reduced sedimentation and less runoff and thus a cleansing of the system by degradational processes. "Valley tides" result when the cave system beneath the valley is unable to carry the excess runoff (due to blockage by sediments and debris) and flood water resurges from caves higher on the slopes of the valley and flows down valley as a flash flood. Tremendous impact results from such devastating runoff events, and data indicate that good land-use practices can reverse the process.

Huppert et al. (1983) studied a Melrose subdivision west of Cookeville, Putnam County, Tennessee. The subdivision, located in a dry valley drained by several large sinkholes, is situated on the Mississippian Warsaw formation. Natural drainage interference by developers (e.g., grading, paving, vegetation removal, plugging several small sinkholes with construction debris, rock, dirt) resulted in increased runoff and slower drainage. Flooding of several homes resulted during heavy precipitation events and wells were drilled to drain excess water (a stopgap measure). The authors suggest that continued de-velopment of the area may result in more frequent flooding, flooding at other sinks, complete plugging of the wells, or pollution problems. Crawford (1981) and Thrailkill et al. (1983) also provide additional information concerning karst-related problems for urban areas located in the environs of sinkholes.

Construction of highways has been shown to have deleterious effects on cave water quality as well as springs (Garton 1983, Werner 1983b; see also Werner 1983a and Holler 1985). In eastern West Virginia, highway construc-tion in areas underlain by sinkholes and caves has resulted in two fish hatch-eries being affected by sediments washed into the recharge inlets of karst spring systems at Bowden National Fish Hatchery, Randolph County, the most productive fish hatchery in the United States (Garton 1983), and Edray Fish Hatchery, Pocahontas County (Werner 1983b). Garton (1983) reports that as construction proceeded (1972), pits and caves were uncovered, re-sulting in large quantities of clay and silt being washed into the caves, which drained directly into South Spring, one of two springs supplying water to the Bowden Hatchery. Numerous fish kills occurred through 1975; one incident resulted in the asphyxiation of 150 000 trout due to the buildup of silt on their gills. Fish poisoning occurred when diesel fuels were spilled at the con-struction site and washed into the caves. In addition to these construction effects, reduced flows of area wells and of South Spring resulted from the beheading of conduits. These are only examples of many problems encoun-tered in construction on karst terrains; many of them could be alleviated if, for example, geophysical studies other than boreholes were conducted.

Solution of carbonate rocks and removal of dissolved minerals by circu-lating groundwater result in subterranean cavities that weaken the structure of the overlying rock. Although collapse may be a natural result, activities of man, such as heavy structural loading and excessive pumping of water from underlying aquifers, accelerate subsidence (ground–surface collapse). Thrailkill et al. (1983) reported that in the inner Bluegrass karst region of central Kentucky two major problems result from urban development in and near sinkholes: flooding (see above) and soil subsidence (caused by increased runoff), which occurs within sinkholes and in areas overlying the sinkhole drain. In Alabama, an estimated 4000 sinkholes are believed to have formed since the turn of the century, most resulting from a decline in the subterranean water level due to groundwater withdrawals (Newton 1977). Spigner and Graves (1983) pointed out numerous problems associated with well drilling, particularly in the Ironton area, Jefferson County, Alabama. Subsidence is troublesome because potentially dangerous ground collapse can occur without warning close to or away from the well site, either during well development or after the well is completed. Brook and Allison (1983) reported that one possible consequence of a recent increase in the use of groundwater for irrigation in the Ocala aquifer underlying the Doughterty Plain in south-western Georgia is accelerated sinkhole development due to a lowering of the regional piezometric surface or to the formation of depression cones at irrigation wells. They developed ground subsidence susceptibility maps using sinkhole and bedrock fracture data. Ogden and Reger (1977) concluded from studies in Monroe County, West Virginia, that the percentage of the limestone area in sinkholes and the sinkhole density were reliable indicators of areas prone to subsidence. Simply put, areas underlain by the most cavernous rock demonstrate the greatest number of sinkholes. Examples of spectacular sinkhole subsidence include Devil's Millhopper, Alachua County, Florida (not related to man's activities), a 106-m-diameter, 30-m-deep sinkhole formed at Winter Park (see Jammal 1984), Orange County, Florida (in excess of $4 million in damages) during a period of drought on 8 May 1981, and a 40-m-wide × 10-m-deep sinkhole opened in downtown Jacksonville, Florida, in October 1985; others have appeared in Tampa and Brooksville, Florida, and many in Shelby County, Alabama (see Beck et al. 1984, Lamoreaux 1984). Also, it is not uncommon for lakes or sinkhole ponds to empty literally overnight.

In karst regions where dams are proposed to store surface water, there is concern as to whether the reservoirs will hold water. Leakage often occurs as the hydrostatic head opens previously plugged conduits in the rock or water enters various caves and other solution openings that had been dry in the previously unsaturated zone. Grouting is a common technique to insure the impermeability of the rock (see LeGrand 1973 and Eraso 1981). A classic example of a leaking reservoir in a karst area is the Hales-Bar Dam on the Tennessee River. Burwell and Moneymaker (1950) reported that the dam was built without sufficient exploration or foundation excavation, and without remedial treatment of an extensively cavernous limestone foundation. The project, estimated at $3 million and scheduled to be completed in two years, actually cost $11.5 million and required eight years for completion due primarily to leakage through the foundation rock.

Nickajack Cave, Marion County, Tennessee, is one of the most famous southeastern caves. Its entrance is 42 m wide × 15 m high and the cave serves as the type locality for a number of troglobitic organisms (e.g., the isopods *Caecidotea nickajackensis* and *C. richardsonae*, the amphipod *Crangonyx antennatus*, and the crayfish *Cambarus (A.) hamulatus*; see Cope and Packard 1881). Yet, in 1967 the cave was partially flooded by an impoundment of the Tennessee River, resulting in Nickajack Lake (Fig. 21). Only about 100 m of this extensive cave is not inundated (passages extend beneath Jackson County, Alabama, and narrowly miss Dade County, Georgia).

When dams do successfully impound water, much alteration of surface and subsurface habitats is inevitable. Lisowski and Poulson (1981) present an excellent account of the changes and adverse effects Lock and Dam No. 6 on the Green River have had on the Mammoth Cave system.

There are increasing signs that heavy use of caves may be significantly altering the environment and thus the community structure of numerous caves in the southeast. In addition to the dumping of calcium hydroxide (used calcium carbide, a fuel used for carbide lights; see Lavoie 1980b, 1981c), accumulations of litter (bottles, cans, blankets, mattresses, feces, etc.) are characteristic features of the scenery in heavily traveled caves. Disturbance of hibernating bats or of maternity colonies is probably the greatest impact cavers have on cave communities (effectively reducing the energy input into aquatic systems). Collection by overzealous individuals has grossly reduced numbers of cavernicoles. George (1972b, p. 48) reports that Under The Road Cave in Breckenridge County, Kentucky, formerly supported a large popu-

FIGURE 21. Nickajack Cave and Nickajack Lake, Marion County, Tennessee.

lation of the troglobitic fish, *Amblyopsis spelaea* ("40–60 in each pool"), but that due to over-collecting, "now only 3 or 4 fish can be seen."

Permanent closure of caves is an event observed in all karst regions. Entrance access is halted more commonly either by landowners who have lost their tolerance for inconsiderate cavers or by bulldozers and concrete or asphalt in urban areas. A "closed" cave simply means that it is no longer available for visitation and study, and it may well mean extinction for certain cave fauna, particularly trogloxenes. *Procambarus (Leconticambarus) milleri*, a troglobitic crayfish, is known only from a well at a nursery and garden store in Miami, Florida; the fate of this site is tenuous, at best.

RECOMMENDATIONS FOR FUTURE RESEARCH AND MANAGEMENT

Very simply stated, research in caves and springs is still hovering around the basic descriptive, alpha level. This does not infer that more in-depth studies have not occurred (see Culver 1982), yet a survey of the literature forces one to realize that information available on the structure and function of cave and spring ecosystems is very limited.

Caves and springs are relatively simple ecosystems and as such are excellent natural laboratories. Simplicity exists because caves are characterized by few, yet predictable, variables, such as limited autotrophy, small species pools, simplified food webs, and minimal physicochemical fluctuations. As Culver (1982) indicates, most ecological models that have been tested involve more complex systems or include organisms with elaborate life histories. Some of the more critical model assumptions, such as near-equilibrium conditions, are more likely to be met in caves and springs than in more complex epigean systems. Community trophic dynamics as well as cave energetics are virtually unstudied and only a few attempts have been made to decipher them in spring ecosystems.

Protection, management, and preservation of cave and spring environments (see Table 6) are imperative (see Poulson 1976). Because the surface environment supplies energy to hypogean systems (input from streams, migration of facultative organisms, etc.), management must include surface habitat protection as well. Aquatic cave ecosystems tend to be more sensitive than terrestrial ecosystems because (1) they are more restricted in number and in area; (2) they are highly vulnerable to localized perturbations such as pollutant inputs; (3) the organisms have low reproductive potential involving infrequent breeding intervals, and in general, greater longevity; and (4) they depend on outside sources of energy for maintaining community structure and function.

Many southeastern states (Alabama, Florida, Georgia, Kentucky, Tennessee, Texas, Virginia, and West Virginia) have adopted cave protection laws, and a federal law was recently passed by Congress. As in surface environments, the number of endangered and threatened species in caves and

TABLE 6   Public and Preserved Lands of the Southeastern United States with Caves and Springs

ALABAMA
Monte Sano State Park
Russell Cave National Monument

ARKANSAS
Buffalo River State Park
Devil's Den State Park
Hot Springs National Park
Mammoth Springs State Park
Ouachita National Forest
Ozark National Forest
Petit Jean State Park
Withrow Springs State Park

FLORIDA
Apalachicola National Forest
Devil's Millhopper State Geological
   Site
Falling Waters State Recreation Area
Florida Caverns
Ichetucknee Springs State Park
Manatee Springs State Park
National Bridge Monument
Ocala National Forest
Peacock Springs Preserve
Wakulla Springs, Registered Natural
   Landmark
Wekiwa Springs State Park

GEORGIA
Chattahoochee National Forest
Cloudland Canyon State Park

KENTUCKY
Barren River State Resort Park
Bernheium Forest
Carter Caves State Resort Park
Daniel Boone National Forest
General Butler State Resort Park
Kentucky Ridge State Forest
Mammoth Cave National Park
Natural Bridge State Resort Park
Otter Creek Park

LOUISIANA
Kisatchie National Forest

MISSISSIPPI
Holly Springs National Forest
Natchez Trace Parkway

NORTH CAROLINA
Boone's Cave State Park
Chimney Rock Park
Great Smokey Mountain National
   Park
Mount Jefferson State Park
Nantahala National Forest
Pisgah National Forest

SOUTH CAROLINA
Santee State Park

TENNESSEE
Carter Caves State Natural Area
Cedars of Lebanon State Park and
   Forest
Cherokee National Forest
Chickasaw State Park and Forest
Chuck Swan Forest and Wildlife
   Management Area
Edgar Evans State Park
Franklin State Forest
Great Smokey Mountain National
   Park
Land Between the Lakes
Pickett State Park
Savage Gulf State Natural Area

VIRGINIA
Cave Spring Recreational Area
Cumberland Gap National History
   Park
George Washington National Forest
Jefferson National Forest
Natural Tunnel State Park
Shenandoah National Park

WEST VIRGINIA
Coopers Rock State Forest
Jefferson National Forest
Monongahela National Forest

springs is increasing (see Tables 7 and 8). Over-collecting in various caves (some by biospeleologists!) has led to gross reductions of populations of various taxa.

A wide variety of questions remains to be answered concerning cave and spring ecosystems, particularly with regard to evolution and ecology of troglobites. Some obvious ones are How are available energy and nutrients partitioned among trophic components? What are the energy budgets? How do such parameters as competition, predation, and parasitism (collectively or individually) regulate community dynamics? What is the role of stream drift in shaping communities and as a dispersal mechanism? What are the roles of resistance and resilience in predictability and stability? Can we stereotype all cavernicoles as K-strategists? Does species richness reflect only a balance between colonization and extinction? What are the dynamics of hot springs communities?

Future research in caves and springs will continue for intrinsic academic reasons. In addition, it is important that long-term, continuous data sets (for various physicochemical parameters as well as for populations and communities) are accumulated in order that adequate baselines are established. These natural laboratories should be protected, but they should be utilized by students of genetics, evolution, ecology, hydrology, palaeontology, ethology, natural history, systematics, etc.

Hazards posed by groundwater (spring and cave water) contamination are numerous and complex. They vary according to the nature of the compounds (their concentration in the aquifer and their persistence in the environment), the degree of exposure to them, and the volume discharged. Too often we are concerned primarily with the impact on humans; yet sometimes these delicate spring–cave ecosystems are severely impacted or even destroyed. Additionally and unfortunately, and partly because of their chronic nature, pollution sources are not easily observed, and often their effects do not become evident until potential catastrophe is well underway. Simply stated, groundwater contamination is just that: It occurs underground, out of sight, out of mind. The tangible effects (often irreversible) of contamination too often are recognized some time after the causal incident; often the precise source of contamination is difficult to determine.

Evidence of spring and cave water pollution in the southeastern United States is accumulating at alarming rates. Examples of destroyed or grossly altered caves are no longer isolated incidents; collapse of surface areas (subsidence) due to a lowering of the buoyant water table resulting in large sinkholes (decline of piezometric surface and increase in water consumption) is becoming a frequent occurrence in northern and central Florida and parts of Alabama. These problems and their causes and sources, must be studied closely. Purchase, protection, and sound management of these nonrenewable resources, such as the Peacock Spring Preserve in Suwannee County, Florida, are of utmost importance. Education and restriction of entry into caves always or at critical times (hibernacular or maternity caves) are two means of protecting subterranean communities; other methods certainly exist. Greater effort must be made to determine how best to use, manage, and protect these

TABLE 7 Rare Species Associated with Caves, Springs, and Seeps

Species	Description	Status
FLORIDA		
Blechnum occidentale	Sinkhole fern	E
Drosera intermedia	Water sundew	T
Aphaostracon asthenes	Blue Spring aphaostracon	RS
A. charlarogyrus	Loose-coiled snail	E
A. monas	Wekiwa Spring aphaostracon	RS
A. pycnus	Thick-shelled aphaostracon	RS
A. xynoelictus	Fenney Spring aphaostracon	RS
Cincinnatia helicogyra	Helicoid spring snail	RS
C. mica	Sand grain snail	RS
C. ponderosa	Ponterous Spring snail	RS
C. vanhyningi	Seminole Spring snail	RS
C. wekiwae	Wekiwa Spring snail	RS
Somatochlora provocans	Kite emerald dragonfly	T
Tachopteryx thoreyi	Gray petaltail dragonfly	R
Agarodes (Psiloneura) libalis	Spring-loving psiloneuran	R
Crangonyx grandimanus	Florida cave amphipod	SC
C. hobbsi	Hobbs' cave amphipod	SC
Caecidotea hobbsi	Hobbs' cave isopod	SC
Palaemonetes cummingi	Squirrel Chimney Cave shrimp	RS
Cambarus (Jugicambarus) cryptodytes	Doughtery Plain Cave crayfish	SC
Procambarus (Lonnbergius) acherontis	Orlando Cave crayfish	RS
P. (Leconticambarus) milleri	Miller's Cave crayfish	SC
P. (Ortomannicus) erythrops	Red-eyed cave crayfish	SC
P. (O.) franzi	Orange Lake cave crayfish	SC
P. (O.) horsti	Horst's cave crayfish	SC
P. (O.) lucifugus	Light-fleeing cave crayfish	SC
P. (O.) orcinus	Woodville Cave crayfish	SC
P. (O.) pallidus	Pallid Cave crayfish	SC
Troglocambarus maclanei	McLane's cave crayfish	SC
Alligator mississippiensis	Ameican alligator	T
Haideotriton wallacei	Georgia blind salamander	RS
Trichecus manatus latirostris	West Indian manatee	T
KENTUCKY		
Drosera brevifolia	Dwarf sundew	E
Drosera intermedia	Sundew	E
Lycopodium appressum	S. bog clubmoss	E
Caecidotea barri	Clifton Cave isopod	RS
Palaemonetes ganteri	Kentucky cave shrimp	E
Orconctes australis	Crayfish	T
O. i. inermis	Crayfish	T
O. pellucidus	Crayfish	SC

TABLE 7   (*Continued*)

Species	Description	Status
Amblyopsis spelaea	N. cavefish	SC
Typhlichthys subterraneus	S. cavefish	SC
MISSISSIPPI		
Lindera subcoriacea	Bog spice bush	CI
Sarracenia leucophylla	Crimson pitcher plant	R/T
S. parpurea	Side-saddle pitcher plant	CI
S. rubra wherryi	Wherry's pitcher plant	CI/ FC
Fallicambarus byersi	Lavender burrowing crayfish	R/T
Gyrinophilus porphyriticus	N. spring salamander	E
Eurycea lucifuga	Cave salamander	E
NORTH CAROLINA		
Stygobromus carolinensis	Carolina seep scud	SC
Caecidotea carolinensis	N.C. cave isopod	SC
Eurycea longicauda	Long-tail salamander	SC
Trichechus manatus	Florida manatee	E
SOUTH CAROLINA		
Utricularia olivacea	Dwarf bladderwort	T
U. floridana	Florida bladderwort	T
Dionaea muscipula	Venus flytrap	T
Sarracenia jonesii	Mountain sweet pitcher plant	E/FC
Trichechus manatus	Florida manatee	E
VIRGINIA		
Thelypteris simulata	Massachusetts fern	T
Diphylleia cymosa	Umbrella leaf	T
Lirceus culveri	Rye Cove cave isopod	T
Antrolana lira	Madison Cave isopod	T
Caecidotea holsingeri	Greenbrier Valley cave isopod	SC
Caecidotea henroti	Henrots' cave isopod	SC
Caecidotea incurva	Incurved cave isopod	SC
Caecidotea pricei	Price's cave isopod	SC
Caecidotea recurvata	Southwestern Virginia cave isopod	SC
Caecidotea vandeli	Vandel's cave isopod	SC
Caecidotea richardsonae	Tennessee Valley cave isopod	U
Lirceus usdagalun	Lee County cave isopod	SC
Stygobromus biggersi	Bigger's cave amphipod	T
Stygobromus hoffmani	Alleghany County cave amphipod	T
Stygobromus abditus	James Cave amphipod	T
Stygobromus pseudospinosus	Luray Caverns amphipod	T
Stygobromus stegerorum	Madison Cave amphipod	T

TABLE 7   (*Continued*)

Species	Description	Status
Crangonyx antennatus	Appalachian Valley cave amphipod	SC
Stygobromus mundus	Bath County cave amphipod	SC
Stygobromus spinosus	Blue Ridge Mountain amphipod	SC
Stygobromus conradi	Brunsville Cove cave amphipod	SC
Stygobromus estesi	Craig County cave amphipod	SC
Stygobromus cumberlandus	Cumberland cave amphipod	SC
Stygobromus ephemerus	Ephemeral cave amphipod	SC
Stygobromus fergusoni	Montgomery County cave amphipod	SC
Stygobromus leensis	Lee County cave amphipod	SC
Stygobromus mackini	Southwestern Virginia cave amphipod	SC
Stygobromus morrisoni	Morrison's cave amphipod	SC
Stygobromus pizzini	Pizzini's groundwater amphipod	SC
Stygobromus tenuis potomacus	Potomac groundwater amphipod	SC
Stygobromus barodyi	Rockbridge County cave amphipod	SC
Stygobromus gracilips	Shenandoah Valley cave amphipod	SC
Stygobromus araeus	Tidewater interstitial amphipod	SC
Stygobromus identatus	Tidewater stygonectid amphipod	SC
Stygobromus interitus	New Castle Murder Hole amphipod	U
Stygobromus phreaticus	Northern Virginia well amphipod	U
Stygobromus kenki	Rock Creek groundwater amphipod	U
Tachopteryx thoreyi	Thorey's grayback dragonfly	SC
Sphalloplana subtilis	Bigger's groundwater planarian	E
Sphalloplana holsingeri	Holsinger's groundwater planarian	E
Sphalloplana virginiana	Rockbridge County cave planarian	E
Sphalloplana consimilis	Powell Valley cave planarian	SC
Semotilus margarita	Pearl dace	SC
Cottus cognatus	Slimy sculpin	SC
WEST VIRGINIA		
Eurycea lucifuga	Cave salamander	E
Gyrinophilus subterraneus	Cave salamander	E

Note. E, endangered; T, threatened; SC, special concern; R, rare; RS, review status; CI, critically imperiled; FC, federal candidate species; U, undetermined.

Source. Data primarily from state Natural Heritage programs.

TABLE 8   Federally Endangered or Threatened Aquatic Species from Southeastern Caves and Springs

Organism	Description	Status
Stygobromus hayi	Hay's spring amphipod	Endangered
Antrolana lira	Madison cave isopod	Threatened
Palaemonias alabamae	Alabama cave shrimp	Endangered
Palaemonetes ganteri	Kentucky cave shrimp	Endangered
Cambarus (Jugicambarus) aculabrum	Cave crayfish	Endangered
Cambarus (Jugicambarus) zophonastes	Cave crayfish	Endangered
Amblyopsis rosae	Ozark blind fish	Threatened
Speoplatyrhinus poulsoni	Alabama cave fish	Endangered
Etheostoma nuchale	Watercress darter	Endangered
Trichechus manatus	West Indian manatee	Endangered

valuable resources. Caves and springs are fragile windows through which one can observe and study unique environments. The vast majority of the caves in the southeastern United States have been only partially explored or mapped and in comparatively few are the biological resources well known. Likewise, springs and seeps of the region have not received adequate attention or protection.

REFERENCES

Adams, C. S., and A. C. Swinnerton. 1937. The solubility of calcium carbonate. *Trans. Am. Geophys. Union* 11:504–508.

Altig, R., and P. H. Ireland. 1984. A key to salamander larvae and larviform adults of the United States and Canada. *Herpetologica* 40(2):212–218.

Anonymous. 1957. Addenda to the caves of West Virginia. *NSS Bull.* 19:28–39.

Anonymous. 1974. Pigeon Mountain and Ellison's Cave System. Dogwood City Grotto of the National Speleological Society, Atlanta.

Anonymous. 1986. How many caves are in Virginia? *Underground Mountaineer* 7(1):3.

Armstrong, J. G., and J. D. Williams. 1971. Cave and spring fishes of the southern bend of the Tennessee River. *J. Tennessee Acad. Sci.* 46(3):107–115.

Atkinson, T. C., and D. I. Smith. 1976. The erosion of limestone. In T. D. Ford and C. H. D. Cullingford (eds.), *The Science of Speleology*. New York: Academic.

Back, W., and B. B. Hanshaw. 1970. Comparison of chemical hydrology of Florida and Yucatan. *J. Hydrol.* 10:330–368.

Barbour, T. 1944. *That Vanishing Eden: A Naturalist's Florida*. Boston: Little, Brown.

Barlow, C. A., and A. E. Ogden. 1982. A statistical comparison of joint, straight cave segment, and photo-lineament orientations. *NSS Bull.* 44(4):107–110.

Barr, T. C., Jr. 1961. Caves of Tennessee. *Bull. Tennessee Div. Geol.* 64:1–567.

Barr, T. C., Jr. 1963. Ecological classification of cavernicoles. *Cave Notes* 5:9–16.

Barr, T. C., Jr. 1967a. Ecological studies in the Mammoth Cave System of Kentucky, I: the biota. *Int. J. Speleol.* 3:147–203.

Barr, T. C., Jr. 1967b. Observations on the ecology of caves. *Am. Natur.* 101:475–492.

Barr, T. C., Jr. 1968. Cave ecology and the evolution of troglobites. *Evol. Biol.* 2:35–102.

Barr, T. C., Jr. 1985. Cave life of Kentucky. Caves and karst of Kentucky. *Kentucky Geological Survey, Special Publication* 12(series XI):146–167.

Barr, T. C., Jr., and R. A. Kuehne. 1971. Ecological studies in the Mammoth Cave system of Kentucky, II: the ecosystem. *Ann. Speleol.* 26:47–96.

Bechler, D. L. 1981. Agonistic behavior in the Amblyopsidae, the spring cave and swamp fishes. *Proceedings of the 8th International Congress of Speleology* 1:68–69.

Bechler, D. L., and A. G. Fernandez. 1981. Preliminary observations on foraging behavior in a hypogean crustacean community. *Proceedings of the 8th International Congress of Speleology* 1:66–67.

Beck, B. F. 1980. An introduction to caves and cave-exploring in Georgia. *Georgia Geologic Survey, Geologic Guide* 5.

Beck, B. F., and D. D. Arden. 1984. Karst hydrogeology and geomorphology of the Dougherty Plain, southwest Georgia. *Southeastern Geological Society, Guidebook* No. 26.

Beck, B. F., R. Ceryak, D. T. Jenkins, T. M. Scott, and D. P. Spangler. 1984. Field guide to some illustrative examples of karst hydrogeology in central and northern Florida. *Florida Sinkhole Research Institute, University of Central Florida, Orlando,* Report No. 84-85-2.

Beck, J. T. 1979. Population interactions between parasitic castrator, *Probopyrus pandalicola* (Isopoda: Bopyridae), and one of its freshwater shrimp hosts, *Palaemonetes paludosus* (Decapoda:Caridea). *Parasitology* 79:431–449.

Beckman, H. C., and N. S. Hinchey. 1944. The large springs of Missouri. *Missouri Geological Survey and Water Resources,* 29(2d series).

Besharse, J. C., and J. R. Holsinger. 1977. *Gyrinophilus subterraneus,* a new troglobitic salamander from southern West Virginia. *Copeia* (1977):624–634.

Black, A. P., E. Brown, and J. M. Pearce. 1953. Salt water intrusion in Florida, 1953. *Florida Water Survey and Research* 9:1–38.

Bogli, A. 1971. Corrosion by mixing of Karst waters. *Trans. Cave Res. Group, Great Britain* 13(2):109–114.

Bogli, A. 1980. *Karst Hydrology and Physical Speleology.* New York: Springer.

Bosnak, A. D., and E. L. Morgan. 1981a. Acute toxicity of cadmium, zinc, and total residual chlorine to epigean and hypogean isopods (Asellidae). *NSS Bull.* 43(1):12–18.

Bosnak, A. D., and E. L. Morgan. 1981b. Comparisons of acute toxicity for Cd, Cr, and Cu between two distinct populations of aquatic hypogean isopods (*Caecidotea* sp.). *Proceedings of the 8th International Congress of Speleology* 1:72–74.

Botosaneanu, L. (ed.). 1986. *Stygofauna Mundi: A Faunistic, Distributional, and Ecological Synthesis of the World Fauna Inhabiting Subterranean Waters.* Leiden, Netherlands: E. J. Brill.

Bousfield, E. L. 1958. Fresh-water amphipod crustaceans of glaciated North America. *Can. Field Natur.* 72(2):55–113.

Bowman, T. E., and B. Sket. 1985. *Remasellus*, a new genus for the troglobitic swimming Florida asellid isopod, *Asellus parvus* Steeves. *Proceedings of the Biological Society of Washington* 98(3):554–560.

Brandon, R. A. 1966. Systematics of the salamander genus *Gyrinophilus*. *Illinois Biol. Monogr.* 35:1–86.

Brandon, R. A. 1971. North American troglobitic salamanders: some aspects of modification in cave habitats, with special reference to *Gyrinophilus palleucus*. *NSS Bull.* 33:1–22.

Bray, L. G. 1969. Some notes on the chemical investigation of cave waters. *Trans. Cave Res. Group* 11(3):165–174.

Bretz, J. H. 1942. Vadose and phreatic features of limestone caverns. *J. Geol.* 50(6):675–811.

Bretz, J. H. 1953. Genetic relations of caves to peneplains and big springs in the Ozarks. *Am. J. Sci.* 251:1–24.

Brock, T. D., F. Passman, and I. Yoder. 1973. Absence of obligately psychrophilic bacteria in constantly cold springs associated with caves in southern Indiana. *Am. Midland Natur.* 90(1):240–246.

Brook, G. A., and T. L. Allison. 1983. Fracture mapping and ground subsidence susceptibility modeling in covered karst terrain: Dougherty County, Georgia. In P. H. Dougherty (ed.), *Environmental Karst*. Cincinnati, OH: GeoSpeleo, pp. 91–108.

Brooks, H. K. 1961. The submarine spring off Crescent Beach, Florida. *Quart. J. Florida Acad. Sci.* 24(2):122–134.

Brown, L. C., and C. W. Broughton. 1981. A survey of pollution indicator bacteria in water wells of Rutherford County, Tennessee. *J. Tennessee Acad. Sci.* 56(3):73–75.

Brucker, R. 1979. New Kentucky junction: Proctor–Mammoth link puts system over 200 miles. *NSS News* 37(10):231–236.

Brune, G. 1975. Major and historical springs of Texas. Texas Water Development Board, Report 189.

Bryan, K. 1924. The Hot Springs of Arkansas. *J. Geol.* 32(6):449–459.

Bryant, W., R. Erisman, and J. Hockersmith. 1970. Sculpin Cave: an unusual Blue Grass Cavern. *Happy Hunting Grounds* 1970:9–12.

Burbank, W. D., J. P. Edwards, and M. P. Burbank. 1948. Toleration of lowered oxygen tension by cave and stream crayfish. *Ecology* 29(3):360–367.

Burwell, E. B., and B. C. Moneymaker. 1950. Geology and dam construction. In S. Paige, Chairman, *Application of Geology to Engineering Practice*. Geological Society of America, Berky Volume, 11–43.

Callahan, J. T. 1964. The yield of sedimentary aquifers of the Coastal Plain, Southeast River Basins. *United States Geological Survey Water Supply Paper*, 1660-W.

Carpenter, J. H. 1970. *Geocentrophora cavernicola* n.sp. (Turbellaria, Aleoeocoela): first cave alleoeocoel. *Trans. Am. Microsc. Soc.* 89:124–133.

Carpenter, J. H. 1982. Observations on the biology of cave planarians of the United States. *Int. J. Speleol.* 12:9–26.

Caumartin, V. 1959. Morphologie et position systematique du *Perabacterium spelei*. *Soc. Bot. Nord France Bull.* 12(1):15–17.

Caumartin, V. 1963. Review of the microbiology of underground environments. *NSS Bull.* 25(1):1–14.

Christiansen, K. 1966. The genus *Arrhopalites* in the U.S. and Canada. *Int. J. Speleol.* 2:43–73.

Christiansen, K. 1982. Zoogeography of cave Collembola east of the Great Plains. *NSS Bull.* 44(2):32–41.

Christiansen, K., and P. Bellinger. 1980. *The Collembola of North America North of the Rio Grande.* Grinnell, IA: Grinnell College.

Christiansen, K., and D. C. Culver. 1968. Geographical variation and evolution in *Pseudosinella hirsuta. Evolution* 22:237–255.

Cliburn, J. W., and A. L. Middleton, Jr. 1983. The vertebrate fauna of Mississippi caves. *NSS Bull.* 45:45–49.

Collins, W. D. 1925. Temperature of water available for industrial use in the United States. *United States Geological Survey Water Supply Paper* 520:97–104.

Collins, W. D., M. D. Foster, F. Reeves, and R. P. Meacham. 1930. Springs of Virginia. Virginia Commission on Conservation and Development, *Water Resources and Power Division Bulletin* 1:1–55.

Cooke, C. W. 1936. Geology of the Coastal Plain of South Carolina. *U.S. Geological Survey Bulletin* 867.

Cooper, J. E. 1968. The salamander *Gyrinophilus palleucus* in Georgia, with notes on Alabama and Tennessee populations. *J. Alabama Acad. Sci.* 39(3):182–185.

Cooper, J. E. 1975. Ecological and behavioral studies in Shelta Cave, Alabama, with emphasis on decapod crustaceans. Thesis, University of Kentucky, Lexington.

Cooper, J. E. 1978. American cave fishes and salamanders. *NSS Bull.* 40(3):89.

Cooper, J. E. 1985a. Revised recovery plan for the Alabama cavefish, *Speoplatyrhinus poulsoni* Cooper and Kuehne 1974. Atlanta, GA: United States Fish and Wildlife Service.

Cooper, J. E. 1985b. Alabama cavefish recovery plan. Rockville, MD: U.S. Fish and Wildlife Service.

Cooper, J. E., and D. B. Beiter. 1972. The southern cave fish, *Typhlichthys subterraneus* (Pisces:Amblyopsidae), in the eastern Mississippian Plateau of Kentucky. *Copeia* 4:879–881.

Cooper, J. E., and M. R. Cooper. 1968. Cave-associated herpetozoa, II: salamanders of the genus *Gyrinophilus* in Alabama caves. *NSS Bull.* 30(2):19–24.

Cooper, J. E., and M. R. Cooper. 1971. Studies of the aquatic ecology of Shelta Cave, Huntsville, Alabama. *ASB Bull.* 18(2):30.

Cooper, J. E., and M. R. Cooper. 1978. Growth, longevity, and reproductive strategies in Shelta Cave crayfishes. *NSS Bull.* 40(3):97 (Abstract).

Cooper, J. E., and R. A. Kuehne. 1974. *Speoplatyrhinus poulsoni*, a new genus and species of subterranean fish from Alabama. *Copeia* 2:486–493.

Cooper, M. R., and J. E. Cooper. 1976. Growth and longevity in cave crayfishes. *ASB Bull.* 23(2):52 (Abstract).

Cope, E. D., and A. S. Packard, Jr. 1881. The fauna of Nickajack Cave. *Am. Natur.* 15:877–882.

Crawford, N. 1981. Karst flooding in urban areas: Bowling Green, Kentucky. *Proceedings of the 8th International Congress of Speleology* 2:763–765.

Crawford, R. L. 1981. A critique of the analogy between caves and islands. *Proceedings of the 8th International Congress of Speleology* 1:295–297.

Cropley, J. B. 1965. Influence of surface conditions on temperatures in large cave systems. *NSS Bull.* 27:1–10.

Cubbon, B. D. 1969. The collection of cave fungi. *Memoirs of the Northern Cavern and Mine Research Society* 1969:85–92.

Culver, D. C. 1970a. Analysis of simple cave communities: niche separation and species packing. *Ecology* 51:949–958.

Culver, D. C. 1970b. Analysis of simple cave communities, I: caves as islands. *Evolution* 24:463–474.

Culver, D. C. 1971a. Analysis of simple cave communities, III: control of abundance. *Am. Midland Natur.* 85:173–187.

Culver, D. C. 1971b. Caves as archipelagoes. *NSS Bull.* 33:91–100.

Culver, D. C. 1973a. Competition in spatially heterogeneous systems: an analysis of simple cave communities. *Ecology* 54:102–110.

Culver, D. C. 1973b. Feeding behavior of the salamander *Gyrinophilus porphyriticus* in caves. *Int. J. Speleol.* 5:369–377.

Culver, D. C. 1976. The evolution of aquatic cave communities. *Am. Natur.* 110(976):945–957.

Culver, D. C. 1981a. Some implications of competition for cave stream communities. *Int. J. Speleol.* 11(1981):49–62.

Culver, D. C. 1981b. The effect of competition on species composition of some cave communities. *Proceedings of the 8th International Congress of Speleology* 1:207–209.

Culver, D. C. 1982. *Cave Life, Evolution and Ecology.* Cambridge, MA: Harvard UP.

Culver, D. C. 1985. Trophic relationships in aquatic cave environments. *Stygologia* 1(1):43–53.

Culver, D. C. 1987. Eye morphometrics of cave and spring populations of *Gammarus minus* (Amphipoda:Gammaricae). *J. Crustacean Biol.* 7(1):136–147.

Culver, D. C., and T. J. Ehlinger. 1980. Effect of microhabitat size and competition size on two cave isopods. *Brimleyana* 4:103–113.

Culver, D. C., and T. J. Ehlinger. 1982. Determinants of body size of two subterranean isopods: *Caecidotea cannulus* and *Caecidotea holsingeri* (Isopoda:Asellidae). *Polski Arch. Hydrobiol.* 29(2):463–470.

Culver, D. C., and T. L. Poulson. 1970. Community boundaries: faunal diversity around a cave entrance. *Ann. Speleol.* 25:853–860.

Culver, D. C., and T. L. Poulson. 1971. Oxygen consumption and activity in closely related amphipod populations from cave and surface habitats. *Am. Midland Natur.* 85(1):74–84.

Culver, D. C., J. R. Holsinger, and R. Baroody. 1973. Toward a predictive cave biogeography: the Greenbrier Valley as a case study. *Evolution* 27:689–695.

Dalby, D. H. 1966. The growth of plants under reduced light. *Studies Speleol.* 1(4):193–203.

Dasher, G. 1983. Cave distribution in West Virginia. *National Speleological Society Convention Guidebook* 23:49–50.

Davies, W. E. 1950. Caves of Maryland. *Maryland Department of Geology, Mines and Water Resources Bulletin 7.*

Davies, W. E. 1965. Caverns of West Virginia with supplement. *West Virginia Geological and Economic Survey, Morgantown, West Virginia,* Vol. 19A.

Davies, W. E., and H. E. Legrand. 1972. Karst of the United States. In M. Herak and V. T. Stringfield (eds.), *Karst: Important Karst Regions of the Northern Hemisphere.* New York: Elsevier, pp. 467–505.

Davis, W. M. 1930. Origin of limestone caverns. *Bull. Geol. Soc. Am.* 41:475–628.

Dearolf, K. 1956. Survey of North American cave vertebrates. *Proceedings of the Pennsylvania Academy of Science* 30:201–206.

Deike, G. H. 1960. Origin and geologic relations of Breathing Cave, Virginia. *NSS Bull.* 22:30–42.

Deloach, N. 1978. *Diving Guide to Underwater Florida.* Jacksonville, FL: New World.

Dickson, G. W. 1975. A preliminary study of heterotrophic microorganisms as factors in substrate selection of troglobitic invertebrates. *NSS Bull.* 37(4):89–93.

Dickson, G. W. 1977. Variation among populations of the troglobitic amphipod crustacean *Crangonyx antennatus* Packard living in different habitats, I: morphology. *Int. J. Speleol.* 9:43–58.

Dickson, G. W., and R. Franz. 1980. Respiration rates, ATP turnover and adenylate energy charge in excised gills of surface and cave crayfish. *Comp. Biochem. Physiol.* 65A:375–379.

Dickson, G. W., and J. R. Holsinger. 1981. Variation among populations of the troglobitic amphipod crustacean *Crangonyx antennatus* Packard (Crangonyctidae) living in different habitats, III: population dynamics and stability. *Int. J. Speleol.* 11:33–48.

Dickson, G. W., and P. W. Kirk, Jr. 1976. Distribution of heterotrophic microorganisms in relation to detritivores in Virginia caves (with supplemental bibliography on cave mycology and microbiology). In B. C. Parker and M. K. Roane (eds.), *The Distributional History of the Biota of the Southern Appalachians,* Part IV, *Algae and Fungi.* Charlottesville: Virginia UP, pp. 205–226.

Dickson, G. W., L. A. Briese, and J. P. Giesy. 1979. Tissue metal concentrations in two crayfish species cohabitating a Tennessee cave stream. *Oecologia* 44:8–12.

Dilamarter, R. R., and S. C. Csallany (eds.). 1986. *Hydrologic Problems in Karst Regions.* Bowling Green: Western Kentucky University.

Dougherty, P. H. 1981. The impact of the agricultural land-use cycle on flood surges and runoff in an Kentucky karst region. *Proceedings of the 8th International Congress of Speleology* pp. 267–269.

Dougherty, P. H. 1983. Valley tides: a water balance analysis of land use response floods in a karst region. In P. H. Dougherty (ed.), *Environmental Karst,* Cincinnati, OH: GeoSpeleo Publication, pp. 25–36.

Douglas, H. H. 1964. *Caves of Virginia.* Falls Church: Virginia Cave Survey.

Drever, J. I. 1982. *The Geochemistry of Natural Waters.* Englewood Cliffs, NJ: Prentice-Hall.

Eraso, A. 1981. New contributions to the problems of dam building in karstic regions. *Proceedings of the 8th International Congress on Speleology* 1:348–350.

Estes, G. 1984. Connate contamination and karstic "pollution conduits" in the Big Sinking Creek oil field. *Cave Research Foundation 1981 Annual Report* 1981:12–14.

Estes, J. A., and J. R. Holsinger. 1976. A second troglobitic species of the genus *Lirceus* (Isopoda, Asellidae) from southwestern Virginia. *Proceedings of the Biological Society of Washington* 89(42):481–490.

Estes, J. A., and J. R. Holsinger. 1982. A comparison of the structure of two populations of the troglobitic isopod crustacean *Lirceus usdagalun* (Asellidae). *Polski Arch. Hydrobiol.* 29(2):453–461.

Exley, S., and N. DeLoach. 1981. The world's largest underwater cave. *Proceedings of the 8th International Congress of Speleology* 1:16–17.

Farlow, T. W. 1987. Nove. 14015. The South Carolina trip. *Der Fledermaus* 15(3):7.

Ferguson, G. E., C. W. Lingham, S. K. Love, and R. O. Vernon. 1947. Springs of Florida. *Florida Geological Survey, Geology Bulletin* 31:1–196.

Fieseler, R. G. 1978. Cave and karst distribution of Texas. In R. G. Fieseler, J. Jasek, and M. Jasek (eds.), *An Introduction to the Caves of Texas. NSS Convention Guidebook No. 19.* Austin, TX: Speleo, pp. 13–53.

Fleming, L. E. 1973. The evolution of the eastern North American isopods of the genus *Asellus* (Crustacea:Asellidae). *Int. J. Speleol.* 5(1973):283–310.

Fletcher, M. W. 1976. Microbial ecology of bat guano. *Annual Report of the Cave Research Foundation* 1975:51–53.

Fliermans, C. B., and E. L. Schmidt. 1977. *Nitrobacter* in Mammoth Cave. *Int. J. Speleol.* 9(1):1–20.

Folkerts, G. W. 1982. The Gulf Coast pitcher plant bogs. *Am. Sci.* 70(3):260–267.

Ford, D. C. 1965. The origin of limestone caves: a model from the Central Mendip Hills, England. *NSS Bull.* 27(4):109–132.

Franklin, E. R. 1978. The structure and dynamics of arthropod communities of bat guano ecosystems. *Bat Res. News* 19(1):23.

Franz, R. (ed.). 1982. *Rare and Endangered Biota of Florida*, Vol. 6, *Invertebrates.* Gainesville: Florida UP.

Franz, R., and H. H. Hobbs, Jr. 1983. *Procambarus (Ortmannicus) leitheuseri*, new species, another troglobitic crayfish (Decapoda:Cambaridae) from peninsular Florida. *Proceedings of the Biological Society of Washington* 96(2):323–332.

Franz, R., and D. S. Lee. 1982. Distribution and evolution of Florida's troglobitic crayfishes. *Bulletin of the Florida State Museum, Biological Science* 28(3):53–78.

Fuller, M. L. 1905. Notes on certain large springs of the Ozark region, Missouri and Arkansas. *Contributions to the hydrology of eastern United States, United States Geological Survey, Water Supply* 145:207–210.

Gardiner, W. W. 1971. A study of pollution levels of selected underground streams, Monroe County. *Speleo Tymes* 3(1):11–16.

Gardner, J. E. 1986. Invertebrate fauna from Missouri: caves and springs. *Missouri Department of Conservation, Natural History* 3:1–72.

Gardner, J. H. 1935. Origin and development of limestone caverns. *Geol. Soc. Am. Bull.* 46:1255–1274.

Garton, E. R., Jr. 1983. The effects of highway construction on the hydrogeologic environment at Bowden, West Virginia. R. R. Dilamarter and S. C. Csallany (eds.), *Hydrologic Problems in Karst Regions* Bowling Green: Western Kentucky University, pp. 439–449.

George, A. I. 1972a. Karst of Meade County, Kentucky. In *Guidebook to the Kentucky Speleo-fest, 1972*. Louisville, KY: Speleopress, pp. 1–10.

George, A. I. 1972b. Under the Road Cave. In *Guidebook to the Kentucky Speleo-fest, 1972*. Louisville, KY: Speleopress, p. 48.

George, A. I. 1985. Caves of Kentucky. *Caves and Karst Kentucky, Kentucky Geological Survey, Special Publication* 12(series XI):18–27.

Ginet, R. 1960. Ecologie, Ethologie et Biologie de *Niphargus* (Amphipodes Gammarides). *Ann. Speleol.* 15:1–254.

Ginet, R. 1969. Rhthme saisonnier des reproduction de *Niphargus. Ann. Speleol.* 24:2,387–397.

Gittleson, S. M., and R. L. Hoover. 1969. Cavernicolous protozoa: review of the literature and new studies in Mammoth Cave, Kentucky. *Ann. Speleol.* 24(4):737–776.

Gittleson, S. M., and R. L. Hoover. 1970. Protozoa of underground waters in caves. *Ann. Speleol.* 25(1):91–106.

Gounot, A. M. 1967. La microflore des limons argileux souterrains: son activite productrice dans la biocoenose cavernicole. *Ann. Speleol.* 22(1):23–146.

Grove, J. L. 1974. Ecology of Blanchard Springs Caverns, Ozark National Forest, Arkansas. Thesis, Memphis State University.

Grund, A. 1903. Die Karsthydrographie, Studien aus Westbosnien. *Geogr. Abh.* 7:1–200.

Gulden, B. 1984. Deep caves of the United States. *NSS News* 42(11):330.

Gulden, B. 1985. The long caves of the United States. *NSS News* 43(1):4–6.

Halliday, W. R. 1954. Ice caves of the United States. *NSS Bull.* 16:3–28.

Halliday, W. R. 1960. Pseudokarst in the United States. *NSS Bull.* 22:109–113.

Hamilton-Smith, E. 1971. The classification of cavernicoles. *NSS Bull.* 33:63–66.

Harris, J. A. 1970. Bat-guano cave environment. *Science* 169:1342–1343.

Hawes, R. S. 1939. The flood factor in the ecology of caves. *J. Anim. Ecol.* 8(1):1–5.

Hazelton, M., and E. A. Glennie. 1962. Cave fauna and flora. In C. H. D. Cullingford (ed.), *British Caving*, 2d ed. London: Routledge & Kegan Paul, Chapter 9, pp. 347–388.

Hempel, J. C. (compiler). 1975. Cave and karst of Monroe County, West Virginia. *Bulletin of the West Virginia Speleological Survey* 4.

Herrick, S. M., and H. E. Legrand. 1964. Solution subsidence of a limestone terrace in southwest Georgia. *Int. Assoc. Sci. Hydrol. Bull.* 9(2):25–36.

Hess, J. W., and W. B. White. 1989. Chemical hydrology. In W. B. White and E. L. White (eds.), *Karst Hydrology: Concepts from the Mammoth Cave Area*. New York: Van Nostrand Reinhold, pp. 145–174.

Hess, J. W., S. G. Wells, and T. A. Brucker. 1974. A survey of springs along the Green and Barren Rivers, Central Kentucky Karst. *NSS Bull.* 36(3):1–7.

Hess, J. W., S. G. Wells, J. F. Quinlan, and W. B. White. 1989. Hydrogeology of the south-central Kentucky karst. In W. B. White, and E. L. White (eds.), *Karst Hydrology: Concepts from the Mammoth Cave Area.* New York: Van Nostrand, pp. 15–63.

Hill, C. A. 1976. *Cave Minerals.* Austin, TX: Speleo.

Hill, C. A., and P. Forti. 1986. *Cave Minerals of the World.* Huntsville, AL: National Speleological Society.

Hill, S. B. 1981. Ecology of bat guano in Tamana Cave, Trinidad, W. I. *Proceedings of the 8th International Congress of Speleology* 1:243–246.

Hobbs, H. H., Jr. 1938. A new crawfish from Florida. *J. Washington Acad. Sci.* 28(2):61–65.

Hobbs, H. H., Jr. 1945. The subspecies and integrades of the Florida burrowing crayfish. *J. Washington Acad. Sci.* 35(8):247–260.

Hobbs, H. H., Jr. 1974. A checklist of the North and Middle American crayfishes (Decapoda:Astacidae and Cambaridae). *Smithsonian Contributions to Zoology* 166:1–161.

Hobbs, H. H., Jr. 1981. The crayfishes of Georgia. *Smithsonian Contributions to Zoology* 318:1–549.

Hobbs, H. H., Jr. 1989. An illustrated checklist of the American crayfishes (Decapoda:Astacidae, Cambaridae, and Parastacidae). *Smithsonian Contribution to Zoology* 480:1–236.

Hobbs, H. H., Jr., and R. Franz. 1986. New troglobitic crayfish with comments on its relationship to epigean and other hypogean crayfishes of Florida. *J. Crustacean Biol.* 6(3):509–519.

Hobbs, H. H., Jr., and D. S. Lee. 1976. A new troglobitic crayfish (Decapoda, Cambaridae) from peninsular Florida. *Proceedings of the Biological Society of Washington* 89(32):383–391.

Hobbs, H. H., Jr., H. H. Hobbs III, and M. A. Daniel. 1977. A review of the troglobitic decapod crustaceans of the Americas. *Smithsonian Contributions to Zoology* 244:1–183.

Hobbs, H. H., III. 1973. The population dynamics of cave crayfishes and their commensal ostracods from southern Indiana. Thesis, Indiana University, Bloomington, 247 pp.

Hobbs, H. H., III. 1985. Bermuda: a unique western Atlantic karst. *Pholeos* 5(2):3–15.

Hobbs, H. H., III. 1987. An Arkansas cave crayfish in peril? *Am. Caves* 2(3):24–25.

Hobbs, H. H., III, and F. M. Bagley. 1989. Shelta Cave Management Plan. *National Speleological Society, Special Publication, Huntsville, AL.*

Hoey, M. A. 1976. Pollution of karst waters in the Bloomington area. *Bloomington Indiana Grotto Newsletter* 12(3):19–21.

Holler, C. O., Jr. 1975. North Carolina Cave Survey. *North Carolina Cave Survey* 1(1):1–80.

Holler, C. O., Jr. 1981. North Carolina's Bat Caves: A significant region of tectonokarst. *Proceedings of the 8th International Congress of Speleology* 1:190–191.

Holler, C. O., Jr. 1985. Letter. *Der Fledermaus* 13(6):2–3.

Holsinger, J. R. 1963. Annotated checklist of the macroscopic troglobites of Virginia with notes on their geographic distribution. *NSS Bull.* 25:23–36.

Holsinger, J. R. 1966. A preliminary study on the effects of organic pollution of Banners Corner Cave, Virginia. *Int. J. Speleol.* 2:76–89.

Holsinger, J. R. 1967. Systematics, speciation, and distribution of the subterranean amphipod genus *Stygonectes* (Gammaridae). *Bulletin of the United States National Museum* 259.

Holsinger, J. R. 1972. The freshwater amphipod crustaceans (Gammaridae) of North America. *Biota of Freshwater Ecosystem Identification Manual No. 5, Environmental Protection Agency.*

Holsinger, J. R. 1975a. Observations on the dispersal of the cavernicolous amphipod crustacean *Crangonyx antennatus* (Gammaridae). *ASB Bull.* 22(2):58.

Holsinger, J. R. 1975b. Description of Virginia Caves. *Bulletin of the Virginia Division of Mineral Resources* 85:1–450.

Holsinger, J. R. 1977. A review of the systematics of the Holarctic amphipod family Crangonyctidae. *Crust. Suppl.* 4:244–281.

Holsinger, J. R. 1978. Systematics of the subterranean amphipod genus *Stygobromus* (Crangonyctidae), II: species of the eastern United States. *Smithsonian Contributions to Zoology* 266:1–144.

Holsinger, J. R. 1982. A preliminary report on the cave fauna of Burnsville Cove, Virginia. *NSS Bull.* 4(3):98–101.

Holsinger, J. R. 1988. Troglobites: the evolution of cave-dwelling organisms. *Am. Sci.* 76:147–153.

Holsinger, J. R., and T. E. Bowman. 1973. A new troglobitic isopod of the genus *Lirceus* (Asellidae) from southwestern Virginia, with notes on its ecology and additional cave records for the genus in the Appalachians. *Int. J. Speleol.* 5:261–271.

Holsinger, J. R., and D. C. Culver. 1970. Morphological variation in *Gammarus minus* Say (Amphipoda, Gammaridae) with emphasis on subterranean forms. *Postilla* 146:1–24.

Holsinger, J. R., and G. W. Dickson. 1978. Burrowing activity in the troglobitic amphipod crustacean *Crangonyx antennatus* Packard (Crangonyctidae). *NSS Bull.* 40(3):85.

Holsinger, J. R., and G. Longley. 1980. The subterranean amphipod crustacean fauna of an artesian well in Texas. *Smithsonian Contributions to Zoology* 308:1–62.

Holsinger, J. R. and S. B. Peck. 1971. The invertebrate cave fauna of Georgia. *NSS Bull.* 33:23–44.

Holsinger, J. R., R. A. Baroody, and D. C. Culver. 1976. The invertebrate cave fauna of West Virginia. *Bulletin of the West Virginia Speleological Survey* 7.

Holt, P. C. 1973. Branchiobdellids (Annelida:Clitellata) from some eastern North American caves, with descriptions of new species of the genus *Cambarincola*. *Int. J. Speleol.* 1973:219–256.

Horst, R. 1972. Bats as primary producers in an ecosystem. *NSS Bull.* 34(2):49–54.

Howard, A. D. 1963. The development of karst features. *NSS Bull.* 25:45–66.

Howell, W. M., and R. D. Caldwell. 1965. *Etheostoma (Oligocephalus) nuchale*, a new darter from a limestone spring in Alabama. *Tulane Studies Zool.* 12(4):101–108.

Hubbs, C. L., and E. R. Allen. 1943. Fishes of Silver Springs, Florida. *Proceedings of the Florida Academy of Science* 6:110–130.

Hubricht, L. 1976. The genus *Fontigens* from Appalachian caves (Hydrobiidae: Mesogastropoda). *The Nautilus* 90(2):86–88.

Hubricht, L. 1979. A new species of *Amnicola* from an Arkansas cave (Hydrobiidae). *The Nautilus* 94(4):142.

Hubricht, L., and J. G. Mackin. 1949. The freshwater isopods of the genus *Lirceus* (Asellota:Asellidae). *Am. Midland Natur.* 42(2):334–349.

Hunter, B. B., and W. J. Thomas. 1975. Isolation and existence of imperfect fungi in caves. *Proceedings of the Pennsylvania Academy of Science* 49:62–66.

Huppert, George, B. Wheeler, and L. Knox. 1983. Suburban expansion and sinkhole flooding: a case study from Tennessee. In P. H. Dougherty (ed.), *Environmental Karst*. Cincinnati, OH: GeoSpeleo Publication, pp. 15–23.

Hutchinson, G. E. 1957. *A Treatice on Limnology*, Vol. 1, *Geography, Physics, and Chemistry*. New York: Wiley.

Hyman, L. H. 1939. North American triclad Turbellaria, X: additional species of cave planarians. *Trans. Am. Microsc. Soc.* 58:276–284.

Hyman, L. H. 1951. North American triclad Turbellaria, XII: synopsis of the known species of freshwater planarians of North America. *Trans. Am. Microsc. Soc.* 70:154–167.

Jacobson, R. L., and D. Langmuir. 1970. The chemical history of some spring waters in carbonate rocks. *Ground Water* 8(3):5–9.

Jammal, S. E. 1984. Maturation of the Winter Park sinkhole. In B. F. Beck (ed.), *Sinkholes: Their Geology, Engineering and Environmental Impact*. Accord, MA: Balkeema, pp. 363–369.

Jegla, T. C. 1964. Studies of the eyestalk, metabolism, and molting and reproductive cycles in a cave crayfish. Thesis, University of Illinois, Urbana.

Jegla, T. C. 1966. Reproductive and molting cycles in cave crayfish. *Biol. Bull.* 130:345–358.

Jegla, T. C. 1969. Cave crayfish: annual periods of molting and reproduction. *Proceedings of the 4th International Congress of Speleology* 4–5:129–133.

Jegla, T. C., and T. L. Poulson. 1970. Circannian rhythms, I: reproduction in the cave crayfish *Orconectes pellucidus inermis*. *Comp. Biochem. Physiol.* 33:347–355.

Jefferson, G. T. 1976. Cave fauna. In T. D. Ford and C. H. D. Cullingford (eds.), *The Science of Speleology*. New York: Academic, pp. 357–421.

Jennings, J. N. 1971. *Karst*. London: MIT Press.

Jennings, J. N. 1985. *Karst Geomorphology*. Oxford, UK: Blackwell.

Johnson, M. L. 1981. Hydrochemical facies: a method to delineate the hydrology of inaccessible features of karst plumbing systems. *Proceedings of the 8th International Congress of Speleology*, 627–629.

Jones, W. K. 1973. Hydrology of limestone karst in Greenbrier County, West Virginia. *Bulletin of the United States Geological Survey* 36.

Jones. W. K. 1981. A karst hydrology study in Monroe County, West Virginia (U.S.A.). *Proceedings of the 8th International Congress of Speleology* 1:345–347.

Kashef, A. I. 1971. On ground water management in coastal aquifers. *Ground Water* 9:12–20.

Kashef, A. I. 1973. Pollution of groundwater by salt invasion. In Editorial Board (eds.), *Groundwater Pollution*. St. Louis, MO: Underwater Research Institute, pp. 72–90.

Kaye, C. A. 1957. The effect of solvent motion on limestone solution. *J. Geol.* 65:35–46.

Keith, J. H., and T. L. Poulson. 1981. Broken-back syndrome in *Amblyopsis spelaea*, Donaldson-Twin Cave, Indiana. *Cave Research Foundation, 1979 Annual Report*, 45–48.

Kenk, R. 1970. Freshwater triclads (Turbellaria) of North America, IV: the polypharngeal species of *Phagocata*. *Smithsonian Contributions to Zoology* 80:1–17.

Kenk, R. 1972. Freshwater planarians (Turbellaria) of North America. *Biota of Freshwater Ecosystems, Identification Manual No. 1*. Washington, DC: Environmental Protection Agency.

Kenk, R. 1973. Freshwater triclads (Turbellaria) of North America, VI: the genus *Dendrocoelopsis*. *Smithsonian Contributions to Zoology* 138:1–16.

Kenk, R. 1979. Freshwater triclads (Turbellaria) of North America, XI: *Phagocata holleri*, new species, from a cave in North Carolina. *Proceedings of the Biological Society of Washington* 92(2):389–393.

Kenk, R. 1984. Freshwater triclads (Turbellaria) of North America, XV: two new subterranean species from the Appalachian region. *Proceedings of the Biological Society of Washington* 97(1):209–216.

Kirk, P. W., Jr. 1973. Distributional survey of heterotrophic microorganisms in Old Mill Cave, Virginia. *NSS Bull.* 35(1):14–15.

Knight, E. L., and B. N. Irby. 1972. Additional information on a new caving area in the United States. *NSS News* 30(11):157.

Knight, E. L., B. N. Irby, and S. Carey. 1974. *Caves of Mississippi*. Hattiesburg, MS: Southern Mississippi Grotto of the National Speleological Society.

Knight, R. L. 1983. Energy basis of ecosystem control at Silver Springs, Florida, In T. D. Fontaine III and S. M. Bartell (eds.), *Dynamics of Lotic Ecosystems*. Ann Arbor, MI: Ann Arbor Science, pp. 161–179.

Kofoid, C. A. 1900. The plankton of Echo River, Mammoth Cave. *Trans. Am. Microsc. Soc.* 21:113–126.

Kohout, F. A., G. W. Leve, F. T. Smith, and F. T. Manheim. 1975. Red Snapper Sink and ground-water flow, offshore northeastern Florida. *International Association of Hydrologists, 12th International Congress, Karst Hydrology (Abstracts and Program), Alabama Geological Survey*, p. 60.

Kostalos, M., and R. L. Seymour. 1976. Role of microbial enriched detritus in the nutrition of *Gammarus minus* (Amphipoda). *Oikos* 27:512–516.

Lamoreaux, P. 1984. Catastrophic subsidence, Shelby County, Alabama. In B. F. Beck (ed.), *Sinkholes: Their Geology, Engineering and Environmental Impact*. Accord, MA: Balkema, pp. 131–136.

Lane, E. 1986. Karst in Florida. *Florida Geological Survey, Special Publication* No. 29.

Lange, A. L. 1953. Speleogens and their meaning. *Western Speleological Institute, Technical Report* 3:40–71.

Langmuir, D. 1971. The geochemistry of some carbonate ground waters in central Pennsylvania. *Geochim. Cosmochim. Acta* 35:1023–1045.

Langmuir, D. 1972. Controls on the amounts of pollutants in subsurface waters. *Earth Miner. Sci.* 42(2):9–13.

Lavoie, K. H. 1980a. Microbial and invertebrate interactions curing dung decomposition. *Annual Report of the Cave Research Foundation* 1980:20–21.

Lavoie, K. H. 1980b. Toxicity of carbide waste to heterotrophic microorganisms in caves. *Microb. Ecol.* 6(2):173–180.

Lavoie, K. H. 1981a. Abiotic effects on the successional decomposition of dung. *Proceedings of the 8th International Congress of Speleology* 1:262–264.

Lavoie, K. H. 1981b. Invertebrate interaction with microbes during the successional decomposition of dung. *Proceedings of the 8th International Congress of Speleology* 1:265–266.

Lavoie, K. H. 1981c. Toxicity of spent carbide waste to microbes in caves. Cave Research Foundation, 1979 Annual Report, pp. 44–45.

Lee, D. S. 1969a. A food study of the salamander *Haideotriton wallacei* Carr. *Herpetologica* 25:175–177.

Lee, D. S. 1969b. Notes on the feeding behavior of cave-dwelling bullfrogs. *Herpetologica* 25(3):211–212.

LeGrand, H. E. 1973. Hydrological and ecological problems of karst regions. *Science* 179(4076):859–864.

Leitheuser, A. T., J. R. Holsinger, R. Olson, N. R. Pace, R. L. Whitman, and T. Whitmore. 1985. Ecological analysis of the Kentucky Cave Shrimp, *Palaemonias ganteri* Hay, at Mammoth Cave National Park (Phase V). *Old Dominion University Research Foundation, Norfolk*.

Lewis, J. J. 1980. A comparison of *Pseudobaicasellus* and *Caecidotea*, with a description of *Caecidotea bowmani*, n. sp. (Crustacea:Isopoda:Asellidae). *Proceedings of the Biological Society of Washington* 93(2):314–326.

Lewis, J. J. 1981. *Caecidotea salemensis* and *C. fustis*, new subterranean asellids from the Salem Plateau (Crustacea:Isopoda:Asellidae). *Proceedings of the Biological Society of Washington* 94(2):579–590.

Lewis, J. J., and T. E. Bowman. 1977. *Caecidotea carolinensis*, n. sp., the first subterranean water slater from North Carolina (Crustacea:Isopoda:Asellidae). *Proceedings of the Biological Society of Washington* 90(4):968–974.

Lewis, J. J., and J. R. Holsinger. 1985. *Caecidotea pheatica*, a new phreatobitic isopod crustacean (Asellidae) from southeastern Virginia. *Proceedings of the Biological Society of Washington* 98(4):1004–1011.

Lisowski, E. A. 1983. Distribution, habitat, and behavior of the Kentucky cave shrimp *Palaemonias ganteri* Hay. *J. Crustacean Biol.* 3(1):88–92.

Lisowski, E. A., and T. L. Poulson. 1981. Impacts of Lock and Dam Six on baselevel ecosystems in Mammoth Cave. *Cave Research Foundation, 1979 Annual Report*, pp. 48–54.

Lovett, T. 1949. Microorganisms in caves. *Bull. Br. Speleol. Assoc.* 2(10):51–52.

MacArthur, R. H., and E. O. Wilson. 1967. *The Theory of Island Biogeography.* Princeton, NJ: Princeton UP.

MacNale, B. 1956. Some interesting facts about caves of South Carolina. *The State,* 9 September, 2C.

Maegerlein, S. 1970. Florida cave diving. *Bloomington, Indiana Grotto Newsletter* 9(2):30–38.

Mallman, W. L., and W. N. Mack. 1961. Biological contamination of groundwater. *Public Health Service Technical Report* No. W61-5:35–43.

Malott, C. A. 1937. Invasion theory of cavern development. *Proceedings of the Geological Society of America for 1937* 41:285–316.

Maness, L. V., and C. O. Holler. 1979. North Carolina coastal plain caves. *NSS Bull.* 41(4):113.

Marshall, N. 1947. The spring run and cave habitats of *Erimystax harperi* (Fouler). *Ecology* 28(1):68–75.

Martel, E. A. 1910. LaTheorie de la "Grundwasser" et les eaux souterraines du karst. *La Geographie* 21:126–130.

Martin, B. 1977. The influence of patterns of guano renewal on bat guano arthropod communities. *Annual Report of the Cave Research Foundation* 1976:36–42.

Mason-Williams, A. 1965. The growth of fungi in caves in Great Britain. *Studies Speleol.* 1(2-3):96–99.

Mason-Williams, A., and K. Benson-Evans. 1958. A preliminary investigation into the bacterial and botanical flora of caves in South Wales. *Cave Research Group, Great Britain*, 8.

Matson, G. C. 1909. Water resources of the Bluegrass region, Kentucky. *United States Geological Survey, Water Supply Papers* 233:42–44.

Matthews, L. E. 1971. Descriptions of Tennessee Caves. *Bulletin of the Tennessee Division of Geology* 69:1–150.

Matthews, R. C., Jr., A. D. Bosnak, D. S. Tennant, and E. L. Morgan. 1975. Assessment of chlorinated stream water on cave crayfish (*Orconectes australis australis*) survival. *ASB Bull.* 22(2):67.

McCallie, S. W. 1908. A preliminary report on the underground waters of Georgia. Georgia Geological Survey Bulletin, 15.

McDaniel, V. R., and J. E. Gardner. 1977. Cave fauna of Arkansas: vertebrate taxa. *Proceedings of the Arkansas Academy of Science* 31:68–71.

McDaniel, V. R., and K. L. Smith. 1976. Cave fauna of Arkansas: selected invertebrate taxa. *Procdings of the Arkansas Academy of Science* 30:57–60.

McDaniel, V. R., K. N. Paige, and C. R. Tumlison. 1979. Cave fauna of Arkansas: additional invertebrate and vertebrate records. *Proceedings of the Arkansas Academy of Science* 33:84–85.

Medville, D., and H. Medville. 1971. Caves of Randolph County. *Bulletin of the West Virginia Speleological Survey*, 1:1–218.

Meinzer, O. E. 1923. The occurrence of ground water in the United States. *United States Geological Survey, Water Supply Papers* 489:1–321.

Meinzer, O. E. 1927a. Geology of large springs. *Bull. Geol. Soc. Am.* 38:213–216.

Meinzer, O. E. 1927b. Large springs in the United States. *United States Geological Survey, Water Supply Papers* 557.

Minckley, W. L. 1963. The ecology of a spring stream Doe Run, Meade County, Kentucky. *Wildlife Monogr.* 11:1–124.

Minckley, W. L., and G. A. Cole. 1963. Ecological and morphological studies on gammarid amphipods (*Gammarus* spp.) in spring-fed streams of northern Kentucky. *Occasional Papers of the C. C. Adams Center for Ecological Studies* 10:1–35.

Minear, R. A., and J. W. Patterson. 1973. Septic tanks and groundwater pollution. In *Groundwater*. St. Louis, MO: Underwater Research Institute, pp. 53–71.

Mitchell, R. W. 1970. Total number and density estimates of some species of cavernicoles inhabiting Fern Cave, Texas. *Ann. Speleol.* 25:73–90.

Modlin, R. F. 1986. *Caecidotea dauphina*, a new subterranean isopod from a barrier island in the northern Gulf of Mexico (Crustacea:Isopoda:Asellidae). *Proceedings of the Biological Society of Washington* 99(2):316–322.

Moneymaker, B. C. 1941. Subriver solution cavities in the Tennessee Valley. *J. Geol.* 49:74–86.

Moore, D. G. 1954. Origin and development of sea caves. *NSS Bull.* 16:71–76.

Moore, G. W. 1952. Speleothem: a new cave term. *NSS News* 10(6):2.

Morehouse, D. F. 1968. Cave development via the sulfuric acid reactions. *NSS Bull.* 30:1–10.

Muxart, R., T. Stchouzkoy, and J. C. Franck. 1968. Observations hydrokarstologiques dans le bassin amont de la seille (Jura). *Proceedings of the 4th International Congress of Speleology* 3:175–180.

Mylroie, J. E. 1978. A functional classification of karst features. *NSS Bull.* 40:3.

Newton, J. G. 1977. Induced sinkholes: a continuing problem along Alabama highways. In J. S. Tolson and F. L. Doyle (eds.), *Karst Hydrogeology*. Huntsville: Alabama UP, pp. 303–304.

Nicholas, B. G. 1960. Checklist of macroscopic troglobitic organisms of the United States. *Am. Midland Natur.* 64(1):123–160.

Odum, H. T. 1953. Factors controlling marine invasion in Florida freshwaters. *Bulletin of Marine Sciences of the Gulf and Carribean* 3:134–156.

Odum, H. T. 1956. Primary production in flowing waters. *Limnol. Oceanogr.* 1:102–117.

Odum, H. R. 1957a. Primary production measurements in eleven Florida springs and a marine turtle grass community. *Limnol. Oceanogr.* 2:85–97.

Odum, H. T. 1957b. Trophic structures and productivity of Silver Springs, Florida. *Ecol. Monogr.* 27(1):55–112.

Ogden, A. E. 1978. Cavernous strata of Arkansas. *NSS Bull.* 40(3):95.

Ogden, A. E. 1980. Pseudokarst caves of Arkansas. *NSS Bull.* 42(2):27.

Ogden, A. E. 1981. Pseudo Karst caves of Arkansas. *Proceedings of the 8th International Congress of Speleology* 2:766–768.

Ogden, A. E. 1982. Karst denudation rates for selected spring basins in West Virginia. *NSS Bull.* 44(1):6–10.

Ogden, A. E., and S. K. Flint. 1976. Road salt contamination of springs and wells in carbonate rocks, Monroe County, West Virginia. *Proceedings of the West Virginia Academy of Science* 48(1):42.

Ogden, A. E., and J. P. Reger. 1977. Morphometric analysis of dolines for predicting ground subsidence, Monroe County, West Virginia. In R. R. Dilamarter and S. C. Csallany (eds.), *Hydrologic Problems in Karst Regions*. Bowling Green: Western Kentucky UP, pp. 130–139.

Ogden, A. E., W. M. Goodman, and S. R. Rothermel. 1981. Speleogenesis of Arkansas Ozark caves. *Proceedings of the 8th International Congress of Speleology* 2:769–771.

Olsen, S. J. 1958. The Wakulla Cave. *Natur. History* 67(7):396–403.

Otvos, E. G., Jr. 1976. "Pseudokarst" and "pseudokarst terrains": problems of terminology. *Bull. Geol. Soc. Am.* 87:1021–1027.

Palmer, A. N. 1981. *A Geological Guide to Mammoth Cave National Park.* Teaneck, NJ: Zephyrus.

Palmer, R. 1985. *The Blue Holes of the Bahamas.* London: Jonathan Cape.

Palmer, R. 1989. *Deep into Blue Holes.* London: Unwin Hyman.

Park, O., and T. C. Barr, Jr. 1961. Some observations on a cave cricket. *Bull. Entomol. Soc. Am.* 7:144(abstract).

Parker, G. G., L. M. Shown, and K. W. Ratzlaff. 1964. Officers Cave, a pseudokarst feature in altered tuff and volcanic ash of the John Day Formation in eastern Oregon. *Geol. Soc. Am.* 75:393–402.

Paul, R. W., Jr., E. F. Benfield, and J. Cairns, Jr. 1983. Dynamics of leaf processing in a medium-sized river. In T. D. Fontaine III and S. M. Bartell (eds.), *Dynamics of Lotic Ecosystems,* Ann Arbor, MI: Ann Arbor Science, pp. 403–423.

Peck, S. B. 1971. The invertebrate fauna of tropical American caves, I: Chilibrillo Cave, Panama. *Ann. Speleol.* 26:423–437.

Peck, S. B. 1973. Feeding efficiency in the cave salamander *Haideotriton wallacei. Int. J. Speleol.* 5:15–19.

Peck, S. B. 1974. The food of the salamanders *Eurycea lucifuga* and *Plethodon glutinosus* in caves. *NSS Bull.* 36(4):7–10.

Peck, S. B., and B. L. Richardson. 1976. Feeding ecology of the salamander *Eurycea lucifugus* in the entrance, twilight, and dark zones of caves. *Ann. Speleol.* 31:175–182.

Peek, H. M. 1951. Cessation of flow of Kissingen Spring in Polk County, Florida. *Florida Geological Survey, Report of Investigations* 7(3):75–82.

Picknett, R. G. 1967. Caver's Perfume. *Cave Research Group, Great Britian* 105:13–14.

Picknett, R. G., L. G. Bray, and R. D. Stenner. 1976. The chemistry of cave waters. In T. D. Ford and C. H. D. Cullingford (eds.), *The Science of Speleology.* New York: Academic, pp. 213–266.

Pohl, E. R. and W. B. White. 1965. Sulfate minerals: their origin in the central Kentucky karst. *Am. Mineral.* 50:1461–1465.

Popenoe, P., F. A. Kohout, and F. T. Manheim. 1984. Seismic-reflection studies of sinkholes and limestone dissolution features on the northeastern Florida shelf. In B. F. Beck (ed.), *Sinkholes: Their Geology, Engineering and Environment Impact.* Accord, MA: Balkema, pp. 43–57.

Poulson, T. L. 1963. Cave adaptation in amblyopsid fishes. *Am. Midland Natur.* 70(2):257–290.

Poulson, T. L. 1964. Animals in aquatic environments: animals in caves. In D. B. Dill (ed.), *Handbook of Physiology: Adaptation to the Environment.* American Physiological Society.

Poulson, T. L. 1967. Comparison of cave stream communities. *Cave Research Foundation, 9th Annual Report,* pp. 33–34.

Poulson, T. L. 1968. Aquatic cave communities. *Cave Research Foundation, 10th Annual Report,* pp. 16–18.

Poulson, T. L. 1972. Bat guano ecosystems. *NSS Bull.* 34(2):55–59.

Poulson, T. L. 1976. Management of biological resources in caves. *Proceedings of the 1st National Cave Management Symposium, Speleobooks, Albuquerque*, pp. 46–52.

Poulson, T. L. 1978. Organization of cave communities: energy availability. *Bull. Ecol. Soc. Am.* 59(2):94–95.

Poulson, T. L. 1979. Community organization. *Annual Report of the Cave Research Foundation*, pp. 41–45.

Poulson, T. L., and W. B. White. 1969. The cave environment. *Science* 165:971–981.

Prager, R. D. 1972. Groundwater pollution of the karst regions of the eastern U.S. *Georgia Underground* 9(2):34.

Price, P. H., J. B. McCue, and H. A. Hoskins. 1936. *Springs of West Virginia.* Morgantown: West Virginia Geological Survey.

Prins, R. 1968. Comparative ecology of the crayfishes *Orconectes rusticus rusticus* and *Cambarus tenebrosus* in Doe Run, Mead County, Kentucky. *Int. Rev. gesamten Hydrobiol.* 53(5):667–714.

Purdue, A. H., and H. D. Miser. 1916. Eureka Springs and Harrison Quadrangles, Arkansas-Missouri. *United States Geological Survey Atlas*, Folio 202, pp. 1–2, Fig. 2.

Pushkar, P. 1985. Sloans Valley Water Study Completed. *Cave Cricket Gazette* 1985:42–43.

Pye, V. I., and R. Patrick. 1983. Ground water contamination in the United States. *Science* 221:713–718.

Pylka, J. M., and R. D. Warren. 1958. A population of *Haideotriton* in Florida. *Copeia* 4:334–336.

Quinlin, J. F. 1983. Groundwater pollution by sewage, creamery waste, and heavy metals in the Horse Cave area, Kentucky. In P. H. Dougherty (ed.), *Environmental Karst*. Cincinnati, OH: GeoSpeleo, p. 52 (abstract).

Reeves, F. 1932. Thermal springs of Virginia. *Virginia Geological Survey Bulletin* 36.

Reichle, D. E., J. D. Palmer, and O. Park. 1965. Persistent locomotor activity in the cave cricket, *Hadenoecus subterraneus*, and its ecological significance. *Am. Midland Natur.* 74:57–66.

Relyea, K., and B. Sutton. 1972. Notes on epigean populations of fishes in subterranean waters of Florida. *Quart. J. Florida Acad. Sci.* 35(suppl. 1):14.

Relyea, K., and B. Sutton. 1973a. Cave dwelling yellow bullheads in Florida. *Florida Sci.* 36(1):31–34.

Relyea, K., and B. Sutton. 1973b. Ecological data for a Florida troglobitic crayfish. *Florida Sci.* 36(204):234–235.

Relyea, K., D. Blody, and K. Bankowski. 1976. A Florida troglobitic crayfish: biogeographic implications. *Florida Sci.* 39(2):71–72.

Rhoades, R., and M. N. Sinacori. 1941. Pattern of groundwater flow and solution. *J. Geol.* 49:785–794.

Richards, A. M. 1971. An ecological study of the cavernicolous fauna of the Nullarbor Plain, Southern Australia. *J. Zool.* 164:1–60.

Robbins, J. W. D., and G. J. Krig. 1973. Groundwater pollution by agriculture. In

Editorial Board (eds.), *Groundwater Pollution*. St. Louis, MO: Underwater Research Institute, pp. 91–114.

Robison, H. W., and K. L. Smith. 1982. The endemic flora and fauna of Arkansas. *Proceedings of the Arkansas Academy of Sciences* 36:52–57.

Rosenau, J. C., G. L. Faulkner, C. W. Hendry, Jr., and R. W. Hull. 1977. Springs of Florida. *Florida Bureau of Geology Bulletin* 31:1–461.

Rutherford, J. M., and R. H. Handley. 1976. The Greenbrier Caverns. *NSS Bull.* 38(3):41–52.

Saunders, J. W. 1973. A reconnaissance of springs and caves along the south side of Green River in the central Kentucky karst. *Wisconsin Speleol.* 12(2):45–50.

Saunders, J. W., R. K. Ortiz, and W. F. Koerschner III. 1981. Major groundwater flow direction in the Sinking Creek and Meadow Creek drainage basins of Giles and Craig Counties, Virginia, USA. *Proceedings of the 8th International Congress of Speleology* 1:398–400.

Schagel, S. R., and C. M. Breder. 1947. A study of the oxygen consumption of blind and eyed characins in light and in darkness. *Zoologica* 32:17–28.

Schindel, G. 1984. Pollution of the Lost River Cave System. *Central Kentucky Cave Survey Bulletin* 1:83–84.

Schram, M. D. 1980. The troglobitic Asellidae (Crustacea:Isopoda) of northwest Arkansas. Thesis, University of Arkansas, Fayetteville.

Scott, W. 1909. An ecological study of the plankton of Shawnee Cave. *Biol. Bull.* 17:386–402.

Senger, C. M. 1980. Relationships between cave morphology and cave climate. *N. Am. Biospeleol. Newslett.* 20:2.

Sevenair, J. P., and D. R. Williamson. 1983. The Caves of Louisiana. *Proceedings of the Louisiana Academy of Sciences* 46:109–113.

Shuster, E. T., and W. B. White. 1971. Seasonal fluctuations in the chemistry of limestone springs: a possible means for characterising carbonate aquifers. *J. Hydrol.* 14:93–128.

Sloan, W. C. 1956. The distribution of aquatic insects in two Florida springs. *Ecology* 37(1):81–98.

Smith, A. R. 1971. Cave and karst regions of Texas. Natural History of Texas Caves, In Lundelius, E. L., and B. H. Slaughter (eds.), *Natural History of Texas Caves*. Dallas, TX: Gulf Natural History.

Smith, D. I., and D. G. Mead. 1962. The solution of limestone. *Proceedings of the Speleological Society of the University of Bristol* 9:188–211.

Smith, D. I., T. C. Atkinson, and D. P. Crew. 1976. The hydrology of limestone terrains. In T. D. Ford and C. H. D. Cullingford (eds.), *The Science of Speleology*. New York: Academic, pp. 179–212.

Smith, K. L. 1977. Biological aspects of *Asellus antricolus* (Creaser) (Isopoda: Asellidae) in an Ozark cave stream. Thesis, Arkansas State University.

Smith, K. L. 1984. The status of *Cambarus zophonastes* Hobbs and Bedinger, an endemic cave crayfish from Arkansas. *Arkansas Natural Heritage Commission, Status Report, Little Rock, Arkansas*.

Spangler, L. E., and J. Thrailkill. 1981. Hydrogeology of northern Fayette County and southern Scott County, Kentucky, USA. *Proceedings of the 8th International Congress of Speleology* 2:553–555.

Spigner, B. C., and S. L. Graves. 1983. Ground-water development problems associated with folded carbonate-rock aquifers in the Irondale area, Alabama. In R. R. Dilamarter and S. C. Csallany (eds.), *Hydrologic Problems in Karst Regions*. Bowling Green: Western Kentucky University, pp. 241–248.

Stearns, N. D., H. T. Stearns, and G. A. Waring. 1937. Thermal springs in the United States. *United States Geological Survey Water Supply Papers* 679-B.

Stecko, D. 1985. Industrial waste dumped at landfill. *Cave Cricket Gazette*, p. 43.

Steeves, H. R., III. 1963. Two new troglobitic asellids from West Virginia. *Am. Midland Natur.* 70(2):462–465.

Steeves, H. R., III. 1966. Evolutionary aspects of the troglobitic asellids of the United States: The *Hobbsi, Stygius* and *Cannulus* Groups. *Am. Midland Natur.* 75(2):392–403.

Stephenson, L. W., and J. O. Veatch. 1915. Underground waters of the Coastal Plain of Georgia and a discussion of the quality of the waters, by R. B. Dole. *United States Geological Survey Water Supply Paper* 341.

Stieglitz, R. D., and W. E. Schuster. 1985. Water quality and land development in a karst area: Door County, Wisconsin. *Wisconsin Acad. Rev.* December:11–13.

Stone, W. C. (ed.) 1989. *The Wakulla Springs Project.* Austin, TX: Raines Graphics.

Stringfield, V. T. 1966. Artesian water in tertiary limestone in the southeastern United States. *United States Geological Survey Professional Paper* 517.

Sun, P. C., J. H. Griner, and J. L. Poole. 1963. Large springs of east Tennessee. *United States Geological Survey, Water Supply Paper* 1755.

Sweeting, M. M. 1973. *Karst Landforms.* New York: Columbia UP.

Swinnerton, A. C. 1932. Origin of limestone caverns. *Bull. Geol. Soc. Am.* 43:662–693.

Teal, J. M. 1957. Community metabolism in a temperate cold spring. *Ecol. Monogr.* 27:283–302.

Thrailkill, J. 1985. The Inner Blue Grass Karst Region. In *Caves and Karst of Kentucky, Kentucky Geological Survey, Special Publication* 12(Series XI):28–62.

Thrailkill, J., W. M. Hopper, M. R. McCann, and J. W. Troester. 1983. Problems associated with urbanization in the inner Bluegrass karst region. In P. H. Dougherty (ed.), *Environmental Karst.* Cincinnati, OH: GeoSpeleo, pp. 51–52.

Tilly, L. J. 1968. The structure and dynamics of Cone Spring. *Ecol. Monogr.* 38(2):169–197.

Tomaselli, R. 1947. Notes sur la vegetation des grottes de l'Herault. *Ann. Speleol.* 1(4):173–185.

Tomaselli, R. 1951. La vegetozione delle Grotte. *Speleon* 2(1):63–68.

Tomaselli, R. 1953. Observations de Biospeleologie vegetale. *Bull. Fed. Speleol. Belgium* 4:28–32.

Tomaselli, R. 1956. Relazione sulla nomenclatura botanica speleologica. *Atti 1st Bot. Lab. Crittogam. Pavia, Series 5* 12(2):203–221.

Troianai, D. 1954. Laconsommation d'oxygene quelque Gammaridae. *Compt. Rend.* 239:1540–1542.

Trudgill, S. 1985. *Limestone Geomorphology*. London: Longman.

U.S. Fish and Wildlife Service. 1984. *Watercress Darter Recovery Plan*. Atlanta, GA: U.S. Fish and Wildlife Services.

VanCouvering, J. A. 1962. Characteristics of large springs in Kentucky. *Kentucky Geological Survey*, Series 10, Information Circular 8.

Vandel, A. 1965. *Biospeleology: The Biology of Cavernicolous Animals*. English Edition Translation by B. E. Freedman. New York: Pergamon.

Van Luik, S. C. 1981. Ecology of crayfishes from West Virginia caves. *Proceedings of the 8th International Congress of Speleology* 2:657–658.

Varnedoe, W. W., Jr. 1973. Alabama caves and caverns. Private publication, Varnedoe, Huntsville, Alabama, p. 1478.

Varnedoe, W. W., Jr. 1980. A history of Alabama caves and caving. *NSS Bull.* 42(2):25(Abstract).

Veitch, J. D. 1967. The Caves of Alabama. A guide for the 1967 Convention of the National Speleological Society. Private Publication, John Veitch, Huntsville Grotto.

Vineyard, J. D. 1981. Guidebook to karst and caves of the Ozark Region of Missouri and Arkansas. In B. F. Beck (ed.), *8th International Congress of Speleology Guidebook to Karst and Caves of Tennessee and Missouri*. Americus: Georgia Southwestern College, pp. 33–68.

Vineyard, J. D., and G. L. Feder. 1974. Springs of Missouri. *Missouri Geological Survey and Water Resources*, WR29.

Wagner, G. H., K. F. Steele, H. C. McDonald, and T. L. Coughlin. 1976. Water quality as related to linears, rock chemistry, and rain water chemistry in a rural carbonate terrain. *J. Environ. Quality* 5(4):444–451.

Walker, B. A. 1961. Isopods in Doe Run. *Trans. Am. Microsc. Soc.* 80:385–390.

Warden, T. 1971. A new caving area in the United States. *NSS News* 29(2):17–19.

Waring, G. A. 1965. Thermal springs of the United States and other countries of the world: a summary. (revised by R. R. Blankenship and R. Bentall). *United States Geological Survey, Professional Paper* 492.

Warren, R. D. 1961. The obligative cavernicoles of Florida. *The Florida Speleological Society, Special Papers* 1:1–10.

Warren, R. D. 1962. A report of the Florida Cave Survey. *Florida Cave Survey, Special Report*.

Watson, T. L. 1924. Thermal springs of the southeast Atlantic states. *J. Geol.* 32(5):373–384.

Weingartner, D. L. 1977. Production and trophic ecology of two crayfish species cohabiting an Indiana cave. Thesis, Michigan State University.

Wells, S. G. 1973. Environmental problems in Indiana's Lost River Karst watershed. *NSS Bull.* 35(1):11–12.

Werner, E. 1983a. Chloride ion variations in some springs of the Greenbrier limestone karst of West Virginia. In R. R. Dilamarter and S. C. Csallany (eds.), *Hydrologic Problems in Karst Regions*. Bowling Green: Western Kentucky University, pp. 357–363.

Werner, E. 1983b. Effects of highways on karst springs: an example from Pocahontas County, West Virginia. In P. H. Dougherty (ed.), *Environmental Karst*. Cincinnati, OH: GeoSpeleo, pp. 3–13.

Wetterhall, W. S. 1965. Reconnaissance of springs and sinks in west-central Florida. *Florida Geological Survey, Report of Investigations*, No. 39.

White, E. L., and B. M. Reich. 1970. Behavior of annual floods in limestone basins in Pennsylvania. *J. Hydrol.* 10:193–198.

White, W. B. 1960. Terminations of passages in Appalachian caves as evidence for a shallow phreatic origin. *NSS Bull.* 22:43–54.

White, W. B. 1976. Cave minerals and speleothems. In T. D. Ford and C. H. D. Cullingford (eds.), *The Science of Speleology*. New York: Academic, pp. 267–327.

White, W. B. 1988. *Geomorphology and Hydrology of Karst Terrains*. New York: Oxford UP.

White, W. B., and J. Longyear. 1962. Some limitations on speleogenetic speculation imposed by the hydraulics of groundwater flow in limestone. *Nittany Grotto Newsletter* 10(9):155–167.

White, W. B., and E. L. White (eds.). 1989. *Karst Hydrology: Concepts from the Mammoth Cave Area*. New York: Van Nostrand Reinhold.

White, W. B., R. A. Watson, E. R. Pohl, and R. Brucker. 1970. The Central Kentucky Karst. *Geograph. Rev.* 60(1):88–115.

Whitford, L. A. 1956. The communities of algae in the springs and spring streams of Florida. *Ecology* 37:433–442.

Whitlach, G. I. 1943. Mineral springs and wells in Tennessee. *J. Tennessee Acad. Sci.* 18(4):305–324.

Whittaker, F. H., and L. G. Hill. 1968. *Proteocephalus chologasteri* sp. n. (Cestoda:Proteocephalidae) from the spring cavefish *Chologaster agassizi* Putman, 1782 (Pisces:Amblyopsidae) of Kentucky. *Proceedings of the Helminthological Society of Washington* 35(1):15–18.

Williams, J. D. 1968. A new species of sculpin, *Cottus pygmaeus*, from a spring in the Alabama River Basin. *Copeia* 1968(2):334–342.

Williams, J. D., and D. A. Etnier. 1982. Description of a new species, *Fundulus julisia*, with a redescription of *Fundulus albolineatus* and a diagnosis of the subgenus *Xenisma* (Teleostei:Cyprinodontidae). *Occasional Papers of the Museum of Natural History, The University of Kansas* 102:1–20.

Williams, J. D., and W. M. Howell. 1979. An albino sculpin from a cave in the New River drainage of West Virginia (Pisces:Cottidae). *Brimleyana* 1:141–146.

Willis, L. D., Jr. 1986. *Recovery Plan for the Ozark Cavefish (Amblyopsis rosae)*. Jackson, MS: United States Fish and Wildlife Service.

Willis, L. D., and A. V. Brown. 1985. Distribution and habitat requirements of the Ozark cavefish, *Amblyopsis rosae*. *Am. Midland Natur.* 114(2):311–317.

Wilmoth, B. M. 1972. Salty groundwater and meteoric flushing of contaminated aquifers in West Virginia. *Ground Water* 10(1):99–106.

Wilson, C. W., Jr. 1980. Geology of Cedars of Lebanon State Park and Forest and vicinity in Wilson County, Tennessee. *Tennessee Division of Geology, State Park Series* 1:1–20.

Wilson, J. 1977. Caves: changing ecosystems? *Studies Speleol.* 3(1):35–38.

Wolfe, T. E. 1964. Cavern development in the Greenbrier series. *NSS Bull.* 26:37–60.

Woods, L. P., and R. F. Inger. 1958. The cave, spring and swamp fishes of the family Amblyopsidae of central and eastern United States. *Am. Midland Natur.* 58(1):232–256.

Wray, D. L. 1952. Some new North American Collembola. *Bull. Brooklyn Entomol. Soc.* 47:95–106.

Yeatman, H. C. 1964. A new cavernicolous cyclopoid copepod from Tennessee and Illinois. *J. Tennessee Acad. Sci.* 39(3):95–98.

Youngsteadt, N. W., and J. O. Youngsteadt. 1978. A survey of some cave invertebrates from northern Arkansas. *Arkansas Cave Studies* 1:1–13.

4 High-Gradient Streams of the Appalachians

J. BRUCE WALLACE

Department of Entomology, University of Georgia, Athens, GA 30602

JACKSON R. WEBSTER

Department of Biology, Virginia Polytechnic Institute and State University, Blacksburg, VA 24061

REX L. LOWE

Department of Biology, Bowling Green State University, Bowling Green, OH 43404

The southern Appalachian region encompasses portions of nine states: West Virginia, Maryland, Virginia, Kentucky, Tennessee, North and South Carolina, Georgia, and Alabama. Within this region of abundant rainfall, extensive forests, and rugged terrain, many streams originate, serving a growing and diverse assemblage of interests, such as municipal, manufacturing, hydroelectric, and recreational needs. This region includes the headwaters, or portions of headwaters, of many rivers, including the Potomac, Ohio, James, Roanoke, Yadkin-Pee Dee, Santee, Savannah, Chattahoochee–Apalachicola, Alabama, and Tennessee. Water originating from these watersheds constitutes an important resource for major metropolitan areas. Thus, management of headwater streams and their watersheds is vital to the rapidly growing population of the region.

The major objectives of this chapter are to (1) synthesize our current knowledge relative to the biotic structure and function of high-gradient streams in the Southeast; (2) identify factors that control or regulate biotic structure and function in these systems; (3) evaluate the impact of various management (and mismanagement) practices on the biota of streams within the region; and, (4) address immediate and long-term research needs for high-gradient streams. We characterize high-gradient streams as those with longitudinal gradients exceeding 0.15% slope within the Blue Ridge and Ridge and Valley provinces. Many streams draining the Piedmont Plateau have reaches with slopes far exceeding 0.15% especially in the vicinity of the fall line, and these systems are addressed elsewhere in this book (see Mulholland and Lenat,

Chapter 5). Although slopes of 0.15% may appear low for high-gradient streams, larger fifth- and sixth-order streams of tortuous whitewater rivers, such as the Chattooga (Georgia–South Carolina) and the Ocoee (Tennessee), have gradients of only 0.5–1.1 %. In contrast, many small headwater (first- and second-order) streams of the southern Appalachians have gradients exceeding 30%.

PHYSIOGRAPHY OF THE REGION

In the Appalachian region, climate, lithology, and relief display considerable variation. Each of these components of the physical environment influences the structure of biotic communities of streams within this region. The southern Appalachian region consists of a series of different geological provinces, including the Blue Ridge, Ridge and Valley, and the Cumberland and Allegheny plateaus (Fig. 1). The Blue Ridge province extends from northern Georgia to southern Pennsylvania and is bordered on the east by the Piedmont Plateau. In most locations the Blue Ridge Escarpment rises abruptly from the lower Piedmont Plateau (Fig. 2). A dense deciduous forest covers much of the Blue Ridge, and in the southern section a coniferous forest consisting of Fraser fir, *Abies fraseri*, and red spruce, *Picea rubens*, constitutes the dominant vegetation near the crests of the higher mountains. Slopes are generally steep, with many exceeding 30% (Fenneman 1938).

The Ridge and Valley province, extending from the Coastal Plain in Alabama to the St. Lawrence Valley in the north, borders the western side of the Blue Ridge province (Fig. 1, Fenneman 1938). The Ridge and Valley province is characterized by long, narrow, rather even-topped ridges with streams in valley floors. Streams have played a major role in shaping the present topography (Fenneman 1938, Hack 1980). Streams in the Ridge and Valley Province follow primarily belts of nonresistant rock and display a distinct trellised pattern.

The Appalachian plateaus (Cumberland in the south and Allegheny in the north) border the western side of the Ridge and Valley province. Overall, the elevations of the Appalachian plateaus decline from the east toward their western side. Some spectacular river gorges, such as that of the New River are found in the Allegheny region. Farther south in extreme southwestern Virginia, eastern Kentucky and Tennessee, and northern Alabama, the Cumberland Mountains and Cumberland Plateau constitute the southern portion of the Appalachian plateaus. The Cumberland mountains represent a narrow region in eastern Kentucky, eastern Tennessee, and southwestern Virginia and they constitute the main divide between the Ohio and Tennessee rivers. The Cumberland Mountain section is analogous to the Allegheny Front farther north (Fenneman 1938). Although rock types in the Cumberland Plateau tend to be more resistant than in the Allegheny Plateau, the boundary between the two areas is somewhat arbitrary (Fenneman 1938).

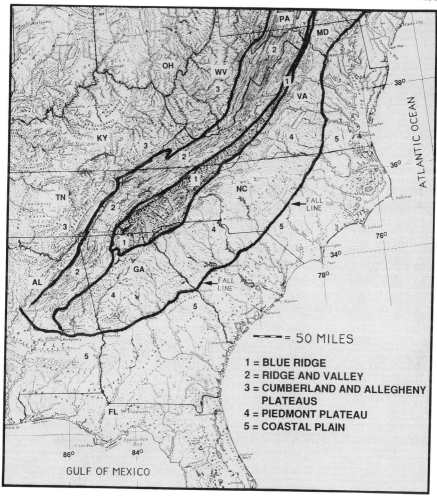

FIGURE 1. Physiographic regions of southeastern United States. (Modified from Fenneman 1938.)

Climate

Summer temperatures in the region are generally cool and the winters are relatively mild (Table 1). Temperatures at a given lattitude are strongly influenced by elevation, as temperature decreases about 1.8 °C for each 305-m increase in altitude. Thus, some of the higher peaks in the southern Appalachians, such as Mount Mitchell (>1800 m), Grandfather Mountain in North Carolina, and Clingman's Dome (>1800 m) in Tennessee may have cooler temperatures than regions found much farther north. Groundwater and springs tend to maintain a rather constant year round temperature close to the mean annual temperature of the region. Depending on elevation, small streams and spring runs where flows are strongly influenced by groundwater

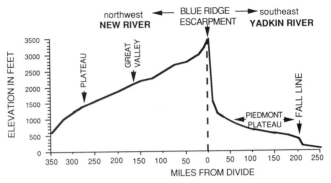

FIGURE 2. Profiles of the New and Yadkin Rivers from their common divide along the Blue Ridge escarpment in western North Carolina. (Redrawn from Hack 1969.)

TABLE 1   Temperature and Precipitation from Various Locations in the Southern Appalachians

Location (State/Local)	Elevation (m)	Mean Annual Temperature (°C)	Mean Monthly Temperature (°C) Jan.	July	Precipitation Mean Annual (cm)	Percentage of Annual as Frozen[a]
Maryland						
Oakland	737.6	8.7	−2.8	19.7	119.7	17.1
West Virginia						
White Sulfur Springs	585.2	11.6	0.2	22.4	97.2	6.7
Virginia						
Big Meadows	1077.5	8.6	−2.1	19.2	126.9	9.6
Floyd	792.5	11.2	0.9	21.3	105.2	4.9
Pennington Gap	460.2	12.4	1.3	22.8	132.2	3.5
North Carolina						
Grandfather Mountain	>1610	5.7	−2.9	16.7	155.5	>10.0
Coweeta Hydrologic Lab.[b]	679.0	13.0	3.2	21.6	181.2	5–10
Coweeta Hydrologic Lab.	1364	—	—	—	238.6	>10.0
Tennessee						
Crossville	551.7	12.1	0.5	22.6	147.1	2.8
Georgia						
Blairsville	584.3	12.6	2.6	22.4	142.5	2.9
Alabama						
Valley Head	317.0	14.3	3.3	24.6	140.2	0.9

[a]Based on a conversion of 10 to 1 (snow to rainfall).

[b]Based on 45–50 years of record; complete temperature records for the high-elevation station are not available

Source. From NOAA (1980) and records of the Coweeta Hydrologic Laboratory.

display a wide range of temperature regimes within an area. Groundwater temperatures closely approximate mean annual temperature of the region (Vannote and Sweeney 1980). Thus, a headwater stream near the summit of some of the high-elevation mountains in North Carolina and Tennessee may accumulate only about 2200 degree days per year (mean temperature × 365) compared to some 5000+ degree days for similar streams in northern Alabama. Downstream, annual temperature variation tends to increase. For example, at the Coweeta Hydrologic Laboratory in the Little Tennessee River headwaters in western North Carolina, high-elevation (>1200 m) springs have annual temperature ranges of about 6.7 to 12.8 °C and accumulate about 3450 annual degree days. Downstream at an elevation of 865 m, a fourth-order reach has an annual range of 0 to 21 °C and accumulates about 4150 annual degree days. Farther downstream the little Tennessee River (elev. = 537 m) has an annual temperature range of 0 to 27 °C and accumulates around 5000 annual degree days (J. B. Wallace and North Carolina Department of Natural Resources, unpublished data).

Rainfall is abundant throughout the region, ranging from about 100 cm/year to over 200 cm/year. The highest rainfall in eastern North America occurs in southwestern North Carolina, western South Carolina, and northeastern Georgia where the boundaries of the three states meet; for example, note the 238 cm/year at the higher elevation site within the Coweeta Hydrologic Laboratory (Table 1). Seasonally, rainfall is fairly evenly distributed throughout the region and in most locations less than 10% of the total precipitation is frozen.

Hydrology

Because of abundant rainfall, most streams are permanent, though some streams, particularly those in limestone areas, may occasionally dry up during extended droughts. Precipitation is fairly evenly distributed throughout the year and seasonal differences in baseflow discharge reflect evapotranspiration of the terrestrial vegetation, with highest baseflow in winter and early spring. Lowest flows typically occur in later summer and early autumn. High discharge associated with storms may occur at anytime of the year.

For streams with forested mountain watersheds, annual water yield (annual stream flow/annual precipitation over the watershed) averages about 40–60%. Most of this occurs as base or delayed flow (Woodruff and Hewlett 1970). Quick flow averages around 4–12% of annual stream flow in the southern Appalachians (Woodruff and Hewlett, 1970).

Logging or other removal of vegetation results in increased stream flow (e.g., Hewlett and Hibbert 1961, Swank et al. 1988). The increased stream flow resulting from forest cutting declines with time as the forest regrows (e.g., Hibbert 1966, Swift and Swank 1981). The type of forest cover can also influence stream flow. For example, converting deciduous hardwoods to pine reduces annual stream flow by 20% in the southern Appalachians (Swank and Douglass 1974).

Geology

The prolonged and diverse mountain formation period in the southern Appalachians has produced bedrock patterns that are complex, both locally and regionally (Hack 1969, King 1977, and Chapter 2, this volume). Most of the interior of the Appalachians in the southeast had been throughly deformed and consolidated by mid-Paleozoic time. The sedimentary Appalachians, to the west of the crystalline Appalachians, were formed during the later Paleozoic phase. The Paleozoic period of Appalachian formation has been followed by perhaps 200 million years of downwasting or erosion (Hack 1969, King 1977). Different rock types forming the Appalachians have resulted in asymmetrical stream and mountain slopes in many areas. The Blue Ridge Escarpment (eastern continental divide, Atlantic and Mississippi drainage system) is clearly asymmetric in North Carolina (Hack 1969, and Fig. 2). The Yadkin River originates on the southeastern side of the escarpment where it has a very steep gradient that decreases greatly once it reaches the Piedmont where the river flows across nonresistent rocks (primarily gneisses and schists). In contrast, the New River flows northwest, crossing extensive areas of rock containing quartz veins, which are more resistant to chemical and mechanical weathering (Hack 1969). Farther downstream, in the vicinity of the Great Valley, the New River crosses large areas of quartzites and sandstones and maintains a steep gradient for a river of large discharge and length. The divide appears to be migrating westward as a result of erosion and uplift (Hack 1969, 1980). This migration is believed to have resulted in a series of stream captures and major modifications of drainage systems (Fenneman, 1938, Ross 1969.

Even large topographic features of the Appalachian Highlands are attributable to differential erosion of rocks of different resistence, and the major drainage systems of the region have become closely adjusted to rock type (Hack 1980). Most areas of high relief and high altitude have been formed on resistent rocks. In contrast, the Cambrian and Ordovician belt, extending from Alabama to the Canadian border on the western side of the Blue Ridge, contains mostly shale and carbonate rocks, which are less resistent to weathering.

Channel slopes in the smaller headwater streams are normally inversely related to discharge, and because discharge and length are directly related, channel slopes are generally inversely proportional to stream lengths. Thus, most longitudinal stream valley profiles assume a concave form (Hack 1980). However, the overall channel slope is strongly influenced by underlying rock type (Hack 1973). According to the hypothesis presented by Hack (1980), profiles of the major Appalachian rivers are explained in part by adjustments as streams cross downstream areas of resistent rock. The resistent downstream bedrock forms local base levels at different altitudes that influence upstream profiles. Major river systems penetrating the Appalachian Highlands show large differences in profile and altitude along reaches of Cambrian and Ordovician rocks, while all have headwater reaches in more resistent rock (Hack

1980). Rock types have important influences on local relief and landscape forms and water chemistry. Hack (1969) summarized some of the characteristics of the predominant rock types of the southern Appalachians as follows:

1. *Granites and other light-colored, course-grained, crystalline rocks.* Are fairly resistant to erosion and water penetration and tend to form areas of steep slopes. These rocks erode primarily by mechanical chipping.
2. *Mica schists.* Tend to occupy areas of low relief in the Blue Ridge, since they are generally very susceptible to chemical and mechanical weathering.
3. *Sandstone and quartzites.* Are resistant to both chemical and mechanical weathering. These rocks tend to form ridges in the Appalachian Valley, are the underlying rock of plateau tops in the Cumberlands, and are usually found in areas of steep slopes. Streams flowing through sandstones and quartzites are lined with cobbles and boulder substrate.
4. *Carbonate rocks.* Include limestone and dolomite, which react with weak acids in rainwater, soil, or groundwater. The limestone, or dolomite, is dissolved and components are transported in solution. Some limestones contain silica and calcium sulfates, which results in higher relief than belts of pure limestone.
5. *Shale.* A nonresistant rock, which is easily penetrated by water, decomposes readily into smaller fragments, and tends to form areas of relatively low relief and thin soils.

Geologic structure influences water chemistry, drainage basin patterns, hydrologic behavior, local substrate, slope, and longitudinal profiles of the streambed. These, in turn, influence the distribution, abundance, and productivity of the stream flora and fauna. Hence, the local geomorphic processes that occurred some 200–300 million years ago exert strong influence on present stream biota and processes.

High-gradient mountain streams of the Blue Ridge, or crystalline Appalachians, are typically dendritic. Stream density may be as high as 6.2 km/km^2 (Harshbarger 1978). In the ridge and valley, or sedimentary Appalachians, streams may have a dendritic pattern, but downstream they tend to follow a somewhat trellised pattern. Gradients may be greater than 20% in some small streams, but vary considerably depending on geological characteristics. In the Ridge and Valley province, adjacent valleys, one principally limestone and the other shale, may have very different stream gradients.

Chemical Environment

Baseflow concentrations of most ions, for example, Cl^-, K^+, Na^+, Ca^{2+}, Mg^{2+}, and SO_4^{2-} are usually low (<1 mg/L) and concentrations of nutrients such as NO_3-N, NH_4-N, and PO_4-P may be very low (0.001–0.004 mg/L for each) in undisturbed streams draining the crystalline Appalachians (Swank and Douglass 1977, Silsbee and Larson 1982). Bedrock geology, elevation, and disturbance history influences nutrient concentrations. The pH of most streams in the crystalline Appalachians is circumneutral (ca. 7.0) and Messer et al. (1986) found that less than 3.2% of stream sites in the Blue Ridge had pH values below 6.4. However, Silsbee and Larson (1982) reported lower pH values (e.g., 5.2–5.5) in streams draining watersheds with the Anakeesta formations (Silsbee and Larson 1982). Anakeesta groups contain pyrites and other sulfides which form dilute sulfuric acid solutions, resulting in lowered pH and alkalinity, and increased conductivity and magnesium concentrations. Similar pyritic rocks occur in other areas of the Appalachians (see Silsbee and Larson 1982). During episodic rain storms or snow melt pH can be depressed rapidly. In Raven Fork, North Carolina, pH occassionally decreased to below 5.0 during storms and, during the same periods, total monomeric aluminum concentrations increased from <50 to >350 μg/L (Jones et al. 1983). Stream pH, alkalinity, Si, K, Na, and turbidity decrease with increasing elevation in the Great Smoky Mountain National Park, whereas nitrate concentrations increased with elevation (Silsbee and Larson 1982). In undisturbed streams in the Coweeta Basin of western North Carolina, highest values for nitrites and sulfates were also found in higher elevation catchments (Swank and Douglass 1977). Clear-cutting and logging also influence nutrient concentrations. Logged catchments in various stages of natural revegetation show elevated nitrate concentrations for at least 10 years following cutting, but appear to return to baseline levels within two decades of regrowth (Swank and Douglass 1977).

Kaufmann et al. (1988) provided an extensive chemical survey of streams draining various geological provinces in the eastern United States. Their results indicate the general tendency for concentrations of many chemicals as well as acid-neutralizing capacity (ANC), conductivity, pH, and total base cations to increase downstream (Table 2). Compared with other Appalachian regions, the southern Blue Ridge has lower ANC, calcium, conductivity, dissolved organic carbon, bicarbonate, nitrate, sulfate and total base cations (Table 2). In contrast to the slow weathering rate of crystalline rock, sedimentary rocks such as limestones have faster weathering rates, which result in streams with higher concentrations of dissolved substances. Alkalinity, pH, and concentrations of nutrients are usually higher in limestone drainages (Table 3).

Many management practices influence water quality of Appalachian streams, such as clear-cutting, surface mining and acid mine drainage, industrial and municipal wastes, riparian disturbances, agricultural practices, impoundments, and the potential problem of acid precipitation, which will be discussed at the end of this chapter. The U.S. Geological Survey Water-Data Reports for the various states are good sources for additional information on stream chemistry in various areas of the Appalachians.

TABLE 2 Means of Selected Chemical Characteristics of Some Appalachian Streams

Chemical Parameter	Units	Ridge and Valley[a]		Southern Blue Ridge[b]		Southern Appalachians[c]	
		Upper Reach	Lower Reach	Upper Reach	Lower Reach	Upper Reach	Lower Reach
ANC[d]	(µeq/L)	644.8	819.9	241.2	257.2	796.0	1001.8
Calcium	(µeq/L)	673.9	810.9	173.4	191.9	641.2	833.1
Conductivity	(µS/cm)	117.6	138.6	29.2	34.7	103.3	129.8
DOC[e]	(mg/L)	1.9	1.6	0.6	0.8	1.6	1.7
Bicarbonate	(µeq/L)	610.5	774.9	214.3	219.2	710.8	928.5
Potassium	(µeq/L)	33.3	31.9	16.6	18.2	31.1	34.2
Ammonium	(µeq/L)	2.1	1.5	0.9	1.0	2.6	2.4
Nitrate	(µeq/L)	92.7	90.6	10.0	12.1	31.5	24.1
pH	(pH units)	6.0	7.2	6.9	7.0	6.4	6.9
Phosphorus[f]	(µM)	0.4	0.3	0.7	1.3	0.4	0.3
Sulfate	(µeq/L)	246.9	278.0	28.5	39.1	155.9	244.0
Base Cations[g]	(µeq/L)	1148.1	1384.1	307.0	338.2	1049.3	1341.0

Note. The data include upper and lower reaches of various streams. Only means for the various streams are given here and in many cases standard deviations within a region are high.

[a]Includes northern and western Virginia, western Maryland, and southern Pennsylvania.

[b]Includes Georgia, South Carolina, Tennessee, and southwestern North Carolina.

[c]Includes southern Ridge and Valley (Alabama, Georgia, Tennessee), Cumberland Plateau (Tennessee, Alabama), central Blue Ridge (North Carolina, Virginia).

[d]ANC, acid-neutralizing capacity.

[e]DOC, dissolved organic carbon.

[f]Total dissolved phosphorus.

[g]Total base cations.

Source. Data from Kaufmann et al. (1988).

141

TABLE 3   Comparison of Nutrient Concentrations of
Stream Water from Control Catchments at the
Coweeta Hydrologic Laboratory in Western North
Carolina with Those of Walker Branch in Eastern
Tennessee

Item	Coweeta[a]	Walker Branch[b]
pH of streamwater	6.64	7.6
NO_3-N	5 ppb	13.0 ppb
NH_4-N	4 ppb	22 ppb
PO_4-P	1–2 ppb	2 ppb
K	0.3–0.4 ppm	0.7 ppm
Na	0.8 ppm	0.6 ppm
Ca	0.5–0.6 ppm	24.5 ppm
Mg	0.3 ppm	13.3 ppm

Note. The underlying rock of the Coweeta Basin is pre-Cam-
brian gneiss and that of Walker Branch is Knox dolomite of
Cambrian and Ordovician age.
[a]Data from Swank and Douglas (1977).
[b]Data from Elwood and Nelson (1972) and Elwood and Henderson (1975).

PLANT COMMUNITIES AND ENERGY SOURCES

In general, high-gradient streams of the southeastern United States support
a reduced flora relative to lentic habitats and low-gradient streams. A majority
of the high-gradient streams occupy watersheds that are not suitable for agriculture and are therefore densely shaded by riparian vegetation (Fig. 3a,
b). A second factor that limits the stream flora is the high current velocity,
which forces most autotrophs to be intimately substrate associated and eliminates many microhabitats such as planktonic or epipelic. Although this results
in reduced standing crops and a limited flora, rheophilous communities often
contain the most characteristic and endemic species in a region (Patrick 1948).

Vascular Plants and Bryophytes

To survive in fast-flowing water, vascular plants must have adventitious roots,
rhizomes (stolons), flexible stems, and streamlined narrow leaves (Westlake
1975). *Podostemum ceratophyllum* Michx. exemplifies the morphology necessary to occupy high-gradient streams. Attaching to rocks with disk-like
processes and giving rise to linearly divided leaves, *Podostemum* is usually
found in clear streams with good aeration. Although *Podostemum* is seldom
reported since it occurs in swift, usually "white water" (Fassett 1966), Meijer
(1975) documented its broad distribution in the southern Appalachian Mountains and suggested that it is an indicator of clean streams in the region.
Podostemum is the dominant macrophyte in the New River and contributes

FIGURE 3. Some typical Appalachian streams. (*A*) Hugh White Creek, a second-order stream at the Coweeta Hydrologic Laboratory in western North Carolina. Note the dense riparian rhododendron. (*B*) A larger, fifth-order stream in the Great Smoky Mountain National Park. Note absence of dense rhododendron canopy.

significantly to the river's organic matter budget (Hill and Webster 1984), generally entering the food chain as an autumnal pulse of rapidly decomposed detritus (Hill and Webster 1982b, 1983).

The water willow, *Justicia americana* (L.) Vahl., is also an important macrophyte of southeastern streams. Unlike *Podostemum*, *Justicia* is rooted in the sediments (Schmalzer et al. 1985), which excludes it from the more turbulent habitats. However, *Justicia* is the dominant emergent plant in the New River, contributing 12% of the aquatic macrophyte biomass (Hill 1981).

Mosses and liverworts are the dominant macrophytes in environments with the highest turbulent flows (Westlake 1975). This situation may be partially due to the fact that bryophytes are able to use free CO_2 as a carbon source (Gessner 1959) and turbulent water insures CO_2 saturation. Messer et al. (1986) and Kaufmann et al. (1988) reported that most small streams in the southern Blue Ridge are probably supersaturated with respect to CO_2. Substrate stability is probably another factor influencing local moss distribution, since mosses tend to be most abundant on bedrock and large boulder substrates (Gurtz and Wallace 1984, Huryn and Wallace 1987). Gilme (1968) surveyed the bryophyte flora of high-gradient Appalachian streams extensively and found four bryophytes to dominate. *Fontinalis dalecarlica* is the most ubiquitous aquatic moss, occurring in first- to third-order streams in depths of 10–100 cm. The gametophyte typically forms mats with "streamers". *Hygroamblystegietum fluviatile* dominates relatively shallow, first- and second-order streams, forming thick mats on submerged rocks, but may occur as a subdominant in larger streams. The distribution of two other potentially dominant bryophytes, *Sciaromium lescurii* and *Scapania undulata*, is poorly known (Glime 1968). An aquatic species of *Fissidens*, a largely terrestrial moss genus, seems to occur in habitats with higher levels of NO_3 and PO_4 as well as CO_2, such as Doe Run in Kentucky (Minckley 1963).

Algae

The algae of high-gradient streams are likewise limited to species that are anchored to stable substrates. Attachment to large, stable objects is of prime importance for the success of this group. Although an algal population may expand to smaller rocks and pebbles during low flows, a spate that causes stones to tumble may remove individuals from all but the largest stones (Minckley and Tindall 1963). The algal flora of the high-gradient streams of the southeastern United States is dominated by filamentous red algae (Rhodophyta), filamentous green algae (Chlorophyta), and diatoms (Bacillariophyta), although other groups are represented in reduced numbers.

Many species of algae appear to be restricted to or at least maintain large populations in this region. Two taxa of red algae, *Nemalionopsis shawii* f. *caroliniana* and *Boldia erythrosiphon*, occur only in streams of the southeast (Howard and Parker 1979, 1980). Camburn and Lowe (1978) described a new diatom from high-gradient streams of the region (*Achnanthes subrostrata* v. *appalachiana* that comprised as much as 73% of the algal community in high-gradient streams in the Great Smokies. There have been a limited number of algal surveys of high-gradient streams of the southeastern United States (Silva and Sharp 1944, Dillard 1969, 1971, Camburn et al. 1978, Lowe and Kociolek 1984). Communities of microalgae are most often dominated by diatoms with a high degree of substrate affinity. Species of *Achnanthes* and *Eunotia*, and to a lesser extent *Meridon*, *Diatoma*, *Gomphonema*, and *Navicula*, dominate the diatom communities of turbulent first- and second-order streams in the southern Appalachian Mountains (Fig. 4a, b) (Kociolek 1982, Keithan and Lowe 1985, Lowe et al. 1986). *Achnanthes* is a genus of tightly

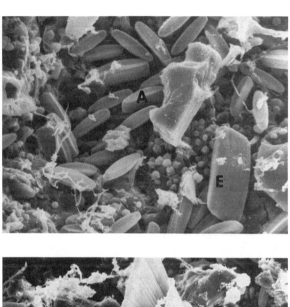

FIGURE 4. Scanning electron micrographs of epilithic periphyton from Camel Hump, Great Smoky Mountain National Park. (*A*) A, *Achnanthes deflexa*; E, *Eunotia rhomboidea* (at 1500× magnification). (*B*) M, *Meridon circulare* (at 700× magnification). (Photos supplied by Elaine Keithan.)

adhering diatoms that may be the most abundant diatom genus in swift streams of the Great Smoky Mountains. Kociolek (1982) observed 21 species of *Achnanthes* in the Smokies. In the two streams most carefully investigated by Kociolek, *Achnanthes* comprised 40–50% of the diatom community. The highly motile and planktonic genera that often dominate low-gradient streams (*Nitzschia*, *Suriella*, *Cyclotella*, and *Stephanodiscus*) are conspicuously sparse in high-gradient streams in the Smokies.

Whitford and Schumacher (1963) collected extensively in high-gradient streams in North Carolina, including the French Broad, New Watauga, Tuckasegee, Cullasaja, and Oconaluftee rivers. Their work provides excellent general information on algal distribution in these streams. The diatoms *Gomphonema parvulum* v. *subelliptica*, *Eunotia alpina*, and *E. lunaris*; the green algae *Oedogonium kurzii* and *Protoderma viride*; the red algae *Compsopogon coeruleus*, *Audouinella violacea*, *Batrachospermum boryanum*, *B. sirodotii*, *Lemania fucina*, and *L. australis*; the chrysophytes *Vaucheria ornithocephala* and *Phaeodermatium rivulare*; and the blue-green algae *Entophysalis lemaniane*, *E. rivularis* and *Phormidium subfuscum* were all recognized as lotic species by Whitford and Schumacher (1963) and were most abundant in swift rapids.

Primary Production

The rate of primary production in high-gradient streams varies with stream order, season, degree of shading, and nutrients. Hornick et al. (1981) estimated gross primary production (GPP as carbon) in a third-order hardwater stream in Virginia as 6.54 g C/m^2 $year^{-1}$ (or as ash free dry mass, ca. 14 g/m^2 $year^{-1}$). They found that GPP in unshaded sites was three times that of shaded stream sites. Keithan and Lowe (1985) found very similar rates of primary productivity (7–9 g C/m^2 $year^{-1}$) in two small streams in the Great Smoky Mountains. Primary production in a heavily shaded, second-order, softwater stream at Coweeta in western North Carolina was only 1.3 g C/m^2 $year^{-1}$ (= ca. 2.9 g AFDM/m^2 $year^{-1}$) (Webster et al. 1983). In contrast, that of a nearby stream draining a clear-cut catchment was 38.9 g C/m^2 $year^{-1}$ (= ca. 86.6 g AFDM/m^2 $year^{-1}$). However, within two years, rapid regrowth of riparian vegetation on the clear-cut catchment resulted in heavy shading, and primary production declined to 3.9 g C/m^2 $year^{-1}$ (= 8.8 g AFDM/m^2 $year^{-1}$) (Webster et al. 1983). In a later study comparing the same streams, periphyton biomass in the clear-cut stream was still higher than in the reference stream; however, in both streams periphyton biomass was unaffected by nutrient additions (Lowe et al 1986). Elwood and Nelson (1972) measured periphyton production with ^{32}P in Walker Branch, Tennessee, as 7.54 g AFDW/m^2 $year^{-1}$, about 2.6 times that found in a similar size softwater stream at Coweeta. Hill and Webster (1982b) found periphyton primary production values ranging from 9.3 to 1,059 mg C/m^2 day^{-1} in the sixth-order, New River, Virginia, with production in the hardwater reaches exceeding that of the softwater reaches by 3–5 times. They attributed differences in production between reaches to greater dissolved inorganic carbon in the hardwater reaches of the river.

Macrophyte production has been studied in two reaches of the New River, Virginia (Hill and Webster 1983, Rodgers et al. 1983), and the Watauga River, Tennessee (Rodgers et al. 1983). In the upper portion of the New River, macrophytes contributed about 63.65 g AFDM/m^2 of river surface area a year to organic matter inputs of the river (Hill and Webster 1983). Macrophytes contributed about 13.1% of the total inputs to the upper reaches of this river. *Podostemum ceratophyllum* contributed about 80% of the total macrophyte inputs into the upper reaches of the New River, with smaller contributions by *Justicia americana* (12.5%) and *Typha latifolia* (6.8%), and minor contributions by *Potamogeton crispus* and *Elodea canadensis* (Hill and Webster 1983). However, in downstream reaches *Justicia americana* and *Typha latifolia* were found to contribute most macrophyte production (Rodgers et al. 1983). *Podostemum ceratophyllum* and *Nitella flexilis* were the most productive macrophytes in the Watauga River (Rodgers et al. 1983).

Allochthonous Energy Sources

Allochthonous organic material, that is, direct litterfall and lateral movement of leaves and wood from riparian forests, is the predominant energy source in high-gradient streams of the southern Appalachians (e.g., Hornick et al. 1981, Webster et al. 1983). Forest litterfall in the region averages about 400 g dry mass/m^2 year^{-1} (Bray and Gorham 1964), and this is probably a good estimate for direct litterfall inputs to headwater streams (Table 4). However, as stream width increases, direct litterfall decreases (e.g., Hornick et al. 1981, Connors and Naiman 1984). Lateral movement is highly variable, depending on such factors as wind patterns, aspect, and bank slope, and its relative contribution to stream inputs varies with stream width (Table 4). Wood generally comprises about 25% of the total input but may approach 50% (Table 5). Logging greatly reduces allochthonous inputs (Webster and Waide 1982), but inputs may be quantitatively near normal within 6–8 years after logging (Webster et al. 1990). However, qualitative differences in litter inputs may persist for many years following logging (Webster et al. 1983, Webster et al. 1990). Inputs from successional vegetation are generally more labile and decay more rapidly in the stream (Webster and Waide 1982, Webster et al. 1983, Benfield et al., 1991). In summary, allochthonous inputs of terrestrial organic matter represent a much larger energy input to small, undisturbed headwater streams of the Appalachians than autochthonous production.

Dissolved organic carbon (DOC) represents another potential energy source to stream ecosystems. External sources of DOC include groundwater inputs and throughfall, while instream sources include leaching from detritus stored in the stream bed as well as dissolved exudates from the biota. Leaching of DOC may be enhanced by microbial activity and macroinvertebrate feeding (Meyer and O'Hop 1983). At the two sites, Coweeta and Walker Branch,

TABLE 4 Litter Inputs to Undisturbed Southern Appalachian Streams.

Stream	Average Stream Width (m)	Litterfall (g/m² year⁻¹)	Lateral movement (g/m year⁻¹)	Lateral movement (g/m² year⁻¹)	Ratio of Litterfall to Lateral Movement
Coweeta WS 7, NC[a]	1.65	259	175	212	1.2
Walker Branch, TN[b]	5.56	372	278	100	3.7
Guys Run, VA[c]	—	347	—	113	3.1
Coweeta WS 14, NC[d]	4.04	415	89	44	9.5
Coweeta WS 18, NC[d]	1.24	482	136	220	2.2

Note. Lateral movement is given in both per unit length of stream and per m² of stream area. The ratio of litterfall to lateral movement is based on inputs per unit area of stream.
[a]Webster and Waide (1982).
[b]Comiskey et al. (1977).
[c]Hornick et al. (1981).
[d]Webster et al. (1990).

TABLE 5 Litter Inputs to Streams at Coweeta Hydrologic Laboratory

Location	Litterfall (g/m² year⁻¹)		Lateral Movement (g/m year⁻¹)	
	Leaf	Wood	Leaf	Wood
Watershed 18 undisturbed, 1st order	482.2	259.7	136.5	10.3
Watershed 14 (undisturbed, 2d order)	415.4	90.9	88.8	9.2
Watershed 6 (successional, 1st order)	332.3	105.6	31.8	4.1
Watershed 7 (successional, 2d order)	354.2	33.9	52.5	24.1

Note. Lateral movement is based on input per linear meter of channel.
Source. From Webster et al. (1990).

where DOC has been studied extensively, concentrations of DOC are low (<1.5 mg/L) during base flow condtions (Meyer et al. 1988, Elwood and Turner 1989). At Walker Branch, concentrations of DOC in streamwater are comparable to those entering via the groundwater at base flow, which suggests either that stream concentrations mirror that of entering groundwater or that a steady state exists between groundwater and in-channel inputs (leaching and exudates) and in-channel uptake, adsorption, and oxidation (Elwood and Turner 1989). In contrast, at Coweeta there are increases in DOC concentration from seeps to downstreams reaches (Meyer and Tate 1983, Meyer et al. 1988, Wallace et al., unpublished). Leaching of organic matter in the stream bed with influence of biological activity appears to be the major source of this downstream increase in DOC (Meyer and O'Hop 1983, Meyer et al. 1988).

Role of Woody Debris

Most smaller first- through third-order streams have low stream power (Leopold et al. 1964), very high channel roughness (Chow 1959), and shallow, narrow channels that are easily obstructed. These features enhance the retention of coarse particulate organic matter (CPOM) such as woody debris and leaves within these channels (Sedell et al., 1978, Naiman and Sedell 1979a, Bilby and Likens 1980, Wallace et al. 1982a, Cummins et al. 1983, Minshall et al. 1983).

In these small headwater streams within forested regions, woody debris not only is a potential energy source, but also serves an important structural role (Swanson et al. 1982, Harmon et al. 1986). These types of streams are common in New England (Bilby and Likens 1980), Oregon (Anderson and Sedell 1979, Naiman and Sedell 1979a, Triska and Cromack 1981, Speaker et al. 1984), the southern Rocky Mountains of New Mexico (Molles 1982), and the southern Appalachians (Wallace et al. 1982a, Webster and Swank 1985a, Golloday et al. 1987, 1989). Woody debris has many roles in high-gradient streams (Harmon et al. 1986), which include contributing to stair-step profiles that result in rapid dissipation of the stream's energy (Bilby and Likens 1980); retention of other particulate organic matter (e.g., Bilby and Likens 1980, Molles 1982, Speaker et al. 1984, Golloday et al. 1987), which may influence both trophic and nutrient dynamics (e.g., Bilby 1981, Molles 1982, Newbold et al. 1982, Melillo et al. 1983, Webster and Swank 1985a, Webster et al., 1990); providing fish habitat (Triska and Cromack 1981, Sedell et al. 1982); and providing a substrate for some stream invertebrates (Anderson et al. 1978), and food for some aquatic invertebrates that may be xylophagous (Anderson et al. 1978, Anderson and Sedell 1979, Dudley and Anderson 1982, Pereira et al. 1982).

Organic Matter Processing

Organic matter exported to downstream reaches consists primarily of fine particulate organic matter (FPOM) and dissolved organic matter (DOM) (Naiman and Sedell 1979b, Webster and Patten 1979, Wallace et al. 1982a, Minshall et al. 1983). Despite the large preponderance of CPOM inputs to small headwater streams at the Coweeta Hydrologic Laboratory in western North Carolina, about 80–95% of the particulate organic matter exported to downstream reaches consists of FPOM. Most of the CPOM export occurs during a few major storms during the year (Cuffney and Wallace 1989, Wallace et al., 1991). Therefore, these small headwater streams function as sites for storage, processing (CPOM to FPOM and DOM), and transport of organic matter (Cuffney et al. 1984, Wallace et al. 1986).

The FPOM and DOM exported from small headwater streams to downstream areas appears to comprise an important energy and nutrient source to downstream microbial flora and fauna. Those fauna adapted for deposit and filter feeding (Short and Maslin 1977, Anderson and Sedell 1979, Wallace and Merritt 1980) may be especially dependant on upstream sources of FPOM. Up to 96% of the annual FPOM flux through downstream segments may originate from upstream sources (Fisher 1977). This longitudinal upstream to downstream linkage has formed the basis for concepts such as nutrient spiraling (Webster and Patten 1979, Newbold et al. 1982) and the River Continuum Concept (Vannote et al. 1980, Minshall et al. 1983).

ANIMAL COMMUNITIES

The streams of the southern Appalachian region contain a diverse fauna of invertebrates, salamanders, and fish. High-gradient streams in the southern Appalachians, ranging from 300 to >2000 m above sea level, are subjected to very different thermal regimes. The wide array of temperatures, combined with diverse stream chemistries, flow, and local geomorphology, influence the species that constitute animal communities.

Invertebrates

The diversity of aquatic invertebrate species in the southern Appalachian Mountains is probably greater than that of any region in North America (Holt 1969, Brigham et al. 1982). Holt (1969) suggested that the area represents an important center for evolution and also an area with many endemic species. This diverse fauna has been attributed to the long-term stability resulting from little major geological change other than climatic trends and fluctuations since the Cretaceous, some 63–135 million years ago (Holt 1969). In their comprehensive treatment of the aquatic insects and oligochaetes of North and South Carolina, Brigham et al. (1982) acknowledged that their treatment of the fauna was incomplete and suggested that for some groups, such as the Chironomidae (Diptera), over 50% of the fauna remains undescribed.

The cool, high-elevation streams may contain taxa that are typical of northern climates and, therefore, do not occur elsewhere in the southeastern region. Detailed systematic treatment of various groups of invertebrates is beyond the scope and objectives of this chapter. Brigham et al. (1982), Merritt and Cummins (1984), and Pennak (1978) should be consulted for general systematic references. We will stress a functional approach built around habitat, trophic organization, and the functional role of invertebrates in streams to bridge the diverse stream habitats of the southern Appalachians.

Cumins proposed a scheme of functional classification based on morphobehavioral mechanisms used to acquire food (Cummins 1973, Cummins and Klug 1979, Merritt and Cummins 1984). These functional feeding groups are as follows:

Scrapers: Animals adapted to graze or scrape materials (periphyton, or attached algae, and its associated microflora) from mineral and organic substrates.

Shredders: Organisms that chew primarily large pieces of decomposing vascular plant tissue (>1 mm diameter) along with its associated microflora and fauna, feed directly on living vascular hydrophytes, or gouge decomposing wood submerged in streams.

Collector–gatherers: Animals that feed primarily on fine pieces of decomposing particulate organic matter (FPOM = <1 mm diameter) deposited within streams.

Collector–filterers: Animals that have specialized anatomical structures (setae, mouthbrushes, fans, etc.) or silk and silk-like secretions that act as sieves to remove particulate matter from suspension (Jorgensen 1966, Wallace and Merritt 1980).

Predators: Those organisms that feed on animal tissue by either engulfing their prey or piercing prey and sucking body contents.

These functional feeding groups refer primarily to *modes* of feeding and not type of food per se. For example, many filter-feeding insects of high-gradient streams are primarily carnivores (Benke and Wallace 1980). Scrapers also consume quantities of what must be characterized as epilithon (Lock 1981) and not solely periphytic algae. Likewise, although shredders may select those leaves that have been "microbially conditioned" by colonizing fungi and bacteria (e.g., Cummins and Klug 1979), they also ingest attached algal cells, protozoans, and various other components of the meiofauna during feeding (Merritt and Cummns 1984). Some shredders apparently obtain very little of their assimilated energy directly from microbial biomass (Cummins and Klug 1979, Findlay et al. 1984), although microbially derived enzymes from endosymbionts or enzymes obtained from microbes ingested with leaf tissue may be important in cellulose hydrolysis (Sinsabaugh, et al. 1985). While it appears valid to separate taxa according to the mechanisms used to obtain foods, many questions remain concerning the sources of protein, carbohydrates, fats, and assimilated energy to each of these functional groups.

A major problem faced by invertebrates in high-gradient streams is maintaining their position in areas of high current velocity. Many taxa have evolved rather elaborate morphological and behavioral adaptations for maintaining their position in such microhabitats (see Hynes 1970, Merritt and Cummins 1984).

Scrapers Scrapers in high-gradient southeastern streams include a rather diverse assemblage of taxa (Table 6). In small woodland streams draining areas of the crystalline rock, scraper production appears to be limited by levels of primary production (e.g., Wallace and Gurtz 1986). Some scrapers, for example, *Goerita semata* Ross (Trichoptera: Limnephilidae), have two-year life cycles in heavily shaded, high-elevation streams. Seasonal differences in larval growth rates of *G. semata* suggest that these are closely linked to seasonal levels of primary production and temperature regimes (Huryn and Wallace 1985). Even in the open, fourth-order, sunlit streams of the Blue Ridge, invertebrate scrapers show considerable temporal segregation in secondary production, and the seasonal distribution of secondary production correlates well with that reported for seasonal changes in periphyton production (Georgian and Wallace 1983). To date, no study has compared seasonal estimates of periphyton production, scraper production, and their bioenergetic efficiencies to estimate the proportion of periphyton production utilized by the scraper guild. Lamberti and Moore (1984) summarized the ecological roles proposed for scrapers, or grazers, in stream ecosystems.

Shredders In areas dominated by crystalline rocks, the predominant shredders in the southern Appalachians include crayfish and aquatic insects (Table 6). Feeding of macroinvertebrates on CPOM, or "shredding," increases at the rate which CPOM is converted to FPOM (Cummins 1973, Petersen and Cummins 1974). Shredder feeding also enhances the conversion of CPOM to dissolved organic matter (Meyer and O'Hop 1983). The generation of large

TABLE 6 Examples of Some High-Gradient Stream Invertebrates Belonging to Various Functional Feeding Groups.

Functional Feeding Group	Dominant Groups	Dominant Families
Scrapers	Ephemeroptera Trichoptera Coleoptera Diptera Gastropoda	Baetidae, Ephemerellidae, and Leptophlebiidae Glossosomatidae, Brachycentridae, and Limnephilidae Psephenidae and Elmidae Blephariceridae, Thaumaleidae, and Chironomidae
Shredders	Plecoptera Trichoptera Diptera Crustacea Mollusca	Pteronarycidae, Peltoperlidae, Nemouridae, and Leuctridae, and Capniidae Lepidostomatidae, Limnephilidae, and Sericostomatidae Tipulidae and some Chironomidae Cambariidae and Isopoda Goniobasis spp. (faculatative scraper-shredders).
Collector–gatherers	Oligochaeta (most) and Collembola Ephemeroptera Plecoptera Coleoptera Diptera Crustacea Mollusca	Heptageniidae, Ephemerellidae, and Leptophlebiidae Taeniopterygidae and Nemouridae Elmidae Psychodidae, Dixidae, Tipulidae, and Chironomidae Copepoda, Ostracoda, Isopoda, and Amphipoda Gastropoda Some Oligoneuridae
Collector–filterers	Ephemeroptera Trichoptera Diptera Mollusca	Philopotamidae, Hydropyschidae, and Brachycentridae Simuliidae, Dixidae, and Chironomidae Sphaeriidae and Pelecypoda
Predators	Turbellaria, Nematoda, and Odonata Plecoptera Megaloptera Trichoptea Diptera Crustacea	Perlidae, Perlodidae, and Chloroperlidae Corydalidae and Sialidae Rhyacophilidae, Molannidae, Leptoceridae and several filtering-collector taxa exploit invertebrate drift Tipulidae, Chironomidae, Ceratopogonidae, Tabanidae, Athericidae, Dolichopodidae, Empididae, and Muscidae Some Decapoda

Note. For detail listing of various genera, consult works such as those of Pennak (1978), Brigham et al. (1982), and Merritt and Cummins (1984).

quantities of small particles, which are more amenable to downstream transport and increase the surface area for microbial colonization, is probably far more important than the shredders' ability to directly degrade organic material by metabolic respiration. Direct metabolic respiration by invertebrate fauna in Bear Brook, New Hampshire, was considered to represent <1% of the annual flux of organic matter through the stream (Fisher and Likens 1973). However, when feeding activities, bioenergetic efficiencies, and secondary production of invertebrates were considered, their overall impact on detritus processing was much greater than 1% in a second-order southern Appalachian stream (Webster 1983). Webster's (1983) model estimated that shredders were responsible for 13% of the leaf litter processing and macroinvertebrates accounted for 27% of the annual particulate organic matter (POM) transport.

Despite the indirect evidence for the role of shredders in processing organic matter in headwater streams there has been little direct evidence to quantify the importance of shredders (Merritt et al. 1984). In a retentive southern Appalachian headwater stream in western North Carolina, the application of an insecticide resulted in massive invertebrate drift and subsequent changes in community structure that eliminated >90% of the insect density and biomass. Elimination of the aquatic insects significantly reduced leaf litter processing rates and export to FPOM to downstream reaches compared to an adjacent, untreated, reference stream (Wallace et al. 1982b, Cuffney et al. 1984). Furthermore, restoration of shredder functional group biomass coincided with restoration of leaf litter processing rates and FPOM export in the treated stream (Wallace et al. 1986). A more recent and expanded experimental manipulation of macroinvertebrate populations in another headwater stream at Coweeta showed that macroinvertebrates accounted for 25 to 28% of annual leaf litter processing (Cuffney et al. 1990) and 65% of the annual FPOM export (Cuffney and Wallace 1989).

Thus, biological processes in small, high-gradient streams, where there is high physical retention of CPOM inputs, favor entrainment by processing CPOM to smaller particles (FPOM) that are more amenable to transport than CPOM (Wallace et al. 1982b, Cuffney et al. 1984). The biota may play a significant role in the upstream to downstream linkage although the instream biota represent a small fraction of total watershed biomass. The above studies were conducted in small first-order streams in the Blue Ridge province and the extent to which these studies apply to larger streams and/or other geographical areas has not been assessed.

Collector–Gatherers   Collector–gatherers are adapted to feeding primarily on small particles (<1 mm diameter) that are deposited on substrate surfaces or in depositional areas of streams. Some typical examples of collector–gatherers found in Appalachian streams are given in Table 6.

The functional role of collector–gatherer invertebrates in high-gradient streams of the Southeast has not been studied directly. Fisher and Gray (1983) provided an excellent account of the role of the collector–gatherers in Sycamore Creek, Arizona. While assimilation efficiencies were low (7–15%), the animals had high ingestion rates in this stream and consumed food equivalent to their own body weight every 4–6 h. Fisher and Gray suggested that egested feces (lower food quality) were rapidly colonized by microbes, which were reingested every 2–3 days on the average. Although growth rates of some collector–gatherers, such as Chironomidae, may be surpisingly high, even in cool, high-elevation Appalachian streams (Huryn and Wallace 1986, Huryn 1990), they are not nearly as rapid as those found in the warm water stream studied by Fisher and Gray (1983). Many of the collector–gatherers in Table 6 contribute to similar processes of FPOM turnover in Appalachian streams. For example, collector-gatherer chironomids alone may consume and egest a large portion of the FPOM stored in headwater Appalachian streams (Schurr and Wallace, unpublished data).

Collector–Filterers There is extensive literature on filter-feeding insects (see Wallace and Merritt 1980, Merritt and Wallace 1981) and other invertebrate filter feeders (Jorgensen 1966, 1975). The animals listed in Table 6 constitute a heterogenous group with respect to feeding, since many of the Hydropsychidae and Brachycentridae (Trichoptera) rely primarily on animal drift (Wallace et al. 1977, Benke and Wallace 1980, Georgian and Wallace 1981, Ross and Wallace 1981, 1983). Although these animals may be filter feeders based on their mode of food capture, they may also be carnivorous, while some taxa, such as Philopotamidae, Simuliidae, Chironomidae (Wallace and Merritt 1980, Merritt and Wallace 1981), and Sphaeriidae (Pennak 1978), exploit minute particles suspended in the water column. With the exception of Brachycentridae, larvae of Trichoptera use a diverse assemblage of woven silken nets to capture particles. Individual pore or mesh sizes of catchnets of various taxa range from <1 µm to 500 × 500 µm. Larger catchnet mesh sizes, such as found within the Arctopsychinae (Hydropsychidae), are located primarily in high-velocity microhabitats, such as swift, moss-covered,rock-face habitats (Gurtz and Wallace 1986, Smith-Cuffney and Wallace 1987, Huryn and Wallace 1987), whereas those with minute mesh openings, such as Philopotamidae, are located in microhabitats of low velocity, for example, on undersides of stones (Wallace et al. 1977, Malas and Wallace, 1977, Georgian and Wallace 1981). Some filter feeders, such as the Philopotamidae and the Simulidae, may actually increase particle sizes by ingesting minute FPOM and egesting compacted fecal particles larger than those originally consumed (Wallace and Malas 1976). Thus, these animals may perform two very important functions: (1) They remove very fine particulate organic matter from suspension (which would otherwise pass through the stream segment), and (2) they defecate larger particles, which are available to a broad spectrum of larger-particle-feeding detritivores.

Studies conducted in high-gradient streams of the Southeast indicate that filter feeders remove a minute fraction of the total particulate organic matter in transport and that their major impact appears to be on seston quality rather than quantity (Benke and Wallace 1980, Georgian and Wallace 1981, Haefner and Wallace 1981a, Ross and Wallace 1983). Newbold et al. (1982) suggested that filter-feeders may shorten nutrient spiralling length when particulate transport is high and there is strong nutrient limitation. However, when there is a high rate of nutrient regeneration, filter feeders probably have little influence on spiralling length.

Predators   Invertebrate predators commonly inhabiting high-gradient streams include several Turbellaria, some Nematoda, Hydracarina, several groups of insects, and some crayfish (Table 6). Brigham et al. (1982) and Merritt and Cummins (1984) should be consulted for specific taxa of aquatic insect predators within the region. Allan (1983) and Peckarsky (1984) reviewed the literature associated with predator–prey relationships in streams. Although there is some experimental evidence that predators can influence the structure of lentic and intertidal communities, there is no strong evidence for predators significantly influencing lotic community structure (Allan 1983). Allan suggested several reasons why benthic communities of streams are not structured by predation. These include the absence of a dominant predator, the presence of many refuges, cryptic coloration, and behavioral adaptations of prey. Another reason that it may be hard to show significant influences on invertebrate community structure as a consequence of predation may be that, on the average, prey standing stock biomass and the generation time of many prey far exceed that of their invertebrate predators in most high-gradient streams of the Southeast. The majority of invertebrate predators have slow growth rates, uni- or semivoltine life cycles and rather low annual production to standing stock biomass ratios ($P/B \le 5$). In addition, these predators generally represent less than 20% of the total invertebrate standing stock biomass, whereas nonpredators comprise over 80% of total biomass and have life cycles that range from a few weeks (e.g., Chironomidae) to a year, with annual P/B's ranging from 5 to >40 (see Production section below). Thus, it is difficult to envision that predators, with high bioenergetic efficiencies (e.g., Lawton 1970, Brown and Fitzpatrick 1978), long life cycles, slow growth rates, and hence, slow population response times, would produce immediate and substantial effects on lotic community structure. Heterogenous substrates, which may provide refuges for benthic prey (Allan 1983), and the selection of a broad spectrum of prey taxa by individual invertebrate predators would further dampen the direct effects of predation in these small headwater streams. Data on prey and predator production, the relative availability of prey production to predators, and the bioenergetics of predators are required before making any definitive conclusions about the influence of predation on lotic communities. Assessments of predator–prey relationships based on either numerical abundances or biomass are tenuous.

Longitudinal and Mesospatial Distributions   Streams display continuous changes in physical and chemical characteristics from headwaters to mouth, which may influence the structure and function of biological communities along this continuum (Vannote et al. 1980). Changes along this continuum may be interrupted or discontinuous due to changes in hydrodynamic characteristics

(Statzner and Higler 1985). From headwaters to mouth, streams display longitudinal shifts in many attributes, including (1) the relative proportion of allochthonous and autochthonous (instream primary production) organic matter contributions; (2) the relative importance of organic matter inputs from upstream sources (3) longitudinal changes in physical characteristics such as retention; and (4) discharge and thermal regimes. These characteristics play an important role in the relative distribution of stream animals that are adapted to utilize various food resources (Vannote et al. 1980). Furthermore, differences in current velocity and retention characteristics may occur over short reaches. Too often biologists fail to recognize that very localized differences in stream geomorphology result in extremely strong influences on benthic community structure. For example, Brussock et al. (1985) discussed the importance of considering channel form in stream ecosystem models.

Secondary Production of Invertebrates   Benke (1984) defined secondary production as "the living organic matter, or biomass, produced by an animal population during an interval of time." Secondary production is thus a measure of the rate at which animal biomass is produced, regardless of its fate (e.g., loss to predators, natural mortality, emergence), and its units are biomass or energy per unit area, per unit time. Most studies of secondary production in stream ecosystems have been limited to a few taxa within a given stream, and there have been few studies that assessed secondary production of the entire macroinvertebrate community within a stream. Secondary production estimates require a knowledge of life cycles or specific growth rates, standing stock densities, and biomass. Thus, such studies are labor intensive. Voltinism and length of immature development have been identified as the two most important factors influencing secondary production of aquatic invertebrates (Benke 1979, Waters 1979). The integration of production, feeding habits, and bioenergetic data can yield a much better understanding of the role of animal populations in ecosystems than either abundances or standing stock biomass (Benke and Wallace 1980, Fisher and Gray 1983, Webster 1983, Benke 1984).

To date, most secondary production studies of invertebrates in high-gradient streams have focused on bivoltine, univoltine, or semivoltine species with clearly discernable life cycles. In some cases, shortcut methods, such as estimating secondary production as the product of standing crop biomass and some production/biomass (P/B) ratio (usually 3.5 to 5; Waters 1977), have been used. The estimates listed in Table 7 include only examples where secondary production was actually measured, and not those based on assumed P/B ratios.

The study of Huryn and Wallace (1987a) adequately assessed secondary production of the entire invertebrate community in high-gradient streams of the Southeast and, to our knowledge, is the only such study for any high-gradient stream. In upper Ball Creek, a high-elevation (1035–1188 m a.s.l.) stream in western North Carolina, total secondary production of invertebrates was about 7.1 g ash free dry mass/m^2 year^{-1} (Fig. 5a–d and Table 8) (Huryn

TABLE 7 Secondary Production of Aquatic Invertebrates in Some Southern Appalachian Streams

Copepoda: *Bryocamptus zschokkei* [a], P = ca. 360, NC
Ephemeroptera: *Baetis*, spp. [b, c, e], P = 15–1,112[1], NC, VA, WV;
  Ephemerella [c], P = 20–71, VA; *Seratella* sp. [d], P = 7–2,476, NC;
  Heterocloeon curiosum [e], P = 540, WV; *Isonychia* sp. [c], P = 34–83, VA;
  Stenonema sp. [c, d], P = 34–205, NC, VA
Odonata: *Lanthus* [c, d], P = 6–53, NC, VA
Plecoptera: *Paracapnia* [c], P = 16–132, VA; *Allocapnia* spp. [d], P = 18–44, NC;
  Amphinemura sp. [d], P = 14–130, NC; *Leuctra* spp. [d], P = 33–416, NC;
  Peltoperlidae [d, f], P = 32–560, NC; *Pteronarcys* [c], P = 36–248, VA;
  Sweltsa spp. [d, g], P = 20–142, NC, TN; *Acroneuria* [c], P = 31–171, VA;
  Isoperla spp. [d], P = 24–159, NC
Megaloptera: *Nigronia* [c], P = 13–87, VA
Trichoptera: *Rhyacophila* spp. [d, h], P = 2–211[2], NC; *Wormaldia* sp. [d, i], P = 2–67, NC; *Diplectrona modesta* [d, g, j, k, l], P = 2–647, GA, NC, TN;
  Parapsyche apicalis [d, i], P = 10–323, NC; *Parapsyche cardis* [d, j, k, l], P = 33–4,274, GA, NC; *Arctopsyche irrorata* [j], P = 604, GA, NC; *Hydropsyche* spp., [i, j], P = 27–175[3], GA, NC; *Cheumatopsyche* spp. [i, e], P = 26–84,654, NC, WV; *Glossosoma* sp. [l], P = 612, NC; *Agapetus* spp. [l], P = 21, NC; *Neophylax* spp. [d, l], P = 27–176, NC; *Goera fuscula* [l], P = 9–16, NC; *Goerita semata* [m], P = 238, NC; *Brachycentrus spinae*, [n], P = 261, NC
Coleoptera: *Psephenus* [d, c], P = 15–180, NC, VA; Elmidae [c, d], P = 1–83, NC, VA.
Diptera: Chironomidae [d, o], P = 689–3,636, NC; *Blepharicera* spp. [l], P = 307–325[4], NC; *Prosimulium* spp. [h], P = 32–167, NC; *Simulium* spp. [e, h], P = 54–25,995, NC, WV; *Tipula* spp. [d], P = 16–896, NC
Decapoda: *Cambarus* spp. [c, p], P = 70–872, NC, VA;
Mollusca: *Leptoxis carinata* [c], P = 1,853–8,194 (AFDM of tissue), VA

Note. All values are in mg AFDM/m[2] year[−1]. Those values initially reported as dry weight have been converted to AFDM based on a 15% ash content for insects and 30% for crayfish. Wet weight values were converted to AFDM based on 20% of wet weight. Following each taxon, references listed by letters in brackets refer to those listed at the bottom of the table. P, production, or range of production measured; localities are listed by state. This table is not complete. For production of additional taxa, consult the references listed below.

References a–m as follows: a = O'Doherty (1985); b = Wallace and Gurtz (1986); c = Miller (1985); d = Huryn and Wallace, 1987a; e = Voshell (1985); f = O'Hop et al. (1984); g = Cushman et al. (1977); h = D. H. Ross and J. B. Wallace (unpublished data); i = Ross and Wallace (1983); j = Benke and Wallace (1980); k = Haefner and Wallace (1981b); l = Georgian and Wallace (1983); m = Huryn and Wallace (1985); n = Ross and Wallace (1981); o = Lugthart et al. in press; and, p = Huryn and Wallace 1987b.

[1] weighted stream production for all substrates for 21 month period.
[2] range for individual species (n = 5–6 species).
[3] range for individual species (n = 4 species).
[4] total production for 3 species.

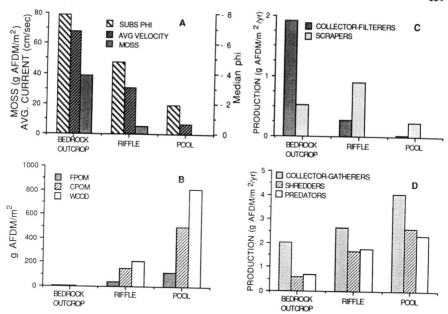

FIGURE 5. Comparison of some habitat characteristics with production of macroinvertebrate functional groups in Upper Ball Creek (WS 27) at the Coweeta Hydrological Laboratory in western North Carolina. (*A*) Moss, water velocity, and substrate particle size as measured for bedrock outcrops, riffles, and pools. (*B*) Average standing crops of fine particulate organic matter (FPOM), coarse particulate organic matter (CPOM) exclusive of wood, and small woody debris. (*C*) Secondary production for collector–filterer and scraper functional feeding groups. (*D*) Secondary production for collector–gatherer, shredder, and predator functional feeding groups. (Data from Huryn and Wallace 1987a, 1988.)

and Wallace 1987a). Primary consumers (collector–gatherers, collector–filterers, shredders, and scrapers) contributed 5.7 g, and engulfing predators about 1.4 g of the total production in upper Ball Creek. Collector–gatherers contributed 51% to total primary consumer production and shredders 26%.

Figures 5a–d also shows the importance of considering substrate and stream geomorphology in assessing secondary productivity of high-gradient streams in the southern Appalachians. Within a given stream reach, distinct differences exist in functional group production for different types of substrates. The physical characteristics of these meso-spatial habitats influence resource availability and mode of resource availability to invertebrate consumers (Huryn and Wallace 1987a, 1988). Within the retentive pool habitats, collector–gatherers and shredders dominated invertebrate production, whereas collector–filterer production was largely restricted to the high-entrainment, low-retention, moss-covered rock face. In the rock-face habitat, moss also facilitates the retention of some fine particulate organic matter for collector–gatherers (see also Lugthart et al. 1990). Therefore, smaller meso-spatial

TABLE 8   Annual Substrate-Specific and Substrate-Weighted (based on proportion of substrate types available) Macroinvertebrate Production, by Functional Feeding Group, Measured in Upper Ball Creek (Watershed 27) at the Coweeta Hydrologic Laboratory, Macon County, North Carolina

	Substrate Type			
Functional Group	Rock Face	Cobble Riffle	Pool	Substrate-weighted (= stream average)
Collector–gatherer	2.03	2.64	4.05	2.93
Collector–filterer	1.92	0.28	0.03	0.57
Shredder	0.58	1.66	2.62	1.48
Scraper	0.52	0.91	0.24	0.68
Engulfing-predator	0.69	1.38	2.23	1.40
Total production	5.74	6.87	9.17	7.06

Note. Rock face substrates = outcrops bedrock with attached moss, cobble riffles = primarily cobble and pebble (16 – to 256-mm-diameter particles), and pools = areas upstreams of debris dams. All values are in g AFDM/m^2 year^{-1} and macroinvertebrates consist of all animals retained by a 230-μm mesh.
Source. From Huryn and Wallace (1987a).

reaches occur within headwater streams which have physical characteristics that resemble various sites of the river continuum of Vannote et al. (1980). In turn, functional group production within specific meso-spatial reaches corresponds to that predicted for various reaches of the entire river continuum (Huryn and Wallace 1987a, 1988).

Chemical characteristics have a pronounced influence on gastropod abundances and production in southern Appalachian streams. At five sites along Guys Run, Virginia, secondary production of invertebrates ranged from 1.8 to 12.5 g DW/m^2 year^{-1}. Those stations in Guys Run with higher pH and hardness generally had the highest production, most of which was attributable to gastropods (Miller 1985). There was no gastropod production in upper, softwater reaches. However, some of the functional group placements, for example, crayfish as 100% predators and gastropods as 100% scrapers in Miller's study are questionable (cf. Elwood et al. 1981, Huryn and Wallace 1987b). The annual P/B ratios of the entire invertebrate community at the two softwater sites studied by Miller also seem to be very low (1.3 and 1.6) compared to those of a cooler softwater stream in western North Carolina (cf. Huryn 1986). Despite the questionable functional group assignments and P/B ratios, the Guys Run study indicates the importance of chemical parameters in gastropod abundance and production. For example, in Walker Branch, Tennessee, a hardwater stream that drains a dolomite watershed, *Goniobasis clavaeformis*, a grazer–shredder, constitutes >95% of the macroinvertebrate biomass (Elwood et al. 1981) whereas at Coweeta in the crystalline Appalachian region of western North Carolina, snails are rare.

In contrast to the low invertebrate secondary production of headwater streams in the Appalachians, some extremely high, habitat-specific production has been reported for Appalachian rivers. On the *Podostemum*-covered, rock outcrop substrates in the New River below Bluestone Dam, West Virginia, Voshell (1985) estimated secondary production of invertebrates as 427.6 g dry mass/m^2 year^{-1}. This appears to be among the highest secondary production known for any stream. The high production is attributable primarily to chironomids and filter-feeding hydropsychid caddisflies and blackflies, which utilize the high-quality seston in the outflow from the reservoir within the favorable environment created by the thick *Podostemum* mat (Table 7) (Voshell 1985).

Vertebrates

Two groups of vertebrates are significant components of high-gradient stream communities of the Southeast: fish and salamanders. Salamanders usually occupy small headwater streams, and fish are found farther downstream. There is seldom much overlap in the distribution of these two groups, though the very large hellbender (*Cryptobranchus alleganiensis*) occurs in large clear streams primarily in the Mississippi drainage (Martof et al. 1980). Salamanders are generally restricted to small streams or the banks of larger streams, probably because of predation by fish.

Salamanders The most common stream salamanders belong to the genus *Desmognathus* (Plethodontidae), though shovel-nosed salamanders (*Leurognathus marmoratus*) and two-lined salamanders (*Eurycea bislineata*) are not uncommon in some streams, and other species may occasionally be found especially in spring seeps. Members of the genus *Desmognathus* range from aquatic to terrestrial. From 3 to 5 species are found in most areas of the southern Appalachian Mountains (Hairston 1986). The most common species are *D. quadramaculatus*, the black-belly salamander; *D. monticola*, seal salamander; and *D. ochrophaeus*, mountain dusky salamander. *Desmognathus quadramaculatus* is the largest and the most aquatic of the three, *D. ochrophaeus* is the smallest and most terrestrial, and *D. monticola* is intermediate (Hairston 1986). Other species of *Desmognathus* may occur in streams in part of the area (e.g., Martof et al. 1980)

Interspecific interactions of this group of *Desmognathus* salamanders have been widely studied (Hairston 1949, 1980, 1981, 1983, 1986, 1987, Organ 1961, Tilley 1968, Krzysik 1979, Keen 1979, 1982, Kleeburger 1984, Southerland 1986a,b). Competition for prey and predation of large salamanders on small salamanders appear to be the major factors determining the within-stream distribution of species. The role of salamanders in stream ecosystems has been much less studied, primarily because of difficulties of accurately determining population abundance and of identifying immature forms.

Spight (1967) estimated a stream population of *Desmognathus fuscus* of 0.4–1.4 individuals/m^2 in a North Carolina Piedmont stream. Using a mean weight of 1.05 g/individual (based on his data), this converts to 0.4–1.5 g/m^2 standing stock. Spight also calculated production of this species of 0.1–0.3 g/m^2 year^{-1} (Table 9). The resultant P/B of 0.2 seems low compared to fish (e.g., Neves and Pardue 1983). Orser and Shure (1972) found densities of *D. fuscus* varying from 0 to 10 individuals/m^2 in small streams near Atlanta, Georgia, using mark/recapture techniques. Based on Surber samples, Woodall (1972) estimated standing crop of *Desmognathous* salamanders (probably primarily *D. monticola*) in a small stream at Coweeta Hydrologic Laboratory in the southern Appalachians as 0.3 g/m^2. In the same stream, Drumheller (1979) used a census method to estimate salamander biomass and found a much lower density but a higher biomass (0.8 g/m^2), since he collected fewer but larger individuals. For various first- to third-order streams in the area, Drumheller found that *Desmognathus* biomass ranged from 0.16 to 0.77 g/m^2 and production ranged from 0.20 to 1.16 g/m^2 year^{-1}. Production and biomass were both greater in first-order than in third-order streams. Drumheller's P/B ratios ranged from 1.5 to 1.8 for *D. quadramaculatus* and *D. ochrophaeus*, and from 0.9 to 1.0 for *D. monticola*. Huryn and Wallace (1987a) report P/B ratios of *Desmognathus* spp. ranging from 1.9 to 2.7. These P/B ratios are similar to or slightly greater than values usually reported for fish.

   Stream salamanders are entirely predaceous and feed largely on terrestrial insects. Drumheller (1979) and others have shown that *Desmognathus* salamanders are size-selective predators, with larger salamanders consuming larger prey. Hairston (1949) reported that *D. quadramaculatus* stomachs contained 65%, *D. monticola* 75%, and *D. ochrophaeus* 100% terrestrial insects. Krzysik (1979) found that adult and larval dipterans were the primary food items of *Desmognathus* salamanders. Lepidoptera larvae, Coleoptera adults, and Plecoptera (adults and nymphs) were also important. More aquatic species consumed a larger proportion of aquatic insects, while more terrestrial species fed extensively on terrestrial invertebrates. In their examination of *Desmognathus* salamander gut contents, W. R. Woodall and J. B. Wallace (unpublished data) found that 54% of the food of adults was terrestrial, primarily Collembola and Hymenoptera. Overall, larval Diptera, primarily Chironomidae, were the most abundant aquatic food items for adult salamanders. Immature salamanders were found to feed much more extensively on aquatic

TABLE 9   Salamander Abundance, Biomass, and Production in Southeastern Headwater Streams

Species	Abundances (No./m^2)	Biomass (g/m^2)	Production (g/m^2 year^{-1})	Reference
D. fuscus	0.4–1.4	0.4–1.5	0.1–0.3	Spight (1967)
D. fuscus	0–10	—	—	Orser and Shure (1972)
D. spp.	—	0.3	—	Woodall (1972)
D. spp.	—	0.16–0.77	0.20–1.16	Drumheller (1979)
D. spp.	2.7	0.113	0.242	Huryn and Wallace (1987a)

organisms: 96% of the food items in immature guts was aquatic. In forest streams gut content composition and composition of the benthic fauna were closely correlated, suggesting that salamanders are opportunistic feeders.

Fish Fish communities of high-gradient Southeastern streams may contain a variety of species (Table 10), but are usually dominated by trout, particularly brook trout (*Salvalinus fontinalis*). There has been little research on species other than trout beyond systematic and zoogeographical studies. There is a distinct zonation of fish species, which can be seen either by sampling along specific streams (e.g., Burton and Odum 1945; Neves and Pardue 1983) or by sampling many streams at different elevations (e.g., Harshbarger, in preparation). This zonation is reflected in species occurrence, biomass, and production (Tables 11 and 12). As noted above, fish are absent from the headwaters of streams. In this shallow water, salamanders are essentially the ecological equivalent of fish. Brook trout are the first to enter. Farther downstream they are joined by sculpins (*Cottus bairdi*), dace (e.g., *Rhinichthys atratulus*), darters (e.g., *Etheostoma flabellare*), and perhaps introduced rainbow (*Oncorhynchus mykiss*) and brown (*S. trutta*) trout. Proceeding downstream, other dace, darters, chubs, shiners, suckers, and other fish are found (Table 11). In larger downstream reaches smallmouth bass *Micropterus d. dolomieu* may constitute an important game fish. The most abundant populations of *M. d. dolomieu* occur where about 40% of the substrate consists of riffles flowing over clean gravel, boulder, or bedrock substrates with a maximum depth >1.2 m and gradients of 0.8–4.8 m/km, and where considerable water willow is present (Burton and Odum 1945, Trautman 1942, 1981).

Of particular interest is the small number of species found near the headwaters of streams in association with trout. Coker (1925) suggested that this situation may exist because of the relative high tolerance of brook trout to the typically lower pH of headwater streams compared to other species (e.g., Table 11) and the possible interacting influences of temperature, current, and dissolved oxygen. Powers (1929) wrote that the commonly observed distribution of brook trout and rainbow trout (brook trout upstream, rainbow trout at lower elevations) might result from the presence or absence of dissolved organic substances. From their analyses of five streams in western Virginia, Burton and Odum (1945) concluded that temperature and stream gradient were the primary factors affecting longitudinal fish zonation.

On a broader scale, the geographic distribution of fish within the southern Appalachians has been extensively affected by anthropogenic disturbances. Presettlement fish distributions reflected drainage basin origins and extensive interbasin exchanges resulting from numerous headwater piracies (e.g., Ross 1969, 1971, Hocutt et al. 1978, Hocutt 1979; Chapter 2, this volume). Logging (e.g., Douglass and Seehorn 1974) and mining (e.g., Hill 1975) have been particularly detrimental to fish, especially brook trout. Introductions, whether purposeful or unintentional (minnow bucket introductions) have greatly modified fish faunas. Some of these factors are illustrated by the history of brook trout in the Great Smoky Mountains National Park (GSMNP).

TABLE 10 Typical Fish of High-Gradient Streams of the Southern Appalachian Mountains.

Species	Common Name	Feeding Habits and References
SALMONIDAE		
Oncorhynchus mykiss	Rainbow trout	Insects and crustaceans; large individuals piscivorus; large indi-
Salmo trutta	Brown trout	viduals as much as 50% of diet may be terrestrial insects (Ricker 1934, Tebo and Hasler 1963, Carlander 1969
Salvelinus fontinalis	Brook trout	Needham 1938, Scott & Crossman 1973)
COTTIDAE		
Cottus bairdi	Mottled sculpin	Benthic aquatic insects (Ricker 1934, Daiber 1956)
CYPRINIDAE		
Rhinichthys atratulus	Blacknose dace	Invertebrates; some plant material (Flemer & Woolcott 1966, Minshall 1967, Tarter 1970)
R. cataractae	Longnose dace	Aquatic insects (Tarter 1970)
Clinostomus funduloides	Rosyside dace	Insects mainly terrestrial (Breder & Crawford 1922, Flemer & Woolcott 1966)
Phoxinus oreas	Mountain redbelly dace	Primarily herbivorous; diatoms and filamentous algae (Flemer & Woolcott 1966)
Campostoma anomalum	Stoneroller	Primarily diatoms on rock surface (Kraatz 1923)
Notropis albeolus	White shiner	Nothing published; probably similar to *N. cornutus* (Lee et al. 1980)
N. cornutus	common shiner	Omnivorous; terrestrial insects, algae, leaves (Breder & Craw-ford 1922, Miller 1964)
N. rubellus	Rosyface shiner	Aquatic and terrestrial insects, (Pfeiffer 1955, Reed 1957, Miller 1964, Gruchy et al. 1973

Species	Common name	Food
N. photogenis	Silver shiner	Immature and adult aquatic insects; turbellarians (Gruchy et al.1973)
N. spectrunculus	Mirror shinner	Nothing published
Nocomis leptocephalus	Bluehead chub	Omnivorous; selects plant food (Flemer & Woolcott 1966)
N. micropogon	River chub	Benthic invertebrates; some plant material (Lachner 1950)
Semotilus atromaculatus	Creek chub	Benthic and terrestrial invertebrates; some plant material (Ricker 1934, Scott & Crossman 1973)

PERCIDAE

Species	Common name	Food
Etheostoma flabellare	Fantail darter	Aquatic insects (Forbes 1880, Turner 1921, Daiber 1956, Karr 1964)
E. longimanum	Longfin darter	Nothing published
E. Kanawhae	Kanawa darter	Nothing published
E. blennioides	Greenside darter	Aquatic insects (Forbes 1880, Turner 1921)

CATOSTOMIDAE

Species	Common name	Food
Moxostoma rhothoecum	Torrent sucker	Plant material (Flemer & Woolcott 1966)
Catostomus commersoni	White sucker	Bottom feeder; algae, mollusks, chironomids (Flemer & Woolcott 1966)
Hypentelium nigricans	Northern hog sucker	Scrapes and turns rocks: insects, crustaceans, diatoms (Flemer & Woolcott 1966)

ICTALURIDAE

Species	Common name	Food
Noturus insignis	Margined madtom	Insects, fish? (Flemer & Woolcott 1966)

Source. From Burton and Odum (1945), Neves and Pardue (1983), and Harshbarger (in preparation).

TABLE 11 Distribution of Fish in Little Stony Creek, Virginia

	Headwaters					Downstream Gradient							Mouth	
Elevation (m)	1120	1067	1021	981	939	922	917	905	838	754	667	603	556	515
pH	5.6	5.6	5.8	5.8	5.9	6.2	6.4	6.6	7.0	7.0	7.1	7.2	7.2	7.4
Temperature (°C)	15	15	16	16	17	18	18	18	18	19	19	20	20	21
Brook trout	X	X	X	X	X	X	X	X	X	X				
Blacknose dace					X	X	X		X			X	X	
Fantail darter							X			X				X
Rainbow trout								X	X	X	X	X	X	X
Mottled sculpin												X	X	X
Stoneroller													X	X
White shiner													X	X
Longnose dace														X
White sucker														X

Source. From Burton and Odum (1945).

TABLE 12   Percent Contribution by Fish Species to Annual Production in Three Sections of Guys Run, Virginia

Species	Stream Section		
	Upper	Middle	Lower
Brook trout	60	61	14
Mottled sculpin	29	18	27
Blacknose dace	10	9	9
Torrent sucker	1	5	4
Bluehead chub	<1	5	37
Fantail darter	<1	<1	
Longnose dace		1	4
Rosyside dace		<1	3
Mountain redbelly dace			<1
Rock bass			<1
Smallmouth bass			<1
Total fish production (g/m^2 year^{-1})	2.84	3.16	3.96
Total fish biomass (g/m^2)	2.14	2.56	4.74

Source. From Neves and Pardue (1983).

It is apparent that the range of brook trout in GSMNP is decreasing and has been decreasing for many years (Jones 1978, Kelley et al. 1980, Bevins et al. 1985). Brook trout are now restricted to small, high-elevation streams. This decline has been attributed to logging and associated habitat deterioration at lower elevations and to competition with introduced rainbow trout. Based on discussions with local fishermen, Powers (1929) found that brook trout had moved upstream prior to introduction of rainbows in 1919 and that the planting of rainbows was a response to this decline. Some of the fishermen attributed this decline to overfishing, but Powers noted that the decline was simultaneous with logging and that the distribution of brook trout was coincident with areas that had not been logged. However, with the regrowth of lower elevation forests, brook trout have not recolonized lower-elevation streams, and, in fact, the brook trout retreat is continuing (Jones 1978). The evidence for competitive displacement by rainbow trout is quite clear (Moore et al. 1983, Larson and Moore 1985). Populations of brook trout will probably persist in the high-elevation streams of GSMNP and throughout the southern Appalachians, though random extinctions of isolated populations in these marginal habitats should be expected (Larson and Moore 1985).

The fish of southern Appalachian streams are primarily insectivorous predators (Table 10). Trout, some of the dace, and some of the chubs are midwater and surface feeders, catching drifting aquatic invertebrates, terrestrial insects, and adult aquatic insects. Sculpins, darters, most chubs and minnows, and some dace feed predominantly on benthic invertebrates, searching on and among the rock and gravel streambed. Several species turn over small rocks in search of prey. Because of limited primarily production in these streams, plant feeders such as *Campostoma* occur only in somewhat larger streams with open canopy and lower gradient. Detritivorous fish are uncommon in high-gradient streams of the region. References to published feeding studies are given by Carlander (1969), Scott and Grossman (1973), and Lee et al. (1980).

Neves and Pardue (1983) estimated production of the fish community at three sites in Guys Run, Virginia (Table 12). Extrapolating from their study and comparing with other components of stream energetics, two conclusions are evident. First, fish predation pressure on stream invertebrates is considerable, but when probable sampling errors are taken into account, predation in not greater than estimated invertebrate production. Second, the fish community as a whole and brook trout in particular depend heavily on allochthonous energy sources. This connection is direct through feeding on terrestrial insects, and indirect through invertebrate dependence on allochthonous detritus. Thus, effects of forest logging and other disturbances on fish populations can be attributed to both lower water quality and modified food resources. Although data for fish production (Table 12), standing stock biomass (Table 13), and invertebrate production (Table 7 and 8) are from different streams, these data have important implications for fisheries biologists because they suggest the following: (1) Most small, high-gradient Appalachian streams, especially those draining areas dominated by crystalline rock, have relatively low levels of invertebrate production. (2) A considerable portion of this production is used by predaceous invertebrates. (3) In small, fishless, headwater streams secondary production of salamanders is similar to production of fish in larger downstream areas. (4) Secondary production of carnivorous vertebrates, especially those that rely on instream food resources, may be strongly influenced by availability of food resources.

REPRESENTATIVE HIGH-GRADIENT STREAMS

There are numerous small first- through fourth-order high-gradient streams in the southern Appalachians that can be considered representative or typical high-gradient streams. Many of these can be found on federal, state, and privately owned lands. In the southern Appalachians, we probably know less about the biotic structure and function of large, high-gradient rivers than smaller streams. This is attributable, at least in part, to the physical difficulty and expense involved in studying these larger rivers. Unfortunately, these rivers will probably come under increasing environmental pressure from municipal, industrial, and recreational interests to meet the growing population needs of the area.

Three rivers in the southern Appalachians were included in the original Wild and Scenic Rivers Act: the Chattooga River in Georgia, North Carolina, and South Carolina; the Obed River in Tennessee; and the headwaters of

TABLE 13 Fish Standing Stock Biomass and Trout Contribution to Total Fish Biomass at Various Elevations in Western North Carolina Streams

Parameter	Elevation (m)							
	>1219	1067–1219	914–1067	762–914	610–762	457–610	<457
Total fish biomass (g/m^2)	1.00	1.55	1.74	2.10	2.10	2.15	1.66
Trout contribution (% of fish biomass)	99.9	86.8	63.5	26.6	33.0	36.3	25.7
Brook trout as (% of total trout numbers)	93.1	30.7	29.7	30.7	7.8	5.8	9.1

Source. From Harshbarger (in preparation).

169

TABLE 14   Some High-Gradient Creeks and Rivers in the Southern Appalachians That Were Included in the Nationwide Rivers Inventory Based on Outstanding Scenic, Recreational, Geologic, and Wildlife values

Alabama

Little Cahaba and Shoal Creek (Bibb and Shelby), Little River (Cherokee and De-kalb), Locust Fork of Black Warrior (Jefferson, Blount, Cullman, Marshall, and Etowah), and Mulberry Fork of Black Warrior (Blount and Cullman)

Georgia

Amicalola Creek (Dawson), Chattahoochee River (Hall, Habersham, and White), Chattooga River* (Rabun and also South Carolina), Chestatee River (Lumpkin), Conasauga River (Gordon, Whitfield, and Murray), Coosawatee River (Gilmer), Etowah (Bartow and Floyd)

Kentucky

South Fork and Little South Fork of Cumberland (McCreary and Wayne), Martins Fork (Harlan and Bell), Rock Creek (McCreary), Tygarts Creek (Carter)

Maryland

North Branch of Potomac, Savage*, and Youghiogheny Rivers* (Garrett)

North Carolina

Big Laurel Creek (Madison), Cane River (Madison), Davidson River (Transylvania), Green River (Polk), Linville River (Burke and Avery), Mitchell River (Surry and Alleghany), Nanahala River (Swain and Macon), Oconaluftee (Swain), Tellico (Cherokee), Tuckasegee (Swain and Jackson), Watauga (Watauga and Avery), and Yadkin (Davidson)

South Carolina

Chauga River (Oconee)

Tennessee

Abrams and Anthony Creeks (Blount), Clear Creek (Morgan, Fentress, and Cumberland), Conasauga River (Bradley and Polk), East Fork of Obed (Fentress and Overton), French Broad (Knox and Sevier), Hiwasee and Ocoee Rivers (Polk), Piney Creek (Rhea), South Fork of Cumberland (Scott), and Tellico (Monroe)

Virginia

Bullpasture and Cowpasture Rivers (Allegheny, Bath, and Highland), Dan River (headwaters, Patrick), South Fork of Holston River (Washington), Jackson River (Allegheny, Bath, and Highland), Back Creek (Bath and Highland), Little River (Pulaski, Montgomery, and Floyd), Maury River (Rockbridge), New River* (Grayson, Carroll, and Pulaski), Cedar Creek (Shenandoah and Frederick), Passage Creek (Shenandoah), South Fork of Shenandoah River (Warren and Page), Stoney Creek* (Giles)

West Virginia

Big Sandy Creek (Preston), Birch River (Nicholas and Braxton), Blackwater River (Tucker), Bluestone River (Mercer and Summers), Buckhanon River (Barbour, Upshur, and Randolph), Cacapon (Morgan, Hampshire, and Hardy), Cheat

TABLE 14   (*Continued*)

River including Dry Fork, Glady Fork, Shavers Fork (Tucker and Randolph), Elk River (Braxton, Webster, and Randolph), Gauley River including Cherry and Cranberry rivers (Greenbrier, Nicholas, Webster, Pocahontas, and Randolph), Greenbrier River (Pocahontas), Meadow River (Nicholas, Fayette, and Greenbrier), Middle Fork River (Barbour, Upshur, and Randolph), North and South Branches of Potomac River (Grant, Pendleton, Hardy, and Highland), Tygart Valley River (Barbour, Taylor, and Marion).

Note. Streams are listed by state and counties in which they are located are in parentheses. The list is not intended to be complete. Those marked with an asterisk were added by the authors. For additional streams and information about each consult National Park Service (1981).

the New River in North Carolina. In addition to these streams, numerous other examples of large high-gradient creeks and rivers can be found in the first Nationwide Rivers Inventory of the National Park Service (1981).

The purpose of the National Park Service's survey was to provide congress, federal, state, and local government agencies, and the private sector with comprehensive consistent data on the nation's free-flowing streams, which could be used to: (1) provide baseline data on the condition and extent of significant free-flowing river resources so that they can be monitored over time; (2) provide informed decisions on river use for recreation, water supply, irrigation, hydroelectric power, flood control, and conservation of scenic and wild rivers; (3) assist and encourage state, local, and private efforts to conserve rivers; and (4) permit comparisons with the National Scenic and Wild River system and identify other rivers that would complete the system. The following criteria were applied to streams selected in the First Nationwide Rivers Inventory: Rivers must appear 25 miles or longer on 1:500 000 scale maps; dammed or channelized rivers were not considered; excessive cultural development such as cities of >10 000 in population, power plants, and active strip mines within 400 m of the bank excluded streams from consideration; the river must have sustained flow; and, the river had to pass field and helicopter video analyses. The list of the nation' rivers did not include rivers presently in, or in consideration for, the National Wild and Scenic River system. Some examples of streams included in the first survey are listed in Table 14.

RESOURCE USE AND MANAGEMENT EFFECTS

A survey of 231 individuals engaged in natural resource management who represented private industry and local, state, and federal governments in a 22-county area of western North Carolina indicate that the most important management priority is the maintenance of high-quality streams (SARRMC 1977). This survey concluded that both the general public and municipal leaders view severe stream degradation as their greatest resource problem.

Identification of the sources, composition, and quantities of wastes and runoff contributing to stream degradation was viewed as the most important research target (SARRMC 1977). Municipalities, manufacturing, recreational, and energy (hydroelectric) interests, combined with forest management (logging) place increasing demands on southern Appalachian streams where heavy rainfall, combined with a dense forest cover, gives rise to many clean streams. Since the late 1970s, there has also been growing public awareness of the potential problems associated with acidic precipitation. Acidic precipitation may present serious problems in the immediate future for streams draining the crystalline regions of the Appalachians. Coal mining also represents an ongoing problem for many Appalachian streams.

Sewage and industrial effluents can produce pronounced effects on water quality and on both microbial and macroinvertebrate community structure within small Appalachian streams. Kondratieff and Simmons (1982) and Kondratieff et al. (1984) reported that the macroinvertebrate community of Cedar Run, Virginia, consists of a diverse assemblage of all functional groups in upstream reference sections compared to a community dominated by collector–gatherers below areas of waste outfall. Farther downstream, collector–filterers constituted the majority of the macroinvertebrate community. Macroinvertebrate scraper biomass was much lower at heavily polluted sites compared to reference stations located upstream and several kilometers downstream of outfalls. The effluents seriously disrupted the integrity of the macroinvertebrate community and microbial community response paralleled that of the macroinvertebrate community. At reference stations in Cedar Run, microbial assemblages were dominated by autotrophs, primarily diatoms, whereas heterotrophic microbiota predominated below waste outfalls (Kondratieff et al. 1984). All of the particulate organic carbon (POC) added by the sewage effluent was removed from the water column within 3 km downstream. Kondratieff and Simmons (1982) suggested that biological uptake by filter-feeding macroinvertebrates and microbial decomposition were responsible for the majority of POC removal. Although organic effluents produce significant alterations in stream biota, the tolerant species may assume an important role in processing these excessive inputs and enhance the self-purification capacity of the stream. Reports of various state agencies represent valuable resources for additional documentation of stream degradation by municipal and industrial waste, for example, The Georgia Water Quality Control Board (GWCB 1970) and the Benthic Macroinvertebrate Ambient Network (BMAN 1985) in North Carolina.

Watershed disturbances such as clear-cutting may produce a multitude of both short- and long-term changes in stream ecosystems. Sedimentation resulting from logging and associated practices such as road building may produce an immediate impact on stream biota (Tebo 1955, Gurtz and Wallace 1984). Clear-cutting increases stream temperatures (Swift 1983, Swift and Messer 1971). Clear-cutting also shifts the relative abundance of allochthonous to autochthonous inputs, which may shift the energy base of the stream (Webster et al. 1983) and be reflected in the invertebrate community (Woodall and Wallace 1972, Webster and Patten 1979, Haefner and Wallace 1981b, Gurtz and Wallace 1984, Wallace and Gurtz 1986). There may also be long-term changes in benthic organic matter storage, especially woody debris (Golloday et al. 1989). Although wood decomposes slowly (Triska and Cromack 1981, Golloday and Webster 1988), there is little input of woody debris other than slash added to the stream during logging, until the forest regenerates. Likens and Bilby (1982) and Hedin et al. (1988) suggested that recovery of streams may lag behind, or be out of phase with that of the terrestrial forest, since large stable debris dams will be reestablished only after mature trees die and fall into the stream channel. Thus, while total allochthonous inputs may be restored to the stream channel within 6–10 years of forest regrowth, the absence of large woody debris dams may influence retention characteristics of the stream for many decades (Webster et al. 1990).

The impact of forest fire on stream biota in the Appalachians is not well known. Most studies of fire have been confined to western North America (Wright and Bailey 1982, Minshall et al. 1989). Some of the immediate effects of severe fire from these western studies appear to resemble those of clear-cutting. These include elevated stream temperatures, increased sediment loads, and, increased nutrients and primary productivity, which may stimulate secondary production of some aquatic insects. However, the increased sediment loads may destroy spawning sites for fish and increased stream temperatures following destruction of the canopy cover may lead to higher incidence of fish disease (Wright and Bailey 1982). Long-term effects may parallel those outlined for clear-cutting (Minshall et al. 1989).

Acid mine drainage is an important source of pollution in coal mining areas of the Appalachians. Groundwater and water percolating through spoil banks may contain sulfuric acid as well as various metalic salts. Iron sulfides, especially pyrite (Fe_2S), react with water in the presence of oxygen to form sulfuric acid. In addition, substrates in stream channels draining mined areas are often coated with reddish-yellow deposits of ferric hydroxide. Typically, streams having acid mine drainage problems have high conductivities, higher iron and sulfate concentrations, and lower pH and alkalinities than streams draining unmined watersheds in the region (Herricks and Cairns 1974).

Between the years 1930 and 1971, some 2017 km^2 of the southern Appalachians were surface mined. About 32–48% of this area was not reclaimed, and abandoned mines represent an ongoing problem (Samuel et al. 1978). Streams draining abandoned mines have reduced invertebrate and fish populations for up to two decades following cessation of mining operations. In some cases, Odonata, Ephemeroptera, and Plecoptera have been completely eliminated, while the Trichoptera, Megaloptera, and Diptera species have been severely affected (Roback and Richardson 1969). Some streams such as Shavers Fork, West Virginia, show increasing degradation since the early 1960s that is associated with increased mining operations. Acid "slugs," released from mined areas during high spring runoff, appear to be the main factor contributing to degradation of Shavers Fork (Samuel et al. 1978). Tarter (1976) noted that coal mining and industrialization, especially chemical plants, have polluted many streams in some regions of West Virginia to the extent that only the most tolerant species of benthic organisms can inhabit them.

The impoundment of many Appalachian streams for hydroelectric and recreational purposes has been a controversial topic since the early 1900s. Obviously, community structure is altered from lotic to lentic conditions within the area of stream inundated. However, in the last two decades it has become increasingly evident that impoundments produce major modifications on the biota below dams. Ward and Stanford (1983) and Ward (1984) addressed some of the changes induced by stream regulation, including altered thermal regimes, which may produce winter-warm or summer-cool conditions below deep release dams creating thermal regimes that are intolerable by many species. Flow regimes may also be very different from that of unregulated streams and eliminate certain species. Dams may alter the ratio of CPOM/FPOM supplied to downstream areas as well as altering the quality of FPOM. Particles suspended in the water column of lake outlets have high nutritive value (Kondratieff and Simmons 1984, Voshell and Parker 1985). Provided species can tolerate the altered thermal regimes, the enriched seston may result in abundant filter-feeder production (Voshell 1985, and Table 7).

The gypsy moth, *Lymantria dispar* (L.), feeds on the foliage of over 600 deciduous and coniferous plants. This forest pest has the potential ability to impact small, high-gradient streams of the Southeast by several mechanisms. Heavy insect defoliation by the fall cankerworm, *Alsophila pometaria* (Harris), has been shown to increase nitrate (NO_3–N) export from southern Appalachian watersheds, which influences watershed nutrient budgets (Swank et al. 1981). Severe defoliation also influences leaf production, frass, and nutrient inputs to the forest floor and stream (Swank et al. 1981), which may increase insolation of the stream and alter the timing, quantity, and quality of stream energy resources. Furthermore, aerial spraying with pesticides to control forest pests may induce catastrophic drift of stream invertebrates (Wallace and Hynes 1975) and, in instances where persistent chlorinated hydrocarbon pesticides were used to control forest pests, benthic populations may be reduced for several years (Ide 1967). Recently, Swift and Cummins (1991) reported that leaf litter from trees previously sprayed with the insecticide Dimilin, which is widely used for gypsy moth control, increased mortality of aquatic shredders feeding upon leaves from sprayed compared to unsprayed plots.

There are currently few data indicating that acidic precipitation has produced significant changes in the biota of southern Appalachian streams. This reflects the absence of long-term data bases and does not imply that this is an area for little concern. Trend analysis of precipitation chemistry does indicate the need for concern (Schertz and Hirsch 1985). Lynch and Dise (1985) suggested that all basins in the Shenandoah National Park (SNP) in Virginia show some signs of acidification by atmospheric deposition. In the SNP, streams draining resistant siliceous bedrocks are viewed as extremely sensitive (alkalinity <20 μeq/L) and those draining granitic rock as having a high sensitivity to acid deposition (alkalinities of 20–100 μeq/L) (Lynch and Dise 1985). However, Messer et al. (1986) examined 54 randomly selected streams in the southern Blue Ridge during the spring and found that despite generally circumneutral pH values, the majority of the streams possessed relatively low acid neutralizing capacity (ANC). Messer et al. found that 6.3% of the combined stream length possessed ANC of <20 μeq/L, while 74.4% were estimated to be <200 μeq/L.

In other regions subject to acidificiation, episodic runoff, especially that associated with snowmelt, may increase the solubility of aluminum in stream water (Hall et al. 1985). The downstream drift rate of some species of invertebrates, especially mayflies, chironomids, and blackflies, increases with either the addition of strong acids (Hall et al. 1980, 1982) or aluminum chloride (Hall et al. 1985). These results are consistent with those from Scandanavia that report altered macroinvertebrate community structure associated with acidified streams (Drablos and Tollan 1980). Community structure of algae in streams of the GSMNP was found to differ among high and low pH sites. The highly acidic sites had high cell biovolume, chlorophyll *a* density, and areal primary productivity compared to sites with higher pH (Mulholland et al. 1986). In contrast, decomposing leaf litter at low-pH sites in the GSMNP had lower microbial production and respiration rates compared to high-pH streams (Palumbo et al. 1987).

In the northeastern United States, Canada, and Scandanavia, where soils and water have little buffering capacity, all trophic levels of aquatic ecosystems (decomposers, primary producers, primary and secondary consumers) appear to be affected by acidification (Haines 1981, Drablos and Tollan 1980). In the southern Appalachians, streams draining high-elevation watersheds with shallow soils underlain by siliceous and granitic bedrock are thought to represent areas most sensitive to acidic precipitation (Record et al. 1982). For example, Silsbee and Larson (1982) found that bacterial densities, pH, alkalinity, turbidity, as well as concentrations of Na, K, and Si decrease with increasing stream elevations in the GSMNP. Bedrock geology had a strong influence on pH, alkalinity, conductivity, hardness, and concentrations of several elements.

RECOMMENDATIONS FOR FUTURE RESEARCH AND MANAGEMENT

Since the early 1900s, there have been numerous conflicts between industries, logging and mining interests, municipalities, private citizens, and the federal government over land and stream use in the southern Appalachians. Mastran and Lowerre (1983) provide an excellent and entertaining historical perspective of these conflicts that have shaped the development of the Appalachian landscape. Unfortunately, many of the conflicts remain unresolved. Many inhabitants of the region voice open opposition to the creation of more wilderness areas for several reasons: (1) the ban on logging in wilderness areas; (2) loss of county tax revenues; (3) exclusion of motorized vehicles; (4) the "invasion" of wilderness areas by "outsiders"; (5) the threat to private ownership within and adjacent to wilderness areas; and, (6) rights of individuals versus rights of the federal government (Mastran and Lowerre 1983). While these issues are not easily resolved, there are others for which solutions do exist.

Clearcut logging, road, residential, and industrial construction, and agricultural practices have and continue to be conducted with little regard for environmental influences on streams that drain Appalachian watersheds. In some areas, coal mining and industrial and municipal pollution represent ongoing problems for Appalachian streams. These operations contribute to both point and nonpoint sources of stream pollution.

In most cases adequate knowledge exists to sharply curtail such stream degradation. The main problems are the absence of adequate laws or the enforcement of existing laws, and the absence of adequate transfer of technology from researchers to those engaged in forest management, agriculture, and industrial, municipal, urban, and rural residential development. Responsibility resides with all aspects of society, including government, researchers, developers, manufacturers, business, and private citizens. For example, early work in the Appalachian region showed that roads contribute to sediment inputs into streams. Proper design of forest access roads can reduce >90% of the sediment input into streams according to Swift (1988), who also noted, "Guidelines for forest road design are available which minimize the impact of construction and use on water quality. The task is to apply these guidelines." Implementation of proper design of spoil banks and adequate reclamation of surface mined sites could also greatly reduce the impact of acid mine drainage on streams in coal mining regions (see Samuel et al. 1978). Unfortunately, as long as government officials, corporation and business interests, and private citizens continue to have strong vested economic considerations in practices that may directly or indirectly contribute to stream degradation, there will continue to be environmental degradation of waterways.

For watershed disturbances such as clear-cutting or other major disturbances where both the physical environment and energy inputs to the stream are altered, any return of the biota to some resemblance of a predisturbance configuration may require long-term (>50 years) studies (Webster and Swank 1985b, Wallace et al. 1986). Extensive long-term data and more knowledge on the exact mechanisms and consequences for ecosystem level processes are required to adequately assess the potential influence of acidic precipitation on stream biota. Unfortunately, we lack sufficient long-term data bases to document the potential subtle effects of acidification, or other pollutants, on stream biota in most streams. It has been almost 30 years since Hynes (1960, p. 174) clearly stated the necessity for long-term studies as follows: "In the past, when damage has been caused or suspected, it has usually been impossible to estimate what was the previous biological condition of the water, except by inference. If records existed the fact of increased or decreased damage could easily and quickly be estimated." The need for such long-term studies appears to be as great today as it was when Hynes stated the problem.

ACKNOWLEDGMENTS

Much of the research presented in this chapter was supported by grants from the National Science Foundation. We thank J. W. Elwood and M. Adams for critical comments on earlier drafts of this chapter.

REFERENCES

Allan, J. D. 1983. Predator–prey relationships in streams. In J. R. Barnes and G. W. Minshall (eds.), *Stream Ecology*. New York: Plenum, pp. 191–229.

Anderson, N. H., and J. R. Sedell. 1979. Detritus processing by macroinvertebrates in stream ecosystems. *Annu. Rev. Entomol.* 24:351–377.

Anderson, N. H., J. R. Sedell, L. M. Roberts, and F. J. Triska. 1978. The role of invertebrates in processing wood debris in coniferous forest streams. *Am. Midland Natur.* 100:64–82.

Benfield, E. F., J. R. Webster, S. W. Golladay, G. T. Peters, and B. Stout. 1991. Effects of forest disturbance on leaf breakdown in southern Appalachian streams. *Verhandlungen Int. Vereingung Theoret. Angew. Limnol.* 24:1687–1690.

Benke, A. C. 1979. A modification of the Hynes method for estimating secondary production with particular significance for multivoltine populations. *Limnol. Oceanogr.* 24:168–171.

Benke, A. C. 1984. Secondary production of aquatic insects. In V. H. Resh and D. M. Rosenberg (eds.), *The Ecology of Aquatic Insects*. New York: Praeger, pp. 289–322.

Benke, A. C., and J. B. Wallace. 1980. Trophic basis of production among netspinning caddisflies in a southern Appalachian stream. *Ecology* 61:108–118.

Bevins, R. D., R. J. Strange, and D. C. Peterson. 1985. Current distribution of the native brook trout in the Appalachian region of Tennessee. *J. Tennessee Acad. Sci.* 60:101–105.

Bilby, R. E. 1981. Role of organic debris dams in regulating the export of dissolved and particulate matter from a forested watershed. *Ecology* 62:1234–1243.

Bilby, R. E., and G. E. Likens. 1980. Importance of organic debris dams in the structure and function of stream ecosystems. *Ecology* 61:1107–1113.

BMAN. 1985. Benthic Macroinvertebrate Ambient Network (BMAN) data review for 1984: Raleigh: North Carolina Department of Natural Resources and Community Development, Division of Environmental Management Water Quality Section, Report No. 85-11.

Bray, J. R., and E. Gorham. 1964. Litter production in forests of the world. *Adv. Ecol. Res.* 2:101–157.

Breder, C. M., Jr., and D. R. Crawford. 1922. The food of certain minnows. *Zoologica* 2:287–327.

Brigham, A. R., W. U. Brigham, and A. Gnilka (eds.). 1982. *Aquatic Insects and Oligochaetes of North and South Carolina*. Mahomet, IL: Midwest Aquatic Enterprises.

Brown, A. V., and L. C. Fitzpatrick. 1978. Life history and population energetics of the dobson fly, *Corydalus cornutus*. Ecology 59:1091–1108.

Brown, G. W., and J. T. Krygier. 1971. Effects of clear-cutting on stream temperature. *Water Resources Res.* 6:1133–1139.

Brussock, P. P. 1986. Macrodistribution and movements of benthic macroinvertebrates in relation to stream geomorphology. Thesis, University of Arkansas, Fayetteville.

Brussock, P. P., A. V. Brown, and J. C. Dixon. 1985. Channel form and stream ecosystem models. *Water Resources Bull.* 21:859–866.

Burton, G. W., and E. P. Odum. 1945. The distribution of stream fish in the vicinity of Mountain Lake, Virginia. *Ecology* 26:182–193.

Camburn, K. E., and R. L. Lowe. 1978. *Achnanthes subrostrata* v. *appalachiana* Camburn and Lowe var. nov., a new diatom from the southern Appalachian Mountains. *Castanea* 43:247–255.

Camburn, K. E., R. L. Lowe, and D. L. Stoneburner. 1978. The haptobenthic diatom flora of Long Branch Creek, South Carolina. *Nova Hedwigia* 30:149–279.

Carlander, K. D. 1969. *Handbook of Freshwater Fishery Biology*, 3d ed., Vol. 1. Ames: Iowa State UP.

Chow, V. T. 1959. *Open Channel Hydraulics*. New York: McGraw-Hill.

Coker, R. E. 1925. Observations of hydrogen-ion concentration and fishes in waters tributary to the Catawba River, North Carolina (with supplementary observations in some waters of Cape Cod, Massachusetts. *Ecology* 6:52–65.

Comisky, C. E., G. S. Henderson, R. H. Gardner, and F. W. Woods. 1977. Patterns of organic matter transport on Walker Branch Watershed. In Correll, D. L. (ed.), *Watershed Research in Eastern North America*. Washington, DC: Smithsonian Institution, pp. 439–469.

Conners, M. E., and R. J. Naiman. 1984. Particulate allochthonous inputs: relationships with stream size in an undisturbed watershed. *Can. J. Fisheries Aquatic Sci.* 41:1473–1484.

Cuffney, T. F., and J. B. Wallace. 1989. Discharge–export relationships in headwater streams: influence of invertebrate manipulations and drought. *J North Am. Benthol. Soc.* 8:331–341.

Cuffney, T. F., J. B. Wallace, and J. R. Webster. 1984. Pesticide manipulation of a headwater stream: invertebrate responses and their significance for ecosystem processes. *Freshwater Invertebrate Biol* 3:153–171.

Cuffney, T. F., J. B. Wallace, and G. J. Lugthart. 1990. Experimental evidence quantifying the role of benthic invertebrates in organic matter dynamics of headwater streams. *Freshwater Biol.* 23:281–299.

Cummins, K. W. 1973. Trophic relations of aquatic insects. *Annu. Rev. Entomol.* 18:183–206.

Cummins, K. W., and M. J. Klug. 1979. Feeding ecology of stream invertebrates. *Annu. Rev. Ecol. Systematics* 10:147–172.

Cummins, K. W., J. R. Sedell, F. J. Swanson, G. W. Minshall, S. G. Fisher, C. E. Cushing, R. C. Petersen, and R. L. Vannote. 1983. Organic matter budgets for stream ecosystems: problems in their evaluation. In J. R. Barnes and G. W. Minshall (eds.), *Stream Ecology*. New York: Plenum, pp. 299–353.

Cushman, R. M., J. W. Elwood, and S. G. Hildebrand. 1977. Life history and production of *Alloperla mediana* and *Diplectrona modesta* in Walker Branch, Tennessee. *Am. Midland Natur.* 98:354–364.

Daiber, F. C. 1956. A comparative analysis of the winter feeding habits of two benthic stream fishes. *Copia* 1956:141–151.

Dillard, G. E. 1969. The benthic algal communities of a North Carolina Piedmont stream. *Nova Hedwigia* 17:7–29.

Dillard, G. E. 1971. An epilithic diatom community of a North Carolina sandhills stream. *Extrait Rev. Algol.* 2:118–127.

Douglass, J. E., and M. E. Seehorn. 1974. Forest management impacts on cold water fisheries. In *Proceedings of Symposium on Trout Habitat Research and Management.* Asheville, NC: Southeast Forest Experiment Station, pp. 32–46.

Drablos, D., and A. Tollan (eds.). 1980. Acid precipitation: effects on forest and fish project. *Proceedings of the International Conference on the Ecological Impact of Acid Precipitation*, Oslo, Norway.

Drumheller, D. 1979. Organization and function of *Desmognathus* salamander communities in southern Appalachian stream ecosystems. Thesis, Clemson University, Clemson, SC.

Dudley, T., and N. H. Anderson. 1982. A survey of invertebrates associated with wood debris in aquatic habitats. *Melanderia* 39:1–21.

Elwood, J. W., and G. S. Henderson. 1975. Hydrologic and chemical budgets at Oak Ridge, Tennessee. In A. D. Hasler (ed.), *Coupling of Land and Water Systems.* New York: Springer, pp. 31–51.

Elwood, J. W., and D. J. Nelson. 1972. Periphyton production and grazing rates in a stream measured with a ^{32}P material balance method. *Oikos* 23:295–303.

Elwood, J. W., and R. R. Turner. 1989. Streams: water chemistry and ecology. In D. W. Johnson and R. I. Van Hook (eds.), *Analysis of Biogeochemical Cycling Processes in Walker Branch Watershed.* New York: Springer, pp. 301–350.

Elwood, J. W., J. D. Newbold, A. F. Trimble, and R. W. Stark. 1981. The limiting role of phosphorous in a woodland stream ecosystem: effects of P enrichment on leaf decomposition and primary producers. *Ecology* 62:146–158.

Fassett, N. C. 1966. *A Manual of Aquatic Plants.* Madison: Wisconsin UP.

Fenneman, N. M. 1938. *Physiography of Eastern United States.* New York: McGraw-Hill.

Findlay, S., J. L. Meyer, and P. J. Smith. 1984. Significance of bacterial biomass in the nutrition of a freshwater isopod (*Lirceus*). *Oecologia* (*Berlin*) 63:38–42.

Fisher, S. G. 1977. Organic matter processing by a stream-segment ecosystem. Fort River, Massachusetts, U.S.A. *Int. Rev. gesamten Hydrobiol.* 62:701–727.

Fisher, S. G., and L. J. Gray. 1983. Secondary production and organic matter processing by collector macroinvertebrates in a desert stream. *Ecology* 64:1217–1224.

Fisher, S. G., and G. E. Likens. 1973. Energy flow in Bear Brook, New Hampshire: an integrative approach to stream ecosystem metabolism. *Ecol. Monogr.* 43:421–439.

Flemer, D. A., and W. S. Woolcott. 1966. Food habits and distribution of the fishes of Tuckaloe Creek, Virginia, with special emphasis on the bluegill, *Lepomis m. macrochirus* Rafinesque. *Chesapeake Sci.* 7:75–89.

Forbes, S. A. 1880. The food of the darters. *Am. Natur.* 14:697–703.

Georgian, T. J., Jr., and J. B. Wallace. 1981. A model of seston capture by net-spinning caddisflies. *Oikos* 36:147–157.

Georgian, T. J., Jr., and J. B. Wallace. 1983. Seasonal production dynamics in a guild of periphyton-grazing insects in a southern Appalachian stream. *Ecology* 64:1236–1248.

Gessner, F. 1959. Hydrobotanik, II: Stoffhaushalt. *Veb. Deutsch. Ver. Wissensch. Berlin* 53:75–92.

Glime, J. M. 1968. Ecological observations on some bryophytes in Appalachian Mountain Streams. *Castanea* 33:300–325.

Golladay, S. W., and J. R. Webster. 1988. Effects of clearcut logging on wood breakdown in Appalachian Mountain streams. *Am. Midland Natur.* 119:143–155.

Golladay, S. W., J. R. Webster, and E. F. Benfield. 1987. Changes in stream morphology and storm transport of organic matter following watershed disturbance. *J. North Am. Benthol. Soc.* 6:1–11.

Golladay, S. W., J. R. Webster, and E. F. Benfield. 1989. Changes in stream benthic organic matter following watershed disturbance. *Holarctic Ecol.* 12:96–105.

Gruchy, C. G., R. H. Bowen, and I. M. Gruchy. 1973. First records of the silver shiner, *Notropis photogenis*, from Canada. *J. Fisheries Res. Board Can.* 30:1379–1382.

Grzenda, A. R., H. P. Nicholson, J. T. Teasley, and J. H. Patric. 1964. DDT residues in mountain stream water as influenced by treatment practices. *J. Econ. Entomol.* 57:615–618.

Gurtz, M. E., and J. B. Wallace. 1984. Substrate-mediated response of stream invertebrates to disturbance. *Ecology* 65:1556–1569.

Gurtz, M. E., and J. B. Wallace. 1986. Substratum production relationships in net-spinning caddisflies (Trichoptera) in disturbed and undisturbed hardwood catchments. *J. North Am. Benthol. Soc.* 5:230–236.

Gurtz, M. E., J. R. Webster, and J. B. Wallace. 1980. Seston dynamics in southern Appalachian streams: effects of clear-cutting. *Can. J. Fisheries Aquatic Sci.* 37:624–631.

GWCB. 1970. *Georgia Water Quality Control Board—Coosa River Basin Study.* Atlanta: Georgia Water Quality Control Board.

Hack, J. T. 1969. The area, its geology: Cenozoic development of the southern Appalachians. In P. C. Holt (ed.), *The Distributional History of the Biota of the Southern Appalachians*, Part I: *Invertebrates*. Research Division Monograph 1. Blacksburg: Virginia Polytechnic Institute, pp. 1–17.

Hack, J. T. 1973. Drainage adjustment in the Appalachians. In M. Morisawa (ed.), *Fluvial Geomorphology*. Publications in Geomorphology. Binghampton: State University of New York, pp. 51–69.

Hack, J. T. 1980. Rock control and tectonism: their importance in shaping the Appalachian Highlands. *United States Geological Survey Professional Paper* 1126-B.

Haefner, J. D., and J. B. Wallace. 1981a. Production and potential seston utilization by *Parapsyche cardis* and *Diplectrona modesta* (Trichoptera: Hydropsychidae) in two streams draining contrasting southern Appalachian watersheds. *Environ. Entomol.* 10:433–441.

Haefner, J. D., and J. B. Wallace. 1981b. Shifts in aquatic insect populations in a first-order southern Appalachian stream following a decade of old field succession. *Can. J. Fisheries Aquatic Sci.* 38:353–359.

Haines, T. A. 1981. Acidic precipitation and its consequences for aquatic ecosystems: a review. *Trans. Am. Fisheries Soc.* 110:669–707.

Hairston, N. G. 1949. The local distribution and ecology of the plethodontid salamanders in the southern Appalachians. *Ecol. Monogr.* 19:47–73.

Hairston, N. G. 1980. Species packing in the salamander genus *Desmognathus*: what are the interspecific interactions involved? *Am. Natur.* 115:354–366.

Hairston, N. G. 1981. An experimental test of a guild: salamander competion. *Ecology* 62:65–72.

Hairston, N. G. 1983. Alpha selection in competing salamanders: experimental verification of an a priori hypothesis. *Am. Natur.* 122:105–113.

Hairston, N. G. 1986. Species packing in *Desmognathous* salamanders: experimental demonstration of predation and competition. *Am. Natur.* 127:266–291.

Hairston, N. G. 1987. Community ecology and salamander guilds. Cambridge, UK: Cambridge UP.

Hall, R. J., G. E. Likens, S. B. Fiance, and G. R. Hendry. 1980. Experimental acidification of a stream in the Hubbard Brook Experimental Forest, New Hampshire. *Ecology* 61:976–989.

Hall, R. J., J. M. Pratt, and G. E. Likens. 1982. Effects of experimental acidification on macroinvertebrate drift diversity in a mountain stream. *Water, Air,* and Soil Pollution 18:273–287.

Hall, R. J., C. T. Driscoll, G. E. Likens, and J. M. Pratt. 1985. Physical, chemical, and biological consequences of episodic aluminum additions to a stream. *Limnol. Oceanogr.* 30:212–220.

Harmon, M. E., J. F. Franklin, F. J. Swanson, P. Sollins, S. V. Gregory, J. D. Lattin, N. H. Anderson, S. P. Cline, N. G. Aumen, J. R. Sedell, G. W. Lienkaemper, K. Cromack, Jr., and K. W. Cummins. 1986. Ecology of coarse woody debris in temperate ecosystems. *Adv. Ecol. Res.* 15:133–302.

Harshbarger, T. J. 1978. Factors affecting regional trout stream productivity. In *Proceedings of the Southeastern Trout Resource: Ecology and Management Symposium.* Asheville, NC: Southeast Forest Experiment Station, pp. 11–27.

Harshbarger, T. J. In preparation. Distribution and biomass of fish with elevation in western North Carolina (a compilation of data from North Carolina Wildlife Resources Commission, 1983). In *Final Report Federal Aid in Fish Restoration, Project F24-S: Survey and Classification of State-Managed Trout Streams, Dist. 7,8, and 9.*

Hedin, L.O., M. S. Mayer, and G. E. Likens. 1988. The effect of deforestation on organic debris dams. *Verhandlungen Int. Vereingung Theoret Angew. Limnol.* 23:1135–1141.

Herricks, E. E., and J. J. Cairns, Jr. 1974. Rehabilitation of streams receiving acid mine drainage. *Virginia Polytechnic Institute and State University Water Resources Center Bulletin* 66.

Hewlett, J. D., and A. R. Hibbert. 1961. Increases in water yield after several types of forest cutting. *Quart. Bull. Int. Assoc. Sci. Hydrol.* 6:5–16.

Hibbert, A. R. 1966. Forest treatment effects on water yield. In W. E. Soper and H. W. Lull (eds.), *Forest Hydrology.* Oxford, UK: Pergamon, pp. 527–543.

Hill, B. H. 1981. Distribution and production of *Justicia americana* in the New River, Virginia. *Castanea* 46:162–169.

Hill, B. H., and J. R. Webster. 1982a. Periphyton production in an Appalachian river. *Hydrobiologia* 97:275–280.

Hill, B. H, and J. R. Webster. 1982b. Aquatic macrophyte breakdown in an Appa-lachian river. *Hydrobiologia* 89:53–59.

Hill, B. H., and J. R. Webster. 1983. Aquatic macrophyte contribution to the New River organic matter budget. In T. D. Fontaine III and S. M. Bartell (eds.), *Dynamics of Lotic Ecosystems*, Ann Arbor, MI: Ann Arbor Science, pp. 273–282.

Hill, B. H., and J. R. Webster. 1984. Productivity of *Podostemum ceratophyllum* in the New River, Virginia. *Am. J. Botany* 71:130–136.

Hill, R. D. 1975. Mining impacts on trout habitat. In *Symposium on Trout Habitat Research and Management Proceedings*. Asheville, NC: Southeast Forest Experi-ment Station, pp. 47–57.

Hobbs, H. H. 1974. A checklist of the North and Middle American crayfishes. (De-capoda: Astocidae and Cambaridae). *Smithsonian Contributions in Zoology* 166.

Hocutt, C. H. 1979. Drainage evolution and fish dispersal in the central Appalachians. *Geol. Soc. Am. Bull.* 90:129–130.

Hocutt, C. H., R. F. Denoncourt, and J. R. Stauffer, Jr. 1978. Fishes of the Greenbrier River West Virginia with drainage history of the central Appalachians. *J. Biogeogr.* 5:59–80.

Holt, P. C. 1969. Epilogue. In P. C. Holt (ed.). *The Distributional History of the Biota of the Southern Appalachians*, Part I: *Invertebrates*. Research Division Mon-ograph 1. Blacksburg: Virginia Polytechnic Institute.

Hornick, L. E., J. R. Webster, and E. F. Benfield. 1981. Periphyton production in an Appalachian Mountain trout stream. *Am. Midland Natur.* 106:22–36.

Howard, R. V., and B. C. Parker. 1979. *Nemalionopsis shawii* forma *caroliniana* (forma nov.) (Rhodophyta: Nemalionales) from the southeastern United States. *Phycologia* 18:330–337.

Howard, R. V., and B. C. Parker. 1980. Revision of *Boldia erythrosiphon* Herndon (Rhodophyta, Bangiales). *Am. J. Botany* 67:413–422.

Huryn, A. D. 1986. Secondary production of the macroinvertebrate community of a high-elevation stream in the southern Appalachian Mountains. Thesis, University of Georgia, Athens.

Huryn, A. D. 1990. Growth and voltinism of lotic midge larvae; patterns across an Appalachian Mountain Basin. *Limnol. Oceanogr.* 35:339–351.

Huryn, A. D., and J. B. Wallace. 1985. Life history and production of *Goerita semata* Ross (Trichoptera: Limnephilidae) in the southern Appalachian Mountains. *Can. J. Zool.* 63:2604–2611.

Huryn, A. D., and J. B. Wallace. 1986. A method for obtaining in situ growth rates of larval Chironomidae (Diptera) and its application to studies of secondary pro-duction. *Limnol. Oceanogr.* 31:216–222.

Huryn, A. D., and J. B. Wallace. 1987a. Local geomorphology as a determinant of macrofaunal production in a mountain stream. *Ecology* 68:1932–1942.

Huryn, A. D., and J. B. Wallace. 1987b. Production and litter processing by crayfish in an Appalachian Mountain stream. *Freshwater Biol.* 18:277–286.

Huryn, A. D., and J. B. Wallace. 1988. Community structure of Trichoptera in a mountain stream: spatial patterns of production and functional organization. *Fresh-water Biol.* 20:141–155.

Hynes, H. B. N. 1960. The biology of polluted waters. Liverpool, UK: Liverpool UP.

Hynes, H. B. N. 1970. The ecology of running waters. Toronto: Toronto UP.

Ide, F. P. 1967. Effects of forest spraying with DDT on aquatic insects of salmon streams in New Brunswick. *J. Fisheries Res. Board Can.* 24:769–805.

Jones, H. C., et al. 1983. Investigations of the cause of fishkills in fish-rearing facilities in the Raven Fork Watershed. *Tennessee Valley Authority, Division of Air and Water Resources, TVA/ONR/WR-83/9.*

Jones, R. D. 1978. Regional distribution trends of the trout resource. *Symposium on Trout Habitat Research and Management Proceedings.* Asheville, NC: Southeast Forest Experimental Station, pp. 1–10.

Jorgensen, C. B. 1966. *Biology of Suspension Feeding.* Oxford, UK: Pergamon.

Jorgensen, C. B. 1975. Comparative physiology of suspension feeding. *Annu. Rev. Physiol.* 37:57–79.

Karr, J. R. 1964. Age, growth, fecundity and food habits of fantail darters in Boone County, Iowa. *Iowa Acad. Sci.* 71:274–280.

Kaufmann, P. R., et al. 1988. Chemical characteristics of streams in the mid-Atlantic and southeastern United States (National Stream Survey–Phase 1), Vol. 2: streams sampled, descriptive statistics, and compendium of physical and chemical data. *United States Environmental Protection Agency Publication EPA/600/3-88/021b.*

Keen, W. H. 1979. Feeding and activity patterns in the salamander *Desmognathous ochrophaeus* (Amphibia, Urodela, Plethodontidae). *J. Herpetol.* 13:461–467.

Keen, W. H. 1982. Habitat selection and interspecific competition in two species of plethodontid salamanders. *Ecology* 63:94–102.

Keithan, E. D., and R. L. Lowe. 1985. Primary productivity and spatial structure of phytolithic growth in streams in the Great Smokey Mountains National Park. *Hydrobiologia* 123:59–67.

Kelly, A. G., J. G. Griffith, and R. D. Jones. 1980. Changes in the distribution of trout in Great Smoky Mountain National Park, 1900–1977. *U. S. Fish and Wildlife Service Technical Paper 102.*

King, P. B. 1977. *The Evolution of North America.* Princeton, NJ: Princeton UP.

Kleeberger, S. R. 1984. A test of competition in two sympatric populations of desmognathine salamanders. *Ecology* 65:1846–1859.

Kociolek, J. P. 1982. The diatom flora (Bacillariophyceae) of two streams in the Great Smoky Mountains National Park. Thesis, Bowling Green State University, Bowling Green, Ohio.

Kondratieff, P. F., and G. M. Simmons, Jr. 1982. Nutrient retention and macroinvertebrate community structure in a small stream receiving sewage effluent. *Archiv Hydrobiol.* 94:83–98.

Kondratieff, P. F., and G. M. Simmons, Jr. 1984. Nutritive quality and size fractions of natural seston in an impounded river. *Archiv Hydrobiol.* 101:401–412.

Kondratieff, P. F., R. A. Matthews, and A. L. Buikema Jr. 1984. A stressed stream ecosystem: macroinvertebrate community integrity and microbial trophic response. *Hydrobiologia* 111:81–91.

Kraatz, W. C. 1923. A study of the food of the minnow *Campostoma anomalum.* *Ohio J. Sci.* 23:265–283.

Krzysik, A. J. 1979. Resource allocation, coexistence, and the niche structure of a streambank salamander community. *Ecol. Monogr.* 49:173–194.

Lachner, E. A. 1950. The comparative food habits of the cyprinid fishes *Nocomis biguttatus* and *Nocomis micropogan* in western New York. *J. Washington Acad. Sci.* 40:229–263.

Lamberti, G. A., and J. W. Moore. 1984. Aquatic insects as primary consumers. In V. H. Resh, and D. M. Rosenberg (eds.), *The Ecology of Aquatic Insects*. New York: Praeger Scientific, pp. 164–195.

Larson, G. L., and S. E. Moore. 1985. Encroachment of exotic rainbow trout into stream populations of native brook trout in the southern Appalachian Mountains. *Trans. Am. Fisheries Soc.* 114:195–203.

Lawton, J. H. 1970. Feeding and food energy assimilation in larvae of the damselfly *Pyrrhosoma nymphula* (Sulz.) (Odonata:Zygoptera). *J. Anim. Ecol.* 39:669–698.

Lee, D. S., C. R. Gilbert, C. H. Hocutt, R. E. Jenkins, D. E. McAllister, and J. R. Stauffer, Jr. 1980. *Atlas of North American Freshwater Fish*. Raleigh: North Carolina State Museum of Natural History.

Leopold, L. B., M. G. Wolman, and J. P. Miller. 1964. Fluvial processes in geomorphology. San Francisco, CA: Freeman.

Likens, G. E., and R. E. Bilby. 1982. Development, maintenance, and role of organic-debris dams in New England streams. In F. J. Swanson, R. J. Janda, T. Dunne, and D. N. Swanson (eds.), *Sediment Budgets and Routing in Forested Drainage Basins*. U.S. Forest Service PNW For. Range Exp. Sta. Gen. Tech. Rept. PNW-141, pp. 122–128.

Lock, M. A. 1981. River epilithon: a light and energy transducer. In M. A. Lock and D. D. Williams (eds.), *Perspectives in Running Water Ecology*. New York: Plenum, pp. 3–40.

Lowe, R. L., and J. P. Kociolek. 1984. New and rare diatoms from Great Smoky Mountains National Park. *Nova Hedwigia* 39:465–476.

Lowe, R. L., S. W. Golladay, and J. R. Webster. 1986. Periphyton response to nutrient manipulation in streams draining clearcut and forested watersheds. *J. North Am. Benthol. Soc.* 5:221–229.

Lugthart, G. J., J. B. Wallace, and A. D. Huryn. 1990. Secondary production of chironomid communities in insecticide-treated and untreated headwater streams. *Freshwater Biol.* 24:417–427.

Lynch, D. D., and N. B. Dise. 1985. Sensitivity of stream basins in the Shenandoah National Park to acid deposition. *United States Geological Survey Water-Resources Investigations Report 85-4115*. Richmond, Virginia. 61p.

Malas, D., and J. B. Wallace. 1977. Strategies for coexistence in three species of net-spinning caddisflies in second-order southern Appalachian streams. *Can. J. Zool.* 55:1829–1840.

Martof, B. S., W. M. Palmer, J. R. Bailey, J. R. Harrison III, and J. Dermid. 1980. Amphibians and reptiles of the Carolinas and Virginia. Chapel Hill: North Carolina UP.

Mastran, S. S., and N. Lowerre. 1983. Mountaineers and rangers: a history of federal forest management in the Southern Appalachians. *USDA, U.S. Forest Service Publ. No. FS-380*. Washington, DC.

Meijer, W. 1975. A note on *Podostemum ceratophyllum* Michx., as an indicator of clean streams in and around the Appalachian Mountains. *Castanea* 41:319–324.

Melillo, J. M., R. J. Naiman, J. D. Aber, and K. N. Eshleman. 1983. The influence of substrate quality and stream size on wood decomposition dynamics. *Oecologia (Berlin)* 58:281–285.

Merritt, R. W., and K. W. Cummins (eds.). 1984. An introduction to the aquatic insects of North America, 2d ed. Dubuque, IA: Kendall/Hunt.

Merritt, R. W., and J. B. Wallace. 1981. Filter-feeding insects. *Sci. Am.* 244:132–144.

Merritt, R. W., K. W. Cummins, and T. M. Burton. 1984. The role of aquatic insects in the processing and cycling of nutrients. In V. H. Resh and D. M. Rosenberg (eds.), *The Ecology of Aquatic Insects*. New York: Praeger, pp. 134–163.

Messer, J. J., et al. 1986. National surface water survey: national stream survey phase 1-pilot study. *United States Environmental Protection Agency Publication EPA/600/4-86/026.*

Meyer, J L., and J. O'Hop. 1983. Leaf-shredding insects as a source of DOC in headwater streams. *Am. Midland Natur.* 109:175–183.

Meyer, J. L., and C. M. Tate. 1983. The effects of watershed disturbance on the dissolved organic carbon dynamics of a stream. *Ecology* 64:33–44.

Meyer, J. L., C. M. Tate, R. T. Edwards, and M. T. Crocker. 1988. The trophic significance of DOC in streams. In W. T. Swank and D. A. Crossley (eds.), *Forest Hydrology and Ecology at Coweeta*. New York: Springer, pp. 269–278.

Miller, C. 1985. Correlates of habitat favourability for benthic macroinvertebrates at five stream sites in an Appalachian Mountain drainage basin. U.S.A. *Freshwater Biol.* 15:709–733.

Miller, R. S. 1964. Behavior and ecology of some North American cyprinid fishes. *Am. Midland Natur.* 72:313–357.

Minckley, W. L. 1963. The ecology of a spring stream Doe Run, Meade County, Kentucky. *Wildlife Monogr.* 11:1–124.

Minckley, W. L., and D. R. Tindall. 1963. Ecology of *Batrachospermum* sp. in Doe Run, Meade County, Kentucky. *Bull. Torrey Botanical Club* 90:391–400.

Minshall, G. W. 1967. Role of allochthonous detritus in the trophic structure of a woodland springbrook community. *Ecology* 48:139–149.

Minshall, G. W., R. C. Peterson, K. W. Cummins, T. L. Bott, J. R. Sedell, C. E. Cushing, and R. L. Vannote. 1983. Interbiome comparisons of stream ecosystem dynamics. *Ecol. Monogr.* 53:1–25.

Minshall, G. W., J. T. Brock, and J. D. Varley. 1989. Wildfires and Yellowstone's stream ecosystems. *BioScience* 39:707–715.

Molles, M. C., Jr. 1982. Trichopteran communities of streams associated with aspen and conifer forests: long-term structural change. *Ecology* 63:1–6.

Moore, S. T., B. Ridley, and G. L. Larson. 1983. Standing crops of brook trout concurrent with removal of rainbow trout from selected streams in the Great Smoky Mountain National Park. *North Am. J. Fisheries Management* 3:72–80.

Mulholland, P. J., J. W. Elwood, A. V. Palumbo, and R. J. Stevenson. 1986. Effect of stream acidification on periphyton composition, chlorophyll, and productivity. *Can. J. Fish. Aquat. Sci.* 43:1846–1858.

Naiman, R. J., and J. R. Sedell. 1979a. Benthic organic matter as a function of stream order in Oregon. *Archiv Hydrobiol.* 87:404–422.

Naiman, R. J., and J. R. Sedell. 1979b. Characterization of particulate organic matter transported by some Cascade Mountain streams. *J. Fisheries Res. Board Can.* 36:17–31.

National Park Service. 1981. The first nation-wide rivers inventory. *The National Park Service, U.S.D.I., Supt. Docs/GPO P.0000-0046.* Washington, DC.

Needham, P. R. 1938. *Trout Streams.* Ithaca, NY: Comstock.

Neves, R. J., and G. B. Pardue. 1983. Abundance and production of fishes in a small Appalachian stream. *Trans. Am. Fisheries Soc.* 112:21–26.

Newbold, J. D., R. V. O'Neill, J. W. Elwood, and W. VanWinkle. 1982. Nutrient spiralling in streams: implications for nutrient limitation and invertebrate activity. *Am. Natur.* 120:628–652.

NOAA. 1980. Climates of the States, Vols. 1 and 2. National Oceanic and Atmospheric Administration. Detroit, MI: Gale Research.

O'Doherty, E. C. 1985. Stream dwelling copepods: their life history and ecological significance. *Linnol. Oceanogr.* 30:554–564.

O'Hop, J, J. B. Wallace, and J. D. Haefner. 1984. Production of a stream shredder, *Peltoperla maria* (Plecoptera: Peltoperlidae) in disturbed and undisturbed hardwood catchments. *Freshwater Biol.* 14:13–21.

Organ, J. A. 1961. Studies of the local distribution, life history, and population dynamics of the salamander genus *Desmognathous* in Virginia. *Ecol. Monogr.* 31:189–226.

Orser, P. W., and D. J. Shure. 1972. Effects of urbanization on the salamander *Desmognathous fuscus fuscus. Ecology* 53:1148–1154.

Palumbo, A. V., P. J. Mulholland, and J. W. Elwood. 1987. Response of microbial communities on leaf material to low pH and to aluminum accumulation. *Can. J. Fish. Aquat. Sci.* 44:1064–1070.

Patrick, R. 1948. Factors effecting the distribution of diatoms. *Botanical Rev.* 14:473–524.

Peckarsky, B. L. 1984. Predatory–prey interactions among aquatic insects. In V. H. Resh and D. M. Rosenberg (eds.), *The Ecology of Aquatic Insects.* New York: Praeger, pp. 196–254.

Pennak, R. L. 1978. Fresh-water invertebrates of the United States, 2d ed. New York: Wiley.

Pereira, C. R. D., N. H. Anderson, and T. Dudley. 1982. Gut content analysis of aquatic insects from wood substrates. *Melanderia* 39:23–33.

Petersen, R. C., and K. W. Cummins. 1974. Leaf processing in a woodland stream. *Freshwater Biol.* 4:343–363.

Pfeiffer, R. A. 1955. Studies on the life history of the rosyface shiner, *Notropis rubellus. Copia* 1955:95–104.

Powers, E. B. 1929. Freshwater studies, I: the relative temperature, oxygen concentration, alkali reserve, the carbon dioxide tension and pH of waters of certain mountain streams at different altitudes in the GSMNP. *Ecology* 10:97–110.

Record, F. A., D. V. Bubenick, and R. J. Kindya. 1982. *Acid Rain Information Book.* Park Ridge, NJ: Noyes Data.

Reed, R. J. 1957. Phases of the life history of the rosyface shiner. *Notropis rubellus*, in northwestern Pennsylvania. *Copia* 1957:286–290.

Ricker, W. E. 1934. An ecological classification of certain Ontario streams. *Univ. Toronto Studies Biol. Ser.* 37:1–144.

Roback, S. S., and J. W. Richardson. 1969. The effects of acid mine drainage on aquatic insects. *Proc. Acad. Natur. Sci. Philadelphia* 121:81–107.

Rodgers, J. H., Jr., M. E. McKevitt, D. O. Hammerlund, K. L. Dickson, and J. Cairns, Jr. 1983. Primary production and decomposition of submergent and emergent aquatic plants of two Appalachian rivers. In T. D. Fontaine III and S. M. Bartell (eds.). *Dynamics of Lotic Ecosystems.* Ann Arbor, MI: Ann Arbor Science, pp. 283–301.

Ross, D. H., and J. B. Wallace. 1981. Production of *Brachycentrus spinae* Ross (Trichoptera: Brachycentridae) and its role in seston dynamics of a southern Appalachian stream. *Environ. Entomol.* 10:240–246.

Ross, D. H., and J. B. Wallace. 1983. Longitudinal patterns of production, food consumption, and seston utilization by net-spinning caddisflies (Trichoptera) in a southern Appalachian stream (USA). *Holarctic Ecol.* 6:270–284.

Ross, R. D. 1969. Drainage evolution and fish distribution problems in the southern Appalachians of Virginia. In P. C. Holt (ed.), *The Distributional History of the Biota of the Southern Appalachians*, Part I: *Invertebrates.* Research Division Monograph 1. Blacksburg: Virginia Polytechnic Institute, pp. 277–292.

Ross, R. D. 1971. The drainage history of the Tennessee River. In P. C. Holt (ed.), *The Distributional History of the Biota of the Southern Appalachians*, Part III. Research Division Monograph 4, Blacksburg: Virginia Polytechnic Institute, pp. 11–42.

Samuel, D. E., J. R. Stauffer, and C. H. Hocutt (eds). 1978. Surface mining and fish/wildlife needs in the eastern United States. Fish and Wildlife Service, Office of Biological Services (FWS/OBS-78/81).

SARRMC, 1977. Western North Carolina priorities for natural resources research: a systems analysis of problems, opportunities, and information needs for management and research. The Southern Appalachian Research/Resource Management Cooperative, Raleigh, North Carolina, May 1977.

Schertz, T. L., and R. M. Hirsch. 1985. Trend analysis of weekly acid rain data: 1978–1983. *United States Geological Survey Water Resources Investigations Report 85-4211, Austin, Texas.*

Schmalzer, P. A., T. S. Patrick, and H. R. DeSelm. 1985. Vascular flora of the Obed Wild and Scenic River, Tennessee. *Castanea* 50:71–88.

Scott, W. B., and E. J. Crossman. 1973. *Freshwater Fishes of Canada.* Ottawa: Fisheries Research Board of Canada.

Sedell, J. R., R. J. Naiman, K. W. Cummins, G. W. Minshall, and R. L. Vannote. 1978. Transport of particulate organic matter in streams as a function of physical processes. *Int. Vereinigung Theoret. Angew. Limnol. Verhandlungen* 20:1366–1375.

Sedell, J. R., F. H. Everest, and F. J. Swanson. 1982. Fish habitat and streamside management: past and present. In *Proceedings of the Society of American Foresters Annual Meeting.* Bethesda, MD: Society of American Forestry, pp. 244–255.

Short, R. A., and P. E. Maslin. 1977. Processing of leaf litter by a stream detritivore: effect on nutrient availability to collectors. *Ecology* 58:935–938.

Silsbee, D. G., and G. L. Larson. 1982. Water quality of streams in the Great Smoky Mountains National Park. *Hydrobiologia* 89:97–115.

Silva, H., and A. J. Sharp. 1944. Some algae of the southern Appalachians. *J. Tennessee Acad. Sci.* 19:337–347.

Sinsabaugh, R. L., A. E. Linkins, and E. F. Benfield. 1985. Cellulose digestion and assimilation by three leaf-shredding insects. *Ecology* 66:1464–1471.

Smith-Cuffney, F. L., and J. B. Wallace. 1987. The influence of microhabitat on availability of drifting invertebrate prey to a net-spinning caddisfly. *Freshwater Biol.* 17:91–98.

Southerland, M. T. 1986a. Behavioral interactions among species of the salamander genus *Desmognathous*. *Ecology* 67:175–181.

Southerland, M. T. 1986b. Coexistence of three congeneric salamanders: the importance of habitat and body size. *Ecology* 67:721–728.

Speaker, R. W., K. W. Moore, and S. V. Gregory. 1984. Analysis of the process of retention of organic matter in stream ecosystems. *Int. Vereinigung Theoret. Angew. Limnol. Verhandlungen* 22: 1835–1841.

Spight, T. M. 1967. Population structure and biomass production by a stream salamander. *Am. Midland Natur.* 78:437–447.

Statzner, B., and B. Higler. 1985. Questions and comments on the river continuum concept. *Can. J. Fisheries Aquatic Sci.* 42:1038–1044.

Swank, W. T., and J. E. Douglass. 1974. Streamflow greatly reduced by converting deciduous hardwood stands to pine. *Science* 185:857–859.

Swank, W. T., and J. E. Douglass. 1977. Nutrient budgets for undisturbed and manipulated hardwood forest ecosystems in the mountains of North Carolina. In D. L. Correll (ed.), *Watershed Research in Eastern North America*. Washington, DC: Smithsonian, pp. 343–362.

Swank, W. T., J. B. Waide, D. A. Crossley, Jr., and R. L. Todd. 1981. Insect defoliation enhances nitrate export from forest ecosystems. *Oecologia* (*Berlin*) 51:297–299.

Swank, W. T., L. W. Swift, Jr., and J. E. Douglass. 1988. Streamflow changes associated with forest cutting, species conversions, and natural disturbances. In W. T. Swank and D. A. Crossley (eds.), *Forest Hydrology and Ecology at Coweeta*. New York: Springer, pp. 297–312.

Swanson, F. J., S. V. Gregory, J. R. Sedell, and A. G. Campbell, 1982. Land–water interactions: the riparian zone. In R. L. Edmonds (ed.), *Analysis of Coniferous Forest Ecosystems in the Western United States*. Stroudsburg, PA: Hutchinson Ross, pp. 233–266.

Swift, L. W., Jr. 1983. Duration of stream temperature increases following forest cutting in the southern Appalachian Mountains. In *Proceedings of the International Symposium on Hydrometerology, Denver, Colorado, USA, 13–17 June 1982*. Bethesda, MD: American Water Resources Association, pp. 273–275.

Swift, L. W., Jr. 1988. Forest access roads: design, maintenance, and soil loss. In W. T. Swank and D. A. Crossley (eds.), *Forest Hydrology and Ecology at Coweeta*. New York: Springer, pp. 313–324.

Swift, L. W., Jr., and J. B. Messer. 1971. Forest cuttings raise temperatures of small streams in the southern Appalachians. *J. Soil Water Conservation* 26:111–116.

Swift, L. W., Jr., and W. T. Swank. 1981. Long-term responses of streamflow following clearcutting and regrowth. *Hydrol. Sci. Bull.* 26:245–256.

Swift, M. C., and K. W. Cummins. 1991. Long-term effects of Dimilin on stream metabolism. *Verhandlungen Int. Vereingung Theoret. Angew. Limnol.* 24: not published yet.

Tarter, D. C. 1970. Food and feeding habits of the western blacknose dace, *Rhinichthys atratulus melengris* Agassiz, in Doe Run, Meade County, Kentucky. *Am. Midland Natur.* 83:134–159.

Tarter, D. C. 1976. *Limnology in West Virginia: A Lecture and Laboratory Manual.* Huntington, WV: Marshall University Bookstore.

Tebo, L. B., Jr. 1955. Effects of siltation, resulting from improper logging, on the bottom fauna of a small trout stream in the southern Appalachians. *Progressive Fish-Culturist* 55:64–70.

Tebo, L. B., Jr., and W. W. Hassler. 1963. Food of brook, brown, and rainbow trout from streams in western North Carolina. *J. Elisha Mitchell Soc.* 79:44–53.

Tilley, S. G. 1968. Size-fecundity relationships and their evolutionary implications in five desmognathine salamanders. *Evolution* 22:806–816.

Trautman, M. B. 1942. Fish distribution and abundance correlated with stream gradients as a consideration in stocking programs. *Transactions 7th North American Wildlife Conference, Washington, DC*, 211–223.

Trautman, M. B. 1981. *The Fishes of Ohio.* Columbus: Ohio State UP.

Triska, F. J., and K. Cromack, Jr. 1981. The role of woody debris in forests and streams. In R. H. Waring (ed.), *Forest: Fresh Perspectives from Ecosystem Analysis.* Corvallis: Oregon State UP, pp. 171–190.

Turner, C. L. 1921. Food of the common Ohio darters. *Ohio J. Sci.* 22:41–62.

Vannote, R. L. 1978. A geometric model describing a quasi-equilibrium of energy flow in populations of stream insects. *Proc. Nat. Acad. Sci. (USA)* 75:381–384.

Vannote, R. L., and B. W. Sweeney. 1980. Geographic analysis of thermal equilibria: a conceptual model for evaluating the effect of natural and modified thermal regimes on aquatic insect communities. *Am. Natur.* 115:667–695.

Vannote, R. L., G. W. Minshall, K. W. Cummins, J. R. Sedell, and C. E. Cushing. 1980. The river continuum concept. *Can. J. Fisheries Aquatic Sci.* 37:130–137.

Voshell, J. R., Jr. 1985. Trophic basis of production for macroinvertebrates in the New River below Bluestone Dam. Mimeographed report. Blacksburg: Department of Entomology, Virginia Polytechnic Institute and State University.

Voshell, J. R., Jr., and C. R. Parker. 1985. Quantity and quality of seston in an impounded and free-flowing river in Virginia, U.S.A. *Hydrobiologia* 122:271–280.

Wallace, J. B., and M. E. Gurtz. 1986. Response of *Baetis* mayflies (Ephemeroptera) to catchment logging. *Am. Midland Natur.* 115:25–41.

Wallace, J. B., and D. Malas. 1976. The fine structure of capture nets of larval Philopotamidae, with special emphasis on *Dolophilodes distinctus*. *Can. J. Zool.* 54:1788–1802.

Wallace, J. B., and R. W. Merritt. 1980. Filter-feeding ecology of aquatic insects. *Annu. Rev. Entomol.* 25:103–132.

Wallace, J. B., J. R. Webster, and W. R. Woodall. 1977. The role of filter feeders in flowing waters. *Archiv Hydrobiol* 79:506–532.

Wallace, J. B., D. H. Ross, and J. L. Meyer. 1982a. Seston and dissolved organic carbon dynamics in a southern Appalachian stream. *Ecology* 63:824–838.

Wallace, J. B., J. R. Webster, and T. F. Cuffney. 1982b. Stream detritus dynamics: regulation by invertebrate consumers. *Oecologia (Berlin)* 53:197–200.

Wallace, J. B., D. S. Vogel, and T. F. Cuffey. 1986. Recovery of a headwater stream from an insecticide-induced community disturbance. *J. North Am. Benthol. Soc.* 5:115–126.

Wallace, J. B., M. E. Gurtz, and F. Smith-Cuffney. 1988. Long-term comparisons of insect abundances in disturbed and undisturbed Appalachian headwater streams. *Verhandlungen Int. Vereingung Theoret. Angew. Limnol.* 23:1224–1231.

Wallace, J. B., T. F. Cuffney, B. S. Goldowitz, K. Chung, and G. J. Lugthart. 1991. Long-term studies of the influence of invertebrate manipulations and drought on particulate organic matter export from headwater streams. *Verhandlungen Int. Vereingung Theoret. Angew. Limnol.* 24:1676–1680.

Wallace, R. R., and H. B. N. Hynes. 1975. The catastrophic drift of stream insects after treatments with methoxychlor (1,1,1-trichloro-2,2-bis(*p*-methoxyphenyl) ethane). *Environ. Pollution* 8:255–268.

Ward, J. V. 1984. Ecological perspectives in the management of aquatic insect habitat. In V. H. Resh and D. M. Rosenberg (eds.), *The Ecology of Aquatic Insects*. New York: Praeger Scientific, pp. 558–577.

Ward, J. V., and J. A. Stanford. 1983. The serial discontinuity concept of lotic ecosystems. In T. D. Fontaine III and S. M. Bartell (eds.), *Dynamics of Lotic Ecosystems*. Ann Arbor, MI: Ann Arbor Science, pp. 29–42.

Waters, T. F. 1977. Secondary production in inland waters. *Adv. Ecol. Res.* 10:91–164.

Waters, T. F. 1979. Influence of benthos life history upon the estimation of secondary production. *J. Fisheries Res. Board Can.* 36:1425–1430.

Webster, J. R. 1983. The role of benthic macroinvertebrates in detritus dynamics of streams: a computer simulation. *Ecol. Monogr.* 53:383–404.

Webster, J. R., and S. W. Golladay. 1984. Seston transport in streams at Coweeta Hydrologic Laboratory, North Carolina, USA. *Verhandlungen Int. Vereingung Theoret. Angew. Limnol.* 22:1911–1919.

Webster, J. R., and B. C. Patten. 1979. Effects of watershed perturbation on stream potassium and calcium dynamics. *Ecol. Monogr.* 49:51–72.

Webster, J. R., and W. T. Swank. 1985a. Within-stream factors affecting nutrient transport from forested and logged watersheds. In B. G. Blackmon (ed.), *Proceedings of Forestry and Water Quality: A Mid-South Symposium, Little Rock, Arkansas, May 8–9, 1985*. Fayetteville: Cooperative Extension Service, University of Arkansas, pp. 18–41.

Webster, J. R., and W. T. Swank. 1985b. Stream research at Coweeta Hydrologic Laboratory. In *Proceedings Speciality Conference Hydraulics and Hydrology in the Small Computer Age. Hydraulics Division, ASCE, Lake Buena Vista, Florida, Aug. 12–17, 1985*.

Webster, J. R., and J. B. Waide. 1982. Effects of forest clearcutting on leaf breakdown in a southern Appalachian stream. *Freshwater Biol.* 12:331–344.

Webster, J. R., M. E. Gurtz, J. J. Hains, J. L. Meyer, W. T. Swank, J. B. Waide, and J. B. Wallace. 1983. Stability of stream ecosystems. In J. R. Barnes and G. W. Minshall (eds.), *Stream Ecology*. New York: Plenum, pp. 355–395.

Webster, J. R., S. W. Golladay, E. F. Benfield, D. J. D'Angelo, and G. T. Peters. 1990. Effects of forest disturbance on particulate organic matter budgets of small streams. *J. North Am. Benthol. Soc.* 9:120–140.

Westlake, D. F. 1975. Macrophytes. In B. A. Whitton (ed.), *River Ecology*. Berkeley: California UP, pp. 106–128.

Whitford, L. A., and G. J. Schumacher. 1963. Communities of algae in North Carolina streams and their seasonal relations. *Hydrobiologia* 22:133–196.

Winterbourn, M. J., J. S. Rounick, and B. Cowie. 1981. Are New Zealand stream ecosystems really different? *New Zealand J. Marine Freshwater Res.* 15:321–328.

Woodall, W. R., Jr. 1972. Nutrient pathways in small mountain streams. Thesis, University of Georgia, Athens.

Woodall, W. R., Jr., and J. B. Wallace. 1972. The benthic fauna in four small southern Appalachian streams. *Am. Midland Natur.* 88:393–407.

Woodruff, J. F., and J. D. Hewlett. 1970. Predicting and mapping the average hydrologic response for the eastern United States. *Water Resources Res.* 6:1312–1326.

Wright, H. A., and A. W. Bailey. 1982. *Fire Ecology: United States and Southern Canada*. New York: Wiley.

5 Streams of the Southeastern Piedmont, Atlantic Drainage

PATRICK J. MULHOLLAND

Environmental Sciences Division, Oak Ridge National Laboratory, Oak Ridge, TN

DAVID R. LENAT

North Carolina Division of Environmental Management, Water Quality Section, Raleigh, NC

This chapter discusses the physical, chemical, and biological characteristics of small to medium-size streams and rivers (<sixth order) in the Piedmont geophysical province of the southeastern United States (Virginia, North Carolina, South Carolina, and Georgia). The Piedmont is bounded on the east and south by the Atlantic Coastal Plain province, and on the west and north by the Blue Ridge province (Fig. 1). The boundary between the Piedmont and Coastal Plain, known as the Fall Line, is so named for the falls and rapids found in rivers that cross this zone. The topography of the Piedmont is characterized by rolling hills, broad and narrow ridges, and deeply eroded valleys. Stream elevations range from approximately 50 to 600 m above sea level.

Climate

The southeastern Piedmont has a humid, warm temperate climate. Mean monthly air temperatures range from January lows of 5.5 °C at Atlanta, Georgia (in the southern portion of the province) and 2.9 °C at Richmond, Virginia (in the northern portion), to July highs of 26 °C in the south and 25 °C in the north (National Oceanic and Atmospheric Administration Climatic Data Center, Asheville, NC). The average annual precipitation across the province increases from north to south, ranging from 100 to 120 cm in the Virginia Piedmont to 120 to 150 cm in the Georgia Piedmont (Cherry 1961, Powell and Abe 1985). Precipitation is relatively evenly distributed throughout the year, although there are some minor regional differences. Monthly averages are greatest in the winter and spring months in the south, and are greatest in the spring and summer months in the north. Autumn is generally the driest period across the Piedmont.

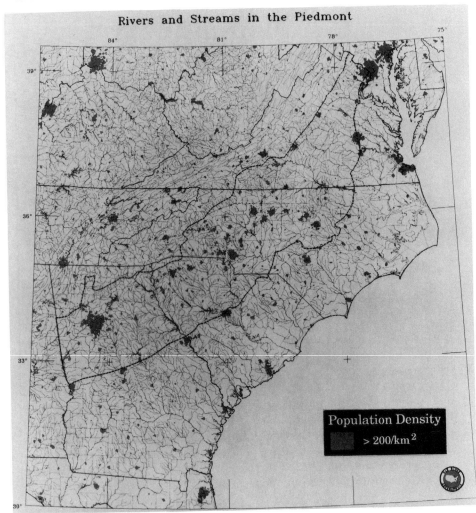

FIGURE 1. Principal drainages and regions of population density exceeding 200/km²
in the southeastern United States. The Piedmont province is outlined by the heavy
dashed line.

Average annual runoff in the Piedmont ranges from approximately 30 to
50% of precipitation, with values of 35–40% being most common (Cherry
1961, Simmons and Heath 1979). Seasonal differences in runoff are large
because high evapotranspiration rates during the growing season considerably
reduce runoff (particularly during the months of June through September).

Geology and Soils

The geology of much of the southeastern Piedmont is dominated by relatively
insoluble igneous and metamorphic rocks of late Precambrian and Paleozoic
age. It therefore more closely resembles the geology of the Blue Ridge prov-

ince than the younger sedimentary geology of the Coastal Plain (Stose et al. 1939, North Carolina Department of Conservation and Development 1958, Overstreet and Bell 1965). The igneous rocks are primarily granite, hornblende and diorite gneiss, gabbro, and basalt, with small amounts of pyroxenite and peridotite. The metamorphic rocks include biotite gneiss and shist, quartzite, slate, and marble. The geology of much of the southeastern Piedmont can be generalized as several northeast-trending belts, distinguished from each other primarily by differing grades of metamorphism (Stuckey and Conrad 1958, Patterson and Padgett 1984). The least metamorphosed of these, the Carolina Slate Belt, occurs generally in the eastern portions of the Piedmont and yields groundwaters (and consequently surface waters) with the highest levels of total dissolved solids, hardness, and pH.

Soils of the Piedmont province are primarily saprolite, derived from a long history of in-place weathering of the underlying rock uninterrupted by glacial or tectonic activity. Aeolean deposits are rare. Soils tend to be deep and highly weathered, the result of relatively warm temperatures, high annual precipitation, and the lack of glaciation. Clay content is high, and surface organic layers are much thinner than in northeastern and north-central portions of the United States. The iron oxide content of Piedmont soils is quite high, producing the orange-red color frequently observed along road cuts and other excavations. In general, Piedmont soils are acidic and have relatively low cation exchange capacity.

Thick deposits of alluvium, resulting from past periods of heavy erosion in upslope areas, often occur along streams. Erosional processes may sort the transported and deposited materials by size and density. Floodplain soils may differ substantially from upslope soils in grain size and organic content. These differences can have large effects on the hydraulic characteristics and chemistry of groundwater and surface water in the floodplain and stream channels.

Land Use

Almost all Piedmont streams have been substantially affected by historical land use changes. The Piedmont region of the Southeast was one of the first areas to be heavily impacted by European settlers. Widespread forest clearing and farming began in the 18th century, continued throughout the 19th century, and peaked in the late 19th and early 20th centuries. During the 19th century, the typical farming practice in the region was to clear a patch of forest, cultivate cotton or corn for several years, and then abandon the field because of high rates of erosion and depletion of soil fertility (Glenn 1911). During the 20th century, much farmland reverted to forest as agriculture moved westward, often in response to exhaustion of soil fertility after years of intensive farming and soil erosion.

Soil erosion during the 19th and early 20th centuries was extremely high in the southeastern Piedmont due to the relatively high annual rainfall and occasional intense storms, the thick, fine-grained soils, moderate to steep slopes, and poor land-use practices following clearing of the native forest (Trimble 1975a, Meade 1976). Trimble has reported human-induced soil losses of more than 25 cm in parts of the Santee, Savannah, and Altamaha river basins in South Carolina and Georgia. The eroded soil greatly increased sediment loads in Piedmont streams and rivers, resulting in extensive deposits of sand and silt on the floodplains (averaging up to 120 cm in depth in small valleys (Happ 1945)). The sediment deposited in the channels and floodplains of headwater streams continues to serve as a source of sediment to larger streams and rivers (Trimble 1975b). Consequently, despite a reduction in soil erosion rates during the last 50 years or so (due to decreases in row cropping, improved soil conservation techniques, and forest regrowth), the rate of sediment transport in Piedmont rivers and larger streams has generally remained high (Meade and Trimble 1974). Only very recently has there been a report of a significant reduction in sediment transport in a large Piedmont basin (Patterson and Cooney 1986).

Land-use trends in the Piedmont over the past 50 years include native forest regrowth, intensive forestry (primarily pine plantations), and urbanization. Field crops and livestock continue to be important, although they have declined from their peak in the early 20th century. In many areas, regrowth of the native forest, following abandonment of agricultural lands, has been slow. Productivity is often low as a result of the depletion of soil nutrients and organic matter during the preceding periods of high erosion rates. Nonetheless, regrowth of native forests has improved stream habitat through reduced sedimentation, bank stabilization, lower temperature variation, and increased input of organic debris (Cook et al. 1983). Establishment of pine plantations in areas that were formerly farmed has also resulted in some improvement in stream habitat, although stream water tends to be more acidic and organic debris inputs (particularly wood) are lower than in streams in native forest. In addition, pine plantations that are fertilized can result in high nutrient inputs to adjacent streams.

Urbanization and the consequent loss of forest, direct alteration of stream channel morphology, and increased input of nutrients, oxygen-demanding organics, and toxic substances can be particularly detrimental to stream habitat. In addition, certain activities that accompany urbanization (construction, improper operation of sanitary landfills, etc.) result in extremely high rates of erosion and sedimentation in Piedmont streams (U.S. Department of Agriculture 1977). Further, urbanization in the Southeast is highly concentrated in the Piedmont region, with the major metropolitan areas of Richmond, Virginia; Greensboro/Winston-Salem/High Point, Raleigh/Durham, and Charlotte in North Carolina; Greenville/Spartanburg and Columbia in South Carolina; and Atlanta, Georgia (Fig. 1).

PHYSICAL AND CHEMICAL CHARACTERISTICS

This section gives a brief overview of the physical and chemical characteristics of streams in the southeastern Piedmont. We present characteristics of streams minimally impacted by human activities, as well as those altered by such disturbances. Although undisturbed Piedmont streams do exist, the physical and chemical conditions of most of the streams in the southeastern Piedmont are currently affected in some way by humans. Even among those streams that are not currently impacted by human activities, most continue to reflect some residual impact from a long history of intensive farming and erosion in their basins.

Channel Morphology, Sediments, and Floodplains

Piedmont streams are quite variable in channel shape, streambed composition, and gradient. In his report describing the natural ecosystems of Georgia, Wharton (1978) gave the following description of Piedmont streams:

> Alluvial streams in the Piedmont have three general appearances: shoals, sometimes with white water and falls; the gently meandering slower runs; and the strongly meandering slow water. Narrow floodplains, usually less than one-half mile in width, may border one or both sides of the stream. Shoals which are degrading tend to alternate with floodplains or bottomlands (swamps) which are generally maintained by aggrading.

Wharton also stated that gradients in shoals areas are typically 2–20 m/km, but can be much higher over short reaches. The streambed composition in shoals areas is dominated by cobble and bedrock, whereas pools and slow-flowing reaches have sand and silt bottoms.

The interaction between the stream and floodplain can be an important mechanism regulating the physical, chemical, and biological properties of streams (Meyer and Edwards 1989). Flooding of low-lying areas that border the stream channel reduces water velocities and sediment transport. This results in temporary water storage, which smooths the hydrograph and reduces the magnitude of peak flow downstream (Wharton 1970). Floodplain swamp forests are common and extensive along many Piedmont streams, particularly in the lower Piedmont along low-gradient sections of streams of third order and greater. Nelson et al. (1957) reported that 9% of the forest land in the Georgia Piedmont was bottomland hardwoods bordering streams and rivers. The floodplain swamp bordering the Flint River in Georgia was shown to act as a filter for various pollutants transported in river flow (Georgia Environmental Protection Department 1971). In a study of the effect of floodplain forests on chemical fluxes in Coastal Plain streams, Kuenzler et al. (1980) and Yarbro et al. (1984) reported that uptake by floodplain vegetation and soils during flooding reduced fluxes of stream nutrients (nitrogen and phosphorus). The floodplain can also be an important source of organic carbon to the main channel, thereby supporting high rates of secondary productivity there (Mulholland 1981, Benke et al. 1984, Meyer and Edwards 1989). Although Piedmont floodplains are flooded less frequently and for shorter duration than are those of the Coastal Plain, they probably also serve as sinks for plant nutrients and sources of dissolved organic carbon during flooding.

Disturbance Effects   Human activities have had a profound impact on the morphology of most Piedmont streams. We have directly altered the physical characteristics, and consequently the biological habitat, of many Piedmont streams by channelization and removal of streamside vegetation. Channelization usually results in reduced channel sinuosity and increased slope, thereby increasing the stream gradient and erosion of the bed or banks (Nunnally 1985). Often the stream bank is also stabilized with rock riprap or concrete. In extreme cases, the streams are converted into concrete culverts.

As discussed above, the conversion of forested watersheds to agricultural and urban uses has also had an indirect impact on the morphology of most Piedmont streams. A recent review of channel alteration resulting from agricultural activities in the southeastern Piedmont (Nunnally 1985) showed that high rates of erosion followed the conversion of forests to agricultural uses, resulting in large loads of sand and silt in runoff and rapid sedimentation in headwater channels. The shifting, fine-grain sediments formed a poor substratum for many benthic organisms, and tributary streams became shallower and wider and flooded more frequently. In recent decades, erosion rates have declined with declining agriculture, improved conservation practices, and reforestation. In many headwater streams, the accumulated sediments are now being eroded, resulting in more entrenched and narrower channels.

The morphology of some Piedmont streams and rivers also has been greatly altered by impoundment. Construction of reservoirs is a relatively recent disturbance, with most of the activity occurring since 1940. Reservoirs are ecosystems with features very different from streams and rivers and are the subject of separate chapters in this volume (see Chapter 11).

Streamflow and Water Temperature

Despite the lack of large seasonal differences in precipitation, Piedmont streams exhibit a large seasonal variation in flow. High rates of evapotranspiration during the growing season deplete soil moisture content and reduce groundwater input to streams. Consequently, average streamflows are generally much lower during this period compared with winter and early spring. However, thunderstorms during the summer months can be intense and produce sharp increases in storm flow although the storm hydrograph is usually short. Occasionally, severe storms derived from hurricanes occur in late summer and fall, and such storms may cause extremely large and relatively long storm hydrographs. During winter and early spring, evapotranspiration is very low and groundwater discharge is usually considerably higher, resulting in higher baseflow.

During summer droughts small streams often cease flowing, and aquatic biota must find refuge in moist sediments, groundwater, or permanent pools. Even streams as large as New Hope Creek near Durham, North Carolina (a third- to fourth-order stream with an average width of 10 m and an average depth of 0.5 m) cease flowing during summer (Hall 1972). In small first- and second-order streams, permanent pools can be scarce or nonexistent over long reaches. In large streams (third-order and above), the duration of no-flow periods is shorter, and pools are usually present to serve as refugia for aquatic biota. However, dissolved oxygen concentrations often become very low in the pools during these periods.

Watershed geology and soil development can greatly influence streamflow. In a study of Piedmont streams in Virginia, Lynch et al. (1987) reported that streams that drained areas of older, more deeply weathered, crystalline rock generally flowed continuously, whereas streams that drained younger, less weathered igneous and sedimentary rock of Triassic age commonly ceased flowing during long drought periods. These differences reflect the greater storage of groundwater in the older, deeper soils.

The hydrographs of many Piedmont streams have been altered because of human activities. The conversion of forest to agricultural and urban uses increases the frequency and magnitude of runoff events (Nunnally 1985). This is primarily the result of stream channelization and the extensive impervious surfaces in urban areas. The hydrographs of streams in agricultural areas also exhibit rapid response to rainfall due to channelization, tributary drainage ditches, and the somewhat reduced infiltration capacity of cultivated soils. Channelization also causes higher peaks in storm hydrographs in higher-order systems downstream.

Piedmont streams typically have broad annual temperature ranges, and substantial variation in diel temperature in the summer can be observed even in forested areas. For example, Reice (1977) reported water temperatures ranging from 0 °C in winter to 25 °C in summer in New Hope Creek, North Carolina, with diel variations of as much as 5 °C in summer. Nelson and Scott (1962) reported average monthly values ranging from 5.8 °C in January to 27.3 °C in August in the Middle Oconee River, Georgia. The large annual and diel variation in water temperature results in dominance by eurythermal biota, which are also capable of withstanding the periods of relatively low concentration of dissolved oxygen during the warmer months.

Chemistry

The baseflow chemistry of Piedmont streams can vary substantially, depending on the underlying geology and the extent of human disturbance. In the first part of this section, we discuss the influence of geology and hydrology on stream water chemistry. In the second part, we discuss anthropogenic effects.

Influence of Geology Of the three geophysical provinces that make up the southeastern Atlantic drainage, the Piedmont has streams that are generally the highest in total dissolved constituents, hardness, and pH, although still quite low compared to other regions of the United States, particularly those with carbonate geology. An excellent summary of the baseflow and stormflow chemistry of Piedmont streams compared with that of streams of the Blue Ridge and Coastal Plain provinces of North Carolina is presented in Simmons and Heath (1979). They report that Piedmont streams in North Carolina can be divided roughly into two groups, depending on the underlying geology. Streams in areas underlain by rocks that are highly resistant to weathering (such as granite, gneiss, and schist—zone I) generally have lower concentrations of base cations (particularly calcium), anions (particularly chloride and bicarbonate), and total dissolved solids, and slightly lower pH compared with streams in areas underlain by somewhat more weatherable metamorphosed volcanic rock of the diorite group (zone II) (Table 1). Most of the streams in the western portion of the North Carolina Piedmont, as well as those of the Blue Ridge, are in the former category. The diorite group dominates the eastern North Carolina Piedmont, including the Carolina Slate Belt (extending into South Carolina as well) and the Durham and Wadesboro Triassic basins. This pattern of higher concentrations of dissolved constituents in the eastern and some central portions of the North Carolina Piedmont, compared with the portion adjacent to the Blue Ridge, is probably also true for the southeastern Piedmont as a whole.

Table 2 summarizes water chemistry data for four streams draining unmanaged forested areas and having no point sources of pollution. Although these streams occur in areas of zone II geology and therefore have somewhat higher concentrations of many dissolved constituents than do streams of zone I, they are nonetheless examples of relatively "pristine" Piedmont streams.

TABLE 1   Mean Concentrations of Dissolved Constituents at Baseflow and Stormflow in North Carolina Piedmont Streams with No Concentrated Sources of Pollution

Parameter	Zone I (mg/L)		Zone II (mg/L)	
	Baseflow	Stormflow	Baseflow	Stormflow
Sodium	1.5	0.8	5.9	1.8
Potassium	1.0	0.6	1.0	1.1
Calcium	1.3	1.3	5.6	2.6
Magnesium	0.6	0.4	2.4	1.0
Bicarbonate	7.4	5.1	34	5.0
Sulfate	2.2	2.2	2.6	6.8
Chloride	0.9	0.7	5.4	2.1
Silica	8.1	6.6	2.1	5.8
Nitrate nitrogen	0.08	0.17	0.16	0.07
Ammonia nitrogen	<0.01	0.01	0.01	0.01
Organic nitrogen	0.11	0.13	0.22	0.48
Total phosphorus	0.01	0.01	0.02	0.03
Total dissolved solids	19	15	61	24

Note. Data are presented for streams draining areas of weathering-resistant granite group geology (zone I) and diorite group geology (zone II).

Source. Simmons and Heath (1979).

TABLE 2 Some Chemical Characteristics of Relatively Undisturbed Piedmont Streams in Georgia and North Carolina

Parameter	Cable Branch	Falling Creek	Cane Creek	Dutchmans Creek	Smith Creek
Watershed (km^2)	0.47	187.1	1.66	[a]	16.1
Conductance (mS/cm)	56.9	121	34	[b]	60
	(10–95)	(28–222)	(27–76)	(28–55)	(40–85)
pH	6.9	7.2	6.3	[b]	6.5
	(6.0–7.9)	(5.3–8.9)	(5.9–7.3)	(6.1–7.3)	(5.7–7.2)
Alkalinity (mg/L)	21	50	9	[b]	[c]
	(4–43)	(15–80)	(3–39)	(3–20)	[c]
Hardness (mg/L)	18	44	12	[b]	[c]
	(6–36)	(17–62)	(9–28)	(7–14)	[c]
NO$_3$ + NO$_2$ (mg/L)	0.23	0.05	0.05	[b]	0.08
	(0.02–0.44)	(0–0.21)	(0.01–0.14)	(0.04–0.28)	(<0.01–0.24)
Total P (mg/L)	0.09	0.03	0.05	[b]	0.09
	(0.02–0.34)	(0–0.25)	(0.01–0.16)	(0–0.17)	(<0.01–0.46)

Note. Location of streams: Cable Branch, Green County, Georgia; Falling Creek, near Juliette, Georgia; Cane Creek, near Buckhorn, North Carolina; Dutchmans Creek, near Uwharrie, North Carolina; Smith Creek, Franklin County, North Carolina. Median or mean values are presented with range given in parentheses.

[a] Watershed area not available. First-order stream.

[b] Median value not available.

[c] Values not available.

Source. Georgia Department of Natural Resources (1983); Buell and Grams (1985); Crawford and Lenat (1989); and J. K. Crawford, U.S. Geological Survey, Raleigh, North Carolina, unpublished data.

Influence of Hydrology   Piedmont stream chemistry can also vary depending on the hydrologic condition (Table 1). In relatively undisturbed streams, many solutes are diluted during periods of high flow. Rainwater generally has lower concentrations of dissolved substances than does groundwater. Concentrations of some solutes, noteably sulfate, nitrate, organic carbon, and phosphorus usually increase during storms. Increases in concentrations of these solutes during storms are probably due to high concentrations in rain (primarily sulfate and nitrate), mobilization from litter and surface soil horizons by lateral flow, particularly in riparian zones, and flushing from areas within the stream channel that are dry or isolated pools during baseflow periods. Iron is another exception to the usual dilution pattern during stormflow. In much of the North Carolina Piedmont, iron concentrations increase by up to an order of magnitude during stormflow, with mean concentrations of 5 mg/L in geochemical zone II streams (Simmons and Heath 1979).

In contrast to many dissolved species, concentrations of particulate materials, both inorganic and organic, increase greatly during storms, often peaking at or slightly before the peak in water discharge. The increase in concentration of particulate materials is largely the result of entrainment of streambed sediments (as water velocity increases and the stream expands laterally) and input of particulates from ephemeral channels. However, a portion of the increase in particulates may be the result of precipitation/flocculation of iron when soil waters rich in reduced iron are discharged to the stream.

Disturbance Effects   Human impacts on water chemistry of Piedmont streams are widespread and substantial. In their baseline chemistry study of Piedmont streams in North Carolina, Simmons and Heath (1979) could find only 20 perennial streams (all with drainage basins $\geq 80\%$ forested) that showed no discernible human effects on water quality. The largest of these, Jacob Fork near Ramsey, North Carolina, drained 66 km^2, but most drained <10 km^2. Although the intensities of agriculture and soil erosion have declined considerably since peaks in the early 20th century, both are still present, and increased use of fertilizers and herbicides has added new stresses to streams draining agricultural watersheds. Urbanization and industrialization have increased greatly during the past 25 years. Although sewage treatment practices are much improved as well, the increased population and industrial development have resulted in more and larger point-source discharges of pollutants to Piedmont streams and rivers.

The effect of agriculture and urbanization on the quality of water in Piedmont streams in Georgia has been the subject of several studies (Benke et al. 1981, Buell and Grams 1985, Georgia Department of Natural Resources 1983). Agriculture usually results in greater concentrations of most dissolved constituents, particularly nitrogen and phosphorus, and increased sediment transport and sedimentation. Results of a statewide water quality assessment in North Carolina (North Carolina Division of Environmental Management 1978) indicated that 70% of the erosion in North Carolina occurred on agricultural lands, with a disproportionate amount of the erosion observed in the Piedmont area. Only 16% of the cropland was adequately protected against erosion. In a more recent study of water quality and biota of agricultural, urban, and largely forested streams in Piedmont, North Carolina, Crawford and Lenat (1989) reported that the major impact of agriculture was higher concentrations of nutrients, whereas the greatest impact of urbanization was greatly increased suspended sediment loads. Urbanization can also result in low concentrations of dissolved oxygen and increased concentrations of toxic metals and nutrients in streams (Klein 1985). Finally, sediment transport in Piedmont streams during storms, even those draining watersheds that are not currently farmed or urbanized, can be very high as a result of the heavy erosion in the past. Even in reforested areas, many stream valleys are still heavily sedimented and serve as a sediment source during storms.

PLANT COMMUNITIES AND ENERGY SOURCES

Surprisingly few studies of the plant communities and energy sources of Piedmont streams have been performed. This probably reflects the disturbed condition of many Piedmont streams and the reluctance of researchers to study altered ecosystems. Below, we present data for a few streams and rivers and discuss the probable influences of anthropogenic disturbances.

Algae

Comprehensive studies of the algal communities of Piedmont streams are scarce. Some studies have been performed by state agencies as a part of overall water quality assessments of streams affected by human activities. These studies often consist of quantifying the algae in one-time grab samples of stream water or placing artificial substrata in the stream (often for 1–3 weeks) to determine rates of periphyton colonization. There are problems with both of these techniques. The types and relative abundance of algae found in stream water may not be representative of the distribution or density of algae growing on the stream bottom. Collection of algae on artificial substrata probably gives a more accurate sample of stream algal types and biomass, although the type of artificial substratum used and the length of the colonization period may influence the results. This technique is biased toward the early successional species, particularly diatoms, and may miss much of the natural community that requires longer periods to become established.

The periphyton of Piedmont streams is often sparse because of the heavy silt loads and shifting sand bottoms. The suspended sediment greatly limits light penetration and consequently the development of benthic algae. Suspended sediment also abrades stationary substrata, inhibiting growth of periphyton mats and filamentous forms. For example, periods of persistent turbidity in the Cape Fear River at Lillington, North Carolina, and in the Pee Dee River at Rockingham, North Carolina, resulted in reduced periphyton growth (D. R. Lenat, personal observation). Stream bottoms dominated by deposits of sand and silt offer a poor substratum for periphyton because the individual grains are alternately buried and exposed during periods of unstable streamflow. Sizeable periphyton accumulations generally occur in these streams only during extended periods of stable flow.

Many Piedmont streams have elevated nutrient concentrations resulting from point-source and non-point-source inputs from urban and agricultural areas. These streams can support large algal growths, particularly during periods of stable streamflow when scouring effects are minimal. In addition, some Piedmont streams do not have a full riparian canopy and are therefore exposed to higher light levels and water temperatures than are forested streams. These anthropogenic impacts strongly influence the composition of the algal community as well as the algal biomass. High light levels and nutrient concentrations may favor the filamentous algal forms over unicellular forms and green algae over chrysophytes and some diatoms.

Periphyton biomass and species composition in Piedmont streams vary seasonally. In streams in heavily forested areas, biomass peaks often occur in spring when light reaching the stream surface is maximum. In streams lacking a vegetation canopy or in those whose width precludes a complete canopy cover, seasonal patterns in flow often control periphyton abundance. In such streams, low stable flows in summer and autumn allow maximum accumulations of periphyton. However, summer spates resulting from thunderstorm activity also can inhibit accumulation of periphyton.

Selected Studies   Cook et al. (1983) conducted a 2-year study comparing algal biomass and community composition in four Piedmont streams in northeastern Georgia: a stream in a relatively undisturbed natural forest, and streams draining a managed forest, an agricultural watershed, and an urban watershed. Artificial substrata (glass slides) were used to determine chlorophyll a, organic content, and diatom species composition. Average chlorophyll a densities varied from 0.4 mg/m^2 in the managed forest stream to 1.4 mg/m^2 in the natural forest stream. The natural forest stream exhibited peaks in chlorophyll a in late winter and spring (11–15 mg/m^2). In contrast, chlorophyll a densities in the agricultural stream generally peaked in summer (5–7 mg/m^2). The low chlorophyll a density in the managed forest stream was probably the result of high sediment loads stemming from recent extensive clear-cutting. The low chlorophyll a density in the agricultural stream, compared with the natural forest stream, probably reflected lower algal biomass on the unstable, shifting sand bottom, which would reduce the rate of colonization on the artificial substrata.

In terms of community composition, the greatest number of diatom taxa were found in the agricultural stream and the least in the urban stream. The natural forest stream had the least even diatom distribution and was almost completely dominated by *Achnanthes lanceolata*. The urban stream had the most even distribution of diatom taxa, with subdominants accounting for almost half the occurrences. The study by Cook et al. (1983) did not include data on other groups of algae in these streams.

The Georgia Department of Natural Resources also used glass slide techniques in conducting a study of the periphyton of the South River and Flint River in the Georgia Piedmont (Georgia Department of Natural Resources 1983). In this study, periphyton were grouped by family; sheath bacteria and fungi were also quantified. Diatoms were the most abundant component of the periphyton in both rivers, constituting nearly 90% of total cell numbers. Although green and blue-green algae accounted for only about 5% of the cells on the glass slides, they could be more abundant on natural substrates having more mature algal communities.

A study of periphyton abundance in a small Piedmont river in Virginia (Mechums River, Albemarle County) was conducted by Tett et al. (1978). Chlorophyll a densities on the natural substrata during the period May through December were generally 50–150 mg/m^2 of stream bottom, with midsummer and autumn peaks of approximately 200 mg/m^2. The autumn peak in chlorophyll a was attributed to an accumulation of green algae, blue-green algae, and diatoms, particularly *Melosira* sp. (Kelly et al. 1974, Tett et al. 1978). Tett and coworkers concluded that most of the temporal variation in chlorophyll a density in the Mechums River was the result of streamflow variation, with high flows dislodging both unicellular and filamentous algae that had accumulated on substrata during periods of low, stable flow. In such streams, with their large accumulations of unstable sand and silt sediments, the frequency of storms has a particularly strong influence on periphyton abundance.

Macrophytes

In general, macrophytes have been rarely reported in Piedmont streams. Hobbs (1981) indicates that there are no vascular plants in shaded portions of Piedmont streams, but *Vallisneria* may grow in areas exposed to direct sunlight. M. G. Kelly (University of Virginia) has conducted extensive studies of primary production in Piedmont streams and rivers and believes that the lack of vascular plants is the result of unstable sediments, moderate to high gradients, and the large variations in streamflow typical of most Piedmont streams (M. G. Kelly, personal communication). Kelly has not observed macrophytes even in the open-canopy streams in portions of the Virginia Piedmont where light should not be limiting (M. G. Kelly, personal communication).

An exception to the general lack of vascular plants in Piedmont streams is *Podostemum ceratophyllum* (river weed). *Podostemum* grows attached to rock surfaces and is therefore not dependent on stable fine-grain substrata. Nelson and Scott (1962) reported that it dominated the benthic flora of a rock outcrop reach of the Middle Oconee River, Georgia. Using a harvest technique, they found that *Podostemum* had a net annual productivity of 1023 g AFDM (ash-free dry mass)/m^2. Its productivity was greatest during moderate and stable streamflow, when the streambed was completely flooded but water velocities were not great. In the Middle Oconee River, *Podostemum* also served as an important substratum and refuge for clinging invertebrates, such as *Simulium* pupae and *Calopsectra* (*Tarytarsus*) larvae. Nelson and Scott concluded that much of the *Podostemum* was probably not used directly as a food resource by invertebrates, but entered the detrital food chain after being dislodged from rock surfaces during high flow or drying out when exposed to air during low flow. Approximately one-half of the total plant detritus on the bottom of this reach of the Middle Oconee was *Podostemum*.

Podostemum has also been reported to occur in the Mechums River in Virginia (Tett et al. 1978). It was most abundant there in May and June and declined during July and August. New growth appeared in the autumn, but the plant disappeared completely during the winter months. At its peak, *Podostemum* covered approximately 20–30% of the rock surfaces, which constituted about one-quarter of the streambed. The chlorophyll *a* density of *Podostemum* at its peak in the Mechums River was <10 mg/m^2 and was small compared with that of the periphyton during this time (50–150 mg/m^2). However, more recent Mechums River observations by M. G. Kelly (personal communication) indicated much greater coverage by *Podostemum* during stable streamflow conditions in summer with nearly all rock surfaces covered.

Lush growths of *Podostemum* have also been reported in the South Anna River in Virginia during summer and autumn, but it was absent from the North Anna River (Parker and Voshell 1983). No reasons were given for the lack of *Podostemum* in the North Anna, although it could be related to the existence of an upstream impoundment on the river. *Podostemum* has been observed in the North Carolina Piedmont as well, being most abundant in larger streams and rivers. However, it is absent from many Piedmont streams (D. R. Lenat, personal observation), presumably due to the lack of exposed rock substrata resulting from watershed erosion and sedimentation in stream channels. In addition, *Podostemum* may require relatively stable streamflows to accumulate significant biomass, and at high nutrient levels, filamentous green algae may outcompete it for rock surface space (M. G. Kelly, personal communication).

Aquatic bryophytes are also uncommon in Piedmont streams, in contrast to streams of the Blue Ridge province where they are common. The paucity of bryophytes may be the result of the high silt loads in Piedmont streams. The warmer temperatures of Piedmont streams may result in increased growth rates of algae competing for space or of heterotrophic microbes decomposing moss tissue.

Macrophytes may be more abundant in larger rivers of the Piedmont, particularly along river margins where fine-grain sediments are relatively stable. J. J. Haines (personal communication, unpublished manuscript) has noted growths of several macrophytes including *Potamogeton*, *Callitriche*, and *Najas*, as well as *Podostemum* in free-flowing sections of the Savannah River. Haines has also observed the aquatic moss, *Fontinalis*, and large growths of the macroalga, *Nitella*, in some areas of the Savannah River. The observations of Haines suggest that the larger rivers of the Piedmont support a greater variety of plant forms than the smaller streams. This is likely because of the presence of different substrate types, greater stability of fine-grain sediments, and greater light availability in the larger rivers compared with small streams.

Primary Production, Respiration, and Allochthonous Energy Sources

There have been a few studies of primary production and total respiration in Piedmont streams using the diurnal dissolved oxygen change technique introduced by Odum (1956). A summary of the results of these studies is presented in Table 3. In the Middle Oconee River and New Hope Creek, total respiration substantially exceeded gross primary production, indicating that allochthonous inputs of organic matter were important in supporting heterotrophic metabolism. However, in some of the Virginia rivers, total respiration only slightly exceeded gross primary production on many dates, indicating that autochthonous production probably supported most of the heterotrophic community metabolism in these stream reaches. There is some evidence that larger rivers are more metabolically active than smaller rivers and streams. For example, the Rappahannock River 4 km above Fredricksburg, Virginia, was considerably deeper (average depth of 2 m compared with <0.5 m) and had substantially greater flow (22 m^3/s at baseflow versus

TABLE 3 Rates of Gross Primary Production (GPP) and Total Respiration (TR) Measured Using the Diurnal Dissolved Oxygen Change Technique in Some Piedmont Streams and Rivers

Stream	GPP	TR	Period	Reference[a]
Middle Oconee River	0.23	0.88	August	1
near Athens, GA	0.28	1.10	September	1
	0.07	0.45	December	1
New Hope Creek, near Durham, NC	0.81	1.33	Annual average	2
Rappahannock River near Fredricksburg, VA	2.4–9.6	4.0–13.6	April–June	3
Mechums River near Crozet, VA	0.4–4.1	1.2–5.4	April–August	3
South Fork Rivanna River upstream of Charlottesville, VA	0.3–3.1	1.2–4.4	May–July	3
Rivanna River downstream of Charlottesville, VA	0.7–3.0	2.5–8.3	July–October	3

Note. All values are given in g O_2/m^2 day^{-1}.
[a]*References*: 1, Nelson and Scott (1962); 2, Hall (1972); 3, Kelly et al. (1975).

< 3 m³/s) than the other rivers reported in Table 3. Although the productivity and respiration of all the streams and rivers presented in Table 3 probably reflect to some extent the influence of point- and non-point-source pollution, the Rappahannock was not appreciably higher in nitrate or phosphate concentrations (Kelly et al. 1975). The greater rates of metabolism measured for the Rappahannock are probably the result of the contribution of a significant water column community as well as a benthic community of algae and bacteria.

There have been very few direct measurements of allochthonous input of organic matter to Piedmont streams. Hall (1972) reported a leaf input of 476 g dry mass/m² year⁻¹ to New Hope Creek, North Carolina, and S. R. Reice (University of North Carolina, personal communication) measured leaf input of 896 g/m² year⁻¹ to a tributary of New Hope Creek. Leaf-fall in temperate deciduous forests is generally in the range of 350–550 g dry mass/m² year⁻¹ (Bray and Gorham 1964). Monk et al. (1970) reported leaf-fall of 445 g dry mass/m² year⁻¹ in a Piedmont oak–hickory forest in Georgia. For small streams (≤ third order) in forested areas, annual leaf input, on a per unit area basis, will usually exceed the average annual forest value because of lateral transport of leaves into the stream from its banks. The allochthonous input is distinctly seasonal for Piedmont streams in deciduous forests, with peak input occurring during the latter part of October to mid-November. Allochthonous inputs to streams with sparse riparian vegetation, such as many agricultural streams, are considerably lower than inputs to forested streams.

ANIMAL COMMUNITIES

Macroinvertebrates

Prior to 1980 the study of benthic macroinvertebrates in Piedmont streams had been impeded by the lack of a comprehensive taxonomic guide. The publication of *Aquatic Insects and Oligochaetes of North and South Carolina* (Brigham et al. 1982) established a sound, uniform taxonomic base for macroinvertebrate investigations. Furthermore, some sections of this guide (Ephemeroptera, Plecoptera, Trichoptera) have tabulated the major species according to their distribution in the Blue Ridge, Piedmont, and Coastal Plain. More recent studies continue to update the list of known Piedmont invertebrates (especially in the Chironomidae), but the guide by Brigham et al. (1982) serves as a common starting point for all investigators. Several informal working groups in the Southeast (Carolinas Area Benthologists, Southeastern Water Pollution Biologists Association) circulate information that supplements the North Carolina and South Carolina species lists. Other invertebrate groups (e.g., meiofauna) are poorly known in Piedmont streams.

There are many parameters that can be used to characterize the benthic macroinvertebrate fauna of streams: for example, measurements of taxa richness, various measures of abundance or production, and information on feeding type and life cycles. This section summarizes the available data on Piedmont streams and makes some comparisons with streams in the Blue Ridge and Coastal Plain provinces. Data are included both for streams that have been minimally impacted by humans and for altered streams.

Natural Communities

Taxa Richness   The total taxa richness of Piedmont streams is roughly equivalent to that of Blue Ridge or Coastal Plain streams. However, comparisons of taxa richness of major taxonomic groups (especially orders of aquatic insects) in third- to fifth-order streams in North Carolina demonstrates some differences (Table 4). Piedmont streams generally are lower in taxa richness for Ephemeroptera, Plecoptera, and Trichoptera, but higher in taxa richness for Coleoptera, Odonata, Mollusca, Crustacea, and other taxa (Turbellaria, Hirudinea, Hemiptera, etc). The major difference between Piedmont and Coastal Plain streams is the lower number of Ephemeroptera in the latter.

Taxa richness is thought to be a function of stream size (Harrell and Dorris 1968, Minshall et al. 1985); however, investigations by the North Carolina Division of Environmental Management (unpublished data) indicate that small Piedmont streams do not have appreciably lower taxa richness unless they become intermittent. Data on Piedmont streams in North Carolina indicate that there is only a 20% increase in taxa richness between second-order and fifth-order streams (D. R. Lenat and D. L. Penrose, unpublished data). However, the taxa richness of small first-order streams is quite variable.

TABLE 4   Taxa Richness, by Group, During Summer for Unstressed Streams in Different Geophysical Provinces of North Carolina

Group	Blue Ridge ($n = 13$)	Piedmont ($n = 2$)	Coastal Plain ($n = 8$)
Ephemeroptera	22.0	15.5	10.0
Plecoptera	4.3	3.0	2.6[a]
Trichoptera	17.5	15.5	13.8[a]
Coleoptera	6.2[a]	8.0	9.3
Odonata	4.5	9.0	10.1
Megaloptera	2.0	2.0	2.0
Diptera:Chironomidae	22.8	23.0	23.4
Diptera:Miscellaneous	4.7	2.0	4.4
Oligochaeta	3.4	4.0	3.3
Crustacea	1.0	1.5	2.5
Mollusca	4.3	7.0	4.1
Other	1.9	5.0	4.4
Total	94.7	96.5	89.3

[a]Highly variable.

Source. Data from Penrose et al. (1980); Lenat (1983, unpublished data).

In general, the macroinvertebrate communities of undisturbed Piedmont streams are numerically dominated by Ephemeroptera, as are the streams in the Blue Ridge province (Table 5). However, Piedmont streams usually have fewer Plecoptera and more Diptera than the Blue Ridge streams. The lower abundance of Plecoptera in Piedmont streams is probably due to higher water temperatures. However, there are some Piedmont streams that do not follow these general patterns. Streams in the North Carolina Slate Belt were found to be strongly dominated by filter-feeding Trichoptera (Table 5). Furthermore, one stream in Union County (unnamed tributary of Lanes Creek) was dominated by Coleoptera (*Stenelmis*) and Diptera (mainly *Microtendipes*). The reasons for these latter differences among Piedmont streams are not yet evident.

The distribution of mollusks, especially Unionidae, in Piedmont streams and rivers is not well documented. In low-alkalinity streams, the number of mollusks is limited (D. R. Lenat, personal observation), presumably because of their low calcium concentrations or anthropogenic disturbance. In higher alkalinity streams, mussels can be abundant. There has been much recent interest in the mollusk fauna of North and South Carolina. Intensive surveys have been conducted by the North Carolina Wildlife Resources Commission (John Alderman, personal communication) and by the U.S. Fish and Wildlife Service (Eugene Keferl, Brunswick College, Brunswick, GA, personal communication). These studies have indicated that high mollusk diversity may still be found in some Piedmont streams and rivers. Although widespread point- and non-point-source pollution has reduced the diversity of Mollusca in most Piedmont streams, there appear to be a limited number of "refugia" which should be afforded special protection (John Alderman, personal com-

TABLE 5 Dominance (% abundance) of Major Invertebrate Groups in Streams of the Blue Ridge and Piedmont Provinces

		Piedmont		
Group	Blue Ridge ($n = 10$)	"Typical" ($n = 4$)	Carolina Slate Belt ($n = 2$)	Unnamed Tributary of Lanes Creek ($n = 1$)
Ephemeroptera	42 (26–61)	42 (38–46)	(12–13)	6
Plecoptera	16 (7–30)	8 (2–14)	(3–5)	3
Trichoptera	19 (11–34)	14 (10–16)	(50–70)	14
Coleoptera	6 (1–17)	5 (1–9)	(3–5)	29
Diptera	14 (7–23)	27 (25–29)	(8–20)	45
Other	3 (1–6)	4 (1–5)	(2–9)	3

Note. Piedmont streams are subdivided into "typical" streams, Carolina Slate Belt streams, and an additional unclassified stream (Lanes Creek). Values are averages with range given in parentheses.

Source. Data are from Penrose et al. (1980) for the Blue Ridge streams and Lenat (1983, unpublished data) for the Piedmont streams.

munication). North Carolina Piedmont streams with unusual mussel species include the headwaters of the Tar River, the headwaters of the Neuse River (especially the Flat River), the Little River (Neuse River basin) and the Little and Uwharrie Rivers (Yadkin River basin). These systems are all located within the Carolina Slate Belt. M. G. Kelly (personal communication) has observed that mussels are abundant in the Rappahannock and James rivers of Virginia as well.

Pleurocerid snails (especially, *Elimia* spp.) may become very abundant in some Piedmont streams. Krieger and Burbanck (1976) indicated that *Elimia* in one Piedmont catchment was abundant in larger streams but often absent from the smallest streams. They found that its distribution was independent of dissolved oxygen, pH, temperature and turbidity. The most important factors were substrate (cobble/boulder) and vegetation (presence of *Podostemum*).

One exotic mollusk, *Corbicula fluminea*, has invaded most Piedmont river basins. It has been suggested that this "asiatic clam" may compete with native mussels, but there has been little good evidence to substantiate this theory. *Corbicula* is more tolerant of low dissolved oxygen levels than most mussels, and replacement of unionids by *Corbicula* may reflect a decline in water quality, rather than competition. Occasional "mussel kills" in areas colonized by *Corbicula* have been reported across the United States, including the Tar River, North Carolina. The cause of this phenomenon is still unknown.

Seasonality   Investigations of seasonal fluctuations in the abundance of macroinvertebrates in Piedmont streams have shown the expected spring and autumn peaks in both density (Stoneburner and Smock 1979, Reisen and Prins 1972) and taxa richness (Lenat 1988). Seasonal changes in dominance have not been well studied in the Piedmont. One investigation indicated relatively stable dominance patterns in some Piedmont streams, particularly in comparison with polluted sites (Lenat 1984). Most Plecoptera are present primarily during the winter–spring period, and this group is often the largest source of seasonal variability in Piedmont streams. Trichoptera (especially Hydropsychidae) often have a distinct autumn maximum, and some Simuliidae may develop very large winter populations.

Feeding Types   There have been few studies of feeding types in Piedmont streams. Limited data, mostly unpublished, indicate that filter-feeders and collector–gatherers are dominant in most Piedmont streams. Streams with open canopies may develop larger populations of scraper–grazers in response to increased periphyton production. Unlike many streams in the Blue Ridge, shredders rarely constitute a large proportion of the numerical standing crop in Piedmont streams, although some stoneflies (especially *Taeniopteryx, Strophopteryx, Allocapnia*, and *Amphinemura*) and Tipulidae can make up a large proportion of the autumn and winter biomass. Piedmont streams usually support a diverse predator community (about 25% of the macroinvertebrate species), but few of these species are abundant. The larger Odonata and

Megaloptera often make up a large share of the total biomass. Crayfish (usually classified as scavengers) also may be important in terms of biomass, although not in terms of numerical abundance.

Life Cycles   Detailed studies of the life cycles of aquatic invertebrates in Piedmont streams have been limited to Trichoptera (Parker and Voshell, 1982, Freeman and Wallace 1984) and Ephemeroptera (Kondratieff and Voshell 1980, 1981). These investigations were generally limited to the most abundant species. All Trichoptera (six species of Hydropsychidae and two species of Philopotamidae) were trivoltine or bivoltine. The trivoltine species had two rapidly growing spring and summer generations and one slower-growing autumn–winter generation. Of the mayflies studied by Kondratieff and Voshell, two were trivoltine, three were bivoltine, and five were univoltine.

Lenat (1988) analyzed the life cycles of invertebrates in a Piedmont stream in North Carolina (Little River) from monthly larval sampling. This coarse method of analysis may overlook or combine some of the more rapid summer generations, but it allows a rough determination of overall trends. Life cycles were determined for 145 of the more common species; 98 were univoltine/semivoltine, 34 were bivoltine, and 13 were polyvoltine. Although the majority of the species present were univoltine; most of the numerically dominant species were bivoltine or polyvoltine. The data suggested that temporal separation has been an important means of reducing competition between macroinvertebrate species in Piedmont streams.

Production and Biomass   There have been few estimates of production for benthic macroinvertebrates in Piedmont streams. Generally, estimates are available only for the filter-feeding community (i.e., certain Trichoptera and Simuliidae). Nelson and Scott (1962) estimated that filter-feeders in a rock outcrop of the Middle Oconee River, Georgia, annually produced at least 20 g AFDM/m^2. Similarly, Freeman and Wallace (1984) reported that filter-feeding Hydropsychidae and Simuliidae in Rose Creek, Oconee County, Georgia, together had an annual production of 10 g AFDM/m^2. They speculated that sediment scouring in this sandy stream may have limited hydropsychid production during winter and spring. However, production of the dominant species (*Hydropsyche betteni*) was eight times greater than the combined production of the six dominant caddisfly species in a Georgia mountain river. This difference was attributed to higher temperature and nutrient levels in the Piedmont stream. Parker and Voshell (1983) reported that annual production of filter-feeding Trichoptera (five Hydropsychidae and two Philopotamidae) varied from 9 to 50 g dry mass/m^2 at free-flowing sites in the North and South Anna rivers of Piedmont Virginia. Most of the species investigated were multivoltine and had production:biomass (*P:B*) ratios of 10–15.

The filter-feeding community often constitutes the dominant fraction (about 50%) of the total biomass of the benthic community in Piedmont streams (D. R. Lenat, unpublished data). Based on this assertion and assuming similar

P:B ratios for the other species, the total annual production by macroinvertebrates in Piedmont streams would be in the range of 20–200 g AFDM/m². Annual production of macroinvertebrates in Piedmont streams is probably greater than that in streams of the Blue Ridge province because of the greater proportion of multivoltine species, the warmer temperatures, and generally higher concentrations of nutrients in Piedmont streams. However, as is the case with most Blue Ridge streams, many Piedmont streams have relatively low alkalinity levels (low calcium concentrations), which should limit total production of invertebrates (Krueger and Waters 1983). It is not clear to what extent higher temperatures and nutrient levels may compensate for low alkalinity by increasing food resources (microbial and algal production) in Piedmont streams compared with Blue Ridge streams.

Our observations of North Carolina streams suggest that the average size of benthic macroinvertebrates is smaller in Piedmont streams than in mountain streams. The association of smaller body size with higher temperatures has been well documented (Kinne 1963) and may be the reason for the difference in average size. Therefore, it is expected that invertebrates in Piedmont streams will have higher *P:B* ratios than those in colder mountain streams.

Detrital Processing   The importance of macroinvertebrates in detrital processing in Piedmont streams is not clear. Reice (1978) reported no relationship between leaf processing and the density or taxa richness of the invertebrate community in a third-order Piedmont stream in North Carolina (Hew Hope Creek). Shredders were largely absent. However, this study was performed in June, a period when shredders are not usually abundant in Piedmont streams. Later work by Reice (1983) in the same stream did indicate a relationship between the rate of leaf decomposition and colonization of leaf packs by shredders in autumn. Anderson and Sedell (1979) found that the life cycles of shredders are keyed to the predictable timing of detrital inputs, that is, to inputs of leaves during autumn in forested Piedmont streams. Personal observations (D. R. Lenat) of other Piedmont streams in North Carolina have indicated that shredders (Taeniopterygidae, Nemouridae, Capniidae, and Tipulidae) are abundant in autumn and winter. Other shredders (Amphipoda, Gastropoda, and Astacidae) may be abundant in some Piedmont streams. Further experiments, similar to those of Wallace et al. (1982), should be conducted to evaluate the significance of shredders to the decomposition of leaves in Piedmont streams.

Drift   Drift studies in South Carolina Piedmont streams (Reisen and Prins 1972, Stoneburner and Smock 1979) indicated that total drift densities are relatively constant throughout the year, but have early spring and late summer peaks. *Baetis* spp. and Simuliidae were the most abundant taxa in drift samples (Stoneburner and Smock 1979), and drift appeared to be related to seasonal growth and emergence patterns. Walton (1980), in a lab study of predator/prey interactions for Piedmont stream species, demonstrated that *Acroneuria* induced prey drift, but the effect was not very strong.

Biotic vs Abiotic Control of Community Structure   Peckarsky (1983), borrowing from the work of Menge (1976), has suggested that streams can be placed on a gradient from harsh to benign environmental conditions. Physical/chemical factors control the composition of the stream invertebrate community in harsh environments, while biotic factors (competition/predation) control community structure in benign environments. Where do Piedmont streams fit on this gradient?

Factors that control community structure in Piedmont streams have been studied by very few investigators. Only in one stream (New Hope Creek, North Carolina) has there been an attempt to compare the relative importance of biotic vs abiotic controls (Reice 1977, 1978, 1981, 1983, Diamond and Reice 1985). These studies suggested that abiotic factors (substrate composition, frequency of flooding) were of much greater importance than either competition or vertebrate predation. McAuliffe (1984) has suggested that Reice's sample sizes (625 cm^2) were too large to examine potential competition between invertebrate species. He indicated that sample sizes of 2–10 cm^2 were required to test for competition effects. Similarly, Minshall and Petersen (1985) indicated that stream "islands" (= individual rocks) reach equilibrium levels of competitors quickly, even in seemingly variable environments. Although Reice's work has contributed substantially to our understanding of factors controlling invertebrate community structure, further experiments may be required to examine the importance of competition in regulating invertebrate communities in Piedmont streams.

New Hope Creek is located in the North Carolina Slate Belt, an ecoregion characterized by rocky streams and poorly drained soils. Because these clay soils promote overland runoff (and low groundwater storage) Slate Belt streams have both very high flows after rainfall events and very low flows during droughts. These extreme fluctuations in flow tend to create "harsh" environmental conditions. Furthermore, the relatively impervious stream bed does not allow an extensive hyporheic refuge during floods or droughts. Streams in another geologic area, the Triassic Basin, have similar flow regimes. Other geologic zones of the Piedmont, however, have more "benign" stream flow characteristics due to differences in soil characteristics (e.g., depth, hydraulic conductivity).

Competition is expected to be an important factor in the control of community structure when space becomes a limiting factor, either through the presence of fixed retreats, or through high densities of mobile invertebrates. Personal observations (D. R. Lenat) indicate that the following taxa might be important competitors in some Piedmont Streams:

Filter-feeders: Hydropsychids (especially *Hydropsyche betteni* and *Cheumatopsyche*), some Chironomidae (especially *Microtendipes*).

Grazers: Pleurocerids (especially *Elimia*), *Ferrisia*, *Petrophila*, many Chironomidae, *Neophylax*, Baetids. Streams with large numbers of *Elimia* often have relatively low densities of grazing Chironomidae, suggesting a competitive interaction.

Collector–gatherers: Heptageniids (especially *Stenonema*, *Stenacron* and *Heptagenia*).

Both vertebrate and invertebrate predators are abundant in some Piedmont streams. Salamanders may be abundant in the smallest streams, while benthic insectivores (darters, madtoms) also occur in high densities in some streams. Important invertebrate predators include Plecoptera (especially *Acroneuria abnormis*), Megaloptera (*Corydalus/Nigronia*), many Odonata (especially in soft sediments and near banks), and some Chironmidae.

Natural Piedmont streams include examples of both harsh and benign environments. To estimate the importance of biotic vs. abiotic factors in controlling community structure in these systems, the following items should be evaluated:

1. Stream size. Small headwater systems have a greater tendency to dry up during drought, limiting the distribution of important keystone species.
2. Soils and underlying rock type. Effects flow regime, substrate type, and depth of hyporheic zone.
3. Abundance of invertebrate competitors.
4. Abundance of invertebrate predators.
5. Abundance of vertebrate predators, especially benthic insectivores. Salamanders in the smallest streams, darter and madtoms in larger streams.
6. Amount of disturbance in catchment.

Disturbance Effects   The macroinvertebrate communities of many Piedmont streams are influenced by human disturbances, including point-source discharges of pollutants, non-point-source runoff from urban and agricultural areas, and channel alteration. While point-source discharges can result in large alterations in the macroinvertebrate communities at the site of the discharge and for some distance downstream, non-point-source discharges are generally a greater problem for the Piedmont region as a whole. Research on the effect of non-point-source runoff in Piedmont streams has been conducted by several state agencies in North Carolina and Georgia (Lenat et al. 1979, Penrose et al. 1980, Cook et al. 1983, Georgia Department of Natural Resources 1983). These investigations have examined the effects of urban runoff, agriculture, silviculture, and construction.

Urban Runoff   There is abundant information on the effects of urban runoff on the macroinvertebrate fauna of Piedmont streams (Lenat and Eagleson 1981, Duda et al. 1979, 1982, Benke et al. 1981, Cook et al. 1983, Georgia Department of Natural Resources 1983, Crawford and Lenat 1989). Urban streams are often among the most heavily impacted of Piedmont streams and usually have a wide assortment of pollutants discharged to them from both point and nonpoint sources. A review of "priority pollutants" in storm-water runoff indicated that an alarming number of toxic materials, particularly heavy metals, may enter urban streams (Cole et al. 1984).

Studies of stream fauna in many large Piedmont cities have invariably indicated poor water quality. The effects of nonpoint urban runoff are often more severe than the effects of poorly managed point-source discharges. This can greatly complicate analysis of the effects of discharges originating in large cities. Some investigators have speculated that urban runoff (i.e., runoff from impervious surfaces) may not be solely responsible for the pollution observed in urban streams, since many small (unpermitted) point-source discharges are often observed in these areas (Duda et al. 1982). However, investigations of Marsh Creek in Raleigh, North Carolina (Crawford and Lenat 1989), have shown the same type of severe pollution with no indication of these small point-source discharges.

Taxa richness (especially for Ephemeroptera, Plecoptera, Trichoptera, and Coleoptera) is drastically reduced in urban streams (50–80%) relative to control sites. Diptera and Oligochaeta become dominant, constituting >90% of the total density. The most abundant Chironomidae usually belong to the "toxic assemblage" described by Winner et al. (1980) and Simpson and Bode (1980). These species are known to be tolerant of many heavy metals. Total macroinvertebrate abundance also is reduced, unless there is contamination from sewer overflows and/or cross-connections between storm sewers and sanitary sewers. This situation may produce a classic organic pollution assemblage, with low taxa richness and high abundance of certain species (Duda et al. 1982). Such buildups of organic pollution indicator species are highly unstable and may be washed out during large storms.

Studies have indicated that urban runoff will result in noticeable changes in macroinvertebrate communities if impervious surfaces exceed 15% of the watershed (Klein 1979) or if the amount of developed land is >30% of the watershed (Benke et al. 1981). In small towns (populations <20 000), Lenat and Eagleson (1981) found that sediment impacts were of greater importance than toxic or organic pollutants. Similarly, Benke et al. (1981) found that in watersheds with housing densities of 50–200/km^2, streams had higher taxa richness than in watersheds with greater housing densities.

Agriculture Macroinvertebrate communities in streams draining agricultural areas can be affected by increased sedimentation (from erosion), nutrients (from fertilizers), organic inputs (especially animal wastes), and toxic materials (pesticides). Several studies have addressed the effects of agricultural runoff on the macroinvertebrates of Piedmont streams (Penrose et al. 1980, Cook et al. 1983, Georgia Department of Natural Resources 1983, Lenat 1984, Crawford and Lenat 1989). Some streams were found to be only slightly affected by agricultural runoff, whereas severe problems have been noted in others. Lenat (1984) has suggested that much of this variation can be attributed to the effectiveness of erosion control programs. Comparisons of streams in control watersheds, "well-managed" watersheds, and "poorly managed" watersheds indicated that good erosion control practices reduced the impact of agricultural runoff by a least 50%. This implies that most of the damage observed in Piedmont agricultural streams is associated with sediment loads and that much of it is preventable through erosion control practices.

Reductions in total taxa richness of only 12–22% have been observed in most agricultural streams, but reductions of up to 50% have been observed in some. The greatest reductions in taxa richness were observed for Ephemeroptera, Plecoptera, and Trichoptera. However, the magnitude of these changes is generally much less than the changes caused by urban runoff. In some agricultural streams, the disappearance of intolerant species is offset by the appearance of more tolerant species. Total abundance of macroinvertebrates in agricultural streams is often 1.5–3 times greater than that observed in forested streams. High nutrient inputs in agricultural streams stimulate periphyton growth (Cook et al. 1983, Georgia Department of Natural Resources 1983), thereby increasing food resources for macroinvertebrates. Species that increase in abundance in agricultural streams include some Ephemeroptera (Baetidae, *Ephemerella catawba*), Trichoptera (tolerant Hydropsychidae), Simuliidae, and many Chironomidae. These groups may have very rapid changes in abundance due to either short life cycles or washout during high flows. Lenat (1984) found a gradient in stability in comparing control sites (most stable), well-managed sites, and poorly managed sites (least stable).

Species that often do well in agricultural streams include filter-feeders, collector–gatherers, and scrapers. High nutrient inputs in agricultural streams stimulate periphyton growth, which in turn directly benefit scrapers. Periphyton growths that break off can be washed downstream, benefitting both filter-feeders and collector–gatherers. Schlosser and Karr (1981) found that maximum concentrations of suspended solids during low-flow periods coincided with high productivity of benthic algae, suggesting that suspended material could be a high-quality food source for invertebrate fauna.

Other Anthropogenic Disturbances   Silviculture, construction, and channelization are directly related to the introduction of inorganic sediment into streams and to alteration of both channel substratum and flow regime. Generally, chemical changes in water quality due to these disturbances are minor. Canopy removal may result in changes in the food web resulting from the increased periphyton production and reduced allochthonous organic inputs. Both silviculture and construction usually involve increased sedimentation in streams over a short period (1–2 years) prior to earth stabilization and revegetation. Recovery should occur when the sediment inputs return to low levels and the sediment accumulated on the streambed is washed out during storms. In high-gradient mountain streams, recovery may occur in less than 1 year (Tebo 1955), but slower recovery may be expected in low-gradient Piedmont streams. Channelization, because of the direct disturbance to the streambed and periodic maintenance activities, often results in more long-lasting effects on stream macroinvertebrates.

Cook et al. (1983) and the Georgia Department of Natural Resources (1983) reported massive sedimentation from the construction of logging roads but few other effects from logging on a Piedmont stream in Georgia. Boschung and O'Neil (1981) found that a small (18 ha), properly conducted clear-cutting had no effect on the macroinvertebrates of a Piedmont stream in Alabama.

The effects of construction, particularly road construction, on stream macroinvertebrates have been investigated by Reed (1977) and Penrose et al. (1980). Road construction can produce very high erosion rates. Although structures for controlling erosion are now used in most road-building projects, they are rarely properly maintained throughout the construction period. Many studies report a 40–60% reduction in the abundance of macroinvertebrates. Taxa richness is not as severely affected, with reductions of the order of 25% observed. Dominance often shifts to Diptera (especially Orthocladiinae) and Oligochaeta (Lumbriculidae).

Although sediment inputs usually result in a large decrease in macroinvertebrate density, occasionally a benthic community develops with a density much greater than that at control sites. This phenomenon occurs during periods of low flow, when minimal resuspension of the sand and silt substratum allows development of large periphyton growths (Lenat et al. 1981). These communities, growing on the temporarily stable substratum, are characterized by large numbers of small Chironomidae, Naididae, and Ephemeroptera (Baetidae, *Ephemerella catawba* gr.). However, during periods of high flow, they may be decimated when covered by sediment following resuspension of the sand and silt substratum. Thus, the frequency of storms may have a much greater impact on the macroinvertebrate communities of sediment-impacted streams than on undisturbed streams.

Vertebrates

Amphibians and Reptiles Few amphibians and reptiles are found in lotic habitats in larger Piedmont streams and rivers. However, in very small springs and seeps, some amphibians (particularly salamanders) may be very abundant. It is likely that predation by fish limits the distribution of many of these species in larger streams and rivers.

Amphibians and reptiles tend to be associated with the terrestrial–aquatic interface in streams and rivers. Consequently, backwater and floodplain pools in Piedmont streams and rivers constitute an important habitat for these organisms (particularly turtles). The confluence of a large, slow-moving river and a small tributary stream can also be a particularly rich habitat for frogs, turtles, salamanders, and snakes (A. Braswell, North Carolina Museum of Natural History, personal communication).

The taxonomy and distribution of amphibians and reptiles in the Piedmont region is reasonably well known. Accounts of individual species may be found in the *Catalogue of American Amphibians and Reptiles* (American Society of Ichthyologists and Herpetologists 1963– .). A less formal description of species in Virginia and North and South Carolina is given in Martof et al. (1980).

Fish

Natural Communities   Despite several works on the taxonomy and distribution of Piedmont fish (Fowler 1945, Dahlberg and Scott 1971, Lee et al.
1980, Menhinick 1991), there have been few published studies of fish community structure in Piedmont streams. However, there are several investigations in progress (P. Angermeir, Virginia Polytechnic Institute and State
University, personal communication; V. Schneider, North Carolina Division
of Environmental Management, personal communication) whose objectives
are to generate baseline data for an "index of biotic integrity" (IBI) for
Piedmont streams like that developed by Karr (1981) for midwestern streams.
Fausch et al. (1984) have shown that this index can be used for a wide variety
of streams, but some initial research is required to establish the expected
levels of IBI parameters for undisturbed streams in each river system (e.g.,
species richness, percentage of insectivorous cyprinids, percentage of omnivores, and number of darters, suckers, sunfish, and "intolerant" species).

It has been well established that species richness of stream fish communities
is positively correlated with stream size or stream order (Sheldon 1968). This
pattern has been demonstrated for streams in the Roanoke River basin in
the Virginia Piedmont (Hambrick 1973). Collections made in the Broad River
and Catawba River basins of the North Carolina Piedmont yielded an average
of 8–9 species in small unpolluted streams, increasing to 12–14 species in
larger streams and rivers (Louder 1964, Messer et al. 1965). In Piedmont
sections of the Tar River and Neuse River basins, small unpolluted streams
averaged 18–19 species, increasing to 26 species in larger streams (Bayless
and Smith 1962, 1964).

Species richness may also be a function of the specific river basin. In an
extensive survey of North Carolina streams conducted during the 1960s, the
North Carolina Wildlife Resources Commission, Division of Inland Fisheries,
collected data indicating that fish species richness in Piedmont streams was
usually much lower than that observed in Coastal Plain streams (North Carolina Wildlife Resources Commission, unpublished data). The higher species
richness in Piedmont sections of the Neuse River and Tar River basins compared with that in the Broad River and Catawba River basins presented above
may be the result of the former rivers being located primarily in the Coastal
Plain, with only small sections of their headwaters in the Piedmont.

Two investigations of fish community structure in North Carolina Piedmont
streams have been conducted. Lemly (1985) found that an introduced species,
the green sunfish (*Lepomis cyanellus*), was usually dominant in first-order
Piedmont streams but much less abundant in second- and third-order streams.
Removal experiments indicated that green sunfish suppressed native fish populations in a first-order stream, apparently through predation on juveniles.
Lemly also reported that the population structure in undisturbed streams was
stable over 2 years.

Gatz (1979) investigated niche overlap in fish communities in North Carolina Piedmont streams. Species richness, determined from repeated collections at three sites, varied from 14 to 25 species. It is not clear to what extent

differences in species richness reflect differences in stream size, habitat type, and river basin. However, the average niche overlap was relatively constant for all streams, suggesting that the degree of species packing was equal for all sites. Gatz hypothesized that consistent patterns exist in Piedmont stream communities and that co-occurring species are not randomly associated.

Some Piedmont streams support excellent populations of game fish, with redbreast sunfish (*Lepomis auritus*), bluegill (*Lepomis macrochirus*), warmouth (*Lepomis gulosus*), and largemouth bass (*Micropterus salmoides*) being the species preferred by anglers. However, sedimentation and turbidity are widespread problems in most Piedmont streams, and most fishermen concentrate their efforts on farm ponds and reservoirs.

Most nongame fish in Piedmont streams have fairly limited distributions, often being abundant in only one or two states. The most widespread species are *Anguilla rostrata*, *Hybognathus regius*, *Nocomis leptocephalus*, *Notemigonus crysoleucas*, *Notropis hudsoni*, *N. niveus*, *N. procne*, *Semotilus atromaculatus*, *Erimyzon oblongus*, *Moxostoma anisurum*, *Ictalurus natalis*, *I. platycephalus*, *Noturus insignis*, and *Etheostoma olmstedi*. The catostomids are spring spawners; adults are usually observed only in spring, but juveniles are year-round residents.

There are fewer darters in Piedmont streams than in mountain streams. It is not clear if this reflects a difference in habitat or a difference in the level of disturbance. "Shiners" (mostly *Notropis* spp.) are usually abundant in Piedmont streams, but few species encompass the entire Piedmont region. The most widespread species (*Notemigonus crysoleucas*) has been spread by "bait-bucket" introductions. Most states have 6–8 common shiners in their Piedmont region, although over 25 shiners are common in the Piedmont of at least one state in the southeast (Lee et al. 1980).

Disturbance Effects   Few studies have investigated the effects of pollution on Piedmont stream fisheries. The studies by the North Carolina Wildlife Resources Commission have usually suggested an association between non-point-source runoff and a reduction in fish abundance and/or species richness. Studies by Karr et al. (1985) in midwestern streams have indicated that agriculture has the greatest impact on small headwater streams. In contrast, studies of Piedmont streams have indicated more severe reductions in the fish communities of streams in urban watersheds than in agricultural streams (Cook et al. 1983, Georgia Department of Natural Resources 1983, Crawford and Lenat 1989, North Carolina Wildlife Resources Commission, unpublished data). Some smaller urban streams may be totally devoid of fish.

UNIQUE BIOTA AND SPECIAL FEATURES

Threatened and Endangered Species

One plant species, *Harperella* sp., a mollusk, the Tar River spiny mussel (*Elliptio steinstanana*), and one fish species, the Cape Fear shiner (*Notropis mekistocholas*), are the only species listed as threatened or endangered by the U.S. Fish and Wildlife Department as of 1988 and known to occur in Piedmont streams of the Southeast. *Harperella* has been reported in sections of the Tar, Deep, and Rocky rivers in North Carolina. *E. steinstanana* is found in Swift Creek and other streams in the upper sections of the Tar River drainage. *N. mekistocholas* is found in sections of the Deep and Rocky rivers. It is anticipated that a number of additional species will be added to the Federal list.

There are a number of state-listed threatened and endangered species or species of special concern found in Piedmont rivers and streams of North Carolina, South Carolina, and Georgia. Several species of mollusks and fish and one crustacean, listed as threatened, endangered, or of special concern by North Carolina, are known to occur in North Carolina Piedmont streams and rivers (Table 6). Although many Oligochaeta and Diptera appear to have a very limited distribution, historical difficulties with taxonomy and inadequate collection have precluded the listing of any species in these groups. This may change in the near future with expansion of the data base through

TABLE 6   Threatened (t) and Endangered (e) Species and Species of Special Concern (sc) Reported in Piedmont Rivers and Streams of North Carolina

Species	River drainage
MOLLUSCA	
Alasmidonta (Prolasmidonta) heterodon (e)	Upper Neuse
Fusconaia masoni (t)	Little, Uwharrie, Tar
Elliptio lanceolata (sc)	Little, Uwharrie, Tar
Strophitus undulatus (sc)	Little, Tar
Lampsilis cariosa (sc)	Little, Tar
Lioplax subcarinata (e)	Tar
CRUSTACEA	
Orconectes n. sp. (sc)	Neuse
FISH	
Exoglossum maxillinqua (cutlips minnow) (sc)	Roanoke-Dan
Hybopsis n. sp. (sc)	Broad, Catawba
Notropis mekistocholas (Cape Fear shiner) (e)	Cape Fear
Moxostoma ariommum (bigeye jumprock) (sc)	Roanoke-Dan
Noturus gilberti (orangefin madtom) (t)	Roanoke-Dan
Ambloplites cavifrons (Roanoke bass) (sc)	Tar, Neuse, Roanoke-Dan
Micropterus punctulatus (spotted bass) (sc)	Cape Fear
Etheostoma collis (Carolina darter) (t)	Cape Fear, Neuse, Tar
Etheostoma podostemone (riverweed darter) (sc)	Roanoke-Dan
Etheostoma thalassinum (seagreen darter) (sc)	Broad, Catawba

Source. North Carolina Division of Environmental Management, unpublished; Bailey (1977); Fuller (1977); Fuller et al. (1979).

ongoing collections by the North Carolina Division of Environmental Management and other groups.

The rocky shoals spider lily (*Hymenocallis coronaria*) is listed by the State of South Carolina as threatened and is currently a candidate for federal listing. It may be found in rocky shoals of Piedmont rivers. In addition, the Carolina darter (*Etheostoma collis collis*) is listed by South Carolina as threatened and may be found in Piedmont rivers and streams.

The Georgia Department of Natural Resources has conducted extensive studies of potential threatened and endangered species. Those species listed by the state as being threatened or of special concern that are known to occur in Piedmont rivers and streams in Georgia are given in Table 7.

State listings of threatened or endangered species may have been updated by the time of this printing. Current listings should be obtained from appropriate state agencies.

Wild and Scenic Rivers

No Piedmont rivers have been included in the National Wild and Scenic Rivers System recently established by Congressional mandate (PL 90-542). However, the entire Ogeechee River of Piedmont and Coastal Plain Georgia has been designated for study for possible inclusion in the system. Designation as a National Wild and Scenic River affords considerable protection from federal or federally sponsored projects that might affect river quality. In addition, efforts are made to secure public ownership of the river corridor or to control activities by private landowners within the corridor.

Numerous Piedmont rivers and streams have been listed in the National Park Services's Nationwide Rivers Inventory (National Park Service 1982)

TABLE 7 Invertebrate and Fish Species Listed by Georgia as Threatened (t) or of Special Concern (sc)

Species	River Drainage
INVERTEBRATES	
Marstonia agarhecta (t)	Ocmulgee
Notogillia sathon (t)	Ocmulgee
Spilochlamys furgida (t)	Ocmulgee
Amblema boykiniana (t)	Chattahootchie and Flint
FISH	
Notropis caeruleus (blue shiner) (t)	Coosawattee
Percina callitaenia (bluestripe shiner) (t)	Chattahootchie and Flint
Noturus munitus (frecklebelly madtom) (t)	Etowah
Percina antesella (amber darter) (t)	Etowah
Percina aurolineata (goldline darter) (t)	Coosawattee
Percina lenticula (freckled darter) (t)	Etowah
Micropterus sp. cf. *M. coosae* (shoal bass) (sc)	Chattahootchie and Flint

Source. Georgia Department of Natural Resources (1981).

(Table 8). This inventory established a list of rivers (or river segments) for potential designation as National Wild and Scenic Rivers. The major criteria used to select streams and rivers for this list were (1) that the river appear on 1:500 000 scale maps and be largely unchannelized and free flowing; (2) that cities (>10 000 population), power plants, and active strip mines be no closer than 0.4 km from the river bank; and (3) that there be sustained flow. The purpose of the inventory was primarily to identify a balanced representation (in terms of physiographic provinces) of the most significant river segments in the United States and to stimulate governmental and private actions that will ensure conservation of, and public access to, these rivers.

TABLE 8   Piedmont Streams and Rivers Listed in The Nationwide Rivers Inventory

Virginia	North Carolina	South Carolina	Georgia
Appomattox River	Ararat River	Broad River	Alcovy River
Big Otter River	Barnes Creek	Catawba River	Amicalola Creek
Bull River	Cane Creek	Chauga River	Apalachee River
Catoctin Creek	Cape Fear River	Savannah River	Big Cedar Creek
Covington River	Caraway River	Turkey Creek	Little Cedar Creek
Cub Creek	Cedar Creek	Tyger River	Rier and Big
Dan River	Dan River	North Tyger River	Blood Mtn Creek
Deep Creek	Deep River	Wateree River	Brier Creek
Falling River	Downing Creek		Big Brier Creek
Goose Creek	Dutchman's Creek		Broad River
James River	Eno River		Middle Fork
Hardware River	Fisher River		Chattahoochee River
Hazel River	Fishing Creek		Chestatee Creek
Meherrin River	Haw River		Coosawattee River
North Anna River	Henry Fork of		Dicks Creek
Nottoway River	Catawba River		Etowah River
Potomac River	Johns River		Flat Shoals Creek
Rappahannock River	Lower Little River		Flint River
Rapidan River	Lumber Creek		Middle Oconee River
Rivanna River	Mayo River		Murder Creek
Roanoke River	Mitchell River		Ocmulgee River
Rucker River	Mountain Creek		Oconee River
Rush River	Neuse River		Savannah River
Slate River	Pee Dee River		Sope Creek
South Anna River	Rock Creek		South River
Tye River	Rocky River		Sweetwater Creek
Willis River	Smith River		Talking Rock River
	Tar River		Towaliga River
	Uwharrie River		
	Warrior Fork of		
	Catawba River		
	Yadkin River		

Source. National Park Service (1982).

MANAGEMENT AND FUTURE RESEARCH NEEDS

As discussed above, most Piedmont streams are disturbed to varying degrees from high rates of erosion and sedimentation in the past plus current water quality and sedimentation problems associated with runoff from agricultural and urban areas. Perhaps because of their disturbed condition, Piedmont streams have not been studied as intensively as have streams in the Blue Ridge province. In particular, algal and microbial communities, meiofauna, and autotrophic and heterotrophic metabolism have received little attention. Fish community composition and production have also been neglected. Although some studies of the macroinvertebrates in Piedmont streams have been made, production studies are uncommon. Finally, as is the general situation in lotic ecology, the meiofaunal community and its role in energy conversion and nutrient cycling in Piedmont streams are largely unknown.

Expanding urban areas threaten increasing numbers of Piedmont streams with point-source and non-point-source discharges of toxic materials. Channelization and removal of vegetation from riparian areas also have reduced the quality of stream habitat in the Piedmont region, although government-sponsored channelization programs have been greatly curtailed. It is imperative that urban expansion be planned and managed so as to ensure the least impact on streams. Good erosion control practices must be used during construction activity, and stream channels should be disturbed to the least extent possible. Natural riparian vegetation should be maintained as buffer strips along streams. Undisturbed, natural corridors along streams, serving as greenbelts and parks, can enhance the quality of urban life. Floodplains must be given protection from commercial development through zoning or other governmental regulation, because of their importance in maintaining high water quality and providing rich biological habitat when flooded. Installation of detention basins and better control of the numerous unpermitted discharges and spills are needed to prevent further deterioration of urban streams and larger rivers downstream from urban areas.

There is little information on the extent of toxic contaminants associated with agricultural runoff. The existing information suggests that better implementation of present erosion control technologies and strategies would greatly improve water quality in Piedmont streams in agricultural areas. However, encouragement and/or enforcement of better erosion control practices is a complex political issue. The technical side of this problem has largely been resolved.

Historically, there have been few problems with water withdrawals in the Piedmont area (i.e., there has been a plentiful water supply). However, the increased demand for water in some drainages now requires that water withdrawals be regulated. Prior to water withdrawal regulation in the western states, some streams and rivers went dry. Some drainages in the southeastern Piedmont are now approaching this condition. Further, planners and managers should seek to maintain the periodicity of entire natural hydrographs, not just low-flow requirements, to maintain biotic communities.

As a result of improved erosion control techniques and forest regrowth, many Piedmont streams are slowly recovering from past disturbances. The deposits of sand and silt in the stream channels are eroding, uncovering previously buried gravels (Nunnally 1985). We must protect the relatively natural Piedmont streams that still exist as they continue to recover from the effects of historical sedimentation. These streams are a valuable natural and scientific resource in a region that is rapidly becoming more heavily populated with increasing demands on its freshwater resources.

Perhaps the most compelling argument for restoring or preserving the integrity of streams in the southeastern Piedmont is that they will be the principal source of drinking water for an expanding human population. The plant and animal communities present in the streams can provide an indication of water quality and the extent to which these streams can be utilized for drinking water, recreation, and fishing.

A final note on future management of Piedmont streams concerns the impoundment of streams and rivers. The construction of reservoirs for maintaining a stable water supply and for recreational purposes has resulted in the elimination of many stream and river reaches over the past several decades. Although reservoirs can often provide an improved drinking water supply (in terms of reduced seasonal variability in quantity) and a more extensive open-water recreational resource than streams or rivers, planners should also consider the unique resource of streams and rivers in their free-flowing condition. Streams and rivers provide recreational activities of a different type (fishing, canoeing, rafting, etc.), as well as a unique and rich biotic habitat formed by the floodplain-riparian zone. This zone, formed by the interface between terrestrial and stream/river ecosystems, often provides the richest component of both terrestrial and aquatic habitats and should be given special consideration in decisions regarding reservoir construction as well as urban and agricultural development.

In this chapter we have attempted to summarize the published literature on the biotic communities of Piedmont streams and to provide additional insights from our personal observations as well as those of several scientists that have studied these systems. We have surely missed some information published as state agency reports and university theses. A more thorough search and synopsis of this type of material would certainly be helpful.

ACKNOWLEDGMENTS

In preparing this chapter, we have been aided by numerous people in state agencies in Georgia, South Carolina, North Carolina, and Virginia, and by staff of the U.S. Geological Survey and the National Park Service. In particular, we thank Kent Crawford, Dennis Lynch, and Glenn Patterson (U.S. Geological Survey in Raleigh, North Carolina; Richmond, Virginia; and Columbia, South Carolina, respectively). We also thank Mahlon G. Kelly (University of Virginia) for sharing his insights on algae and macrophyte distribution. The chapter benefitted from critical reviews by Mahlon Kelly, Michael Sale (Environmental Sciences Division, Oak Ridge National Laboratory), Seth R. Reice (University of North Carolina), and John J. Haines (U.S.A.E. Waterways Experiment Station, Calhoun Falls, South Carolina).

This study was partially supported by the Office of Health and Environmental Research, U.S. Department of Energy, under Contract DE-AC05-840R21400 with Martin Marietta Energy Systems, Inc. Publication No. 3285, Environmental Sciences Division, Oak Ridge National Laboratory.

REFERENCES

American Society of Ichthyologists and Herpetologists. 1963–. Herpetological Catalogue Committee, W. J. Reimer (ed.), *Catalogue of American Amphibians and Reptiles* (loose leaf). Bethesda, MD: Am. Soc. Ichthyol. Herpetol.

Anderson, N. H., and J. R. Sedell. 1979. Detritus processing by macroinvertebrates in stream ecosystems. *Annu. Rev. Entomol.* 24:351–377.

Bailey, J. R. 1977. Freshwater fishes. In J. E. Cooper, S. S. Robinson, and J. B. Funderburg (eds.), *Endangered and Threatened Plants and Animals of North Carolina*. Raleigh: North Carolina State Museum of Natural History, pp. 265–298.

Bayless, J., and W. B. Smith. 1962. *Survey and Classification of the Neuse River and Tributaries*. Raleigh: North Carolina Wildlife Resources Commission.

Bayless, J., and W. B. Smith. 1964. *Survey and Classification of the Tar River and Tributaries*. Raleigh: North Carolina Wildlife Resources Commission.

Benke, A. C., G. E. Willke, F. K. Parrish, and D. L. Stites. 1981. *Effects of Urbanization of Stream Ecosystems*, ERC 08-81. Atlanta: Georgia Institute of Technology.

Benke, A. C., T. C. van Arsdall, Jr., D. M. Gillespie, and F. K. Parrish. 1984. Invertebrate productivity in a subtropical blackwater river: the importance of habitat and life history. *Ecol. Monogr.* 54:25–63.

Boschung, H., and P. O'Neil. 1981. The effects of forest clear-cutting on fishes and macroinvertebrates in an Alabama stream. In L. A. Krumholz (ed.), *The Warmwater Streams Symposium*. Bethesda, MD: South. Div., Am. Fish. Soc., pp. 200–217.

Bray, J. R., and E. Gorham. 1964. Litter production in forests of the world. *Adv. Ecol. Res.* 2:101–157.

Brigham, A. R., W. U. Brigham, and A. Gnilka. 1982. *Aquatic Insects and Oligochaetes of North and South Carolina*. Mahomet, IL: Midwest Aquatic Enterprises.

Buell, G. R., and S. C. Grams. 1985. *The Hydrologic Bench-Mark Program: A Standard to Evaluate Time-Series Trends in Selected Water-Quality Constituents for Streams in Georgia*, USGS Water Resour. Invest. Rep. 84-4318. Doraville, GA: U.S. Geological Survey.

Cherry, R. N. 1961. *Chemical Quality of Water of Georgia Streams, 1957–58*, Geol. Surv. Bull. No. 69. Atlanta: Georgia State Division of Conservation.

Cole, R. H., R. E. Frederick, R. P. Healy, and R. G. Rolan. 1984. Preliminary findings of the priority pollutant monitoring project of the Nationwide Urban Runoff Program. *J. Water Pollut. Control Fed.* 56:898–908.

Cook, W. L., F. Parrish, J. D. Satterfield, W. G. Nolan, and P. E. Gafney. 1983. *Biological and Chemical Assessment of Nonpoint Source Pollution in Georgia:*

Ridge-Valley and Sea Island Streams. Atlanta: Department of Biology, Georgia State University.

Crawford, J. K. and D. R. Lenat. 1989. *Effects of Land Use on the Water Quality and Biota of Three Streams in the Piedmont Province of North Carolina,* Water Resour. Invest. Rep. 89-4007. Raleigh, NC: U.S. Geological Survey.

Dahlberg, M. D., and D. C. Scott. 1971. The freshwater fishes of Georgia. Sapelo Island Marine Institute Contribution No. 212. *Bull. Ga. Acad. Sci.* 29:2–64.

Diamond, J. M., and S. R. Reice. 1985. Effects of selective taxa removal on lotic macroinvertebrate colonization in a wooded piedmont USA stream. *J. Freshwater Ecol.* 3:193–201.

Duda, A. M., D. R. Lenat, and D. L. Penrose. 1979. Water quality degradation in urban streams of the Southeast: will non-point source controls make any difference? In *International Symposium on Urban Storm Runoff.* Lexington: University of Kentucky, pp. 151–159.

Duda, A. M., D. R. Lenat, and D. L. Penrose. 1982. Water quality in urban streams— what we can expect. *J. Water Pollut. Control Fed.* 54:1139–1147.

Fausch, K. D., J. R. Karr, and P. R. Yant. 1984. Regional application of an index of biotic integrity based on stream fish communities. *Trans. Am. Fish. Soc.* 113:39–55.

Fowler, H. W. 1945. A study of the fishes of the southern piedmont and coastal plain. *Monogr. Acad. Nat. Sci. Philadelphia* 7:1–408.

Freeman, M. C., and J. B. Wallace. 1984. Production of net-spinning caddisflies (Hydropsychidae) and black flies (Simuliidae) on rock outcrop substrate in a small southeastern piedmont stream. *Hydrobiologia* 112:3–15.

Fuller, S. L. H. 1977. Freshwater and terrestrial mollusks. In J. E. Cooper, S. S. Robinson, and J. B. Funderburg (eds.), *Endangered and Threatened Plants and Animals of North Carolina.* Raleigh: North Carolina State Museum of Natural History, pp. 143–194.

Fuller, S. L. H., F. Grimm, T. L. Laavy, H. J. Porter, and A. H. Shoemaker. 1979. Status report: freshwater and terrestrial mollusks. In D. M. Forsythe and W. B. Ezell, Jr. (eds.), *Proceedings of the First South Carolina Endangered Species Symposium.* Columbia: South Carolina Wildlife and Marine Resources Department, pp. 55–81.

Gatz, A. J. 1979. Community organization in fishes as indicated by morphological features. *Ecology* 60:711–718.

Georgia Department of Natural Resources. 1981. *Statewide Endangered Wildlife Program, Annual Progress Report, July 1, 1980–June 30, 1981.* Atlanta: GDNR.

Georgia Department of Natural Resources. 1983. *Georgia Nonpoint Source Impact Assessment Study* (final report from CTA, Inc.). Atlanta: GDNR.

Georgia Environmental Protection Department. 1971. *Flint River Water Quality Study, Atlanta-Griffin.* Atlanta: Water Quality Control Section, Environmental Protection Division, State of Georgia.

Glenn, L. C. 1911. *Denudation and Erosion in the Southern Appalachian Region and the Monongahela Basin,* USGS Prof. Pap. No. 72. Reston, VA: U.S. Geological Survey.

Hall, C. A. S. 1972. Migration and metabolism in a temperate stream ecosystem. *Ecology* 53:585–604.

Hambrick, P. S. 1973. Composition, longitudinal distribution, and zoogeography of the fish fauna of Back Creek, Blackwater River, and Pigg River, tributaries of the

Roanoke River in south-central Virginia. MS Thesis, Virginia Polytechnic Institute and State University, Blacksburg.

Happ, S. C. 1945. Sedimentation in South Carolina Piedmont valleys. *Am. J. Sci.* 243:113–126.

Harrell, R. C., and T. C. Dorris. 1968. Stream order, morphometry, physico-chemical conditions, and community structure of benthic macroinvertebrates in an intermittent stream system. *Am. Midl. Nat.* 80:220–251.

Hobbs, H. H., Jr. 1981. The crayfishes of Georgia. *Smithson. Contrib. Zool.* 318:1–549.

Karr, J. R. 1981. Assessment of biotic integrity using fish communities. *Fisheries* 6:21–27.

Karr, J. R., L. A. Toth, and D. A. Dudley. 1985. Fish communities of midwestern rivers: a history of degradation. *BioScience* 35:90–95.

Kelly, M. G., G. M. Hornberger, and B. J. Cosby. 1974. Continuous automated measurement of rates of photosynthesis and respiration in an undisturbed river community. *Limnol. Oceanogr.* 19:305–312.

Kelly, M. G., G. M. Hornberger, and B. J. Cosby. 1975. *A Method for Monitoring Eutrophication in Rivers*, Project Completion Report for OWRR Contract No. 14-31-0001-4231. Charlottesville: Department of Environmental Sciences, University of Virginia.

Kinne, O. 1963. The effects of temperature and salinity on marine and brackish water animals. *Oceanogr. Mar. Biol.* 1:301–340.

Klein, R. D. 1979. Urbanization and stream quality impairment. *Water Res. Bull.* 15:948–963.

Klein, R. D. 1985. *Effects of Urbanization upon Aquatic Resources*. Annapolis: Maryland Department of Natural Resources.

Kondratieff, B. C., and J. R. Voshell, Jr. 1980. Life history and ecology of *Stenonema modestum* (Banks) (Ephemeroptera: Heptageniidae) in Virginia, USA. *Aquat. Insects* 2:177–189.

Kondratieff, B. C., and J. R. Voshell, Jr. 1981. Seasonal distribution of mayflies (Ephemeroptera) in two piedmont rivers in Virginia. *Entomol. News* 92:189–195.

Krieger, K. A., and W. D. Burbanck. 1976. Distribution and dispersal mechanisms of *Oxytrema* (= *Goniobasis*) *suturalis* Haldeman (Gastropoda: Pleuroceridae) in the Yellow River, Georgia, USA. *Am. Midl. Nat.* 95:49–63.

Krueger, C. C., and T. F. Waters. 1983. Annual production of macroinvertebrates in three streams of different water quality. *Ecology* 64:840–850.

Kuenzler, E. J., P. J. Mulholland, L. A. Yarbro, and L. A. Smock. 1980. *Distributions and Budgets of Carbon, Phosphorus, Iron, and Manganese in Floodplain Swamp Ecosystem*, Rep. No. 157. Raleigh: University of North Carolina, Water Resources Research Institute.

Lee, D. S., C. R. Gilbert, C. H. Hocutt, R. E. Jenkins, D. E. McAllister, and J. R. Stauffer. 1980. *Atlas of North American Freshwater Fishes*. Raleigh: North Carolina State Museum of Natural History.

Lemly, A. D. 1985. Suppression of native fish species by green sunfish in first-order streams of Piedmont, North Carolina. *Trans. Am. Fish. Soc.* 114:705–712.

Lenat, D. R. 1983. Benthic macroinvertebrates of Cane Creek, North Carolina, and comparisons with other southeastern streams. *Brimleyana* 9:53–68.

Lenat, D. R. 1984. Agriculture and stream water quality: a biological evaluation of erosion control practices. *Environ. Manage.* 8:333–344.

Lenat, D. R. 1988. The macroinvertebrate fauna of the Little River, North Carolina: taxa list and seasonal trends. *Arch. Hydrobiol.* 110:19–43.

Lenat, D. R., and K. W. Eagleson. 1981. *Ecological Effects of Urban Runoff on North Carolina Streams*, Biol. Ser. No. 104. Raleigh: North Carolina Division of Environmental Management.

Lenat, D. R., D. L. Penrose, and K. W. Eagleson. 1979. *Biological Evaluation of Nonpoint Source Pollution in North Carolina Streams and Rivers*, Biol. Ser. No. 102. Raleigh: North Carolina Division of Environmental Management.

Lenat, D. R., D. L. Penrose, and K. W. Eagleson. 1981. Variable effects of sediment addition on stream benthos. *Hydrobiologia* 79:187–194.

Louder, D. E. 1964. *Survey and Classification of the Catawba River and Tributaries*. Raleigh: North Carolina Wildlife Resources Commission.

Lynch, D. D., E. H. Nuckels, and C. Zenone. 1987. *Low Flow Characteristics and Chemical Quality of Streams in the Culpepper Geologic Basin, Virginia and Maryland*, USGS Misc. Invest. Ser., MAP-I-1313-H. Richmond, VA: U.S. Geological Survey.

Martof, B. S., W. M. Palmer, J. R. Bailey, and J. R. Harrison, III. 1980. *Amphibians and Reptiles of the Carolinas and Virginia*. Chapel Hill: University of North Carolina Press.

McAuliffe, J. R. 1984. Competition for space, disturbance, and the structure of a benthic stream community. *Ecology* 65:894–908.

Meade, R. H. 1976. Sediment problems in the Savannah River basin. In B. L. Dillman and J. M. Stepp (eds.), *The Future of the Savannah River*. Clemson, SC: Clemson University Water Resources Research Institute, pp. 105–129.

Meade, R. H., and S. W. Trimble. 1974. Changes in sediment loads in rivers of the Atlantic drainage of the United States since 1900. *IAHS-AISH Publ.* 113:99–104.

Menge, B. A. 1976. Organization of the New England rocky intertidal community: role of predation, competition, and environmental heteogeity. *Ecol. Monogr.* 46:355–393.

Menhinick, E. F. 1991. *The Freshwater Fishes of North Carolina*. North Carolina Wildlife Resources, Commission, Raleigh, NC, 227 pp.

Messer, J. B., J. R. Davis, T. E. Crowell, and W. C. Carnes. 1965. *Survey and Classification of the Broad River and its Tributaries, North Carolina*. Raleigh: North Carolina Wildlife Resources Commission.

Meyer, J. L., and R. T. Edwards. 1989. Ecosystem metabolism and turnover of organic carbon along a blackwater river continuum. *Ecology* 71:668–677.

Minshall, G. W., and R. C. Petersen, Jr. 1985. Towards a theory of macroinvertebrate community structure in stream ecosystems. *Arch. Hydrobiol.* 104:49–76.

Minshall, G. W., R. C. Petersen, Jr., and C. F. Nimz. 1985. Species richness in streams of different size from the same drainage basin. *Am. Nat.* 125:16–38.

Monk, C. D., G. I. Child, and S. A. Nicholson. 1970. Biomass, litter, and leaf surface area estimates of an oak-hickory forest. *Oikos* 21:138–141.

Mulholland, P. J. 1981. Organic carbon flow in a swamp-stream ecosystem. *Ecol. Monogr.* 51:307–322.

National Park Service. 1982. *The Nationwide Rivers Inventory*. Washington, DC: U.S. Department of the Interior.

Nelson, D. J., and D. C. Scott. 1962. Role of detritus in the productivity of a rock-outcrop community in a Piedmont stream. *Limnol. Oceanogr.* 7:396–413.

Nelson, T. C., R. D. Ross, and G. D. Walker. 1957. *The Extent of Moist Sites in the Georgia Piedmont and Their Forest Associations*, Res. Note No. 102. Asheville, NC: Southeast Forestry Experiment Station, U.S. Forest Service.

North Carolina Department of Conservation and Development. 1958. *Geologic Map of North Carolina*. Raleigh: NCDCD.

North Carolina Division of Environmental Management. 1978. *208 Phase I Results*. Raleigh: NCDEM.

Nunnally, N. R. 1985. Application of fluvial relationships to planning and design of channel modifications. *Environ. Manage.* 9:415–426.

Odum, H. P. 1956. Primary production of flowing waters. *Limnol. Oceanogr.* 2:85–97.

Overstreet, W. C., and H. Bell. 1965. *The Crystalline Rocks of South Carolina*, USGS Water-Supply Bull. No. 1183. Columbia, SC: U.S. Geological Survey.

Parker, C. R., and J. R. Voshell, Jr. 1982. Life histories of some filter-feeding Trichoptera in Virginia. *Can. J. Zool.* 60:1732–1742.

Parker, C. R., and J. R. Voshell, Jr. 1983. Production of filter-feeding Trichoptera in an impounded and a free-flowing river. *Can. J. Zool.* 61:70–87.

Patterson, G. G., and T. W. Cooney. 1986. Sediment transport and deposition in Lakes Marion and Moultrie, South Carolina. In *Proceedings of the 3rd International Symposium on River Sedimentation*, Jackson, Mississippi, March 31–April 4, 1986.

Patterson, G. G., and G. G. Padgett. 1984. *Quality of Water from Bedrock Aquifers in the South Carolina Piedmont*, USGS Water Resour. Invest. Rep. 84-4028. Columbia, SC: U.S. Geological Survey.

Peckarsky, B. L. 1983. Biotic interaction or abiotic limitations? A model of lotic community structure. In T. D. Fontaine, III and S. M. Bartell (eds.), *Dynamics of Lotic Ecosystems*. Ann Arbor, MI: Ann Arbor Science, pp. 303–324.

Penrose, D. L., D. R. Lenat, and K. W. Eagleson. 1980. *Biological Evaluation of Water Quality in North Carolina Streams and Rivers*, Biol. Ser. No. 103. Raleigh: North Carolina Division of Environmental Management.

Powell, J. D., and J. M. Abe. 1985. *Availability and Quality of Ground Water in the Piedmont Province of Virginia*, USGS Water Resour. Invest. Rep. 85-4235. Richmond, VA: U.S. Geological Survey.

Reed, J. R., Jr. 1977. Stream community response to road construction sediments. *Bull.—Va. Water Resour. Res. Cent.* 97:1–61.

Reice, S. R. 1977. The role of animal associations and current velocity in sediment-specific leaf litter decomposition. *Oikos* 29:357–365.

Reice, S. R. 1978. The role of detritivore selectivity in species specific litter decomposition in a woodland stream. *Verh. Int. Ver. Theor. Angew. Limnol.* 20:1396–1400.

Reice, S. R. 1981. Interspecific associations in a woodland stream. *Can. J. Fish. Aquat. Sci.* 38:1271–1280.

Reice, S. R. 1983. Predation and substratum: factors in lotic community structure. In T. D. Fontaine and S. M. Bartell (eds.), *Dynamics of Lotic Ecosystems*. Ann Arbor, MI: Ann Arbor Science, pp. 325–346.

Reisen, W. K., and R. Prins. 1972. Some ecological relationships of the invertebrate drift in Praters Creek, Pickens County, South Carolina. *Ecology* 53:876–884.

Schlosser, I. J., and J. R. Karr. 1981. Water quality in agricultural watersheds: impact of riparian vegetation during base flow. *Water Res. Bull.* 17:233–240.

Sheldon, A. L. 1968. Species diversity and longitudinal succession in stream fishes. *Ecology* 49:193–198.

Simmons, C. E., and R. C. Heath. 1979. *Water-Quality Characteristics of Streams in Forested and Rural Areas of North Carolina*, USGS Water Resour. Invest. Rep. 79-108. Raleigh, NC: U.S. Geological Survey.

Simpson, K. W., and R. W. Bode. 1980. Common larvae of Chironomidae (Diptera) from New York State streams and rivers. *Bull. N. Y. State Mus.* 439:1–105.

Stoneburner, D. L., and L. A. Smock. 1979. Seasonal fluctuations of macroinvertebrate drift in a South Carolina piedmont stream. *Hydrobiologia* 63:49–56.

Stose, G. W., C. Cooke, C. Wythe, G. W. Crickmay, and C. Butts. 1939. *Geologic Map of Georgia*. Atlanta: Georgia Department of Mines, Mining and Geology.

Stuckey, J. L., and S. G. Conrad. 1958. *Explanatory Text for Geologic Map of North Carolina* Bull. No. 71. Raleigh: North Carolina Department of Conservation and Development.

Tebo, L. B., Jr. 1955. Effects of siltation, resulting from improper logging on the bottom fauna of a small trout stream in the southern Appalachians. *Prog. Fish-Cult.* 17:64–70.

Tett, P., C. Gallegos, M. G. Kelly, G. M. Hornberger, and B. J. Cosby. 1978. Relationships among substrate, flow, and benthic microalgal pigment density in the Mechums River, Virginia. *Limnol. Oceanogr.* 23:785–797.

Trimble, S. W. 1975a. A volumetric estimate of man-induced soil erosion on the southern Piedmont Plateau. In *Present and Prospective Technology for Predicting Sediment Yields and Sources*, ARS-S-40. Washington, DC: U.S. Agricultural Research Service, pp. 142–152.

Trimble, S. W. 1975b. Denudation studies: can we assume stream steady state? *Science* 188:1207–1208.

U.S. Department of Agriculture. 1977. *Erosion and Sediment Inventory, North Carolina*. Raleigh, NC: Soil Conservation Service, U.S. Department of Agriculture.

Wallace, J. B., J. R. Webster, and T. F. Cuffney. 1982. Stream detritus dynamics: regulation by invertebrate consumers. *Oecologia* 53:197–200.

Walton, O. E., Jr. 1980. Invertebrate drift from predator-prey associations. *Ecology* 61:1486–1498.

Wharton, C. H. 1970. *The Southern River Swamp—A Multiple Use Environment*. Atlanta: Georgia State University.

Wharton, C. H. 1978. *The Natural Environments of Georgia*. Atlanta: Georgia Geologic and Water Resources Division and Department of Natural Resources.

Winner, R. W., M. W. Boesel, and M. P. Farrell. 1980. Insect community structure as an index of heavy metal pollution in lotic ecosystems. *Can. J. Fish. Aquat. Sci.* 37:647–655.

Yarbro, L. A., E. J. Kuenzler, P. J. Mulholland, and R. P. Sniffen. 1984. Effects of stream channelization on exports of nitrogen and phosphorus from North Carolina Coastal Plain watersheds. *Environ. Manage.* 8:151–160.

6 Medium–Low-Gradient Streams of the Gulf Coastal Plain

JAMES D. FELLEY

Department of Biological and Environmental Sciences, McNeese State University, Lake Charles, LA 70609

PHYSICAL ENVIRONMENT

Geology

The geology traversed by the streams covered in this chapter is of relatively recent origin (Thornbury 1965). Medium-low gradient streams of the Gulf coastal plain primarily flow through Tertiary formations identified by Thornbury. These portions of the Gulf coastal plain were largely formed during the Oligocene. Large amounts of water, now locked up as ice, were then in liquid form (Vail and Hardenbol 1979), and oceans intruded almost to the fall line (a demarcation between the metamorphosed geologic regions of the interior— the present-day Piedmont—and sediments deposited by Cretaceous seas). When the Oligocene oceans retreated, they left behind the relatively flat expanses of clayey and sandy soils that now comprise most of the coastal plains. Murray (1961) has presented a compendium of coastal plain environments in North America. Riggs (1984) discusses post-Oligocene sea-level changes and their geological effects. Summaries of the geology of the region may also be found in Swift et al. (1986) and Conner and Suttkus (1986). Other than clays and sands deposited during the Oligocene, special geological features of the Gulf coastal plain include deposits of loess in western Mississippi, and uplifted coralline beds in peninsular Florida and in portions of the Florida panhandle (Bernard and LeBlanc 1965).

Subsequent changes in sea level have left their mark (Vail and Hardenbol 1979). Evidence of Pleistocene sea level changes can be found in some areas (Cooke 1939, Flint 1957, MacNeil 1949). At least three sets of Pleistocene terraces in Louisiana mark sea-level changes (Saucier and Fleetwood 1970).

James D. Felley's present address is the Office of Information Resource Management, Room 2310 A&I Building, Smithsonian Institution, 900 Jefferson Dr., S.W., Washington, DC 20560.

On the Florida peninsula, soils associated with contours less than 30 m are likely of marine Pleistocene origin. Opdyke et al. (1984) demonstrated that dissolution of limestone bedrock of the area has caused epeirogenic uplift of more than 30 m. Other soils of the Gulf coastal plain may be of Eolian origin (e.g., loess soils of Mississippi) (Flint 1957, Snowdon and Priddy 1968).

Poorly consolidated soils of Pleistocene or Pliocene origin were easily eroded (Grissinger et al. 1982, Holland 1944), and now even headwater streams have relatively low gradients (though see Beck 1973). Most medium-low gradient streams of the Gulf coastal plain flow over sand or sand/clay substrates. A few streams have gravel bottoms (e.g., Bayou Pierre in Mississippi) and at elevations less than 10 m, many streams in Louisiana have muddy bottoms. Streams in the Ochlockonee and Aucilla drainages have eroded their beds down to basement limestone and many are partially subterranean (Sellards 1917). Exposed limestone may be seen in the Chipola river drainage. Many streams in Florida limestone areas are spring-fed (Rosenau et al. 1977). Streams of the Florida panhandle west of the Apalachicola continue to erode their beds in response to uplift of the coastal plain in the region (Price and Whetstone 1977).

Climate

The Gulf coastal plain is affected by two general types of rainfall patterns: Weather fronts passing through in winter and spring bring most of the year's rain; in summer and fall, these fronts do not reach as far south as the coastal plain. In summer and fall, southerly winds bring rainfall in the form of thermal thundershowers [50–70 inches per year over a particular area (Geraghty et al. 1973)]. Hurricanes are unpredictable but important sources of rainfall in late summer and fall. Muller (1977) provided a detailed summary of weather patterns that affect the Gulf coastal area. Runoff tends to be highest in winter and spring, when rainfall is highest and temperature and evapotranspiration are lowest. Conversely, runoff is lowest in summer and fall (Geraghty et al. 1973, Gosselink et al. 1979). In the Florida peninsula and parts of the Florida panhandle, frontal rain events are much less important, and summer and fall thundershowers account for most of the yearly rain (Fernald and Patton 1984). Fernald and Patton (1984) and Geraghty et al. (1973) show that stream flow in these regions tends to be highest in summer and fall.

Hydrology

The streams considered here are categorized as warmwater streams, which tend to have low gradients, moderate to high discharges, low turbulence, and rubble-sand-mud substrates (Winger 1981). Average discharge rates for selected streams of the Gulf coastal plain are listed in Table 1. Variation in flow among these streams is most closely tied to drainage area ($r = .923$, $p > .001$).

TABLE 1   Drainage Areas and Yearly Averages of Physicochemical Variables for Selected Stream Systems of the Gulf Coastal Plain

Drainage	Area (km²)	Flow (m³/s)	Temperature (°C)	Conductivity (µS/cm)	Dissolved Oxygen (mg/L)	pH	Hardness (mg/L)	Phosphate (mg/L)
Calcasieu[a]	4 403	71.1	19.0	184	7.6	6.8	17.2	.08
Amite[a]	133	1.5	20.0	181	8.5	6.2	12.9	.10
Tickfaw[a]	247	6.0	18.4	219	8.4	6.4	13.7	.11
Tangipahoa[a]	1 673	24.0	19.4	188	8.4	6.2	13.1	.15
Perdido[b]	2 396	21.8	19.7	22	8.4	4.7	3.6	.10
Escambia[b]	10 878	185.1	20.3	79	8.1	6.4	6.6	.13
Blackwater[b]	2 227	30.0	17.8	25	9.0	5.0	6.6	—
Yellow[b]	3 626	65.0	18.2	54	8.3	6.3	20.7	.08
Choctawhatchee[b]	12 033	204.8	20.8	98	7.5	6.6	37.9	.10
Ochlockonee[b]	5 957	45.7	19.8	121	7.5	6.1	25.8	.64
Aucilla[b]	2 279	15.7	19.4	116	7.3	5.9	55.2	.17
S. Withlacoochee[b]	5 180	32.0	22.4	264	5.4	7.6	125.4	.12
Hillsborough[b]	1 787	16.8	22.8	287	5.8	7.1	138.8	2.30
Peace[b]	5 957	32.7	22.2	316	7.8	6.9	117.6	8.30

[a]Data from STORET computer library, made available by Louisiana Department of Environment Quality.
[b]Data from Bass (1984).

There is strong seasonal variation in discharge within streams. Evapotranspiration and rainfall in the two seasons described above produce a period of low flow from June through October and a period of high flow from November through May (Felley and Felley 1987, Finger and Stewart 1987, Ross et al. 1987). Lowest flows are in August, September, and October; the highest flows are in January, February, and March, except in regions of Florida where summer thunderstorms are important contributors to total yearly rainfall (Beck 1973, Fernald and Patton 1984, Geraghty et al. 1973). Ross and Baker (1983) and Ross et al. (1987) presented discharge data over a period of years for tributaries of the Pascagoula River. Figure 1 illustrates the yearly flow regime of the Calcasieu River at a point where it is a sixth-order stream. Data for these streams show that differences between highest and lowest flows may be quite dramatic, high flows being up to 10 times greater than low flows.

Within streams, there is also much spatial variation in discharge-related variables. This variation is tied to stream order (Strahler 1957). Data collected by myself and S. M. Felley in 1984 showed that in headwater streams of the Calcasieu drainage (stream orders 1–2), flow (as measured by current speed), and variation in flow were greatest in the wet season; flow in these small, runoff-fed streams was negligible during the dry season (Fig. 2). Moore (1970) found that after high rainfall, temporary ponds may form in uppermost reaches of such small streams. By contrast, streams of orders 3–5 in the Calcasieu drainage had more spatial variation in current speeds in the dry season when

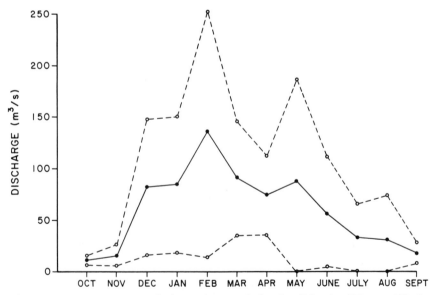

FIGURE 1. Average stream discharge by month for the Calcasieu River at Kinder, Louisiana (stream order 6). Raw data are monthly averages from 1975–1985 (U.S. Geological Survey stream gauging station). The solid line joins discharge means for the 10-year period; upper and lower dashed lines delimit one standard deviation.

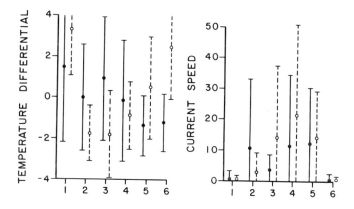

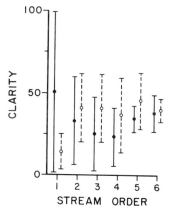

FIGURE 2. Means and one standard deviation of physical variables in streams of the Calcasieu drainage, Louisiana, by stream order. Solid lines and circles represent wet season (Dec–May) values; dashed lines and open circles represent dry season (June–Nov) values. Temperature differential (°C) is the difference between a location's temperature and the average drainage temperature in that season. Current speed was measured in cm/s, clarity as Secchi disk depth in cm. Sample sizes for particular stream orders (in the wet and dry season) were as follows: stream order 1 (7, 7); 2 (15, 7); 3 (14, 9); 4 (24, 21); 5 (8, 10); 6 (10, 7).

many of these reaches were characterized by pool–riffle sequences. In the wet season, high water obliterated pools and riffles, and flow was relatively constant. Variation in water temperature (Fig. 2) also followed a similar pattern: Upstream areas were most variable in the wet season, and downstream areas were most variable in the dry season.

Suspended load is related to discharge, and differs among streams (Beck 1965) due to erosion of different soil types and to different agricultural practices in these drainages. Within a stream system, there is also temporal and spatial variation in suspended load. Figure 2 illustrates variation in clarity in streams of different orders within the Calcasieu drainage. Table 2 illustrates the high variability of suspended load (turbidity) in five Louisiana streams during both wet and dry seasons. Most coefficients of variation were greater than 100%. These examples indicate that in Gulf coastal streams, suspended load is highest and most variable in the wet season (due to high and periodic rainfall and runoff).

Thus, spatial and temporal variability in hydrological variables is related to patterns of rainfall and evapotranspiration. In most systems, flow is low in late summer and fall, when upstream areas dry up or become pool–riffle sequences. Temperatures increase downstream. In the wet season, suspended load is high, temperatures are higher upstream (as a consequence of runoff), and pool–riffle sequences are obliterated by high water.

Chemicals

Streams of the Gulf coastal plain tend to be extremely low in dissolved substances, reflecting their watershed geochemistries (Shoup 1947). Averages for a number of streams in Florida and Louisiana (Table 1) demonstrate the low pH, conductivity, hardness, and nutrient levels of these streams. Exceptions are streams of the Florida peninsula that drain limestone deposits high in phosphate. These streams tend to have high levels of pH, dissolved substances, and nutrients. Oxygen levels in streams of the Gulf coastal plain tend to be relatively high, normally not dropping below 70% saturation all year long. Oxygen depletion may occur in low-order streams during low flow periods, in streams receiving municipal/industrial effluents, in the freshwater portions of tidal streams (Felley 1987), and in spring-fed streams (Rosenau et al. 1977). Data on physicochemistry may be found in Grady et al. (1983) for Bayou Sara, Louisiana; Beck (1973) for the Blackwater river system of Florida; and Bass and Cox (1985) for Florida streams.

Seasonal variation was apparent in water chemistry of Louisiana streams (Table 2). Conductivity and pH levels were lower in the wet season, higher in the dry season. Nitrate levels were higher during the wet season, presumably due to high runoff and low assimilation by primary producers during the cooler winter season. Lower nitrate levels during the dry season likely reflected high primary productivity that removed nitrates from the water column. By contrast, phosphate in these streams showed no seasonal variation and was found in low amounts. Levels of these two nutrients were highly variable in both seasons (Table 2).

Variation within streams of different orders in the Calcasieu drainage indicated that small streams (orders 1–3) were more variable in pH and oxygen saturation levels than larger streams in the same drainage. This variation was more pronounced in the dry season than in the wet season (Fig. 3). Conductivity was highly variable between Louisiana streams, but no pattern of spatial variation was evident within the Calcasieu river system (Fig. 3).

TABLE 2 Seasonal Means and Coefficients of Variation of Physicochemical Variables in Five Louisiana Rivers

River	Season	pH	Temperature	O$_2$ (%)	Turbidity	Conductivity	Nitrate	Phosphate
Calcasieu	Wet	6.4 (5.9)	12.8 (42.0)	75 (20.0)	26.5 (133.3)	78 (151.6)	.10 (100.0)	.05 (55.8)
	Dry	6.6 (6.8)	22.3 (19.1)	72 (15.9)	19.8 (61.7)	121 (151.4)	.10 (71.1)	.06 (90.6)
Amite	Wet	6.4 (7.6)	14.0 (40.0)	88 (15.0)	37.0 (163.3)	66 (139.0)	.23 (110.4)	.09 (81.1)
	Dry	6.6 (5.7)	22.4 (20.6)	88 (14.4)	13.1 (81.0)	65 (94.5)	.14 (59.0)	.07 (110.0)
Tickfaw	Wet	6.3 (9.9)	14.2 (38.5)	88 (17.4)	25.5 (70.5)	62 (90.6)	.30 (52.1)	.13 (85.8)
	Dry	6.6 (9.1)	23.0 (23.0)	88 (16.3)	13.5 (61.0)	72 (75.7)	.17 (58.4)	.09 (62.8)
Tangipahoa	Wet	6.5 (5.2)	15.8 (32.0)	97 (12.3)	32.4 (117.7)	67 (67.9)	.35 (30.8)	.13 (91.0)
	Dry	6.6 (4.9)	23.3 (18.6)	98 (13.2)	14.4 (195.3)	66 (38.9)	.29 (89.5)	.12 (144.4)
Tchefuncte	Wet	6.3 (7.7)	14.3 (36.9)	88 (15.0)	23.7 (175.6)	65 (102.2)	.31 (55.2)	.10 (53.8)
	Dry	6.4 (7.6)	21.6 (20.8)	88 (13.8)	22.7 (210.0)	79 (116.0)	.18 (62.2)	.10 (64.6)

Note. The wet season is from December through May, the dry season from June through November. Dissolved oxygen is expressed as percent saturation, turbidity in JTUs. Other units are as in Table 1.

Source. From the Louisiana Department of Water Quality records (1975–1985) (Louisiana Department of Environmental Quality, 1984).

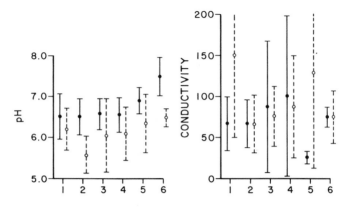

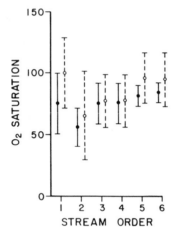

FIGURE 3. Means and one standard deviation of chemical variables in streams of the Calcasieu drainage, Louisiana, by stream order. Solid lines and circles represent wet season (Dec–May) values; dashed lines and open circles represents dry season (June–Nov) values. Conductivity was measured as μS/cm, oxygen saturation as a percentage. Sample sizes for stream orders (in the wet and dry seasons) were as follows: stream order 1 (7, 7); 2 (15, 7); 3 (14, 9); 4 (24, 21); 5 (8, 10); 6 (10, 7).

Drainages and Watershed Characteristics

There are recognized physicochemical differences among Gulf coastal drainages. Beck (1965) divided Florida waterways (including drainages covered here) into a number of categories, according to stream size and chemistry. Beck grouped the Florida streams covered in this chapter into the following categories: *sand-bottomed streams* (most of the streams considered here), *calcareous streams* (Aucilla, Peace, Hillsborough), and *larger rivers* (Escambia and Choctawhatchee). These streams are most easily divisible by their sizes and by their chemistries. Bass (1984) used Beck's (1965) scheme to characterize Florida waterways and gave drainage means for various physical and chemical variables. Bass and Cox (1985) discuss some other categorizations of Florida streams.

Beck's scheme does not adequately characterize streams of alluvial areas below 10 m elevation. These streams are often muddy-bottomed with banks lined by bald cypress (*Taxodium distichum*) and water tupelo (*Nyssa aquatica*). These trees form "living levees" which retard bank erosion. As a consequence, stream beds tend to be deep and steep-sided. Examples include the streams flowing into the downstream reaches of the Calcasieu river and estuary (Felley and Felley 1987, Felley 1987).

I identified patterns of environmental differences among drainages across the Gulf coastal plain, using principal components analysis. Data for this analysis were drainage means [from Bass (1984) for Florida streams; from Louisiana Department of Water Quality for five Louisiana streams]. Three separate locations in the Calcasieu drainage were also included.

A major trend differentiating Gulf-coastal streams relates to water chemistry (Fig. 4). The streams most differentiated from the others on principal component 1 were streams of the Florida peninsula that drain calcareous

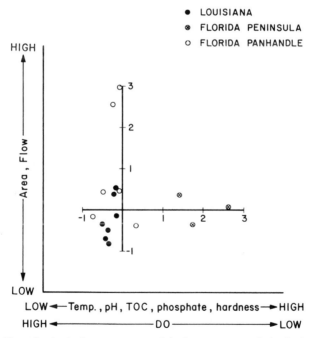

FIGURE 4. Plot of principal components of drainage means of physical and chemical variables. The *x* axis represents principal component 1 (relating to between-stream differences in temperature, pH, total organic carbon, phosphate and hardness). Principal component 2 (*y* axis) identifies drainage differences on the basis of drainage area and discharge.

deposits high in phosphates (Hillsborough, Peace, southern Withlacoochee). This division corresponds to a division between all other streams and the calcareous stream category of Beck (1965). Calcareous streams have high pH and high alkalinity, and tend to be quite productive. The other streams analyzed share similar chemical properties, being low in pH, dissolved solids, and so forth. Their chemical makeups are also quite similar to those of streams in the Atlantic coastal plain (Beck et al. 1974) (see Chapter 8).

Another trend differentiating streams of the Gulf coastal plain related to their drainage areas, and thus their average yearly flows. On principal component 2, the Escambia and Choctawhatchee systems are clearly separated (scores >2) from the others included in this study. In their downstream reaches, these two systems represent Beck's large rivers. Finally, smaller streams with low pH and dissolved solids represent Beck's sand-bottomed streams. These streams have scores between −1.0 and 1.0 on both principal components 1 and 2.

For much of the Gulf coastal plain, the typical stream seems to be a sand-bottomed stream. A few streams are markedly different on the basis of water chemistry or drainage size. There are some differences between the principal components results and Beck's (1965) classification. For example, Beck considered the southern Withlacoochee River to be a sand-bottomed stream, while Fig. 4 indicates that it is a calcareous stream. Despite minor differences such as this, a principal components analysis of stream means for environmental variables demonstrated that three of the stream-type categories identified by Beck can be used to classify the systems included in this analysis.

PLANT COMMUNITIES

Submergent Plants

Submergent plants are not much in evidence in most streams of the Gulf coastal plain, especially in upstream reaches. Much of the primary production that feeds these streams occurs in the riparian forests or swamp areas bordering the streams (see below). De la Cruz and Post (1977) found that gross primary production of a second order stream was from 4 to 28 g O_2/m^2 day^{-1}, which accounted for only one-fourth to one-half of the respiration in the stream. Primary production within the stream may be more important in downstream areas (Vannote et al. 1980).

Occurrence of submerged plants seems related to stream order. Moore (1970) found that temporary ponds (of the type found in headwaters) harbored large populations of *Eudorina* sp. (Volvocaceae), desmids (*Xanthidium* sp. and *Staurastrum* sp.), as well as *Spirogyra* sp. Organic detritus was thick in such pools. In the Calcasieu drainage, producers in low-order streams include algae and periphyton growing upon dead branches and snags in the water, as well as localized mats of diatoms in shallow, still areas (J. Felley personal observation). O'Quinn and Sullivan (1983) found encrusting diatoms to be an important community of primary producers in a small, calcareous Mississippi stream. Other unicellular producers have been found to be rare in running waters. In those environments covered in this chapter, Pecora (1973) found phytoflagellates only in canals, sloughs, and larger, slow-moving streams. Prescott (1942) noted that flowing waters of forested regions were poor habitats for algae. Table 3 lists dominant unicellular forms.

Table 3 also lists the vascular submerged species found in selected streams of the Gulf coastal plain. See Godfrey and Wooten (1979, 1981) for details of these species. Vegetated areas are not common in most of these streams, perhaps due to high variability in flow and turbidity, and to the presence of unstable sandy bottoms. In the Calcasieu drainage, vascular submergents tend to be abundant only in the lower reaches (stream orders 5–6) and in canals and oxbows. These are slow-flowing areas where clay or silt bottoms dominate. In southern Mississippi, *Sparganium americanum* may be common in some areas (S. T. Ross, personal communication). In Florida drainages surveyed by Bass and Hitt (1977, 1978) and Bass et al. (1979, 1980), vegetated areas were also sparse, accounting for only 5–6% of their benthic sample locations. In a small stream of the Calcasieu drainage, Carver (1975) found no submergent vascular plants at any of his benthic sampling locations. By contrast, sampling site descriptions from small tributaries of the Hillsborough and Peace rivers show that submergent vegetation may be quite dense in these drainages (Cowell and Carew 1976, Robertson and Piwowar 1985). In the environments found in Gulf coastal streams, primary producers may be mostly limited to substrates presented by snags, and to quiet water situations found in downstream areas, sloughs, and canals.

Emergent Plants

In the Gulf coastal plain, the important producers contributing to medium-low gradient streams are terrestrial plants surrounding headwaters (Gemborys and Hodgkins 1971, Finger and Stewart 1987), and plants growing in floodplains. Allochthonous inputs of fixed carbon from such communities have been studied by de la Cruz and Post (1977) and Post and de la Cruz (1977). Wharton et al. (1982) included an account of emergent plant communities of southeastern streams and floodplains, and Wooten (1986) examined edaphic features important to species of *Sagittaria*. In the Calcasieu drainage, some type of emergent vegetation was present at more than 30% of Felley and Felley's (1987) sampling locations. Table 3 lists important taxa found by various researchers. Godfrey and Wooten (1979, 1981) describe many of these species. Many species of emergent plants can grow or survive in flooded conditions as well as on saturated or drying soil. Thus, these plants are more resistant to the water level and flow fluctuations that characterize medium-low gradient streams of the Gulf coastal plain. Tree species such as bald cypress (*Taxodium distichum*) and water tupelo (*Nyssa aquatica*) grow along the edges of low gradient streams in areas which may be dry for only a short time each year, or only intermittently from year to year (Wharton et al. 1982). As mentioned above, these trees may stabilize the banks of such low-gradient streams.

TABLE 3 Dominant Macrophytes (Submergent and Emergent) and Unicellular Plants in Gulf Coastal Plain Streams

River System	Macrophytes	
	Submergent	Emergent
Calcasieu[1]	Najas quadelupensis (water nymph) Ceratophyllum demersum (coontail) Cabomba caroliniana (cabomba) Utricularia vulgaris (bladderwort)	Taxodium distichum (baldcypress) Cladium jamaicense (sawgrass) Pontedoria cordata (pickerelweed) Alternanthera philoxeroides (alligatorweed) Brasenia schreberi (watershield) Nymphea odorata (white water lily) Cephalanthus occidentalis (buttonbush) Eichornia crassipes (water hyacinth)
Blackwater[2]	Potamogeton sp. (pondweed) Vallisneria americana (eelgrass) Mayaca fluviatilis (bogmoss) Bacopa monnieri (water hyssop) Utricularia vulgaris (bladderwort)	Sagittaria sp. (arrowhead) Cladium jamaicense (sawgrass) Rynchospora sp. (spike rush) Orontium aquaticum (golden club) Xyris sp. (yellow-eyed grass) Pontedoria cordata (pickerelweed) Juncus repens (rush) Nuphar ulvacea (black nuphar) Nymphea sp. (water lily) Hypericum fasciculatum (St. John's wort)
Black Creek (MS)[3]		Sparganium americanum (burr-reed)
Yellow River[4]	Vallisneria sp. (eelgrass)	Typha sp. (cattail) Cladium jamaicense (sawgrass)

Location	Taxa
Hilsborough,[5,6]	*Hypericum americanum* (St. John's wort) *Orontium* sp. (golden club) *Nuphar ulvacea* (black nuphar) *Vallisneria neotropicalis* (eelgrass) *Potamogeton illinoiensis* (pondweed) *Egeria densa* (waterweed) *Hydrilla verticillata* (hydrilla) *Eichoria crassipes* (water hyacinth)
Peace[7]	*Sagittaria latifolia* (arrowhead) *Hydrochloa carolinensis* (hydrochloa)

Unicellular forms

Location	Taxa
Small Creek in MS[8]	*Achnanthes minutissima* (diatom) *Cymbella turgida* (diatom) *Epithemia sorex* (diatom) *Gomphonema angustatum* (diatom) *Navicula cryptocephala* (diatom) *N. menisculus* (diatom) *N. minima* (diatom) *Nitzschia dissipata* (diatom) *N. palea* (diatom)
Small Creek in MS[8]	
Various LA drainages[9]	*Chlamydomonas pertusa* (phytoflagellate) *Euglena proxima* (phytoflagellate) *Trachelomonas gibberosa* (phytoflagellate)

References. 1, J. Felley (personal observation); 2, Bass and Hitt (1977); 3, Ross and Baker (1983); 4, Bass et al. (1979); 5, Cowell and Carew (1976); 6, Beck and Cowell (1976); 7, Robertson and Piwowar (1985); 8, O'Quinn and Sullivan (1983); 9, Pecora (1973).

ANIMAL COMMUNITIES

Invertebrates

The Gulf coastal plain is a region of high diversity for aquatic invertebrates, and biogeographic history accounts for much variation in invertebrate faunas among drainages of the Gulf coastal plain. Berner (1950) tabulated the distributions of mayfly genera in the southeastern United States, showing that the coastal plain has more genera than any other physiographic region of the southeastern United States. Beck (1980) discussed the biogeography of chironomids in the southeast. Barr and Chapin (1988) characterized Louisiana water beetles phylogenetically and biogeographically. Penn (1959) listed distributions and ecological characteristics of Louisiana crayfish species. Franz and Lee (1982) discussed the crayfishes found in underground streams of Florida's calcareous regions.

Within a stream system, invertebrate community characteristics change from headwater to downstream environments, as predicted by Vannote et al. (1980). Table 4 lists the dominant taxa in different habitats of small, medium-sized, and larger streams. In extreme headwaters, invertebrates are abundant in temporary ponds and include rotifers, copepods (primarily *Diaptomus* spp.), and cladocerans (primarily *Ceriodaphnia quadrangulata*) (Moore 1970). Larger arthropods are also found, including amphipods, isopods, odonates, and culicids. White (1985) discussed the invertebrate community of an intermittently flooded wetland area. The trophic base of these areas is mostly detritus.

In all habitats of permanent streams, oligochaetes and chironomids tend to be the dominant taxa (Bass and Hitt 1977, 1978, Bass et al. 1979, 1980, Carver 1975). Ephemeropterans, ceratopogonids and gastropods are also abundant. D. G. Bass, Jr. (personal communication) felt that stream invertebrate sampling techniques are biased against crayfish species, which may be important contributors to invertebrate biomass in these streams. The importance of other invertebrate taxa differs among habitats in medium-low gradient streams of the Gulf coastal plain.

In sand-bottomed streams (i.e., most upstream areas and small streams), riffle beetles (Elmidae) and trichopterans are abundant, and their importance decreases downstream. These forms are most often associated with snags and woody debris. Beck (1965) found that elmids were generally restricted to sand-bottomed streams. He also found that the ephemeropteran genus *Stenonema* and the trichopteran genus *Cheumatopsyche* were important in such streams. Peters and Jones (1973) described in some detail the invertebrate fauna of the Blackwater River, as well as the habitat preferences of particular species. They identified the nymphs of two mayflies (*Dolania americana* and *Homoeoneuria dolani*) as being specially adapted to living in the shifting substrates of sand-bottomed streams. The natural history of *D. americana* was further described by Peters and Peters (1977) and Tsui and Hubbard (1979). Berner (1950) gave ecological descriptions of various Florida ephemeropteran species. He found only nymphs of *Callibaetis floridanus* and *Caenis*

TABLE 4 Relative Abundances of Benthic Invertebrate Taxa in Various Stream Sizes and Habitats

Taxon	Feeding Group	Stream Size			Bottom Type			
		Large	Medium	Small	Sand	Sand/Litter	Mud/Litter	Vegetation
Oligochaeta	Substrate, debris feeding	+++	+++	+++	++	+++	+++	++
Isopoda	Scavenger, debris feeding	++	+	+++		+	++	+++
Amphipoda	Omnivorous	++	+		+	+	++	++
Hydracarina	Piercer	+	+			+	++	+
Ephemeroptera	Collector–gatherer, scraper	++	++	++	++	++	++	+++
Anisoptera	Engulfer	+	+		+	+	+	+
Megaloptera	Engulfer	+	+		+	+	+	
Trichoptera	Shredder, collector–gatherer	+	++	++	+	+	++	++
Coleoptera (Elmidae)	Collector–gatherer, scraper	+	+	++	+	+	++	+
Chaoborinae	Piercer	++					++	
Chironomidae	Collector–gatherer, piercer	+++	+++	+++	+++	+++	+++	+++
Ceratopogonidae	Engulfer	++	++	++	++	++	++	++
Gastropoda	Scraper	++	++	++	+	++	++	++
Pelecypoda	Filterer	++	++		++	++	++	++

Note. + + +, abundant; + +, common; +, occasional.

Source. From data in Bass and Hitt (1977, 1978), Bass et al. (1979, 1980), and Carver (1975). Feeding groups are according to Merritt and Cummins (1978) and Pennak (1978).

diminuta in intermittent headwater streams. These two species were ubiquitous in the Florida drainages considered here, as were *Stenonema smithae* and *Hexagenia munda marilandica*. Mayfly nymphs characteristic of sand-bottomed streams with little vegetation included *Blasturus intermedius*, *Paraleptophlebia bradleyi*, and *Habrophlebiodes brunneipinnis*. Ephemeroptera characteristic of streams with vegetation were *Siphlopectron speciosum*, *Ephemerella trilineata*, *Acentrella ephippiata*, and *Baetis spinosus*.

Forms more characteristic of downstream reaches (Beck's larger rivers) include isopods, amphipods, phantom midge larvae (Chaoborinae), and pelecypods (Bass and Hitt 1978, Bass et al. 1980). Beck (1965) identified only the odonate genus *Argia* as restricted to larger rivers. Among the ephemeropterans, Berner (1950) found (in addition to the ubiquitous species listed above) *Stenonema exiguum* and *S. proximum* in larger rivers. Beck and Cowell (1976) reported on life history aspects of a freshwater shrimp (*Palaemonetes paludosus*) common in Florida rivers. The dominant pelecypod species of Florida streams was the introduced Asiatic *Corbicula* sp. (Bass and Hitt 1977, 1978, Bass et al. 1979, 1980). Beck (1965) found that the only taxa seemingly restricted to calcareous streams were the gastropod genera *Goniobasis* and *Campeloma*.

Average seasonal invertebrate biomass varies between drainages. Data in Bass and Hitt (1977, 1978) and Bass et al. (1979, 1980) showed differences in benthic invertebrate production of four Florida drainages (Escambia, Blackwater, Yellow, and Choctawhatchee). Averaged over all seasons and all habitat types, the Blackwater drainage produced 0.6 g/m^2, Escambia 3.4 g/m^2, Yellow 9.4 g/m^2, and Choctawhatchee 39.2 g/m^2 (wet weight in all cases).

There is also substantial variation in invertebrate productivity within each drainage, associated with different habitat types. The most productive habitats are those with vegetation or find sand/mud substrates with detritus (producing 50–60 g/m^2, wet weight). Benke et al. (1984) also found this in a blackwater river of the Atlantic coastal plain. These habitats have slow currents and a large trophic base (both detritus and primary producers). Thus, such habitats are open to invertebrate species lacking special adaptations to current, and to species from a variety of feeding groups (Merritt and Cummins 1978). Gregg and Rose (1982) showed that macrophytes interact with the environment, providing more area for periphyton growth, and creating microhabitats with low current speeds, finer substrates, and more detritus.

Sand substrates with litter were less productive (21.0 g/m^2) and had fewer species (Bass and Hitt 1977, 1978, Bass et al. 1979, 1980). Bare sand substrates had the least biomass (9.0 g/m^2) and the fewest individuals of invertebrate taxa, in the four Florida drainages. Clean sand is normally quite low in benthos (Bass and Cox 1985). The sand and sand/litter habitats are those most typical of small streams of the Gulf coastal plain from the Florida panhandle to Louisiana.

Snags and woody debris represent a relatively poorly studied habitat type that accounts for much invertebrate production in sand-bottomed streams. Benke et al. (1984) demonstrated that the invertebrate community living on snags was extremely important in terms of total invertebrate production in a southeastern blackwater stream. However, snag habitat was not sampled or quantified in the studies on the four Florida drainages or in Carver's (1975) study. This habitat deserves intensive scrutiny, because Benke et al. (1985) have indicated that it is of great importance to higher level consumers (fishes). Other studies have documented the high invertebrate production associated with snag habitat [Ager et al. (1985) in the Apalachicola River, Smock et al. (1985) and Thorp et al. (1985) in Atlantic coastal drainages].

Invertebrate trophic groupings (Merritt and Cummins 1978) varied spatially, reflecting stream order position (Table 4). Upstream communities are dominated by collector/gatherers and scrapers. These groupings are also important downstream, but predators (piercers, engulfers) are also more in evidence.

Data in Bass and Hitt (1977, 1978), Bass et al. (1978, 1980), and Carver (1975) suggest that invertebrate biomass varies seasonally in Gulf coastal streams. Small and medium-sized streams (orders 1–4) have biomass minima in summer months, while larger streams (orders 5–6) have peaks of biomass in summer. This may reflect the different trophic bases of these stream sizes. The important resource base in small and medium-sized Gulf coastal streams is detritus entering streams, mostly in fall, winter, and spring (de la Cruz and Post 1977, Post and de la Cruz 1977). Consumers in downstream areas may be more dependent on primary producers within the stream itself (Vannote et al. 1980). The growing season for these producers would be spring and summer.

Invertebrate taxa other than oligochaetes and arthropods may also be important in Gulf Coast drainages. Everitt (1975) found several species of ectoprocts to be abundant in small acidic streams of Louisiana, usually found in association with macrophytes, which the ectoprocts used as attachment sites. Moore (1953) and Poirrier (1969) reported on sponges of Louisiana streams. They found *Spongilla lacustris*, *S. fragilis*, and *Heteromeyenia ryderi* in sand-bottomed streams.

In summary, invertebrate productivity varies spatially, being higher in areas with vegetation and detritus, lower in areas with clean sand bottoms. Invertebrate communities on woody snags may account for a large portion of stream invertebrate production. Sand-bottomed streams may have production minima in the low flow season (perhaps due to a lesser input of detritus in this season). Important taxa in sand-bottomed streams include elmids, chironomids, ephemeropterans, and oligochaetes. Larger rivers (with more autochthonous primary production) tend to have peaks of invertebrate production in the low flow season of summer and fall. Important taxa are all those of the sand-bottomed streams (except for elmids), as well as a number of predators. Perhaps higher productivity of downstream areas allows larger predator populations than are found in environments upstream.

Vertebrates

Fishes   Variation among stream fish faunas is tied to the geologic and zoogeographic histories of these streams. Conner and Suttkus (1986), Gilbert

(1987), and Swift et al. (1986) presented detailed analyses of fish zoogeography in the waterways considered here. Table 5 lists the abundant fish species of Gulf coastal streams. Included are the freshwater forms; estuarine and marine species may penetrate downstream areas, especially in low water seasons (Felley 1987). Lee et al. (1980) provide summaries and references for taxonomies and natural histories of all these species. In general, streams west of the Suwannee drainage are primarily inhabited by minnows (Cyprinidae), sunfishes (Centrarchidae), and darters (Percidae), as well as suckers (Catostomidae). Predatory fish are not important numerically in most cases, but do comprise an important amount of the biomass (Bass and Hitt 1977, 1978, Bass et al. 1979, 1980, Bass and Cox 1985). The most important predatory species are spotted gar (*Lepisosteus oculatus*), bowfin (*Amia calva*), and largemouth and spotted bass (*Micropterus salmoides* and *M. punctulatus*). Streams of the Florida peninsula are largely lacking in minnows, suckers, and darters. For example, the ichthyofauna of the Peace, Hillsborough, and southern Withlacoochee rivers are dominated by sunfishes (Bass 1984). Gilbert (1987) presented a detailed discussion of zoogeographic patterns of peninsular Florida drainages.

There are relatively few endemic freshwater fish species limited to the medium-low gradient streams of the Gulf coastal plain. Many of the important species listed in Table 5 are quite widely distributed. Species limited to particular drainages covered by this chapter include a number of different shiners (*Notropis*), topminnows (*Fundulus*), and darters (*Etheostoma*). In general, these genera are considered to be ones that speciate readily, and many of the species are limited to only a few drainages.

There have been a number of investigations on the natural history and reproductive biology of some of the characteristic fishes of small Gulf coastal streams. These species include the longnose shiner *Notropis longirostris* (Heins and Clemmer 1975, 1976), Sabine shiner *N. sabinae* (Heins 1981), weed shiner *N. texanus* (Bresnick and Heins 1977, Heins and Davis 1984, Heins and Rabito 1988), blacktail shiner *N. venustus* (Heins and Baker 1987), naked sand darter *Ammocrypta beani* (Heins and Rooks 1984), Florida sand darter *A. bifascia* (Heins 1985), gulf darter *Etheostoma swaini* (Ruple et al. 1984), and saddleback darter *Percina vigil* (Heins and Baker 1989). Most of these species have a protracted breeding season when many egg clutches are produced, lasting from late spring through early fall (the season of lowest stream flow for most of the drainages in which these species live). Heins and Clemmer (1976) felt that such a long spawning season is adaptive in a highly variable environment, since it assures that no overwhelming fraction of the propagules will be destroyed by some dry period or sudden spate. Heins and Baker (1987) and Heins and Rabito (1986) further discussed life history strategies of particular species in these variable environments. They concluded that the high variability of Gulf coastal streams translates into strong selective pressures that affect life history traits in the species they studied.

Baker and Ross (1981), Ross et al. (1987), and Felley and Felley (1987) examined habitat use of fishes in Gulf coastal streams. Species in these as-

TABLE 5 Fish Species Characteristic of Small to Medium-Sized Streams of the Gulf Coastal Plain

Species	Stream Order		Current		Cover, Vegetation	
	Low	High	None	Some	Much	Little
Forms Found Throughout the Area Under Consideration						
Amia calva (bowfin)		+	+		+	
Anguilla rostrata (American eel)		+	+		+	
Esox americanus (grass pickerel)	+	+	+	+	+	
E. niger (chain pickerel)	+	+	+		+	
Notemigonus crysoleucas (golden shiner)	+	+	+		+	
Notropis chalybaeus (ironcolor shiner)	+		+	+	+	+
N. emiliae (pugnose minnow)	+	+	+	+	+	
N. maculatus (taillight shiner)	+	+	+		+	
Aphredoderus sayanus (pirate perch)	+	+	+		+	
Fundulus chrysotus (golden topminnow)	+	+	+		+	
Gambusia affinis (mosquitofish)	+	+	+		+	
Labidesthes sicculus (brook silverside)	+	+	+		+	+
Elassoma zonatum (banded pygmy sunfish)	+	+	+		+	
Lepomis gulosus (warmouth)	+	+	+		+	
L. macrochirus (bluegill)	+	+	+		+	
L. microlophus (redear sunfish)	+	+	+		+	
L. punctatus (spotted sunfish)	+	+	+		+	
Micropterus salmoides (largemouth bass)	+	+	+		+	+
Etheostoma fusiforme (swamp darter)		+	+	+	+	
Forms Widely Distributed, but Absent from Florida Peninsula						
Ichthyomyzon castaneus (chestnut lamprey)	+	+	+	+	+	+
I. gagei (southern brook lamprey)	+	+	+	+	+	+
Lepisosteus oculatus (spotted gar)		+		+	+	
Notropis texanus (weed shiner)	+	+		+	+	+

TABLE 5 (Continued)

Species	Stream Order		Current			Cover, Vegetation	
	Low	High	None	Some		Much	Little
N. venustus (blacktail shiner)	+	+		+			+
Pimephales vigilax (bullhead minnow)	+	+	+	+			+
Minytrema melanops (spotted sucker)	+	+	+			+	+
Moxostoma poecilurum (blacktail redhorse)	+	+	+	+		+	+
Fundulus notatus (blackstripe topminnow)	+	+	+			+	
F. olivaceus (blackspotted topminnow)	+	+	+	+		+	
Lepomis megalotis (longear sunfish)	+	+	+	+		+	+
Micropterus punctulatus (spotted bass)	+	+		+		+	+
Pomoxis annularis (white crappie)	+	+	+			+	

Forms Found in Various Coastal Drainages from Florida to East Louisiana

Species	Stream Order		Current			Cover, Vegetation	
	Low	High	None	Some		Much	Little
Ericymba buccata (silverjaw minnow)	+		+	+			+
Notropis hypselopterus (sailfin shiner)	+	+	+	+		+	
N. longirostris (longnose shiner)	+		+	+			+
N. roseipinnis (cherryfin shiner)	+		+	+			+
N. signipinnis (flagfin shiner)	+		+	+		+	+
N. welaka (bluenose shiner)		+	+			+	
Erimyzon tenuis (sharpfin chubsucker)	+	+	+	+		+	
Noturus leptacanthus (speckled madtom)	+	+		+		+	+
Fundulus euryzonus (broadstripe topminnow)	+		+	+		+	
F. notti (starhead topminnow)	+	+	+	+		+	
Ambloplites ariommus (shadow bass)	+	+		+		+	+
Ammocrypta beani (naked sand darter)	+	+		+			+
A. bifascia (Florida sand darter)	+	+	+	+			+
Etheostoma davisoni (Choctawhatchee darter)	+	+		+		+	
E. edwini (brown darter)	+	+		+		+	
Percina nigrofasciata (blackbanded darter)	+	+		+		+	

Forms Found in Western Louisiana and in Mississippi River Tributaries

Species						
Hybognathus nuchalis (central silvery minnow)		+			+	+
Notropis fumeus (ribbon shiner)		+			+	+
N. sabinae (Sabine shiner)	+	+		+	+	
N. umbratilis (redfin shiner)	+	+	+	+	+	
N. volucellus (mimic shiner)			+	+	+	
Ammocrypta vivax (scaly sand darter)	+	+		+	+	
Etheostoma chlorosomum (bluntnose darter)	+	+	+	+	+	
E. gracile (slough darter)	+	+	+	+	+	
E. histrio (harlequin darter)	+	+	+	+	+	
E. proeliare (cypress darter)	+	+		+	+	
E. swaini (Gulf darter)	+	+	+	+	+	
Percina sciera (dusky darter)	+	+	+	+	+	

Forms Found Only in Florida (Peninsula and Panhandle)

Species						
Lepisosteus platyrhincus (Florida gar)	+	+	+	+	
Fundulus escambiae (eastern starhead topminnow)	+	+	+	+	
F. lineolatus (lined topminnow)	+	+	+	+	
Leptolucania ommata (pygmy killifish)	+	+		+	
Acantharchus pomotis (mud sunfish)	+	+		+	
Elassoma evergladei (Everglades pygmy sunfish)	+	+		+	
E. okefenokee (Okefenokee pygmy sunfish)	+	+		+	
Enneacanthus chaetodon (blackbanded sunfish)	+	+	+	+	
E. gloriosus (bluespotted sunfish)	+	+		+	
E. obesus (banded sunfish)	+	+		+	
Lepomis auritus (redbreast sunfish)	+	+		+	+
Micropterus notius (Suwannee bass)	+	+		+	

Note. This is not an exhaustive list. Species are divided according to geographic distribution within the Gulf coastal plain, and by ecological preferences for stream order (low order = headwaters, high = downstream), current condition ("some" current corresponds to riffle habitat, there are essentially no "fast" current habitats within these streams), and amount of cover and/or vegetation.

Source. Information on particular species is from Baker and Ross (1981), Felley and Felley (1987), Lee *et al.* (1980), Page (1983), Pflieger (1975), and Ross et al. (1987).

253

semblages tend to be ecologically differentiated in terms of their uses of different stream orders, current speeds, and amounts of debris, cover, and vegetation. Table 5 categorizes fish species according to their preferences for differing states of these three environmental variables. Baker and Ross (1981) also found that water column position tended to differentiate cyprinids of a small coastal plain stream. Habitat use by species is relatively constant from year to year (Ross et al. 1987).

There may be large seasonal differences in habitat use by species. Felley and Felley (1987) showed that in the dry season, fish species tended to be more restricted in their use of habitats, and tended to depend more heavily on detritus as a food source. In the wet season, most species were found in a wide range of different habitats, and tended to feed more on invertebrates. Ross and Baker (1983) showed that several species feed heavily in flooded areas during the wet season, and that amount of flooding in the wet season may partly determine reproductive success in the following dry season. Finger and Stewart (1987) documented the importance of floods for fishes of medium-low gradient streams of southeastern Missouri. For fishes in these streams, the wet season seems to be a "season of plenty," while the dry season is a lean season (Schoener 1982).

Stream physicochemistry is closely related to fish production in medium-low gradient streams of the Gulf coastal plain. Kautz (1981) predicted fish biomass in various Florida streams based the streams' conductivities, total dissolved solids, and total nitrogen values. Bass and Hitt (1977, 1978) and Bass et al. (1979, 1980) investigated fish productivity in streams of the Florida panhandle. Using electrofishing techniques, they found that these streams produced between 5 and 14 kg of fish per hour of electrofishing (a standard method of determining fish biomass). These productivity estimates correspond to the relative invertebrate productivities of these streams (see previous section). The Blackwater River, producing an average of 0.6 g benthic invertebrates/m^2, had an average of 5.12 kg of fish per hour of sampling. The Choctawhatchee River system had the highest average value for benthic invertebrates (39.2 g/m^2) and for fish biomass (13.9 kg/h).

Reptiles and Amphibians   Table 6 lists a number of reptiles and amphibians characteristic of medium-low gradient streams of the Gulf coastal plain, along with their ecological preferences for stream and river habitats (most often flowing water with little vegetation) or swamp–lake environments with abundant macrophytes (such environments are found in the floodplains and backwaters of these streams). Species found more often in streams include waterdogs (*Necturus* spp.), salamanders characteristic of small streams and springs (*Pseudotriton* spp., *Eurycea bislineata*), musk turtles (*Sternotherus* spp.), and map turtles (*Graptemys* spp.). A number of giant salamanders (*Siren* spp., *Amphiuma* spp.) are more likely to be found in vegetated backwaters and pools, as are most aquatic frog species (*Rana* spp.), mud turtles (*Kinosternon* spp.), and species of water snakes (*Nerodia* spp.). Several turtle species (*Pseu-*

TABLE 6 Amphibians and Reptiles Characteristic of Small to Medium-Sized Streams of the Gulf Coastal Plain

Species	Stream	Floodplain
Amphibians		
SALAMANDERS		
Gulf Coast waterdog (*Necturus beyeri*)*	+	
Alabama waterdog (*N. alabamensis*)	+	
Greater siren (*Siren lacertina*)		+
Lesser siren (*S. intermedia*)*		+
Dwarf siren (*Pseudobranchus striatus*)*		+
Two-toed amphiuma (*Amphiuma means*)		+
Three-toed amphiuma (*A. tridactlyum*)		+
Southern dusky salamander (*Desmognathus auriculatus*)	+	+
Mud salamander (*Pseudotriton montanus*)*	+	
Red salamander (*P. ruber*)	+	
Two-lined salamander (*Eurycea bislineata*)	+	+
Three-lined salamander (*E. longicauda*)	+	+
Dwarf salamander (*E. quadridigitata*)		+
Eastern newt (*Notopthalmus viridescens*)*		+
FROGS		
Cricket frog (*Acris gryllus*)*	+	+
Bullfrog (*Rana catesbiana*)		+
River frog (*R. hecksheri*)		+
Pig frog (*R. grylio*)		+
Bronze frog (*R. clamitans*)	+	+
Southern leopard frog (*R. sphenocephala*)		+
Reptiles		
ALLIGATORS AND CROCODILES		
American alligator (*Alligator mississippiensis*)	+	+
TURTLES		
Snapping turtle (*Chelydra serpentina*)*	+	+
Alligator snapping turtle (*Macroclemys temminckii*)	+	+
Stinkpot (*Sternotherus odoratus*)	+	+
Razorback musk turtle (*S. carinatus*)	+	+
Loggerhead mud turtle (*S. minor*)*	+	
Striped mud turtle (*Kinosternon baurii*)		+
Eastern mud turtle (*K. subrubrum*)*		+
False map turtle (*Graptemys pseudogeographica*)*	+	
Barbour's map turtle (*G. barbouri*)	+	
Yellow-blotched map turtle (*G. flavimaculata*)	+	
Mississippi map turtle (*G. kohni*)	+	
Black-knobbed map turtle (*G. nigrinoda*)*	+	
Ringed map turtle (*G. oculifera*)	+	
Alabama map turtle (*G. pulchra*)	+	
Slider (*Pseudemys scripta*)*	+	+

TABLE 6   (*Continued*)

Species	Stream	Floodplain
River cooter (*P. concinna*)*	+	+
Cooter (*P. floridana*)*	+	+
Florida softshell (*Trionyx ferox*)	+	+
Smooth softshell (*T. muticus*)*	+	+
Spiny softshell (*T. spinifer*)*	+	+
SNAKES		
Mud snake (*Farancia abacura*)*	+	+
Rainbow snake (*F. erytrogramma*)*	+	+
Green water snake (*Nerodia cyclopion*)*		+
Plainbelly water snake (*N. erythrogaster*)*	+	+
Southern water snake (*N. fasciata*)*	+	+
Diamondback water snake (*N. rhombifera*)		+
Brown water snake (*N. taxispilota*)	+	+
Graham's crayfish snake (*Regina grahami*)		+
Glossy crayfish snake (*R. rigida*)*		+
Cottonmouth (*Agkistrodon piscivorous*)*	+	+

Note. This is not an exhaustive list. The list does not include species that are found in these streams during only one period in their lives (e.g., terrestrial amphibians that may breed in these habitats). Forms are characterized ecologically, based on whether they are more likely to be found within the stream, or in floodplain pools, or both.
*Species with more than one subspecies in covered areas.

Source. Ecological information summarized from Cochran and Goin (1970), Conant (1975), Mount (1975); nomenclature follows Collins et al. (1982).

demys spp. and *Trionyx* spp.) may be found in either type of environment, as may mud snakes (*Farancia*) and some water snakes.

Most of these reptiles and amphibians are widely distributed and many are represented by several subspecies in the geographical area covered by this chapter. Auffenberg and Milstead (1965) discussed the effect of Pleistocene sea-level changes on speciation and zoogeography of reptiles in the southeastern United States. In particular, they hypothesized that the Gulf Coast served as a "corridor" of temperate climatic conditions that allowed exchange of species between the Florida peninsula and Texas. Only a few species are restricted to particular drainages (as are some species of fishes, for example). In particular, a number of map turtles (Barbour's, yellow-blotched, black-knobbed, Alabama) are confined to particular drainages, and the false map turtle has evolved into several subspecies confined to different drainages. Much like fishes, these map turtles are ecologically confined to rivers and streams, which may permit isolation and speciation.

Few researchers have concentrated on the ecology of reptiles or amphibians in Gulf coastal streams. Shively and Jackson (1985) investigated the ecology of the Sabine map turtle (*Graptemys ouachitensis sabinensis*) in Whisky Chitto

Creek, a tributary of the Calcasieu River. They found that this species is limited to areas with many snags, which provide sunning spots and substrate for growth of green algae, a food of this turtle.

REPRESENTATIVE AQUATIC COMMUNITIES

Medium-low gradient streams of the Gulf coastal plain share a number of physicochemical characteristics. All are warm and, except for extreme head-waters, water flows year-round. Stream beds typically consist of sand or sand and clay, and pH and dissolved solids tend to be quite low. In most such streams, oxygen levels are high, typically staying near 80% saturation. Low levels of dissolved oxygen may occur in spring-fed streams, since groundwater is characterized by low dissolved oxygen. Periodic low flow in tidally affected streams may result in transient low oxygen (Felley 1987).

Seasonal variability in these streams relates to flow regimes. All streams experience a dry and a wet season, and for most streams considered here, the wet season is winter and spring. During the wet period, flow is high and streams tend to flood, while flow drops off in summer and fall. This change in flow regime affects temperature, pH, concentrations of dissolved solids, and oxygen levels, all of which show seasonal differences reflecting seasonal flow patterns.

There is spatial variation in stream physicochemistry, tied to stream order. In general, and as predicted by Vannote et al. (1980), headwaters are more variable in flow and physicochemical variables. Downstream areas tend to integrate the effects of many headwater streams and are less variable.

The most important primary producers of these streams are the terrestrial and emergent plants of the stream edges and surrounding floodplains. Detritus generated by these communities is the principal food base of most streams. Primary production within the stream itself seems mostly tied to algae and aufwuchs that grow on snags. Submerged macrophytes are important in the most downstream portions of these drainages, in canals and floodplain ponds— all areas where current is low. Where they occur, submerged macrophytes create habitats rich in animal biomass and diversity. However, the shifting sand bottom and high flow variability of typical streams apparently limit the distribution of submerged macrophytes.

The importance of snags and woody debris has been demonstrated for aquatic invertebrates, fishes, and reptiles. Invertebrate diversity and biomass is high on snags (Ager et al. 1985, Benke et al. 1984). Fish depend on these invertebrates for food (Benke et al. 1985) and some species seem to prefer snags and cover for protection from predators. At least one reptile species (Sabine map turtle) is partially limited in its distribution by the occurrence of snags.

Seasonal changes in flow rates are reflected in the biology of particular fish species. Some forms take advantage of spring floods to forage in sur-rounding floodplains (Ross and Baker 1983, Finger and Stewart 1987). In the dry season, species tend to be more limited in their distributions; Felley and Felley (1987) found that species were constrained in their distribution to "exclusive environments," perhaps areas where they were safest from predators or could avoid competitors. They also found that food seemed most limiting in the dry season.

There are important geographical differences from the typical ecosystem outlined above, which essentially represents a sand-bottomed stream (Beck 1965). Larger rivers (here including only the Escambia and Choctawhatchee rivers in their downstream reaches) may have higher levels of primary production within the stream. Invertebrate production is highest in summer and fall. Judging from upstream–downstream comparisons in the Calcasieu drainage, larger rivers should be much less physicochemically variable than upstream, sand-bottomed reaches. Calcareous streams of the Florida peninsula are under a different hydrological cycle, because the low flow season is winter and spring. These streams have high levels of dissolved substances and nutrients and more submergent macrophytes than do either sand-bottomed streams or large rivers. There are some invertebrates limited to these streams (primarily gastropods), and fishes are mostly sunfishes and topminnows.

RESOURCE USE AND MANAGEMENT IMPACTS

Past and Future Effects

Originally, the stream systems discussed in this chapter drained pine and mixed pine–hardwood woodlands. In the 19th century, much of this forest was logged, and crops (cotton, corn, etc.) were planted on the cleared land (Hilliard 1984). This was doubtless a time when waterway siltation increased. Much of the planted area has now been returned to forest, in the guise of pine plantations. Thus, in some sense the streams covered in this chapter now drain landscapes more similar to those before human exploitation.

In the 20th century, industrialization and population growth have introduced a greater diversity of impacts than existed previously. Today, human impacts include pollution of waterways by human wastes, by pesticides, and by industrial wastes (such as those originating from pulp mills), effects of forestry and agricultural practices, and effects related to direct modification of waterways.

Municipal effluents (sewage) produce lowered oxygen levels, nutrient enrichment (addition of nitrates and phosphates), and raised bacterial counts. The Louisiana Department of Environmental Quality (1984) lists various waterways, streams, and stream sections that receive high levels of municipal effluents. Point sources of municipal sewage affect all the Louisiana streams discussed in this chapter. The effects of sewage on stream systems are well known (Krenkel and Novotny 1980). In nutrient-poor streams with normally high oxygen levels, the effects of sewage on communities could be expected to be quite dramatic. Invertebrate community composition in sand-bottomed streams is changed by organic pollution. Beck (1954) illustrated a method of using presence/absence of particular invertebrate species to indicate organic pollution. However, nutrient enrichment alone may not always produce a change in community composition. Lackey and Morgan (1960) and Lackey and Putnam (1965) investigated the effect of phosphorus enrichment (through phosphate mining) on microbial and blue-green algal populations. Comparing the phosphate-poor Santa Fe River with the Peace River (phosphate-rich), they found populations in the Peace River to be hundreds of times greater. Yet the species were no different in the two streams, suggesting that while prokaryote community structure would be affected by phosphate enrichment, species composition would not be so affected.

Industrial wastes affect many Gulf coastal streams. The Louisiana Department of Environmental Quality (DEQ) monitors substances that identify industrial effluents and agricultural runoff, including heavy metals (arsenic, cadmium, copper, mercury), and PCBs. DEQ data (1975 to present) on Louisiana streams covered in this chapter indicate levels of these substances ranging from 0 to 20 $\mu g/L$. Lowest levels were found in the Tangipahoa and Tickfaw rivers, two recreationally important waterways. Similar levels were recorded from the Florida rivers considered here (data from Florida Department of Environmental Regulation, mid-1970s–present). Lafleur (1956) investigated the effects of pulp mills on invertebrate communities in the Calcasieu drainage. He found that mayflies were potential indicators of pulp mill pollution, because they seemed intolerant of the physicochemical conditions produced by these mills. Herrman (1981) discussed the physicochemical effects of pulp mills in certain warmwater streams; these effects may be generalized to Gulf coastal streams. Some Florida waterways have been heavily affected by municipal and industrial pollution. Bass (1984) considered the small Fenholloway River to be "ruined." He reviewed and summarized studies on pollution and pollution abatement programs on the Escambia River, a system he felt was improving due to these programs.

Agricultural and silvicultural practices affect these waterways. Clear-cutting of pine may result in siltation of adjacent waterways (Gibbons and Salo 1973). Siltation and community responses to it have apparently been little studied in Gulf coastal regions. Some effects may be quite local. For example, flushing of rice fields in summer produces pulses of turbidity and oxygen demand in the lower Calcasieu drainage (during a period of the year when turbidity is usually low and primary production high). The Louisiana DEQ Office of Water Resources (1984) felt that agricultural/silvicultural effects were a significant source of water-quality problems along some streams, but were often minor in comparison to other insults acting upon many Louisiana streams.

Some impacts affect small streams directly, as these waterways become impounded or otherwise managed. Woodruff et al. (1963) listed a series of channel control or impoundment facilities that would affect the streams of the Florida panhandle (from the Perdido River to the Santa Fe River). These included four proposed reservoirs on the Ochlockonee River, one on the

Choctawhatchee and one on the Yellow River. Channelization of waterways involves a series of changes, generally deleterious to native biota (see Henegar and Harman 1971). Habitat diversity, which is of demonstrated importance to fishes, is decreased (Gorman and Karr 1978). The waterway is so constructed as to produce pulses of high water as runoff is flushed away. Woody structure (of importance to these ecosystems) is removed. All of these effects reduce biotic diversity in these streams and add physicochemical stresses uncharacteristic of small southeastern streams. These may present conditions to which local species are poorly adapted. In the Calcasieu drainage, fishes inhabiting channelized waterways are forms more characteristic of physicochemically stressful prairie streams (streams with widely variable flow and temperatures and little cover (Matthews and Hill 1980)), including the red shiner (*Notropis lutrensis*), black bullhead (*Ictalurus melas*), mosquitofish (*Gambusia affinis*), and green sunfish (*Lepomis cyanellus*).

There are a number of endangered species, threatened species, and species of special concern inhabiting the streams covered in this chapter. Endangered and threatened species include the American alligator and two darters confined to one or two small drainage systems (U.S. Fish and Wildlife Service 1986). The Okaloosa darter (*Etheostoma okaloosae*) is confined to a few small streams draining into Choctawhatchee Bay, Florida (Gilbert 1978), and the bayou darter (*E. rubrum*) is found only in the Bayou Pierre system, Mississippi (Suttkus and Clemmer 1977). The recently described blackmouth shiner, *Notropis melanostomus*, is found only in the Blackwater and Yellow river systems; its life history is being intensively studied by the Florida Game and Freshwater Fish Commission (Bortone 1989). Several species of map turtles and the alligator snapping turtle are being considered for designation as endangered species. Beccasio et al. (1982) provide an ecological inventory (incorporating references pertinent to endangered or threatened species) covering some areas included in this chapter. In particular, their inventory covers Louisiana streams draining into Lake Pontchartrain, and streams of the Florida panhandle.

A special problem is associated with streams of the Florida peninsula and to a lesser extent other streams covered here. Introduced fish species (Courtenay and Stauffer 1984) are actively displacing native species in this area. Wilson and Porras (1983) discuss some of the problems facing south Florida's herpetofauna.

Unique Resources and Management Problems

The sand-bottomed streams of the Gulf coastal plain represent a valued recreational resource. Portions of the Tangipahoa and Tickfaw rivers (two Louisiana scenic waterways) are popular with canoeists from the Baton Rouge and New Orleans areas. In southwestern Louisiana, Whisky Chitto Creek is also a popular recreational waterway. In Mississippi, portions of Black Creek have been designated as wild and scenic rivers. In Florida, the Blackwater River and Yellow River are both protected by national forests or other federal lands, and both have been proposed for consideration as national scenic rivers. Bass (1984), reviewed literature on these streams, noting that their popularity with fishermen and canoeists had produced accumulations of trash, beer cans, and so on. Waste left by campers and fishermen may not have much effect at the ecosystem level, but it represents a management problem for those responsible for maintaining the waterways for the public. A more serious problem is the clearing of snags and logs to "improve" these waterways for boaters.

The major importance of snags and other woody material in these streams must be understood and accepted. The typical small stream ecosystem of the Gulf coastal plain revolves around debris and large woody snags. The high productivity of this habitat was demonstrated in the Apalachicola River by Ager et al. (1985). Angermeier and Karr (1984) investigated the effects of changing amounts of structure (snags and debris) in a midwestern stream. Benke et al. (1984, 1985) and Thorp et al. (1985) demonstrated the importance of snag habitat in waterways of the Atlantic coastal plain. Thus, a major management aim should be to retain and enhance snag habitat, because the animal community and a large segment of the plant community depend on these snags. This will require rethinking of flood management and navigation improvement programs. These programs usually include removal of snags, since snags tend to impede flow (Marzolf, 1978).

The major importance of seasonal floods to these streams must also be understood and accepted. The biota of these streams have evolved in an environment that floods seasonally. Alterations that reduce flooding may be expected to cause ecosystem-level changes in these streams. Use of medium-low gradient streams and associated floodplains should be approached with the attitude that seasonal flooding is normal and proper. Such an attitude translates into land use policies that forbid encroachment onto floodplains and that promote reduction of soil erosion and consequent catastrophic floods. Massive projects to prevent floods are inappropriate in these streams. In the medium- and low-gradient streams covered in this chapter, flood control (if absolutely necessary) may be better effected by constructing check dams in extreme headwaters. Ebert and Knight (1981) outlined such a flood-control program, along with a fish habitat enhancement program, in a Mississippi watershed.

Unique Natural and Induced Impacts

Important stressors in the small stream ecosystems covered in this chapter have already been outlined. Year-to-year variation in rainfall and in seasonal flooding may be counted a stress. This variability may determine reproductive success of some species (Ross and Baker 1983). For fish species, the dry season may represent a period of shortage, for example, of food and appropriate habitat (Felley and Felley 1987). However, the typical communities that inhabit these streams are (by definition) composed of species that are able to maintain populations in the face of such events. Human-induced impacts present these communities with events to which many species are not adapted. Pollutants (e.g., heavy metals and organic chemicals), high nutrient levels, and low oxygen levels may all induce changes in the species composition of these communities. Removal of substrates for attachment of epiphytic plants and logging or clearing of streamside communities both deprive these streams of necessary primary production. Channelization produces conditions that may favor wholly different communities. Human impacts, as opposed to natural ones, may produce conditions that can wipe out a species in a drainage. Thus, a darter species that inhabits a single drainage is not likely to be vulnerable to short-term climatic events, but is vulnerable to short-term environmental changes due to human activities.

RECOMMENDATIONS FOR FURTHER RESEARCH AND MANAGEMENT

Both research and management should, in the future, concentrate on those aspects that make Gulf coastal streams unique. These include the low dissolved solids and low pH condition of these streams. What effect does adding dissolved substances have on the biota? Are any species limited by these physicochemical conditions? Which ones? Such "sensitive" species would be most vulnerable to changes in the environment. Heins and Baker (1987) and Heins and Rabito (1988) have shown that life history and population structure of certain fishes respond to differences in the physicochemical environments of these streams. How does the stream community interact with the floodplain community? Apparently, some species are intimately tied to both communities, while others are not (Ross and Baker 1983, Finger and Stewart 1987). What are the important producers within these streams, and are they limited by hard substrate (snags) as so many studies show (Ager et al. 1985, Benke et al. 1984, 1985, Thorp et al. 1985)? What are the relative importances of within-stream producers and allochthonous inputs of fixed carbon? What are the patterns of spatial and temporal variation in the importance of these two sources of primary production? How do different trophic levels respond to the low-water/high-water cycle? These are ecosystem levels that require the intensive study of a representative Gulf coastal drainage. A relatively large drainage, such as the Pascagoula system, should be intensively investigated with a view to answering such questions as those outlined above.

Each of these questions has management implications. Management recommendations at this stage must be relatively crude and perhaps obvious: Do not channelize, leave woody material and snags within the stream, do not build in the floodplain, restrict municipal/industrial discharges and enforce adherence to adequate treatment level to maintain stream quality. These recommendations would benefit any stream ecosystem. Study of medium-low-gradient streams at the ecosystem level will allow more subtle management decisions.

ACKNOWLEDGMENTS

I thank Susan M. Felley, and the people below, for reading and commenting on various aspects of this manuscript. Kenneth O'Hara of the Louisiana Department of Environmental Quality helped greatly in obtaining water chemistry data for Louisiana and Florida streams. D. Gray Bass also provided data on Florida streams. Carter R. Gilbert, David C. Heins, and Stephen T. Ross, provided unpublished data and copies of works in progress. Dugan S. Sabins, Lynn Wellman, Robert Maples, and Mark Wygoda gave advice on particular sections of the manuscript.

REFERENCES

Ager, L. A., C. Mesing, and M. Hill. 1985. *Fishery Study, Apalachicola River Maintenance Dredging Disposal Site Evaluation Program*. Florida Game and Freshwater Fish Commission Report, prepared for U.S. Army Corps of Engineers, Mobile District.

Angermeier, P. L., and J. R. Karr. 1984. Relationships between woody debris and fish habitat in a small warmwater stream. *Trans. Am. Fish. Soc.* 113:716–726.

Auffenberg, W., and W. W. Milstead. 1965. Reptiles in the Quaternary of North America. In H. E. Wright and D. G. Frey (eds.), *The Quaternary of the United States*. Princeton, NJ: Princeton University Press, pp. 557–568.

Baker, J. A., and S. T. Ross. 1981. Spatial and temporal resource utilization by southeastern cyprinids. *Copeia* 1981:178–189.

Barr, C. B., and J. B. Chapin. 1988. The aquatic Dryopoidea of Louisiana (Coleoptera: Psephenidae, Dryopidae, Elmidae). *Tulane Stud. Zool. Bot.* 21:1–164.

Bass, D. G., Jr. 1984. *Rivers of Florida and Their Fishes*, Study III. Dingell-Johnson Project F-36. Florida Game and Freshwater Fish Commission Report. Florida Game and Freshwater Fish Commission, Tallahassee, Florida.

Bass, D. G., Jr., and D. T. Cox. 1985. River habitat and fishery resources of Florida. In W. Seaman, Jr. (ed.), *Florida Aquatic Habitat and Fishery Resources*. Kissimmee, FL: Florida Chapter American Fisheries Society, pp. 121–187.

Bass, D. G., Jr., and V. G. Hitt. 1977. *Ecology of the Blackwater River System, Florida*. Florida Game and Freshwater Fish Commission Report. Florida Game and Freshwater Fish Commission, Tallahassee, Florida.

Bass, D. G., Jr., and V. G. Hitt. 1978. *Sport Fishery Ecology of the Escambia River, Florida*. Florida Game and Freshwater Fish Commission Report. Florida Game and Freshwater Fish Commission, Tallahassee, Florida.

Bass, D. G., Jr., D. M. Yeager, and V. G. Hitt. 1979. *Ecology of the Yellow River System, Florida*. Florida Game and Freshwater Fish Commission Report. Florida Game and Freshwater Fish Commission, Tallahassee, Florida.

Bass, D. G., Jr., D. M. Yeager, and V. G. Hitt. 1980. *Ecology of the Choctawhatchee River System, Florida*. Florida Game and Freshwater Fish Commission Report. Florida Game and Freshwater Fish Commission, Tallahassee, Florida.

Beccasio, A. D., N. Fotheringham, A. E. Redfield, R. L. Frew, W. L. Levitan, J. E. Smith, and J. O. Woodrow, Jr. 1982. Gulf coast ecological inventory: user's guide and information base. *U.S., Fish Wildl. Serv., Biol. Serv. Program* FWS/OBS-82/55.

Beck, J. T., and B. C. Cowell. 1976. Life history and ecology of the freshwater caridean shrimp, *Palaemonetes paludosus* (Gibbes). *Am. Midl. Nat.* 96:52–65.

Beck, K. C., J. H. Reuter, and E. M. Perdue. 1974. Organic and inorganic geochemistry of some coastal plain rivers of the southeastern United States. *Geochim. Cosmochim. Acta* 38:341–364.

Beck, W. M., Jr. 1954. Studies of stream pollution biology. I. A simplified ecological classification of organisms. *J. Fla. Acad. Sci.* 17:211–227.

Beck, W. M., Jr. 1965. The streams of Florida. *Bull. Fla. State Mus.* 10:91–126.

Beck, W. M., Jr. 1973. Chemical and physical aspects of the Blackwater River in northwestern Florida. In W. L. Peters and J. G. Peters (eds.), *Proceedings of the First International Conference on Ephemeroptera*. Leiden, The Netherlands: E. J. Brill, pp. 231–241.

Beck, W. M., Jr. 1980. Interesting new chironomid records for the southern United States (Diptera: Chironomidae). *J. Ga. Entomol. Soc.* 15:69–73.

Benke, A. C., T. C. van Arsdall, Jr., D. M. Gillespie, and F. K. Parrish. 1984. Invertebrate productivity in a subtropical blackwater river: the importance of habitat and life history. *Ecol. Monogr.* 54:25–63.

Benke, A. C., R. L. Henry, III, D. M. Gillespie, and R. J. Hunter. 1985. Importance of snag habitat for animal production in southeastern streams. *Fisheries* 10:8–13.

Bernard, H. A., and R. J. LeBlanc. 1965. Resumé of the Quaternary geology of the northwestern Gulf of Mexico province. In H. E. Wright and D. G. Frey (eds.), *The Quaternary of the United States*. Princeton, NJ: Princeton University Press, pp. 137–185.

Berner, L. 1950. The mayflies of Florida. *Univ. Fla. Publ., Biol. Sci. Ser.* 4:1–267.

Bortone, S. A. 1989. *Notropis melanostomus*, a new species of cyprinid fish from the Blackwater-Yellow river drainage of northwest Florida. *Copeia* 1989:737–741.

Bresnick, G. I., and D. C. Heins. 1977. The age and growth of the weed shiner, *Notropis texanus* (Girard). *Am. Midl. Nat.* 98:495–499.

Carver, D. C. 1975. Life history of the spotted bass, *Micropterus punctulatus* (Rafinesque) in Six-mile Creek, Louisiana. *La. Dep. Wildl. Fish., Fish. Div. Bull.* 13:1–103.

Cochran, D. M., and C. J. Goin. 1970. *The New Field Book of Reptiles and Amphibians*. New York: Putnam's.

Collins, J. T., R. Conant, J. E. Huheey, J. L. Knight, E. M. Rundquist, and H. M. Smith. 1982. Standard common and current scientific names for North American amphibians and reptiles. *Soc. Study Amphibians Reptiles, Herpetol. Circ.* 12.

Conant, R. 1975. *A Field Guide to Reptiles and Amphibians of Eastern and Central North America*. Boston, MA: Houghton Mifflin.

Conner, J. V., and R. D. Suttkus. 1986. Zoogeography of freshwater fishes of the western Gulf slope of North America. In C. H. Hocutt and E. O. Wiley (eds.), *Zoogeography of North American Freshwater Fishes*. New York: Wiley, pp. 413–456.

Cooke, C. W. 1939. Scenery of Florida interpreted by a geologist. *Fla. Geol. Surv.* 17:1–118.

Courtenay, D. J., Jr., and J. R. Stauffer, Jr. (eds.). 1984. *Distribution, Biology, and Management of Exotic Fishes*. Baltimore, MD: Johns Hopkins University Press.

Cowell, B. C., and W. C. Carew. 1976. Seasonal and diel periodicity in the drift of aquatic insects in a subtropical Florida stream. *Freshwater Biol.* 6:587–594.

de la Cruz, A. A., and H. A. Post. 1977. Production and transport of organic matter in a woodland stream. *Arch. Hydrobiol.* 80:227–238.

Ebert, D. J., and L. A. Knight, Jr. 1981. Management of warmwater stream systems in the Holly Springs National Forest, Mississippi. In L. A. Krumholz (ed.), *Warmwater Streams Symposium, American Fisheries Society*. Lawrence, KS: Allen Press, pp. 382–387.

Everitt, B. 1975. Freshwater Ectoprocta: distribution and ecology of five species in southeastern Louisiana. *Trans. Am. Microsc. Soc.* 94:130–134.

Felley, J. D. 1987. Nekton assemblages of three tributaries to the Calcasieu Estuary, Louisiana. *Estuaries* 10:321–329.

Felley, J. D., and S. M. Felley. 1987. Relationships between habitat selection by individuals of a species and patterns of habitat segregation among species: fishes of the Calcasieu drainage. In W. J. Matthews and D. C. Heins (eds.), *Community and Evolutionary Ecology of North American Stream Fishes*. Norman: Oklahoma University Press, pp. 61–68.

Fernald, E. A., and D. J. Patton (eds.). 1984. *Water Resources Atlas of Florida*. Gainesville: Florida State University.

Finger, T. R., and E. M. Stewart. 1987. Response of fishes to flooding regime in lowland hardwoods wetlands. In W. J. Matthews and D. C. Heins (eds.), *Community and Evolutionary Ecology of North American Stream Fishes*. Norman: Oklahoma University Press, pp. 86–92.

Flint, R. F. 1957. *Glacial and Pleistocene Geology*. New York: Wiley.

Franz, R., and D. S. Lee. 1982. Distribution and evolution of Florida's troglobitic crayfishes. *Bull. Fla. State Mus., Biol. Sci.* 28:53–78.

Gemborys, S. R., and E. J. Hodgkins. 1971. Forests of small stream bottoms in the coastal plain of southwestern Alabama. *Ecology* 52:70–84.

Geraghty, J. J., D. W. Miller, F. van der Leeden, and F. L. Troise. 1973. *Water Atlas of the United States*. Port Washington, NY: Water Information Center Publication.

Gibbons, D. R., and E. O. Salo. 1973. An annotated bibliography of the effects of logging on fish of the western United States. *USDA For. Serv. Gen. Tech. Rep.* PNW-10.

Gilbert, C. R. (ed.). 1978. *Rare and Endangered Biota of Florida*, Vol. 4. Gainesville: University Presses of Florida.

Gilbert, C. R. 1987. Zoogeography of the freshwater fish fauna of southern Georgia and peninsular Florida. *Brimleyana* 13:25–54.

Godfrey, R. K., and J. W. Wooten. 1979. *Aquatic and Wetland Plants of Southeastern United States: Monocotyledons*. Athens: University of Georgia Press.

Godfrey, R. K., and J. W. Wooten. 1981. *Aquatic and Wetland Plants of Southeastern United States: Dicotyledons*. Athens: University of Georgia Press.

Gorman, O. T., and J. R. Karr. 1978. Habitat structure and stream fish communities. *Ecology* 59:507–515.

Gosselink, J. G., C. L. Cordes, and J. W. Parsons. 1979. An ecological characterization study of the Chenier Plain coastal ecosystem of Louisiana and Texas. *U.S. Fish Wildl. Serv., Off. Biol. Serv.* FWS/OBS-78/9–78/11.

Grady, J. M., R. C. Cashner, and J. S. Rogers. 1983. Fishes of the Bayou Sara drainage, Louisiana and Mississippi with a discriminant functions analysis of factors influencing species distribution. *Tulane Stud. Zool. Bot.* 24:83–100.

Gregg, W. W., and F. C. Rose. 1982. The effects of aquatic macrophytes on the stream microenvironment. *Aquat. Bot.* 14:309–324.

Grissinger, E. H., J. B. Murphey, and W. C. Little. 1982. Late-Quaternary valley-fill deposits in north-central Mississippi. *Southeast. Geol.* 23:147–162.

Heins, D. C. 1981. Life history pattern of *Notropis sabinae* (Pisces: Cyprinidae) in the lower Sabine River drainage of Louisiana and Texas. *Tulane Stud. Zool. Bot.* 22:67–84.

Heins, D. C. 1985. Life history traits of the Florida sand darter *Ammocrypta bifascia*, and comparisons with the naked sand darter *Ammocrypta beani*. *Am. Midl. Nat.* 113:209–216.

Heins, D. C., and J. A. Baker. 1987. Analysis of factors associated with intraspecific variation in propagule size of a stream-dwelling fish. In W. J. Matthews and D. C. Heins (eds.), *Community and Evolutionary Ecology of North American Stream Fishes* Norman: Oklahoma University Press, pp. 223–231.

Heins, D. C., and J. A. Baker. 1989. Growth, population structure and reproduction of the percid fish *Percina vigil*. *Copeia* 1989:727–736.

Heins, D. C., and G. H. Clemmer. 1975. Ecology, foods and feeding of the longnose shiner, *Notropis longirostris* (Hay) in Mississippi. *Am. Midl. Nat.* 94:284–295.

Heins, D. C., and G. H. Clemmer. 1976. The reproductive biology, age and growth of the North American cyprinid fish *Notropis longirostris* (Hay). *J. Fish Biol.* 8:365–379.

Heins, D. C., and D. Davis. 1984. The reproductive season of the weed shiner, *Notropis texanus* (Pisces, Cyprinidae) in southeastern Mississippi. *Southwest. Nat.* 29:133–140.

Heins, D. C., and F. G. Rabito, Jr. 1986. Spawning performance in North American minnows: direct evidence of the occurrence of multiple clutches in the genus *Notropis*. *J. Fish Biol.* 28:343–357.

Heins, D. C., and F. G. Rabito, Jr. 1988. Reproductive traits in populations of the weed shiner, *Notropis texanus*, from the Gulf coastal plain. *Southwest. Nat.* 33:147–156.

Heins, D. C., and J. R. Rooks. 1984. Life history of the naked sand darter, *Ammocrypta beani*, in southeastern Mississippi. In D. G. Lindquist and L. M. Page (eds.), *Environmental Biology of Darters*. The Hague, Netherlands: D. W. Junk Publishers, pp. 61–69.

Henegar, D. L., and K. W. Harman. 1971. A review of references to channelization and its environmental impact. In E. Schneburger and J. L. Junk (eds.), *Stream Channelization, a Symposium*, Spec. Publ. No. 2. Am. Fish. Soc., North Cent. Div., pp. 79–83.

Herrman, R. B. 1981. Studies of warmwater fish communities exposed to treated pulp mill wastes. In L. A. Krumholz (ed.), *Warmwater Streams Symposium, American Fisheries Society*. Lawrence, KS: Allen Press, pp. 127–141.

Hilliard, S. B. 1984. *Atlas of Antebellum Southern Agriculture*. Baton Rouge: Louisiana State University Press.

Holland, W. C. 1944. Physiographic divisions of the Quaternary lowlands of Louisiana. *Proc. La. Acad. Sci.* 8:11–24.

Kautz, R. S. 1981. Fish populations and water quality in north Florida rivers. *Proc. Annu. Conf. Southeast. Assoc. Fish Wildl. Agencies* 35:495–507.

Krenkel, P. A., and V. Novotny. 1980. *Water Quality Management.* New York: Academic Press.

Lackey, J. B., and G. B. Morgan. 1960. Chemical microbiotic relationships in certain Florida surface water supplies of flowing waters in Florida. *Q. J. Fla. Acad. Sci.* 23:289–301.

Lackey, J. B., and H. D. Putnam. 1965. Ability of streams to assimilate wastes. *Q. J. Fla. Acad. Sci.* 28:303–317.

Lafleur, R. A. 1956. A biological and chemical survey of the Calcasieu River. MS Thesis, Louisiana State University, Baton Rouge.

Lee, D. S., C. R. Gilbert, C. H. Hocutt, R. E. Jenkins, D. E. McAllister, and J. R. Stauffer, Jr. 1980. *Atlas of North American Freshwater Fishes.* Raleigh: North Carolina State Museum of Natural History.

Louisiana Department of Environmental Quality. 1984. *Louisiana Water Quality Report.* Baton Rouge: LA DEQ Office of Water Resources.

MacNeil, F. S. 1949. Pleistocene shorelines in Florida and Georgia. *U.S., Geol. Surv.-Prof. Pap.* 21-F:95–106.

Marzolf, G. R., 1978. *The Potential Effects of Clearing and Snagging on Stream Ecosystems,* FWS-OBS-78/14. Washington, DC: U.S. Fish and Wildlife Service, Office of Biological Services.

Matthews, W. J., and L. G. Hill. 1980. Habitat partitioning in the fish community of a southwestern river. *Southwest. Nat.* 25:51–66.

Merritt, R. W., and K. W. Cummins. 1978. *An Introduction to the Aquatic Insects of North America.* Dubuque, IA: Kendall/Hunt.

Moore, W. G. 1953. Louisiana freshwater sponges, with ecological observations on certain sponges of the New Orleans area. *Trans. Am. Microsc. Soc.* 72:24–32.

Moore, W. G. 1970. Limnological studies of temporary ponds in southeastern Louisiana. *Southwest. Nat.* 15:83–110.

Mount, R. H. 1975. *The Reptiles and Amphibians of Alabama.* Auburn, AL.: Auburn Printing Company.

Muller, R. A. 1977. A comparative climatology for environmental baseline analysis: New Orleans. *J. Appl. Meteorol.* 16:20–35.

Murray, G. E. 1961. *Geology of the Atlantic and Gulf Coastal Province of North America.* New York: Harper.

Opdyke, N. D., D. P. Spangler, D. L. Smith, D. S. Jones, and R. C. Lindquist. 1984. Origin of the epeirogenic uplift of Pliocene-Pleistocene beach ridges in Florida and development of the Florida karst. *Geology* 12:226–228.

O'Quinn, R., and M. J. Sullivan. 1983. Community structure dynamics of epilithic and epiphytic diatoms in a Mississippi stream. *J. Phycol.* 19:123–128.

Page, L. M. 1983. *Handbook of Darters.* Neptune City, NJ: TFH Publications.

Pecora, R. A. 1973. A report on the algal flora of southwestern Louisiana: Phytoflagellates. *Proc. La. Acad. Sci.* 36:76–82.

Penn, H. G. 1959. An illustrated key to the crawfish of Louisiana with a summary of their distribution within the state. *Tulane Stud. Zool.* 7:1–20.

Pennak, R. W. 1978. *Freshwater Invertebrates of the United States*. New York: Ronald Press.

Peters, W. L., and J. Jones. 1973. Historical and biological aspects of the Blackwater River in northwestern Florida. In W. L. Peters and J. G. Peters (eds.), *Proceedings of the First International Conference on Ephemeroptera*. Leiden, The Netherlands: E. J. Brill, pp. 242–253.

Peters, W. L., and J. G. Peters. 1977. Adult life and emergence of *Dolania americana* in northwestern Florida (Ephemeroptera: Behningiidae). *Int. Rev. Gesamten Hydrobiol.* 62:409–438.

Pflieger, W. L. 1975. *The Fishes of Missouri*. Jefferson City: Missouri Department of Conservation.

Poirrier, M. A. 1969. Some freshwater sponge hosts of Louisiana and Texas spongilla-flies, with new locality records. *Am. Midl. Nat.* 81:573–574.

Post, H. A., and A. A. de la Cruz. 1977. Litterfall, litter decomposition and flux of particulate organic material in a coastal plain stream. *Hydrobiologia* 55:201–207.

Prescott, G. W. 1942. The freshwater algae of southern United States. II. The algae of Louisiana, with description of some new forms and notes on distribution. *Trans. Am. Microsc. Soc.* 61:109–119.

Price, R. C., and K. N. Whetstone. 1977. Lateral stream migration as evidence for regional geologic structure in the eastern Gulf coastal plain. *Southeast. Geol.* 12:129–147.

Riggs, S. R. 1984. Paleoceanographic model of Neogene phosphate deposition, U.S. Atlantic continental margin. *Science* 223:123–131.

Robertson, D. J., and K. Piwowar. 1985. Comparison of four samplers for evaluating macroinvertebrates of a sandy Gulf coast plain stream. *J. Freshwater Ecol.* 3:223–231.

Rosenau, J. C., G. L. Faulkner, C. W. Hendry, Jr., and R. W. Hull. 1977. Springs of Florida. *Geol. Surv. Bull.* (*U.S.*) 31.

Ross, S. T., and J. A. Baker. 1983. The response of fishes to periodic spring floods in a southeastern stream. *Am. Mid. Nat.* 109:1–14.

Ross, S. T., J. A. Baker, and K. E. Clark. 1987. Microhabitat partitioning of southeastern stream fishes: temporal and spatial predictability. In W. J. Matthews and D. C. Heins (eds.), *Community and Evolutionary Ecology of North American Stream Fishes*. Norman: Oklahoma University Press, pp. 42–51.

Ruple, D. L., R. H. McMichael, and R. H. Baker. 1984. Life history of the gulf darter, *Etheostoma swaini* (Pisces: Percidae). *Environ. Biol. Fishes* 11:121–130.

Saucier, R. T., and A. R. Fleetwood. 1970. Origin and chronological significance of late Quaternary terraces, Ouachita River, Arkansas and Louisiana. *Geol. Soc. Am. Bull.* 81:869–890.

Schoener, T. W. 1982. The controversy over interspecific competition. *Am. Sci.* 70:586–595.

Sellards, E. H. 1917. Geology between the Ocklocknee and Aucilla rivers in Florida. *Fla. State Geol. Surv.* 9:85–139.

Shively, S. H., and J. F. Jackson. 1985. Factors limiting the upstream distribution of the Sabine map turtle. *Am. Midl. Nat.* 114:292–303.

Shoup, C. S. 1947. Geochemical interpretation of water analyses from Tennessee waters. *Trans. Am. Fish. Soc.* 74:223–239.

Smock, L. A., E. Gilinsky, and D. L. Stoneburner. 1985. Macroinvertebrate production in a southeastern United States blackwater stream. *Ecology* 66:1491–1503.

Snowdon, J. O., Jr., and R. R. Priddy. 1968. Geology of Mississippi loess. *Bull.— Miss., Geol., Econ. Topogr. Surv.* 3:13–203.

Strahler, A. N. 1957. Quantitative analysis of watershed geomorphology. *Trans., Am. Geophys. Union* 38:913–920.

Suttkus, R. C., and G. H. Clemmer. 1977. A status report on the bayou darter, *Etheostoma rubrum*, and the Bayou Pierre system. *Southeast. Fishes Counc. Proc.* 1:1–4.

Swift, C. C., C. R. Gilbert, S. A. Bortone, G. H. Burgess, and R. W. Yerger. 1986. Zoogeography of freshwater fishes of the southeastern United States: Savannah River to Lake Pontchartrain. In C. H. Hocutt and E. O. Wiley (eds.), *Zoogeography of North American Freshwater Fishes*. New York: Wiley, pp. 213–265.

Thornbury, W. D. 1965. *Regional Geomorphology of the United States*. New York: Wiley.

Thorp, J. H., E. M. McEwan, M. F. Flynn, and F. R. Hauer. 1985. Invertebrate colonization of submerged wood in a cypress-tupelo swamp and blackwater stream. *Am. Midl. Nat.* 113:56–68.

Tsui, P. T. P., and M. D. Hubbard. 1979. Feeding habits of the predaceous nymphs of *Dolania americana* in northwestern Florida (Ephemeroptera: Behningiidae). *Hydrobiologia* 67:119–123.

U.S. Fish and Wildlife Service. 1986. *Endangered and Threatened Wildlife and Plants*, 50 CFR 17.11 and 17.12. Washington, DC: U.S. Fish Wildl. Serv., Dep. of the Interior.

Vail, P., and J. Hardenbol. 1979. Sea level changes during the Tertiary. *Oceanus* 22:71–79.

Vannote, R. L., W. Minshall, K. W. Cummins, J. R. Sedell, and C. E. Cushing. 1980. The river continuum concept. *Can. J. Fish. Aquat. Sci.* 37:130–137.

Wharton, C. H., W. M. Kitchens, E. C. Pendleton, and T. W. Sipe. 1982. *The Ecology of Bottomland Hardwood Swamps of the Southeast: A Community Profile*, FWS/OBS-81/37. Washington, DC: U.S. Fish Wildl. Serv., Biol. Serv. Program.

White, D. C. 1985. Lowland and hardwood wetland invertebrate community and production in Missouri. *Arch. Hydrobiol.* 103:509–533.

Wilson, L. D., and L. Porras. 1983. *The Ecological Impact of Man on the South Florida Herpetofauna*. Lawrence: University of Kansas Publications, Museum of Natural History.

Winger, P. V. 1981. Physical and chemical characteristics of warmwater streams: a review. In L. A. Krumholz (ed.), *The Warmwater Streams Symposium, American Fisheries Society*. Lawrence, KS: Allen Press, pp. 32–44.

Woodruff, J. W., Jr., H. A. Morris, T. A. Adams, C. W. Chapman, H. W. Chapman, W. A. Gresh, J. H. Hammond, W. E. Hiatt, L. S. Moody, and R. G. Price. 1963. *Plan for Development of the Land and Water Resources of the Southeast River Basins*. Atlanta, GA: U.S. Study Commission, Southeast River Basins.

Wooten, J. W. 1986. The edaphic factors associated with eleven species of *Sagittaria* (Alismataceae). *Aquat. Bot.* 24:35–41.

7 Coastal Plain Blackwater Streams

LEONARD A. SMOCK,
ELLEN GILINSKY

Department of Biology, Virginia Commonwealth University, Richmond,
VA 23284

Blackwater streams and rivers are an important and diverse aquatic habitat of the southeastern United States, occurring throughout the Coastal Plain (Fig. 1). They are far more common than other types of streams that occur in this physiographic province, such as the whitewater streams of Florida or the alluvial rivers that flow through the Coastal Plain. The ecological characteristics of blackwater streams are predominantly controlled by the physiography of the area. The flat topography, mostly sandy soils, and close coupling of stream channels with often extensive, productive floodplains make the physical, chemical, and biological characteristics of these streams quite different from those of higher-gradient, rocky streams of the Piedmont and mountains.

In this chapter the physical, chemical, and biological characteristics of Coastal Plain blackwater streams of stream order six or less are discussed. The coverage focuses initially on the abiotic and biotic characteristics of relatively unperturbed streams and then reviews anthropogenic influences and resource management problems associated with blackwater streams. Although floodplains are an integral component of these streams and rivers, they are discussed only as they directly affect channel characteristics. An excellent review of the ecology of southeastern floodplains is provided by Wharton et al. (1982).

PHYSICAL ENVIRONMENT

Climate

The climate of the Coastal Plain ranges from humid continental in the north to humid subtropic in the south. Hot summers and mild winters are characteristic. Mean monthly low air temperatures are 3–7 °C and mean monthly high temperatures are 25–27 °C (Park 1979); average annual temperatures are 12–22 °C.

Average annual precipitation ranges from about 102 cm in Virginia to about 152 cm along the Gulf Coast (Park 1979), nearly all of it occurring as rainfall.

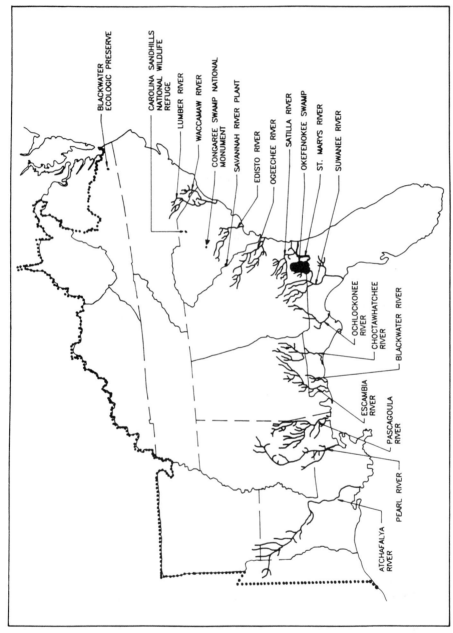

FIGURE 1. Major southeastern Coastal Plain blackwater river systems and important preserves for the biota of blackwater streams.

Precipitation is fairly evenly distributed over the entire year, autumn being the driest period. There are, however, large seasonal differences in runoff, with the lowest discharge occurring during the summer and early autumn because of high rates of evapotranspiration. Tropical storms and hurricanes annually cause short periods of high precipitation and increased runoff from late summer to late autumn.

Whereas precipitation tends to decrease inland from the coast, average annual runoff per unit area increases inland due to greater permeability of inland soils. In South Carolina, precipitation decreases from 140 to 114 cm from the coast to the fall line, whereas runoff per unit area increases from 33 to 46 cm (Lovingood and Purvis 1975). Runoff thus averages about 24 and 40% of precipitation along the coast and fall line, respectively. Watersheds closer to the coast therefore receive about 10% more precipitation but yield about 25% less runoff than inland watersheds.

Geology

The geology of the Coastal Plain is characterized by sedimentary rocks set in an area of low relief with only a few belts of rolling hills. Drainage is much slower than in the mountains and Piedmont; extensive floodplains often develop even along headwater streams. These factors are of considerable importance to the physical, chemical, and biological aspects of streams flowing through this area.

The Atlantic Coastal Plain often is divided into the Upper and Lower Coastal Plain. The Upper Coastal Plain occurs from the fall line, where metamorphic rock of the Piedmont dips beneath the sedimentary rock of the Coastal Plain, coastward to the innermost seacut terraces (Bloxham 1976). The Lower Coastal Plain extends from the terraces to the coast. Soil and drainage conditions differ between the two areas. Drainage is more rapid in the Upper Coastal Plain; in the poorly drained soils of the Lower Coastal Plain drainage is predominantly overland (Bergeaux 1969).

Soils are primarily leached, acidic sands and clays deposited during Pleistocene rises in sea level, along with fluvial and aeolian sands from the Holocene (Sumsion 1970). Tertiary clay, sand, and limestone predominate, but sand, clay, and shell beds of Quaternary age occur near the coast. A band of Cretaceous sand, clay, and marl underlain by a layer of quartz sand is particularly prominent inland in North and South Carolina. A section of the Cretaceous formations in these two states, known as the Sandhills, consists of a layer of surficial quartz sand that is up to 45 m (150 ft) thick (Simmons and Heath 1979).

Hydrology and water chemistry are strongly influenced by the different sediments. Limestone and shell beds are relatively soluble, and thus runoff from these areas has higher concentrations of dissolved solids. Dissolved ion concentrations are lower in runoff from the less soluble Tertiary sands and clays and much lower in runoff originating from the Cretaceous sediments. This is especially true in the Sandhills, where the high permeability of the soils causes a high proportion of precipitation to infiltrate through relatively insoluble quartz sand (Simmons and Heath 1979).

Channel Morphology

A factor of overriding importance to the ecology of Coastal Plain streams is their low slope, which affects sediment characteristics, water velocity and chemistry, and ultimately the biological characteristics of the streams. Slopes of headwater streams are usually less than 0.1% (i.e., a 1-m drop every kilometer) and around 0.02% for middle-order streams. Streams in the Lower Coastal Plain typically have lower slopes than streams in the Upper Coastal Plain.

Channels normally consist of long reaches, often with numerous pools; few streams have any riffles except for local areas in the Sandhills. Channels are not deeply entrenched and often are braided throughout their floodplains (Winner and Simmons 1977). Sediments range from highly organic soils, predominating in the Lower Coastal Plain, to loose, shifting sand, predominating in the Upper Coastal Plain and in larger streams and rivers. Because of the nature of these sediments, channel roughness and heterogeneity are low and the substrate is unstable, being prone to frequent disturbance during storms.

Debris dams and snags, resulting from the input of wood primarily through bank undercutting, are prominent and important features of the channels. Wood abundance, and hence the availability of dams and snags, increases from lower- to higher-order streams on the Coastal Plain, which is the opposite of the trend usually found in higher-gradient river systems (Benke and Wallace 1990). Debris dams covered only 1–3% of stream surface in two first-order streams (Smock et al. 1989); snags provided 4–36% of the available surface area in six second- to sixth-order rivers (Benke et al. 1984, Wallace and Benke 1984, Smock et al. 1985). Primarily because they are the major stable substrate in channels, dams and snags have a major role in the ecology of Coastal Plain streams and rivers. They are important substrates for invertebrates and provide refuge and food for fish. Their detritus-retention capability is higher than that of the relatively homogenous and often unstable sediment, and thus dams have an important influence on stream energetics and material spiraling. They also greatly affect the hydrodynamic nature of streams by increasing channel roughness, decreasing water velocity, causing more frequent and longer flooding of floodplains, and rerouting headwater channels through the floodplain.

Hydrology

Southeastern blackwater streams typically exhibit large seasonal variation in flow (Fig. 2). The period of high flow begins with the onset of winter rains in November and December, causing a large increase in aquatic habitat through floodplain inundation. Initial inundation of floodplains provides a water storage function for river systems; upon inundation, much of the total river

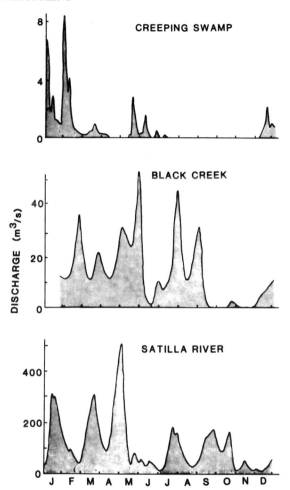

FIGURE 2. Annual discharge patterns for three blackwater streams. (Modified from Kuenzler et al. 1977, Findlay et al. 1986b, and Benke et al. 1979.)

discharge flows through the floodplain (Pernik and Roberts 1985). Low-flow periods occur during summer and autumn and are caused mainly by high rates of evapotranspiration rather than lower precipitation. For example, in a stream–swamp system in North Carolina, 61% of total annual precipitation was lost to evapotranspiration, whereas only 37 and 2% were lost as stream flow and net deep aquifer recharge, respectively (Winner and Simmons 1977).

Streams on the Upper Coastal Plain have less variability in flow than do Lower Coastal Plain and Piedmont steams because many of the former streams have cut into permeable sand and gravel aquifers that provide a more dependable water source than does rainfall (Bloxham 1976). This is especially true of streams in the Sandhills. Lower-order streams depend more on precipitation than do higher-order streams and rivers for their flow, and thus have a more variable discharge pattern. Even larger streams, however, can experience extremes in flow. For example, discharge over one year in the sixth-order Satilla River, Georgia, ranged from about 10 m³/s to over 500 m³/s; flow ranged from 0.03 m³/s to over 50 m³/s in fourth-order Black Creek, Georgia (Fig. 2).

Many headwater streams in the Lower Coastal Plain tend to dry in the summer except for isolated pools. These pools become sites of intense biological activity and serve as refugia for aquatic organisms until flow in channels is restored. The length of the dry period can be highly variable even for a given stream. For example, the no-flow period in Creeping Swamp, a small headwater stream in North Carolina, ranged from 20 to 130 consecutive days over a seven-year period (Sniffen 1981). Flow typically resumed in December and continued in the channel until the following June, the floodplain being inundated from December to April. Over a one-year period, 89% of the total aquatic area in this stream–swamp system was inundated floodplain.

Water Chemistry

The chemical characteristics of blackwater streams, although quite variable, are different from whitewater streams in a number of respects. In particular, a high dissolved organic carbon concentration imparts a dark color (hence the name blackwater) and high acidity. Buffering capacity and nutrient concentrations are low. Because they lack the power of alluvial brownwater rivers that arise in the mountains and Piedmont, blackwater rivers carry a low suspended solids load.

Southeastern blackwater streams can be broadly separated into two categories based on the type of area they drain (Beck 1965, Noltemeir et al. 1986). Peat-draining streams originate in swamps, bogs, and marshes and predominate in the Lower Coastal Plain. Streams that drain mineral soils predominate in the Upper Coastal Plain and especially the Sandhills region. Dissolved organic carbon concentrations, and thus color and acidity, are lower in these latter streams than in peat-draining streams. Alkalinity, hardness,and concentrations of dissolved oxygen, cations, and nutrients tend to be higher in Upper Coastal Plain streams.

The extent of the influence of geology on water chemistry can be seen from differences in mean concentrations of filterable constituents from unpolluted streams draining three geochemical zones in North Carolina (Table 1). Highest concentrations occur in areas draining the more soluble Tertiary and Quaternary limestone and shell sediments (zone III), whereas lowest concentrations occur in the Sandhills (zone I) because of the low solubility of the quartz sand that predominates there.

Although blackwater streams are considered to be acidic systems, pH varies widely between streams and in some cases can be neutral to slightly alkaline if the streams receive a significant amount of calcium input from groundwater. Acidity tends to decrease with increasing stream size. The pH of most peat-draining streams is 3.5–6; pH in mineral soil-draining streams typically is 4–7. Variability in pH due to local geology and hydrology can be evident even

TABLE 1 Mean Concentrations (mg/L) of Filterable (<0.45 μm) Constituents at Base Flow in Relatively Unpolluted North Carolina Blackwater Streams Flowing in Three Geochemical Zones

Parameter	Zone I	Zone II	Zone III
Calcium	0.6	0.6	2.3
Magnesium	0.4	0.4	0.8
Sodium	1.3	2.3	3.6
Potassium	0.5	0.5	0.9
Silica	4.8	5.9	7.2
Bicarbonate	3.5	1.0	4.0
Chloride	2.4	3.4	5.3
Sulfate	2.2	7.6	9.2
Nitrate–nitrogen	0.11	0.02	0.01
Ammonia–nitrogen	0.01	0.00	0.01
Organic nitrogen	0.22	0.35	0.45
Total phosphorus	0.01	0.02	0.01
Total dissolved solids	13	22	32

Note. Zone I, Sandhills; Zone II, Black Creek and Pee Dee Formations of Cretaceous age; Zone III, Castle Hayne and Yorktown formations and surficial Quaternary sediments.

Source. From Simmons and Heath (1979).

in a given river system. For example, the Santa Fe River in Florida arises from swampy tributaries as a blackwater stream with an annual mean pH of 5.3. Downstream it receives swamp drainage during high flow but alkaline groundwater input during low flow, resulting in a pH of 6.4. The lower river receives inputs from artesian springs, raising the pH to 7.4 (Wharton et al. 1982).

Dissolved oxygen concentrations follow a typical seasonal pattern. Low concentrations in the summer and autumn are caused by low-flow conditions and lowered solubility because of warmer water temperatures. An autumn minimum often coincides with leaf fall and the related increase in microbial respiration. Increased flow and colder temperatures in the winter and spring raise oxygen concentrations. Conditions of oxygen stress rarely occur in unpolluted streams, concentrations greater than 3 mg/L having been measured throughout the year even in intermittent stream pools (Kuenzler et al. 1980). Oxygen concentrations, however, often are low to anoxic just several centimeters below the sediment surface in many small streams (Strommer and Smock 1989, Metzler and Smock 1990).

Dissolved organic carbon concentrations in blackwater streams usually are 5–50 mg/L, whereas concentrations rarely are greater than 3 mg/L in higher-gradient streams of the Southeast. There are pronounced seasonal variations in concentrations and export of dissolved organic carbon. Export depends on the frequency and duration of floodplain inundation and on the timing and magnitude of autumn floods (Kuenzler et al. 1977, Mulholland and Kuenzler 1979, Meyer 1986).

Whereas most rivers of the world have a ratio of dissolved inorganic to organic constituents of about 10:1, the ratio in blackwater rivers of the southeastern United States is about 1:1 (Wharton and Brinson 1979). Most organic matter is in a dissolved rather than particulate form (Benke and Meyer 1988). The relative contribution of dissolved material to total organic concentrations is greater than in whitewater systems (Schlesinger and Melack 1981, Meybeck 1982). This relative abundance of dissolved organics, and the related low pH, is in distinct contrast with the majority of nonblackwater river systems in the world. The dissolved organics are predominately humic and fulvic acids mainly derived from the leaching of swamp soils (Reuter and Perdue 1972, 1977, Beck et al. 1974). The high acidity of these compounds imparts hydrogen ions to the water and hence causes low pH. Dissolved organics also are important to the chemistry of many cations through complexation reactions.

The inorganic chemistry of many blackwater streams is based on a sodium sulfate rather than a calcium carbonate regulating system (Blood 1980, 1986). Bicarbonate, a major ion in most streams worldwide, occurs in low concentrations in blackwater streams; buffering capacity thus is low. The dominance of inorganic ions by sodium and chloride suggests that rainfall is important in determining the distributions and concentration of major elements in many of these streams (Beck et al. 1974).

Concentrations of both dissolved and particulate-associated inorganic ions are low because of the absence of readily soluble minerals in the highly leached soils (Kennedy 1964, Neiheisel and Weaver 1967, Windom et al. 1971, Beck et al. 1974, Bloxham 1976). Concentrations of dissolved solids, and thus conductivity, tend to increase from the Upper to the Lower Coastal Plain. Conductivity increases from a mean of 21 μS/cm to 64 μS/cm in streams in the Upper and Lower South Carolina Coastal Plain, respectively (Gardner 1983). Storms are particularly important to the transport of dissolved and particulate matter. The total element transport rates and the chemical forms of metals and nutrients in transport change considerably between base flow and storm flow conditions (Simmons and Heath 1979, Mulholland et al. 1981, Smock and Kuenzler 1983).

Nitrogen and phosphorus concentrations are low in pristine southeastern blackwater streams, due to both low concentrations of these nutrients in drainage basin soils and their retention in the floodplain. It is well documented that floodplains serve as a sink and alter the form of these nutrients (e.g., Kitchens et al. 1975, Brinson et al. 1980, Yarbro 1983). Nitrogen removal occurs primarily through denitrification (Bradshaw 1977), and phosphorus removal is due to uptake by algae and sorption onto clays and silts (Yarbro 1983).

PLANT COMMUNITIES AND ENERGY SOURCES

There is a growing body of information on plant communities of the floodplain forests associated with Coastal Plain blackwater streams and the importance of these areas as an energy source to the streams. However, very little is known about the plant communities and autochthonous energy sources within stream channels. Whereas physical characteristics of streams limit development of extensive algal and macrophyte communities, there is some evidence that these communities can at times be an important aspect of the ecology of blackwater streams.

Algae

The algal community of southeastern blackwater streams is dominated by diatoms on sand sediment and by filamentous algae on stable substrates in both the stream and floodplain. Woodson and Wilson (1973) provide one of the more complete lists of algal species inhabiting blackwater streams. The dominant periphytic diatoms in a first-order stream in Virginia were *Eunotia pectinalis*, *Pinnularia major*, *Gomphonema constrictum*, and *Frustulia* sp. (B. Rosen and L. A. Smock, unpublished data). *E. pectinalis* also was a dominant species in two other Virginia streams (Woodson and Wilson 1973) and in North Carolina Coastal Plain streams (Whitford and Schumacher 1963). It probably is common in lotic systems throughout the Southeast, since it is a cosmopolitan species in acidic, low-conductivity streams (Sheath and Burkholder 1985).

Filamentous algal species can be prominent in patches on the floodplain, from where they are occasionally transported to stream channels during floods. Extensive growth of filamentous species often occurs in channels and floodplains inundated by beaver activity. Dominant species include the desmids *Hyalotheca dissiliens* and *Desmidium aptogonum*, filamentous greens such as *Spirogyra* spp., *Oedogonium* spp., *Bulbochaete* spp., *Rhizoclonium* sp., and *Ulothrix* sp., the Cyanophyceae *Lyngbya* sp. and *Nodularia* sp., and the diatom *E. pectinalis* (Kuenzler et al. 1980, Atchue et al. 1983).

Quantitative studies on densities and biomass of periphytic algae in southeastern blackwater streams are rare, and only a few estimates of chlorophyll concentrations (Kuenzler et al. 1977, Kondratieff and Kondratieff 1984) or productivity (discussed below) have been reported. Greatest algal growth occurs during the late winter and early spring months, when large patches of filamentous species often are noticeable in floodplains (Wharton and Brinson 1979, Mulholland 1981). Phytoplankton densities are low in smaller streams (Herlong and Mallin 1985), and probably in larger streams as well. Highest densities of phytoplankton occur during summer in slow flowing areas of streams (Kuenzler et al. 1977, Herlong and Mallin 1985).

Low light intensity no doubt is an important factor limiting algal growth in blackwater streams and rivers. High concentrations of dissolved organic carbon result in rapid light attenuation, so that even in larger streams with no overhanging canopy very little light can reach the channel substrate. The late winter–early spring algal bloom occurs during the time that flooded conditions coincide with an open canopy, allowing high light penetration, particularly in the floodplain. Nutrient limitation, often important in lentic waters, probably is not an important factor in these streams (Brown 1981).

Another important factor limiting algal growth is the lack of stable substrates for attachment. Algae growing on sand sediment are subjected to frequent disturbance by increased flow. This relegates much of the algal growth to snags, which are especially important as substrates for algae in uncanopied, higher-order streams.

Macrophytes

Macrophytes rarely occur in smaller streams, but can be locally abundant in some of the larger streams and rivers. Their significance to the ecology of blackwater streams has been relatively unstudied, with the exception of several studies that included an examination of macrophyte–macroinvertebrate relationships (Smock et al. 1985, Wallace and O'Hop 1985).

Typical species include *Sparganium americanum*, a species tolerant of low light conditions and found in fully canopied, second-order Cedar Creek in the Congaree Swamp National Monument, South Carolina, and *Vallisneria americana* and *Potamogeton epihydrus*, two species common in Upper Three Runs Creek on the Savannah River Plant in South Carolina (Morse et al. 1980). Bryophytes, including especially the moss *Brachelyma subulatum* and the liverwort *Porella pinnata*, are occasionally abundant on old submerged snags and tree trunks.

Large beds of macrophytes often occur in the backwaters of large, uncanopied rivers (Dennis 1973, Twilley et al. 1985, Wallace and O'Hop 1985). For example, above-ground production of the water lily *Nuphar luteum* in the sixth-order Ogeechee River, Georgia, was high in local beds (575 g dry mass/m^2 of river surface area over a 179-day period). Much of this production entered the food chain through grazing by water lily beetles (*Pyrrhalta nymphaea*) (Wallace and O'Hop 1985).

Light intensity is an important limiting factor to the growth of macrophytes in blackwater streams, because of both canopy development over smaller streams during the macrophyte growing season and light attenuation in larger rivers. Discharge patterns of these streams are probably also important. W. R. English (personal communication) suggests that the highly developed macrophyte beds in Upper Three Runs Creek, South Carolina, are a function of that stream's more constant discharge versus the more fluctuating discharge patterns of Cedar Creek and the Ogeechee River, both of which have less developed macrophyte growth.

Primary Production and Allochthonous Energy Sources

Rates of primary production, based on whole-system oxygen metabolism studies, show that blackwater streams and rivers have low primary productivity and are highly heterotrophic systems (Table 2). All of the low-order streams are highly heterotrophic, photosynthesis to respiration (*P/R*) ratios typically ranging from 0.2 to 0.7 over a year and annual average *P/R*'s being in the lower part of this range. *P/R* ratios increase with increasing stream

TABLE 2 Rates of Gross Primary Production (GPP) and Respiration (R) Measured in Southeastern Blackwater Streams and Rivers

Stream	GPP	R	Reference
North Carolina streams	0.3–9.8	0.7–21.5	Hoskin (1959)
North Carolina headwater streams	0.5–5.7	1.8–6.7	Kuenzler et al. (1977)
Creeping Swamp, NC	0.1–3.9	0.005–0.15* 0–4.7**	Mulholland (1981)
Jourdan River, MS	4.0–28.4	7.9–50.1	de la Cruz and Post (1977)
Ogeechee River system, GA	0.02–14.0	0.5–11.8	Edwards and Meyer (1987a), Meyer and Edwards (1990)

Note. All values are reported as g O_2/m^2 day^{-1}. *, Floodplain water column; **, flooded floodplain floor.

size, but even mid-order rivers, which according to the river continuum concept (Vannote et al. 1980) should tend toward autotrophy, are highly heterotrophic (Meyer and Edwards 1990). For example, the Ogeechee River had an annual average P/R of 0.25, with P/R never exceeding 0.6 except in July when it reached 1.3 during low-flow conditions (Edwards and Meyer 1987a).

A number of factors interact to cause these heterotrophic conditions. Primary production is low because of the factors noted above. In addition, detritus transported from floodplains to channels contributes to high respiration rates by providing large quantities of allochthonous organic carbon used by bacteria as a growth substrate (Findlay et al. 1986a,b, Meyer et al. 1987). The high bacterial biomass found in larger rivers also seems to originate predominantly in floodplains (Edwards and Meyer 1986).

The low productivity and P/R ratios suggest the importance of allochthonous organic matter inputs to the energy budgets of southeastern blackwater streams. Detrital dynamics are influenced by channel morphology and hydrologic characteristics of streams, and these factors in turn influence the inputs, processing, storage, and transport of organic matter within the floodplain–channel system. Linkages between floodplains and channels are particularly important, especially in larger river systems with extensive floodplains. Timing and magnitude of hydrologic events determine if floodplains act as a sink or a source of organic matter (Shure et al. 1986). Studies on the Ogeechee River floodplain show it to be a major energy source to the river. Inputs from the floodplain are seven times greater than annual average in-channel primary production (Edwards and Meyer 1987a, Cuffney 1988). Much of the autumnal leaf fall to floodplains is transported directly to rivers (Cuffney 1988) or first processed and then transported to channels as smaller particles (Smock 1990). Inputs to channels subsidize microbial and macroinvertebrate communities (Edwards 1987, Edwards and Meyer 1987b, Wallace et al. 1987). Floodplains thus serve as the functional headwaters of river systems that arise predominately on the Coastal Plain.

Studies on annual direct litter fall into southeastern blackwater streams indicate levels of inputs similar to or slightly higher than those in temperate deciduous forests (Brinson 1977, Post and de la Cruz 1977, Mulholland 1981, Elder and Cairns 1982, Cuffney 1988, Smock 1990). However, total organic matter inputs per unit length of channel are much higher than in other streams because of inputs from floodplains. The magnitude of inputs from floodplains is dependent on interactions between floodplain topography and elevation, hydrologic events, and seasonal changes in the availability of detritus in floodplains.

Detritus processing in blackwater streams has been the subject of a number of studies (Post and de la Cruz 1977, Hauer et al. 1986, McArthur et al. 1986, Cuffney and Wallace 1987, Allred and Giesy 1988, Metzler and Smock 1990). Processing rates are highly dependent on hydrologic conditions and the location within the stream where processing occurs, leaves buried in the sediment being processed at a slower rate than leaves on the sediment surface because of low, frequently anoxic, oxygen concentrations and lower availability of leaves to leaf-shredding macroinvertebrates (Mayack et al. 1989, Metzler and Smock 1990). Processing rates also decrease rapidly as the period of flowing water in channels or floodplains decrease (Day et al. 1988, Walters and Smock, 1991). One reason for this trend is the effect that intermittent flow has on shredding macroinvertebrates. Only perennially flowing, sand-bottomed blackwater streams support those shredders, such as craneflies and some caddisflies, with high leaf consumption rates. Leaf processing is relatively rapid in these streams.

In intermittent-flowing streams, isopods and amphipods are the predominant shredders (Cuffney and Wallace 1987, Gladden and Smock, 1990). These species are, however, facultative shredders, commonly feeding on fine particulate organic matter, algae, and the surface microbial matrix on leaves rather than actually shredding leaves. Leaf processing in these streams thus is highly dependent on microbial activity and to a lesser extent on physical abrasion, both of which are negatively affected by progressively shorter periods of inundation.

Detritus storage and transport also are important to the energetics of blackwater streams. The quantity of stored organic matter tends to be greater than in comparably sized higher-gradient streams primarily because of higher settling and burial rates. Differences in storage are especially pronounced if a whole-system view is taken, recognizing the highly coupled nature of channels and floodplains and thereby including detritus stored in floodplains as being within the functional "stream system."

The majority of detritus stored in sand-bottomed stream channels is buried. Microbial activity associated with this material may be limited in small streams because of low oxygen concentrations in the sediment. In larger rivers, subsurface oxygen concentrations are higher because of higher water exchange rates with the surface; bacterial activity on detritus thus occurs deep in the sediments (Meyer 1988).

Deep accumulations of peat-like sediment occur in streams with very low flow. In faster-flowing streams with sand sediments, the majority of detritus on the channel surface is stored in debris dams. Debris dam abundance is especially influenced by age of the riparian forest. Because most forests of the Coastal Plain are timbered before attaining a senescent stage, the input of large logs to streams is limited, decreasing the number of dams. The effectiveness of debris dams in detritus retention also can be limited because of the erodable nature of the sediment. New channels can easily be cut around and under logs, thereby making the debris dams leaky in terms of detritus retention.

Detritus retention on the channel surface of sand-bottomed streams is low compared to high-gradient streams because of the combined effects of a lack of a rocky substrate, lower retention capabilities of debris dams, and the flushing effects of storms. Rising flow associated with even small storms is efficient at flushing detritus from the channel surface (Leff and McArthur 1988, Smock et al. 1989). During major storms, for example as associated with rainfall from tropical storms and hurricanes, large quantities of detritus are removed from both surface and subsurface storage. Such events have long-term effects on stream metabolism and productivity.

ANIMAL COMMUNITIES

Invertebrates

Prior to 1975, published information on invertebrates inhabiting southeastern blackwater streams was limited almost entirely to studies of Florida streams by William Beck and William Peters and colleagues at Florida A & M University (e.g., Beck 1965, Peters and Jones 1973, Beck and Beck 1974). Since that time there has been a continuing increase in the number of studies describing the taxonomic composition, abundance, and function of invertebrates in the streams, rivers, and floodplains throughout the entire southeastern United States. These studies are providing important insights into the ecological functioning of Coastal Plain river systems, aquatic ecological theory in general, and the best management practices for these river systems.

Most recent research has been on macroinvertebrates; little is known about zooplankton and protozoan communities in blackwater systems. Protozoans may be at least locally important to the functioning of stream communities. High protozoan densities, primarily of microflagellates and lesser numbers of ciliates, have been reported in the Ogeechee River system, especially in backwater areas (Carlough 1989, Carlough and Meyer 1990). Net protozoan production was 600 μg C/L day^{-1} (Carlough and Meyer 1989) and may be a key factor in the trophic dynamics of this river through controlling bacterial production and by supporting a large fraction of the abundant micro-filter-feeding macroinvertebrates in the river (Findlay et al. 1986a, Wallace et al.

1987). Further study of the spatial distribution and functional importance of protozoans is warranted.

Densities of zooplankton are typically low in blackwater rivers. In Black Creek, South Carolina, densities were 48–2093 individuals/m³ of water during the spring and summer (Herlong and Mallin 1985). The zooplankton community consisted of species of rotifers, copepods, and cladocerans, such as *Bosmina longirostris*, *Alona*, and *Alonella*. Harpacticoid copepods, including *Acanthocyclops vernalis* and *Attheyella illinoisensis*, can also be an important component of the invertebrates within the sandy substrates of blackwater streams (Strommer and Smock 1989) and in floodplains (Gladden and Smock, 1990).

Recent studies on macroinvertebrates have put to rest the classic but unwarranted view of blackwater streams as having low macroinvertebrate taxonomic richness and productivity. Species richness can be very high, often being comparable to that found in streams located in other geophysical provinces in the Southeast (Lenat 1983). Over 240 taxa have been collected from a first-order stream in Virginia (L. A. Smock and J. E. Gladden, unpublished data), 170 taxa were collected from second-order Cedar Creek in South Carolina (Smock and Gilinsky 1982), and over 550 species of aquatic insects were collected around fourth-order Upper Three Runs Creek (Morse et al. 1980, 1983, Kondratieff and Kondratieff 1984). The latter number reflects the intensive work on adult insects at that site by J. C. Morse and colleagues. While it includes species occurring in other areas of the drainage basin, it also reflects the more heterogeneous nature of the stream due to the occurrence of large beds of macrophytes that are not found at the other sites.

Although the number of taxa collected at the three sites noted above differ, relative differences in the number of species in each insect order are approximately the same as those shown for Upper Three Runs Creek in Fig. 3. Diptera typically compose the greatest proportion of the taxa; Trichoptera also are well represented. Many of the insect species are endemic to the region. For example, of the approximately 450 species of Trichoptera found in southeastern Coastal Plain streams, about 75 species are endemic (Holzenthal and Hamilton 1984). Of these, 45% are Hydroptilidae, 18 are Leptoceridae, and 14% are Hydropsychidae. Most of the endemic species are southern disjuncts of typically northern species. Considerable speciation and endemism also have occurred in the Mollusca throughout the Coastal Plain; numerous studies on molluscan taxonomy and biogeography have been published (e.g., Goodrich 1942, Clench and Turner 1956, Heard 1970, Johnson 1970, Burch 1973, 1982, Sepkowski and Rex 1974).

Only a few species of Mollusca are normally present in the smaller and more acidic streams. *Physa pumilia* was the only mollusk found by Beck (1965) in swamp streams in Florida, and only 10 of 242 and 4 of 170 macroinvertebrate taxa in Virginia and South Carolina streams, respectively, were mollusks (Smock and Gilinsky 1982, L. A. Smock and J. E. Gladden, unpublished data). The most abundant mollusks are the Sphaeriidae, or fingernail clams, which have a greater tolerance to high acidity than do most

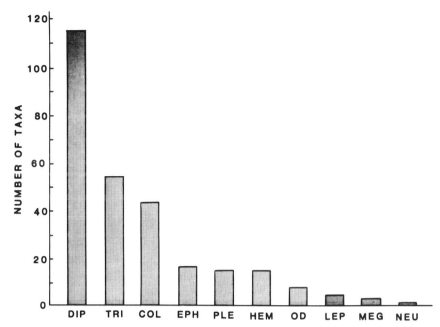

FIGURE 3. Number of taxa of different insect orders present in Upper Three Runs Creek, South Carolina. (Adapted from Morse et al. 1980, 1983.) DIP, Diptera; TRI, Trichoptera; COL, Coleoptera; EPH, Ephemeroptera; PLE, Plecoptera; HEM, Hemiptera; OD, Odonata; LEP, Lepidoptera; MEG, Megaloptera; NEU, Neuroptera.

other mollusks (Pennak 1978). Mollusk abundance increases as stream size increases and is correlated with the decrease in acidity and higher calcium concentrations that often occur with increasing stream size. *Corbicula fluminea*, the introduced Asiatic clam, is abundant in some rivers such as the Ogeechee, but it is not found in other rivers most likely because of low calcium concentrations (Benke and Meyer 1988).

Information on life histories of invertebrates from blackwater streams is minimal. The mayfly *Dolania americana* (Behningiidae) has received much attention because of its "threatened" status in Florida and because, unlike most other mayflies, it is predaceous and lives in sand substrate (Peters and Peters 1977, Tsui and Hubbard 1979, Finn and Herlong 1980, Fink 1986, Peters et al. 1987). Life histories for a variety of species have been presented by Benke et al. (1984), Smock (1988), and Roeding and Smock (1989). Morse (Morse et al. 1980, 1983) lists adult flight periods for most of the 550 species collected from Upper Three Runs Creek, and Soponis (1983) reports on the emergence patterns of five species of the chironomid genus *Polypedilum*.

Many species of aquatic invertebrates inhabiting southeastern blackwater streams are multivoltine (having multiple generations per year) because of the generally warm year-long water temperatures. This is particularly true of many blackflies (Simuliidae) and midge flies (Chironomidae). Cohort production intervals, or larval development times, of less than two weeks for blackflies and chironomids have been measured at typical summer water temperatures (Hauer and Benke 1987, Stites and Benke 1989); even larger taxa, such as some species of mayflies, have development times of less than a month (Benke and Jacobi 1986). Such short development times are a critical aspect of the production ecology of Coastal Plain macroinvertebrate communities since they indicate rapid turnover times and hence high annual production. A key question that remains unanswered is whether rapid development rates are found throughout the Coastal Plain, especially given the often low levels of high-quality food found in smaller streams during warmer months.

Besides the effects of warm water temperature, the distribution, abundance and production of macroinvertebrates in blackwater streams is controlled in large part by stream hydrology and morphology. In particular, the predominance of sandy, unstable sediment in the channel, combined with frequent periods of high, substrate-scouring flow, can limit the development of macroinvertebrate communities. This was shown in Virginia where high discharge during a spate caused extensive scouring of the sand substrate of a headwater stream, resulting in the loss of 42% of the individuals and 75% of the biomass of the invertebrate community (Strommer and Smock 1989). Similarly, extremely high discharge associated with the storm surge from Hurricane Hugo in 1989 scoured the sediment of a small coastal stream in South Carolina, decreasing invertebrate numbers and biomass by 97% (L. A. Smock, unpublished data). The effects of scouring can be quickly offset, however, by recolonization from upstream and subsurface areas and in particular because of the short generation times of many of the smaller species.

Conditions of substrate-limited growth are likely on the few stable substrates in channels, including debris dams, snags, and macrophytes. Macroinvertebrate colonization of new snags is rapid (Thorp et al. 1985) and species richness is higher on snags and in debris dams than in other habitats. Highest biomass and production also occur in these habitats, production being 3–7 times higher on snags than in sandy or silty stream sediments (Benke et al. 1984, Wallace and Benke 1984, Smock et al. 1985, Smock and Roeding 1986, Benke and Meyer 1988). Annual production ranged up to 72 g/m^2 on snags in the Satilla River and 36 g/m^2 in debris dams in a headwater stream. This high production is critical to the trophic structure of Coastal Plain streams and rivers.

Abundance of macroinvertebrates also is higher on macrophytes (Smock et al. 1985) and in macrophyte beds (W. R. English, personal communication) than in sandy sediment. Macrophytes stabilize sediment and are an important substrate and, upon their death, food for invertebrates. D. R. Lenat (personal communication) considers the annual cycle of water lily abundance in many Coastal Plain rivers to be the major factor influencing seasonal variation in macroinvertebrate abundance.

Production in sandy and silty sediments is roughly equivalent. Species composition, however, differs considerably between these two sediments.

Oligochaetes and isopods are abundant in the silty areas, whereas chironomids and nematodes predominate in sandy areas. Sand sediment also can have an abundant and active subsurface invertebrate community.

Total macroinvertebrate production in a stream cannot be assessed simply by measuring production in specific habitats since all habitats are not equally abundant. Estimates of production for a stretch of stream, calculated by adjusting habitat-specific production according to habitat availability, show that the majority of production in Coastal Plain streams occurs in sediments, even though production on snags and in debris dams is higher than in sediments on a per unit surface areas basis. The large areas of sediment compared to that of snags and dams, and the predominance of species with very short development times in the sediments causes the trophic dynamics of Coastal Plain streams to be highly influenced by the sediment-dwelling community.

Annual macroinvertebrate production increases with increasing stream size (Table 3). Published estimates range from 2.1 g/m² in intermittent-flowing Creeping Swamp, North Carolina, to over 33 g/m² in the sixth-order Satilla River. The latter value is high compared to production in higher-order rivers; however, production in headwater blackwater streams is low compared to published estimates of 5–22 g/m² for higher-gradient headwaters. Lower production in blackwater headwater streams may be a function of a number of factors, including especially high mortality of organisms in sand sediment during spates as well as the low retention of detritus on the sediment surface (resulting in low food availability). However, the supposed differences in production may not be real, since no comparative studies on production in low- and high-gradient streams have been performed using the same sampling and analytical techniques and encompassing all potential habitats in the streams.

A stream system approach needs to be taken in assessing macroinvertebrate production in Coastal Plain waterways, encompassing channel surface areas as well as the often extensive subsurface and floodplain areas. Production in sandy subsurface areas can be far higher than that on the surface, for example,

TABLE 3 Annual Macroinvertebrate Production in Blackwater Streams

Site	Q	P	Comment	Reference
Creeping Swamp, NC	?	2.1	Channel surface	Sniffen (1981)
		5.1	Floodplain	Sniffen (1981)
Buzzards Branch, VA	0.8	5.6	Channel surface	Smock et al. (in press)
		1.7	Floodplain	Gladden and Smock (1990)
Cedar Creek, SC	1.2	3.1–4.1	—	Smock et al. (1985)
Satilla River, GA	44.7	16.8	Upstream site	Benke et al. (1984)
	87.2	33.5	Downstream site	Benke et al. (1984)

Note. In all studies habitat-specific production was adjusted according to the relative abundance of the major habitats in the stream. Q = annual mean discharge (m³/s); P = production (g dry mass/m² year⁻¹).

5.6 g/m² on the surface and over 10 g/m² in the subsurface of a Virginia stream (Smock et al. 1992). Production in floodplains also can be an important component of total production of the stream systems. Given the extensive size of many floodplains relative to sizes of stream channels, production in floodplains per linear meter of channel usually constitutes the major proportion of total system production. The extent of linkages of invertebrate communities and trophic dynamics in general between channels and floodplains is unknown but is an important issue for evaluating the ecological importance of floodplains. More work in this area is needed.

Downstream drift of macroinvertebrates is another important aspect of the trophic dynamics of at least the larger blackwater rivers given the importance of macroinvertebrates to the diets of many gamefish. Densities of drifting macroinvertebrates can be high, the majority of drifting organisms originating from snags (Benke et al. 1986). Highest densities in the drift occur in the late spring and summer (Soponis 1980, Soponis and Russell 1984, Specht et al. 1984, Benke et al. 1986) when fish feeding rates are highest, although Cowell and Carew (1976) found greatest drift during the winter and early spring in Florida. Drift of invertebrates from snags is an important food source for fish in large rivers (Benke et al. 1985).

The relative abundance of taxa of each of the functional feeding groups in southeastern blackwater streams differs somewhat from that proposed by the river continuum concept (Vannote et al. 1980) for river systems in general. Whereas collector–gatherers predominate in streams of all orders (Fig. 4; also see Benke and Meyer 1988), collector–filters are abundant even in some headwater streams. Scrapers are not common in these river systems.

A wide variety of species in the collector–gatherer feeding guild are found in blackwater streams, the dominant species changing over short geographical distances and with changing environmental conditions. Their abundance is linked with their ability to exploit the high levels of fine particulate organic matter found in most habitats in blackwater streams. Chironomids usually

FIGURE 4. Relative abundance of macroinvertebrate functional-feeding groups in second-order Cedar Creek (Smock et al. 1985), fourth-order Upper Three Runs Creek (Morse et al. 1983), and the sixth-order Satilla River (Benke et al. 1984). Shred, shredder.

are the most abundant collector–gatherers in sandy streams, some of the more commonly encountered genera being *Corynoneura*, *Lopescladius*, *Phaenopsectra*, *Polypedilum*, *Rheosmittia*, and *Tanytarsus*. The oligochaete *Barbidrilus paucisetus* (Enchytraeidae) can be a dominant member of the sand community of larger rivers (Sites 1987). A different set of species dominates the collector–gatherer group in backwater areas with more organic matter in the sediment; genera such as the oligochaete *Limnodrilus* and the chironomids *Chironomus*, *Dicrotendipes*, *Diplocladius*, *Parametriocnemus* and *Polypedilum* often predominant. Snags and dams harbor the majority of the larger species of collector–gatherers, especially mayflies (*Stenonema* and species of Ephemerellidae).

High densities of collector–gatherers occasionally occur deep within sand sediment. Small species, with rapid development rates and the ability to exploit interstitial areas among sand grains, are most abundant. Dissolved oxygen concentration limits invertebrate vertical distribution and abundance. Some chironomids and enchytraeid and lumbriculid oligochaetes are especially common in the well-aerated Ogeechee River sediments. Little decrease in their densities occurs down to 20 cm (Gillespie et al. 1985, Stites 1986). Abundance is lower in the sediment subsurface of smaller streams where oxygen concentrations drop to near zero only centimeters below the surface during summer (Strommer and Smock 1989).

Most filter-feeders occur on snags, although fingernail clams (*Sphaerium* and *Pisidium*) and the tube-building caddisfly *Phylocentropus* (Wallace et al. 1976, Benke et al. 1984) are occasionally abundant in sediments. The predominant filterers in headwater streams are fine-particle feeders such as blackflies, Tanytarsini chironomids, and the hydropsychid caddisfly *Macrostemum*. As stream size increases, there is an addition of those species that feed on larger particles, such as caddisflies of the genus *Hydropsyche* (Benke et al. 1984, Wallace and Benke 1984). Variation in blackfly production, and possibly production of other microfilterers, is linked to floodplain inundation patterns and the export of high-quality food to stream channels (Benke and Parsons 1990). Reservoirs built on blackwater streams change the filter-feeding community from fine-particle feeders upstream of the reservoir to large-particle feeders below the dam. Downstream organisms feed on the abundant zooplankton originating from the reservoir (Herlong and Mallin 1985).

Scraping macroinvertebrates, which in higher-gradient streams are important algal grazers, are not abundant in Coastal Plain river systems (Fig. 4), including even mid-order rivers where theoretically they should be at a maximum. Lack of stable substrates and the resulting low primary production keep densities low. However, those few streams where a rocky substrate exists, such as local gravel riffles in the Sandhills, support a diverse and occasionally abundant scraper community (D. R. Lenat, personal communication).

Shredders, feeding on large pieces of leaf litter and wood, are not often abundant in blackwater streams (Fig. 4), although their abundance can be high in debris dams. Shredders accounted for less than 10% of all macroinvertebrates in a first-order Florida stream (Scheiring 1985) and less than 5%

of macroinvertebrate numbers and production in second- to fourth-order streams (Morse et al. 1983, Smock et al. 1985, Scheiring 1985). However, they did account for 33% of invertebrate production on the channel surface of a first-order stream (L. A. Smock, unpublished data). Low abundance is related to low leaf litter retention on sediments. The most abundant shredders such as the caddisfly *Pycnopsyche* and the cranefly *Tipula* occur primarily in debris dams (Roeding and Smock 1989). Their abundance probably is food and substrate limited. The caddisfly *Agarodes* (Sericostomatidae) and the isopods *Caecidotea* and *Lirceus* are the only shredders commonly found in sandy and peaty sediments, respectively. The amphipod *Gammarus tigrinus* is a particularly abundant facultative shredder in some intermittent-flowing streams close to estuarine areas. Its abundance is related to its tolerance of the occasionally high salinities that occur in these streams during strong off-shore storms.

Macroinvertebrate production in blackwater streams is based on allochthonous detritus, in particular on fine particulate organic matter (FPOM, 0.045–1 mm). Smock and Roeding (1986) found that FPOM supported 47–64% of macroinvertebrate production in second-order Cedar Creek, the proportion being greater in low-flow, swampy areas compared to sand-bottomed areas of the stream (Fig. 5). Amorphous detritus and associated bacteria, much of which originates on the floodplain, are an important link between terrestrial primary production and invertebrate production in blackwater streams and rivers, for example, accounting for the majority of production by invertebrate microfilterers and collector–gatherers of the Ogeechee River (Edwards 1987, Wallace et al. 1987).

Autochthonous energy sources were an important aspect of the trophic dynamics of Cedar Creek (Fig. 5), algae supporting a large proportion of the production of some abundant filter feeders. The majority of ingested algae was of the same species that were seasonally abundant in the Cedar Creek floodplain. Occasional floods that flush algae from floodplains may provide a nutritionally important food source for filter feeders in these streams. Algae

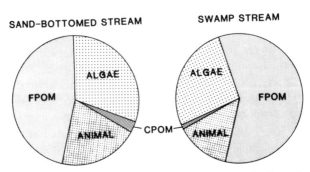

FIGURE 5. Relative contribution of food types to the production of macroinvertebrates at two sites on Cedar Creek, South Carolina (Smock and Roeding 1986). CPOM, coarse particulate organic matter; FPOM, fine particulate organic matter.

also were an important component of the diet for some species in the Ogee-chee River (Wallace et al. 1987).

Because of low abundance of shredders, little macroinvertebrate production is supported directly by CPOM in most blackwater streams (Fig. 5). Only permanent-flowing, first-order streams have at least moderate levels of shredder abundance and thus leaf processing by macroinvertebrates. However, even in these streams, common shredders such as *Pycnopsyche* and *Agarodes* rely heavily on wood as a food resource. The proportion of wood ingested increases as leaves become less abundant throughout the late winter and spring because of low leaf retention in channels (Roeding and Smock 1989). Some shredders also may rely heavily on algae as a food resource in larger rivers (Wallace et al. 1987).

Leaf processing in floodplains, and the subsequent movement of the resulting FPOM to stream channels during floods, probably accounts for much of the FPOM used by the majority of macroinvertebrates. The often abundant isopods, amphipods, and terrestrial shredders in floodplains likely are responsible for much of the leaf shredding in these river systems; hence, floodplains must be considered an integral component of river system functioning.

Fish

Southeastern blackwater streams have an abundant and diverse fish community, especially compared to Piedmont and mountain streams. From 30 to 55 species of fish were found in first- to fourth-order streams on the Savannah River Plant, a total of 64 species being collected from five streams at that site (Dahlberg and Scott 1971, McFarlane 1976, Meffe and Sheldon 1988). Forty-two species occurred in a small Mississippi stream (Ross and Baker 1983) and 37 species were found in two undisturbed headwater streams in eastern North Carolina (Pardue et al. 1975, Huish and Pardue 1978). Besides being diverse, the fish communities of blackwater drainage basins also have complex zoogeographic histories and affinities with interesting patterns of endemism and subspeciation (Hocutt et al. 1986, Swift et al. 1986).

Species composition of fish assemblages is largely determined by hydrologic and habitat structure characteristics, including water depth and flow, extent of available cover (roots, snags, undercuts), and substrate type (Meffe and Sheldon 1988, 1990). Most species are typical of slow-flowing, deep water habitats (Table 4). Dominant species often differ between streams because of differences such as stream size and magnitude and permanency of flow. In a study by Pardue et al. (1975), redfin pickerel (27%), brown bullhead (17%), and creek chubsucker (15%) were the most abundant species by weight in Duke Swamp Creek, whereas American eel (31%), redfin pickerel (21%), yellow bullhead (15%), and flier (12%) were the most abundant species in nearby Hoggard Mill Creek. The ratio of forage fish to carnivorous fish in those streams was low, ranging from 1.0 to 2.8 (Huish and Pardue 1978).

Coastal Plain streams and their floodplains are important spawning and nursery areas for many fish species, including many anadromous species and

TABLE 4   Fish Species Commonly Found in Eastern North Carolina Blackwater Streams

Common Name	Species
Bowfin	*Amia calva*
American eel	*Anquilla rostrata*
Herring	*Alosa* spp.
Eastern mudminnow	*Umbra pygmaea*
Redfin pickerel	*Esox a. americana*
Chain pickerel	*Esox lucius*
Golden shiner	*Notemiqonus crysoleucus*
Ironcolor shiner	*Notropis chalybaeus*
Creek chubsucker	*Erimyzon oblonqus*
Yellow bullhead	*Ictalurus natalis*
Brown bullhead	*Ictalurus nebulosus*
Tadpole madtom	*Noturus gyrinus*
Swampfish	*Choloqaster cornuta*
Pirate perch	*Aphredoderus sayanus*
Mud sunfish	*Acantharcus pomotis*
Flier	*Centrarchus macropterus*
Blue-spotted sunfish	*Enneacanthus gloriosus*
Swamp darter	*Etheostoma fusiforme*
Sawcheck darter	*Etheostoma serriferum*
Warmouth bass	*Lepomis gulosus*

Source. Adapted from Pardue et al. (1975) and Huish and Pardue (1978).

the catadromous American eel. Studies have documented the widespread use of blackwater streams by anadromous culpeids for spawning and as nursery areas for larval and juvenile fish (e.g., Gasaway 1973, Davis and Cheek 1966, Baker 1968, Pate 1972, Frankensteen 1976). Highest numbers of fish occur from April to June, the time for most spawning runs and highest water and thus greatest area of inundated floodplain. Many species of fish move onto floodplains to forage, and some species spawn on the floodplain floor. Damming of streams can be locally catastrophic to migrating fish, preventing upstream migration and thus eliminating species from their historical spawning areas.

Studies on relatively undisturbed headwater streams in eastern North Carolina indicate that these streams typically support a fish standing stock biomass of 50–370 kg/ha (Table 5 and Bayless and Smith 1964). Standing stock estimates are quite variable between streams, being especially affected by variations in water level. The relatively high estimate of 1112 kg/ha for Potecasi Creek (Table 5) was the result of sampling after a prolonged dry period that concentrated fish into a small area (Huish and Pardue 1978).

Blackwater streams and rivers compose a large portion of the public fishing waters in the southeastern United States. They are excellent gamefish waters. All but the smallest streams typically support 60–200 kg of gamefish/ha (Table 5). From 50 to 70% of the total fish standing stock biomass usually is composed

TABLE 5   Fish Standing Stock Biomass (kg/ha) in Coastal Plain Blackwater Streams and Rivers

Site	Standing Crop		Reference
	Total	Gamefish	
Eastern NC (28 streams)	174	86	Tarplee et al. (1971)
Eastern NC (30 streams)	—	143	Bayless and Smith (1964)
Duke Swamp Creek, NC	280	61	Huish and Pardue (1978)
Hoggard Mill Creek, NC	308	226	Huish and Pardue (1978)
Potecasi Creek, NC	1112	706	Huish and Pardue (1978)
Satilla River, GA	—	179	Holder and Ruebsamen (1976)

of gamefish. In terms of numbers of gamefish, Bayless and Smith (1964) found a mean of 1211 gamefish larger than 15.2 cm (6″) per ha in their study of 30 undisturbed headwater to mid-order streams in North Carolina. The most commonly caught gamefish include redfin and chain pickerel, warmouth, redbreast sunfish (*Lepomis auritus*), bluegill (*Lepomis macrochirus*), pumpkinseed (*Lepomis gibbosus*), and largemouth bass (*Micropterus salmoides*). In larger blackwater rivers, channel catfish (*Ictalurus punctatus*) and various species of bullheads are typically sought by fishermen.

Because of the close coupling of channels and floodplains, fish probably heavily utilize terrestrial organisms and food resources originating in the floodplain. Many fish move onto floodplains as soon as they are flooded (Ross and Baker 1983), possibly in response to large numbers of litter-dwelling terrestrial invertebrates trapped by rising water. As the period of floodplain inundation lengthens, an aquatic invertebrate community develops that, along with terrestrial insects caught in the surface film, supports the fish community.

Snags are a particularly important aspect of the ecology of fish in larger blackwater rivers. They add physical complexity to channels, thus serving as cover for fish. In addition, Benke (Benke et al. 1979, 1985) has shown their importance to fish feeding ecology. He suggests that two major food chains occur in the Satilla River. Small fish, such as shiners, silversides, minnows, darters, and mosquitofish, rely heavily on sediment-dwelling chironomids and terrestrial insects caught at the water's surface. However, up to 90% of the food items, and 60% of prey biomass, of most larger fish are invertebrates, especially chironomids and caddisflies, originating on snags. This was especially true for the diets of sunfish and pirate perch. Many prey were probably caught as they drifted downstream after leaving snags. At the top of each of these food chains were piscivorous largemouth bass and chain and redfin pickerel. Aquatic insects in general, and particularly insects dwelling on snags, thus are an integral aspect of the Satilla River food web and no doubt of other blackwater rivers.

UNIQUE BIOTA AND SPECIAL FEATURES

Threatened and Endangered Species

No species occurring in Coastal Plain streams are included on the U. S. Fish and Wildlife Service's list of endangered and threatened wildlife and plants. A number of species, however, are listed as endangered, threatened, or of special concern by the states in this region. The species included in the lists for North and South Carolina are noted in Appendix 1. They include several species of Unionidae mussels, five species of crayfish, various fish, and an insect. Other species are included in lists for other states (e.g., Williams et al. 1989).

Many species listed for North and South Carolina occur in the closely associated Lumber and Waccamaw River systems. A number of aquatic species are endemic to one or both of these systems; Bailey (1977) estimated that nearly 20% of the fish in the Lumber River system are of unusual interest because of their limited distribution. Appendix 1 does not include those species of fish, mussels, and a snail that are endemic to Lake Waccamaw; further sampling throughout this river system may determine that some of these species are not confined to the lake, but also occur in riverine habitats. Considering the number of species listed in Appendix 1 and the number of endemic species that occur in these two river systems, greater scientific investigation and special consideration in terms of further development in their watersheds is warranted.

The Sandhills area is another area of special concern. The Sandhills Game Management Area in North Carolina and the Carolina Sandhills National Wildlife Refuge in South Carolina (Fig. 1) have been established to preserve the flora and fauna of this distinctive habitat. A listing of the fish that occur in the Refuge and a discussion of the status of some of the species of special interest that occur there are given by Olmsted and Cloutman (1979). Other areas such as the Congaree Swamp National Monument and the Savannah River Plant in South Carolina, the Blackwater Ecologic Preserve in Virginia, the Blackwater River State Forest in Florida, and the Okeefenokee Swamp also serve as important preserves for the biota of blackwater streams and associated swamps.

Wild and Scenic Rivers

Many blackwater streams and rivers are noted in the Nationwide Rivers Inventory (Appendix 2). They are recognized as significant free-flowing areas that could be designated as Wild and Scenic Rivers. It is our hope that a number of these ecologically and esthetically significant rivers come under federal and/or state protection. Upper Three Runs Creek on the Savannah River Plant has been designated by the U.S. Geological Survey as a National Hydrologic Benchmark Stream, providing long-term hydrologic data for this well-studied, ecologically significant stream.

RESOURCE USE AND MANAGEMENT EFFECTS

Much of the Coastal Plain is experiencing some form of agricultural or urban use and nearly all of this area has at one time been utilized, and altered, by man. The ecology of blackwater streams thus must be examined both in terms of present day and historical human impacts.

The Coastal Plain is an agriculture-dominated area with extensive tracts of both traditional and forestry crops. For example, in South Carolina about 45% of the Lower Coastal Plain and 30% of the Upper Coastal Plain is used for nonforestry agriculture. Forests and wetlands compose about 79% of the Chowan River basin in Virginia and North Carolina; the remainder of the basin is used primarily for agriculture (Fig. 6). Much of the forested areas throughout the Coastal Plain are managed for forest products. Impacts on blackwater streams therefore are mainly agriculture-associated problems.

Urbanization is not as great a problem as it is on the Piedmont since there are few major population centers on the Coastal Plain. Because of this, the demands on surface water for drinking water and for industrial uses are relatively light compared to the Piedmont, although agricultural demands for irrigation water can be locally significant. Historically, the major water resource problem for the Coastal Plain has not been the retention of surface water in reservoirs for drinking water or for flood control, but rather the draining of water from land to allow agricultural use. This is usually accomplished through channelization projects.

The relatively low urbanization also has meant that streams and rivers of the area have not experienced many of the man-induced problems that have occurred in the Piedmont. Whereas many Piedmont streams have been dammed, few reservoirs have been built on blackwater streams, mainly because groundwater has traditionally supplied the required drinking water and because the flat topography of the area is not suitable for the construction of deep, long-lived reservoirs. Where reservoirs have been built, their biological impact has

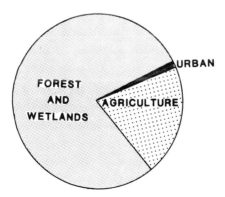

FIGURE 6. Land use in the Chowan River basin in Virginia and North Carolina. (Modified from Virginia Water Control Board 1985.)

included prevention of upstream migrations of anadromous fish (Huish and Pardue 1978) and changes in invertebrate community composition (Herlong and Mallin 1985).

In addition, fewer Coastal Plain waterways have been significantly degraded by sewage inputs than have Piedmont waterways. Where sewage inputs have occurred, the assimilation capacities of the streams often have been sufficient to minimize and localize the problem. Similarly, the effects of nonpoint runoff, at times significant, seems to be lower in the Coastal Plain than in the Piedmont area (Cook et al. 1983, Georgia Department of Natural Resources 1983, Lenat and Eagleson 1981). With few heavy industries in the area, there have been fewer instances of stream pollution due to industrial wastes than on the Piedmont. However, significant impact from heated water discharges to streams on the Savannah River Plant have been well documented (e.g., McFarlane 1976, Kondratieff and Kondratieff 1984, Sadowski and Matthews 1984, Specht et al. 1984, Poff and Matthews 1985, 1986, Hauer et al. 1986), and water-quality degradation due to discharges from the paper-products industry have occurred.

Agricultural activities have a variety of effects on Coastal Plain streams. The hydrology of streams is often altered through channelization projects that are necessary for farming in some areas. Dissolved solids concentrations are increased through nonpoint runoff of fertilizers, pesticides, animal wastes, and septic tank effluent (Simmons and Heath 1979). Also, increased sediment load and particulate-associated ion concentrations occur in streams because of erosion from cleared and cultivated land. Alterations in the benthic macroinvertebrate community occur because of these water-quality problems (Lenat 1984). A key factor in reducing degradation of these streams is maintenance of the riparian vegetation (Lowrance et al. 1984), so that the vegetation can effectively filter nutrients and soil from the water before reaching the waterways.

Effects on water quality occur even if a small percentage of a watershed is used for agriculture. Concentrations of calcium, sodium, bicarbonate, sulfate, and chloride increase in streams flowing through areas where agricultural activity exceeds 10% of land use (Simmons and Heath 1979). Fertilizers and liming are the major sources of calcium, magnesium, potassium, chloride, and bicarbonate for South Carolina streams (Gardner 1983).

Nitrogen and phosphorus concentrations in streams are increased by agricultural activity (e.g., Gambrell et al. 1974, Simmons and Heath 1979, Yarbro et al. 1984). Base and storm flow concentrations of both of these nutrients increase proportionally with increasing agricultural activity in the watershed (Fig. 7). Few studies have documented the biological effects of nonpoint runoff of nutrients on the biology of blackwater streams. In the Chowan River, a combination of nonpoint and point-source inputs have been at least partially responsible for frequent blue-green algal blooms (Witherspoon et al. 1979, Duda 1982, Paerl 1982). About 73 and 89% of the total phosphorus and nitrogen inputs, respectively, to the Chowan River watershed are from nonpoint sources, the majority of these inputs being from agricultural

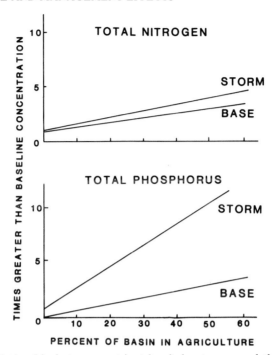

FIGURE 7. Relationship between nutrient levels in streams and the percentage of the watershed used for agriculture. The vertical axis was calculated as the nutrient concentration in the stream divided by the pristine, baseline concentration. (Modified from Simmons and Heath 1979.)

land versus forests and wetlands (Fig. 8). Adherence to best-management practices for agricultural land effectively decreases this runoff.

The effects of the timber industry on blackwater streams have not been sufficiently investigated. Conversion of land to monoculture pine stands or selective cutting of certain species alters the type of detrital inputs to streams, potentially causing ecological changes. The effects of clear-cutting, although well studied in high-gradient systems, are not clear for the low-gradient streams of the Coastal Plain. A small-scale (16 ha) clear-cutting operation in Georgia did not alter the water chemistry or biology of a stream (Cook et al. 1983). Effects could be extensive, however, if large portions of floodplains are included in timbering operations.

Channelization projects are common on the Coastal Plain. Their goal is to increase discharge of floodwaters and to lower groundwater levels, thereby making land adjacent to channelized streams more suitable for agriculture. In the past, channelization usually included the widening, deepening, and straightening of channels and the removal of snags. Timber was often cleared from a 5- to 10-m-wide strip along one or both banks; spoil from the channel was placed on the banks, creating a levee. Today, most projects are limited to removing snags and excess accumulated sediment.

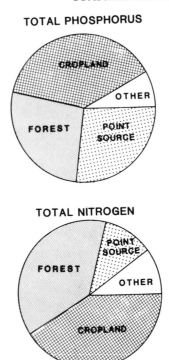

FIGURE 8. Sources of point and nonpoint nutrient inputs to the Chowan River in Virginia and North Carolina. (Virginia Water Control Board 1985.)

The main hydrologic effect of classic channelization is to isolate the channel from the floodplain, which results in a reduction of the frequency and magnitude of floodplain inundation. The relative contribution of groundwater to a stream's water budget increases, while that of overland flow decreases. Both base flow and peak rates of storm flow increase (Winner and Simmons 1977), with the result that many streams that were seasonally flowing are converted to perennially flowing systems.

The increased flow and reduced interaction of the floodplain and channel due to channelization result in a number of changes in water chemistry. Dissolved oxygen, turbidity, conductivity, alkalinity, and pH increase (Kuenzler et al. 1977, Huish and Pardue 1978). Organic matter concentrations and color decrease because of decreased floodplain inundation. Water temperature can increase several degrees during the summer if the forest canopy is removed (Tarplee et al. 1971), although increased input of usually cooler groundwater can ameliorate this effect. Concentrations and export of nutrients and many major elements increase, especially during spates, due to increased discharge and erosion of soils, faster groundwater drawdown (decreasing nutrient retention in soils), and increased inputs from land turned to agricultural use (O'Rear 1975, Kuenzler et al. 1977, Yarbro et al. 1984).

Biological effects of channelization include increased periphyton growth (Kuenzler et al. 1977) and decreased macroinvertebrate density and diversity (Chapin 1975, Huish and Pardue 1978). Occasional high densities of fast-growing macroinvertebrates, such as chironomids and oligochaetes, can develop during prolonged low-flow conditions, but they are highly susceptible to spates because of the unstable nature of sand sediment (Chapin 1975).

The effects of channelization on fish communities have been well studied. Channelization results in decreased diversity, mean size, density, and standing crop of fish (Bayless and Smith 1964, Tarplee et al. 1971, Huish and Pardue 1978). For example, both the number of gamefish greater than 15 cm and their standing crop were more than 80% lower in channelized versus natural streams (Bayless and Smith 1964). Whereas Tarplee et al. (1971) found that fish species diversity returned to pre-channelization levels after about 15 years, Bayless and Smith (1964) noted no significant return of numbers or standing crop of fish to natural levels 40 years after channelization. Effects of channelization probably are long term.

The major reasons for the adverse effects of channelization on fish include reduced habitat heterogeneity and hydrologic stability, higher flow rates, decreased food resources, and warmer summer water temperature. Frankensteen (1976) noted that spawning success of blueback herring (*Alosa aestivalis*) was lower in channelized streams because their eggs, which are adhesive and normally attach to instream vegetation and debris, are flushed out of the system before hatching. The draining of floodplains also adversely affects spawning of hickory shad (*Alosa mediocris*) and some species of river herring since they spawn in floodplain channels and pools.

Another management technique often practiced in the Southeast is snagging, in which large tree trunks and other obstructions are removed from river channels for navigation purposes. Unfortunately, this practice has significant hydrologic and energetic effects on river systems, especially by increasing transport rates of nutrients out of the system, altering river–floodplain interactions, and decreasing macroinvertebrate and fish production (Wallace and Benke 1984). Benke et al. (1979, 1985) suggest that extensive snagging in the Satilla River could reduce production of major fish species by at least 50%. Production of some species such as sunfish may be reduced by more than 70% because of their heavy reliance on macroinvertebrates originating on snags for their food.

As a final note, it should be remembered that there are few, if any, remaining watersheds on the Coastal Plain that have not been significantly impacted by man. Because of this, the ecological characteristics of streams in this area may not be similar to their pristine condition. Unfortunately, an accurate historical perspective is not easily obtained, since many potential effects have been obscured by time. For example, snagging operations have occurred in the Southeast at least since the early 1800s; over 15 000 snags have been removed from the lower 260 km of the Satilla River (J. R. Sedell, personal communication, cited in Wallace and Benke 1984). Our present concept of this river system does not include the ecological changes caused by this impact. Similarly, timbering practices probably have reduced the input of wood into streams, while the effects of fire control on nutrient input to waterways is unclear (Richter et al. 1982, Wilbur and Christensen 1983).

FUTURE MANAGEMENT AND RESEARCH NEEDS

A variety of aspects concerning the management and ecology of blackwater systems are unstudied or at least poorly understood. Of prime importance is the necessity of having ecologists, resource managers, and land-use planners recognize that floodplains are an integral component of these river ecosystems. The broad ecologic and economic significance of southeastern riverine floodplains, including their value in terms of hydrology, water quality, system productivity, education and recreation has often been noted (Wharton 1970, Wharton et al. 1982). However, many specific ecological- and management-oriented questions remain unanswered.

In particular, studies on floodplain–channel linkages are required, which examine such aspects as energy and nutrient exchanges and the extent of floodplain use by river-dwelling organisms. For example, floodplains are generally considered to be energy exporters; however, the dynamics and actual significance of this exported material to river ecosystems is not well documented. Neither are the effects of floodplain disturbances on energy export dynamics or river trophic dynamics understood. Also, whereas many studies have indicated that some species of fish utilize floodplains for foraging and spawning, little quantitative information exists detailing the factors governing floodplain use and the actual energetic value of the floodplain to these species.

Other subjects that need to be addressed include the ecological differences between floodplains on headwater streams and those associated with large rivers. Similarly, is there an ecological or hydrological necessity to preserve continuous tracts of undisturbed floodplains versus having large but fragmented and partially developed floodplains? Until such specific information is obtained, the effects of floodplain clearing, draining, and development can only be surmised in general terms, decreasing the effectiveness of protection and planning efforts for these critical areas.

Another research need is an integrated analysis of trophic linkages within these streams and rivers, encompassing allochthonous inputs and primary producers through top carnivores. As noted above, trophic linkages between the channel and the floodplain must be included in this analysis as well as the often ignored microbial, protozoan, and meiofanua communities. Also, almost no information exists on the importance of vertebrates other than fish (e.g., salamanders and birds) on the food webs of blackwater streams. It is only with this type of holistic information that the full ramifications of watershed and/or river management projects can be understood.

An additional important area of research and management involves the biological effects of agricultural runoff. Information is required on such aspects as the effects of increased sedimentation from soil erosion, the effects of increased nutrient concentrations on primary production and the potential acute and chronic effects of pesticide runoff. Research, development, and implementation of best-management practices, such as buffer strips, conservation tillage, and use of winter cover crops to reduce nonpoint runoff, are necessary to limit potentially adverse environmental effects. Given the close proximity of Coastal Plain streams to the Atlantic and Gulf estuaries, these problems have a scope broader than that of freshwater systems alone.

A future problem, which may already be upon us, is the effects of acid deposition. The acidity of rainwater has been increasing throughout the southeastern United States (Haines 1980). Changes in water quality and the ecological characteristics of blackwater streams could be rapid and severe given the naturally low buffering capacity of these streams. Little research on this problem has been initiated. A reverse aspect of the acidity problem is that many of the endemic, rare, and endangered species on the Coastal Plain occur in streams with naturally low pH. Since agricultural impacts usually raise pH, some consideration, and possibly research, on maintaining low pH in these streams may be necessary (Morgan 1984).

Blackwater streams and rivers have an aesthetic beauty that is quite different from the high-gradient trout streams of the mountains. In addition, they support a diverse flora and fauna that in many areas of the Coastal Plain is threatened by human activities. The establishment of protected areas or refuges in critical areas should be of high priority for government agencies and land-use planners and managers. The Lumber, Waccamaw, Black, and South River watersheds in North and South Carolina are prime examples of critical areas. They support a relatively large number of endemic and endangered and threatened species and have a high diversity of aquatic habitats that support a high diversity of all forms of aquatic life. There is almost no federal or state protection of any areas within these drainage basins. Other such critical areas exist throughout the southeastern Coastal Plain. Until their significance is recognized and the areas receive some form of government or private protection, unplanned development will continue to irreversibly consume ecologically, hydrologically, economically, and aesthetically important tracts of land throughout the Southeast.

ACKNOWLEDGMENTS

Much of the unpublished data referred to was collected through support by the National Science Foundation (BSR-8315763 and BSR-8614828) and the National Park Service (CX 5000-0-0946). We thank Art Benke, Dave Lenat, and Judy Meyer for their critical reviews and insightful comments.

REFERENCES

Allred, P. M., and J. P. Giesy. 1988. Use of in situ microcosms to study mass loss and chemical composition of leaf litter being processed in a blackwater stream. *Arch. Hydrobiol.* 114:231–250.

Atchue, J. A., III, F. P. Day, Jr., and H. G. Marshall. 1983. Algal dynamics and nitrogen and phosphorus cycling in a cypress stand in the seasonally flooded Great Dismal Swamp. *Hydrobiologia* 106:115–122.

Baker, W. D. 1968. *A Reconnaissance of Anadromous Fish Runs into the Inland Fishing Waters of North Carolina*, Project AFS-3. Raleigh: North Carolina Wildlife Research Commission, Federal Aid in Fish Restoration.

Bailey, J. R. 1977. Freshwater fishes. In J. E. Cooper, S. S. Robinson, and J. B. Funderburg (eds.), *Endangered and Threatened Plants and Animals of North Carolina*. Raleigh: North Carolina State Museum of Natural History, pp. 265–298.

Bayless, J., and W. B. Smith. 1964. The effects of channelization upon the fish populations of lotic waters in eastern North Carolina. *Proc. Annu. Conf. Southeast. Assoc. Game Fish Comm.* 18:230–238.

Beck, K. C., J. H. Reuter, and E. M. Perdue. 1974. Organic and inorganic geochemistry of some coastal plain rivers of the southeastern United States. *Geochim. Cosmochim. Acta* 38:341–364.

Beck, W. M., Jr. 1965. The streams of Florida. *Bull. Fla. State Mus., Biol. Sci.* 10:91–126.

Beck, W. M., Jr., and E. C. Beck. 1974. The Blackwater River basin and the Chironomidae of Florida. *Entomol. Tidskr., Suppl.* 95:18–20.

Benke, A. C., and D. I. Jacobi. 1986. Growth rates of mayflies in a subtropical river and their implications for secondary production. *J. North Am. Benthol. Soc.* 5:173–190.

Benke, A. C., and J. L. Meyer. 1988. Structure and function of a blackwater river in the southeastern U.S.A. *Verh. —Int. Vere. Theor. Angew. Limnol.* 23:1209–1218.

Benke, A. C., and K. A. Parsons. 1990. Modelling black fly production dynamics in blackwater streams. *Freshwater Biol.* 24:167–180.

Benke, A. C., and J. B. Wallace. 1990. Wood dynamics in Coastal Plain blackwater streams. *Can. J. Fish. Aquat. Sci.* 47:92–99.

Benke, A. C., D. M. Gillespie, F. K. Parrish, T. C. van Arsdall, Jr., R. J. Hunter, and R. L. Henry, III. 1979. *Biological Basis for Assessing Impacts of Channel Modifications: Invertebrate Production, Drift and Fish Feeding in a Southeastern Blackwater River*, Environ. Resour. Cent. Publ. No. ERCO6-79. Atlanta: Georgia Institute of Technology.

Benke, A. C., T. C. van Arsdall, Jr., D. M. Gillespie, and F. K. Parrish. 1984. Invertebrate productivity in a subtropical blackwater river: the importance of habitat and life history. *Ecol. Monogr.* 54:25–63.

Benke, A. C., R. L. Henry, III, D. M. Gillespie, and R. J. Hunter. 1985. Importance of snag habitat for animal production in southeastern streams. *Fisheries* 10:8–13.

Benke, A. C., R. J. Hunter, and F. K. Parrish. 1986. Invertebrate drift dynamics in a subtropical blackwater river. *J. North Am. Benthol. Soc.* 5:173–190.

Bergeaux, P. J. 1969. *Soils in Georgia*, Bull. No. 662. Athens: Cooperative Extension Service, College of Agriculture, University of Georgia.

Blood, E. R. 1980. Surface water hydrology and biogeochemistry of the Okefenokee Swamp watershed. Ph.D. Dissertation, University of Georgia, Athens.

Blood, E. R. 1986. Biogeochemistry of blackwater Coastal Plain streams. *Assoc. Southeast. Biol. Bull.* 33:60.

Bloxham, W. M. 1976. *Low-Flow Characteristics of Streams in the Inner Coastal Plain of South Carolina*, Rep. No. 5. Columbia: South Carolina Water Resources Commission.

Bradshaw, H. D. 1977. Nitrogen cycling in an alluvial swamp forest. MS Thesis, East Carolina University, Greenville, NC.

Brinson, M. M. 1977. Decomposition and nutrient exchange of litter in an alluvial swamp forest. *Ecology* 58:601–609.

Brinson, M. M., H. D. Bradshaw, R. N. Holmes, and J. B. Elkins, Jr. 1980. Litterfall, stemflow, and throughfall nutrient fluxes in an alluvial swamp forest. *Ecology* 61:827–835.

Brown, D. C. 1981. The interactive effects of temperature and nutrients on periphyton assemblages from a low pH South Carolina stream. Ph.D. Dissertation, University of South Carolina, Columbia.

Burch, J. B. 1973. Freshwater Unionacean Clams (Mollusca: Pelecypoda) of North America. Biota of Freshwater Ecosystems Identification Manual Number 11. *Water Pollut. Control Res. Ser.* 18050 ELD03/73.

Burch, J. B. 1982. *Freshwater Snails (Mollusca: Gastropoda) of North America*, EPA-600/3-82-026. Cincinnati, OH: U.S. Environmental Protection Agency.

Carlough, L. A. 1989. Fluctuations in the community composition of water-column protozoa in two southeastern blackwater rivers (Georgia, USA). *Hydrobiologia* 185:55–62.

Carlough, L. A., and J. L. Meyer. 1989. Protozoans in two southeastern blackwater rivers and their importance to trophic transfer. *Limnol. Oceanogr.* 34:163–177.

Carlough, L. A., and J. L. Meyer. 1990. Rates of protozoan bacterivory in three habitats of a southeastern blackwater river. *J. North Am. Benthol. Soc.* 9:45–53.

Chapin, J. W. 1975. A study of benthic macroinvertebrates in channelized and un-channelized Coastal Plain stream-swamps. MS Thesis, East Carolina University, Greenville.

Clench, W. J., and R. D. Turner. 1956. Freshwater mollusks of Alabama, Georgia, and Florida from the Escambia to the Suwannee River. *Bull., Fla. State Mus. Biol. Sci.* 1:97–239.

Cook, W. L., F. Parrish, J. D. Satterfield, W. G. Nolan, and P. E. Gaffney. 1983. *Biological and Chemical Assessment of Nonpoint Source Pollution in Georgia: Ridge-Valley and Sea Island Streams*. Atlanta: Department of Biology, Georgia State University.

Cooper, J. E., S. S. Robinson, and J. B. Funderburg (eds.). 1977. *Endangered and Threatened Plants and Animals of North Carolina*. Raleigh: North Carolina State Museum of Natural History.

Cowell, B. C., and W. C. Carew. 1976. Seasonal and diel periodicity in the drift of aquatic insects in a subtropical Florida stream. *Freshwater Biol.* 6:587–594.

Cuffney, T. F. 1988. Input, movement and exchange of organic matter within a subtropical coastal blackwater river-floodplain system. *Freshwater Biol.* 19:305–320.

Cuffney, T. F., and J. B. Wallace. 1987. Leaf litter processing in Coastal Plain streams and floodplains of southeastern Georgia, U.S.A. *Arch. Hydrobiol., Suppl.* 76:1–24.

Dahlberg, M. D., and D. C. Scott. 1971. The freshwater fishes of Georgia. *Bull. Ga. Acad. Sci.* 29:1–64.

Davis, J. R., and R. P. Cheek. 1966. Distribution, food habits, and growth of young culpeids, Cape Fear River system, North Carolina. *Proc. Annu. Conf. Southeast. Assoc. Game Fish Comm.* 20:250–259.

Day, F. P., Jr., S. K. West, and E. G. Tupacz. 1988. The influence of groundwater dynamics in a periodically flooded ecosystem, the Great Dismal Swamp. *Wetlands* 8:1–13.

de la Cruz, A. A., and H. A. Post. 1977. Production and transport of organic matter in a woodland stream. *Arch. Hydrobiol.* 80:227–238.

Dennis, W. M. 1973. A synecological study of the Santee Swamp, Sumter County, South Carolina. MS Thesis, University of South Carolina, Columbia.

Duda, A. M. 1982. Municipal point source and agricultural nonpoint source contributions to coastal eutrophication. *Water Resour. Bull.* 18:397–407.

Edwards, R. T. 1987. Sestonic bacteria as a food source for filtering invertebrates in two southeastern blackwater rivers. *Limnol. Oceanogr.* 32:221–234.

Edwards, R. T., and J. L. Meyer. 1986. Production and turnover of planktonic bacteria in two southeastern blackwater rivers. *Appl. Environ. Microbiol.* 52:1317–1323.

Edwards, R. T., and J. L. Meyer. 1987a. Metabolism of a sub-tropical low-gradient blackwater river. *Freshwater Biol.* 17:251–263.

Edwards, R. T., and J. L. Meyer. 1987b. Bacteria as a food source for black fly larvae in a blackwater river. *J. North Am. Benthol. Soc.* 6:241–250.

Elder, J. F., and D. J. Cairns. 1982. Production and decomposition of forest litterfall in the Appalachicola River floodplain, Florida. *U.S., Geol. Surv., Rep.* 82-252.

Findlay, S., L. Carlough, M. T. Crocker, H. K. Gill, J. L. Meyer, and P. J. Smith. 1986a. Bacterial growth on macrophyte leachate and fate of bacterial production. *Limnol. Oceanogr.* 31:1335–1341.

Findlay, S., J. L. Meyer, and R. Risley. 1986b. Benthic bacterial biomass and production in two blackwater rivers. *Can. J. Fish. Aquat. Sci.* 43:1271–1276.

Fink, T. J. 1986. The reproductive life history of the predaceous, sand-burrowing mayfly *Dolania americana* (Ephemeroptera: Behningiidae). Ph.D. Dissertation, Florida State University, Tallahassee.

Finn, P. L., and D. D. Herlong. 1980. New distributional record of *Dolania americana* (Ephemeroptera: Behningiidae). *Entomol. News* 91:102–104.

Forsythe, D. M., and W. B. Ezell, Jr. (eds.). 1979. *Proceedings of the First South Carolina Endangered Species Symposium.* Columbia: South Carolina Wildlife and Marine Resources Department.

Frankensteen, E. D. 1976. Genus *Alosa* in a channelized and an unchannelized creek of the Tar River basin, North Carolina. MA Thesis, East Carolina University, Greenville, NC.

Gambrell, R. P., J. W. Gilliam, and S. B. Weed. 1974. *The Fate of Fertilizer Nutrients as Related to Water Quality in the North Carolina Coastal Plain*, Rep. No. 93. Raleigh: North Carolina, Water Resources Research Institute.

Gardner, L. R. 1983. Element mass balances for South Carolina Coastal Plain watersheds. In R. Lowrance, R. Todd, L. Asmussen, and R. Leonard (eds.), *Nutrient Cycling in Agricultural Ecosystems*, Spec. Publ. No. 23. Athens: University of Georgia, College of Agriculture Experiment Stations, pp. 263–279.

Gasaway, R. D. 1973. *Study of Fish Movements from Tributary Streams into the Suwannee River*, Annu. Prog. Rep., Statewide Fish. Invest., F-21-5, Study VI, Job

2. Atlanta: Georgia Game and Fish Commission, Department of Natural Resources.

Georgia Department of Natural Resources. 1983. *Georgia Nonpoint Source Impact Assessment Study*. Atlanta: Environmental Protection Division, Department of Natural Resources, State of Georgia.

Gillespie, D. M., D. L. Stites, and A. C. Benke. 1985. An inexpensive sampler for use in sandy substrates. *Freshwater Invertebr. Biol.* 5:147–151.

Gladden, J. E., and L. A. Smock. 1990. Macroinvertebrate distribution and production on the floodplains of two lowland headwater streams. *Freshwater Biol.* 24:533–545.

Goodrich, C. 1942. The Pleuroceridae of the Atlantic Coastal Plain. *Occas. Pap., Mus. Zool., Univ. Mich.* 456:1–6.

Haines, B. 1980. Acid precipitation in southeastern United States: a brief review. *Ga. J. Sci.* 37:185–191.

Hauer, F. R., and A. C. Benke. 1987. Influence of temperature and river hydrograph on black fly growth rates in a subtropical blackwater river. *J. North Am. Benthol. Soc.* 6:251–261.

Hauer, F. R., N. L. Poff, and P. L. Firth. 1986. Leaf litter decomposition across broad thermal gradients in southeastern coastal plain streams and swamps. *J. Freshwater Ecol.* 3:545–552.

Heard, W. H. 1970. Eastern freshwater mollusks (II). The south Atlantic and Gulf drainages. *Malacologia* 10:23–31.

Herlong, D. D., and M. A. Mallin. 1985. The benthos-plankton relationship upstream and downstream of a blackwater impoundment. *J. Freshwater Ecol.* 3:47–59.

Hocutt, C. H., R. E. Jenkins, and J. R. Stauffer, Jr. 1986. Zoogeography of the fishes of the central Appalachians and central Atlantic Coastal Plain. In C. H. Hocutt and E. O. Wiley (eds.), *The Zoogeography of North American Fishes*. New York: Wiley, pp. 161–211.

Holder, D. R., and R. Ruebsamen. 1976. *A Comparison of the Fisheries of the Upper and Lower Satilla River*. Atlanta: Georgia Department of Natural Resources, Game and Fish Division.

Holzenthal, R. W., and S. W. Hamilton. 1984. Trichopteran diversity and endemism in the Southeastern Coastal Plain. *Program Abstracts, 33rd Annual Meeting of the North American Benthological Society*, Raleigh, NC.

Hoskin, C. M. 1959. Studies of oxygen metabolism of streams of North Carolina. *Inst. Publ. Mar. Sci., Univ. Tex.* 6:186–192.

Huish, M. T., and G. B. Pardue. 1978. *Ecological Studies of One Channelized and Two Unchannelized Wooded Coastal Swamp Streams in North Carolina*, FWS/OBS-78/85. Harpers Ferry, WV: U.S. Fish and Wildlife Service, Office of Biological Services.

Johnson, R. I. 1970. The systematics and zoogeography of the Unionidae (Mollusca: Bivalvia) of the southern Atlantic Slope region. *Bull., Mus. Comp. Zool.* 140:263–449.

Kennedy, V. C. 1964. Sediment transport by Georgia streams. *U.S., Geol. Surv., Water Supply Pap.* 1668:1–100.

Kitchens, W. M., Jr., J. M. Dean, L. H. Stevenson, and J. H. Cooper. 1975. The Santee Swamp as a nutrient sink. In F. G. Howell, J. B. Gentry, and M. H. Smith (eds.), *Mineral Cycling in Southeastern Ecosystems*. ERDA Symp. Ser., CONF-740513. Springfield, VA: U.S. Energy Research and Development Administration, pp. 349–366.

Kondratieff, B. C., and P. Kondratieff. 1984. *A Lower Food Chain Community Study: Thermal Effects and Post-Thermal Recovery in the Streams and Swamps of the Savannah River Plant*, Rep. ECS-SR-15. Aiken, SC: Environmental & Chemical Sciences, Inc.

Kuenzler, E. J., P. J. Mulholland, L. A. Ruley, and R. P. Sniffen. 1977. *Water Quality in North Carolina Coastal Plain Streams and Effects of Channelization*, Res. Rep. No. 127. Raleigh: North Carolina Water Resources Research Institute.

Kuenzler, E. J., P. J. Mulholland, L. A. Yarbro, and L. A. Smock. 1980. *Distributions and Budgets of Carbon, Phosphorus, Iron and Manganese in a Floodplain Swamp Ecosystem*. Res. Rep. No. 157. Raleigh: North Carolina Water Resources Research Institute.

Leff, L. G., and J. V. McArthur. 1988. Seston and dissolved organic carbon transport during storm flows in a natural and a disturbed Coastal Plain stream. *J. Freshwater Ecol.* 4:271–278.

Lenat, D. R. 1983. Benthic macroinvertebrates of Cane Creek, North Carolina, and comparisons with other southeastern streams. *Brimleyana* 9:53–68.

Lenat, D. R. 1984. Agriculture and stream water quality: a biological evaluation of erosion control practices. *Environ. Manage.* 8:333–344.

Lenat, D. R., and K. Eagleson. 1981. *Ecological Effects of Urban Runoff on North Carolina Streams*, Biol. Ser. No. 104. Raleigh: North Carolina Department of Natural Resources and Community Development, Division of Environmental Management.

Lovingood, P. E., and J. C. Purvis. 1975. *The Nature of Precipitation, South Carolina 1941–1970*. Columbia: South Carolina Crop and Livestock Reporting Service.

Lowrance, R., R. Todd, J. Fail, Jr., O. Hendrickson, Jr., R. Leonard, and L. Asmussen. 1984. Riparian forests as nutrient filters in agricultural watersheds. *BioScience* 34:374–377.

Mayack, D. T., J. H. Thorp, and M. Cothran. 1989. Effects of burial and floodplain retention on stream processing of allochthonous litter. *Oikos* 54:378–388.

McArthur, J. V., L. G. Leff, D. A. Kovacic, and J. Jaroscak. 1986. Green leaf decomposition in coastal plain streams. *J. Freshwater Ecol.* 3:553–558.

McFarlane, R. W. 1976. Fish diversity in adjacent ambient, thermal, and post-thermal freshwater streams. In G. W. Esch and R. W. McFarlane (eds.), *Thermal Ecology II*, CONF-750425. Springfield, VA: Technical Information Center, U.S. Energy Research and Development Administration, pp. 268–271.

Meffe, G. K., and A. L. Sheldon. 1988. The influence of habitat structure on fish assemblage composition in southeastern blackwater streams. *Am. Midl. Nat.* 120:225–240.

Meffe, G. T., and A. L. Sheldon. 1990. Post-defaunation recovery of fish assemblages in southeastern blackwater streams. *Ecology* 71:657–667.

Metzler, G. M., and L. A. Smock. 1990. Storage and dynamics of subsurface detritus in a sand-bottomed stream. *Can. J. Fish. Aquat. Sci.* 47:588–594.

Meybeck, M. 1982. Carbon, nitrogen, and phosphorus transport by world rivers. *Am. J. Sci.* 282:401–450.

Meyer, J. L. 1986. Dissolved organic carbon dynamics in two subtropical blackwater rivers. *Arch. Hydrobiol.* 108:119–134.

Meyer, J. L. 1988. Benthic bacterial biomass and production in a blackwater river. *Verh.—Int. Vere. Theore. Angew. Limnol.* 23:1832–1838.

Meyer, J. L., and R. T. Edwards. 1990. Ecosystem metabolism and turnover of organic carbon along a blackwater river continuum. *Ecology* 71:668–677.

Meyer, J. L., R. T. Edwards, and R. Risley. 1987. Bacterial growth on dissolved organic carbon from a blackwater river. *Microb. Ecol.* 13:13–29.

Morgan, M. D. 1984. Acidification of headwater streams in the New Jersey Pine Barrens: a re-evaluation. *Limnol. Oceanogr.* 19:1259–1266.

Morse, J. C., J. W. Chapin, D. D. Herlong, and R. S. Harvey. 1980. Aquatic insects of Upper Three Runs Creek, Savannah River Plant, South Carolina. Part I: Orders other than Diptera. *J. Ga. Entomol. Soc.* 15:73–101.

Morse, J. C., J. W. Chapin, D. D. Herlong, and R. S. Harvey. 1983. Aquatic insects of Upper Three Runs Creek, Savannah River Plant, South Carolina. Part II: Diptera. *J. Ga. Entomol. Soc.* 18:303–316.

Mulholland, P. J. 1981. Organic carbon flow in a swamp-stream ecosystem. *Ecol. Monogr.* 51:307–322.

Mulholland, P. J., and E. J. Kuenzler. 1979. Organic carbon export from upland and forested wetland watersheds. *Limnol. Oceanogr.* 24:960–966.

Mulholland, P. J., L. A. Yarbro, R. P. Sniffen, and E. J. Kuenzler. 1981. Effects of floods on nutrient and metal concentrations in a Coastal Plain stream. *Water Resour. Res.* 17:758–764.

National Park Service. 1982. *The Nationwide Rivers Inventory*. Washington, DC: U.S. Department of the Interior.

Neiheisel, J., and C. E. Weaver. 1967. Transport and deposition of clay minerals, Southeastern United States. *J. Sediment. Petrol.* 37:1084–1116.

Noltemeier, D. D., M. M. Brinson, D. Holbert, and M. N. Jones. 1986. Variations in water chemistry in blackwater streams in the Coastal Plain of North Carolina. *Assoc. Southeast. Biol. Bull.* 33:59–60.

Olmsted, L. L., and D. G. Cloutman. 1979. Abundance, distribution, and status of the fishes of the Carolina Sandhills National Wildlife Refuge, Chesterfield County, South Carolina. In D. M. Forsythe and W. B. Ezell, Jr. (eds.), *Proceedings of the First South Carolina Endangered Species Symposium*. Columbia: South Carolina Wildlife and Marine Resources Department, pp. 126–129.

O'Rear, C. W., Jr. 1975. *The Effects of Stream Channelization on the Distribution of Nutrients and Metals*, Rep. No. 108. Raleigh: North Carolina Water Resources Research Institute.

Paerl, H. 1982. *Environmental Factors Promoting and Regulating N_2 Fixing Blue-Green Algal Blooms in the Chowan River, N. C.*, Rep. No. 176. Raleigh: North Carolina Water Resources Research Institute.

Pardue, G. B., M. T. Huish, and H. R. Perry, Jr. 1975. *Ecological Studies of Two Swamp Watersheds in Northeastern North Carolina*, Rep. No. 105. Raleigh: North Carolina Water Resources Research Institute.

Park, A. D. 1979. *Groundwater in the Coastal Plains Region: A Status Report and Handbook*. Charleston, SC: Coastal Plains Regional Commission.

Pate, P. P., Jr. 1972. Life history aspects of the hickory shad, *Alosa medicris* (Mitchell), in the Neuse River, North Carolina. MS Thesis, North Carolina State University, Raleigh.

Pennak, R. W. 1978. *Fresh-Water Invertebrates of the United States.* New York: Wiley (Interscience).

Pernick, M., and P. J. W. Roberts. 1985. *Mixing in a River-Floodplain System*, Rep. No. 85-107. Atlanta: Georgia Institute of Technology.

Peters, J. G., W. L. Peters, and T. J. Fink. 1987. Seasonal synchronization of emergence in *Dolania americana* (Ephemeroptera:Behningiidae). *Can. J. Zool.* 65:3177–3185.

Peters, W. L., and J. Jones. 1973. Historical and biological aspects of the Blackwater River in northwestern Florida. In W. L. Peters and J. G. Peters (eds.), *Proceedings of the First International Conference on Ephemeroptera*. Leiden, The Netherlands: E. J. Brill, pp. 242–253.

Peters, W. L., and J. G. Peters. 1977. Adult life and emergence of *Dolania americana* in northwestern Florida (Ephemeroptera: Behningiidae). *Int. Rev. Gesamten Hydrobiol.* 62:409–438.

Poff, N. L., and R. A. Matthews. 1985. The replacement of *Stenonema* spp. by *Caenis diminuta* Walker as the numerical dominant in the mayfly assemblage of a thermally-stressed stream. *J. Freshwater Ecol.* 3:19–26.

Poff, N. L., and R. A. Matthews. 1986. Benthic macroinvertebrate community structural and functional group response to thermal enhancement in the Savannah River and a coastal plain tributary. *Arch. Hydrobiol.* 106:119–137.

Post, H. A., and A. A. de la Cruz. 1977. Litterfal, litter decomposition, and flux of particulate organic material in a Coastal Plain stream. *Hydrobiologia* 55:201–207.

Reuter, J. H., and E. M. Perdue. 1972. *Chemical Characterization of Dissolved Organic Matter and Its Influence on the Chemistry of River Water*, Publ. No. ERC-0372. Atlanta: Environmental Resources Center, Georgia Institute of Technology.

Reuter, J. H., and E. M. Perdue. 1977. Importance of heavy metal-organic matter interactions in natural waters. *Geochi. Cosmochim. Acta* 41:325–334.

Richter, D. D., C. W. Ralston, and W. R. Harms. 1982. Prescribed fire: effects on water quality and forest nutrient cycling. *Science* 215:661–663.

Roeding, C. E., and L. A. Smock. 1989. Ecology of macroinvertebrate shredders in a low-gradient sandy-bottomed stream. *J. North Am. Benthol. Soc.* 8:149–161.

Ross, S. T., and J. A. Baker. 1983. The response of fishes to periodic spring floods in a southeastern stream. *Am. Midl. Nat.* 109:1–14.

Sadowski, P. W., and R. A. Matthews. 1984. *Effect of Thermal Effluents from the Savannah River Plant on Leaf Decomposition Rates in Onsite Creeks and the Savannah River*, U.S. Department of Energy Rep. DP-84-399. Aiken, SC: E. I. du Pont de Nemours and Company, Savannah River Laboratory.

Scheiring, J. F. 1985. Longitudinal and seasonal patterns of insect trophic structure in a Florida Sand-hill stream. *J. Kans. Entomol. Soc.* 58:207–219.

Schlesinger, W. H., and J. M. Melack. 1981. Transport of organic carbon in the world's rivers. *Tellus* 33:172–187.

Sepkowski, J. J., and M. A. Rex. 1974. Distribution of freshwater mussels: coastal rivers as biogeographic islands. *Syst. Zool.* 23:165–188.

Sheath, R. G., and J. M. Burkholder. 1985. Characteristics of softwater streams in Rhode Island. II. Composition and seasonal dynamics of macroalgal communities. *Hydrobiologia* 128:109–118.

Shure, D. J., M. R. Gottschalk, and K. A. Parsons. 1986. Litter decomposition processes in a floodplain forest. *Am. Midl. Nat.* 115:314–327.

Simmons, C. E., and R. C. Heath. 1979. *Water-Quality Characteristics of Streams in Forested and Rural Areas of North Carolina*, Water Resour. Invest. Rep. 79-108. Raleigh, NC: U. S. Geological Survey.

Smock, L. A. 1988. Life histories, abundance and distribution of some macroinvertebrates from a South Carolina, USA coastal plain stream. *Hydrobiologia* 157:193–208.

Smock, L. A. 1990. Spatial and temporal variation in organic matter storage in low-gradient, headwater streams. *Arch. Hydrobiol.* 118:169–184.

Smock, L. A., and E. Gilinsky. 1982. *Benthic Macroinvertebrate Communities of a Floodplain Creek in the Congaree Swamp National Monument*, Final Rep., Contract No. CX5000-0-0945. Atlanta, GA: National Park Service.

Smock, L. A., and E. J. Kuenzler. 1983. Seasonal changes in the forms and species of iron and manganese in a seasonally-inundated floodplain swamp. *Water Res.* 17:1287–1294.

Smock, L. A., and C. E. Roeding. 1986. The trophic basis of production of the macroinvertebrate community of a southeastern U. S. A. blackwater stream. *Holarctic Ecol.* 9:165–174.

Smock, L. A., E. Gilinsky, and D. L. Stoneburner. 1985. Macroinvertebrate production in a southeastern United States blackwater stream. *Ecology* 66:1491–1503.

Smock, L. A., G. M. Metzler, and J. E. Gladden. 1989. Role of debris dams in the structure and functioning of low-gradient headwater streams. *Ecology* 70:764–775.

Smock, L. A., J. E. Gladden, J. L. Riekenberg, L. C. Smith, and C. R. Black. 1992. Lotic macroinvertebrate production in three dimensions: channel surface, hyporheic, and floodplain environments. *Ecology* 73 (in press).

Sniffen, R. P. 1981. Benthic invertebrate production during seasonal inundation of a floodplain swamp. Ph.D. Dissertation, University of North Carolina, Chapel Hill.

Soponis, A. R. 1980. Taxonomic composition of Chironomidae (Diptera) in a sand-bottomed stream of northern Florida. In D. A. Murray (ed.), *Chironomidae: Ecology, Systematics, Cytology and Physiology*. Oxford, England: Pergamon, pp. 163–169.

Soponis, A. R. 1983. Emergence of *Polypedilum* (Chironomidae) in a sand-bottomed stream of northern Florida. *Mem. Am. Entomol. Soc.* 34:309–313.

Soponis, A. R., and C. L. Russell. 1984. Larval drift of Chironomidae (Diptera) in a north Florida stream. *Aquat. Insects* 6:191–199.

Specht, W., H. J. Kania, and W. Painter. 1984. *Annual Report on the Savannah River Aquatic Ecology Program*, Vol. II, Rep. ECS-SR-9. Aiken, SC: Environmental & Chemical Sciences, Inc.

Stites, D. L. 1986. Secondary production and productivity in the sediments of blackwater rivers. Ph.D. Dissertation, Emory University, Atlanta, GA.

Stites, D. L. 1987. Population and production dynamics of an enchytraeid worm in a subtropical blackwater river. *Can. J. Fish. Aquat. Sci.* 44:1469–1474.

Stites, D. L., and A. C. Benke. 1989. Rapid growth rates of chironomids in three habitats of a subtropical blackwater river and their implications for $P:B$ ratios. *Limnol. Oceanogr.* 34:1278–1289.

Strommer, J. L., and L. A. Smock. 1989. Vertical distribution and abundance of invertebrates within the sandy substrate of a low-gradient headwater stream. *Freshwater Biol.* 22:263–274.

Sumsion, C. T. 1970. Geology and ground-water resources of Pitt County, North Carolina. *U.S., Geol. Surv., Ground Water Bull.* 18.

Swift, C. C., C. R. Gilbert, S. A. Bortone, G. H. Burgess, and R. W. Yerger. 1986. Zoogeography of the freshwater fishes of the southeastern United States: Savannah River to Lake Pontchartrain. In C. H. Hocutt and E. O. Wiley (eds.), *The Zoogeography of North American Freshwater Fishes.* New York: Wiley, pp. 213–265.

Tarplee, W. H., Jr., D. E. Louder, and A. J. Baker. 1971. *Evaluation of the Effects of Channelization on Fish Populations in North Carolina's Coastal Plain Streams.* Raleigh: North Carolina Wildlife Resources Commission.

Thorp, J. H., E. M. McEwan, M. F. Flynn, and F. R. Hauer. 1985. Invertebrate colonization of submerged wood in a cypress-tupelo swamp and blackwater stream. *Am. Midl. Nat.* 113:56–68.

Tsui, P. T. P., and M. D. Hubbard. 1979. Feeding habits of the predaceous nymphs of *Dolania americana* in northwestern Florida (Ephemeroptera:Behningiidae). *Hydrobiologia* 67:119–123.

Twilley, R. T., L. R. Blanton, M. M. Brinson, and G. J. Davis. 1985. Biomass production and nutrient cycling in aquatic macrophyte communities of the Chowan River, North Carolina. *Aquat. Bot.* 22:231–252.

Vannote, R. L., G. W. Minshall, K. W. Cummins, J. R. Sedell, and C. E. Cushing. 1980. The river continuum concept. *Can. J. Fish. Aquat. Sci.* 37:130–137.

Virginia Water Control Board. 1985. *Chowan Basin Nutrient Control Plan for Virginia.* Inf. Bull. No. 561. Richmond: Virginia Water Control Board.

Wallace, J. B., and A. C. Benke. 1984. Quantification of wood habitat in subtropical Coastal Plain streams. *Can. J. Fish. Aquat. Sci.* 41:1643–1652.

Wallace, J. B., and J. O'Hop. 1985. Life on a fast pad: waterlily leaf beetle impact on water lilies. *Ecology* 66:1534–1544.

Wallace, J. B., W. R. Woodall, and A. A. Staats. 1976. The larval dwelling-tube, capture net and food of *Phylocentropus placidus* (Trichoptera:Polycentropodidae). *Ann. Entomol. Soc. Am.* 69:151–154.

Wallace, J. B., A. C. Benke, A. H. Lingle, and K. Parsons. 1987. Trophic pathways of macroinvertebrate primary consumers in subtropical blackwater streams. *Arch. Hydrobiol., Suppl.* 74:423–451.

Walters, K. H., and L. A. Smock. 1991. Cellulase activity of leaf litter and stream-dwelling, shredder macroinvertebrates. *Hydrobiologia* 220:29–35.

Wharton, C. H. 1970. *The Southern River Swamp—A Multiple Use Environment.* Georgia State University, Atlanta: School of Business Administration.

Wharton, C. H., and M. M. Brinson. 1979. Characteristics of southeastern river system. In R. R. Johnson and J. F. McCormick (eds.), *Strategies for Protection and Management of Floodplain Wetlands and Other Riparian Ecosystems. Gen. Tech. Rep. WO—U.S., For. Ser. [Wash. Off.]* GTR-WO-12.

Wharton, C. H., W. M. Kitchens, E. C. Pendleton, and T. W. Sipe. 1982. *The Ecology of Bottomland Hardwood Swamps of the Southeast: A Community Profile*, FWS/ OBS-81/37. Washington, DC: U.S. Department of the Interior, Fish and Wildlife Service.

Whitford, L. A., and G. L. Schumacher. 1963. Communities of algae in North Carolina streams, and their seasonal relationships. *Hydrobiologia* 22:133–195.

Wilbur, R. C., and N. L. Christensen. 1983. Effects of fire on nutrient availability in a North Carolina Coastal Plain pocosin. *Am. Midl. Nat.* 110:54–61.

Williams, J. E., J. W. Johnson, D. A. Hendrickson, S. Contreras-Balderas, J. D. Williams, M. Navarro-Mendoza, D. E. McAllister, and J. E. Deacon. 1989. Fishes of North America endangered, threatened, or of special concern: 1989. *Fisheries* 14:2–20.

Windom, H. L., K. C. Beck, and R. Smith. 1971. Transport of trace metals to the Atlantic Ocean by three southeastern streams. *Southeast. Geol.* 12:169–181.

Winner, M. D., Jr., and C. E. Simmons. 1977. *Hydrology of the Creeping Swamp Watershed, North Carolina, with Reference to Potential Effects of Stream Channelization*, Water Resour. Invest. Rep. 77-26. Raleigh, NC: U.S. Geological Survey.

Witherspoon, A. M., C. Balducci, D. C. Boody, and J. Overton. 1979. *Response of Phytoplankton to Water Quality in the Chowan River System*, Rep. No. 129. Raleigh: North Carolina University, Water Resources Research Institute.

Woodson, B. R., Jr., and W. Wilson, Jr. 1973. A systematic and ecological survey of algae in two streams of Isle of Wight County, Virginia. *Castanea* 38:1–18.

Yarbro, L. A. 1983. The influence of hydrologic variations on phosphorus cycling and retention in a swamp stream ecosystem. In T. D. Fontaine, III and S. M. Bartell (eds.), *Dynamics of Lotic Ecosystems*. Ann Arbor, MI: Ann Arbor Science, pp. 223–245.

Yarbro, L. A., E. J. Kuenzler, P. J. Mulholland, and R. P. Sniffen. 1984. Effects of stream channelization on exports of nitrogen and phosphorus from North Carolina Coastal Plain watersheds. *Environ. Manage.* 8:151–160.

APPENDIX 1

Species designated as endangered (E), threatened (T), or of special concern (SC) that occur in blackwater streams of North and South Carolina

Species	Status
North Carolina	
MUSSELS	
Fusconaia mansoni (Atlantic pigtoe)	T
CRAYFISH	
Procambarus lepidodactylus (Pee Dee lotic crayfish)	T
Orconectes virginiensis (Chowan River crayfish)	SC
FISH	
Etheostoma mariae (Pinewoods darter)	SC
Hybopsis n. sp. II (Undescribed chub)	SC
Noturus n. sp. (Undescribed madtom)	SC
Semotilus lumbee (Sandhills chub)	SC
South Carolina	
MUSSELS	
Fusconaia mansoni (Atlantic pigtoe)	E
Lampsilis ochracea (Tidewater mucket)	SC
CRAYFISH	
Procambarus echinatus	SC
Procambarus hirsutus	SC
Procambarus chacei	SC
INSECTS	
Dolania americana	SC
FISH	
Semotilus lumbee (Sandhills chub)	T
Noturus sp. (Broadtail madtom)	T

Source. From Cooper et al. (1977) and Forsythe and Ezell (1979).

APPENDIX 2

Blackwater streams and rivers of Virginia, North Carolina, South Carolina, and Georgia listed in the Nationwide Rivers Inventory

Virginia	North Carolina	South Carolina	Georgia
Blackwater River	Bennet's Creek	Ashepoo Creek	Alapaha River
Chickahominy River	Big Swamp	Jones Swamp Creek	Canoochee River
Dragon Run	Black River	Black River	Ebenezer Creek
Northwest River	Cashie River	Combahee River	Kinchafoonee River
Nottoway River	Chowan River	Coosawatchie River	Little Ohoopee River
Poropotank River	Colly Creek	Edisto River	Muckalee Creek
Yarmouth Creek	Fishing Creek	South Fork River	Ochlockonee River
	Goshen Swamp	Four Hole Swamp	Ohoopee River
	Great Coharie Swamp	Little Pee Dee River	Satilla River
	Little Coharrie Creek	Little Salkehatchie River	
	Little Marsh Swamp	Lumber River	
	Lumber River	Lynches River	
	Drowning Creek	New River	
	Moores Creek	Great Swamp	
	Northeast Cape Fear River	North Fork Edisto River	
	Potecasi Creek	Pee Dee River	
	Ramsey Creek	Salkehatchie River	
	Six Runs Creek	Waccamaw River	
	South River		
	Town Creek		
	Upper Little River		

Source. National Park Service (1982). Rivers in other states on the Coastal Plain are also included in this inventory.

8 Medium-sized Rivers of the Atlantic Coastal Plain

GREG C. GARMAN

Department of Biology, Virginia Commonwealth University, Richmond, VA 23284

LARRY A. NIELSEN

Department of Fisheries and Wildlife Sciences, Virginia Polytechnic Institute and State University, Blacksburg, VA 24061

In this chapter we discuss major rivers of the Southeast Atlantic slope, most of which begin in the mountains of the Appalachian Region, and follow relatively straight and narrow watersheds to the sea. The scope of this chapter will be limited to river mainstems within the Piedmont and Coastal Plain physiographic provinces. The almost complete lack of data on ecological processes (production, nutrient cycling, etc.) for the rivers being considered necessitates a largely descriptive approach to this chapter. For some ecological components, even descriptive information is unavailable. An attempt to identify broad ecological patterns for rivers of the region will be presented.

The Atlantic Coastal Plain occupies a broad band of land from New England south to Florida and west to Texas. The region extends approximately 300 km inland in the southern Atlantic area, bounded on the west by the fall line, a distinct increase in land elevation. The area is underlaid by metamorphic rocks and is covered by horizontal layers of Cretaceous, Tertiary, and Quarternary sedimentary deposits. The Coastal Plain was submerged ocean bottom through the Eocene; the eastern extension of the plain remains submerged today as the continental shelf. The geological structure of the area is highly complex, with four distinct zones: embayed, Cape Fear Arch, Sea Islands Downwarp, and Peninsular Arch regions (Hunt 1974).

The Piedmont province is the easternmost extension of the Appalachian Region. It lies between the Blue Ridge Mountains on the west and the Atlantic Coastal Plain on the east. Its width gradually increases from north to south, eventually reaching about 200 km in Georgia. The same rock structures that underlie the Coastal Plain also form the Piedmont, but without the overlying sedimentary deposits. These metamorphic structures have produced highly complex geological patterns, resulting in a broad array of soil types. Mountains arose in the Piedmont during the Paleozoic. Since then, the mountains have largely eroded away, supplying the sediments that now cover the Coastal Plain (Hunt 1974).

The erosional and sedimentary relationships of these regions have influenced both their physical and cultural characteristics. The topography grades from the undulating hills of the Piedmont to the uniformly flat landscape of the Coastal Plain. Sections of the Piedmont less resistant to erosion have provided sites for major river impoundments, now supplying electricity and water for the growing human population. River mouths are drowned valleys (rather than alluvial deltas), providing some of the world's premier harbors (Hunt 1974). Most major cities are located at these harbors or along the fall line, where rivers were narrow and readily crossed in colonial times. As a consequence, the central portions of the Coastal Plain and much of the Piedmont are sparsely populated (Hunt 1974).

Soils in the region are old and deeply weathered. The characteristic ultisols and spodosols are generally low in organic matter and high in clay and sand; vertisols and inceptisols are common in specific areas and in floodplains (Hunt 1974, Bailey 1978). Land use is predominantly rural and equally divided between forest and farmland. Soil erosion problems are not considered severe, but total erosion for the region is above the national average of 4.8 tons per acre per year (Council on Environmental Quality (CEQ) 1981).

The physiographic provinces covered by this review fall within two water resource regions, the Mid-Atlantic and South Atlantic-Gulf (Cushman et al. 1980). The regions are rich in water resources, with abundant surface and groundwater flows and relatively low withdrawal rates (CEQ 1981). For example, the region holds two million hectares of fishable freshwater, the highest among 18 water resources regions (Cushman et al. 1980). Total annual runoff ranges from 8 to 15 cm (Mannering 1982).

The human population of the region is growing rapidly, increasing by 60% between 1940 and 1976 (CEQ 1981). As a consequence, prime farmlands have decreased by 5% in the last decade, and saltwater intrusion in coastal areas due to pumping of aquifers is becoming more common (Cushman et al. 1980). Suspended solids remain the major water pollution problem, followed in importance by bacterial contamination (Soil Conservation Service (SCS) 1980).

The major biotic regions are classified as the Southeastern Mixed Forest province and the Outer Coastal Plain Forest province (Bailey 1978). Climax vegetation is dominated by broadleaf deciduous and needleleaf evergreen trees in the Mixed Forest province and by evergreen oaks, laurels, and magnolias in the Coastal Plain province. Floodplain forests are the dominant riparian plant communities throughout the region (Wharton et al. 1982). The biota of the region is exceptionally diverse, with over 1000 species of nondomestic vertebrates (CEQ 1981). In general, however, only a few areas are protected in national park lands or other forms of ecological preserves.

SOUTHEASTERN RIVERS: AN OVERVIEW

Collectively, the river mainstems of the southeastern Atlantic slope incorporate diversities of climate, physiography, biota, and human influence that defy a simple scheme of river classification and may, in some cases, obscure broad geographic patterns in the characteristics of these rivers. Some patterns are apparent, however, and may be used to organize an overview of the large number of lotic systems included within this chapter's mandate.

Several abiotic river parameters (Table 1) exhibit a pattern with respect to latitude, suggesting a classification of southeastern rivers into four groups. The James River and the westernmost tributary of the Chowan River (Roanoke River) are characterized by relatively high gradient, rocky substrate, and clear, slightly alkaline water. Water temperature in this first river group may fluctuate more than 30 C° seasonally, ranging from 1 to 32 °C in the James River (Woolcott 1974, Garman and Smock 1988).

The second, and largest, group of Coastal Plain and Piedmont rivers includes the Tar, Neuse, and Cape Fear of North Carolina, most South Carolina rivers, and the Savannah, Ogeechee, and Ocmulgee of Georgia (Table 1). The gradient of these rivers ranges between 0.4 and 0.2 m/km, pH is neutral to slightly acidic, and perhaps more significant, a sand/silt substrate predominates. Within Piedmont sections of these rivers, water is frequently turbid due to suspended clay particles; in Coastal Plain sections, water may be stained by humic compounds. Dissolved organic carbon comprised the major organic component transported by the lower Savannah River (Cudney and Wallace 1980). In the same study, over 90% of the suspended seston was fine organic particulates (<1 mm), with total organic seston concentrations ranging from approximately 1.0 mg/L during fall months, to 3.5 mg/L in winter months.

The Altamaha, Satilla, and St. Marys rivers drain only the Coastal Plain of Georgia and northeast Florida and, as a consequence, gradient, pH, and alkalinity are further reduced in this third grouping of rivers. Seasonal temperature fluctuations are also reduced, with water temperatures rarely dropping below 10 °C in the Satilla during winter months (Benke et al. 1979). Rivers are darkly stained (5–50 mg/L dissolved organic carbon), and benthic particulate organic matter may range up to 4 kg AFDM/m² (Smock and Gilinsky, Chapter 7). Substrate in rivers of the third category is fine-grained and unstable. As a result, submerged woody debris is a major macroinvertebrate habitat in blackwater rivers like the Satilla, Altamaha and St. Marys (Wallace and Benke 1984, Benke et al. 1985).

A final group of exclusively Floridian rivers is represented here by the St. Johns—a wide, shallow river with almost imperceptible gradient (0.02 m/km), and highly buffered, alkaline water. Here, because of relatively recent geologic events, the distinction between river and estuary is often obscure, and the physical character of such rivers is vastly different from that of rivers in the three previous groupings.

TABLE 1 Selected Physicochemical Parameters for Rivers of the Southeastern Atlantic Slope

River System	Watershed Area (km²)	Total Length (km)	Discharge (m³/s)	Gradient (m/km)	Physiographic Province	Substrate	pH	Alkalinity (mg/L CaCO₃)
James	26 000	536	303	0.5	P/CP	R	7.6	31–65
Chowan	12 652	81	—	0.1	CP	D	—	—
Tar	7 900	288	—	0.5	P/CP	S	—	—
Neuse	15 876	397	—	0.4	P/CP	S	—	—
Cape Fear	—	272	—	—	P/CP	—	—	7–32
Waccamaw	4 000	225	—	0.1	CP	S	7.0	—
Savannah	27 400	476	302	0.4	P/CP	S	6.8	20
Ogeechee	14 109	392	175	0.2	P/CP	S	6.2	30
Ocmulgee	15 750	388	160	0.2	P/CP	S	6.5	18
Altamaha	7 500	220	200	0.1	CP	S/D	6.8	7
Satilla	10 401	362	45	0.1	CP	S/D	5.1	4
St. Marys	4 992	193	20	0.1	CP	S/D	4.9	20
St. Johns	21 409	500	421	0.02	CP	S/D	7.4	500

Note. Unless otherwise stated, values are annual means. P, Piedmont; CP, Coastal Plain; R, rock; D, detritus; S, sand.

Source. From various sources cited in the tex.

Primary Producers

Several genera of emergent aquatic macrophytes, including *Nuphar, Insticia, Bidens, Typha,* and *Sagittaria,* may be encountered in southeastern rivers. Macrophyte biomass ranged from 200 to 1200 g dry mass/m^2 in the lower Chowan River; lowest seasonal biomass occurred after the first frost in October (Blanton et al. 1975). In another study of Chowan River macrophytes (Twilley et al. 1985), annual net primary production of *Justicia americana* and *Nuphar luteum* was 173 and 222 g dry mass/m^2, respectively. High macrophyte production rates, together with the influences of herbivory may result in significant fluxes of nutrients such as phosphorous from the sediment to the water column in some southeastern rivers (Twilley 1976).

Aquatic macrophytes are most abundant in low-gradient rivers with a fine-grained organic substrate. In fact, sedimentation of fine particulates, and hence the suitability of the substrate for further colonization, is significantly enhanced by the presence of macrophyte stands (Blanton et al. 1975). Emergent aquatic plants may be absent from large stretches of rivers with rock or sand substrate. Where abundant, the importance of living emergent and submergent plants to the lotic community is generally that of providing a substrate for periphyton and physical "structure" for fish and invertebrates, rather than as a food source (Barber and Kevern 1973).

Consumers

Macroinvertebrates   The work of Benke et al. (1979, 1985) on the Satilla River of Georgia represents one of the few detailed studies of aquatic invertebrate ecology available for a major southeastern river. The Satilla is a low-gradient system characterized by high levels of fine particulate and dissolved organic matter, and a sand substrate. Filter-feeding insects dominated the benthic fauna, utilizing the abundant fine (<1 mm) particulate organics in the seston. Common genera included *Simulium, Polypedilum,* and *Parakiefferiella* (Diptera), and *Hydropsyche* (Trichoptera); *Corydalus* (Megaloptera) and *Boyeria* (Odonata) were major invertebrate predator taxa. Oligochaetes (e.g., *Limnodrilus*) also were abundant in sand/silt habitats.

Due to the instability and homogeneity of the sand substrate (Rabeni and Minshall 1977), most invertebrates in the Satilla River utilized "snag" habitat (large woody debris). Of several habitat types identified, snags exhibited the highest taxonomic richness (40 genera of macroinvertebrates) and were the most productive (57–72 g dry mass/m^2 year^{-1}). Annual production values within this range are among the highest reported for lotic aquatic systems (Waters 1977, Stites and Benke 1989).

Annual production of six species of filtering caddisfly larvae, including the genera *Hydropsyche, Macronema,* and *Chimarra,* in the lower Savannah River was also very high, ranging from 13 to 23 g dry mass/m^2 (Cudney and Wallace 1980). As in the Satilla, woody debris, especially submerged roots and branches of *Salix nigra,* was the most highly colonized substrate type in the Savannah River.

If these rivers can be considered models for other southeastern river systems, especially within the Coastal Plain, some generalizations are possible. Macroinvertebrate communities in such rivers are energetically dependent on allochthonous detrital inputs from the adjacent flood plain, and production rates of certain functional groups (e.g., filterers) may be high. In addition, woody debris is a critical structural component of these rivers, it is the major substrate used by the fauna, and its availability may limit the abundance and biomass of many macroinvertebrate groups (van Arsdall 1977, Cudney and Wallace 1980).

Unfortunately, few detailed studies of aquatic communities are currently available for southeastern rivers that may differ significantly from this model (e.g., James, St. Johns).

Fishes   The fish fauna of most southeastern rivers is dominated by the family Centrarchidae, with the number of species within this family increasing from north to south (Table 2). The only exception to this pattern is found in the most northerly river, the James, where the dominant family is Cyprinidae, and the numerically dominant genus is *Notropis*. The genus *Notropis* is also numerically dominant in the Waccamaw and the Pee Dee rivers, and again in the Ogeechee. The total number of families represented in collections also increases, from 11 to 23, in a southerly direction (Table 2). Similar latitudinal patterns in taxonomic diversity between temperate and subtropical regions have been observed for most other faunal groups (Karr 1971, Stanton 1979), and may be largely explained by increased resource complexity and a more benign physical environment in subtropical regions (MacArthur 1972). An increase in fish species richness and endemism from north to south for entire drainages is described by Hocutt et al. (1986) for the southeast region. The species richness data in Table 2, however, are primarily based on collections of river mainstems only and do not reflect this pattern.

Fish biomass ranges widely both between and within river systems, often being highest in slough or backwater areas (Coomer and Holder 1980, Wiltz 1984). Total fish biomass for the Piedmont James River ranged between 70.0 and 8.1 kg wet mass/ha; dominant fishes, based on biomass, were smallmouth bass, redbreast sunfish (*L. auritus*), and bull chub (*Nocomis raneyi*) (Garman and Smock 1988). In the Ocmulgee River over 71% of the biomass was contributed by catostomids, especially *Moxostroma anisurum* (Coomer and Holder 1980). The coastal shiner, *Notropis petersoni*, is probably the most widespread cyprinid of southeastern Coastal Plain rivers (Davis and Louder 1971), whereas *Lepomis auritis* and *L. macrochirus* are the dominant centrarchid species (Coomer and Holder 1980). No estimates of annual production of fish are available for any of the rivers under consideration.

The Roanoke River, a tributary to the Chowan River, has the greatest number of endemic fishes, including the Roanoke logperch (*Percina rex*), a hog sucker (*Hypentelium roanokense*), two *Moxostoma* species, *Etheostoma podostemone*, and one undescribed subspecies each of the genera *Noturus* and *Percina*. The near-endemic Roanoke bass (*Ambloplites cavifrons*) is found

TABLE 2 Fish Fauna of Rivers of the Southeastern Atlantic Slope

River System	Species Richness	Number of Families	Dominant Family (number of species)	Dominant Genus	Endemic Taxa (number)	Mean Fish Biomass (kg/ha)
James	72	11	Cyprinidae (18)	*Notropis*	3	41
Chowan	32	15	Centrarchidae (9)	*Lepomis*	—	—
Waccamaw	49	15	Centrarchidae (11)	*Notropis*	5	—
Pee Dee	35	15	Centrarchidae (10)	*Notropis*	2[c]	—
Santee	48	15	Centrarchidae (9)	*Lepomis*	—	—
Lynches	35	13	Centrarchidae (13)	*Lepomis*	—	—
Savannah	88[a]	19	Centrarchidae (14)	*Lepomis*	2[b]	121
Ogeechee	56	19	Centrarchidae (14)	*Notropis*	2[b]	255
Ocmulgee	56	19	Centrarchidae (14)	*Moxostoma*	—	295
Altamaha	42	18	Centrarchidae (12)	*Minytrema*	2(4)[b]	386
Satilla	46	17	Centrarchidae (15)	*Lepomis*	—	179
St. Johns	75[a]	23	Centrarchidae (15)	*Lepomis*	3	—

Note. Dominance is numerical. Species richness values are cumulative totals from published collection records for the respective river mainstem, where available, inclusive of diadromous and euryhaline taxa. As such, the values represent the true number of mainstem river species only to the extent that collections were representative and sampling methods were nonselective. Clearly, in many cases collections underrepresent the total number of species present. Some species lists for these rivers (e.g., Hocutt et al. 1986) are inclusive of the entire watershed and not strictly relevant to the scope of this chapter.

[a]Entire watershed including tributaries (Hocutt et al. 1986, Swift et al. 1986).

[b]*Etheostonia hopkinsi* and *E. inscriptum* are shared by the Altamaha, Ogeechee, and Savannah drainages.

[c]*Semotilus lumbee* is shared with the Cape Fear drainage.

Source. From various sources cited in the text.

in the Neuse and Tar Rivers of North Carolina (Cashner and Jenkins 1982, Garman 1988). Several of the above species are associated with the Roanoke river mainstem. Endemism is also well developed in the Waccamaw system, where at least five fish species or subspecies are unique, or shared with one other drainage (Shute et al. 1981). The endemism of the Waccamaw system is unusual for Coastal Plain rivers (Jenkins et al. 1972), and may be explained by recent geological events that greatly reduced the size of the drainage system. Most of the endemic forms, including *Fundulus waccamensis*, *Etheostoma perlongum*, and *Menidia extensa*, are associated solely with Lake Waccamaw, whereas only one undescribed species of madtom (*Noturus* sp.) is found in the main-stem river. A final endemic form, *Elassoma* sp., has been collected from several tributaries to the Waccamaw River (Shute et al. 1981), and is also known from the Savannah drainage (Hocutt et al. 1986).

Anadromous and catadromous fishes may be important seasonal components of the fauna of most southeastern rivers, including the James, Neuse, Tar, Cape Fear, Santee, Cooper, Altamaha, and Ogeechee rivers (Jensen 1974, Hottell et al. 1983). The clupeids *Alosa sapidissima*, *A. aestivalis*, *A. pseudoharengus*, and *A. mediocris* spawn in most of the above rivers, as does the striped bass. The Atlantic and shortnose sturgeons are also widespread, although the latter is considered endangered (Hottell et al. 1983). Adult *Anguilla rostrata* were collected from all major rivers within the region.

For a more detailed description of fish zoogeography for the southeastern Atlantic slope refer to Hocutt and Wiley (1986).

As with the macroinvertebrate fauna, large woody debris (snags) represents an important structural component of the fish habitat in southeastern rivers (Angermeier and Karr 1984). In the Satilla River, Georgia, four dominant fishes, including *Lepomis auritus*, *L. macrochirus*, *L. gulosus*, and *L. punctatus*, obtained at least 60% of their prey biomass from snag habitats (Benke et al. 1985). Similar results were obtained from the Ogeechee River, Georgia (Benke et al. 1985).

Resource Use

The rivers of the southeastern Coastal Plain remain in a relatively undisturbed condition (Table 3). Even the most highly developed of these, the James, with several major population centers adjacent, is characterized by good water quality (Weand and Grizzard 1986), and great recreational potential. Whereas most rivers experience some impact from adjacent agricultural and silvicultural activities, a few such as the Waccamaw, Ogeechee, and Satilla flow through undeveloped areas (Table 3) and may be considered near pristine (Hottell et al. 1983).

The Cooper and Santee rivers of South Carolina are notable exceptions to the above generalization. Both rivers were dammed early in this century, and were joined by a 12-km diversion canal for hydroelectric purposes in 1942 (see Chapter 15). The Santee–Cooper project resulted in a 200-fold increase in the annual flow of the Cooper and a concomitant siltation problem at the

TABLE 3 Selected Human Impacts for Rivers of the Southeastern Atlantic Slope

River System	Riparian Land Use	Major Pollutant(s)	Active Power Generating Plants (number)	Dams (number)	Major Impoundments (number)
James	A/U	Kepone, sewage	7	12	0
Chowan	A/S	Agricult. runoff	0	0	0
Tar	A/S	—	1	1	0
Neuse	A/S	—	0	3	0
Cape Fear	A/S	—	0	3	0
Waccamaw	F	—	0	0	0
Pee Dee	A/S	—	6	4	4
Santee	A/S	—	1	1	1
Cooper	A/S	—	1	1	1
Savannah	F	Thermal	4	4	3
Ogeechee	F	—	0	0	0
Ocmulgee	A/S	—	2	2	1
Altamaha	F/S	Papermill effluent	1	0	0
Satilla	F/A	—	0	0	0
St. Johns	A/U	Agricultural runoff	0	0	0

Note. A, agriculture; S, silviculture; F, forest; U, urban.

Source. From various sources cited in the text.

river mouth. A rediversion project to partially restore flow to the Santee River has recently been completed. Although greatly disturbed, this system continues to support large fisheries for striped bass and other species; the effects of rediversion, if any, on these fisheries are as yet unknown (Meador et al. 1984).

Most of the states of the region recognize the resource potential of these rivers and associated tributaries, and several states have implemented programs for river protection. Florida and South Carolina have each placed sections of small rivers, including the Myakka, Wekiva, Congaree, and Chattooga, under special status. Virginia currently protects 360 km of streams and rivers, including a section of the upper James and several major tributaries; approximately 60 other sites are under consideration for "scenic" status in Virginia. North Carolina has a water resources plan that proposed 120 water-related natural areas within Piedmont and Coastal Plain provinces. Several river systems discussed in this chapter, including Cape Fear, Neuse, Tar and Waccamaw, and Pee Dee, are identified as potential "Natural and Scenic Rivers" by North Carolina.

The recreational and commercial fishery potential of the major southeastern rivers is considerable (Table 4). Except for the Piedmont James River, where smallmouth bass is the dominant gamefish, largemouth bass, American shad, and striped bass are the major species sought by recreational anglers (Hornsby and Hall 1981). The Cooper/Santee River system, South Carolina, supports heavy sport fishing pressure for black crappie (*Pomoxis nigromaculatus*) and largemouth bass, averaging 66 hours of effort per hectare during

TABLE 4   Fisheries of Rivers of the Southeastern Atlantic Slope

River System	Major Fisheries		Recreational Angler Effort	Recreational Angler Harvest
	Commercial	Recreational		
James	Clupeids	*Micropterus* spp.	—	—
Santee	Blueback herring	Black crappie	96 000 (h/year)	26 (kg/ha)
Cooper	Blueback herring	Largemouth bass	173 000 (h/year)	34 (kg/ha)
Savannah	Catfish	Striped bass	65 000 (trips/year)	—
Ogeechee	American shad	Largemouth bass	40 000 (h/year)	15 (kg/ha)
Ogeechee	—	American shad	2 225[a] (h/year)	1 681 (kg/year)
Ocmulgee	—	*Lepomis* spp.	282 (h/ha)	92 (kg/ha)
Satilla	—	*Lepomis* spp.	175 (h/ha)	19 (kg/ha)
St. Johns	Catfish	Largemouth bass	240 000 (h/year)	180 000 (fish/year)

[a]Season limited to four weeks during the spring.

Source. From various sources cited in the text.

the year. Angler harvest in these rivers averages 32 kg/ha (Christie and Curtis 1983). In a recent review of recreational creel surveys, Schmitt and Hornsby (1985) found that the Altamaha, Satilla, Savannah, and Ogeechee River fisheries of Georgia were dominated by redbreast sunfish, largemouth bass, and catfishes, averaging 17, 14, and 26% by weight, respectively. In one of these rivers, the Altamaha, recreational anglers fish over 100 000 h/year and average 1.4 fish/h, resulting in a total annual harvest of 26 000 kg (Hottell et al. 1983).

Clupeid fishes (*Alosa* spp.) dominate the commercial fisheries of major southeastern rivers. In 1981, approximately 100 000 kg of blueback herring were harvested from the Santee River, South Carolina (Meador et al. 1984). Commercial landings of American and hickory shad during the years 1973–1976 averaged 140, 16, and 77 kg ($\times 1000$) for the rivers of North Carolina, South Carolina, and Georgia, respectively (Holder and Hall 1976).

SPECIFIC RIVER SYSTEMS

Because of the large number of rivers involved, it is not possible for this chapter to present, in detail, a discussion of structure and function for each. Our approach is to identify a few river main stems—the James, Savannah, and St. Johns—which represent three of the four categories of southeastern rivers presented earlier. Since the remaining group of rivers is the subject of a previous chapter (Smock and Gilinsky, Chapter 7), a description of a single representative blackwater system will not be included here. Other criteria for selection, such as the availability of information, were somewhat arbitrary. With a few exceptions, the largest river main stems of the region are also the best studied.

James River

Formed by the confluence of the Jackson and Cowpasture rivers at an elevation of 328 m, the James River drains 25% (26 000 km²) of Virginia, and flows east for a distance of 536 km to empty into the Chesapeake Bay at Hampton Roads. The James River basin, while largely rural, contains three major population centers—Lynchburg, Richmond, and Newport News—and approximately 20 industrial facilities that may potentially compromise water quality (Weand and Grizzard 1986). The climate of the basin is characterized by mean air temperatures ranging from 3 to 26 °C, and by 103 cm of precipitation in an average year.

At the fall line in Richmond, the James River descends 30 m in 15 km, forming the boundary between Piedmont and Coastal Plain regions. For approximately 368 km above Richmond, the James River of the Piedmont (i.e., nontidal) has a moderate gradient, averaging 0.5 m/km, and river discharge is subject to extreme fluctuations as influenced by precipitation events (Woolcott 1974). Substrate of the upper James is typically gravel, boulders, and bedrock. Except for channel margins, the substrate is relatively free of silt or particulate organic matter; large woody debris (snags) is relatively uncommon. A narrow corridor of dense riparian vegetation, dominated by sycamore (*Planatus occidentalis*), American elm (*Ulmus americana*), red maple (*Acer rubrum*), and redbud (*Cercis canadensis*) shades channel margins and usually provides a buffer between the river and open agricultural land beyond.

The geological formations that underlie Piedmont sections of the river range in age from the Paleozoic to the Triassic, and encompass both sedimentary and metamorphic rocks (Hunt 1974). Several smaller rivers, including the Slate and Rivanna, are important tributaries of the Piedmont James.

The ecology of the James River has been poorly studied, and this neglect is particularly evident for the Piedmont. Only one extensive faunal survey (Woolcott 1974, White et al. 1977), and several surveys of limited scope (e.g., Raney 1950), are published; estimates of ecological processes (e.g., primary and secondary production) are totally lacking for any biotic component of the James River system within the Piedmont. An extensive multiyear investigation of the nontidal James River mainstem (Garman and Smock 1988) is expected to yield useful information on fish, macroinvertebrates, periphyton, and physicochemistry.

The lower 168 km of the river is influenced by tides but remains freshwater to the City of Hopewell, some 100 km above Chesapeake Bay. This freshwater tidal portion of the river lies entirely within the Coastal Plain and is characterized by a wide (200 m) meandering channel, low gradient, numerous oxbows and islands, and a substrate of silt and fine particulate organic matter. Partially submerged woody debris is found occasionally along the river margin. Major James River tributaries within the Coastal Plain include the Chickahominy and Appomattox rivers.

The sediments over which the lower James flows reflect a recent geologic past, and are composed of unconsolidated sands and gravels (Jensen 1974). Riparian vegetation is dense, with sycamore, slippery elm (*Ulmus rubra*), and pawpaw (*Asimina triloba*) dominating the floodplain canopy, and herbs (e.g., *Smilax spp.*) dominating the understory (D. Young, Virginia Commonwealth University, personal communication).

Moderate alkalinity (36–61 mg/L $CaCO_3$) and conductivity (110–340 µS), and a mean pH slightly above neutral (7.6) characterize the James within the Piedmont (Woolcott 1974). Dissolved organic acids are not present (i.e., water is not "stained"), but suspended solids levels are often high (>100 mg/L) following rainfall. Water in the uppermost section of the James is often stained by unknown agent(s) entering the Jackson River at Covington, Virginia (G. Garman, personal observation). Nitrate and phosphate nutrient concentrations are relatively low, ≤1.0 and 0.8 mg/L, respectively, but are probably enhanced by agricultural runoff. Dissolved oxygen values measured at Bremo Bluff (Fluvanna Co., Virginia) ranged from 8 mg/L in July to 14 mg/L in December (Woolcott 1974); biochemical oxygen demand (BOD) at river kilometer 224 was consistently below 1.0 mg/L (Weand and Grizzard 1986). Annual mean fecal coliform counts (cells/mL) for seven sites in the James

River above Richmond (Virginia State Water Control Board, unpublished data) ranged from 125 to 960, being greatest just below Lynchburg, Virginia.

A slightly older (1969) study of James River water quality just above Richmond (Jensen 1974) largely corroborates Woolcott's findings, although summer dissolved oxygen levels were substantially lower (6.0 mg/L), and nitrate values were higher, up to 4.0 mg/L. Concentrations of several unspecified metals were within the range expected for natural surface waters. In a recent review of James River water quality (Weand and Grizzard 1986), copper and zinc levels occasionally exceeded water quality criteria, although, with the exception of trace amounts of DDT at a single site, organic contaminants were not detected.

Several chemical parameters, including pH, conductivity, and alkalinity, remain largely unchanged as the James flows from Piedmont to Coastal Plain regions (Jensen 1974). Prior to 1974, however, high BOD, caused by domestic and industrial effluent from Richmond, severely lowered dissolved oxygen values in the James from Drewry's Bluff to the City of Hopewell (Bishop 1985). Following construction of a new wastewater treatment facility at Richmond in 1974, five-day BOD values in the vicinity of Hopewell (river km 120) were consistently below 4 mg/L, a significant improvement over pre-1974 levels (Bishop 1985, Wootton 1985, Weand and Grizzard 1986).

Primary Producers

Periphyton   Forty-four genera of algae, representing five divisions, were collected in one year from natural substrates of the Piedmont James in the vicinity of Bremo Bluff by Woolcott (1974). The genera *Navicula*, *Fragilaria* (Bacillariophyceae), *Cladophora* (Chlorophyta), and *Vaucheria* (Chrysophyceae) were most common on both artificial and natural substrates. Channel margin habitats (silt) yielded a slightly greater number of genera (40 vs. 36) than did rubble substrate in mid-channel. But, since the most abundant taxa (e.g., Bacillariophyceae, *Cladophora* sp.) are epilithic forms (Hynes 1970), it is possible that overall periphyton biomass was greatest on rubble substrate. While diatom density fluctuated seasonally, this group numerically dominated the periphyton community throughout the year.

Heated effluent from a nearby electrical generating facility was probably responsible for dense mats of *Oscillatoria* (Cyanophyta) observed on substrates within the thermal plume. The few other limited descriptions of algal taxa within the James River basin (Woodsen 1959, Woodsen and Prescott 1961) are similar to the findings of the Bremo Bluff study.

Apparently, no published estimates of periphyton standing stock or primary production are available for the James River above Richmond. Periphyton (epilithic) standing stock (mg/m^2 total chlorophyll) ranged from 5 to 34 during the period October–February 1988 (P. Silverman, Virginia Commonwealth University, unpublished data). Based on current conceptual models of river ecology (e.g., Vannote et al. 1980) and the abundance of "grazer" macroinvertebrate fauna in this section of the James River, it is likely that primary production by epilithic diatoms is an important component of the trophic base for the river above the fall line.

Phytoplankton   Unlike the Piedmont James, where samples indicated the almost total absence of planktonic algae (Woolcott 1974), there is a gradual, downstream increase in numbers of phytoplankton within the tidal, freshwater (Coastal Plain) part of the James River (Jensen 1974, Weand and Grizzard 1986). In the summer of 1970, Jensen (1974) collected 27 genera of phytoplankton from the river at Chesterfield (river km 150), representing five divisions. Abundant genera included *Synedra*, *Navicula*, and *Melosira* (Bacillariophycea) and, in general, diatoms dominated the summer phytoplankton community. While no estimates of production are available, the concentration of chlorophyll *a* in surface waters of the lower James during 1984 and 1985 was <10 mg/L immediately below the fall line, but increased sharply to 50 mg/L some 50 km below Richmond at Hopewell (Weand and Grizzard 1986).

Consumers

Zooplankton   Sixteen genera of planktonic microcrustaceans were collected by Jensen (1974) and Bender and Huggett (1985) in samples of the James within the Coastal Plain. Two groups, the copepod *Cyclops* spp. and the cladoceran genera *Bosmina* and *Daphnia*, were dominant. Total numbers of zooplankton during 1970–1971 ranged from 2 to 12 individual organisms per liter. Within the Piedmont, copepods were encountered only rarely in samples of James River water by Woolcott (1974). No information is available on non-planktonic microinvertebrate groups (e.g., meiofauna) in either section of the James, although these organisms may be ecologically important.

Macroinvertebrates   As with other components of the Piedmont James River, the Bremo Bluff study of Woolcott (1974) provides a useful picture of the macroinvertebrate community. Two types of natural substrates, fine sediment in channel margins and cobbles in mid-channel, were sampled every other month over a two-year period (1972–1974); in addition, artificial substrates were employed on a limited basis. Combining substrate types and procedures, 196 genera, representing seven phyla, were identified from the river. The insect order Diptera contained the greatest number of genera (53), followed by Trichoptera (28), Coleoptera (24), Odonta (21), and Ephemeroptera (19). The geographically widespread insect genera *Stenonema*, *Chimarra*, *Cheumatopsyche*, *Hydropsyche*, *Stenelmis*, and *Orthocladius* dominated the macroinvertebrate fauna.

The amphipod *Gammarus fasciatus* and the gastropod *Somatogyrus* spp. also were frequently encountered, in addition to the crayfish taxa *Cambarus* sp. and *Orconectes* sp. Values for taxonomic richness and total number of organisms were substantially greater in channel margin habitats than in mid-channel habitats, regardless of year or sample date. Twenty-one macroinvertebrate genera were collected exclusively from river margins; only four genera were unique to rubble habitat, perhaps suggesting greater microhabitat complexity in the former (Rabeni and Minshall 1977). More recent studies of the nontidal James River main stem at several locations (Garman and Smock 1988) have identified 101 aquatic insect genera, exclusive of chironomids. Dominant insect groups were Diptera (Chironomidae), Ephemeroptera (Heptageniidae), and Coleoptera (Elmidae). Mollusks of the families Hydrobiidae, Pleuroceridae (*Goniobasis* spp.), Sphaeridae, and Corbiculidae (the introduced *Corbicula fluminae*) dominated macroinvertebrate biomass at most Piedmont sites.

From a trophic function standpoint, most common taxa are classified as collectors of fine particulate organic matter (<1 mm), or as scrapers (grazers) of epilithic periphyton (Merritt and Cummins 1984). The importance of these two functional groups would be expected in a shallow, medium-sized river like the Piedmont James (Vannote et al. 1980), and suggests that secondary production is based both on inputs of fine detritus and on production of periphyton. In addition to the above functional groups, insect predators (e.g., *Gomphus*, *Rhyacophila*, *Corydalus*) were also common in samples.

At least one freshwater mussel, *Fusconaia collina* (Bivalvia: Unionidae), is known only from the James River (Boss and Clench 1967, Johnson and Boss 1984). The Asiatic freshwater clam *Corbicula fluminea* was first sampled from the James River in 1971 (Diaz 1974) and has since become established in both Coastal Plain and Piedmont sections, often dominating the fauna of silt and sand substrates (Sickel 1980).

Total macroinvertebrate biomass for four sites within the Piedmont James ranged up to 453 g dry mass/m², due primarily to the mollusks *Corbicula* and *Elliptio*; values <10 g dry mass/m² were, however, more typical (Garman and Smock 1988). Again, no estimates of production are available for any macroinvertebrate group or species of the Piedmont James River.

The macroinvertebrate community of the tidal freshwater James, as described by Sickel (1980) and Jensen (1974), is dominated by two taxonomic groups—oligochaete worms and chironomid larvae—resulting in a community very different from that above Richmond. The widely distributed genus *Limnodrilus* (Hynes 1970) alone represented 40% (numbers) of the benthic fauna in samples taken by Jensen (1974). The mayflies *Caenis* and *Ephemerella*, and the caddisflies *Oecetis* and *Protoptila*, all widespread and tolerant forms, were encountered above and below the fall line (Sickel 1980).

The comparatively lower macroinvertebrate taxonomic richness of the Coastal Plain James River (75 species vs. 196 genera at Bremo Bluff) and the dominance of burrowing tubificids in the lower James might be expected due to a shift in substrate composition to fine organic and inorganic particulates. Inputs of organic and chemical pollutants, largely abated after Jensen's (1974) study, may also have reduced richness to levels typical of stressed environments. No recent data are available for comparison.

Fishes   Based on several collections (Jensen 1974, Woolcott 1974, Garman and Smock 1988), 72 fish species are known from the James River main stem (Table 2). Collections of the Piedmont James fish fauna, made by Woolcott

(1974) at Bremo Bluff, included 50 species representing 11 families—over half of the species known to occur in the entire drainage. All but one of the species (*Anguilla rostrata*) are largely confined to freshwater (Darlington 1957, Berra 1981). Cyprinids dominated the fish assemblage both in number of species and number of individuals, with *Notropis analostanus*, *N. ardens*, and *N. hudsonius* being most abundant. The centrarchid fishes *Lepomis auritus* and *L. macrochirus*, and the catfish *Ictalurus punctatus* also were common. Relative biomass (g/collection) was greatest for *Cyprinus carpio*, *A. rostrata*, and *Catostomus commersoni*.

Diets of the seven most abundant species collected by Woolcott included a wide range of food types, but most fish taxa were benthic invertivores, consuming aquatic insects and the amphipod *Gammarus*. The satinfin shiner (*N. analostanus*) appeared to be the only true omnivore, consuming a broad spectrum of epilithic algae and terrrestrial and aquatic invertebrates. The numerical dominance of *N. analostanus* in the community may be, in part, due to its role as a trophic generalist (Dill 1983). The longnose gar (*Lepisosteus osseus*) was exclusively piscivorous, whereas the other major fish predator, smallmouth bass (*Micropterus dolomieui*), specialized on crayfish, especially during winter months. In apparent contrast, the numerical percentage of crayfish and fish in the diets of 340 juvenile and adult James River smallmouth bass during nonwinter months of 1987 averaged 7.9 and 4.2%, respectively (Garman and Smock 1988). Percent contribution of terrestrial-source insects (e.g., Hymenoptera, Homoptera) in the same study was substantial, ranging up to 88% (Garman 1990).

Although three fish species are endemic to the James drainage, no endemics were collected by Woolcott at Bremo Bluff. Two endemic forms, *Etheostoma longimanum* and *Percina notogramma montuosa*, however, are reported to occur in the main stem James within the Piedmont (Hogarth and Woolcott 1966, Jenkins and Burkhead 1975). The third endemic species, *Notropis semperasper*, is restricted to the Ridge and Valley physiographic province. The darter *Percina roanoka* and the redhorse *Moxostoma cervinum*, previously restricted to the Roanoke and/or Tar rivers, have recently become established in the James above the fall line as a result of introductions (Jenkins and Burkhead 1975). The range of the bridle shiner (*Notropis bifrenatus*) has extended southward to include the James River (Jenkins and Zorach 1970). Several other non-native, or possibly non-native, forms, including *Nocomis raneyi*, *Noturus gilberti*, *Cyprinus carpio*, *Notropis cerasinus*, and *Notropis telescopus*, are known to occur in the James River (Stauffer et al. 1982, Hocutt et al. 1986). In addition, the nonnative game fishes *Micropterus dolomieui*, *M. punctulatus*, *Lepomis cyannelus*, *L. microlophus*, *Ambloplites rupestris*, *Perca flavescens*, *Stizostedion vitreum*, *Pylodictis olivaris*, and *Esox masquinongy* have been introduced, presumably for their sport fishery potential (Hocutt et al. 1986, Garman and Smock 1988). Recent (1989) collections from the James River in and above the fall line (G. C. Garman, unpublished data) included white crappie (*Pomoxis annularis*) and threadfin shad (*Dorosoma petenense*), heretofore not recorded for the main stem James River.

Compared to the Piedmont, fewer species of fish (41) comprised the resident Coastal Plain assemblage as described by Jensen (1974), although recruitment of several estuarine forms (e.g., *Menidia, Trinectes, Mugil, Fundulus*) increases to 17 the number of families represented. Eleven of the 41 species collected were rare in Jensen's samples, occurring only once or twice in 23 samples.

Compared to the Piedmont fish fauna, the cyprinid genus *Notropis* was replaced in the lower James by *Morone* and *Ictalurus* as numerically dominant taxa; several *Lepomis* spp. were abundant both above and below the fall line. The appearance of planktivores (*Dorosoma*) and detritivores (*Mugil*) below the fall line may reflect a food resource no longer limited to benthic macroinvertebrates and periphyton.

In addition to resident fishes, anadromous clupeids (*Alosa aestivalis, A. mediocris, A. pseudoharengus, A. sapidissima*) and striped bass (*Morone saxatilis*) were collected in the tidal freshwater James by Jensen (1974) and by G. C. Garman (unpublished data) within the fall line to William's Dam.

Resource Use   The James River receives input from at least fifteen major municipal and industrial discharges, the most important being the cities of Lynchburg, Richmond, and Hopewell (Table 3). Prior to construction of a new waste treatment facility at Richmond in 1974, the tidal, freshwater river was characterized by low dissolved oxygen levels and eutrophication (Jensen 1974). Water quality is presently considered "good," however, along the entire length of the James (Bishop 1985). Copper and zinc were the only trace metals to occasionally exceed water quality standards in a recent study of the lower James (Weand and Grizzard 1986). In the same study, BOD values ranged between 1.0 and 3.9 mg/L, significantly lower than pre-1974 levels (Jensen 1974).

The effects of thermal pollution from three electric power generating stations were studied by White et al. (1977), and Maurakis and Woolcott (1984). No major impacts on the aquatic biota were observed, although fish distributions were possibly affected (Woolcott 1985).

Dams have dramatically reduced the historical range of anadromous fishes in the James River. Twelve dams presently restrict upstream movement of clupeid species and of striped bass between Richmond and Lynchburg, a distance of 222 km (Mudre et al. 1985). Plans to modify or breach five dams at Richmond are being hampered by political resistance and estimated cost ($2.5–7.5 million); the two lowermost dams, however, were breached in early 1989. None of the above dams create a major impoundment of the James River mainstem.

The discovery in 1975 of high levels of the pesticide Kepone in samples of James River sediment and biota resulted in one of the most publicized environmental problems of the last decade. Kepone entered the lower James River at Hopewell where, between 1966 and 1975, 1.5×10^6 kg of the material was produced. Total contamination of river sediments was estimated at 14 000 kg of Kepone in 1976 (Cutshall et al. 1981), but levels in the sediment have been declining since production was halted. The affinity of Kepone for organic sediment, together with low solubility in natural waters, probably reduced exposure of biota to the contaminant. As a result, the major effect of Kepone may have been economic rather than ecological, due to a recently lifted fishing ban on the lower James River and estuary (Diaz 1989).

The recreational fishery of the Piedmont James River (Table 4) is of considerable economic importance, and is similar in character to the fishery of the smaller Shenandoah River in western Virginia (Kauffman 1980). While harvest and catch composition data are not presently available, the fishery is dominated by smallmouth bass (*Micropterus dolomieui*) and other centrarchids (*Lepomis* spp.), and pressure on this component of the fishery is increasing.

Savannah River

The Savannah River rises in the mountains of Georgia, South Carolina, and North Carolina at an elevation of 1676 m. A network of tributaries, including the Chattooga, Tallulah, Chauga, and Keowee rivers unite to form the Tugaloo and Seneca rivers, which merge to form the Savannah near Anderson, South Carolina. For most of its length, the Savannah forms the border between Georgia and South Carolina. Augusta, Georgia, lies on the fall line, separating Piedmont and Coastal Plain sections of the river. Above Augusta, the river is impounded for most of its 193 km by Clark Hill, Russell, and Hartwell reservoirs. The river is also impounded for navigation at the New Savannah Bluff Lock and Dam at Augusta. Below this point, the river flows freely for 291 km to the sea. Savannah, Georgia, the other major riverbank city, is situated at river km 35.

The large watershed area encompasses a long, narrow corridor equally divided between Georgia and South Carolina. The watershed is primarily wooded; within the Coastal Plain, principal forest species change from loblolly pine (*Pinus taeda*) and slash pine (*P. elliottii*) to longleaf pine (*P. australis*) and slash pine in a downstream direction (Schmitt and Hornsby 1985). Climate consists of long, humid summers and short, mild winters, with average rainfall of 116 cm per year and average air temperature of 18.6 °C (Schmitt and Hornsby 1985).

The Savannah River in the Coastal Plain is a large, low-gradient, warmwater river (Table 1). Mean annual discharge from the river varies greatly, with a 140-fold range over the past 50 years. Range in discharge has declined in recent years with the operation of the upstream impoundments (Patrick et al. 1967). The U.S. Army Corps of Engineers attempts to maintain a minimum streamflow of 173 m^3/s to allow commercial boat navigation (Specht et al. 1985). The negligible gradient in the Coastal Plain produces a sinuous channel; 24% of the river length is in oxbows (Schmitt and Hornsby 1985), and swamps adjoin the channel throughout its length. The substrate is 90% sand, with small quantities of mud/silt and larger rock (Patrick et al. 1967). Turbidity is high during high flows, but the water clears substantially during low flow summer months. The Broad River, entering the Savannah above Clark Hill Dam, is a major contributor of suspended sediment to the river (Gordon and Wallace 1975).

Because of substantial flows and strong tidal influence, the river never stratifies. River water temperatures are similar throughout the Coastal Plain. Midsummer temperatures can exceed 30 °C and midwinter temperatures seldom fall below 8 °C (Patrick et al. 1967, Stickney and Miller 1974). A strong salinity gradient occurs in the tidal portion of the river near Savannah, with ranges as high as 15 ppt from surface to bottom (Stickney and Miller 1974). The river is low in alkalinity and other dissolved nutrients, and maintains a neutral to slightly acidic character. Tributary streams differ widely in physical and chemical character, and include streams of both low and high alkalinity (Specht et al. 1985).

Riparian vegetation consists mainly of hardwood trees characteristic of river swamps and bottomland forests. Main components of the riparian forests are black gum (*Nyssa sylvatica*), tupelo gum (*N. aquatica*), sweetgum (*Liquidambar styraciflua*), swamp hickory (*Carya cordiformis*), sycamore (*Platanus occidentalis*), cypress (*Taxodium distichum*), white ash (*Fraxinus americana*), red maple (*Acer rubrum*), live oak (*Quercus virginiana*), water oak (*Q. nigra*), and laurel oak (*Q. laurifolia*) (Patrick et al. 1967, Schmitt and Hornsby 1985).

Riparian vegetation varies due to the meandering nature of the river. Outside bends of oxbows generally have mature pine–hardwood stands, while inside bends and sand bars characteristically include stands of pioneer grasses and the willow *Salix* spp. (Schmitt and Hornsby 1985). Outside bends often develop undercut banks, which eventually collapse. Fallen trees and accumulated organic debris are the major substrates for aquatic invertebrates and cover for fishes.

Primary Producers The lower river has abundant stands of alligatorweed (*Alternanthera philoxeroides*), which can impede navigation (Schmitt and Hornsby 1985). The lowest 45 km of river is estuarine marsh, dominated by cordgrass (*Spartina* spp.) and rushes (*Juncus* spp.).

The algae of the Savannah River were sampled exhaustively during the 1950s between river km 214 and 280 (Patrick et al. 1967). Over 400 species of algae were reported; an average of 91 species was collected in each sample. Diatoms dominated the flora at all times, followed in diversity and abundance by green algae (Chlorophyta). Highest algal densities occurred during fall sampling.

Abundant diatom genera included *Melosira, Eunotia, Achanthes, Gomphonema*, and *Gyrosigma*. The most abundant genera of green algae included *Vaucheria, Stigeoclonium, Oedgonium*, and *Spirogyra*. Blue green algae were represented by *Oscillatoria, Lyngbya*, and *Anabaena*. Red algae were uncommon relative to other groups, but typical genera included *Compospogon* and *Batrachospermum* (Patrick et al. 1967).

Additional sampling of phytoplankton revealed essentially complete overlap between the attached and drifting fauna. The only exception was *Asterionella formosa*, a lentic form collected below Clark Hill reservoir. The presence of Clark Hill Dam greatly influenced the density of phytoplankton; comparison of collections before and after the operation of the dam showed substantial increases in algal cell density, presumably due to transport of lentic forms through the dam (Patrick et al. 1967).

Specht et al. (1985) investigated periphyton biomass in the vicinity of the Savannah River Plant (river km 220–240) during 1982–1983. Biomass and chlorophyll *a* concentrations peaked in August at 5500 and 95 mg/m^2, respectively. Diatoms dominated the periphyton assemblage, along with green and red algae at specific locations.

Consumers

Macroinvertebrates   Patrick et al. (1967) reported that the protozoan community for the Savannah River was typical of similar sites elsewhere. Ciliates, flagellates, and sarcodinids were all represented among the more than 400 species collected. Protozoans were present in great numbers in the dense beds of *Vaucheria* growing in shallow waters and in other protected microhabitats. As with drifting phytoplankton, number of species increased greatly after Clark Hill Dam was completed in 1952.

Limited information exists for a variety of invertebrate phyla. Of three species of Porifera, the sponge *Spongilla fragilis* was the most abundant (Patrick et al. 1967). Among 16 annelid taxa collected by Patrick et al. (1967), only the oligochaete *Limnodrilus hoffmeisteri* was common. Oligochaetes comprised 50% of the drifting invertebrates reported in upstream areas (Matter et al. 1983) and presumably are also important components of the downstream fauna. Four bryozoans were present, dominated by *Fredericella sultana* (Patrick et al. 1967).

The molluscan fauna, especially the Unionidae, has been described in a variety of studies. Johnson (1970) listed 18 unionid species in the Savannah River system. The most common were *Elliptio complanata*, *E. ictarina*, *Lampsilis cariosa*, and *L. radiata splendida* (Britton and Fuller 1979). In addition, the introduced Asiatic clam, *Corbicula*, is very abundant throughout the river. Six species of Sphaeriidae are present in the river system (Britton and Fuller 1979). Patrick et al. (1967) also list six gastropods, the most common of which was *Physa heterostropha*.

Few crustacean species have been reported for the Savannah. Collections by Patrick et al. (1967) reported ten species, the most abundant of which were the amphipod *Hyalella azteca* and the decapods *Palaemonetes paludosus* and *Procambarus pubescens*. Hobbs et al. (1976) reported a much more diverse decapod fauna for swamps, ponds, and streams adjacent to the river, including 15 species in 5 genera. Although many of these species prefer lentic habitats, several presumably are also present in the Savannah River itself. In collections of macroinvertebrates just below the city of Savannah, Stickney and Miller (1974) noted a low diversity and density of crustaceans, relative to similar estuarine areas; densities of commercial species of *Penaeus* and *Callinectes* were particularly low.

The aquatic insect fauna, in contrast, is particularly well developed. Patrick et al. (1967) listed 67 dipterans and 202 other species from the Savannah. Coleoptera, Ephemeroptera, and Odonata were most diverse, but all major aquatic insect orders were represented. Artificial substrates placed in the river in 1982–1983 developed assemblages dominated by Diptera (74% by mass), Trichoptera (18%), and Ephemeroptera (6%) (Specht et al. 1985). Feeding strategies of the abundant insects indicated a community composed of collectors (74%), engulfers (18%), shredders (4%), scrapers (2%), and piercers (1%) (Specht et al. 1985). Gordon and Wallace (1975) intensively studied the Hydropsychidae within the Coastal Plain area of the Savannah River and found the group dominated by *Hydropsyche* and *Cheumatopsyche*.

Fishes Fishes of the Savannah River represent a diverse riverine fauna with 88 reported species (Table 2). Sampling at river km 252 (McFarlane et al. 1979) revealed a fish community dominated by piscivore (28%) and catostomid omnivore (55%) species. Schmitt and Hornsby (1985) reported that the most common fishes in mainstem habitats were spotted sucker (*Minytrema melanops*, 21% by mass), carp (*Cyprinus carpio*, 15%), silver redhorse (*Moxostoma anisurum*, 15%), and channel catfish (*Ictalurus punctatus*, 8%); total fish standing crop averaged 121 kg/ha. In slackwater areas, biomass was 846 kg/ha, comprised largely of bowfin (*Amia calva*, 23%), gizzard shad (*Dorosoma cepedianum*, 18%), spotted sucker (16%), carp (13%), and crappie (*Pomoxis* spp., 7%). Most notable among recent collections in the Savannah River drainage is the discovery of the mountain mullet (*Agonostomus monticola*) in 1982; the species had not previously been reported outside of Mexico and the West Indies (Schmitt and Hornsby 1985).

Six major anadromous species utilize the Savannah River (Rulifson and Huish 1982). The shortnose sturgeon (*Acipenser brevirostrum*) is considered endangered within Georgia, and recent collections have identified shortnose sturgeon larvae between river km 242 and 253 (Schmitt and Hornsby 1985). Atlantic sturgeon (*Acipenser oxyrhynchus*), American shad (*Alosa sapidissima*), hickory shad (*Alosa mediocris*), and blueback herring (*Alosa aestivalis*) either are stable in abundance or have unknown status.

The most important anadromous species is the striped bass, considered to be declining in abundance in the river because of numerous environmental perturbations (Rulifson and Huish 1982). The Savannah River striped bass, in contrast to most riverine populations, is not anadromous. Spawning occurs throughout the river, but is concentrated in the vicinity of Savannah City (Gilbert et al. 1985). High oceanic temperatures adjacent to the mouth of the Savannah are presumed to inhibit movement of adults and juveniles into the ocean after spawning (Dudley et al. 1977). Because upstream river temperatures are considerably lower than those in the ocean, striped bass move upstream as far as Augusta, where dams prevent further migration (Dudley and McGahee 1983).

Modification of the river at Savannah City to allow navigation presents potential problems for striped bass (Fig. 1). The river splits into two channels at Savannah. The Back River is a naturally shallow channel and is relatively undeveloped; the Front River, the larger channel, has been dredged for navigation and is polluted by riverside industry. A tidal gate constructed across the Back River in the late 1970s traps tidal water moving upstream. The water is diverted through an artificial channel into the Front River to augment navigation flows. Striped bass spawning, most of which occurs in the Back River, is not affected by the tidal gate. Eggs and larvae, however, which previously drifted downstream through the Back River, now are diverted to the Front River, where they are subjected to poor water quality and physical disturbance by boats (Dudley and Black 1978, Gilbert et al. 1985).

Resource Use   The Savannah River serves a variety of human needs (Table 3). In 1980, upstream reservoirs produced 1.6 billion kilowatt-hours of energy, primarily from Hartwell Reservoir (Matter et al. 1983). The river serves as municipal water supply to many shoreline communities, including the city of Savannah. The river provides cooling water for industrial plants along the entire river, including the nuclear power reactors of the Savannah River Plant. Water quality is "drinkable" above Augusta, drops to "fishable" for the 110 km below Augusta, and returns to "drinkable." At Savannah, water quality is reduced to "navigation" (Schmitt and Hornsby 1985, Winger et al. 1990).

The U.S. Army Corps of Engineers maintains the river for navigation throughout the Coastal Plain. The New Savannah Bluff Lock and Dam, constructed in 1937, provides limited navigation to Augusta, and dredging below Augusta has provided a 2.7-m-deep channel downstream to Savannah since 1965. Below Savannah, the Corps maintains a deepwater port for ocean-going vessels. Modifications of the natural channel for navigation and flood control have reduced the total channel length by 13% (Schmitt and Hornsby 1985).

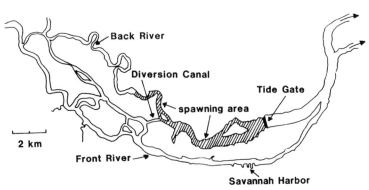

FIGURE 1. Multiple stream channels of the Savannah River near Savannah, Georgia. A tide gate constructed on the Back River diverts inflowing tidal water into the Front River through an upstream diversion canal. (After Gilbert et al. (1985).)

Recreational use of the river is high (Table 4; Schmitt and Hornsby 1985). In 1980, approximately 18.7 million visitors used the river and its reservoirs. The recreational fishery in the Coastal Plain has produced up to 65 000 trips annually (78.8 trips/ha). The fishery depends largely on centrarchid species (45% of catch) and American shad (11%). A small commercial fishery for ictalurids exists in the river (Schmitt and Hornsby 1985). The Savannah River provides the striped bass broadstock for Georgia state hatcheries.

The Savannah River Plant occupies the South Carolina bank of the Savannah River between river kms 227 and 252. The facility is a U.S. Department of Energy National Environmental Research Park performing biological studies and analysis of the ecological impacts of nuclear power generation (Morse et al. 1980). The 750-km² park was closed to the public in 1952 and has been the site of intense ecological study for over three decades. Most of the available knowledge about the Savannah River, including that cited here, is derived from research conducted in association with the Savannah River Plant through the Savannah River Ecology Laboratory, Aiken, South Carolina. The park contains Upper Three Runs Creek, a U.S. Geological Survey National Hydrologic Benchmark Stream. Gibbons et al. (1980) summarized thermal ecology research on the property, and Morse et al. (1980, 1983) described the insect fauna.

St. Johns River

The St. Johns River arises from extensive marshes in east-central Florida at an elevation of approximately 8 m. The St. Johns flows northward for 460 km, turning eastward at Jacksonville, Florida, and flowing 40 km to the Atlantic Ocean. The 24 424-km² watershed lies entirely in Florida, covering the northeastern quadrant of the state. It is the third largest river in Florida (Cox et al. 1976). The watershed contains several moderate-sized urban areas, including Orlando, Deland, Ocala, and Gainesville; the basin's 1985 population was nearly 2 million people (St. Johns River Water Management District (SJRWMD) 1984).

Four distinct river sections comprise the St. Johns system. The upper basin extends from the headwaters to Lake Harney. The upper basin begins in dense marshland which gradually develops into shallow lakes and multiple channels (Williams and Bruger 1972). The natural topography is a broad floodplain, but agricultural canalization and drainage has reduced the natural floodplain by about 60% (SJRWMD 1984). The middle river, from Lake Harney to the confluence with the Oklawaha River, contains a well-defined channel. Most of the middle river basin lies within the Ocala National Forest. The middle river broadens at several locations, forming Lakes Harney, Jessup, Monroe, and George. The river from Lake Harney to the mouth is maintained for navigation by dredging and channelization (Williams and Bruger 1972).

The lower river basin, from the Oklawaha confluence to the Atlantic, is mixed urban and forest land. Freshwater tidal influence occurs as far upstream as Lake George at river km 170 (SJRWMD 1984). The lower river is often wider than 1.5 km, but is extremely shallow. The fourth major section is the Oklawaha River, the largest tributary with a 7040-km^2 drainage basin (SJRWMD 1984). The Oklawaha system contains several large shallow lakes including Lake Apopka, a 13 000-ha hypereutrophic basin averaging 2 m deep.

The St. Johns basin has a humid subtropical climate, with a mean annual air temperture of 20 °C and mean annual rainfall of 143.3 cm. Rainfall in recent decades, however, has been significantly lower than historical levels (Lowe and Gerry 1984). Brief storms during June–August deliver over half of the annual rainfall, producing high river flows in summer and early fall and low flows during the remainder of the year (Leed and Bélanger 1981).

The St. Johns is a warmwater river with extremely low gradient (Table 1). Gradient of the river averages 0.002% (McElroy 1977); consequently, prevailing winds and tidal movements often determine the direction of water flow (Kautz 1981). Water flows towards the ocean only about 75% of the time, with reverse flows as high as 85 m^3/s at Deland (Cox et al. 1976). Because of agricultural drainage and low rainfall, ranges in discharge have increased in recent decades, and average surface elevation has decreased by 0.74 m (Lowe and Gerry 1984). Substrate is mud to sand. Water temperatures are characteristically warm, annually ranging from 9 to 30 °C in the lower river (DeMort and Bowman 1985) and 14 to 30 °C in the upper river (Cox and Moody 1981). High salinities occur throughout the lower river, with levels exceeding 16 ppt at Jacksonville, and occasional pockets of high salinity occur where upstream salt springs enter the river (Williams and Bruger 1972). The river does not stratify, so vertical temperature and salinity gradients are negligible (DeMort and Bowman 1985). The St. Johns is less turbid than other northern Florida rivers, with photic depth near Jacksonville between 1.5 and 3.4 m (Kautz 1981, DeMort and Bowman 1985).

The chemical character of the water is complex. In general, high productivity in the middle and lower river is indicated by conductivity and total dissolved solids concentrations at least double those of other northern Florida rivers (Kautz 1981). The river pH is neutral to slightly basic (Cox and Moody 1981). Nitrate and phosphate levels are highly variable, owing to the local input of organic materials, especially nitrates from blackwater tributaries (DeMort and Bowman 1985). Iron levels generally exceed state standards for drinking water, reaching 600 mg/L in some locations; such concentrations probably occur because of massive inputs of groundwater after agricultural use, low pH in association with organic acids, low dissolved oxygen concentrations, and high turbulence during high flows (Leed and Bélanger 1981). Zinc and copper concentrations are low and are correlated with proportion of urban land use in the watershed (Leed and Bélanger 1981).

Primary Producers Aquatic macrophytes are abundant and diverse throughout the floodplain. Cox et al. (1976) recorded more than 300 species of vascular plants in extensive river, wetland, and floodplain surveys. Freshwater zones are dominated by *Eichornia, Typhus, Pontederia, Myriophyllum*, and *Vallisneria*; saline zones are dominated by *Spartina, Ulva*, and *Enteromorpha*

(DeMort and Bowman 1985). Most primary production in the upper river is macrophytic, presumably because interactions among solar radiation, water color, and high iron content depress dissolved phosphate concentrations (Bélanger 1982).

Marshland vegetation in the upper river occurs in three distinct biotypes (Cox et al. 1976). The headwater biotype is dominated by true marsh plants, including maidencane (*Panicum hemitomon*), sawgrass (*Cladium iamaicense*), and arrowhead (*Sagittaria lampifolium*). The second biotype includes most areas of the river and is dominated by plants preferring drier ground, including dog fennel (*Eupatorium capillifolium*), cordgrass (*Spartina bakeri*), and wax myrtle (*Myrica cerifera*). The third biotype occurs in areas of high salt concentrations and includes salt-tolerant grasses and sedges. The lower rainfall and extensive floodplain drainage of recent years has caused shifts in vegetation towards dryland species, such as willow (*Salix caroliniana*) (Hall and Williams 1984).

The exotic water hyacinth (*Eichornia crassipes*) was first introduced into the United States in 1890 on the St. Johns River at Palatka (Zeigler and McGehee 1977). The water hyacinth forms dense floating mats which at times have entirely covered the mainstem and tributaries for great distances. The U.S. Army Corps of Engineers has been attempting control of water hyacinth since 1900. Current control mechanisms emphasize selective spraying with 2,4-D in areas of high priority and likely success. Management plans restrict pesticide spraying in habitats used by nursing manatees, spawning American shad, and spawning largemouth bass. Phytoplankton accounts for most primary production in the lower river, presumably because higher turbidity prohibits macrophyte production (DeMort and Bowman 1985). Phytoplankton densities peaked during June–August at 25–30 mg chlorophyll/m^3 (6 cells/L) and again in December–January at about half those levels (DeMort and Bowman 1985). Annual chlorophyll averaged 6–10 mg/m^3. These investigators considered phosphorus to be limiting production, based on N:P ratios of 19.5:1 in the water. Blue-green algae (Cyanophyta) are generally dominant in the upper and middle river, with total densities reaching 20–300 million cells/L during blooms (Cox and Moody 1981). The plankton community near Jacksonville is dominated by diatoms (80% of species), with smaller representation of Rhodophyta (7%), Chlorophyta (6%), and Cyanophyta (7%). The ten most abundant algal species were *Skeletonema costatum, Chaetoceros decipiens, Rhizosolenia alata, Nitzschia seriata, Melosira italica, Chaetoceros debile, Coscinodiscus lineatus, Thalassionema nitzschiodes, Thalassiothrix fraunfeddii,* and *Gyrosigma* spp. (DeMort and Bowman 1985).

Consumers The St. Johns River fauna reflects the physical history of peninsular Florida. The St. Johns has been affected repeatedly by massive water level changes on a geological time scale (Burgess and Franz 1978). The latest event, the Wicomico, occurred during the late Pleistocene–early Pliocene when peninsular Florida was covered by seawater at a level 28–30 m higher than today. Most of the St. Johns basin was inundated or connected to the

Suwanee or Sante Fe (a Suwanee tributary) drainages emptying into the Gulf of Mexico. When the water receded, the St. Johns basin developed many endemic taxa and also maintained taxa that have disappeared from adjacent watersheds. The Oklawaha River and Black Creek drainages, both in the western St. Johns drainage, contain most of these special forms (Burgess and Franz 1978).

Zooplankton   The only comprehensive surveys of the invertebrate fauna were conducted by Cox et al. (1976) and Cox and Moody (1981) at stations throughout the upper and middle St. Johns River. Copepods dominated the zooplankton, typically representing more than 50%, by number, of net samples. Zooplankton densities ranged from approximately 1 to 50 organisms/L.

Macroinvertebrates   Cox et al. (1976) listed 68 macroinvertebrate taxa collected in an earlier study of the St. Johns River. Mean invertebrate densities in Ekman dredge samples ranged from 1500 to 14 000 organisms/m^2; samples typically were dominated by chironomids and oligochaetes (Cox and Moody 1981).

Taxa of zoogeographic interest have been reported by Burgess and Franz (1978). The Black Creek drainage has three endemic chironomid taxa and a single endemic crayfish (*Procambarus pictus*). The Oklawaha–St. Johns system has an additional endemic crayfish (*P. deodyte*).

Endemism is most fully developed in the molluscan fauna. The Oklawaha–St. Johns River contains 14 endemic gastropod taxa and shares 4 other species with the Santa Fe or Suwanee rivers. These species are in the family Hydrobiidae, and include the genera *Spilochlamys*, *Cincinnatia*, and *Amnicola*. Johnson (1970) reported a rich unionid fauna, which included *Elliptio dariensis*, *E. icterina*, *E. buckleyi*, *E. jayensis*, *Uniomerus tetralamis*, *Anodonta couperiana*, *Carunculina parva*, *Villosa vibex*, and *V. amygdata*. The Asiatic clam (*Corbicula*) is also common, but declined in abundance following dieoffs in the early 1980s (Morrison 1973, Cox and Moody 1981).

Fishes   The St. Johns River contains a diverse fish fauna for a river of its size (Table 2). Tagatz (1967) reported 55 freshwater and 115 euryhaline species, not including marine species found only at the mouth. Euryhaline species occurred at great distances upstream, presumably because of extended tidal influence and upstream refugia provided by salt springs. The most diverse families were Clupeidae (10 species), Cyprinodontidae (14), Centrarchidae (15), Sciaenidae (9), and Gobiidae (12). The fish community contained a higher proportion of predatory fishes (40–98%) (Cox and Moody 1981) than other northern Florida rivers.

Several St. Johns fishes have unusual geographic distribution patterns. The Lake Eustis minnow (*Cyprinodon hubbsi*), the Florida largemouth bass (*Micropterus salmoides floridianus*), and a subspecies of the pugnose minnow (*Notropis emiliae peninsularis*) are endemic to the St. Johns River (Tagatz 1967, Burgess and Franz 1978). The dusky shiner (*Notropis cummingsae*), bluenose shiner (*N. welaka*), snail bullhead (*Ictalurus brunneus*), and tesse-lated darter (*Etheostoma olmstedi*) occur in the St. Johns, but not in adjacent watersheds of the Atlantic Coastal Plain (Burgess and Franz 1978).

Six anadromous fishes inhabit the St. Johns River (Rulifson and Huish 1982). The striped bass population is presumed to be nonmigratory, and is increasing in abundance due to stocking for recreational angling and the elimination of the commercial fishery. Atlantic sturgeon are captured occa-sionally in the commercial fishery for American shad at the mouth of the river. Shortnose sturgeon, uncommon in the river, are listed as threatened by the state of Florida. Hickory shad and blueback herring occur throughout the river at low, and probably declining, densities.

The American shad is the most abundant and economically important anadromous species in the river. The population, estimated to have had a 1.4-million kg biomass in the 1950s, has supported a commercial fishery since 1858 and currently supports a sport fishery as well (Walburg 1960, Williams and Bruger 1972). The commercial fishery captured approximately 46 metric tons of American shad in 1984–85, valued at \$36 000 (Hale et al. 1985). Because this is the southern limit of their range, American shad have been extensively studied in the St. Johns. Adults enter the river in the late winter and migrate to spawning grounds between river km 259 and 370, losing 60–70% body weight during the migration (Davis 1980, Glebe and Leggett 1981). Spawning fish in the St. Johns are smaller, mature at an earlier age, and have higher fecundity than American shad in northern rivers, and all adults die after spawning once (Leggett and Carscadden 1978). Juveniles move down-river during the year as water temperatures increase, and leave the river in November and December (Williams and Bruger 1972).

Resource Use   The St. Johns River and associated waterways have been used extensively and modified highly through time (Table 3). Presently, the largest use of water is for irrigation, with 1.1 million m^3 extracted per day on the river (SJRWMD 1984). Via Lake Washington, the river supplies drinking water for municipalities in south Brevard County. The basin contains several population areas listed among the 20 fastest growing areas in the nation. Human population density is expected to increase by 46% between 1985 and 2010, indicating the extent of increasing water demand in the coming gen-eration (SJRWMD 1984).

Lake Apopka, the source of the Oklawaha River, illustrates the vulnera-bility of Florida rivers to natural and human disturbance (SJRWMD 1984). The lake maintained good water quality and productive fisheries through 1947, despite extensive inputs of domestic sewage and agricultural wastes. A hur-ricane in 1947 destroyed the lake-wide aquatic macrophyte community, which has never recovered. Dense algal blooms developed and the fish community shifted from mostly sport and commercial fishes to the planktivorous gizzard shad (*Dorosoma cepedianum*), which comprised 80% of the biomass by 1957. Although point pollution has been abated, nonpoint pollution continues to sustain eutrophication of the lake.

Modification of drainage patterns for agricultural benefit began at the turn of the century (Fig. 2). From 1910 to 1940, extensive floodplain areas were drained by canals and levees, so that water normally entering the St. Johns was removed for irrigation and eventually discharged into the Indian River (McElroy 1977). These actions have reduced the historical floodplain by 60%, mostly in the upper river basin (SJRWMD 1984). In 1976, management of the entire basin was consolidated under the St. Johns River Water Management District. Strategic plans now de-emphasize structural modifications and emphasize achieving flood control and water supply objectives through wetland preservation and restoration, relying on the natural water storage capacity of the floodplain (SJRWMD 1984).

The U.S. Army Corps of Engineers maintains a navigation channel from the mouth to Lake Harney via dredging and water hyacinth treatment (SJRWMD 1984). A 4-m-deep channel is maintained from Jacksonville upstream to Sanford and a 1.6-m-deep channel continues to Lake Harney.

Waterborne recreation is important in the basin and was estimated to involve 500 million recreation trips in 1975 (Table 4; SJRWMD 1984). Fishing, which ranks fourth among waterborne recreation, focuses on American shad and largemouth bass (*Micropterus salmoides*). The American shad sport fishery in the middle river captures about 180 000 fish annually, based on an effort of 40 000 angler days (Rulifson and Huish 1982). The St. Johns River enjoys a national reputation for excellent largemouth bass fishing; over half

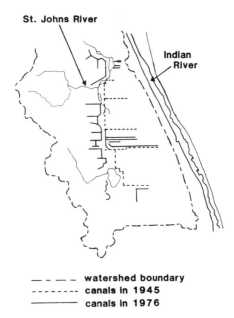

St. Johns River

Indian River

— — — — watershed boundary
--------- canals in 1945
——————— canals in 1976

FIGURE 2. Canalization of the upper portion of the St. Johns River watershed since 1900. Water from several canals on the eastern side of the watershed is now diverted into the Indian River. (After McElroy (1977).)

of the tournament bass fishing in Florida occurs in the St. Johns basin (Chapman and Fish 1983). Fish kills occur regularly on the river, usually due to naturally occurring low dissolved oxygen levels (Cox and Moody 1981).

The commercial fishery of the St. Johns produced 2670 metric tons during the 1984–85 season, valued at $2.3 million dockside (Hale et al. 1985). The commercial fishery for American shad operates via gill nets in Jacksonville harbor and haul seines between Palatka and Welaka (Walburg 1960). Declining catches in recent years have presumably occurred because of overfishing of female shad, reduced water flow rates caused by canalization and diversion, and obstructions to upstream migration (Williams and Bruger 1972). A commercial catfish fishery operates using trap nets of various kinds throughout the first 230 km of river. The fishery in 1984–85 captured 1322 metric tons of catfish valued at $1.4 million (Hale et al. 1985).

SUMMARY

If a single major point arises from this chapter, it is that the rivers of the southeastern Atlantic slope are a diverse lot. A hypothetical straight line between the two rivers that bound the region, the James and St. Johns, represents a distance of approximately 800 km. Such a line would include subtropical rivers of dark, acidic water over shifting sand (e.g., Altamaha), rivers characterized by temperate climate, clear water, and rocky structure (e.g., James), and highly alkaline rivers of almost imperceptible gradient (e.g., St. Johns). Some southeastern rivers have been dramatically altered by human activities (e.g., Santee and Cooper), while others remain in a near-pristine condition (e.g., Ogeechee and Satilla). Several mainstems (e.g., Roanoke and Waccamaw) boast unique fauna assemblages; in others, endemism is poorly developed (e.g., Savannah). It is unlikely that any line of the same length, superimposed over any other region of North America, would encompass a similar diversity of lotic aquatic systems.

This chapter also serves to highlight the paucity of biological data for southeastern river mainstems. Major ecological processes (e.g., primary and secondary production) have been estimated for only a few of the rivers in question; the biota of most are not adequately described. In light of the recreational importance of these river mainstems, and their potential for the study of biodiversity, the lack of published information is surprising, but may be explained, in part, by the logistic difficulties of sampling the large river mainstems of any region.

From a resource management standpoint, the diversity of these systems, and the lack of information about them, may represent something of a conflict. Namely, effective management of any ecological system is dependent on an understanding of that system, but knowledge possessed for one southeastern river may not be appropriately applied to another. Several recent and ongoing studies of southeastern river ecology may, at least partially, correct the situation.

REFERENCES

Angermeier, P. L., and J. R. Karr. 1984. Relationships between woody debris and fish habitat in a small warmwater stream. *Trans. Am. Fish. Soc.* 113:716–726.

Bailey, R. G. 1978. *Descriptions of the Ecoregions of the United States.* Forest Service, Ogden, UT: U.S. Department of Agriculture.

Barber, W. E., and N. R. Kevern. 1973. Ecological factors influencing macroinvertebrate standing crop distribution. *Hydrobiology* 43:53–75.

Bélanger, T. V. 1982. Factors affecting measured phytoplankton production in the Upper St. Johns River. *Fla. Sci.* 45(Suppl.):36.

Bender, M. E., and R. Huggett. 1985. Fate and effects of Kepone in the James River, Virginia. pp. 157–172. In E. Hodgson (ed.), *Reviews in Environmental Toxicology.* New York: Elsevier.

Benke, A. C., D. Gillespie, M. Parrish, F. Van Arsdale, T. Hunter, and R. Henry. 1979. *Biological Basis for Assessing Impacts of Channel Modification: Invertebrate Production, Drift and Fish Feeding in a Southeastern Blackwater River*, Rep. No. 06 79.187. Atlanta: Environ. Res. Cent., Georgia Institute of Technology.

Benke, A. C., R. L. Henry, D. Gillespie, and R. Hunter. 1985. Importance of snag habitat for animal production in southeastern streams. *Fisheries* 10:8–14.

Berra, T. M. 1981. *An Atlas of Distribution of the Freshwater Fish Families of the World.* Lincoln: University of Nebraska Press.

Bishop, J. 1985. Human ecology of the James River. *Va. J. Sci.* 36:14–18.

Blanton, L. R., M. Brinson, R. Twilley, and G. Davis. 1975. Seasonal biomass distribution of aquatic macrophytes in the lower Chowan River. *Assoc. Southeast. Biol. Bull.* 22:42.

Boss, K., and W. Clench. 1967. Notes on *Pleurobema collina* from the James River, Virginia. *Occas. Pap. Mollusks Mus. Comp. Zool. Harv. Univ.* 3:45–52.

Britton, J. C., and S. L. H. Fuller. 1979. *The Freshwater Bivalve Mollusca (Unionidae, Spaeriidae, Corbiculidae) of the Savannah River Plant, South Carolina*, SRO-NERP-3. Aiken, SC: Savannah River Ecology Laboratory.

Burgess, G. H., and R. Franz. 1978. Zoogeography of the aquatic fauna of the St. Johns River system with comments on adjacent peninsula faunas. *Am. Midl. Nat.* 100:160–170.

Cashner, R. C., and R. E. Jenkins. 1982. Systematics of the Roanoke bass, *Ambloplites cavifrons. Copeia* 1982:581–594.

Chapman, P., and W. V. Fish. 1983. Largemouth bass tournament catch results in Florida. *Proc. Annu. Conf. Southeast. Assoc. Fish Wildl. Agencies* 37:495–505.

Christie, R. W., and T. A. Curtis. 1983. Establishment of bluefin killifish (*Lucania goodei*) in Cooper River, North Carolina. *Ga. J. Sci.* 41:91–92.

Coomer, C., and D. R. Holder. 1980. A fisheries survey of the Ocmulgee River. *Ga., Dep. Nat. Resour., Rep.*

Council on Environmental Quality (CEQ). 1981. *Environmental Trends.* Washington, DC: Council on Environmental Quality.

Cox, D. T., and H. L. Moody. 1981. Fisheries of the St. Johns River. Completion Report, Federal Aid in Fish Restoration Project F-33. Tallahassee: Florida Game and Fresh Water Fish Commission.

Cox, D. T., H. L. Moody, E. D. Vosatka, and L. Hartzog. 1976. *Upper St. Johns River Study, 1971–1976*, Completion Report, Federal Aid to Fish Restoration Project F-25. Tallahassee: Florida Game and Fresh Water Fish Commission.

Cudney, M. D., and J. B. Wallace. 1980. Life cycles, microdistribution and production dynamics of six species of net-spinning Caddis Flies in a large southeastern (U.S.A.) river. *Holarctic Ecol.* 3:169–182.

Cushman, R. M., S. B. Gough, M. S. Moran, and R. B. Craig. 1980. *Sourcebook of Hydrologic and Ecological Features*. Ann Arbor, MI. Ann Arbor Science.

Cutshall, N., I. Larsen, and M. Nichols. 1981. Man-made radionuclides confirm rapid burial of Kepone in James River, Virginia sediments. *Science* 213:440–442.

Darlington, P. J. 1957. *Zoogeography*. New York: Wiley.

Davis, J. R., and D. E. Louder. 1971. Life history and ecology of the cyprinid fish *Notropis petersoni* in North Carolina waters. *Trans. Am. Fish. Soc.* 100:726–733.

Davis, S. M. 1980. American shad movement, weight loss and length frequencies before and after spawning in the St. John River, Florida. *Copeia* 1980(4):889–892.

DeMort, C. and R. D. Bowman. 1985. Seasonal cycle of phytoplankton populations and total chlorophyll of the lower St. Johns River Estuary, Florida. *Fla. Sci.* 48:96–107.

Diaz, R. J. 1974. Asiatic clam, *Corbicula manilensis*, new record in the tidal James River, Virginia. *Chesapeake Sci.* 15:118–120.

Diaz, R. J. 1989. Pollution and tidal benthic communities of the James River Estuary, Virginia. *Hydrobiologia* 180:195–211.

Dill, L. M. 1983. Adaptive flexibility in the foraging behavior of fishes. *Can. J. Fish. Aquat. Sci.* 40:398–408.

Dudley, R. G., and K. N. Black. 1978. Distribution of striped bass eggs and larvae in the Savannah River estuary. *Proc. Annu. Conf. Southeast. Assoc. Fish Wildl. Agencies* 32:561–570.

Dudley, R. G., and T. G. McGahee. 1983. Winter and altered spring movements of striped bass in the Savannah River, Georgia. *Fish. Bull.* 81:420–426.

Dudley, R. G., A. W. Mullis, and J. W. Terrill. 1977. Movements of adult striped bass (*Morone saxatilis*) in the Savannah River, Georgia. *Trans. Am. Fish. Soc.* 106:314–322.

Garman, G. C. 1988. *Factors Affecting the Distribution of Ambloplites cavifrons in Virginia*, Final Rep. F-72-R1. Virginia Department of Game Inland Fish, Richmond.

Garman, G. C. 1990. Use of terrestrial arthropod prey by a stream-dwelling cyprinid fish. *Environ. Biol. Fishes* 30:325–331.

Garman, G. C., and L. A. Smock. 1988. *Annual Report of the James River Mainstem Investigation*, F-74-R1. Virginia Department of Game Inland Fish, Richmond.

Gibbons, J. W., P. R. Sharitz, and I. L. Brisbin, Jr. 1980. Thermal ecology research of the Savannah River Plant: a review. *Nucl. Saf.* 21:367–379.

Gilbert, R. J., S. Larson, and A. Wentworth. 1985. *The Relative Importance of the Lower Savannah River as a Striped Bass Spawning Area*, Final Rep., Project F-39. Atlanta: Federal Aid to Fish Restoration, Georgia Department of Natural Resources.

Glebe, B. D., and W. C. Leggett. 1981. Latitudinal differences in energy allocation and use during the freshwater migrations of American shad (*Alosa sapidissima*) and their life history consequences. *Can. J. Fish. Aquat. Sci.* 38:806–820.

Gordon, A. E., and J. B. Wallace. 1975. Distribution of the family Hydropsychidae in the Savannah River basin of North Carolina, South Carolina and Georgia. *Hydrobiologia* 46:405–423.

Hale, M. M., J. E. Crumpton, and D. J. Renfro. 1985. *1984–85 Commercial Fisheries Investigations Report.* Tallahassee: Florida Game and Fresh Water Fish Commission.

Hall, G. B., and M. M. Williams. 1984. Historical changes in the vegetation of the upper St. Johns River in response to hydroperiod alteration. *Fla. Sci.* 47(Suppl.):27.

Hobbs, H. H., III, J. H. Thorp, and G. E. Anderson. 1976. *The Freshwater Decapods and Crustaceans (Palaemonidae, Cambaridae) of the Savannah River Plant, South Carolina.* Aikens, SC: Savannah River Ecology Laboratory.

Hocutt, C. H., and E. O. Wiley (eds.). 1986. *The Zoogeography of North American Freshwater Fishes.* New York: Wiley.

Hocutt, C. H., R. E. Jenkins, and J. R. Stauffer. 1986. Zoogeography of the fishes of the Central Appalachians and Central Atlantic Coastal Plain. In C. H. Hocutt and E. O. Wiley (eds.), *The Zoogeography of North American Freshwater Fishes.* New York: Wiley.

Hogarth, W. T., and W. S. Woolcott. 1966. The mountain stripeback darter, *Percina notogramma montuosa*, n. ssp. from the upper James River, Virginia. *Chesapeake Sci.* 7:101–109.

Holder, D., and C. Hall. 1976. The shad sport fishery of the Ogeechee River in 1975. *Ga. Game Fish Commi. Rep.*

Hornsby, J. H., and C. Hall. 1981. Impact of supplemental stocking of striped bass fingerlings in the Ogeechee River. *Ga. Dep. Nat. Resour. Rep.* p. 62.

Hottell, H. E., D. R. Holder, and C. E. Coomer. 1983. A fisheries survey of the Altamaha River. *Ga. Dep. Nat. Resour. Rep.* p. 68.

Hunt, C. R. 1974. *Natural Regions of the United States and Canada.* San Francisco, CA: Freeman.

Hynes, H. B. N. 1970. *The Ecology of Running Waters.* Toronto: University of Toronto Press.

Jenkins, R. E., and N. M. Burkhead. 1975. Distribution and aspects of life history and morphology of the cyprinid fish *Notropis semperasper* endemic to the Upper James River drainage, Virginia. *Chesapeake Sci.* 16:178–191.

Jenkins, R. E., and T. Zorach. 1970. Zoogeography and characters of the American cyprinid fish *Notropis bifrenatus*. *Chesapeake Sci.* 11:174–182.

Jenkins, R. E., E. A. Lachner, and F. J. Schwartz. 1972. Fishes of the central Appalachian drainages: their distribution and dispersal. pp. 43–117 In P. C. Holt (ed.), *The Distributional History of the Biota of the Southern Appalachians*, Part III, Res. Div. Monogr. 4. Blacksburg: Virginia Polytech. Inst. State University.

Jensen, L. D. 1974. *Environmental Responses to Thermal Discharges from the Chesterfield Station, James River, Virginia*, Rep. RP-49. Electric Power Research Institute.

Johnson, R. I. 1970. The systemmatic and zoogeography of the Unionidae (Mollusca: Bivalvia) of the southern Atlantic slope region. *Bull. Mus. Comp. Zool.* 140:263–450.

Johnson, R. I., and K. Boss. 1984. *Fusconaia collina* from the James River, Virginia, an additional note. *Occas. Pap. Mollusks Mus. Comp. Zool. Harv. Univ.* 4:319–320.

Karr, J. R. 1971. Structure of avian communities in selected Panama and Illinois habitats. *Ecol. Monogr.* 41:207–233.

Kauffman, J. 1980. The effects of a mercury-induced consumption ban on angling pressure. *Fisheries* 5(1):10–12.

Kautz, R. S. 1981. Fish populations and water quality in north Florida rivers. *Proc. Annu. Conf. Southeast. Assoc. Fish Wildl. Agencies* 35:495–507.

Leed, J. A., and T. V. Bélanger. 1981. Selected trace metals in the upper St. Johns River and their land use relationships. *Fla. Sci.* 44:136–150.

Leggett, W. C., and J. E. Carscadden. 1978. Latitudinal variation in reproductive characteristics of American shad (*Alosa sapidissima*): Evidence for population specific life history strategies in fish. *J. Fish. Res. Board Can.* 35:1469–1478.

Lowe, E. F., and L. R. Gerry. 1984. Rainfall and surface water deficits in the upper St. Johns River basin, Florida. *Fla. Sci.* 47(Suppl.):37.

MacArthur, R. H. 1972. *Geographical Ecology*. New York: Harper & Row.

Mannering, J. V. 1982. Soil and water management and conservation. In V. J. Kilmer (ed.), *Handbook of Soils and Climate in Agriculture*. Boca Raton, FL: CRC Press, pp. 349–370.

Matter, W. J., P. L. Hudson, and G. E. Sand. 1983. Invertebrate drift and particulate organic material transport in the Savannah River below Lake Hartwell during a peak power generation cycle. In T. O. Fontaine and S. M. Bartell (eds.), *Dynamics of Lotic Ecosystems*. Ann Arbor, MI: Ann Arbor Science.

Maurakis, E. G., and W. S. Woolcott. 1984. Seasonal occurrence patterns of fishes in a thermally enriched stream. *Va. J. Sci.* 35:5–12.

McElroy, W. J. 1977. Hydrologic considerations of the Upper St. Johns River Basin, Florida. *Water Resour. Bull.* 13:1153–1164.

McFarlane, R. W., R. A. Frietsche, and R. D. Miracle. 1979. Community structure and differential impingement of Savannah River fishes. *Proc. Annu. Conf. Southeast. Assoc. Fish Wildl. Agencies* 33:628–638.

Meador, M. R., A. G. Eversole, and J. S. Bulak. 1984. Utilization of portions of the Santee River system by spawning blueback herring. *North Am. J. Fish. Manage.* 4:155–163.

Merritt, R. W., and K. W. Cummins. 1984. *An Introduction to the Aquatic Insects of North America*. Dubuque, IO: Kendall/Hunt.

Morrison, J. P. E. 1973. Sympatric species of *Elliptio* living in the St. Johns River, Florida. *Bull. Am. Malacol. Union* 38:14.

Morse, J. C., J. W. Chapin, D. D. Herlong, and R. S. Harvey. 1980. Aquatic insects of Upper Three Runs Creek, Savannah River Plant, South Carolina. Part I: Orders other than Diptera. *J. Ga. Entomol. Soc.* 15:73–101.

Morse, J. C., J. W. Chapin, D. D. Herlong, and R. S. Harvey. 1983. Aquatic insects of Upper Three Runs Creek, Savannah River Plant, South Carolina. Part II: Diptera. *J. Ga. Entomol. Soc.* 18:303–316.

Mudre, J. M., J. J. Ney, and R. J. Neves. 1985. *An Analysis of the Impediments to Spawning Migrations of Anadromous Fish in Virginia, Rivers*. Blacksburg: Dep. Fish. Wildl. Sci., Virginia Tech.

Patrick, R. J., J. Cairns, Jr., and S. Roback. 1967. An ecosystematic study of the fauna and flora of the Savannah River. *Proc. Acad. Nat. Sci. Philadelphia* 118:109–407.

Rabeni, C. F., and G. W. Minshall. 1977. Factors affecting microdistribution of stream benthic insects. *Oikos* 29:33–43.

Raney, E. C. 1950. Freshwater fishes. pp. 151–194. In *The James River Basin, Past Present and Future*. Virginia Academy of Science, Richmond.

Rulifson, R. A., and M. T. Huish. 1982. *Anadromous Fish in the Southeastern United States and Recommendations for Development of a Management Plan*, Final Rep., Contract No. 14-16-0004-80-077. Atlanta, GA: U.S. Fish Wildl. Serv.

St. Johns River Water Management District. (SJRWMD). 1984. *Strategic Plan—St. Johns River System*. Jacksonville, FL. SJRWMD.

Schmitt, D. N., and J. H. Hornsby. 1985. *A Fisheries Survey of the Savannah River*. Atlanta: Georgia Fish and Game Commission.

Shute, J. R., P. W. Shute, and D. G. Lindquist. 1981. Fishes of the Waccamaw River drainage, North Carolina and South Carolina, USA. *Brimleyana* 6:1–24.

Sickel, J. B. 1980. Correlation of unionid mussels with bottom sediment composition in the Altamaha River, Georgia. *Bull. Am. Malacol. Union* 46:10–13.

Soil Conservation Service (SCS). 1980. *America's Soil and Water: Conditions and Trends*. Washington, DC: U.S. Department of Agriculture, Soil Conservation Service.

Specht, W., H. J. Kania, and W. Painter. 1985. *Annual Report on the Savannah River Aquatic Ecology Program, September 1982–August 1983*, Vol. II. Aiken, SC: Environmental & Chemical Sciences, Inc.

Stanton, N. L. 1979. Patterns of species diversity in temperate and tropical litter mites. *Ecology* 60:295–304.

Stauffer, J. R., B. Burr, C. Hocutt, and R. Jenkins. 1982. Checklist of the fishes of the central and northern Appalachian mountains. *Proc. Biol. Soc. Wash.* 95:27–47.

Stickney, R. R., and D. Miller. 1974. Chemistry and biology of the lower Savannah River, Georgia, United States. *J. Water Pollut. Control Fed.* 46:2316–2326.

Stites, D. L., and A. C. Benke. 1989. Rapid growth rates of chironomids in three habitats of a subtropical black water river and their implications for P:B ratios. *Limnol. Oceanogr.* 34:1278–1289.

Swift, C. C., C. R. Gilbert, S. A. Bortone, G. H. Burgess, and R. W. Yerger. 1986. Zoogeography of the freshwater fishes of the southeastern United States: Savannah River to Lake Pontchartrain. In C. H. Hocutt and E. O. Wiley (eds.), *The Zoogeography of North American Freshwater Fishes*. New York: Wiley.

Tagatz, M. E. 1967. Fishes of the St. Johns River, Florida. *Q. J. Fla. Acad. Sci.* 30:25–50.

Twilley, R. 1976. Phosphorous cycling in *Nuphar lateum* communities in the lower Chowan River, North Carolina. MS Thesis, East Carolina University, Greenville.

Twilley, R., L. Blanton, M. Brinson, and G. Davis. 1985. Biomass, production and nutrient cycling in aquatic macrophyte communities of the Chowan River, North Carolina. *Aquat. Bot.* 22:231–252.

van Arsdall, T. C., Jr. 1977. Production and colonization of the snag habitat in a southeastern river. MS Thesis, Georgia Institute of Technology, Atlanta.

Vannote, R. L., G. W. Minshall, K. W. Cummins, J. R. Sedell, and C. E. Cushing. 1980. The river continuum concept. *Can. J. Fish. Aquat. Sci.* 37:130–137.

Walburg, C. H. 1960. Abundance and life histories of shad. St. Johns River, Florida. *Fish. Bull.* 60:487–501.

Wallace, J. B., and A. C. Benke. 1984. Quantification of wood habitat in subtropical coastal plain streams. *Can. J. Fish. Aquat. Sci.* 41:1643–1652.

Waters, T. F. 1977. Secondary production in inland waters. *Adv. Ecol. Res.* 10:91–164.

Weand, B. L., and T. J. Grizzard. 1986. *Water Quality Review—James River.* Blacksburg: Virginia Tech.

Wharton, C. H., W. M. Kitchens, and T. W. Sipe. 1982. *The Ecology of Bottomland Hardwood Swamps of the Southeast: A Community Profile*, FWS/OBS-81/37. Washington, DC: U.S. Fish Wildl. Serv.

White, J. W., W. S. Woolcott, and W. Kirk. 1977. A study of the fish community in the vicinity of a thermal discharge in the James River, Virginia. *Chesapeake Sci.* 18:161–171.

Williams, R. O., and G. E. Bruger. 1972. *Investigations on American Shad in the St. Johns River*, Tech. Ser. No. 66. Florida Department of Natural Resources, Marine Research Laboratory.

Wiltz, J. W. 1984. Developmental stages of the highfin carpsucker (*Carpoides velifer*), and its separation from other catastomids of the Altamaha River, Georgia. *Ga. J. Sci.* 42:97–108.

Winger, P. V., D. P. Schultz, and W. W. Johnson. 1990. Environmental contaminant concentrations in biota from the lower Savannah River, Georgia and South Carolina. *Arch. Environ. Contam. Toxicol.* 19:117.

Woodsen, B. R. 1959. A study of the Chlorophyta of the James River Basin, Virginia. *Va. J. Sci.* 11:70–82.

Woodsen, B. R., and G. W. Prescott. 1961. Algae of the James River basin, Virginia. *Trans. Am. Microsc. Soc.* 80:166–175.

Woolcott, W. S. 1974. *The Effects of Thermal Loading by the Bremo Power Station on a Piedmont Section of the James River*, Vols. I and II. Richmond: Virginia Institute of Scientific Research, University of Richmond.

Woolcott, W. S. 1985. James River fisheries and thermal effects. *Va. J. Sci.* 36:26–29.

Wootton, E. T. 1985. Modification of the James River. *Va. J. Sci.* 36:19–22.

Zeigler, C. F., and J. T. McGehee. 1977. Water hyacinth control plan for the St. Johns River. *J. Aquat. Plant Manage.* 15:10–12.

9 Medium-sized Rivers of the Gulf Coastal Plain

ROBERT J. LIVINGSTON

Center for Aquatic Research and Resource Management, Florida State
University, Tallahassee, FL 32306-2043

Rivers of the gulf coastal plain are not well understood. Although numerous
studies may be available for a given system, most such analyses are of limited
scope with little in the way of long-term data or interdisciplinary support.
This problem is exacerbated by the sheer complexity of most river systems,
which comprise multifold habitats and diverse processes that are spatially and
temporally variable. In addition, each such system represents a somewhat
unique situation with respect to how its individual parts and processes interact
through time, so generalization without systematic comparison is not feasible.
Without such understanding, realistic and effective resource management is
not possible, which would explain observed losses of productivity in such
systems in recent times.

Vannote et al. (1980) and Vannote (1981) have constructed a model of
river habitat gradients from the headwaters to the estuary. The model is
process-oriented and is based on organic input and consumer response along
gradients of "fluvial geomorphic processes." The approach emphasizes structural and functional characteristics such as the biological response to a hi-
erarchy of drainage facets as part of a homeostatic process within limits
imposed by the physical system. Thus, strongly heterotrophic headwater streams
(low-order) grade into mid-order streams dominated by autochthonous (macrophyte) energy sources, which form the basis for organisms adapted to
process coarse particulate material. On large rivers (orders 9–12), the system
returns to heterotrophy based on fine particulates buttressed by periodic input
from bordering flood plains and swamp (bottomland) hardwood forests. Water
flow supports the biotic continuum as an important part of the energy loading
system. Equilibrium in space and time is based on continuous species replacement for maximum use of available energy inputs. One problem with
this model is that, within a given river system, there is a complex mosaic of
distinct microhabitats that lead to highly evolved species assemblages that do
not necessarily fit the concept of a stable continuum of energy processing
units. Low-order streams along the main river channel can be extremely

variable in terms of energy processing and microhabitat distribution. For instance, in the Choctawhatchee River system in north Florida, there is a basic dichotomy of ordering between the tributaries and the main stem that depends largely on hydrological conditions, wetlands configuration, and differences in microhabitat distribution (Livingston et al. 1991). Tributaries in this system depend on factors such as wetland type, stream flow, and the physiographic aspects of the drainage system. In terms of biotic complexity, the tributaries are highly developed and ecologically complex. In short, there is a lateral (spatial) component to river biological diversity that is as important, in terms of the definition of the aquatic system, as the longitudinal distribution along the main stem. This aspect of river analysis has not been well developed in the river continuum models and deserves attention.

The highly complex within-system variation is not necessarily in equilibrium. Pulsed floods periodically destabilize the system, moving nutrients and particulates in an episodic fashion that often destabilizes local habitats. Such instability may enhance river production in southeastern rivers although the within- and between-system variability continues to hinder the development of a predictive model that applies to the entire spectrum of highly individualized river systems in the gulf coastal plain.

In sum, rivers along the gulf coastal plain are diverse and remain poorly understood, both individually and as a major habitat form. This review will address what is currently known about such systems and how such knowledge can be applied to what is now recognized as an endangered habitat.

PHYSICAL ENVIRONMENT

The primary rivers of the gulf coastal plain (Fig. 1) include 33 major drainage basins that eventually discharge into the Gulf of Mexico. Stream flow of such rivers varies as a function of precipitation level, gradient, evapotranspiration rates, size of the drainage basin, runoff and infiltration rates, water consumption, and general runoff characteristics. Physical alterations (locks, dams, dredged channels) and freshwater diversion have changed many of these rivers. Such changes have affected natural ecological processes in the riverine habitats in various ways (Darnell 1976) to the extent that any generalizations concerning form and process of present-day conditions should be qualified relative to the extent of anthropogenic impact. Total drainage area of these rivers approximates 4 080 822 km^2 (1 575 607 mi^2) (National Oceanic and Atmospheric Administration 1985). Peak discharge in the northeastern gulf rivers tends to occur in winter and spring months; low flows in such systems often occur during summer–early fall periods. Such flows are somewhat different in rivers along the Florida peninsula where peaks often occur in summer–fall months. As a physical phenomenon, each river is closely associated with an estuarine environment that is often dominated physically by flow rates, nutrient transport, and the transfer of a broad spectrum of pollutants from point and nonpoint sources. In this way, the estuary is a physical ex-

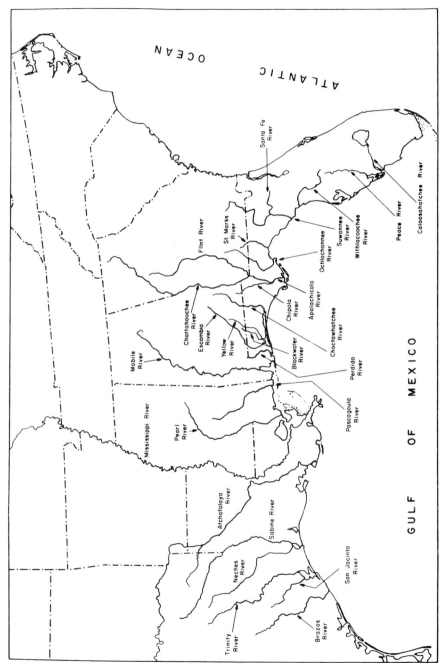

FIGURE 1. Major rivers of the Gulf Coastal Plain.

353

tension of the contributing river drainage and, as such, should be included as a functional part of the river system even though it is rarely studied as such.

Peninsular Florida

Peninsular Florida along the southwest gulf coast is characterized by Miocene limestones, marls, silty sands, and clays. Central areas have phosphatic sandy clay and limestone of various types, and the northwest gulf coast is characterized by Eocene shallow marine limestone and crystalline dolomite (north) (Fernald 1981). Most of the Florida peninsula received deposits of marine carbonates from ancient shallow seas. This region is marked by featureless lowlands that are lined by barrier islands, swamps, and marshes. The karst drainage, characterized by underlying limestone rock, has led to various stream–sinkhole systems (often spring-fed) with poorly defined drainage areas flowing through subsurface channels, swamps, and marshes. Such drainage basins, together with a series of distinct river channels, characterize the freshwater flows along the west coast of Florida. Annual temperatures vary seasonally in the northern portion of Florida (south temperate climate). Farther south, seasonal differences are smaller in the subtropical climate. Average stream temperatures exceed 24 °C in southern Florida with gradients northward (17.8–20.0 °C in north Florida). Annual precipitation along the west coast of Florida averages 127 to 165 cm (50–65 in.).

The main rivers along the peninsular west coast include the Caloosahatchee, Peace, Withlacoochee, and Suwannee. The Caloosahatchee is no longer a river: it has been canalized into a cross-state ditch. Locks and water control structures have been built into the system from Lake Okeechobee to its estuary near Fort Myers on the Gulf Coast. Horel (1960) indicated that a sports fishery in the post-channelized Caloosahatchee was not evident, with relatively poor habitat conditions. The current condition of the river remains unknown. The Peace River, 167 km long, receives water from various point and nonpoint sources of pollution before it ends in the Charlotte Harbor system. It is a warm, hard-water stream with its headwaters in areas lined with rich phosphate deposits. Except for the heavily polluted Alafia River, the Peace River has the highest phosphorous levels of all of Florida's rivers. Periodic mortalities of organisms caused by a series of phosphate spills have been reported. A detailed analysis of the long-term water quality conditions of the Peace River has been given by Fraser (1991).

Between the Peace and Suwannee Rivers are more than a dozen coastal rivers, including several spring-fed systems. Impoundments or water diversions of various kinds affect about one-third of the rivers in this region. Estevez et al. (1991) have described the unique ecological conditions of many of these rivers. The Manatee River is impounded and is seriously degraded. The Little Manatee River suffers from erosion and metal (Ag) contamination. Phosphate mining and runoff from agriculture and urban development have adversely affected many of the streams along the west coast of Florida. Several of these streams, however, such as the Myakka and the Chassahowitzka, remain relatively pristine.

The Withlacoochee River originates in the Green Swamp and represents one of the most ecologically diverse river corridors in west-central Florida. The Withlacoochee has both blackwater and spring-fed habitats and a diverse floodplain forest, and is hydrologically connected to two large lake systems (the Tsala-Apopka and Panasoffkee). Lake Rousseau, near the mouth of the river, is one of the oldest impoundments in Florida. The river is wide in most places downstream of Panasoffkee, with a relatively constant current and substantial input from springs. Water quality is good in some portions, although there are reported impacts due to the citrus processing industry, domestic wastes, and agricultural runoff.

The Suwannee River, 394 km long, has its origin in the Okefenokee Swamp in Georgia and includes a drainage basin of about 25 000 km^2. The Suwanee is the second largest river in Florida in terms of flow rate (about 305 m^3/s). Tributaries include the Withlacoochee (not the above-mentioned stream, which is entirely separate), the Alapaha, and the Santa Fe rivers. The Suwannee is known to have variable water quality along its length with the most important determining features of such quality being the Okefenokee Swamp, discharges from phosphate mining activities, and groundwater input (springs) along the upper and mid-reaches of the river. The upper river is very acidic and low in nutrients and turbidity. Swift Creek, which drains phosphate mining lands, is characterized by high nitrogen, phosphorus, and fluoride. Changes in variables such as pH are thought to be important to the distribution of organisms within the Suwannee basin (Florida Department of Environmental Regulation, unpublished data). Farther down the river, groundwater has a major impact on water quality. Color levels, high in the upper river, decrease, while the pH increases to neutrality and dissolved oxygen recovers. Generally, the various biological components (e.g., periphyton, benthic macroinvertebrates, fishes) tend to increase downstream in numbers, species richness, and species diversity, which could reflect, in part, the changes in water quality features, such as pH (Florida Department of Environmental Regulation, unpublished data). The Suwannee and Santa Fe rivers have a high diversity of habitats, which includes extensive limestone shoals. Algal growth is particularly affected by phosphorus input at Swift Creek. The Suwannee River remains highly productive although various anthropogenic activities are currently having localized impacts (e.g., the confluences of Swift Creek and the Withlacoochee) along the length of this important system.

A series of small streams, including the Fenholloway, Econfina, Aucilla, and St. Marks rivers, are located in the northeastern gulf portion of Florida. The Fenholloway is a damaged system because of the long-term use of this stream as a receptable for about 190 million liters per day of pulp mill wastes. The Econfina, on the other hand, is a relatively natural and unpolluted stream with a rich fauna for a stream of its size (R. J. Livingston, unpublished data). The Aucilla River arises in Georgia and flows 111 km to the Gulf. It has spring-fed tributaries, the Little Aucilla and the Wacissa. A product of the karst topography of the region, portions of the Aucilla flow underground. The Wacissa, for instance, has its origin in a series of 11 springs and is unpolluted, having naturally acidic waters. The Aucilla is still in relatively natural condition.

Eastern Gulf Coast

The coastal plain region from the Florida panhandle to Louisiana is characterized by Cenozoic formations consisting of sand, clay, mud, or calcareous ooze as deposits from ancient seas (Adams et al. 1926). Pleistocene marine clays and sands overlie the older (Paleozoic, Mesozoic, Cenozoic) formations along the coast. Estuarine and fluvial deposits extend up the various major river valleys. Overall, the gulf coastal plain is covered by a thick layer of clastic (erosion-produced) sediments and nonclastic, limestone sediments. Such formations are complicated by considerable Pleistocene fluctuations of sea level. The mouths of various gulf rivers, such as the Choctawhatchee and Escambia, are narrow, drowned Pleistocene floodplains. Sediments of the Terrace I are thought to have provided sands for the barrier islands along the gulf coast. Climate throughout this region is mild, with mean annual temperatures approximating 18.9 °C. Annual rainfall in the region averages 150 cm (range: 137–162 cm). Dry periods usually occur in the fall, and the wettest periods during the summer–early fall months. Hurricanes (winds exceeding 64 knots) commonly occur in this region. Major watersheds in the northeastern gulf region include the Ochlockonee, Apalachicola, Choctawhatchee, Escambia, Mobile, Pascagoula, and Pearl.

The Ochlockonee River flows for about 257 km through Georgia and Florida and is dammed. The sandy-bottomed system is in a relatively natural state with high phosphorus levels and generally good water quality (Florida Department of Environmental Regulation, unpublished data). Recent high concentrations of mercury in the fishes of the impoundment (Lake Talquin) indicate possible problems in the upper Ochlockonee drainage area. The Apalachicola drainage basin, located in Florida, Georgia, and Alabama, is the largest in Florida in terms of flow rates, which average 690 m^3/s (24 642 ft^3/s). The Apalachicola River accounts for about 35% of the total freshwater flow on the west coast of Florida. It is typical of a series of alluvial rivers along the northern gulf coast where extensive portions of the drainages are located outside of Florida. Most of the Apalachicola River basin, formed by the convergence of the Chattahoochee and Flint Rivers, is located in Georgia. Water quality is relatively good in the Apalachicola River basin. However, physical changes in the form of dredging, diking, and channelization along its 171 km (107 mi) length have led to serious habitat problems (Leitman et al. 1991). The headwaters of the Apalachicola are formed by the moderately polluted reservoir called Lake Seminole. Environmental consequences of the physical changes and pollution of the upper portions of the system include a loss of habitat quality and serious reduction of the anadromous fisheries along the river. Despite such changes, the Apalachicola River has a rich ichthyofauna with the largest assemblage of freshwater fish species in Florida.

The Choctawhatchee River basin, arising in Alabama, resembles the Apalachicola in its extensive floodplain forests. This system has the second largest alluvial floodplain in Florida (28 000 ha) (Wharton et al. 1982) and is a highly turbid stream with a series of productive tributaries. In terms of flow rate, the Choctawhatchee is a major river (around 200 m^3/s), which remains in relatively natural condition (Livingston et al. 1991). This river has a largely intact wetlands system and good water quality. The Florida portion of the Choctawhatchee basin remains largely undeveloped, with forestry and agriculture as the major land uses. Like the Apalachicola River, considerable portions of the freshwater wetlands system have been purchased by the state of Florida as part of a progressive program designed to preserve environmentally sensitive habitats. The sports fishery of the Choctawhatchee River system is better developed than that of less mineralized streams of northwest Florida, such as the Blackwater and Yellow rivers. According to Bass and Cox (1985), the fauna of the Choctawhatchee better resembles that of the Escambia River to the west than that of the Apalachicola. At present, however, there is no overall management plan for the Choctawhatchee drainage system.

The Pensacola Bay drainage, consisting of the Escambia, Blackwater, and Yellow rivers, has been influenced by a variety of point and nonpoint sources of pollution (Olinger et al. 1975). With headwaters in Alabama, the Escambia River is the fourth largest of Florida streams in terms of flow rate. There are periodic fish kills in the lower reaches of the river due to hypoxic conditions (Bass and Cox 1985). The Blackwater River, on the other hand, is a relatively unpolluted stream that drains the sand and gravel aquifer of the region. Protection is afforded this basin by the establishment of the Blackwater State Forest in Florida and the Conecuh National Forest in Alabama. Water quality is relatively good in the Blackwater, with a heavily forested watershed protecting the sandy bottomed system. In the upper reaches of the system, some of the instream vegetative cover in the form of logs, fallen trees, and snags has been removed, which is a possible reason for observed low levels of biological productivity in the Blackwater area. The Yellow and Shoal rivers are relatively natural streams that represent cool-temperature, sandy-bottomed environments. Although low in nutrients, these rivers have a relatively species-rich ichthyofauna (Bass and Cox 1985), with a moderate fish biomass and a low level of sports fishing potential.

The Perdido River forms the northwestern boundary of Florida, running between Florida and Alabama. Forestry and agriculture dominate land use in the Florida portion of the stream. This system remains relatively unpolluted, with high levels of dissolved oxygen and low pH. The lower portions of the Perdido are characterized by largely undisturbed marshes and swamps. This cool-water, acidic stream has low nutrient levels, a characteristic of various rivers in northwestern Florida. There is little in the way of an active management plan for the Perdido River, a situation that is general throughout northwest Florida. Overall, in terms of water and habitat quality, the streams of northwest Florida are in better shape than those of the peninsula (Wolfe et al. 1988).

The Mobile River drainage basin includes the Mobile, Tombigbee–Black Warrior, and Alabama–Coosa–Talapossa river systems along with numerous tributary streams and lakes. The total drainage area prior to the construction of the Tennessee–Tombigbee Waterway was 113 001 km² (43 650 mi²), which included about 64% of Alabama and portions of Georgia, Mississippi, and Tennessee. Annual mean discharge of the gauged streams in the Mobile River system approximates 2050 m³/s (72 351 ft³/s), although there is considerable variation due to seasonal changes in precipitation. Peak discharges occur from February to April as a rule, with lowest flow rates from June to November. Flow rates along the Mobile River are regulated by upstream reservoirs on the Etowah, Coosa, and Tallapossa rivers and to a degree by the locks and dams on the Tombigbee River. The lower Mobile system is characterized by tributaries such as the Chickasaw Creek drainage. Natural flows and water quality in such areas are affected by physical alterations and municipal/industrial usage. Conditions in the Mobile Bay system are dominated by the Mobile River.

The Escatawpa–Pascagoula River basins drain areas of Mississippi and Alabama (about 2800 km²). This drainage area empties into the Mississippi Sound. As with other such areas in the northeastern gulf coastal region, the area is warm and humid with mean January temperatures between 10.6 and 11.7 °C and mean July temperatures of 27.2 °C (R. J. Livingston, unpublished data). These rivers are characterized by muddy bottoms with the clay–silt fractions ending up in the adjoining estuarine system. Although the area is not heavily populated, the Escatawpa and Pascagoula rivers have historically been adversely affected by agricultural and industrial activities and dredging. Stream discharge into the Mississippi Sound area in the vicinity of Pascagoula averages 433 m³/s.

Western Gulf Coast

West of the Atchafalaya–Mississippi drainage basin, the primary rivers are in south Texas. The region, subtropical and semiarid, is a flat to gently rolling coastal plain (Norwine 1981). This subtropical steppe has mean annual temperatures of about 24 °C, with mean summer maximum temperatures of 29.4 °C and mean winter (January) temperatures of 15.6 °C. Mean annual rainfall approximates 38–72 cm. Such precipitation is monsoonal: 65% of the total comes during a single period, with minima during spring and midsummer. Except for rivers and man-made ponds and reservoirs, there are few permanent bodies of fresh water in this region. Major drainages in this region include the Sabine, Neches, Trinity, San Jacinto, Brazos, Colorado, and Rio Grande.

Pleistocene climates and hydrology strongly influenced modern-day river basin morphology (Wharton et al. 1982). The various alluvial rivers in the Texas region are small in terms of discharge volume relative to the size of the floodplain. Greater rainfall and river discharge 10 000 to 18 000 years ago, together with high sediment delivery, account for this discrepancy; runoff subsided considerably 10 000 to 12 000 years ago. Steplike terraces formed as a result of recent filling of the ancient floodplain with sediments. The most recent (Holocene) terrace (known as "first bottoms") is underlain by the next lowest terrace (Terrace I), which is distinguishable in rivers such as the Pearl, Pascagoula, Sabine, Trinity, and Brazos.

PLANT COMMUNITIES

The composition and distribution of emergent and submergent aquatic vegetation in coastal plain river systems are dependent on complex interactions of various factors: climate, hydroperiod, water quality, soils/sediment characteristics, watershed properties, and biological interactions. The meandering gulf coastal plain rivers and streams are a product of the gentle slopes, sediment deposition patterns, and precipitation and evapo-transpiration conditions. Various forms of these rivers have been described by Wharton et al. (1982) and Clewell (1991). Spring-fed rivers are often clear and alkaline, and are derived from underground aquifers. Examples are found in north Florida (St. Marks, Suwannee). Submerged aquatic vegetation thrives in such situations. Bog streams, located in the Florida panhandle, are characterized by steady lateral seepage, and such streams support swamp-type vegetation. Bog-fed streams have intermittent discharge (following rainfall) with little or no sediment load. Such streams flood rapidly and drain slowly because of dense emergent vegetation. Alluvial rivers originate in Piedmont areas, forming extensive swamps at the juncture of the Piedmont and coastal plain. Rivers such as the Apalachicola, Choctawhatchee, and Mobile have hydroperiods that reflect flow from distant rainfall and many tributaries. Floodplain interactions strongly modify discharge patterns, with wetlands vegetation playing an important role in defining the hydrology of such streams. Blackwater streams, on the other hand, originate in the coastal plain and depend on local rainfall for their discharge rates. Groundwater seepage is an important component of such discharge. Although the waters of blackwater streams are often clear, (i.e., low in turbidity and suspended matter), they are usually highly colored by various organic compounds such as humic and fulvic acid complexes and lactic acid components. Blackwater streams may act as tributaries to alluvial rivers. Blackwater rivers usually are more acidic, with higher concentrations of dissolved organics than alluvial streams, which are higher in nitrogen, phosphorus, calcium, and magnesium. Such differences contribute to the various forms of aquatic plants (emergent and submergent) found in such systems.

Clewell (1991) has outlined the existence of 10 vegetation types (8 wetland, 2 mesic), with an emphasis on several critical environmental conditions that determine riverine vegetation:

1. Cypress–tupelo swamp
   River swamp
   Strand
   Head
2. Bottomland hardwood forest

3. Hydric hammock
4. Mesic hammock
5. Mesic evergreen hammock
6. Bay swamp
   Bayhead
   Hillside seepage swamp
   Floodplain seepage swamp
7. Thicket
   Willow pointbar thicket
   Strand thicket
8. Mangrove wood
9. Marsh
   Brackish marsh
   Tidal river marsh
   Floodplain marsh
   Slough
10. Aquatic macrophyte assemblage

These vegetation types are distinguished by physiognomy, life-forms of the principle species, and species composition. The subdivisions of cypress–tupelo swamps, bay swamps, thickets, and marshes are based largely on habitat differences and secondarily on species composition. The critical environmental conditions that are thought to play a role in the determination of the plant associations include a number of variables (Wharton et al. 1982, Clewell 1991): hydrology, organic matter, fire, salinity, and biological factors such as predation and competition.

Wetlands vegetation of various types is associated with the coastal plains river basins. Such floodplain vegetation depends on river type (spring-fed, alluvial, blackwater), flow characteristics, soil composition, water quality, and nutrient distribution. In Florida, the major bottomland hardwood forests are found along the alluvial rivers of northwest Florida (Apalachicola, Choctawhatchee) (Fernald and Patten 1984). Along spring-fed streams and rivers, such plants are accustomed to less changeable water levels than those of the periodically flooded alluvial plains. Backwater river swamps, often anaerobic, are populated by distinct plant forms. All of these are highly productive (considerable water and nutrients) and form the basis for complex aquatic and terrestrial food webs. Florida has various other forms of wetlands such as gum ponds and swamps, hydric hammocks, (red cedar, cabbage palm), herb bogs, and savannahs (scattered trees, wildflowers). In peninsular Florida, there are extensive freshwater marshes dominated by sedges such as *Cladium jamaicensis*.

Detailed lists of species found within such areas across the gulf coastal plain are provided by Niering (1985). Emergent vegetation is dominated by narrow- and broad-leaved plants, such as cattails, bulrushes, and emergent sedges and grasses. Floating plants, such as duckweed, water lettuce, water hyacinth, and alligator weed, are also common in such areas. Submergent vegetation of the riverine areas include such dominants as *Elodea*, water celery, pondweeds, milfoils, and bladderworts. Many of these species are seed-bearing aquatic submergents. Because of the environmental vagaries of flowing streams, rooted plants often are not widely distributed in such systems. Such plants require good anchorage and strong stems and leaves. Usually these plants are more prevalent in impounded areas such as bayous, swamps, and ponds. These aquatic plants, emergent, floating, and submergent, con-tribute to the high productivity of the river basins of the southeastern coastal plains region.

Some elements of the emergent, floodplain vegetation of gulf coastal rivers are quite distinctive. Shrub swamps, low-growing evergreen or deciduous forms or pocosins, are primarily located in northwest Florida along the gulf coast. Such areas are dominated by loblolly bay, redbay, and sweetbay forms, in addition to titi, waxmyrtle, and zenobia. Cypress swamps, dominated by bald cypress and pond cypress, are common along the gulf coast to east Texas. Southern bottomland hardwood swamps are found along river floodplains from east Texas to Florida. Such river swamps occur mainly in the coastal plain along continuously flooded areas dominated by bald cypress, water tupelo, Ogeechee tupelo, and various other forms.

The autochthonous, microalgal components of gulf coastal plains rivers are not well studied or understood. However, microalgae of various types con-tribute to the biota of the relatively slow-moving rivers and streams of the gulf coastal plain, which harbor a rich diversity of epiphytic microphytes (Livingston et al. 1991). Such flora is often dominated by diatoms (Patrick and Reimer 1966) and is well developed in areas where the water currents are not swift. In acidic dystrophic waters, high in humates and low in dissolved oxygen and nutrients, the microalgal flora is often specialized, with some species being aerophilous. Current characteristics have a strong influence on the diatom composition. In fast-moving streams, rheophilic types predomi-nate. Usually plankton development is scarce in open-water situations or in streams and rivers of even moderate current strength, with benthic flora limited to edges, the stream bed, or associated pools. Muddy or highly turbid rivers are often associated with relatively impoverished diatom development. Larger rivers, however, may have a well-developed plankton flora. Such associations are often developed along specific stretches of the river and are often derived from benthic or epiphytic populations. Overall, the microflora of the slow-moving gulf coastal rivers, though not well studied in terms of distribution and ecology, may contribute an extremely important dimension to the biological organization of such areas.

ANIMAL ASSEMBLAGES

Within a given drainage system, there is considerable variation in the animal assemblages, depending on stream order, the dimensions of the river basin, and specific changes in key ecological features in time and space. Niering (1985) gives a comprehensive review of the various animal forms taken in river–wetlands areas. Developmental stages of aquatic insects are dominant forms in numerical abundance and species richness at various trophic levels. The adult stages of such insects are usually terrestrial (Resh and Rosenberg 1984). Planktonic forms include protozoans and rotifers, which feed primarily on the microorganisms, blue-green and green algae, diatoms, and flagellates that are considered to be typical river plankters. Organisms found in the substrate include tendipedid larvae, annelid worms, mollusks, and insects. Small-scale habitat changes along a given stream or river can lead to extreme differences in population distributions and community types. Wharton et al. (1982) showed how specific habitat changes lead to complex community types. Nektonic forms include various vertebrates, with fishes taking a predominant role. Certain forms remain in specific habitats throughout their life history, whereas others may be migratory, making use of various parts of the river system as a function of life history stages. Dominant types include the percids, clupeids, topminnows, catfishes, carp, suckers, centrarchids, and gars. Anadromous forms include sturgeon, shad, and striped bass.

There is no simple way to organize the animal associations found in the Gulf Coastal Plains rivers. The river continuum model (Vannote et al. 1980, Vannote 1981) is useful as an organizing principle, but, as pointed out by various authors (Wharton et al. 1982), such generalizations often are not applicable to specific ecological situations. Many so-called detrital-based food webs may involve zooplankton "grazing" links, even in coastal blackwater streams where phytoplankton (diatom) blooms may be locally important. There is a lack of systematic multidisciplinary studies in gulf coastal plain rivers, so precise associations of habitat and animal distributions are not well developed. Comparative studies among different river systems are even less well drawn; so, outside of specific faunal distributions, few generalizations can be made concerning process-oriented functions in such stream and river systems.

Livingston et al. (1991) compared the main stem and the tributaries of the Choctowhatchee river system in terms of overall habitat distribution and biological organization. The study focused on the distribution of habitats and organisms at various levels of biological organization, with an emphasis on a comparison of the main stem of the river with various tributaries. Spatial and temporal hydrological differences between these two general categories contributed to basic differences of certain habitat variables. Such differences accounted for different biological responses. Cumulative species richness of phytoplankton was lowest in the main stem areas and in a tributary that was clear-cut. Infaunal macroinvertebrates were depauperate in numbers and species richness in the main stem of the river relative to the tributaries. Temporal differences in the distribution of epibenthic macroinvertebrates were noted between tributaries and the main channel. Fishes were fewer in number and species richness in the tributaries than the main stem. The fish trophic organization was also quite different in that the main stem was low in the number of levels of carnivory and high in the number of surface feeders relative to tributaries. Smaller streams were more often represented by open water feeders. Fishes in tributaries were less dependent on living and particulate organic matter. Overall, tributaries were highly individualistic in terms of the biological organization, and there was a lateral component to river biological diversity that appeared to be as important, in terms of the definition of the aquatic system, as the longitudinal distribution of organisms along the main stem.

There is extensive scientific literature on lotic systems. Recently, various questions have been asked concerning temporal (long-term) relationships and the need for manipulative field experimentation to explain the relationships of local species assemblages to past and existing biological (interspecific) conditions (Ross et al. 1987). However, most such studies are too short in duration to answer important questions concerning the intricate relationships of the various biotic components of streams and rivers. The role of hydrology in this process is important (Reiter 1986). Stream order, as a descriptive term, may be less important to the determination of system processes than factors, such as mean annual discharge and mean annual range of discharge (Hughes and Omernik 1980). In addition, there may be certain basic differences between northern and southern lotic systems. Winger (1986) reviewed differences between warmwater and coldwater streams, noting that warmwater systems are often characterized by low elevations, more pools and fewer riffles, high turbidity, substrata with small particle size, and more man-made modifications and introduced species. The various rivers of the southeastern United States remain relatively poorly studied from the standpoint of ecosystem ecology. River management, based on ecosystem-level research, is largely lacking.

REPRESENTATIVE SYSTEM: APALACHICOLA–CHATTAHOOCHEE–FLINT RIVER BASIN

The tri-river Apalachicola–Chattahoochee–Flint (ACF) system (Fig. 2) is located in Alabama, Georgia, and Florida. This system represents a broad range of habitats that are characteristic of medium-sized rivers in the Gulf Plain province. In addition, there has been a long-term effort to understand and preserve the Apalachicola River system. Consequently, there is a considerable amount of experience in the application of research results to practical ways of managing a major, river–estuarine resource.

Watershed Characteristics

The ACF system (Fig. 2) drains an area of 50 800 km^2 (19 614 mi^2) (Leitman et al. 1983). The Chattahoochee River is 701 km (436 miles) long and drains an area of 22 714 km^2 (8770 mi^2). The average annual flow rate is 346 m^3/s (12 210 ft^3/s). Maximum basin width is 68 km (43 mi) (U.S. Army Corps of Engineers 1984). The Flint River watershed lies completely within the state

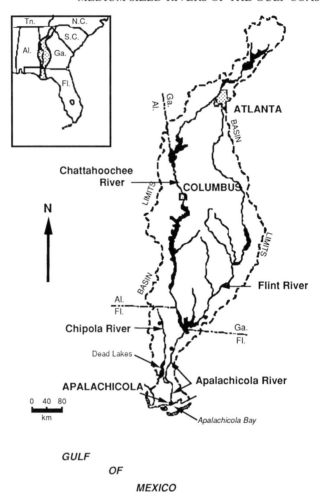

FIGURE 2. The Apalachicola–Chattahoochee–Flint (ACF) drainage basin showing the major rivers in the region and the associated estuary.

of Georgia, where it drains an area of 15 767 km^2 (6088 mi^2). The Flint is about 600 km (375 mi) long with a floodplain approximately 64 km (40 mi) wide. The average annual flow rate is 244 km^3/s (8700 ft^3/s). The Apalachicola River is 171 km (106 mi) long, falling about 12 m from its head at the Jim Woodruff Dam to its mouth at the Apalachicola estuary. This river drains an area of about 6200 km^2 (2393 mi^2), of which about 50% is drained by its tributary, the Chipola River. The average annual flow rate at Chattahoochee, Florida, is 690 m^3/s (24 642 ft^3/s).

The Chattahoochee River originates in north Georgia within the Blue Ridge physiographic province (U.S. Army Corps of Engineers 1984). The upper river is characterized by cold clear water flowing over rock substrate with various rapids, shoals, pools, and falls. Such mountainous areas feature mixed pines and hardwood forests. It then passes through the Red Hills of the Piedmont province to enter the upper coastal plain, where it joins the Flint to form the headwaters of the Apalachicola River. The Chattahoochee thus flows through three physiographic provinces. Within the Piedmont, the gradient decreases as the substrate turns to sand, silt, and gravel. Water temperature and silt load increase. Vegetation here combines Appalachian and coastal plain types within the Piedmont flora, which includes oak/pine associations and hardwood forests. Between Atlanta and Columbus, Georgia, the Piedmont vegetation is dominated by loblolly pine. In the Coastal Plain physiographic province, agriculture is the dominant upland type. Little land in this region remains in its natural condition. In this way, river habitat along the Chattahoochee includes shoal areas with steep gradients and rocky substrates, sandy runs of shallow water of moderate gradient and sand substrate, and deep pools with narrows, almost no gradient, and silt substrate. The various impoundments along the river (Leitman et al. 1991) provide another form of habitat, albeit a highly modified one, that requires constant maintenance.

The Flint River originates in the Piedmont Plateau near Hartsfield International Airport, just south of Atlanta, Georgia. After moving to the south through a swampy region, it flows through a fall area, the transition between the Piedmont and Coastal Plain provinces. This region is characterized by rocks, riffle areas, and a series of rapids and water falls. The upper Flint is in its natural state and has been designated the most scenic stream in the Georgia Piedmont. In the upper coastal plain, the surrounding regions are heavily used for agriculture; numerous sinks and wetlands occur in this region. Bottomland swamps support cypress–tupelo forests with upland slopes supporting magnolia–oak–pine associations. The river is characterized by alternating shallow and deep areas that are relatively short in length. Widths range from 50 to about 135 m, with depths of 1–3 m. The river bed is generally sandy and cut through by a main channel with shallow side areas. Velocities reach 1 m/s. Overall, the Flint River remains relatively undeveloped along its length, although three dams have been constructed and there are anthropogenic alterations in water quality in the urbanized areas (Atlanta, Montezuma, Albany, Bainbridge) (U.S. Army Corps of Engineers 1984).

The Apalachicola River, formed by the confluence of the Flint and Chattahoochee Rivers, lies entirely within the coastal plain. It is subdivided into upper and lower regions: The Marianna lowlands, New Hope Ridge, Tallahassee Hills, and Beacon Slope are part of the Gulf–Atlantic rolling plain, whereas the lower coastal lowlands are part of the Gulf–Atlantic coastal flats (Leitman et al. 1983). Stream forms range from first-order ravine types to higher-order, low-gradient, meandering types. Such streams may range from calcareous and clear in the Marianna lowlands karst plain to highly colored organic (acidic) flows from flatwoods areas. No extensive natural lake systems exist in the valley. The upper river, cutting through Miocene sediments, has a flood plain of 1.5–3 km; this flood plain widens to 3–5 km along middle river regions. The lower river has a floodplain approximating 7 km. Upstream tidal influences on the floodplain do not extend beyond km 40 (from the river mouth). The Chipola River joins the Apalachicola at km 45. The delta area, surrounded by a broad marsh, is about 16 km wide. The lower river lies entirely within the Gulf Coastal Lowlands. Altitudes are less than 15 m in this region, with natural riverbank levels ranging from 15 to 45 m wide, rising 0.5–2.5 m above the floodplain floor (Leitman et al. 1983). The river empties into the Apalachicola estuary, a barrier island lagoonal complex.

On 23 June 1824, the U.S. Congress authorized navigational improvement by the Army Corps of Engineers for a channel 2.7 m deep and 30 m wide along the main channel of the Apalachicola system (Leitman et al. 1983). Later, authorization was made for a channel 30 m wide and 2.7 m deep; such authorization for the tri-river system extends from the Gulf of Mexico to Bainbridge, Georgia (Flint River), and to Columbus, Georgia (Chattahoochee River) (U.S. Army Corps of Engineers 1984). In addition to an active dredging program, a series of dams has been constructed on the Flint and Chattahoochee Rivers (Table 1). Active construction occurred from 1834 to 1975. The series includes 13 dams on the Chattahoochee and two on the Flint, with the Jim Woodruff dam at the confluence of two rivers, for a total of 16 impoundments. The larger dams may influence seasonal, weekly, or daily flows. The upper Chattahoochee River (average annual flow rate, 346 m^3/s) flows through a highly urbanized area (Atlanta), with the lower half flowing through a heavily agricultural area. After it leaves the Atlanta area, the Flint River (average annual flow rate, 244 m^3/s) flows through a largely rural area, with the lower portions of the basin dominated by agricultural use. The Apalachicola River (average annual flow rate, 690 m^3/s) is lightly industrialized at its northern limit, flowing through mostly rural areas in a series of largely natural wetlands systems. The Corps of Engineers has made a series of cutoffs in the Apalachicola River to straighten bends for easier navigation. Such cutoffs have shortened the length of the Apalachicola by 3 km (Leitman et al. 1983). A series of groins was installed along the upper Apalachicola to deepen the channel. The Jim Woodruff Dam at the Georgia/Florida border forms Lake Seminole, the headwaters of the Apalachicola River. Despite continued dredging and channelization of the Apalachicola River, it remains undammed, with a relatively natural wetlands system and no overt water quality problems. As a result, the Apalachicola River is one of the last major coastal plain rivers existing in a relatively natural condition.

Geology

By the Cretaceous period, about 135 million years ago, much of the tri-river valley was submerged under an ancient sea (W. F. Tanner, personal communication). Sea level declined through Cenozoic time (70 million years ago to present), and by middle Miocene, the present upper Apalachicola valley (i.e., Blountstown, Florida) was a deltaic (coastal) area (Olsen 1968). By the Pleistocene epoch, the Apalachicola barrier island chain was located about 22.5 km northeast of Apalachicola, Florida. Present-day structure is about

TABLE 1 Dams Along the Apalachicola–Chattahoochee–Flint System with Notes on the Major Impoundments

Dam	Description
CHATTAHOOCHEE RIVER	
Buford Dam	Flood control, navigation, hydropower, recreation, and water supply
Buford Re-regulation Dam	Not constructed, but funded for planning and engineering; water supply, river recreation
Morgan Falls Dam	Hydropower, maintain minimum downstream flows for water quality and water supply
West Point Dam	Flood control, hydropower, recreation, fish and wildlife development, area redevelopment, streamflow deregulation, and downstream navigation
Langdale Dam	Private dam for hydropower
Riverview Dam	Private dam for hydropower
Bartletts Ferry Dam	Hydropower, municipal and county water supply, and recreation
Goat Rock Dam	Hydropower and recreation
Oliver Dam	Hydropower, industrial and municipal water supply, navigation, and recreation
North Highlands Dam	Hydropower
City Mills Dam	Hydropower
Eagle-Phenix Dam	Hydropower
Walter F. George Lock and Dam	Hydropower, navigation, recreation, and fish and wildlife
George W. Andrews Lock and Dam	Navigation and recreation
FLINT RIVER	
Warwick Dam	Hydropower
Flint River Dam	Hydropower
APALACHICOLA RIVER	
Jim Woodruff Lock and Dam	Navigation and hydropower; secondary benefits include public recreation, stream flow regulation, and fish and wildlife conservation

Source. Leitman et al. (1991).

10 000 years old, with the present gulf barrier island chain position formed about 5000 years ago.

Today, the tri-river drainage system is alluvial, carrying sediment loads that eventually end up in the northeast gulf coastal region. The Flint River arises in Piedmont crystalline formations, by the fall line, and moves through upper Cretaceous sediments as it passes through the coastal plain strata (the so-called Flint River and Clayton formations) (U.S. Army Corps of Engineers 1984). The Chattahoochee originates in Piedmont crystalline formations, which continue to the vicinity of Columbus, Georgia. By the fall line, the river enters the upper Cretaceous sediments with varying strata exposed in the bluffs down-river including Upper Cretaceous, Paleocene, Lower Eocene, Middle Eocene, and Upper Eocene. From here, the transition proceeds from the middle and western Alabama coastal plain to subsurface formations of the eastern coastal plains and Florida. The Apalachicola River lies completely within the coastal plain in a region divided into the Marianna lowlands, the Western highlands, and the Tallahassee Hills. The entire lower portion of the Apalachicola basin lies within the coastal lowlands. The youngest rock formations of this region, Pleistocene and Recent, comprise lower marine and estuarine deposits (sand, clay, silt, or calcareous ooze). Estuarine and fluvial deposits extend up the main river valley, with Pleistocene marine sands and clays along the coast. Suspension of the clays and fine sands of the river bed are the cause of the high turbidity of the river system. The sand/gravel bottom is dynamic, forming a continuously changing series of bars. Frequent dredging of such bars adds to the resuspension of the sands and clays, which contributes to the natural turbidity.

Climate

Temperature in this south temperate area tends to be mild, although elevation and proximity to the coast influence local differences. Mean annual temperatures of the region approximate 20 °C, with peaks usually in August and winter minima from December to February. Mean annual rainfall in the tri-river basin varies from 130 cm in Georgia to 150 cm in Florida (Livingston 1984). Season patterns are similar, with some difference in the actual timing of peaks. Georgia rain tends to form two peaks of equal magnitude in March and July (Meeter et al. 1979). In Florida, the March peak is proportionally lower, with the larger peak in late summer and early fall. Drought periods in both areas occur in mid to late fall. Spectral analyses of long-term data (Meeter et al. 1979) indicate recurrent cycles of peak activity with approximate 6.7-year periods; Georgia rainfall has a slightly different spectrum. Wind direction is generally from the southeast during the spring and from the north or northeast at other times of the year. Strong air masses of continental origin often move through the area during winter. Tropical storms and occasional hurricanes occur during summer and fall months.

Hydrology

The annual mean flow of the Chattahoochee River from 1958 to 1979 was 15% greater than from 1929 to 1957, probably a result of increased rainfall in the three-state drainage area during the later period (Leitman et al. 1983). Monthly river flow is seasonal, with considerable variation in peak flow rates (Meeter et al. 1979). Such peaks usually occur during late winter–early spring periods, at which time rainfall in Georgia is high and evapotranspiration of river–wetland vegetation is low (Livingston and Loucks 1978). Low flows occur from September through November. Statistical analyses showed longer-term cycles of peak flow rates with periods approximating 6–7 years, and indications of even longer-term cycles (Meeter et al. 1979).

Floodplain inundation varies with location along the three rivers according to placement and height of natural levees, flood heights, form and distribution of floodplain vegetation, soil types, and the level of the water table. The surficial sand aquifer of the Apalachicola basin, with a thickness usually less than 35 m, is composed of fine to very fine sand and clayey sand. The exact thickness varies, depending on the distribution of scattered beds of gravel, limestone, marl, and shell. The water table gradient usually is toward the river, with groundwater discharge to the floodplain dependent on local geological and physiographic features. Generally, the aquifer maintains base flows by discharging into the stream. The Floridan aquifer (in Florida) and the Ocala aquifer (in Alabama) are the principle sources of water for the ACF system. Groundwater flow regimes are complex in this region. They are of limited availability in northern portions of the ACF system, but are a major source of water to mid and lower parts of the basin. The extent of the inter-action of surface water and groundwater is not well understood. Groundwater sustains river flow during low flow periods (U.S. Army Corps of Engineers 1984).

Water Quality

Water quality in the groundwater and surface water of the Chipola and Apa-lachicola Rivers is generally good (U.S. Army Corps of Engineers 1984). Nutrient increases have been observed below municipal wastewater treatment plants (Chattahoochee, Blountstown, Marianna), with occasional periods of low dissolved oxygen and increased coliforms from wastewater discharges and natural drainage from riverine wetlands. There is realtively limited agricul-tural and industrial development in the Apalachicola valley. From 1972 to 1974, moderate levels of organochlorine residues in sediments and tissues of organisms were taken from the Apalachicola estuary (Livingston et al. 1978). Recent surveys by the U.S. Geological Survey (H. Mattraw, personal com-munication) indicated relatively low levels of heavy metals and negligible concentrations of herbicides in the Apalachicola River. There were relatively low levels of organochlorine in the water and organisms of the Apalachicola Bay system. Such changes have been attributed to the banning of DDT in 1972 and the flushing action of the river and bay system. However, Winger et al. (1984) found that residues of organochlorine insecticides, polychlori-nated biphenyls, and heavy metals were higher in aquatic biota of the upper Apalachicola River than in the lower river. Such moderately high levels of these pollutants in the upper river were considered to be an early stage of contamination with sources in Georgia. A battery plant on the Chipola River (near Marianna, Florida) released acid wastes and heavy metals (Pb, Al) to the primary drainage. Recent studies (R. J. Livingston, unpublished data) indicate that the problem is local and there appears to be some recovery since termination of plant operations. R. J. Livingston (unpublished data) found that certain agricultural operations, municipal sites, and areas of storm water runoff (septic tanks, roads) caused some contamination of specific areas of the lower Apalachicola basin. This contamination included runoff from a limited agricultural area and forestry operations in wetlands areas above the bay. In general, the Apalachicola/Chipola River and estuary, although not pristine, are relatively free of pollution at this time.

Plant Communities and Nutrient Distribution

The only major alluvial rivers in Florida are the Apalachicola and the Choctawhatchee. The broad floodplains of these systems harbor a great diversity of wetland habitats. Annual winter–spring flooding creates a condition that favors specific assemblages of plants that are among the most productive in the world. The highlands along the upper Apalachicola have been occupied by plants since the late Miocene so that much of the uniqueness of the Apalachicola wetlands flora is attributable to the spatial/temporal continuity of such assemblages (Clewell 1977). Of the 116 plant species along the Apalachicola River, 17 are endangered, 28 threatened, and 30 rare. The Apalachicola watershed contains 9 endemic species. The forested floodplain of the Apalachicola covers about 49 373 ha (122 000 acres), the largest in Florida. The exact distribution of species depends on various factors: soil saturation, dissolved oxygen availability, flood patterns, and soil types (Leitman et al. 1983). Much of the floodplain was logged between 1870 and 1925 (Clewell 1977), and some logging has occurred in recent years. Small parts of the floodplain have been closed for industry, agriculture, and residential development. Overall, however, floodplain vegetation remains intact; regrowth has been rapid and much of the area resembles a mature forest (Leitman et al., 1983).

The Apalachicola forested floodplain is 114 km long and is about 450 km² in area (Wharton et al. 1977). Dominant tree types in terms of basal area include water tupelo, Ogeechee tupelo, bald cypress, Carolina ash, swamp tupelo, and sweetgum (Leitman et al. 1983). Five basic forest types were identified by Leitman et al. (1983) (Table 2): These were determined largely by water depth, duration of inundation and saturation, and water level fluctuations due to flooding patterns. Biomass increased downstream, where the range of river stage fluctuations was one-third that of the upper river. Species-specific differences in distribution followed hydrological conditions along the river. Ogeechee tupelo was uncommon along the upper river, and swamp tupelo was observed only along lower river transects. Sweetbay, cabbage palmetto, and pumpkin ash were found only along the lower river. Lower river forests were dominated by water tupelo, Ogeechee tupelo, bald cypress, Carolina ash, and swamp tupelo (Types C, D, E). Upper river areas had the highest species richness with types such as sugarberry, American hornbeam, and possumhaw found in this area. Sweetgum occurred on higher terraces

TABLE 2 Wetlands Vegetation Types (km²) in the Apalachicola Floodplain

Mapping Category	Upper River	Middle River	Lower River from Wewahitchka to Sumatra	Lower River from Sumatra to Mile 10	Lower River from Mile 10 to Mouth	Total
Pine	136	672	0	204	0	1 010
A/pine	642	1 440	154	464	0	2 710
AB	12 500	32 200	15 800	1 700	48.0	62 300
C	170	1 860	8 310	15 800	6 920	34 100
DE	2 420	2 270	6 240	10 300	456	21 700
Pioneer	0	150	19.2	0	0	169
Marsh	0	0	0	0	9 030	9 030
Open water	2 730	3 110	1 540	2 010	1 260	10 700
Unidentified	1 020	748	81.3	76.8	19.2	1 950
Total	20 600	42 500	32 100	30 600	17 700	144 000

Note. A/pine: sweetgum–sugarberry–wateroak–loblolly pine; AB: water hickory–sweetgum–overcup oak–green ash–sugarberry; C: tupelo–cypress with mixed hardwoods; DE: tupelo–cypress; pioneer: black willow; marsh: sawgrass–bulrushes–big cordgrass–cattail.
Source. After Leitman et al.(1983).

and flats of the upper and middle river. Forest types were more diverse in the upper river because of the broader range of water level fluctuations and topographic relief (relative to the flatter lower river floodplain). Overall, water–tree relationships varied with river location because of the range of water-level fluctuations and topographic relief of the floodplain.

The interaction of the processes of primary production, decomposition, and nutrient dynamics in freshwater wetlands is complex and not well understood (Good et al. 1978). Primary production depends on hydrology, soil conditions, community types, life history, and extrinsic biological features (de la Cruz 1978). Likewise, decomposition processes are poorly understood, particularly with respect to below-ground processes. Hydrology is an important controlling factor in nutrient dynamics (Valiela and Teal 1978) in addition to the biological activity concerning specific nitrogen and phosphorus compounds. Exchanges among the various interacting components of the freshwater marshes and the role of animals in nutrient cycling are also important, yet poorly understood, components of the puzzle. Microbial activity in such processes is also a crucial factor in how freshwater wetlands are related to nutrient distributions in river-associated areas.

In detailed studies of production and decomposition of forest litter fall in the Apalachicola floodplain, Elder and Cairns (1982) found average litter fall levels of about 800 g/m²/year. Levee vegetation had slightly higher values than swamp vegetation. Leaf fall was maximal from September through December, although diversity of tree types led to a lower peak than might be expected. Dominant forms (tupelo, cypress, ash) accounted for over 50% of the litter fall, but were least productive, on a weight-per-biomass basis, than any of the 12 major leaf producers. Leaf decomposition was highly species-specific, with bald cypress and diamond-leaf oak more resistant than tupelo and sweetgum. Carbon loss from leaves was linear with time, whereas nitrogen and phosphorus leaching was more or less complete within one month. Much of the organic matter may be recycled within the wetlands forest, but annual river flooding provided the mechanism for leaf-litter mobilization from the floodplain. Litter fall mobilization depended to a considerable degree on the timing of the annual flood relative to peak autumn leaf fall. Leaf litter fall in the Apalachicola system was greater than that in almost any warm temperate system and many tropical systems, a positive example of the high productivity of alluvial river wetlands systems.

Various studies in the Apalachicola system have led to the development of a model of linkage between the upland (freshwater) wetlands and the estuary. Complex cyclic interrelationships among variables such as hydrology, wetlands forest production, and nutrient distribution (Fig. 3) form the basis of linkage between such interlocking systems (riverine, estuarine). Nutrient transport in the Apalachicola River system was studied by Mattraw and Elder (1984). About 80% of the Apalachicola stream flow entered the river at its headwaters (from the Flint and Chattahoochee systems) with 11% from the Chipola and less than 10% from groundwater and overland runoff. About 35 000 metric tons of particulate organic carbon derived from litter fall was delivered to the estuary in a year. Such transport was especially sharp during winter–spring floods. Approximately 2.1×10^5 metric tons of total organic carbon, 2.2×10^4 metric tons of nitrogen, and 1.7×10^3 metric tons of phosphorus were delivered to the estuary in one year. Such nutrient inputs were derived from the Apalachicola River floodplain. On a net basis, the Apalachicola floodplain exports greater quantities of carbon (13 g/m^2/year) and phosphorus (0.08 g/m^2/year) than most other watersheds studied in this fashion. The combination of annual spring floods and a productive bottomland hardwood forest in the floodplain was crucial to such nutrient flow patterns. In a coordinated, long-term evaluation of POC input to the estuary, Livingston (1981) showed that the exact timing of flow peaks, together with seasonal changes in the floodplain vegetation, were critical to the long-term changes in POC input to the river–estuary from freshwater wetlands. Such trends are currently being modeled relative to the long-term biological response of the Apalachicola Bay system to riverine inputs.

Animal Associations

There is considerable diversity and variability among animal associations in the Apalachicola River system. In an intensive analysis of first- to third-order streams in the Chipola system, Livingston (1988) found that these portions of the basin are highly stressed by seasonal droughts and floods. During periods of low rainfall, such low-order streams dry up. During winter–spring flooding, these areas are inundated by rapid streams of water that heavily scour the system. Nevertheless, such areas are highly productive, in part as a result of extensive blooms of microalgae (primarily diatoms, such as *Gocconeis*, *Eunotia*, *Nitschia*, and *Gloeocystis*). The infaunal macroinvertebrate fauna, with peak numbers during summer–fall periods, was dominated by

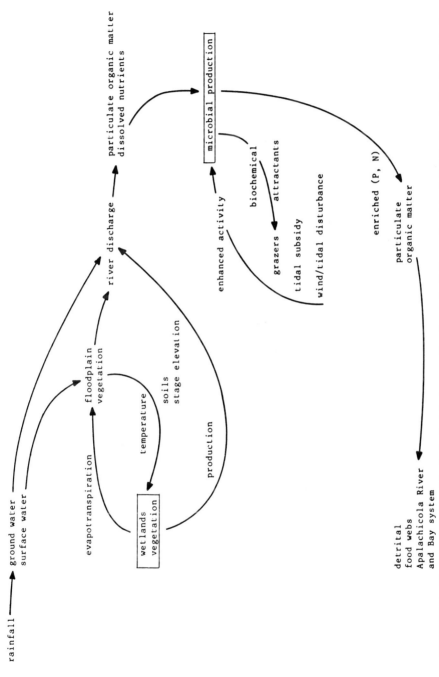

FIGURE 3. Proposed model of hydrological factors, wetlands productivity, and nutrient distribution in the Apalachicola River estuary.

oligochaete worms (*Dero* sp., *Ilyodrilus*) and insect larvae (*Chironomus decorus* group, *Tanytarsus* sp.). Epibenthic macroinvertebrates, peaking during spring–early summer and fall periods, were dominated by insect larvae (*Tanytarsus*, *Chironomus decorus* group). Species richness and diversity in such areas were periodically low during periods of drought, with relatively rapid recovery during periods of increased rainfall. Fishes in such first- to third-order streams were dominated by centrarchids (*Centrarchus*, *Elassoma*, *Lepomis* spp.), topminnows (*Gambusia*), catfish (*Ictalurus*), and *Amia*. Relatively low numbers and biomass of fishes were taken during winter–early spring months.

The trophic organization of the low-order streams was diverse. Among the infaunal and epifaunal macroinvertebrates, omnivores and primary carnivores were prevalent. Browser/scavenger/grazers, engulfers, piercers, and scrapers were represented among the gastropod mollusks, oligochaete worms, crustaceans, arachnids, and insects. Fishes, on the other hand, were dominated by primary and secondary carnivores, which appeared mainly as browsers, scavengers, and grazers. There were seasonal changes in the trophic dominance order of such assemblages, with omnivorous and carnivorous insects as the dominant form in terms of trophic diversity and prevalence.

The main channel of the Apalachicola River has not been well studied. Early analyses showed that chironomids and caddisflies were characteristic bottom-living forms of the upper river. Simplifications of habitat types due to dredging were noted during these and subsequent studies. Cox (1970) and Cox and Auth (1971–1973) indicated that dredging was associated with habitat destruction, simplification of the aquatic fauna, and reduced productivity. More recent work (Ager et al. 1985) showed that long-term dredging of the river has led to habitat alteration, river shortening, and redirected river flow due to the construction of training dikes, maintenance dredging, spoil deposition on the floodplain and along the sides of the river, bendway elimination, and snag removal. Leitman et al. (1991) analyzed the range of waterway activities along the Apalachicola River.

Investigation of mollusks of the Apalachicola River (Heard 1977) showed that this sytem contains the largest number of species of freshwater gastropods and bivalves of all the drainages from the Escambia to the Suwannee River (e.g., the west Floridian molluscan province). The Apalachicola also was a center of endemism for the province. Lotic types, such as pleurocerid snails (*Coniobasis* spp.) and amblemid and unionid clams were prevalent. Several Apalachicola mussels were listed as rare and endangered species, with local extinctions often associated with the appearance of the introduced Asiatic clam, *Corbicula manilensis*, which is prevalent along the river.

Yerger (1977) noted that the deep penetration of the Apalachicola headwater into the southern Apalachian mountains provided a convenient dispersal route for various freshwater fishes. Eighty-six fish species (43 genera, 21 families) have been found in the Apalachicola and its tributaries; this fauna is dominated by primary freshwater types, with some euryhaline marine, diadromous, and secondary freshwater types found as well. Overall, no other river in Florida has so many species of fishes. Various species of fishes are

endemic to the Apalachicola river system and a fourth of such species originated in this river.

Considerable research has been carried out concerning the natural history
of anadromous fishes in the Apalachicola river system. Alabama shad (*Alosa
alabamae*) enter the river in February, reaching peak abundance in mid to
late April. This species remains unutilized as a fishery (Yerger 1977). At one
time, the striped bass (*Morone saxatilis*) was abundant and represented an
important fishery on the river; the Apalachicola was the only river on the
Florida gulf coast to support such a fishery. The endemic Apalachicola population (Barkuloo 1967, 1970) is the last remnant of a native population once
widespread in the rivers of the northern Gulf. Prior to the construction of
the Jim Woodruff Dam, considerable numbers of striped bass supported a
viable sports fishery; drastic declines followed the damming and channelization of the upper river. Native and introduced stock have co-occurred in
recent years although construction activities restricted the distribution to only
17% of their historic spawning grounds along the tri-river system. Such activities have seriously limited population growth (Wooley and Crateau 1983).
Water temperature was an important factor in habitat selection and survival.
The sturgeon (*Acipenser oxyrhynchus*) was once a viable fishery in the Apalachicola; however, by 1970, this commercial resource was no longer viable,
although a remnant, hook-and-line fishery has developed just below the Jim
Woodruff Dam, where the sturgeon are blocked from further migration upstream (Wooley and Crateau 1982).

The Apalachicola floodplain is rich in animal biota. There are 259 species
of vertebrates in addition to the fishes: 44 amphibians, 64 reptiles, 99 breeding
birds, 52 mammals (Means 1977). There is a higher species density of amphibians and reptiles in the upper Apalachicola River basin than anywhere
else in North America north of Mexico. In part, such high species richness
is due to the diversity of the physical habitat (streams of various types, a karst
plain, extensive floodplain wetlands, all in a relatively natural condition).
Many of the vertebrates are endangered, threatened, rare, or of special concern. Endemism is high. Sightings (1951) of the ivory-billed woodpecker and
Bachman's warbler were made in the Chipola valley (Stevenson 1977). All
habitat along the Apalachicola appears highly suitable for such species (Florida Game and Freshwater fish Commission 1982). At the lower end of the
Apalachicola basin, the highest concentration of nesting ospreys in northwest
Florida was observed in some of the most pristine habitat remaining in Florida
(Eichholz 1980). The floodplain forests of the Apalachicola basin present a
highly productive habitat for a diverse group of vertebrate species that in
many ways is as unique as that of the Florida Everglades.

BIOTIC PROCESSES: AN INTEGRATION

Pulsed System

The pulsed flood events of the Apalachicola River provide a continuous
disturbance that maintains the high productivity and diversity of life within

the floodplain (Fig. 4). The continuous flood cycle regulates the chemical properties of the floodplain soils, which, in turn, determine the distribution of vegetation. Feedback, in terms of organic production, is an integral part of the energetics of the system with microhabitat distribution a distinctive result of the specific hydrology of the natural system. Hydrological conditions, as part of a timed sequence of floodplain productivity, also are crucial in the redistribution of dissolved and particulate organic matter and inorganic nutrients to subsidiary portions of the river–estuary. Such conditions, in turn, determine the spatial/temporal patterns of autochthonous primary and secondary production. The continuously changing river levels provide the basis for a moving edge effect between the aquatic and terrestrial systems, which is part of the explanation for the high sustained productivity and diversity of the plant and animal associations along the river. The wetlands plants provide active nutrient and water conservation; such production is both a product of and a stimulus to the microbial activity that forms the vital link between the derived organic constituents and consumers and the mineralization activities by microorganisms that provide the nutrients for plants. All such processes operate within the range of shifting microhabitats and complex trophic responses to the episodic movement of organic by-products through the system. Wharton et al. (1982) give the simplified models of the trophic organization of the aquatic and terrestrial food webs that are to be found in a riverine system such as the Apalachicola. The interaction between the detritus-dominated aquatic regime and the grazing-foraging zone of transition to the terrestrial regime is thus an important part of biological organization that has been observed along the Apalachicola River system.

Coupling with Other Systems

A major effort has been made to evaluate the importance of coupling between the Apalachicola River and its associated estuary (Livingston 1983, 1984). The estuary, although strongly influenced by the offshore gulf and various physical factors associated with wind and tide, is, in an ecological sense, an extension of the river. The limits and boundaries of the estuary are continuously changing relative to daily, seasonal, and interannual periods of river fluctuation. The single most important physical aspect is salinity; variable and fluctuating salinity gradients represent the most important controlling factor of the biological organization of the estuary (Livingston 1984). Without the freeflowing (flooding) river and the modifying effect of the wetlands, which tend to dampen extreme changes in salinity, the estuary would be altered in terms of the biological response to the salinity regime. Turbidity, color, and sedimentation processes in the estuary are also controlled by the extent and patterns of freshwater inflow via the river.

The trophic organization of the Apalachicola estuary follows seasonal and interannual patterns of nutrient input, productivity, and microbial organic transformations (Fig. 4). Such processes are affected by continually changing combinations of autochthonous and allochthonous components that follow

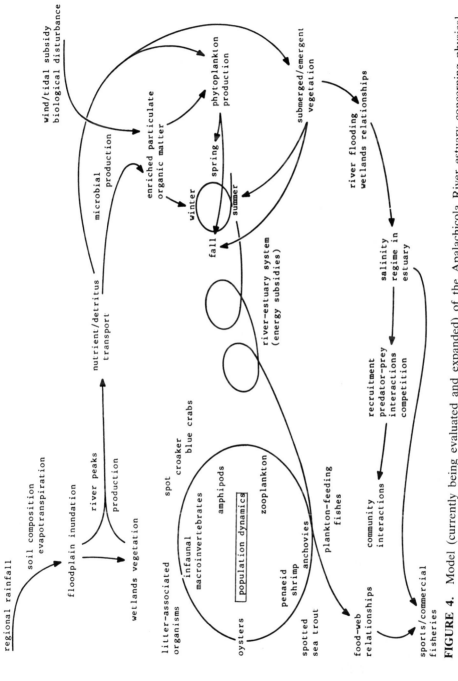

FIGURE 4. Model (currently being evaluated and expanded) of the Apalachicola River estuary concerning physical controls, energy transfers, and biotic response of the aquatic system to varying regimes of freshwater input.

sequenced (seasonal) changes in the system. River input to the estuary is maximal during winter–spring flooding events when particulate organic matter and nutrients, mobilized from the combined floodplains of the tri-river system, form the basis of the detritivore (omnivore) food webs that are predominant in the estuary at this time. The shallow, wind-driven system is a natural habitat for enhanced microbial activity in sediments; microbes, through mineralization, combine the dissolved and particulate organic matter from river floodplains into a nutrient-rich food for various portions of the detrital system. Later in the spring, such matter is combined with phytoplankton production to form the energy basis of the estuarine system at this time. From summer to late fall or early winter, such input is combined with marsh production and submerged aquatic vegetation to fuel the dynamic and highly productive estuarine food webs until the river again floods to renew the cycle. In this way, the importance of the river component to the estuary varies seasonally as well as annually, depending on cyclic patterns of rainfall that are occurring hundreds of miles to the north. It is part of a seasonal succession of nutrient inputs to the river–estuary, which, in turn, form the basis of the overall biological organization of the system in space and time.

One biological component of the coupling of the fresh and brackish water wetlands with river–estuarine productivity involves the periodic movements of diadromous fishes up and down the river. Various marine forms also penetrate the Apalachicola River along its entire length. In addition to receiving and sustaining the continuous waves of migrating offshore organisms that use the estuary as a nursery and feeding ground, the Apalachicola Bay system thus serves a conduit and receptacle for populations that use the river for the same purposes. The extreme variability of such biological processes is the primary reason for the high level of complexity of estuarine systems relative to other areas. Current efforts are now being directed to determine the details of the interactions between the freshwater input to the estuary and the spatial/temporal response of the Apalachicola Bay system. Overall, there are relatively few data concerning the river, per se, since much of the river management is based on the associations with the highly productive Apalachicola estuary. The various data sets (R. J. Livingston, unpublished data) have been organized and models are being constructed for a quantitative determination of the various relationships between the river and the estuary.

RESOURCE USE AND MANAGEMENT ACTIVITIES

The Apalachicola River and Bay system represents an anachronism in today's world of physically altered and often polluted major river systems. The high levels of biological productivity of the Apalachicola basin provide a major source of income for an entire region. Although the anadromous fish resource has been largely destroyed by construction activities along the tri-river system, a commercial catfish fishery still exists on the river. Sports fisheries are also quite strong along the Apalachicola River. Species featured include the white bass (*Morone chrysops*), the hybrid (sunshine) bass (*M. saxatilis* × *M. chry-*

sops), the largemouth bass (*Micropterus salmoides*), various lepomids (*Lepomis* spp.), the yellow perch (*Perca flavescens*), and the Alabama shad (*Alosa alabamae*). In the estuary, over 90% of Florida's commercial oyster (*Crassostrea virginica*) crop is taken. The oyster industry has both historical and cultural significance for the region. The penaeid shrimp nursery of the river–estuary is extensive, providing the basis for the most important (economically) commercial fishery. Blue crabs, black mullet, grouper, whiting, menhaden, flounder, red snapper, white seatrout, and spotted seatrout all contribute to extensive commercial and sports fisheries. A virtually untapped sports fishery exists for species such as tarpon, sheepshead, black drum, cobia, bluefish, red snapper, grouper, and various forms of shark. Along the banks of the river, a lucrative tupelo honey industry thrives, along with worm grunting and other associated by-products of the extensive hunting and fishing activities. In short, the utilization of the natural resource provides the primary industry for the inhabitants of the Apalachicola valley.

Over the past 19 years, a multidisciplinary scientific study has been carried out by various federal, state, and local agencies concerning the ecological relationships of the Apalachicola River estuary. The primary data base was developed by researchers from Florida State University. However, the process did not stop with the research information. A conscious effort was made to apply the scientific data to questions involving management of the extensive Apalachicola resource. The result is one of the most comprehensive management programs ever attempted for a major river–estuary system anywhere in the world. For instance, the following actions have been carried out: The Apalachicola River has been designated as an Outstanding Florida Water by the state of Florida; Apalachicola Bay is a State Aquatic Preserve; much of the estuary has been declared class II waters, a restrictive designation for the protection of oyster-producing waters; the lower river and bay have been designated the largest national estuarine reserve in the United States; the lower valley is part of the Experimental Ecological Reserve System of the National Science Foundation; the lower Apalachicola system has been included in a system of international preserves by the United Nations; St. Vincent Island is a national wildlife refuge; the Apalachicola (Franklin County) region is now an Area of Critical State Concern, with millions of dollars designated by the Florida Legislature for the construction of sewage treatment plants and the implementation of an advanced comprehensive plan for the local (Franklin) county; various portions of the system have been declared private and state preservation areas.

In addition to such management efforts, a major land purchase program by the state of Florida has provided the basis of an unprecedented preservation/management effort. In December 1976, based on the nutrient information that connected the river wetlands with bay productivity, the state of Florida purchased 11 314 ha (28 004 acres) of the bottomland hardwood swamps along the lower Apalachicola River for over $8 million. This action was carried out through the state's Environmentally Endangered Land (EEL) program. In March 1977, Little St. George Island was purchased through the EEL program for $8 838 000. About 526 ha (1300 acres) of Dog Island were purchased by the Nature Conservancy for an island conservation program, which is now being implemented through a cooperative effort with various state agencies. Portions of the eastern end of St. George Island were purchased in 1974 for over $6 500 000 as an inclusion to the existing state park on the island. The Trust for Public Land assisted in the purchase by the state of land on St. George Island (Unit 4) that is close to important oyster-bearing waters in the bay. Wetlands surrounding East Bay have been purchased as part of the establishment of the National Estuarine Sanctuary. Considerable portions of Franklin County, which surrounds the lower river and estuary, have been zoned at low density of one unit per acre or less. In 1986, in the largest single purchase in the history of the state at the time, 25 543 ha (73 000 acres) of Apalachicola and Choctawhatchee River wetlands were purchased through the efforts of the Northwest Florida Water Management District and the Nature Conservancy. This purchase, at a cost of $21 million, represents a major acquisition of bottomland hardwood forests along Florida's major alluvial floodplain. Currently, negotiations are underway for state purchase of most of the remaining Apalachicola River wetlands. If these efforts are successful, this will be the first major river in the United States in which most of the wetlands and a high proportion of the critical coastal fringing area are in public ownership (e.g., are being managed for maintenance of natural productivity). Justification for much of this effort has been based on the scientific data generated by the integrated research program carried out in the Apalachicola valley.

Over the past 5–6 years (1985–1990), there has been a gradual change in the political atmosphere concerning the development and application of research to management problems in the Apalachicola system. This change is perceptible at local, state, and national levels. Local authorities are resisting further purchases of environmentally endangered lands in the basin. Meaningful ecological research in the Apalachicola system has been actively discouraged, especially when it is perceived to interfere with local municipal development. Due to both natural and human activities, the formerly productive oyster industry has been weakened, both as an economic entity and a political force. There is considerable pressure for municipal and agricultural development along the tri-river system, which will increasingly place pressure on freshwater resources. In short, there is a real possibility that the major preservation efforts in the 1970s and early 1980s will be compromised by the growing economic development of the region. The role of research in this area has been reduced accordingly. This follows a pattern that is evident throughout the state (Livingston 1991).

EXTRAPOLATION OF THE APALACHICOLA EXPERIENCE TO OTHER RIVER SYSTEMS

The various environmentally sensitive aspects of medium-sized rivers in the Gulf Coastal Plain are represented by the history of attempts to understand and manage the Apalachicola River resource. Toxic substances and organic enrichment are obvious problems in various areas, and have been well documented. However, the most pervasive impact on such rivers continues to be government-sanctioned and financed physical alterations that include damming, channelizing, diking, dredging, water diversion, and the removal of water for municipal and agricultural uses. In the most comprehensive documentation of such impacts on river–estuary systems, Darnell (1976) pointed out that cumulative, long-term impacts of construction activities along major river systems eventually cause irrevocable changes that adversely affect the natural habitat, productivity, and the unique biological attributes of the systems. Usually, research carried out to document such effects is inadequate in terms of scope and time. The management effort is often hampered by severely limited research information while the scale of the problem is seldom matched by the scale of the research effort. The complexity of system processes cannot be understood when research is limited to virtually inconsequential levels. Even when thorough research is done, the results are often ignored. The question of hydrological changes is crucial to any effective management of river estuaries. Yet, few studies use adequate hydrological information. Few studies have addressed the natural interactions of the various portions of the river system, so that changes in freshwater drainage areas are often translated into losses of associated estuarine productivity. There has been recent scientific interest in larger-scale studies involving riverine ecosystems (Meyer 1990). However, long-term, cumulative impacts of multifold habitat destruction are ignored, even though such impacts may have devastating economic implications (Livingston 1984). Without adequate scientific documentation, realistic management eventually becomes unfeasible when faced with politically expedient destruction of the resource by private concerns and government agencies.

The ultimate question of whether previous attempts of management in the riverine environment of the southeastern United States have been successful is answered by continuing trends. According to Niering (1985), at the time of early settlement of the United States, there were 83 million ha (215 million acres) of wetlands; today only 40 million ha (99 million acres) remain. Fewer than 2 million ha (5 million acres) (less than 20%) of the bottomland hardwood swamps of the Mississippi alluvial plain are left in their original state because of channelization and agriculture. By eliminating the natural flooding process, the productivity of fish populations alone is reduced by 98%. The only realistic way to salvage what remains of the Gulf Coastal Plain rivers is to develop an integrated mechanism for generation of scientific information that can be used for management purposes and to provide a vehicle for such information to be included in the decision-making process. As long as this process is dominated by short-sighted political decisions based on pressure from a few influential special interests, the destruction will continue. Poor planning by state and federal agencies responsible for research in such areas has exacerbated the naturally short-sighted management activities along the Gulf Coast. Without the comprehensive, system-wide research that continues to be shunned by scientists and research agencies, there is scant hope that a reasonable policy will be implemented for Gulf river basins.

REFERENCES

Adams, G. I. C. Butts, L. Stephenson, and W. Cook. 1926. Geology of Alabama. *Ala., Geol. Surv., Spec. Rep.*: 14:1–312.

Ager, L. A., C. Mesing, and M. Hill. 1985. *Fishery Study, Apalachicola Maintenance Dredging Disposal Site Evaluation Program*. Final Rep. Florida Game and Fresh Water Fish Commission. 73 pp.

Barkuloo, J. M. 1967. Florida striped bass, Florida Game and Freshwater Fish Commission. *Fish. Bull.* 4:1–24.

Barkuloo, J. M. 1970. Taxonomic status and reproduction of striped bass (*Morone saxatilis*) in Florida. *Tech. Pap. U.S. Fish Wild. Serv.* 1–16.

Bass, D. G. and D. T. Cox. 1985. River Habitat and Fishery Resources of Florida. In: W. Seaman, Jr. (ed.), *Florida Habitat And Fisheries Resources*. Florida Chapter, American Fisheries Society, pp. 121–187.

Clewell, A. F. 1977. Geobotany of the Apalachicola river region. In R. J. Livingston and E. A. Joyce, Jr. (eds.), *Proceedings of the Conference on the Apalachicola Drainage System*, Flor. Mar. Res. Publ. 26:6–15. Florida Department of Natural Resources, St. Petersburg, Florida.

Clewell, A. F. 1991. Florida rivers: the vegetational mosaic. In R. J. Livingston (ed.), *The Rivers of Florida*. New York: Springer-Verlag, pp. 46–62.

Cox, D. T. 1970. *Annual Progress Report for Investigations Project*. Florida Game and Freshwater Fish Commission (unpublished report).

Cox, D. T., and D. Auth. 1971–1973. *Annual Progress Report for Investigations Project. I. Upper Apalachicola River Study*. Florida Game and Freshwater Fish Commission (unpublished report).

Darnell, R. M. 1976. *Impacts of Construction Activities in Wetlands of the United States*, EPA-6OO/3-76-045. Corvallis, OR: U.S. Environ. Prot. Agency.

de la Cruz, A. A. 1978. Primary production processes: summary and recommendations. In R. E. Good, D. F. Whigham, and R. L. Simpson (eds.), *Freshwater Wetlands*. New York: Academic Press, pp. 79–86.

Eichholz, N. F. 1980. Osprey nest concentrations in northwest Florida. *Fla. Field Nat.* 8:18–19.

Elder, J. F., and D. J. Cairns. 1982. Production and decomposition of forest litter fall on the Apalachicola river flood plain, Florida. *Geol. Surv., Water-Supply Pap. (U.S.)* 2196:1–42.

Estevez, E. D., L. K. Dixon, and M. S. Flannery. 1991. West-coastal rivers of peninsular Florida. In R. J. Livingston (ed.), *The Rivers of Florida*. New York: Springer-Verlag, pp. 183–216.

Fernald, E. A. 1981. *Atlas of Florida*. Tallahassee: Florida State University Foundation.

Fernald, E. A., and D. J. Patten. 1984. *Water Resources Atlas of Florida*. Tallahassee: Florida State University.

Florida Game and Freshwater Fish Commission. 1982. *Lower Apalachicola River E.E.L. Management Plan*.

Fraser, T. F. 1991. The Lower Peace River and Horse Creek: Flow and water quality characteristics, 1976–1986. In R. J. Livingston (ed.), *The Rivers of Florida*. New York: Springer-Verlag, pp. 139–182.

Good, R. E., D. F. Whigham, and R. L. Simpson. 1978. *Freshwater Wetlands*. New York: Academic Press.

Heard, W. H. 1977. Freshwater mollusca of the Apalachicola drainage. In R. J. Livingston and E. A. Joyce, Jr. (eds.), *Proceedings of the Conference on the Apalachicola Drainage System*, Flor. Mar. Res. Publ. 26:20–21. Florida Department of Natural Resources, St. Petersburg, Florida.

Horel, G. J. 1960. Results of Fisheries Investigations and Recommendations for Reestablishment of a Sports Fishery in the Caloosahatchee River, Florida Game and Freshwater Fish Commission, Tallahassee, Florida.

Hughes, R. M. and J. M. Omernik. 1980. Use and Misuse of the Terms Watershed and Stream Order. In L. A. Krumholz (ed.), *The Warmwater Streams Symposium: A National Symposium on Fisheries Aspects of Warmwater Streams*. pp. 320–326.

Leitman, H. M., J. E. Sohm, and M. A. Franklin. 1983. Wetland hydrology and tree distribution of the Apalachicola river floodplain, Florida. *Geol. Surv., Water-Supply Pap. (U.S.)* 2196:1–52.

Leitman, S. F., L. Ager, and C. Mesing. 1991. The Apalachicola experience: environmental effects of physical modifications to a river for navigational purposes. In R. J. Livingston (ed.), *The Rivers of Florida*. New York: Springer-Verlag, pp. 217–241.

Livingston, R. J. 1981. River-derived input of detritus into the Apalachicola estuary. In R. D. Cross and D. L. Williams (eds.), *Proceedings of the National Symposium on Freshwater Inflow to Estuaries*, Publ. FWS/OBS-81/04. Washington, DC: U.S. Fish Wildl. Serv., pp. 320–331.

Livingston, R. J. 1983. *Resource Atlas of the Apalachicola Estuary*, Fla. Sea Grant Coll. Rep. No. 55.

Livingston, R. J. 1984. The ecology of the Apalachicola Bay system: an estuarine profile. *U.S. Fish Wildl. Serv.* FWS/OBS 82/05:1–148.

Livingston, R. J. (1988). *Field Verification of Bioassay Results at Toxic Waste Sites in Three Southeastern Drainage Systems*, Final Rep. Washington, DC: U.S. Environ. Prot. Agency.

Livingston, R. J. (ed.). 1991. *The Rivers of Florida*. New York: Springer-Verlag.

Livingston, R. J., and O. Loucks. 1978. Productivity, trophic interactions, and food web relationships in wetlands and associated systems. In *Wetland Functions and Values: The State of Our Understanding*, Publ. No. 26. St. Petersburg, FL: Am. Water Resour., pp. 101–119.

Livingston, R. J., N. P. Thompson, and D. Meeter. 1978. Long-term variation of organochlorine residues and assemblages of epibenthic organisms in a shallow north Florida (USA) estuary. *Mar. Biol. (Berlin)* 46:355–372.

Livingston, R. J., J. H. Epler, F. Jordan, Jr., W. R. Karsteter, C. C. Koenig, A. K. S. K. Prasad, and G. L. Ray. 1991. Ecology of the Choctawhatchee River system. In R. J. Livingston (ed.), *The Rivers of Florida*. New York: Springer-Verlag, pp. 241–269.

Mattraw, H. C., Jr., and J. F. Elder. 1984. Nutrient and detritus transport in the Apalachicola river, Florida. *Geol. Surv., Water-Supply Pap. (U.S.)* 2196-C:1–62.

Means, D. B. 1977. Aspects of the significance to terrestrial vertebrates of the Apalachicola river drainage basin, Florida. In R. J. Livingston and E. A. Joyce, Jr. (eds.), *Proceedings of the Conference on the Apalachicola Drainage System*, Flor.

Mar. Res. Publ. 26:37–67. Florida Department of Natural Resources, St. Petersburg, Florida.

Meeter, D. A., R. J. Livingston, and G. C. Woodsum. 1979. Long-term climatological cycles and population changes in a river-dominated, estuarine system. In R. J. Livingston (ed.), *Ecological Process in Coastal and Marine Systems*. New York: Plenum, pp. 315–338.

Meyer, J. L. 1990. A blackwater perspective on riverine ecosystems. *BioScience* 40:643–651.

Niering, W. A. 1985. *Wetlands*. Audubon Society Nature Guide, Chanticleer Press, Inc. New York. 638 pp.

Norwine, J. 1981. Precipitation, trends, and variability in the vicinity of the northwest Gulf Coast: 1900–1980. In R. D. Cross and N. G. Benson (eds.), *Proceedings of the National Symposium on Freshwater Inflow to Estuaries*, Vol. II. Washington, DC: U.S. Fish Wildl. Serv., pp. 322–334.

Olsen, S. J. 1968. Miocene vertebrates and north Florida shorelines. *Falarogeo., Palaeoclim., Falaroeocol.* 5:127–134.

Olinger, L. W., P. L. Rogers, R. L. Fore, B. L. Todd, B. L. Mullins, F. T. Blistenfield, and L. A. Wise, II. 1975. Environmental and Recovery Studies of Escambia Bay and Pensacola Bay System, Florida. U.S. Environmental Protection Agency, Atlanta, Georgia.

Patrick, R., and C. W. Reimer. 1966. *The Diatoms of the United States*, 2 vols., No. 13. Philadelphia, PA: Academy of Natural Sciences of Philadelphia.

Reiter, M. A. 1986. Interactions between the hydrodynamics of flowing water and the development of a benthic algal community. *J. Freshwater Ecology* 3:511–517.

Resh, V. H., and D. M. Rosenberg. 1984. *The Ecology of Aquatic Insects*. New York: Praeger.

Ross, S. T., J. A. Baker and K. E. Clark. 1987. Microhabitat partitioning of southeastern stream fishes: Temporal and spatial predictability. In W. J. Mathews and D. C. Heins (eds.), *Community and Evolutionary Ecology of North American Stream Fishes*. Norman, Oklahoma, Univ. Oklahoma Press, pp 42–51.

Stevenson, H. M. 1977. A comparison of the Apalachicola river avifauna above and below Jim Woodruff dam. In R. J. Livingston and E. A. Joyce, Jr. (eds.), *Proceedings of the Conference on the Apalachicola Drainage System*, Flor. Mar. Res. Publ. 26:34–36. Florida Department of Natural Resources, St. Petersburg, Florida.

U.S. Army Corps of Engineers. 1984. *1984 Water Assessment*. Apalachicola-Chattahoochee, Flint River Basin. Mobile, Alabama.

Valiela, I., and J. M. Teal. 1978. Nutrient dynamics: summary and conclusions. In R. E. Good, D. F. Whigham, and R. L. Simpson (eds.), *Freshwater Wetlands*. New York: Academic Press, pp. 259–266.

Vannote, R. L. 1981. The river continuum: a theoretical concept for analysis of river ecosystems. In R. D. Cross and N. G. Benson (eds.), *Proceedings of the National Symposium on Freshwater Inflow to Estuaries*, Vol. II. Washington, DC: U.S. Fish Wildl. Serv., pp. 289–304.

Vannote, R. L., G. W. Minshall, K. W. Cummins, J. R. D. Sedell, and C. E. Cushing. 1980. The river continuum concept. *Can. J. Fish. Aquat. Sci.* 37:130–137.

Wharton, C. M., W. M. Kitchens, E. C. Pendleton, and T. W. Sipe. 1982. *The Ecology of Bottomland Hardwood Swamps of the Southeast: A Community Profile*, FWS/0B5-81/37. Washington, DC: U.S. Fish Wild. Serv.

Winger, P. V. 1986. Forested wetlands of the southeast: Review of major characteristics and role in maintaining water quality. U.S. Fish and Wildlife Service Research Rpt., Pub. 163, Washington, D.C.

Winger, P. V., C. Sieckman, T. W. May, and W. W. Johnson. 1984. Residues of organochlorine insecticides, polychlorinated biphenyls, and heavy metals in biota from the Apalachicola River, Florida, 1978. *J. Assoc. Off. Anal. Chem.* 67:325–333.

Wolfe, S. H., J. A. Reidenauer, and D. B. Means. 1988. An ecological characterization of the Florida panhandle. U.S. Fish and Wildlife Service, Biological Report 88(12), Mineral Management Service, OCS Study/MMS 88-0063, pp 1–277.

Wooley, C. M. and E. J. Crateau. 1982. Observations of Gulf of Mexico sturgeon (*Acipenser oxyrhynchus* desotor) from the Apalachicola River, Florida. *Northeast Gulf Sci.* 5:57–59.

Wooley, C. M. and E. J. Crateau. 1983. Biology, Population estimates, and movement of native and introduced striped bass, Apalachicola River, Florida. *North Am. J. Fish. Man.* 3:383–394.

Yerger, R. W. 1977. Fishes of the Apalachicola River. In R. J. Livingston and E. A. Joyce, Jr. (eds.), *Proceedings of the Conference on the Apalachicola Drainway System*, Flor. Mar. Res. Publ. 26:22–33. Florida Department of Natural Resources, St. Petersburg, Florida

PART III
Lentic Systems

10 Small Impoundments and Ponds

RONALD G. MENZEL

United States Department of Agriculture, Water Quality and
Watershed Research Laboratory, Durant, OK 74701

CHARLES M. COOPER

United States Department of Agriculture, Sedimentation Laboratory,
Oxford, MS 38655

Farm ponds and other small impoundments are a common feature of the landscapes of the Southeastern United States. To the extent that they are small and isolated from other bodies of water, they are more affected by outside influences, both natural and artificial, than are larger impoundments. Therefore, the biotic communities of small impoundments exist in a more variable environment than those of larger impoundments. Furthermore, the management or mismanagement of small impoundments may have great impact on their biotic communities.

The size of impoundments considered in this chapter has arbitrarily been restricted to those smaller than 16 ha (0.16 km^2 or 40 acres) in surface area. The reason for this size limitation is that an inventory of land and water resources compiled by the United States Department of Agriculture (USDA) (1981) separates inland water bodies larger and smaller than 16 ha. Inland water bodies include natural lakes as well as impoundments. Farm ponds were inventoried separately without regard to surface area. Although a few farm ponds in the 12 southeastern states may be larger than 16 ha, their average surface area was only 0.45 ha.

In 1977, there were nearly one million farm ponds in these 12 states. The area covered by farm ponds was 0.28% of the total land area, ranging from 0.07% in Florida to 1.08% in Mississippi (Fig. 1). Similarly, the area in small inland water bodies was 0.47%, ranging from 0.13% in West Virginia to 0.90% in Florida.

The purposes served by small impoundments determine the kind and level of management they receive, possibly setting limits on the biotic communities that develop. Small impoundments are constructed by local governments for community water supplies, recreational areas, or flood control. Occasionally, small impoundments result from road construction or surface mining activities. The majority of farm ponds have been constructed for control of soil

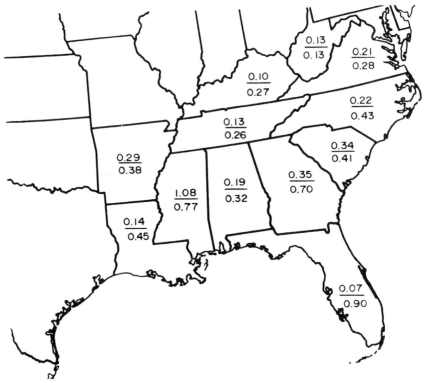

FIGURE 1. Percentage of total land area occupied by farm ponds (upper value) and small inland water bodies <0.16 km² (lower value) in the southeastern United States (U.S. Department of Agriculture, 1981).

erosion and flooding with assistance from the United States Soil Conservation Service. According to the United States Department of Agriculture (1981), 75% of the farm ponds in the 12 southeastern states are also used for recreational fishing or home-use aquaculture. In addition, many are used as sources of water for livestock or irrigation. About 10% of the farm ponds in Georgia were used for irrigation in 1977 (Bruce et al. 1980). The total irrigated acreage and the number of ponds used for irrigation increased severalfold between 1960 and 1977, and is expected to continue increasing across the southeastern United States (Bruce et al. 1980). Management practices, which may include changing the water level, controlling or removing vegetation, harvesting or stocking fish, etc., vary according to the purposes of each impoundment.

PHYSICAL ENVIRONMENT

The physical and chemical characteristics of water and sediments in small impoundments vary considerably in the southeastern United States, depending on the local hydrologic, climatic, and geologic situation. In addition, the inflow of water, sediment, and chemicals changes with different land uses and management practices in the watershed.

A general description of 9 ponds on which extensive biological studies have been made, and which will be referred to frequently in this chapter, is given in Table 1. The pond locations are shown in Fig. 2. These ponds represent a wide variety of small impoundments typical of those in the southeastern United States. Although one pond, Golf Course Pond in Denton County, Texas, is peripheral to the southeastern states, it is included because the very thorough study of its ecosystem allows useful comparisons with biological communities of the other ponds.

Hydrology

Most small impoundments and farm ponds are formed by damming a small, perhaps intermittent, stream. The watershed must be large enough to provide water for the intended uses and to compensate for evaporation and seepage losses. A watershed area at least 10 times the surface area is recommended for farm ponds in most of the southeastern United States (USDA 1982). Minimum water depths of 1.5 m in wet areas and 3 m in dry subhumid areas are recommended.

The average annual runoff over most of the southeastern United States ranges from 38 to 64 cm (U.S. Department of the Interior, 1970). Annual runoff is higher, up to 100 cm, in the southern Appalachians, and lower, down to 12 cm, along the Atlantic Coastal Plain, in Florida, and westward toward Oklahoma and Texas. Annual lake evaporation increases toward the southwest in the region, from less than 100 cm in the West Virginia mountains to 150 cm in Florida and along the Gulf Coast, and 180 cm in eastern Oklahoma and Texas (Water Resources Council, 1968).

Surface runoff is often not the main source of water for small impoundments in the southeastern United States. In the Appalachian and Piedmont areas, runoff is commonly augmented or even exceeded by flow from springs or ground water. For example, springs provided noticeable inflow to 14 of 37 farm ponds in the western Piedmont of Georgia (Gunn 1974). In Florida and in the flood plains of large rivers, excavated ponds may be fed mainly by ground water. Groundwater inflow and seepage losses were major parts of the hydrologic budget for mine reclamation ponds in central Florida (Boody et al. 1985).

Water levels in small impoundments of the southeastern United States are usually fairly constant. Of 37 ponds studied in western Georgia over a 4-year period, 16 maintained a nearly constant water level (Gunn 1974). Fluctuations of up to 30 cm were noted on 12 ponds, and from 35 to 60 cm on 9 ponds. Evaporation during the summer and fall may lower the water level as much as 1 m in Alabama ponds (Boyd 1979). When small impoundments are used for irrigation or other uses with a high demand, water levels may be lowered quickly.

TABLE 1 General Description of Small Impoundments and Farm Ponds Used in Ecosystem Studies

Identification	Location	Purpose	Year Filled	Impoundment Area (m² × 10³)	Volume (m³ × 10³)	Annual Outflow (m³ × 10³)	Watershed Area (m² × 10³)	Description	References
Spruce Knob Lake	Randolph Co., WV	Fishing, recreation	1952	105	220	920	1070	Mixed forest, sandstone	Labaugh 1978
Bays Mountain Lake	Sullivan Co., TN	Nature preserve, former city reservoir	1916	150	500	n.a.[a]	5250	Forested mountaintop	Crowley and Johnson 1982, Johnson and Crowley 1980
Botany Pond	Orange Co., NC	Botanical garden	1966	32	65	n.a.	400	Deciduous forest	Gilinsky 1981, Meyer and McCormick 1971
Risher Pond	Barnwell Co., SC	Abandoned farm pond	1950	13	20	n.a.	n.a.	Mixed forest	Polisini et al. 1970
Morris Pond	Panola Co., MS	Sediment detention	1972	11	17	76	320	55% pasture 45% cultivated	Dendy and Cooper 1984
Lago Pond	Clarke Co., GA	Fishing	1955	12	28	n.a.	n.a.	Pasture, woodlot, residence	Welch 1968
Golf Course Pond	Denton Co., TX	Water hazard, irrigation, former stock tank	1947	10	15	n.a.	100	Fairways, greens	Childress et al. 1981, Kelly et al. 1978
Lake Jasmine	Lee Co., FL	Water level control	1965	12	23	n.a.	n.a.	Residential area	Cassani and Caton 1985
Agrico No. 6	Polk Co., FL	Mine reclamation	1975	44	123	17	90	Pasture	Boody et al. 1985

[a]n.a., not available.

FIGURE 2. Locations of small impoundments and farm ponds used for intensive ecosystem studies.

Inflows of sediment may greatly reduce the capacity of small impoundments. Impoundments with less than 4000 m³ capacity in the southeastern United States lost from 5 to 10% of their capacity per year due to sediment deposition during the first 5 years after impoundment (Dendy and Champion 1978). The loss of capacity was slower for older and larger impoundments, usually less than 1% per year for impoundments older than 20 years or with more than 250 000 m³ capacity.

Most of the inflowing sediment is retained even in small impoundments. Dendy (1982) reported trap efficiencies for 6 small impoundments varying greatly in size, shape, and location in the southeastern United States. They are listed in Table 2 in order of decreasing water residence time, which ranged from 81 to 3.6 days. The trap efficiency decreased slightly in the same order, from 95 to 82%, which indicates that residence times were long enough for all except fine clay particles to be deposited. Fine sediments, generally less than 8 μm, were fairly evenly distributed in the impoundments. Coarse sediments, silt size and larger, were usually concentrated near the upstream end at or near the permanent pool elevation.

TABLE 2 Retention of Sediments in Small Impoundments of the Southeastern United States

Identification	Location	Watershed Area (m² × 10³)	Impoundment Area (m² × 10³)	Volume (m³ × 10³)	Annual Outflow (m³ × 10³)	Sediment Annual Outflow Mg	Annual Deposit Mg	Trap Efficiency (%)
Y-S-122	MS	850	29	49	220	69	552	89
Six Mile No. 6	AR	10720	163	372	3570	266	4766	95
Plum Creek No. 4	KY	3890	56	113	1600	234	2486	91
Salem Fork No. 11-A	WV	750	4	9	390	26	190	88
N. Fk. Broad No. 14	GA	3110	24	28	1820	229	1269	85
Third Creek No. 7A	NC	12540	43	58	5880	426	1941	82

Source. Dendy (1982).

Climate

The climate is moist temperate over most of the region, becoming subtropical in southern Florida and subhumid in eastern Oklahoma and Texas (*Climates of the States* 1974). Precipitation is quite evenly distributed among the seasons, averaging 25–50 cm in winter, 25–40 cm in spring, 25–50 cm in summer, and 20–25 cm in fall over most of the area. However, winter precipitation averages somewhat less than 25 cm in eastern Oklahoma, eastern Texas, and southern Florida. Also, summer rainfall usually exceeds 50 cm along the Gulf Coast and in most of Florida, and fall rainfall is usually more than 25 cm in the coastal and mountain areas.

The surface water temperature in small impoundments approximates average daily air temperatures because of their relatively small volume and shallow depth. Heat exchange occurs quite rapidly. For example, mean temperatures in Morris Pond in north central Mississippi may fluctuate 10 °C in a week (C. M. Cooper, unpublished data, Oxford, MS, 1989). The coolest temperatures in the southeastern United States are found in the West Virginia mountains, where average daily air temperatures range from − 2 °C in January to 20 °C in July (*Climates of the States* 1974). Spruce Knob Lake in Randolph County, West Virginia, had ice cover for only 10 days in the winter of 1973–74, but for nearly 4 months in 1974–75, with ice up to 0.5 m thick (Labaugh 1978). The maximum water temperature was 22 °C in July of each year. In the southern tier of states, from Georgia to eastern Texas, average January air temperatures are 5 °C and higher. Ice may cover small impoundments rarely or occasionally for a few days in this area. The minimum water temperature in mine reclamation ponds in central Florida was 14 °C (Boody et al. 1985). Average July air temperatures are near 30 °C in most of the southern states. Mid-afternoon surface water temperatures may approach 35 °C at this time.

Most small impoundments in this region stratify in response to surface heating during the spring and summer months. Stratification may be quite stable in ponds deeper than 2 m and can lead to anoxia in the bottom water. In fact, all of the impoundments listed in Table 1, except Risher Pond, become anoxic in their deepest parts during the summer. Periodically, summer stratification can break down due to strong winds and rain commonly associated with thunderstorms.

Watershed Characteristics

Characteristics of the watershed can have great impact on small impoundments because they have a large ratio of shoreline length to water surface area. The degree of impact from different land uses and management will vary with their proximity to the impoundment. Trees surrounding an impoundment may provide a significant amount of allochthonous energy input to the aquatic ecosystem through leaf fall. Cropland is generally subject to greater erosion, and thus may provide greater amounts of sediment, nutrients, and pesticides to small impoundments than do grassland or forest. In many cases, livestock have access to farm ponds, disturbing shoreline soils and destroying or consuming vegetation.

There are no comprehensive statistics concerning the watersheds of small impoundments in the southeastern United States. Because of the wide distribution of small impoundments, it is likely that land uses in their watersheds are representative of the whole region. More than half of the region is forest, including grazed forest (USDA 1984). Cultivated cropland occupies 16%, pasture 13%, and other uses 15% of the area. The division of land uses in individual states is shown in Fig. 3.

Gunn (1974) studied vegetation associated with 37 farm ponds in the western Piedmont of Georgia, and noted hardwood or pine adjacent to 22 of them. He noted cultivated fields near 8 ponds and pastures near 5. Livestock had access to 16 ponds for watering. Thus, the land uses near these ponds are typical for the southeastern United States. The watersheds of the 9 study ponds listed in Table 1 include these land uses as well as a residential area and a golf course.

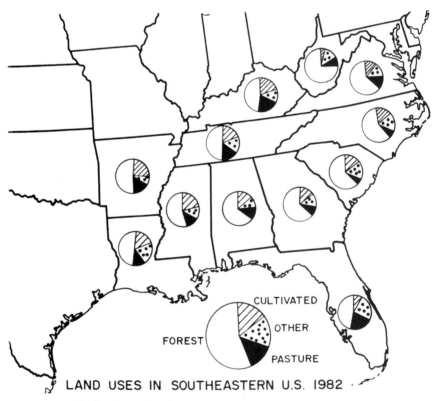

FIGURE 3. Distribution of land uses in the southeastern United States, 1982 (U.S. Department of Agriculture, 1984).

The intensity of management of watersheds and impoundments depends on the land uses and the purposes of the impoundment. Many impoundments are in relatively unmanaged forests. Examples include Spruce Knob Lake in Monongahela National Forest, West Virginia (Labaugh 1978), Risher Pond on the U. S. Atomic Energy Commission Savannah River Plant near Aiken, South Carolina (Polisini et al. 1970), and Bays Mountain Lake, near Kingsport, Tennessee (Johnson and Crowley 1989). Farm ponds are more likely to be in watersheds with more intensively managed pastures and cropland. An example is Morris Pond, which receives drainage from 17.8 ha of pasture and 14.6 ha of cultivated cropland (Dendy and Cooper 1984).

In general, farm ponds studied by Gunn (1974) in the western Georgia piedmont received little management. Most were stocked for fishing, but fishing use declined in the older ponds, possibly because of lack of management. Most dams were initially seeded to grass and were kept clear of woody vegetation. The entire banks were mowed on a few ponds, and limited areas around other ponds were mowed to facilitate access for swimming or boating. Livestock were excluded from about half of the ponds, but had free access to the others.

Chemical Characteristics

Surface waters of the southeastern United States are classified predominantly as soft. Over most of the region, concentrations of dissolved solids are less than 120 mg/L, and the hardness is less than 60 mg/L as $CaCO_3$ (Geraghty et al. 1973). Higher concentrations are found in the limestone regions of the Appalachians, Kentucky, central Tennessee, northern Alabama, central and southern Florida, as well as in the subhumid regions of eastern Oklahoma and Texas. Even in these regions, the concentrations rarely exceed 350 mg/L of dissolved solids and 120 mg/L of hardness.

Surface runoff waters are usually slightly acidic because they contain carbonic acid, which is formed when carbon dioxide is dissolved from the soil and atmosphere. The low base content of most soils in the southeastern United States is insufficient to neutralize this acidity. The pH of impounded water increases somewhat as photosynthesis removes dissolved carbon dioxide from the water. For the impoundments listed in Table 1, pH values ranging from 4.1 to 10.0 were reported. In general, pH values above neutrality were observed during active photosynthesis.

Boyd (1974) reported that many farm ponds in Alabama have hardness less than 20 mg/L. He recommended applying lime for fish production in such ponds. Liming increases the buffering capacity of the water, thus avoiding extreme pH values, and decreases the fixation of phosphorus by acid sediments.

Nutrient concentrations in a number of small impoundments and farm ponds are summarized in Table 3. For comparison, geometric mean concentrations are included from a eutrophication study of European and North American lakes and impoundments (Organization for Economic Co-operation

TABLE 3   Mean Concentrations of Nutrients in Some Small Impoundments and Farm Ponds in the Southeastern United States

Identification	Number of Ponds	Samples per Pond	Sampling Period (months)	Nutrient Concentrations (mg/L)				Trophic Status	References
				Total P	Ortho-P	Total N	Inorganic-N		
Spruce Knob Lake, WV	1	47	27	.087	.001[a]	—	.159[a,b]	Eutrophic	Labaugh 1978
Botany Pond, NC	1	16	6	—	.0015	—	.0004[b]	Oligotrophic	Meyer and McCormick 1971
Risher Pond, SC	1	48	12	.005	—	.10	—	Oligotrophic	Polisini et al. 1970
Morris Pond, MS	1	545	24	.325	.165	—	.280	Eutrophic	Cooper and Knight 1990
Golf Course Pond, TX	1	14	12	—	.10	—	.25[b]	Eutrophic	Kelly et al. 1978
Lake Jasmine, FL	1	36	36	.60	.4	—	1.00	Eutrophic	Cassani and Caton 1985
Mine reclamation, FL	9	4	10	.37	.212	1.28	0.268	7 Eutrophic 2 Mesotrophic	Boody et al. 1985
Wooded ponds, AL	34	1	c	.092	.007	—	1.27	Not determined	Boyd 1976
Pasture ponds, AL	53	1	c	.128	.015	—	.203	Not determined	Boyd 1976
Fertilized ponds, AL	26	1	c	.185	.019	—	.198	Not determined	Boyd 1976
Catfish ponds, MS	25	2	7	.71	.116	4.98	.82	Eutrophic	Tucker and Lloyd 1985
OECD lakes (number of lakes in parentheses)				.042 (101)	.015 (92)	1.23 (57)	.58 (89)	Various	OECD 1982

[a]Summer sampling only.
[b]Nitrate-N form only.
[c]Four of these ponds were sampled 3 additional times each during one summer.

and Development, 1982). Except for total phosphorus, the concentrations are similar in both sets of water bodies. High total phosphorus concentrations in the Florida mine reclamation ponds and in the Mississippi catfish ponds are attributed to phosphate mine wastes and to supplemental feeding, respectively. The concentrations in wooded and pasture ponds in east-central Alabama, which were not fertilized, are probably more typical for the southeastern United States.

Pesticides or their residues occasionally enter small impoundments in runoff. Since persistent organochlorine pesticides are no longer used for field applications, problems with residues have decreased. Major factors controlling the movement of pesticides in runoff are persistence, water solubility, formulation of the pesticide, and timing of a runoff event after application of the pesticide (Wauchope 1978). For the majority of commercial pesticides, total runoff losses are less than 0.5% of the amounts applied, unless severe runoff occurs within 1 or 2 weeks after application. However, wettable-powder formulations of herbicides applied to the soil surface may lose up to 5% because of easy washoff of the powder. Pesticides with solubilities of 10 mg/L or higher are lost mainly in the water phase of runoff so that erosion control practices have little effect on their losses.

PLANT COMMUNITIES

The plant communities of small impoundments consist of submersed, floating, and emergent plants. Submersed plants include phytoplankton, periphyton, and some aquatic macrophytes. Bacteria populate the sediments and water column, and will be considered with submersed plants. Floating plants may be free-floating or may have root systems attached to shallow sediments. Whereas phytoplankton typically contribute most of the primary production in large impounds, the submersed or emergent macrophytes often assume this role in small impoundments, particularly if a relatively large proportion of the pond volume consists of shallow water.

Submersed Plants

The submersed macrophyte, periphyton, and phytoplankton communities are closely associated in a small impoundment. Their major interactions involve competition for light and nutrients, with competition for light probably critical in more cases. Boyd (1975) reported that submersed macrophytes did not grow at depths greater than twice the Secchi depth visibility in ponds at Auburn, Alabama. McVea and Boyd (1975) found that phytoplankton abundance was much reduced in ponds where from 5 to 25% of the surface area was covered with water hyacinth (*Eichhornia crassipes*), a floating macrophyte which provides dense shade. The shading effect of phytoplankton can also suppress the growth of periphyton attached to macrophytes or bottom surfaces (Boyd 1973). Suspended clay particles may have a similar shading effect.

On the other hand, submersed macrophytes that have a sparse growth habit and do not provide dense shade seem to compete ineffectively with phytoplankton. For example, Terrell (1976) found that phytoplankton did not increase in ponds at Fort Gordon, near Augusta, Georgia, when littoral growths of spikerush (*Eleocharis* sp.) and bladderwort (*Utricularia* sp.) were controlled by stocking grass carp (*Ctenopharyngodon idella*). Terrell speculated that these rooted macrophytes were acting as nutrient "pumps," moving nutrients from the sediments into the water column where they could be used by phytoplankton. If so, eliminating the macrophytes eliminated a source of nutrients for phytoplankton. It is also possible that clay turbidity limited the growth of phytoplankton in these ponds (Terrell 1975).

Phytoplankton   Although all of the major groups of algae are found in small impoundments, the species distribution is likely to differ from that of large impoundments. A small impoundment is influenced more by benthic algae and periphyton of the littoral zone, which may detach and become part of the planktonic population. The abundance of truly planktonic forms of algae, such as centric diatoms (Bacillariophyta), may be correspondingly less.

There is often a seasonal progression of phyla in the phytoplankton community, which is usually dominated by diatoms or green algae (Chlorophyta) during winter and spring, and blue-green algae (Cyanophyta) during summer and fall. Such was the case in Golf Course Pond, Denton, Texas, where diatoms, maximal in late spring, were followed by green and then blue-green algae, which dominated in late summer and fall (Kelly et al. 1978). A similar cycle occurred in nine small mine reclamation ponds in Polk County, Florida (Boody et al. 1985). Green and blue-green algae were dominant, with blue-green algae more numerous in the spring and summer (Fig. 4). The greatest number of diatoms were observed in the winter, but they were a small part of the phytoplankton population in the Florida ponds.

Blue-green algae often become dominant in small impoundments when they are enriched by nutrients. Such was the case in Alabama and Georgia ponds that were fertilized with nitrogen and phosphorus to increase fish production (Boyd 1973, Welch 1968). Genera that occasionally formed heavy blooms under these conditions included *Anabaena*, *Anacystis*, *Aphanizomenon*, *Oscillatoria*, *Raphidiopsis*, and *Spirulina*. The dominance of blue-green algae is even more commonly noticed when ponds receive organic manures such as runoff from livestock feeding areas.

On the other hand, blue-green algae were absent from Botany Pond in North Carolina (Meyer and McCormick 1971), which was newly formed and low in nutrients. The plankton community in Botany Pond consisted of a base flora, species present almost continuously throughout the year, and ephemeral flora, species occasionally present in high numbers. The base flora included the green algae *Oocystis lacustris* and *Ankistrodesmus falcatus*, the euglenoid (Euglenophyta) *Trachelomonas volvocina*, and the dinoflagellate (Pyrrhophyta) *Peridinium gatunense*. The ephemeral flora included the diatoms *Cy-*

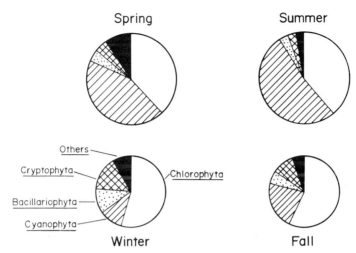

FIGURE 4. Seasonal abundance of phytoplankton phyla in nine small mine reclamation ponds in Polk County, Florida. (Data from Boody et al. (1985).) Area of circles is proportional to cell numbers.

clotella stelligera and *Fragilaria* sp., and the yellow-brown algae (Chrysophta) *Dinobryon bavaricum* and *D. sertularia.*

Macrophytes   Submersed macrophytes are generally found rooted in sediments at a depth of from 1 to 4 m of water, depending on light penetration. This may include most or all of the area of many small impoundments. For example, more than 90% of the bottom area of Lake Jasmine, Florida, with a maximum depth of 3 m was infested with hydrilla (*Hydrilla verticillata*) before an experiment on chemical and biological weed control was undertaken (Cassani and Caton 1985). In Bays Mountain Lake, Tennessee, 55% of the bottom area lies within the 1- to 4-m depth range and supports a sparse stand of American wildcelery (*Vallisneria americana*), which is overgrown by slender naiad (*Najas flexilis*) in late summer (Johnson and Crowley 1980). Although several submersed macrophyte species occurred in the nine mine reclamation ponds in Florida, hydrilla was the most extensive, occupying up to 20% of the bottom area in one pond (Boody et al. 1985).

The floating-leaved or floating macrophytes are not so great a problem in small impoundments as they frequently are in large impoundments. However, if they become established with an adequate nutrient supply, they may rather quickly cover the entire water surface of a small impoundment, greatly reducing photosynthesis and increasing respiration in the water column. The resulting reduction in dissolved oxygen concentration may eliminate fish from the impoundment. Duckweed (*Lemna* spp.), spatterdock (*Nuphar* spp.), and yellow lotus (*Nelumbo* spp.) are widely distributed throughout the southeastern United States. Water hyacinth and alligatorweed (*Alternanthera phi-*

loxeroides) are restricted to the freeze-free zone along the Gulf Coast. Fortunately, the relative isolation of most small impoundments from permanent streams and canals helps to exclude many floating macrophytes.

Microbial Community   Fungi and bacteria are the main decomposers of organic matter in small impoundments. Their activities are evident in the oxygen demand of sediments and hypolimnetic waters. However, there are very few studies describing decomposer populations or their activities in the southeastern United States. Bacterial counts in several West Virginia farm ponds showed that more bacteria were present in the surface few centimeters of sediment than in the 2- to 3-m-deep water column (Wilson et al. 1966). The presence of photosynthetic bacteria in pond sediments was demonstrated by Sylvester (1972). He also showed that more bacteria were present in pond water immediately following runoff events. Whether this reflected the transport of microorganisms in runoff, or transport of nutrients for growth of those already present in the ponds, was not determined.

Respiration rates have been determined in the sediments and water column of fish production ponds in Alabama (Boyd 1979). Benthic respiration rates ranged from 10 to 100 mg O_2/m^2 h^{-1}. The respiration rates of plankton and bacteria in the water ranged from 20 to 700 mg O_2/m^3 h^{-1}. In addition, respiration by catfish (*Ictalurus punctatus*) stocked at a rate of 2500 kg/ha was calculated to be 126 mg O_2/m^2 h^{-1}. Thus, in a heavily stocked and fed pond, more oxygen may be consumed in the water column than in the sediments.

Emergent Plants

Emergent aquatic and semiaquatic plants, such as cattails (*Typha* spp.) and willows (*Salix* spp.), are strikingly visible along the shoreline of many small impoundments, particularly where the water depth is less than 1 m. In addition to serving as an important source of energy materials for the impoundment, emergent plants provide food and habitat for numerous birds and terrestrial animals. Ponds fill with sediments as they age, and the area occupied by emergent plants usually increases.

Gunn (1974) described the changes in littoral vegetation with increasing age of 37 farm ponds in western Georgia. The water sources and general uses of the ponds described are typical of much of the southeastern United States. Pond age ranged from 1 to 95 years (median of 11 years) and pond area from 0.2 to 1.6 ha (median of 0.4 ha). In the first 2 years, the rather sparse vegetation consisted of herbaceous annuals and perennials growing at the pond margins with a few growing in shallow water. Dominant species included sedge (*Carex* sp.), spikerush, rush (*Juncus* sp.), beggartick (*Bidens* sp.), and marsh-purslane (*Ludwigia* sp.). In the next few years, the amount of vegetation increased and included woody species, such as red maple (*Acer rubrum*), hazel alder (*Alnus serrulata*), sweetgum (*Liquidambar styraciflua*), blackgum tupelo (*Nyssa sylvatica*), and black willow (*Salix nigra*). After about 8 years many species of woody plants were well established, with hazel alder along the pond margins being prominent. Nonwoody vegetation was abundant, with sedges and rushes growing along the pond margins on the banks and in the water. Watermilfoil (*Myriophyllum* sp.) and stonewort (*Chara* sp.) occasionally dominated the farther reaches of water.

Ponds from 8 to 24 years old were characterized by the same tree species as younger ponds, but the trees were larger. Black willow and blackgum tupelo were more abundant on the shores as hazel alder became less prominent. The water near the shores was dominated by species of rush, sedge, and watermilfoil. The older ponds in this group had accumulated much sediment and developed some characteristics of swamp ponds. Parrotfeather (*Myriophyllum brasiliense*) and stonewort species frequently dominated large areas of the water with large clumps of common rush, common arrowhead (*Sagittaria latifolia*), and common cattail (*Typha latifolia*) growing near the shores and sometimes across the pond.

Ponds older than 25 years were usually heavily sedimented and shallow with many characteristics of a swamp. The pond margins were virtually covered with woody vegetation. Trees, shrubs, and emergent herbaceous species grew across much of the aquatic area. Pondweed (*Potamogeton* spp.) and parrotfeather often grew completely across the ponds.

The small impoundments listed in Table 1 exhibit trends similar to those observed by Gunn. Botany Pond at 12 years of age had developed a common cattail bed at its shallow end with spikerush clumps in slightly deeper water (Gilinsky 1981). Risher Pond at 18 years of age was completely surrounded with overhanging vegetation, including black willow, tulip poplar (*Liriodendron tulipifera*), and speckled alder (*Alnus rugosa*) (Polisini et al. 1970). Somewhat larger impoundments, such as Spruce Knob Lake and Bays Mountain Lake, were less quickly affected by encroaching vegetation (Labaugh 1978, Johnson and Crowley 1980).

Productivity of Plant Communities

The available measurements indicate that small impoundments of the southeastern United States are quite productive (Table 4). Primary production rates greater than 1 g C/m^2 day^{-1} are typical of eutrophic conditions, whereas rates less than 0.3 g C/m^2 day^{-1} are considered oligotrophic (Wetzel 1975). By these criteria, Spruce Knob Lake and Risher Pond are borderline oligotrophic and the other ponds in Table 4 are strongly eutrophic. The major factors controlling productivity of the submersed plant community are temperature, light intensity, and nutrient concentrations. The low production rate in Risher Pond resulted from very low nutrient concentrations (Table 3). In Spruce Knob Lake, lower temperatures associated with its higher elevation (1173 m) and more northern location limited productivity. Production in Spruce Knob Lake was very low from mid-October through April of each year, when the water temperature did not exceed 10 °C (Labaugh 1978).

High production rates in Lago Pond and Golf Course Pond reflected high nutrient concentrations and warm temperatures. Although nutrient concen-

TABLE 4 Comparison of Primary Production in Small Impoundments and Farm Ponds of the Southeastern United States

Identification	Method(s)	Production Measurement		Comments	References
		Reported Value(s)	Converted Value(s) $(gC/m^2\ day^{-1})$[a]		
Spruce Knob Lake	[14]C fixation	$100\ gC/m^2\ year^{-1}$ $109\ gC/m^2\ year^{-1}$	0.29	Observed over 2 years + 1 summer; hypolimnetic aeration during last 2 summers	Labaugh 1978
Botany Pond	Diurnal O_2	$2.25\ gC/m^2\ day^{-1}$	2.25	Young pond, diverse oligotrophic algal population	Meyer and McCormick 1971
Risher Pond	Light–dark bottle Light–dark bottle	$1.31\ gC/m^2\ day^{-1}$ $0.5\ mg\ O_2/L^1\ day^{-1}$	1.31 0.28	Oligotrophic; only small differences in productivity with depth	Polisini et al. 1970
Lago Pond	Diurnal O_2	$7389\ kcal/m^2\ year^{-1}$	2.02	Fertilized pond; summer blue-green algal bloom, little emergent vegetation	Welch 1968
Golf Course Pond	Diurnal O_2	$22.0\ kcal/m^2\ day^{-1}$	2.20	Significant production from algal mats and attached vegetation	Kelly et al. 1978
Lake Jasmine	Biomass regrowth	$35–42\ g/m^2\ day^{-1}$	1.94	Hydrilla regrowth in exclosures during two 6-month periods	Cassani and Caton 1985

[a]Conversion factors were $2.67\ g\ O_2 = 10\ kcal = 1\ gC$. Fresh weight biomass was converted on the basis of 1 kg fresh wt = 100 g dry wt = 50 gC.

trations were not measured in Lago Pond, it was fertilized regularly for fish production (Welch 1968). High production rates were measured in Botany Pond during the first year after filling (Meyer and McCormick 1971), and may not have been sustainable considering the low levels of soluble nutrients. The extensive nutrient concentration data from Alabama ponds (Table 3) indicate that their productivity would be within the range shown in Table 4.

There are very few data on productivity of emergent plants in small impoundments of the southeastern United States. However, Boyd and Hess (1970) found that average standing crops of cattail shoots were similar in ponds, ditches, and swamps across a large part of this area, with values of 925, 995, and 875 g/m^2, respectively. Assuming that the shoot standing crop represented material regenerated annually, the average cattail productivity in the ponds approximated 0.13 g C/m^2 day^{-1}. This is lower than the productivity of submersed plants shown in Table 4, and may result from low soil fertility. The factor most closely related to standing crop was the dissolved phosphate concentration in the water (Boyd and Hess 1970).

ANIMAL COMMUNITIES

Animal communities can also be divided into two arbitrary groups: those associated with substrate and those that live in the water column. Many animals occupy different habitats during various feeding or life cycle stages. Bottom substrates provide habitat for numerous invertebrates, while macrophytes in the littoral zone provide important habitat for both invertebrates and vertebrates. Pond invertebrates consist mainly of zooplankton and insect larvae, whereas vertebrate communities are composed of fish, reptiles, birds, and other emigrants. Domestic animals, especially livestock, impact the littoral zones of many farm ponds.

Invertebrates

Zooplankton and benthic fauna, the focus of numerous pond studies in the southeastern United States, are composed mainly of primary consumers, an essential component of small impoundment food webs. In addition to consuming bacteria, algae, and debris, they also can assume a secondary consumer role by preying on each other. In turn, they become food for many pond vertebrates. Studies on invertebrates in southeastern ponds have generally focused on specific taxa and their role in aquatic food webs. Only a few studies have evaluated the role of the invertebrate community in energy transfer within pond ecosystems (see Representative Aquatic Communities).

Rotifers and protozoans are sometimes neglected because they generally comprise a small fraction of total pond biomass. However, they are numerically abundant, reproduce rapidly, and may periodically contribute substantially to energy flow in pond ecosystems. Unfortunately, since there has historically been a lack of proper methodology, density estimates are not always comparable and may have large errors. Pandapus Pond, a 3.2-ha impoundment in the Appalachian Highlands near Blackburg, Virginia, is an example of a pond that supported a large rotifer production (660 mg/m^3 year^{-1}), dominated by *Polyarthra vulgaris*, *Keratella cochlearis*, and *Kellicottia bostoniensis* (Knauer and Buikema 1984). Rotifer populations averaged 200/L over an annual cycle in reclaimed lakes of the Florida phosphate mining region (Boody et al. 1985). Prevalent species were *Polyarthra vulgaris*, *Keratella cochlearis*, *K. gracilenta*, and *Brachionus havanaensis*. Rotifers in Morris Pond, in the loess hills of Mississippi, generally were composed of *Polyarthra* sp., *Keratella* sp., and *Filina* sp., but total rotifer counts never reached 200/L (C. M. Cooper, unpublished data, Oxford, MS, 1989). Other miscellaneous invertebrates that inhabit southeastern ponds include freshwater sponges and bryozoans. Although most rotifers and microscopic invertebrates qualify as herbivores, they generally play a minor role in food chains of pond ecosystems.

The Cladocera, mainly consisting of microcrustacea, and the Copepoda are major biomass and energy flow components of farm pond zooplankton communities. Since at least 80% of the energy flow of a pond beyond the primary producer level passes through the detrital food chain, these filter-feeding crustacea constitute one of the major trophic links between primary producers and fish.

Densities of crustacea are cyclic, with limitations imposed by several external factors, including food supply, substrate availability (for littoral species), predation, and interactions between temperature and photoperiod. Crustacean plankton communities for seven impoundments in southeastern West Virginia were grouped by multivariate cluster analysis (Janicki et al. 1979). The four most productive impoundments (based on mean chlorophyll concentration >24 mg/m^3) clustered as a unit, while the three least productive were nearly as different from each other as they were from the main group. Mean crustacean abundances were 432 000 and 182 000 individuals/m^2 in the main cluster and other ponds, respectively. The smallest impoundment (<0.16 km^2) in each group had the highest crustacean abundance (521 000 and 353 000/m^2). *Bosmina longirostris* and *Daphnia parvula* were dominant in the more productive small impoundment, and *D. parvula* and *Mesocyclops edax* were dominant in the less productive one.

Lemly and Dimmick (1982), in a study of littoral zooplankton of three North Carolina lakes, found distinct taxa of Cladocera and Copepoda occupying the littoral zone. These species exhibited seasonal dynamics largely different from limnetic fauna that appeared to be more closely related to intrazooplankton predation than climatic or chemical variables. Copepoda and Cladocera numbers ranged from 100 to 500 000 individuals/m^3.

The dominance of crustacean species in farm pond limnetic zones is often predator controlled (Brooks and Dodson 1965). When predation from planktivorous fish is of low intensity, the small planktonic crustacean herbivores will be competitively eliminated by large forms. However, intense grazing from fish, which is selective for larger Crustacea, allows the smaller zooplankton to achieve dominance.

The benthic macroinvertebrates in small impoundments are composed of heterogeneous taxa and feeding groups. The major taxa include aquatic oligochaetes, crustaceans, and immature insects. Feeding groups include herbivores, detrital feeders, and carnivores. Unlike limnetic plankton, most macroinvertebrates are limited by substrate type. Many, however, are not limited to substrates, but actively feed in the water column.

Benthic organisms are generally more abundant and diverse in shallow water than in deeper water, especially if littoral zones are well developed. Boody et al. (1985) found that immature tubificids, *Chaoborus* sp., *Hyallela azteca, Aulodrilus pigueti, Glyptotendipes* spp., *Chironomus crassicaudatus*, and nematodes were the most commonly collected animals at shallow sites (1–2 m) in reclaimed Florida phosphate mining ponds. *Chaoborus* was the only taxon found in deeper water, where oxygen was depleted in warmer months. *Chaoborus* is important to farm pond ecology because of its ability to exist in turbid and anaerobic conditions. Since it migrates nocturnally into the water column and feeds on plankton, high densities also impact pond plankton structure. Macroinvertebrates greatly increase secondary productivity in ponds. Chironomidae may have six or more life cycles per year in southeastern impoundments, even emerging during warm periods in midwinter (Cooper 1987).

Vertebrates

Five major groups of vertebrates—fish, amphibians, reptiles, birds, and mammals—inhabit or utilize freshwater ponds in the southeastern United States. Fish are the dominant predators in pond ecosystems and more information is available on fish than other pond vertebrates. The paucity of information on amphibians, reptiles, and mammals may be accounted for by the purposes and procedures of observation rather than an actual lack of occurrence. However, it is true that ponds located in single cover watersheds, such as pasture, often have little advantageous riparian habitat for vertebrates. Available literature indicates a limited species diversity in all groups except birds.

Fish, like other consumers, can be divided by feeding groups and habitat preferences, but are often listed by taxonomic group for convenience. Largemouth bass (*Micropterus salmoides*), the top predator in many ponds, eats other fish, amphibians, reptiles, and even small birds and mammals. Green sunfish (*Lepomis cyanellus*), bluegill (*L. macrochirus*), and pumpkinseed (*L. gibbosus*), are smaller sunfishes which may spawn several times each year. Although they serve as a major food source for predaceous fishes, sunfish tend to overpopulate small ponds and produce stunted individuals. Juvenile Centrarchidae, like other immature fish, feed on plankton and detritus. Unlike largemouth bass, which become mainly carnivores as adults, smaller sunfish are omnivores. Pond bottoms are used extensively by bottom-feeding fish, including catfish and suckers. Members of the Cyprinidae and other minnows occupy feeding niches at the water's surface, in the water column, and at the pond bottom. Minnows serve as a major food source for piscivorous fish throughout the year.

Predominant species of fish included bluegill, brown bullhead (*Ictalurus nebulosus*), and golden shiner (*Notemigonus crysoleucas*) in Lullwater Lake, Georgia (Mozley 1968); largemouth bass, redear sunfish (*Lepomis microlophus*), and bluegill in Bays Mountain Lake, Tennessee (Johnson and Crowley 1989); and bluegill, pumpkinseed, yellow perch (*Perca flavescens*), and bass (*Micropterus* spp.) in Botany Pond, North Carolina (Gilinsky 1981). In the nine reclaimed lakes of the Florida phosphate mining region (Boody et al. 1985), the species found in more than half of the lakes were threadfin shad (*Dorosoma petenense*), mosquitofish (*Gambusia affinis*), warmouth (*Lepomis gulosus*), bluegill, largemouth bass, and golden shiner. Estimated fish biomass ranged from 2 to 41 kg/ha in the Florida lakes, but these were considered low estimates. Fish biomass increased with increasing age of these young, 3- to 9-year-old lakes.

Occurrences of other vertebrate wildlife have been recorded infrequently in small southeastern impoundments. Smith et al. (1979) noted only two species each of frogs (*Rana catesbeiana* and *Acris crepitans*) and turtles (*Pseudemys scripta* and *Trionix spinifer*) as inhabitants in Golf Course Pond, Denton, Texas. In Lullwater Lake near Atlanta, Georgia, Mozley (1968) noted two species of water snakes (*Natrix sipedon pleuralis* and *N. septemvittata*) and eight turtles (*Sternotherus odoratus*, *Amyda ferox*, *Pseudemys scripta scripta*, *P. s. elegans*, *Chelydra serpentina*, *Kinosternon subrubrum*, *Chrysemys picta picta*, and *C. p. marginata*). Shallow water, rapid sedimentation, and abundant small fish in this lake may have increased the reptile population. Mozley also noted that many water birds visited the lake and that domestic ducks and geese were permanent residents. At reclaimed mine impoundments in central Florida, Boody et al. (1985) compiled a wildlife species list, including 3 amphibians, 7 reptiles, 55 birds, and 1 mammal that were observed during seasonal visits over one year.

Fish Management

Principles of maintaining balanced fish populations in small impoundments were established by Swingle (1950). A balanced population is defined as one capable of producing satisfactory annual crops of harvestable fish. The important criterion is not the total biomass of fish, but the presence of sufficient individuals of harvestable size. In 55 ponds with balanced populations, the average fish biomass was 346 kg/ha, with 60% of the biomass (207 kg/ha) in harvestable size fish. In 34 ponds with unbalanced populations, there were 440 kg/ha of fish biomass with 25% (108 kg/ha) of harvestable size. To maintain a balanced population, Swingle recommended that the biomass of forage fish should be from 3 to 6 times the biomass of piscivorous fish. Forage fish are comprised of a large number of omnivores that feed largely upon plants, plankton, crustacea, insects, and occasionally small fish. Piscivorous fish cannot grow to normal adult size without appropriate sized prey. Thus, an insufficient number of small forage fish can result in starved populations of piscivores. Adult forage fish, however, must be harvested to reduce the overall grazing pressure which allows an abundant crop of small fish to develop and provide prey for piscivorous fish. A common mistake in fish pond management is to harvest only the piscivorous species from a mixed population. The principal forage and piscivorous fish recognized by Swingle (1950) are channel catfish, bluegill, and other sunfish (*Lepomis* spp.), golden shiner (forage species), largemouth bass, and white crappie (*Pomoxis annularis*) (piscivorous species).

Two examples will illustrate that the structure of fish communities in small impoundments is heavily influenced by stocking and harvesting practices. Lago Pond in Clarke County, Georgia, was fertilized during the spring and summer, and managed for fee fishing (Welch 1968). The average standing crop of fish was 450 kg/ha, and 155 kg/ha was harvested from May 1966 through April 1967. The harvest consisted of 71% (by weight) bluegill, 12.5% largemouth bass, 12.5% redbreast sunfish (*Lepomis auritus*), and 2.2% warmouth. Although biomass estimates were not given for forage and piscivorous fish, the harvest indicates a reasonably well-balanced population with a fair percentage of harvestable size fish.

On the other hand, Golf Course Pond, Denton, Texas, was not managed for fishing. From June 1973 through May 1974 the standing crop of fish averaged 1105 kg/ha, with harvest of 64.5 kg/ha (Jones et al. 1977). The harvest consisted of 83% black bullhead (*Ictalurus melas*), 6.5% largemouth bass, 4.9% white crappie, 3.4% green sunfish, 1.8% bluegill, and 0.2% longear sunfish (*Lepomis megalotis*). Fishing in the pond was prohibited for many years prior to the summer of 1973, which accounts for the larger standing crop. The population appears to be unbalanced with few harvestable size fish.

SMALL POND ECOSYSTEMS

Food wed relationships

The food web of any pond is a reflection of the dynamics of energy flow through the pond ecosystem. Primary consumers in aquatic systems derive energy from phytoplankton, detritus, and bacteria. Food web relationships in Bays Mountain Lake (Fig. 5) are representative of small ponds with energy flowing through numerous primary consumers to fewer secondary consumers (Johnson and Crowley 1989).

Food web relationships in Bays Mountain Lake are based on diet analysis of four major groups of predators in the lake: largemouth bass; bluegill and redear sunfish; odonate larvae (primarily *Tetragoneuria cynosura*, *Celithemis* spp., *Sympetrum vicinum*, and *Enallagma* spp.); and chironomid larvae (Tanypodinae, primarily *Ablabesmyia parajanta* and *Procladius* spp.). While all predators fed on a wide range of prey, invertebrate predators were limited by their size to small prey. Fish ignored smaller organisms to pursue larger, more energetically profitable prey. There was a considerable diet overlap

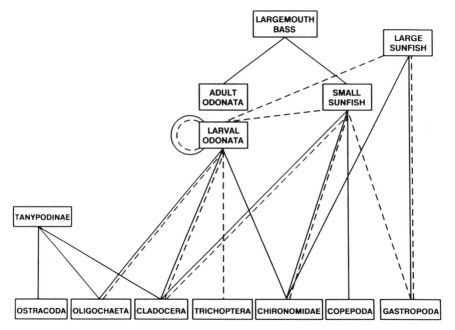

FIGURE 5. Strong links in the food web of the littoral zone of Bays Mountain Lake. —, Prey comprising at least 10% of the diet of predators. - - -, Statistically significant depletion of prey populations by natural densities of predators in *in situ* enclosure/exclosure experiments. (From Johnson and Crowley (1989).)

between predaceous chironomids, larval odonates, and small sunfish, all of which preyed heavily on chironomids and chydorid cladocerans.

Chironomids are important components of the food webs of ponds, with different groups playing varying roles in the food chain. For example, the particular role of chironomid larvae in the food web of Lullwater Lake, a 5.2-ha shallow impoundment near Atlanta, Georgia, was studied by Mozley (1968). Although the lake was rapidly filling with sediment, chironomid populations averaged 10 000 individuals/m² during the study. Major producers in Lullwater Lake included *Gomphonema*, *Gyrosigma*, *Asterionella*, *Closterium*, *Staurastrum*, *Phacus*, and *Ceratium*. *Spirogyra* and *Lyngbya* were prominent in several blooms. The principal fish, in order of decreasing biomass, were bluegill, brown bullhead, and golden shiner.

Four groups of chironomid larvae were distinguished by having different roles in the food web. The numerically abundant *Procladius* spp. (61% of the population) fed on benthic microfauna and the diatom *Gomphonema*. Some *Procladius* were consumed by brown bullhead, but most emerged as adults from the lake. The second group (2% of the population), made up of *Cryptochironomus*, *Clinotanypus*, and *Coelotanypus*, fed largely on oligochaetes and smaller chironomids. This group was consumed by all species of fish, but may be of little importance to them because of their small numbers. The third group (34% of the population) was a heterogeneous assortment, including *Tanytarsini*, *Orthocladiinae*, and a number of *Chironomini*, that fed on algae and detritus in the shallow margins of Lullwater Lake. This group was important prey for all fish in the lake. The fourth group (3% of the population) consisted of larger *Chironomus* larvae which also ingested algae and detritus. Most of this group were consumed by larger bluegill and brown bullhead.

Odonates can also play a major secondary consumer role in pond food chains. For example, investigations in Bays Mountain Lake, which centered on habitat and food web relationships (Crowley and Johnson 1982), showed the average density of odonates ranged from 0.5 to 100 individuals/m^2 in their preferred habitats. Most species showed a definite habitat preference. For example, *Tetragoneuria cynosura*, *Enallagma traviatum*, and *E. signatum* were found in areas with submersed vegetation, *Ischnura posita* and *I. verticalis* with emergent vegetation, while certain species, notably *Enallagma aspersum* and *Plathemis lydia*, were found only in a small adjacent pond where fish were absent. Thus, while specific producers and consumers of importance may be dictated by habitat, abiotic parameters, or geographic location, the generalized pond food web is predictable.

Representative Aquatic Communities

Small pond communities. like any ecosystem, portray individual character-istics that cover a wide range, but an overview of "typical" ponds lends insight into those communities found throughout the southeast. An excellent view of the total pond system structure and function can be obtained by an inte-grated or holistic approach using energy flow. To this end, energy flow in typical small impoundments is summarized in Fig. 6 for Lago Pond in Clarke County, Georgia, and in Fig. 7 for Golf Course Pond in Denton County, Texas. Both ponds are productive, receiving fertilizer directly in the case of Lago Pond, and indirectly from watershed runoff in the case of Golf Course Pond. Their productivity, however, does not appear to be unusually high compared with that of other small impoundments of the southeastern United States (Table 4).

Lago Pond

The transfer of energy through the main food chain components in Lago Pond is depicted in Fig. 6 (Odum 1971), based on the measurements of Welch (1968). The primary producer in this pond was a single species of blue-green algae, probably *Aphanizomenon*. An estimated 26% of the energy fixed by the phytoplankton was ingested by microcrustacea (90% benthic ostracods), larval Diptera (95% Chironomidae) and an energetically insignificant Oli-

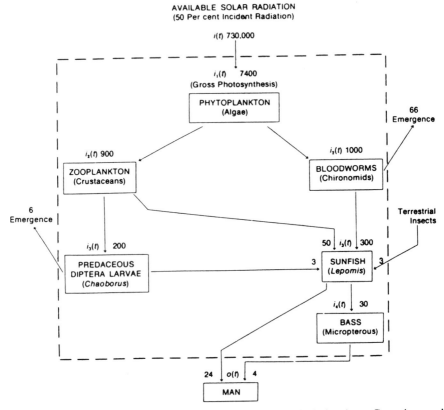

FIGURE 6. Compartment model of the principal food chains in a Georgia pond managed for sports fishing. Estimated energy inputs (i) with respect to time (t) are given in kcal/m^2 year^{-1}. $i_1(t)$, $i_2(t)$, $i_3(t)$, and $i_4(t)$ represent ingested food energy at successive trophic levels; losses during assimilation and respiration are not shown. $o(t)$ is output from the pond in terms of caloric value of fish caught by man. (Data from Welch (1968), with his estimate of assimilated energy at the i_2 level changed to estimated ingested energy on the basis of a 60% assimilation efficiency for zooplankton and 40% for bloodworms.) (From Odum (1971).)

gochaeta population. Average biomass of the benthic community was 19 kcal/m^2, while that of zooplankton was 0.5 kcal/m^2. About 28% of the energy ingested by crustaceans was passed on to predaceous larval Diptera (*Chaoborus*) and sunfish. *Chaoborus* populations were unusually high, averaging 10 000–20 000 larvae/m^2. Only a small percentage of *Chaoborus* was consumed by fish, indicating a high nonpredatory mortality. About 30% of the energy ingested by Chironomidae was available to sunfish. Of the energy ingested by sunfish, about 8% was incorporated by bass, and 7% was harvested by man. Of the energy ingested by bass, 13% was harvested by man. Overall, man harvested 0.4% of the energy fixed by the algae.

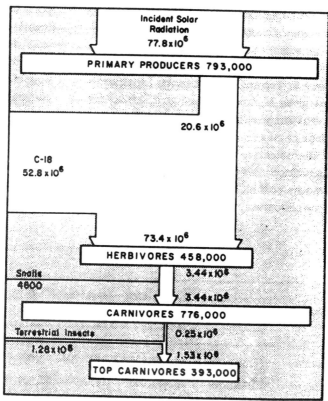

FIGURE 7. Annual energy flow between trophic levels in Golf Course Pond, Denton, Texas. Trophic level values are mean annual biocontent in kcal. Flow values are annual energy flow in kcal/year. Width of trophic levels are proportional to biocontent, and width of flow arrows are proportional to energy flow (From Childress et al. (1981).)

Golf Course Pond

The rates of primary production and energy transfer to herbivores in Golf Course Pond (Fig. 7) do not differ greatly from those in Lago Pond. However, the herbivores in Golf Course Pond derived a large part of their energy from recycled organic matter (C-18, including bacteria, fungi, and detritus). Transfers of energy from herbivores to carnivores and then to top carnivores were about 5 and 7%, respectively, much lower than in Lago Pond. Childress et al. (1981) concluded that much nonutilized organic matter accumulated in the pond sediments.

Community structure in Golf Course Pond differed considerably from that in Lago Pond. Primary production was almost equally divided between phytoplankton, floating algal mats, and attached vegetation. Zooplankton consisted of cladocerans and copepods. Benthic biomass was composed of 59% mayfly (*Brachycercus* sp.), 26% chironomid (*Chironomus decorus, Procladius* sp.), 12% annelids (*Haplotaxis gordioides*), and 1% *Chaoborus* sp. The average biomass of the benthic community was 3.2 kcal/m² and that of zooplankton was 16.2 kcal/m² compared to 19 and 0.5 kcal/m² for the two communities, respectively, in Lago Pond. The relatively sparse benthic community was attributed to fluctuating water level, intense fish predation, and removal of floating algal mats by groundskeepers (Benson et al. 1980). Although the standing crop of fish was high (1105 kg/ha), the fish were mainly small and the total harvest was less than half of that from Lago Pond.

RESOURCE USE AND MANAGEMENT EFFECTS

The majority of farm ponds in the southeastern United States have been constructed within the past 50 years (Hawley 1973). Not only has the number of ponds increased rapidly, but the average size of pond also increased during this period. Many ponds have been constructed for the traditional purposes of livestock watering and soil conservation. The size of ponds is expected to increase as demands for irrigation and commercial fish production increase.

Resource and Management Problems

Small impoundments and ponds ideally should be managed to maintain or increase their intended uses. However, management may be slighted for various reasons. The intended uses may not develop, or the impoundment may prove unsuitable for its intended uses. There may be insufficient funds or trained personnel to implement proper management on a regular basis. Needed information may occasionally be lacking. Conflicts of interest may arise, especially if the impoundment and its watershed affect different interests. Finally, it must be recognized that pond management often has a low priority with landowners.

A major problem affecting all uses is the filling of small impoundments with sediment. The problem may be severe in watersheds where erodible soils are under cultivation. The direct effect is loss of storage capacity in the impoundment, but there may also be important changes in benthic habitat and primary productivity. One result is the encroachment of emergent and woody vegetation into the impoundment. While this may increase overall productivity and use of impoundment by wildlife, it would probably detract from its intended uses.

Some farm ponds are constructed to function as sediment traps, thus improving water quality downstream, and reducing the off-site consequences of soil erosion. Even in such ponds, too rapid filling with sediment shortens the useful life of the pond. Effective measures to reduce soil erosion on the watershed are beneficial. These measures might include changes in land use, construction of terraces, and conservation tillage methods.

Turbidity due to suspended clay limits primary production and detracts from the appearance of many impoundments in the southeastern United

States. The problem is common with Piedmont and Coastal Plain soils. Persistently turbid ponds discussed in this chapter include Lullwater Lake near Atlanta, Georgia (Mozley 1968), the Fort Gordon ponds near Augusta, Georgia (Terrell 1975), and Golf Course Pond in Denton, Texas (Kelly et al. 1978). Gunn (1974) noted that livestock wading caused turbidity in a number of small Georgia ponds. In some ponds, such as Morris Pond, Panola County, Mississippi, most of the suspended sediment settles out during seasons of low runoff, leading to periods of rapid phytoplankton growth and decay (Cooper and Knight 1990). Thus, suspended sediment may mask a eutrophic condition.

The turbidity of small impoundments can be reduced by chemical or biological flocculation of the clay (Boyd 1979, Avnimelech and Menzel 1984). Materials that have been used with varying levels of success include alum, gypsum, hydrated lime, fertilizers, hay, and manure. A lasting effect can be obtained only if sources of inflowing or resuspended sediment are controlled.

The matter of pesticide pollution in small impoundments requires continuing attention. Although persistent organochlorine insecticides are no longer generally used, various nonpersistent pesticides are increasingly used in agriculture and forestry. Herbicides are essential in conservation tillage farming, and are being used more often in forest management (Bouchard et al. 1985, Neary et al. 1985). The biological impact of pesticide runoff into small impoundments may be temporarily severe. Numerous instances of fish kills from insecticides have been reported (U.S. Environmental Protection Agency, 1972), and there may be unnoticed instances of plant inhibition from herbicides. For example, Tucker (1987) found that oxygen production by phytoplankton was reduced at a herbicide concentration of 25 mg/m^3 propanil.

Many small impoundments of the southeastern United States are eutrophic, as indicated by their nutrient concentrations (Table 3) and primary productivity (Table 4). This condition may be beneficial rather than detrimental for some intended uses of impoundments. However, if the pond is used for recreation or as a source of water for livestock, excessive growths of algae or macrophytes may become a nuisance. Biological, chemical, and mechanical methods have been used to control or remove excessive growths, but these sometimes introduce new problems.

Once established, a eutrophic condition is very difficult to ameliorate. Nevertheless, treatments are possible on small impoundments that could not be attempted on larger ones. Control of vegetation is more feasible in a small pond than in a large reservoir. Ponds could be drained and bottom sediments allowed to dry, when they could even be removed for fertilizing upland areas. The ponds or exposed sediments could be amended to reduce nutrient availability, if desired.

The need for management to maintain balanced fish populations for recreational fishing was mentioned above. (Management of ponds for commercial fish production is outside the scope of this chapter.) Population management requires knowledge and effort to determine the biomass ratio of forage fish and piscivorous fish in a pond, and to bring unbalanced populations into proper balance. Professional advice on this aspect of pond management is usually available through state fish and wildlife services.

Future concerns relate to the treatment of small ponds that no longer serve their purposes. Other than abandonment, the alternatives are to drain ponds, converting the land to other uses, or to renovate them. Pond renovation may require removing vegetation, excavating accumulated sediment, or raising the level of embankments. Usually it is more economical to build a new pond if another site is available.

Natural and Human-Induced Stresses

The number of small impoundments and farm ponds in the southeastern United States will continue to increase to meet the demands of an increasing population for agricultural production and recreational opportunities. Each new impoundment will be constructed for one or more specific purposes, such as irrigation, livestock watering, aquaculture, aesthetic value, or recreational fishing. The biological community of the impoundment will replace the former community of the impounded site.

Small impoundments impact directly on a rather small fraction of the total landscape. Nevertheless, construction of small impoundments has inundated nearly 0.5% of the land surface in the southeastern United States. Many of the inundated areas were previously wooded, and near a permanent stream or spring water source. Primary productivity of such areas would generally be high before inundation, and capable of supporting large, diverse wildlife populations. In agricultural areas, farm ponds have usually been placed on low-lying sites that were not agriculturally productive, although they may have been biologically very productive.

Although inundation may decrease total biological productivity, it provides greater habitat diversity. Reduced productivity of the inundated area would also be partially offset by more uniform water supply and increased productivity in the littoral zone of most impoundments. Productivity might be drastically reduced in those impoundments where water level fluctuates greatly due to irrigation, evaporation, or seepage, thus preventing the establishment of much shoreline vegetation.

RECOMMENDATIONS FOR FUTURE RESEARCH AND MANAGEMENT

An improved understanding of small impoundment ecosystems is needed for many reasons. This understanding could help improve the productivity of both recreational fish ponds and commercial aquaculture units. The capability to predict the effects of sediment accumulation, nutrient enrichment, and pesticide pollution in small impoundments would also be improved. Such information would allow better management practices to be developed to promote the intended uses of small impoundments and ponds.

Studies are needed on holistic ecosystem interrelationships in small impoundments and ponds. Ecosystems may adapt to disturbing factors, such as sediment accumulation, nutrient enrichment, and pesticide pollution, by one organism or group of organisms replacing another in certain essential roles. The study of such changes would be easier in a small pond ecosystem than in a larger impoundment, but might provide information that could be applied to the larger impoundments.

Information is also needed to manage water quality problems that are poorly controlled now. Current treatments for the control of turbidity from suspended sediment are expensive and have temporary effect. Research on bioflocculation of clays, perhaps in combination with chemical treatments, may give a more permanent effect. Much also remains to be learned about the entrapment and cycling of nutrients within small impoundments. Ammonium and nitrite released from enriched bottom sediments may become toxic to fish. Microbial processes involved in the production of ammonium and nitrite are well known, but microbial populations of small ponds have been studied very little.

Methods for increasing the use of good pond management practices need to be investigated. Although most ponds are built for specific purposes, routine management of the ponds is often neglected. The predominant reasons (economic, lack of information, lack of incentive, etc.) should be determined. Because of the abundance of impoundments and farm ponds, the potential return from good management practices is very great.

REFERENCES

Avnimelech, Y., and R. G. Menzel. 1984. Coflocculation of clay and algae to clarify turbid impoundments. *J. Soil Water Conserv.* 39:200–203.

Benson, D. J., L. C. Fitzpatrick, and W. D. Pearson. 1980. Production and energy flow in the benthic community of a Texas pond. *Hydrobiologia* 74:81–93.

Boody, O. C., C. D. Pollman, G. H. Tourtellotte, R. E. Dickinson, and A. N. Arcuri. 1985. *Ecological Considerations of Reclaimed Lakes in Central Florida's Phosphate Region*, Publ. 03-018-030. Bartow: Florida Inst. Phosphate Res.

Bouchard, D. C., T. L. Lavy, and E. R. Lawson. 1985. Mobility and persistence of hexazinone in a forest watershed. *J. Environ. Qual.* 14:229–233.

Boyd, C. E. 1973. Summer algal communities and primary productivity in fish ponds. *Hydrobiologia* 41:357–390.

Boyd, C. E. 1974. Lime requirements of Alabama fish ponds. *Ala., Agric. Exp. Stn., Bull.* 459.

Boyd, C. E. 1975. Competition for light by aquatic plants in fish ponds. *Circ.—Ala., Agric. Exp. Stn.* 215.

Boyd, C. E. 1976. Water chemistry and plankton in unfertilized ponds in pastures and in woods. *Trans. Am. Fish. Soc.* 105:634–636.

Boyd, C. E. 1979. *Water Quality in Warmwater Fish Ponds*. Auburn, AL: Auburn University.

Boyd, C. E., and L. W. Hess. 1970. Factors influencing shoot production and mineral nutrient levels in *Typha latifolia*. *Ecology* 51:296–300.

Brooks, J. L., and S. I. Dodson. 1965. Predation, body size, and composition of plankton. *Science* 150:28–35.

Bruce, R. R., J. L. Chesness, T. C. Keisling, J. E. Pallas, Jr., D. A. Smittle, J. R. Stansell, and A. W. Thomas. 1980. *Irrigation of Crops in the Southeastern United States: Principles and Practice*, USDA Agric. Rev. Man., South. Ser., No. 9. New Orleans: LA: U.S. Department of Agriculture.

Cassani, J. R., and W. E. Caton. 1985. Effects of chemical and biological weed control on the ecology of a south Florida pond. *J. Aquat. Plant Manage.* 23:51–58.

Childress, W. M., L. C. Fitzpatrick, and W. D. Pearson. 1981. Trophic structure and energy flow in a Texas pond. *Hydrobiologia* 76:135–143.

Climates of the States. 1974. Vol. 1, *Eastern States Plus Puerto Rico and the United States Virgin Islands.* Port Washington, NY: Water Information Center.

Cooper, C. M. 1987. Benthos in Bear Creek, Mississippi: effects of habitat variation and agricultural sediments. *J. Freshwater Ecol.* 4:101–113.

Cooper, C. M., and S. S. Knight. 1990. Nutrient trapping efficiency of a small sediment detention reservoir. *Agric. Water Manage.* 18:149–158.

Crowley, P. H., and D. M. Johnson. 1982. Habitat and seasonality as niche axes in an odonate community. *Ecology* 63:1064–1077.

Dendy, F. E. 1982. Distribution of sediment deposits in small reservoirs. *Trans. ASAE* 25:100–104.

Dendy, F. E., and W. A. Champion. 1978. Sediment deposition in United States reservoirs: Summary of data reported through 1975. *Misc. Publ.—U.S. Dep. Agric.* 1362.

Dendy, F. E., and C. M. Cooper. 1984. Sediment trap efficiency of a small reservoir. *J. Soil Water Conserv.* 39:278–280.

Geraghty, J. J., D. W. Miller, F. Van der Leeden, and F. L. Troise. 1973. *Water Atlas of the United States.* Port Washington, NY: Water Information Center.

Gilinsky, E. 1981. The role of predation and spatial heterogeneity in determining community structure: the experimental manipulation of a pond system. Thesis, University of North Carolina, Raleigh. (University Microfilms 8125579, Ann Arbor, MI).

Gunn, W. D. 1974. Aquatic vascular flora of some artificial ponds of the western Piedmont of Georgia. Thesis, University of Georgia, Athens. (University Microfilms 75-8145, Ann Arbor, MI).

Hawley, A. J. 1973. Farm ponds in the United States: A new resource for farmers. In W. C. Ackermann, G. F. White, and E. B. Worthington (eds.), *Man-Made Lakes: Their Problems and Environmental Effects.* Washington, DC: Am. Geophys. Union, pp. 746–749.

Janicki, A. J., J. DeCosta, and J. Davis. 1979. The summer crustacean plankton communities of seven small impoundments. *Hydrobiologia* 64:123–129.

Johnson, D. M., and P. H. Crowley. 1980. Habitat and seasonal segregation among coexisting odonate larvae. *Odonatologica* 9:297–308.

Johnson, D. M., and P. H. Crowley. 1989. A ten year study of the odonate assemblage of Bays Mountain Lake, Tennessee. *Adv. Odonatol.* 4:27–43.

Jones, F. V., W. D. Pearson, and L. C. Fitzpatrick. 1977. Yield estimates derived from active and passive creel surveys of a small pond fishery. *Tex. J. Sci.* 29:41–48.

Kelly, M. H., L. C. Fitzpatrick, and W. D. Pearson. 1978. Phytoplankton dynamics, primary productivity, and community metabolism in a north-central Texas pond. *Hydrobiologia* 58:245–260.

Knauer, G. W., and A. L. Buikema, Jr. 1984. Rotifer production in a small impoundment. *Verh.—Int. Ver. Theo. Angew. Limnol.* 22:1475–1481.

Labaugh, J. W. 1978. Hydrologic and total phosphorus budgets, chlorophyll concentrations, and primary production in Spruce Knob Lake, West Virginia, an artificially aerated impoundment. Thesis, West Virginia University, Morgantown. (University Microfilms, 7900886, Ann Arbor, MI).

Lemly, A. D., and J. F. Dimmick. 1982. Structure and dynamics of zooplankton communities in the littoral zone of some North Carolina lakes. *Hydrobiologia* 88:299–307.

McVea, C., and C. E. Boyd. 1975. Effects of waterhyacinth cover on water chemistry, phytoplankton, and fish in ponds. *J. Environ. Qual.* 4:375–378.

Meyer, K. A., and J. F. McCormick. 1971. Seasonal fluctuation of phytoplankton composition, diversity, and production in a freshwater lake. *J. Elisha Mitchell Sci. Soc.* 87:127–138.

Mozley, S. C. 1968. The integrative roles of the chironomid (*Diptera: Chironomidae*) larvae in the trophic web of a shallow five hectare lake in the Piedmont region of Georgia. Thesis, Emory University, Atlanta, GA. (University Microfilms 69-5240, Ann Arbor, MI).

Neary, D. G., P. B. Bush, J. E. Douglass, and R. L. Todd. 1985. Picloram movement in an Appalachian hardwood forest watershed. *J. Environ. Qual.* 14:585–592.

Odum, E. P. 1971. *Fundamentals of Ecology*. Philadelphia, PA: Saunders.

Organization for Economic Co-operation and Development. 1982. *Eutrophication of Waters: Monitoring, Assessment, and Control*. OECD, Paris.

Polisini, J. M., C. E. Boyd, and B. Didgeon. 1970. Nutrient limiting factors in an oligotrophic South Carolina pond. *Oikos* 21:344–347.

Smith, G. A., L. C. Fitzpatrick, and W. D. Pearson. 1979. Structure and dynamics of the zooplankton community in a small north-central Texas pond ecosystem. *Southwest. Nat.* 24:1–16.

Swingle, H. S. 1950. Relationships and dynamics of balanced and unbalanced fish populations. *Ala., Agric. Exp. Stn., Bull.* 274.

Sylvester, M. A. 1972. Some studies on the microflora and the chemical aspects of farm ponds. Thesis, West Virginia University, Morgantown. (University Microfilms 72-26895, Ann Arbor, MI).

Terrell, T. T. 1975. The impact of macrophyte control by the white amur (*Ctenopharyngodon idella*). *Verh.—Int. Ver. Theo. Angew. Limnol.* 19:2510–2514.

Terrell, T. T. 1976. Environmental impact of the white amur (*Ctenopharyngodon idella* Val.) on some Georgia ponds. Thesis, University of Georgia, Athens. (University Microfilms 76-29568, Ann Arbor, MI).

Tucker, C. S. 1987. Short-term effects of propanil on oxygen production by plankton communities from catfish ponds. *Bull. Environ. Contam. Toxicol.* 39:245–250.

Tucker, C. S., and S. W. Lloyd. 1985. Water quality in streams and channel catfish (*Ictalurus punctatus*) ponds in west-central Mississippi. *Miss., Agric. For. Exp. Stn., Tech. Bull.* 129.

U.S. Department of Agriculture. 1981. *Soil, Water, and Related Resources in the United States: Status, Condition, and Trends.* Soil and Water Resources Conservation Act, 1980 Appraisal. Part I. Washington, DC: U.S. Department of Agriculture.

U.S. Department of Agriculture. 1982. Ponds—planning, design, construction. *U.S., Dep. Agric., Agric. Handb.* 590.

U.S. Department of Agriculture. 1984. *Agricultural Statistics 1984.* Washington, DC: U.S. Govt. Printing Office.

U.S. Department of the Interior. 1970. *The National Atlas of the United States of America.* Washington, DC: U.S. Department of the Interior, Geological Survey.

U.S. Environmental Protection Agency. 1972. *Pesticide Usage and Its Impact on the Environment in the Southeast*, Pestic. Study Ser. No. 8. Washington, DC: U.S. Govt. Printing Office.

Water Resources Council. 1968. *The Nation's Water Resources*, Parts 1–7. Washington, DC: U.S. Govt. Printing Office.

Wauchope, R. D. 1978. The pesticide content of surface water draining from agricultural fields—a review. *J. Environ. Qual.* 7:459–472.

Welch, H. E. 1968. Energy flow through the major macroscopic components of an aquatic ecosystem. Thesis, University of Georgia, Athens. (University of Microfilms 69-3488, Ann Arbor, MI).

Wetzel, R. G. 1975. *Limnology.* Philadelphia, PA: Saunders.

Wilson, H. A., T. Miller, and R. Thomas. 1966. Some microbiological, chemical, and physical investigations of farm ponds. *Bull.—W. Va., Agric. Exp. Stn.* 522T.

11 Reservoirs

D. M. SOBALLE and B. L. KIMMEL

Environmental Sciences Division, Oak Ridge National Laboratory, Oak Ridge, TN

R. H. KENNEDY and R. F. GAUGUSH

U. S. Army Corps of Engineers, Waterways Experiment Station, Vicksburg, MS

River impoundments are major aquatic ecosystems of the southeastern United States and they provide an economic, recreational, and environmental resource of increasing value. Impoundments are among the oldest of human engineering works (Baxter 1977) and in the southeast (as elsewhere) reservoirs provide benefits from flood control, hydropower generation, navigation, and water supply. The ecological impacts of river impoundment have been both dramatic and far-reaching, but economics and topography have ultimately directed the construction and operation of these projects, while their ecological consequences have received less consideration.

Dam construction in the southeastern United States proliferated during the last 50 years, with most of this activity attributable to the Tennessee Valley Authority (TVA) and the U. S. Army Corps of Engineers (USCE). Today, however, nearly all suitable sites for major reservoirs in the Southeast have been utilized and dam construction has slowed dramatically (Fig. 1). As a result, the need and opportunity to understand the impacts of new impoundments in this region have declined, but the demand for scientific guidance in managing or manipulating existing reservoirs to maximize their benefits for multiple uses has continued to increase.

Our scientific understanding of reservoirs, as functional ecosystems, is undeveloped relative to that of natural lakes. Early researchers took a comparatively narrow view of reservoir ecology and focused on problems of fish production and the downstream impacts of reservoir discharges on river biota (with a strong emphasis on downstream fisheries and aquatic invertebrates). The great variability of reservoirs, both within individual impoundments and among reservoir types, was not generally appreciated, and only recently have

D. M. Soballe's and R. F. Gaugush's current address is U.S. Fish and Wildlife Service, 575 Lester Ave., Onalaska, Wi 54650.

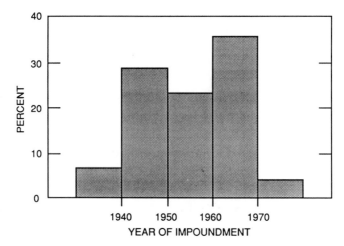

FIGURE 1. Distribution of impoundment dates for major reservoirs operated by the Tennessee Valley Authority and the U.S. Army Corps of Engineers in the southeastern United States. Reservoirs impounded prior to 1940 or after 1970 are pooled. (Based on data reported by Leidy and Jenkins (1977) and Placke (1983).)

researchers and managers recognized that the limnological "dogma" derived from a restricted group of natural lakes is not wholly applicable to the full range of river impoundments.

Recent reservoir studies have built upon earlier reservoir research and have rapidly expanded beyond it. As a result, an approach to reservoir ecology that considers the special limnological features of reservoirs, their functioning as integral components of larger river–reservoir ecosystems, and the full range of their abiotic and biotic interactions is emerging, but many fundamental questions of reservoir ecology remain unanswered.

In this chapter we discuss features and processes that we believe dominate the ecological structure and functioning of southeastern reservoirs, and we outline concepts that we feel are central to an understanding of reservoir ecology. We do not discuss "off-channel" reservoirs that are formed primarily by excavations, by levee systems, or by outlet and shoreline modifications on existing natural lakes. These other "artificial lakes" are in many cases similar to river impoundments, but their inclusion here would expand this chapter inappropriately.

THE PHYSICAL SETTING OF RESERVOIRS IN THE SOUTHEASTERN UNITED STATES

The geographic distribution of reservoirs in the Southeast reflects a complex interaction between topography, climate, economics, and the need to regulate or modify the movement of water in major river basins. While climate and water management goals have guided the design of reservoirs and reservoir systems, economics and topography have ultimately directed their construction and operation. For example, engineering considerations require reservoirs to have a low ratio of dam volume to water storage volume (Fair et al. 1966) and thus reservoirs are most often built in narrow, steeply sloped river reaches that also have broad, branching upstream valleys. Other site requirements include (1) topographic relief that provides a favorable reservoir surface area to volume ratio (i.e., sufficient storage volume without excessive shallow areas), (2) an alignment that does not allow a strong influence of prevailing winds (to avoid both extreme shoreline erosion and wind-driven run-ups of surface water that could overtop the dam), and (3) subsurface geology that affords a stable construction site with low seepage. For economic reasons, the basin and surrounding areas should provide a source of readily available construction materials and should be located to minimize the impacts of construction on populated areas or existing structures.

Approximately 144 major reservoirs have been built on the rivers of the Southeast (Table 1) and they are widely distributed. These impoundments are located in three main drainages of the region (Figs. 2 and 3). Not included in this inventory are small, privately owned impoundments, locks and dams constructed primarily for navigation, and the newly completed impoundments associated with the Tennessee–Tombigbee Waterway (see chapter 15). Of the reservoirs included, 59% are operated by the U.S. Corps of Engineers and 17% by the Tennessee Valley Authority. The remaining reservoirs (24%) are operated by state, local, and private agencies. Hydropower reservoirs in North Carolina and South Carolina operated by the Duke Power Company are included in this latter category.

Southeastern reservoirs can be separated into two major geographic groups: (1) reservoirs located west of the broad alluvial plain of the Mississippi River, and (2) a widely dispersed group located in highland regions east of the Mississippi River (Fig. 4). Conspicuously lacking are reservoirs in low-lying coastal regions and in the lower Mississippi River Valley.

Reservoirs in the western group are located primarily in the Ozark–Ouachita Highlands physical subdivision (as described by Hammond (1964)) of northwestern Arkansas and the gently rolling region along the Louisiana–Texas border (Fig. 4). The northern reservoirs in the western group are primarily mainstem impoundments that provide flood control and navigation along the White and Arkansas rivers. Included in this group are series of storage reservoirs built for flood control on several tributaries of the Red River and along headwater streams of the Ouachita River basin. Large storage reservoirs on the Sabine and Neches rivers are the major systems in the southern half of the western group.

Reservoirs east of the Mississippi River serve a variety of water control functions and provide diverse aquatic habitats. Reservoirs in this group can be broadly subdivided as those on rivers that flow north to the Ohio River and those that impound rivers that flow either east to the Atlantic Ocean or south to the Gulf of Mexico. The boundary separating these subgroups is defined by the Appalachian Highlands subdivision, which extends south from western Virginia and West Virginia along the Tennessee–North Carolina border and into northern Georgia and Alabama (Fig. 4). Reservoirs in the

TABLE 1 Major Reservoirs of the Southeastern United States and Their Primary Tributaries

Reservoir	Tributary	Agency[a]
ALABAMA		
Bankhead	Black Warrior	USCE
Bartletts Ferry (Lake Harding)	Chattahoochee	Other
Claiborne	Alabama	USCE
Coffeeville	Tombigbee	USCE
Demopolis	Tombigbee	USCE
Gainesville	Tombigbee	USCE
Guntersville	Tennessee	TVA
Holt	Black Warrior	USCE
Jones Bluff	Alabama	USCE
Jordan	Coosa	Other
Lay	Coosa	Other
Lewis Smith	Sipsey	Other
Martin	Tallapoosa	Other
Millers Ferry	Alabama	USCE
Mitchell	Coosa	Other
Warrior	Black Warrior	USCE
Weiss	Coosa	Other
Wheeler	Tennessee	TVA
Wilson	Tennessee	TVA
ARKANSAS		
Beaver	White	USCE
Blue Mountain	Petit Jean	USCE
Bull Shoals	White	USCE
Dardanelle	Arkansas	USCE
DeGray	Caddo	USCE
DeQueen	Rolling Fork	USCE
Dierks	Saline	USCE
Gillham	Cossatot	USCE
Greers Ferry	Little Red	USCE
Greeson	Little Missouri	USCE
Millwood	Little	USCE
Nimrod	Fourche La Fave	USCE
Norfolk	North Fork	USCE
Ouachita	Ouachita	USCE
Ozark	Arkansas	USCE
GEORGIA		
Allatoona	Etowah	USCE
Blue Ridge	Oconee	TVA
Carters	Cossawattee	USCE
Chatuge	Hiwassee	TVA
George W. Andres	Chattahoochee	USCE
Hartwell	Savannah	USCE
J. Strom Thurmond (Clarks Hill)	Savannah	USCE

GEORGIA (*Continued*)	Tributary	Agency[a]
Nottely	Hiwassee	TVA
Richard B. Russell	Savannah	USCE
Seminole	Apalachicola	USCE
Sidney Lanier	Chattahoochee	USCE
Sinclair	Oconee	Other
Walter F. George	Chattahoochee	USCE
West Point	Chattahoochee	USCE
KENTUCKY		
Barkley	Cumberland	USCE
Barren River	Barren	USCE
Buckhorn	Kentucky	USCE
Carr Fork	Carr Fork	USCE
Cave Run	Licking	USCE
Cumberland	Cumberland	USCE
Dewey	Johns Creek	USCE
Fishtrap	Levisa Fork	USCE
Grayson	Little Sandy	USCE
Green River	Green	USCE
Kentucky	Tennessee	USCE
Nolin River	Nolin	TVA
Paintsville	Big Sandy	USCE
Red River	Red	USCE
Rough River	Rough	USCE
Taylorsville	Salt	USCE
Yatesville	Big Sandy	USCE
LOUISIANA		
Bodcau	Bayou Bodcau	USCE
Caddo	Willow Pass	USCE
Wallace	Cypress Bayou	USCE
MISSISSIPPI		
Arkabutla	Coldwater	USCE
Enid	Yacona	USCE
Grenada	Yalobusha	USCE
Okatibbee	Chichasawhay	USCE
Ross Barnett	Pearl	Other
Sardis	Little Tallahatchie	USCE
NORTH CAROLINA		
Apalachia	Hiwassee	TVA
B. Everett Jordan	Haw	USCE
Badin	Pee Dee	Other
Blewett Falls	Pee Dee	Other
Cheoah	Little Tennessee	TVA
Falls	Neuse	USCE
Fontana	Little Tennessee	TVA
Hickory	Catawba	Other

(*Continues on next page*)

TABLE 1 *(Continued)*

Reservoir	Tributary	Agency[a]
NORTH CAROLINA *(Continued)*		
High Rock	Pee Dee	Other
Hiwassee	Hiwassee	TVA
James	Catawba	Other
Lookout	Catawba	Other
Nantahala	Nantahala	TVA
Normon	Catawba	Other
Rhodhiss	Catawba	Other
Roanoke Rapids	Roanoke	Other
Santeetlah	Cheoah	TVA
Thorpe	Tuckasegee	TVA
W. Kerr Scott	Yadkin	USCE
SOUTH CAROLINA		
Catawba	Wateree	Other
Greenwood	Saluda	Other
Jocassee	Seneca	Other
Keowee	Seneca	Other
Marion	Santee	Other
Moultrie	Santee	Other
Murray	Saluda	Other
Wateree	Wateree	Other
TENNESSEE		
Boone	Watauga	TVA
Calderwood	Little Tennessee	TVA
Center Hill	Caney Fork	USCE
Cheatham	Cumberland	USCE
Cherokee	Holston	TVA
Chickamauga	Tennessee	TVA
Chilhowee	Little Tennessee	TVA
Cordell Hull	Cumberland	USCE
Dale Hollow	Obey	USCE
Douglas	French Broad	TVA
Fort Loudon	Tennessee	TVA
Ft. Patrick Henry	Watauga	TVA
Great Falls	Caney Fork	Other
Hiwassee	Hiwassee	TVA
J. Percy Priest	Stones	USCE
Melton Hill	Clinch	TVA
Nickajack	Tennessee	TVA
Norris	Clinch	TVA
Oconee	Oconee	TVA
Old Hickory	Cumberland	USCE
Pickwick	Tennessee	TVA
South Holston	South Fork Holston	TVA

TENNESSEE (*Continued*)

Tellico	Little Tennessee	TVA
Tims Ford	Elk River	TVA
Watts Bar	Tennessee	TVA
Watauga	Watauga	TVA

TEXAS

Lake "O" The Pines	Cypress	USCE
Sam Rayburn	Angleina	USCE
Texarkana	Sulphur	USCE
Toledo Bend	Sabine	Other

VIRGINIA

Gaston	Roanoke	Other
John H. Kerr	Roanoke	USCE
John W. Flannagan	Pound	USCE
Leesville	Roanoke	Other
Philipott	Smith	USCE
Smith	Roanoke	Other

WEST VIRGINIA

Bluestone	New	USCE
East Lynn	East Fork Twelve-Pole	USCE
Summersville	Gauley	USCE
Sutton	Elk	USCE
Tygart	Tygart	USCE

[a]USCE, U.S. Army Corp of Engineers; TVA, Tennessee Valley Authority.

Ohio River drainage are located in the Appalachian Highlands and Eastern Interior Uplands and Basins subdivisions in Tennessee, Kentucky, and West Virginia. Reservoirs of the Ohio basin can be further grouped as those of the Tennessee River basin (including the Cumberland River) and those of several smaller tributary basins draining north directly to the Ohio River, including the Green, Kentucky, Licking, and Big Sandy rivers in Kentucky and West Virginia, where several tributary reservoirs have been constructed to provide flood-control storage for the Ohio River. Reservoirs in the Tennessee River basin are operated primarily by the TVA and include main-stem impoundments on the Tennessee River and tributary reservoirs on the Clinch, French Broad, Holston, Little Tennessee, Hiwassee, Elk, and Duck River subbasins. As in the Tennessee basin, the Cumberland River has a series of main-stem impoundments and several major tributary reservoirs. Reservoirs in the Cumberland basin are operated by the USCE.

Numerous dams have been built on rivers that drain the eastern slopes of the Appalachian Highlands subdivision and flow across the Gulf–Atlantic Rolling Plain subdivision to either the Atlantic Ocean or the Gulf of Mexico (Fig. 4). Major basins impounded in this region include those of the Mobile

FIGURE 2. Geographic distribution of major reservoirs in the southeastern United States.

(including the basins of the Tombigbee, Black Warrior, and Alabama rivers), the Apalachicola (including the Chattahoochee River), and the Savannah rivers. The Santee and Pee Dee, Wateree, Seneca, and Saluda rivers in the Carolinas are smaller impounded rivers in this area. As a group, the river–reservoir systems of the Gulf–Atlantic Rolling Plain subdivision provide flood control, hydropower, and navigational access to the interior of the region.

A third, smaller subgroup of reservoirs in the southeast includes four USCE impoundments in northern Mississippi. These reservoirs provide flood protection for the eastern Mississippi alluvial plain and are located on westerly flowing tributaries of the Yazoo River. Further reservoir development in this area has been restricted by the flat topography and, as a result, flooding continues to interfere with agriculture activities in this subregion.

Impoundments in the southeast are most prevalent in regions that have appropriate topographic relief and high runoff or high flood potential. High

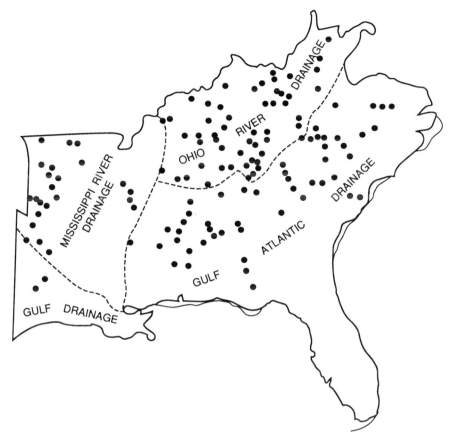

FIGURE 3. Geographic distribution of reservoirs within the major drainage basins of the southeastern United States.

annual runoff (i.e., greater than 50 cm per year) occurs along the Appalachian Mountain ridge in eastern Tennessee and Kentucky, across the Eastern Interior Upland and Basins subdivision of central Tennessee and northern Alabama, and in upland areas of western Arkansas (Fig. 5). Despite high annual runoff, the flat topography of the Gulf coastal plain discourages reservoir construction in that area.

The distribution of reservoirs in the northern and western portions of the region (Fig. 6) corresponds reasonably well to areas with high flood potential. One index of flood potential is the volume of flow in the "10-year" flood. This is the maximal runoff from a standard watershed (800 km^2) that occurs with an average frequency of once every 10 years (U.S. Department of the Interior 1964). Highest values of this flood-potential index are found across a broad area extending from northwestern Arkansas southeasterly to central Alabama (Fig. 6). Although flood potential is high throughout much of the

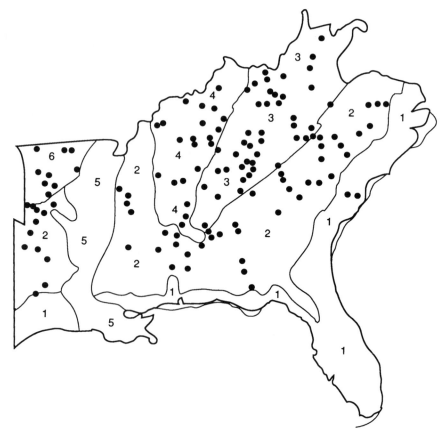

FIGURE 4. Geographic distribution of reservoirs within major physical subdivisions of the southeastern United States: 1, Gulf–Atlantic Coastal Flats; 2, Gulf–Atlantic Rolling Plain; 3, Appalachian highlands; 4, Eastern Interior Uplands and Basins; 5, Lower Mississippi Alluvial Plain; 6, Ozark–Ouachita Highlands. (Based on Hammond (1964).)

Mississippi River alluvial plain, the low topographic relief in this area has favored the construction of levees and floodgates instead of reservoirs for the control of flood waters.

GENERAL LIMNOLOGY OF SOUTHEASTERN RESERVOIRS

As expected from their broad geographic distribution, reservoirs of the southeastern United States span a wide range of hydrological and limnological conditions (Tables 2 and 3). Consequently, generalizations about these varied and variable systems are imprecise. However, based on data from 116 southeastern reservoirs sampled in the National Eutrophication Survey (U.S. En-

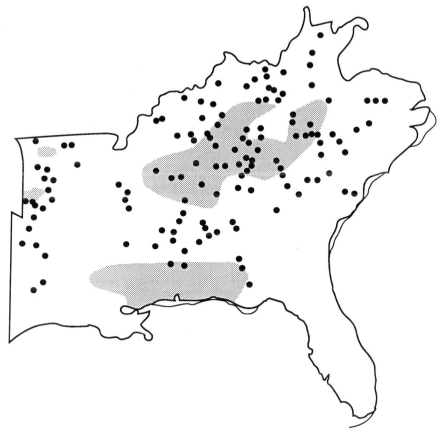

FIGURE 5. Comparison of reservoir locations and distribution of average annual runoff in the southeastern United States. Shaded areas experience runoff in excess of 50 cm per year. (Based on data reported by the U.S. Department of the Interior (1964).)

vironmental Protection Agency (USEPA) 1975, 1978), southeastern reservoirs are relatively large (median surface area = 52 km^2), deep (mean depth = 7.7 m, range = 1.3–41 m), complex morphologically (shape), and strongly influenced by their hydrology.

River–Lake Hybrids

The damming of southeastern rivers has created man-made lakes that are similar in many respects to natural lakes. For example, the limnology of many natural lakes is dominated by vertical (surface to bottom) gradients of light availability, temperature, dissolved and particulate materials, biological production, and decomposition (Hutchinson 1957, Wetzel 1983), and many south-

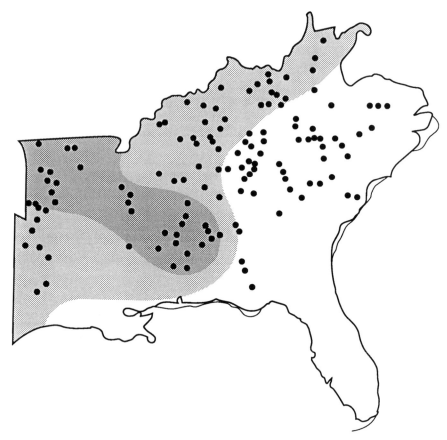

FIGURE 6. Comparison of reservoir locations and distribution of flooding potential for a 10-year flood, based on expected discharge from standard drainages of 800 km². Potential flood discharges in excess of 600 and 900 m³/s are represented by light and dark shading, respectively. (Based on data reported by the U.S. Department of the Interior (1964).)

eastern reservoirs have similar vertical gradients. However, these impoundments also retain important river-like (lotic) characteristics. For example, rivers are dominated by longitudinal (upstream to downstream) gradients in channel morphology, flow velocity, water temperature, bottom substrate type, and biotic community composition (e.g., Hynes 1970, 1975, Cummins 1974, 1979, Vannote et al. 1980, Minshall et al. 1983; see Wallace et al., Chapter 4, this volume), and southeastern reservoirs typically have longitudinal gradients of this type as well. Thus, reservoirs blend together the properties of rivers and lakes, and numerous aquatic ecologists have suggested that river impoundments are intermediate ecologically between rivers and lakes (Regier and Henderson 1973, Carline 1986) and should be viewed as "river–lake

TABLE 2   Morphologic and Hydrologic Characteristics of U.S. Corps of Engineers Reservoirs in the South

Variable	N	Mean	Median
Drainage area (km²)	52	24 682.4	2930.6
Surface area (km²)	52	85.7	52.5
DA/SA ratio[a]	52	439.1	75.6
Shoreline length (km)	52	459.6	277.6
Shoreline development	52	14.3	12.0
Volume (m³ × 10⁶)	52	1 110.6	301.6
Maximum depth (m)	52	31.7	24.8
Mean depth (m)	52	9.5	7.9
Thermocline depth (m)	35	7.8	7.6
Outlet depth (m)	52	17.8	14.0
Pool fluctuation (m)	52	6.3	4.6
Annual discharge (m³ × 10⁶)	51	9 406.8	1567.9
Water residence time (days)	51	143	82

[a]Ratio of drainage basin area to reservoir surface area.

Source. Based on data reported by Leidy and Jenkins (1977).

hybrids" (e.g., Lind 1971, Henderson et al. 1973, Ackermann et al. 1973, Ryder et al. 1974, Margalef 1975, Ryder 1978, Goldman and Kimmel 1978, Thornton et al. 1981, Jenkins 1982, Benson 1982, Kimmel and Groeger 1984, Soballe and Bachmann 1984, Soballe and Threlkeld 1985, Soballe and Kimmel 1987, Kimmel et al 1990). Accumulating evidence supports this view of river impoundments, and we believe that the river–lake hybrid concept of reservoirs is central to an understanding of the ecological structure and functioning of these important southeastern ecosystems.

The fundamental biological, chemical, and physical processes in reservoirs are the same as in rivers and lakes; therefore, there is a large common ground upon which to build an understanding of all three system types. River im-

TABLE 3   Limnological Characteristics of Selected Southeastern Reservoirs

Variable	N	Mean	Median
Phosphorus load (g/m² year⁻¹)	30	6.41	2.72
Nitrogen load (g/m² year⁻¹)	30	66.3	27.2
Total phosphorus (mg/L)	45	0.042	0.31
Total nitrogen (mg/L)	26	0.594	0.525
N/P ratio	26	58.9	50.3
Dissolved solids (mg/L)	51	102.9	60.0
Chlorophyll (μg/L)	45	7.1	6.2
Secchi disk (m)	45	1.6	1.4

Source. Based on data reported by Walker (1981), Leidy and Jenkins (1977), and Placke (1983) for Corps of Engineers and Tennessee Valley Authority reservoirs.

poundments, as artificial and manipulated systems, exaggerate some features and processes while diminishing others relative to conditions found in most natural lakes and rivers. The degree to which the ecological behavior of southeastern reservoirs deviates from the majority of rivers and natural lakes can often be linked to the engineering requirements and operational practices associated with these river-impoundment projects (Thornton 1984, Kennedy et al. 1985).

Basin Morphometry

Morphometry (shape) is a significant factor in the ecology of lakes, streams, and impoundments. The typical dendritic (tree-like) outline of many southeastern reservoirs results directly from the engineering criteria used for site selection; e.g., dams are more often constructed in the upper reaches of drainage basins because of the favorable topography, and impoundment of river water in the more dissected terrain of headwater areas produces deep reservoirs with irregular shorelines. Also, because dam construction is very recent, there has been little time for erosion to smooth these jagged outlines, and, as a consequence, southeastern impoundments are characterized by numerous coves, islands, and embayments. This is illustrated by the morphometry of USCE reservoirs in the southeast. A common index of shoreline irregularity is the shoreline development ratio, which is the ratio of the actual shoreline length to the circumference of a circle that has the same area as the lake surface; a perfectly circular lake has the minimum possible shoreline development value of 1. USCE reservoirs in the Southeast have a median shoreline development ratio of 12.0. By comparison, most natural lakes have shoreline development indices that are less than three. For example, kettle lakes of glacial origin in Wisconsin and other midwestern states have shoreline development ratios that lie mostly between 1.5 and 2.5, and the maximum value for a natural Wisconsin lake is only 3.18 (Hutchinson 1957). The shoreline lengths of the reservoirs in the Southeast (median shoreline length = 278 km; maximum = 1745 km, Cumberland Lake) also suggest their large size and irregular shape.

The complex shape of southeastern reservoirs enhances (by a factor of 4 to 8 over most natural lakes) the direct connection between the aquatic and terrestrial components of these ecosystems. This enhanced degree of land–water interaction is of major ecological significance. The view of lakes as isolated microcosms (Forbes 1887) has given way to the understanding that lakes and rivers are significantly affected by the water, material, and organisms that are transported into them from their surrounding watersheds (Hynes 1975, Likens 1984). In addition to direct shoreline contact, southeastern reservoirs also have extensive upstream watersheds that are roughly twice the size of watersheds that surround natural lakes of comparable lake surface area. Again, the data from 52 southeastern USCE reservoirs provide an example. In this group of impoundments, drainage areas range from 508 km^2 for Grayson Lake (a small tributary reservoir located on the Little Sandy River in northeast Kentucky) to 397 995 km^2 for Ozark Lake (a main-stem impoundment of the Arkansas River; this drainage area includes the watersheds of upstream reservoirs). The median drainage area of USCE reservoirs in the southeast is 2931 km^2 and the median drainage area:reservoir surface area ratio is 76. A review of similar data for natural lakes sampled during the U.S. Environmental Protection Agency's National Eutrophication Survey (as summarized by Walker (1981)) yields a ratio of only 33.

Morphometric data for southeastern reservoirs also strongly suggest that the ecological functioning of these impoundments should, in regard to terrestrial influences, be intermediate between rivers and natural lakes. External influences should be far more significant in these reservoirs than in most natural lakes, but somewhat less important than in rivers which have the strongest link between the aquatic and terrestrial components.

Operational Practices

The operation of the dam can strongly influence the effect of a reservoir on the downstream river and can also alter the ecological structure and functioning within the reservoir itself. Dams are built to regulate water and, unlike most natural lakes and rivers, reservoir water levels and releases are controlled to meet specific project objectives. The most ecologically significant aspects of reservoir operation are (1) the quantity of water released, (2) the timing of releases, and (3) the depth(s) from which the water is released.

The quantity and timing of reservoir releases are driven by the design purposes of the project (e.g., hydropower generation, navigation, flood control) and result in downstream flow patterns that seldom resemble the natural regime that existed before impoundment. In some extreme cases, impounded water is routed via penstocks to generation stations located far downstream or in adjacent drainages, and extensive reaches of streambed directly below the dam are dewatered completely. In other situations, daily cycles of hydropower generation result in alternate drying and flooding of the downstream streambed. Although reservoir operations often serve to dampen long-term (seasonal) fluctuations in streamflow, they may also greatly enhance short-term (e.g., daily) changes in flow. The effects of altered flow regimes on the biotic communities downstream of impoundments have often been severe and are well documented. We will discuss these effects only briefly in this chapter, and we refer the reader to existing reviews on this subject for more detailed information (e.g. Ward and Stanford 1979, Baxter 1977).

Controlled reservoir releases also have significant effects within the impoundment itself. There is significant seasonal variation in runoff in the southeast, and a temporal mismatch exists between reservoir inflow volume (highest in winter and spring) and water demand (highest in late summer and fall). As a result, water levels within southeastern reservoirs are often subject to water-level changes that are far greater than in natural lakes. Water-level variations of several meters are not uncommon, and the greatest fluctuations occur in tributary reservoirs (mean fluctuations: USCE reservoirs = 4.6 m, TVA tributary reservoirs = 5.6 m, TVA main-stem reservoirs = 2.2 m). In the TVA system, short-term (weekly) and low-amplitude (approximately 1 m) fluctuations are used to limit mosquito reproduction, whereas more long-term (seasonal) fluctuations with greater amplitudes result from seasonal drawdown for flood control and hydropower demands. Seasonal demands for peak hydropower generation (i.e., to provide air conditioning in summer) in the Southeast occur typically during minimal streamflow, and continued discharge for hydropower generation without replacement by stream inflow can drop the water level in tributary reservoirs more than 5 m during late summer. This lowering of the water level exposes the otherwise extensive littoral zones of these impoundments to periodic (seasonal) drying and hence the vascular plant beds that would normally occupy this zone cannot develop. Water-level changes in main-stem impoundments normally are less severe because these reservoirs are located in less steeply sloping terrain (greater volume of water per depth increment) and have more extensive upstream watersheds (less seasonal variation in water supply), and (perhaps most importantly) navigational requirements necessitate that water levels be held relatively constant.

Discharge Depth and Stratification

Natural lakes usually discharge by simple overflow of surface water, whereas reservoir releases are controlled by outlet structures which usually can discharge water from one or more subsurface depths. This is an important distinction, because southeastern reservoirs tend to be deep and seasonally stratified, and selective subsurface discharge from these reservoirs couples conditions in the receiving stream to vertical gradients in temperature and water quality in the upstream impoundment.

The formation of strong vertical gradients of dissolved and suspended material in lakes and reservoirs is a consequence of density stratification (Hutchinson 1957, Wetzel 1983). Heating of the water surface by solar radiation, combined with wind mixing of the near-surface water, is the most common cause of this stratification. Heated, less dense, surface water floats on the cold, denser water underneath, thus producing a stable layering that cannot be fully disrupted by wind and convection unless the lake is shallow (or the wind is very strong). A denser layer of water (the hypolimnion or bottom layer) usually persists beneath the well-mixed near-surface layer (epilimnion) from late spring to mid fall in the Southeast. The boundary region between the warm and cool layers is termed the metalimnion and a seemingly small temperature difference (less than 1 °c/m) in the metalimnion can effectively prevent mixing between the epilimnion and hypolimnion (Fig. 7a).

Deep release of hypolimnetic water is common in southeastern hydropower reservoirs, and the 52 USCE reservoirs mentioned previously provide an illustration. Mean and median discharge depths in this group are 17.8 and 14.0 m, respectively. For these same systems, the depth of the thermocline averages 7.0 m, so that, on average, these systems release cool hypolimnetic water downstream. The thermal and chemical effects (or benefits) of these deep releases are of major ecological significance to the receiving streams.

Hypolimnetic discharge also affects the releasing reservoir. The particulate and dissolved nutrients that normally accumulate in the bottom waters of natural lakes are, in deep-discharge impoundments, released downstream. One result is that the releasing reservoir may become depleted in nutrients and its productivity may be reduced, while downstream rivers and reservoirs become enriched (Wright 1967, Martin and Stroud 1973, Elser and Kimmel 1985).

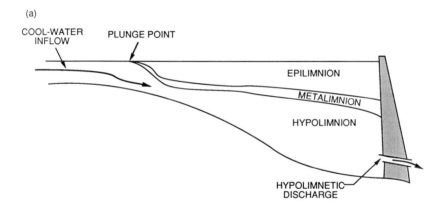

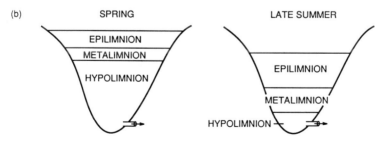

FIGURE 7. (*a*) Summer thermal stratification in a southeastern U.S. reservoir. (*b*) Spring to late-summer changes in patterns of reservoir thermal stratification resulting from hypolimnetic discharge and reduced water level.

In tributary reservoirs, where summer hydropower releases greatly exceed inflows, the release of bottom water lowers the water level while expanding the vertical extent of the epilimnion. This happens because an approximately constant water volume in the epilimnion is being drawn down into a narrowing basin (Fig. 7b). As this happens, the stability of the vertical stratification weakens and the impoundment becomes less resistant to wind mixing. Increased mixing can then enhance reservoir productivity as plant nutrients released by microbial activity in near-bottom waters are transported into the illuminated surface layer where they can be assimilated by photosynthetic algae.

In thermally stratified reservoirs that receive cold inflows from upstream and that do not experience severe drawdown (e.g., downstream impoundments in a series of hydropower reservoirs), the situation is reversed. Cool-water discharges in this situation are replenished by cold-water inputs from upstream and strong vertical stratification is maintained (Fig. 7a).

Density Currents

As river water enters an impoundment, it may not mix immediately or completely with the water mass within the reservoir, but rather can maintain its identity because of density differences (usually caused by differences in temperature) between the inflowing river water and the reservoir water (Fig. 8). Three basic types of water movement result: (1) the river water is less dense (warmer) and flows across the reservoir surface (overflow, Fig. 8a), (2) the river water is denser (cooler) and flows along the reservoir bottom (underflow, Fig. 8b), or (3) the river water flows at an intermediate depth in the reservoir (interflow, Fig. 8c). In all cases, horizontal momentum allows the inflowing river water to push back the lake water temporarily, but as the reservoir cross section (width and depth) increases, the inflowing water slows and then either sinks (plunges) to the bottom (or to a depth of equal density) or is lifted by buoyant forces and flows across the reservoir surface. The point at which gravity or buoyancy overcomes horizontal momentum is called the plunge point or plunge line (although this terminology is inappropriate for overflows).

Density flows depend on the pattern of density stratification in the impoundment, the density and volume of the inflow, and dam operations (i.e., volume, timing, and depth of releases). Density currents control the location and extent of mixing between the lake water and the river inflow, and this in turn controls the transport and fate of materials (e.g., nutrients, contaminants, dissolved oxygen, suspended particles) transported into the impoundment. For example, in a stratified reservoir that receives cold inflow from upstream and that discharges from the hypolimnion, river water entering the impoundment may move along the bottom as a density flow and be discharged downstream before it can mix with the epilimnion. Nutrients transported into the reservoir by the river would contribute little to productivity within this impoundment and the main interaction is expected to be between the river inflow and the reservoir sediments. Density flows are not unique to impound-

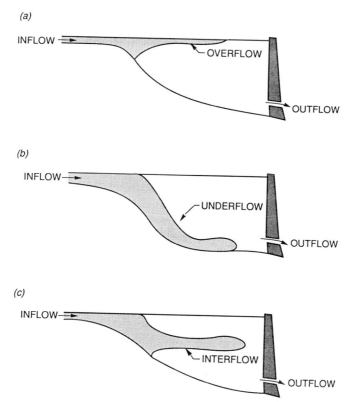

FIGURE 8. Types of density flows in reservoirs. Often the inflowing river and reservoir waters differ in temperature and, therefore, in density. If the river inflow is warmer than the reservoir water, the less dense river water will spread over the reservoir surface as an overflow (*a*). If the river inflow is cooler and denser than the entire reservoir water mass, the inflowing river water plunges from the surface and flows along the reservoir bottom as an underflow (*b*). If the river inflow is of an intermediate temperature and density, it plunges from the surface and proceeds downstream at the depth at which the river and reservoir water densities are equal. (After Wunderlich (1971).)

ments and have been observed occasionally in some large natural lakes (Hutchinson 1957), but the combination of a cool river inflow and a hypolimnetic outflow make density flows commonplace in reservoirs.

Longitudinal Zonation

The morphology and hydrodynamics of reservoirs result in longitudinal gradients in physical and chemical characteristics that coincide with, or result in, longitudinal changes in ecological structure and functioning. Within most southeastern reservoirs, there is an upstream-to-downstream decline in the influence of river inflow and a concomitant change from riverine to lacustrine (lake-like) conditions. This change corresponds to a downstream widening and deepening of the reservoir basin (increasing the reservoir cross section) and a slowing of flow velocity (factors that control the location of the plunge line). The change from a riverine to a lacustrine environment, coupled with the elongate morphology that is typical of southeastern reservoirs, can be idealized into a three-zone model (Fig. 9) that has been suggested in various forms by a number of researchers (e.g., Baxter 1977, Thornton et al. 1981, Kimmel and Groeger 1984, Kimmel et al. 1990). This model illustrates the internal river-lake hybrid nature of reservoirs, with riverine features dominating the upstream zone and lake-like characteristics found near the dam.

Because the longitudinal zonation of reservoirs is flow-induced, it is also temporally and spatially dynamic. Density flows can cause a sharp transition (e.g., via plunging) from riverine to lake-like conditions and the riverine, transition, and lacustrine zones of southeastern reservoirs can expand and contract in response to long-term (annual, seasonal) and short-term (storm

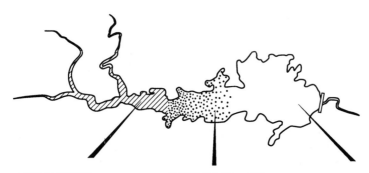

RIVERINE ZONE	TRANSITIONAL ZONE	LACUSTRINE ZONE
• Narrow, channelized basin (small basin cross-section)	• Broader, deeper basin	• Broad, deep, lake-like basin (large basin cross-section)
• Relatively high flow	• Reduced flow	• Little flow
• High suspended solids, turbid, low light available, $Z_p < Z_m$	• Reduced suspended solids, less turbid, light availability increased	• Relatively clear, light more available at depth, $Z_p > Z_m$
• Nutrient supply by advection, relatively high nutrients	• Advective nutrient supply reduced	• Nutrient supply by internal recycling, relatively low nutrients
• Light-limited phytoplankton production	• Phytoplankton production/m^3 relatively high	• Nutrient-limited phytoplankton production
• Cell losses primarily by sedimentation	• Cell losses by sedimentation and grazing	• Cell losses primarily by grazing
• Organic matter supply primarily allochthonous, $P < R$	• Intermediate	• Organic matter supply primarily autochthonous, $P > R$
• More "eutrophic"	• Intermediate	• More "oligotrophic"

FIGURE 9. Longitudinal zonation in environmental factors controlling light and nutrient availability for phytoplankton production, algal productivity and standing crop, organic matter supply, and trophic status in a river impoundment. (Modified from Kimmel and Groeger (1984).)

event, daily, hourly) changes in watershed runoff or river inflow, in thermal stratification, and in reservoir operations. Depending on watershed characteristics and river inflow velocity, all three zones may not always be distinguishable within a particular impoundment. Under low-flow conditions, both the riverine and transitional zones may be compressed into a small, uplake portion of the reservoir; consequently, most of the reservoir will exhibit lacustrine conditions. With increased watershed runoff and river inflow, water residence time is reduced and the riverine and transitional zones expand downstream. Under high-flow conditions, often accompanied by increased turbidity resulting from river-borne sediments, the riverine zone may extend throughout most of the reservoir.

Reservoir Classification and Water Residence Time

Because reservoirs span such a wide range of conditions, our understanding of these systems can be aided by classifying them into more homogeneous subgroups. Several classifications are possible: (1) by project purpose or operational practices (e.g., flood control, water supply, hydropower production, navigation, recreation); (2) by type of release or discharge structure (e.g., surface overflow, near-surface discharge, near-bottom or hypolimnetic discharge, multiport release, pumpback); (3) by location in the drainage basin (e.g., tributary or mainstem); and (4) by water residence time. Our approach to southeastern reservoirs emphasizes the latter two classifications.

Water residence time is a useful index of reservoir functioning because it reflects the extent of river influence on the physical and biological processes of reservoirs and is also related to important morphometric features (reservoir depth, surface area, drainage area). Reservoirs with short residence time (< 30–40 days) tend to be river-like in their ecological structure and function, whereas long-residence time reservoirs (>100 days) are often very similar to natural lakes. In this scheme, impoundments occupy intermediate (and overlapping) positions between rivers and natural lakes on a continuum of aquatic ecosystems that is ordered by water residence time (Fig. 10). This continuum has proven to be a useful index of aquatic ecosystem function (Soballe and Kimmel 1987).

A theoretical water residence time is calculated by dividing reservoir volume by total annual outflow. The resulting value corresponds to the actual water residence time only if water progresses steadily through the reservoir as a defined water mass; but regardless of the actual flow pattern, theoretical residence time is a useful index of water renewal and system structure.

Most southeastern reservoirs have relatively short water residence times as a result of their large watersheds and large hydraulic (water) loadings. The average water residence time of the 116 southeastern reservoirs included in the National Eutrophication Survey is relatively short (140 days or 0.38 years), and more than half the reservoirs have a residence time less than 100 days (0.27 years) (Fig. 11). This implies that large "riverine" reservoirs are most common in the Southeast. Similarly the median theoretical residence time for

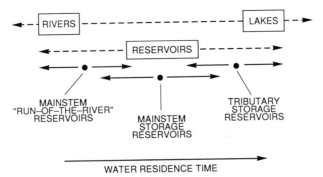

FIGURE 10. Reservoirs occupy an intermediate position between rivers and natural lakes along a continuum of aquatic ecosystem types. The water residence time and the extent of riverine influence determine the relative positions of the various types of impoundments (main-stem run-of-the-river, main-stem storage, tributary storage) along the river-lake continuum. (Modified from Kimmel and Groeger (1984).)

the group of 51 USCE reservoirs discussed earlier is only 81 days (0.22 years) and, thus, for half of these reservoirs, the total reservoir volume is displaced at least 4.5 times each year. In contrast, the median residence time of natural lakes surveyed by Walker (1981) was 270 days (0.74 years = flushing rate of 1.35 times per year). As a result of operational, morphological, and stream-order differences, the mean water residence time for tributary reser-

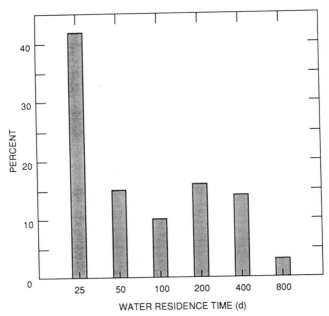

FIGURE 11. Frequency distribution of the average water residence time of 116 major reservoirs located in the southeastern United States.

voirs (263 days, 0.72 years) is significantly longer than for mainstream impoundments (51 days, 0.14 years; Table 4).

As with all the previously discussed characteristics of southeastern reservoirs, the mean water residence time varies greatly (from 1 day to 3 years). This may help explain the wide variations in the structure and function of these systems. Reservoirs with water residence times ranging from about 30 to 100 days represent the transition from riverine to lacustrine-type reservoirs (Soballe and Kimmel 1987). Impoundments in this range should vary most dramatically among themselves, and within individual impoundments they should exhibit marked temporal and longitudinal variations.

Location within the drainage basin is also a useful classification criterion for southeastern impoundments, because location conveys important information about reservoir morphometry and likely operational practices. We can classify southeastern reservoirs into two large groups: (1) *tributary reservoirs*, impoundments located on low-order streams (order < 7–8, see Chapter 3 for definition) with drainage areas less than about 10 000 km^2, which provide off-channel storage and temporal redistribution of seasonally variable flows; and (2) *main-stem reservoirs*, which impound larger rivers (stream order > 7–8, drainage area > 10 000 km^2) and provide direct control of flow in major river basins. There are major differences between these two types of reservoirs (e.g., see Table 4).

TABLE 4 Summary of Differences in Limnological Characteristics of Tributary and Main-Stem Reservoirs

	Tributary		Main Stem	
Characteristic	N	Mean	N	Mean
Annual runoff (cm)	29	51.6	16	43.9
Drawdown (m)	21	5.6	6	2.2
Water residence time (years)	29	0.72	16	0.14
Secchi depth (m)	29	1.8	16	1.3
Mean depth (m)	29	12.0	12	9.1
Surface area (km^2)	29	97.4	12	110.5
Volume ($\times$ 10^6 m^3)	29	1061.8	12	1153.7
Inorganic nitrogen (mg/L)	18	0.173	8	0.352
Total nitrogen (mg/L)	18	0.559	8	0.672
Total nitrogen load	22	26.4	8	164.1
Organic nitrogen load				
(percentage of total nitrogen load)	22	65.0	8	52.0
Total phosphorus (mg/L)	29	0.037	16	0.052
Total phosphorus load (g/m^2 year^{-1})	22	2.3	8	16.6
Chlorophyll *a* (µg/L)	29	7.7	16	5.9

Note. All differences listed are signficant (probability > t < .05).

Source. Based on data reported by Walker (1982) and Placke (1983).

Because tributary reservoirs are located higher in the drainage basin and often in areas of greater topographic relief than main-stem reservoirs, they usually have higher depth:surface area ratios, longer water residence times, lower nutrient loading rates (per unit area), and more pronounced (10–15 m) water level fluctuations as a result of flood control, hydroelectric generation, or flow augmentation operations. The relatively long water-residence time in tributary reservoirs results in lacustrine conditions. But unlike typical natural lakes, many tributary reservoirs are designed to store seasonal flood waters and their water levels must be lowered prior to the high-flow period in the spring. The water-level fluctuations that result from this operational practice prevent the development of shoreline (littoral zone) biotic communities, and, consequently, the food webs of tributary systems often are based primarily on open-water (pelagic) production.

Main-stem reservoirs are located lower in drainage basins and are more influenced by river inflows. They span a wide range of residence times, but generally have shorter water residence times than tributary systems. Relative to tributary reservoirs, main-stem systems also have less dramatic (3–4 m) water level fluctuations (often due to river regulation by upstream impoundments and to navigational requirements) and have expanded riverine and transitional zones (Fig. 9). During high flows, riverine conditions may dominate the entire mainstem reservoir pool and, in "run-of-the-river" main-stem impoundments (average water residence time < 30 days), riverine conditions are the normal situation. The larger upstream drainage areas of main-stem reservoirs, relative to tributary systems, contribute greater nutrient loads (concomitant with high hydrologic loading), and water residence times are sometimes long enough for phytoplankton to grow and accumulate in response to these higher nutrient levels. Littoral communities may also develop in main-stem reservoirs (due to relatively stable water levels) and this may further enhance biological productivity.

The differences between main-stem and tributary impoundments in the Southeast are illustrated by data for COE and TVA reservoirs (Table 4). Like natural lakes, the tributary reservoirs stratify thermally during most of the growing season (late spring to mid fall); however, many main-stem impoundments show intermittent stratification or fail to stratify as a result of rapid flushing and the influence of upstream reservoirs. Both tributary and main-stem reservoirs have complex shorelines and extensive overbank (submerged floodplain) areas that form extensive littoral zones. However, in the tributary reservoirs, winter drawdown for flood control and summer drawdown for hydropower result in an annual (or semiannual) drying and freezing of the littoral area, which suppresses colonization by aquatic plants.

There are strong connections between water residence time, reservoir location, and operational practices because stream size, flow volume, and basin morphometry are interrelated (e.g., Leopold et al. 1964). Thus, water residence time and reservoir location are both useful indices, but at a minimum, information on water residence time is essential for understanding the ecological structure and functioning of impoundments.

Reservoirs in Series

The effect of reservoir releases on downstream water temperature, flow regimes, water quality, and biota have all been studied (Young et al. 1972, Ward 1976, Ward and Stanford 1979, 1981), but relatively little is known of the limnological and ecological interactions that occur in multiple-impoundment series. Water release patterns are a central feature in the ecology of river–reservoir systems and, when reservoirs are constructed in series, this maximizes the potential for controlling the timing, volume, and quality of discharged water. In particular, the hypolimnetic release of water and nutrients from deep-discharge impoundments in a series may amplify the influence of upstream impoundments on the physical, chemical, and biological processes in reservoirs downstream (Neel 1963, Wright 1967, Gloss et al. 1981, Paulson and Baker 1981, Evans and Paulson 1983, Elser and Kimmel 1985, Kimmel et al. 1990).

The ecological interactions that occur among reservoirs in series will likely be site-specific and vary greatly. Basin morphology, reservoir placement, and the design of the reservoir release structures are major determinants of these interactions. For example, bottom release from a deep, stratified, and anoxic impoundment into a relatively shallow downstream system will create conditions that differ markedly from those in a series of shallow, surface-release impoundments. Likewise, if all reservoirs in a series are designed for deep-water discharge, then bottom water that is released from an upstream reservoir in the series may travel through warmer downstream impoundments by underflow, and this can enhance the transport of materials (e.g., nutrients and contaminants) through the series relative to a reservoir series with surface releases. The overall pattern of interactions will depend upon the structure and functioning of each impoundment in the series and the relationship of each impoundment to the reservoirs above and below it in the river–reservoir system.

It is important that the ecology of reservoir series be understood, because dam construction has converted nearly all the major rivers in the Southeast into series of regulated pools. We estimate that over 100 major reservoirs (about 75% of the total number of major impoundments) are arranged in series in the Southeast. The most extreme case is in the Tennessee River basin where 22 major impoundments have been built in series. Eight of these impoundments form a series on the mainstem of the Tennessee River. Side-chains of two-, four-, and five-impoundment series (Clinch, Tennessee, and Little Tennessee basins) join the top of this main-stem chain for a maximum chain length of 13 reservoirs. The Santee River basin (with its Catawba River tributary) in the Carolinas has 14 impoundments and a maximum series length of 10 reservoirs. The Mobile River basin has 19 impoundments in series (outside the Tombigbee subbasin), but, as with the Arkansas River with 11 impoundments, many of these are primarily navigational (lock and dam) "run-of-the-river" structures with water residence times less than 10 days.

We know that the ecological effects of multiple-impoundment projects on southeastern rivers have been profound. The current challenge is to understand the ecological structure and functioning of these impoundments as parts of a larger, interconnected system, so that their ecological impacts and benefits can be scientifically managed.

BIOTIC COMMUNITIES

The transfer of nutrients and energy among biotic components of southeastern reservoirs is a major functional aspect of these ecosystems (Fig. 12) and is closely tied to physical structure. The physical structure of southeastern impoundments varies widely, both within and among reservoirs, and, as might be expected, the interrelationships among the biotic components of these systems also display temporal and spatial heterogeneity. The within-reservoir (autochthonous) photosynthetic activity of suspended algae (phytoplankton), attached algae (periphyton), and macrophytes (macroscopic algae and aquatic vascular plants) supplies most of the organic matter to the food webs of southeastern impoundments. External inputs of organic matter from the watershed may supplement significantly the internal production in some reservoir systems, but apparently only in extremely turbid waters do these external (allochthonous) sources of organic matter have major importance (Adams et al. 1983, Kimmel et al. 1990).

Primary Producers

Autochthonous production and the overall standing crop of photosynthetic organisms (primary producers) in southeastern reservoirs are governed by light availability, nutrient concentrations, substrate (for rooted or attached plants), water level fluctuations, and water residence time (especially for phytoplankton). These environmental factors, operating in concert, also determine the relative contributions of the various producer types (phytoplankton, attached algae, and aquatic macrophytes) to overall system production. Phytoplankton production dominates in most southeastern impoundments because changing water levels inhibit the development of littoral macrophyte and periphyton communities. However, in reservoirs having relatively stable water levels, expanding stands of rooted and floating macrophytes (e.g., eurasian milfoil or *Hydrilla*) can become a nuisance.

Southeastern reservoirs are relatively nutrient-rich and moderately productive. In 30 reservoirs surveyed, the median total concentrations of P and N were 0.031 mg P/L and 0.525 mg N/L. Phytoplankton productivity in southeastern reservoirs (100–1500 mg C/m^2 day^{-1}; Adams et al. 1983) spans the range observed in natural lakes (cf. Wetzel 1983, Kimmel et al. 1990).

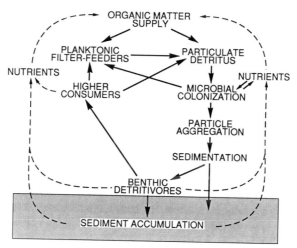

FIGURE 12. Energy and nutrient flow in a reservoir ecosystem. Allochthonous (watershed) inputs of dissolved and particulate organic matter directly enter the "detritus pathway," and support the growth of bacteria, fungi, and detritivores. Some of the autochthonous (in-lake) primary production enters the "grazer pathway" and provides food for herbivores and, ultimately, for higher consumers. However, much of the phytoplankton production may also enter the detritus pathway via phytoplankton mortality and support both planktonic and benthic detritus feeders. Adsorption of dissolved organic compounds, particle aggregation, and microbial colonization produce microheterotroph–detrital aggregates, which, by virtue of their particle size, are available to planktonic filter feeders. Sedimentation of detrital aggregates and zooplankton fecal pellets provides a major energy source to the benthic detritivores, which are preyed upon by higher consumers. Nutrient regeneration occurs at virtually every level of the food web, and only a small fraction of the organic matter produced ultimately accumulates as permanent bottom sediment. (Modified from Goldman and Kimmel (1978).)

Because they have a high ratio of drainage area to lake-surface area, southeastern reservoirs have annual nutrient loads that are high in comparison to most natural lakes (Table 3). Nutrient loads (product of inflow volume and concentration) and concentrations in mainstem impoundments are significantly higher than in tributary reservoirs (Tables 3 and 4). Nitrogen and phosphorus areal loadings (i.e., expressed per unit of reservoir surface area as g/m^2 year^{-1}) are approximately 6 and 7 times higher, respectively, to mainstem than to tributary reservoirs in the southeast (Table 4). These differences are attributable to the larger drainage areas of main-stem reservoirs, and to the more numerous point and nonpoint nutrient sources along the larger streams which flow into main-stem impoundments.

Phosphorus is an essential plant nutrient that, in many freshwater systems, is in short supply relative to the demands of phytoplankton growth. As a result, average phytoplankton standing crop usually correlates with total phos-

phorus concentration unless factors other than phosphorus are more important in controlling phytoplankton abundance. Phytoplankton forms the base
of the food chain in many aquatic systems, and Hoyer and Jones (1983)
demonstrated a correlation between phytoplankton standing crop (chlorophyll
concentration) and fish production in midwestern reservoirs. The planktonic
chlorophyll–total phosphorus relationship (Fig. 13) in 45 southeastern reservoirs is similar to that reported from other areas. The mean and median
concentrations of planktonic chlorophyll for these impoundments are 7.1 and
6.2 mg/m³, respectively, and are similar to values reported for reservoirs in
other regions of the United States, but are significantly lower than for most
natural lakes (Walker 1981, Jones and Bachmann 1976). Walker (1984, 1985)
attributed lower chlorophyll values in reservoirs to the influence of high
concentrations of suspended inorganic material in some reservoirs. However,

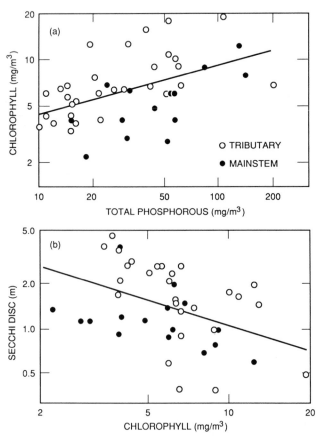

FIGURE 13. Relationships of chlorophyll and total phosphorus concentrations and
Secchi disk transparency and chlorophyll concentrations for 45 tributary (○) and main-
stem (●) reservoirs operated by the Tennessee valley Authority and the U.S. Army
Corps of Engineers in the southeastern United States.

Soballe and Kimmel (1987) showed that this difference is also correlated with the shorter water residence times of impoundments relative to most natural lakes.

The chlorophyll–phosphorus data also show that, as a group, main-stem reservoirs in the Southeast have lower chlorophyll concentrations per unit phosphorus than do tributary reservoirs. Again, this is likely attributable to the shorter water residence times and higher turbidities (reduced light availability for algal growth) in main-stem reservoirs. The latter suggestion is supported, in part, by a comparison of Secchi disk and chlorophyll relationships for the two classes of reservoirs (Fig. 13b). On average, main-stem reservoirs have lower transparency per unit of chlorophyll than do tributary reservoirs, which suggests a significant influence of nonalgal material on light penetration on the main-stem systems.

As with other physical–chemical variables, the relative importance of turbidity in reservoirs varies widely in the Southeast. Although high turbidity and light limitation are major determinants of productivity and abundance in many southeastern reservoirs, particularly in the Gulf–Atlantic Rolling Plain subdivision (Fig. 4), reservoirs in the Ozark and Appalachian Highlands subdivisions have relatively high transparencies.

The phytoplankton communities of southeastern impoundments are also linked to both the upstream and downstream reaches of the river–reservoir system. Larger rivers in the Southeast transport a significant load of suspended algae (Briggs and Ficke 1977) and these organisms can contribute to (or be derived from) impoundments on these rivers (Soballe and Kimmel 1987).

In rapidly flushed reservoirs, planktonic primary producers are quickly washed downstream. In these situations, attached plants (periphyton and aquatic macrophytes) that can maintain their position may have increased significance to total system productivity. In nutrient-poor systems, rooted aquatic macrophytes may have a competitive advantage for nutrients over phytoplankton, and macrophytes differ greatly from phytoplankton in their relationship to the food web. For example, many of the nutrients and much of the organic matter contained in macrophyte biomass may be unavailable to other portions of the food web until the plants die in late summer. Macrophytes provide not only raw organic matter, but also localized concentrations of food organisms that, in comparison to the dispersed food resources found in the planktonic environment, may be more readily utilized by fish and other consumers. For example, Cowell and Hudson (1967) and Claflin (1968) found in two Missouri River reservoirs that summer densities of macroinvertebrates (a major food source for fish) were 4–11 times greater on submersed timber than on adjacent bottom sediments. Ploskey (1986) further showed that in these same systems the abundance of benthic invertebrates was significantly correlated with the abundance (biomass) of attached algae. Rooted aquatic plants also provide important habitat for fish (especially fry) and fish-food organisms. However, if overly abundant, rooted macrophytes can be a highly visible nuisance that degrades the recreational and aesthetic value of the water resource.

Although attached plants are an important producer component of many aquatic systems, water-level fluctuations and high turbidity restrict macrophyte and periphyton growth in many southeastern reservoirs. Substrates located below the 1% light level are generally unsuitable for periphyton colonization, and periphyton desiccate and die if exposed to the air for several days (Ploskey 1986, Benson and Cowell 1967, Claflin 1968). Several weeks, or even months, of inundation are usually required for recovery (e.g., Benson 1973). Likewise, many rooted macrophytes cannot tolerate short-term fluctuations in water depth and, although the warm water temperatures of many southeastern impoundments favor macrophyte growth, cold water discharges from upstream reservoirs may inhibit their growth.

Aquatic macrophytes and attached algae attain maximal abundance in southeastern reservoirs that have relatively clear water and stable water levels. For example, main-stem TVA reservoirs are maintained at a relatively constant water level for navigational purposes, and shallow embayment areas of these impoundments are often infested with *Myriophyllum, Najas, Potamogeton*, and floating algal mats composed of *Oedogonium, Mougeotia*, and *Lyngbya* (Cox 1984). Nonetheless, only in the most plant-infested reservoirs do attached plants cover more than 15% of the water surface. In the Tennessee River system, for example, they generally cover less than 5% of the surface area (e.g., Placke 1983).

The production rates of submersed macrophytes, periphyton, and phytoplankton are all in the same range (typically $1-2$ g C/m^2 day^{-1}; Wetzel 1983) on an areal basis. But because their areal extent is restricted, attached plants in southeastern reservoirs are generally minor contributors to overall ecosystem production. Increased surface area for periphyton colonization and growth occurs in some southeastern reservoirs because standing timber was left to provide game-fish habitat (e.g., DeGray, Richard B. Russell, and West Point reservoirs). In such systems, periphyton may be an important contributor to total system productivity, but unfortunately there are no data to evaluate this on a regional scale.

Longitudinal changes in reservoir basin morphology and flow velocity result in longitudinal differences in the factors that control primary productivity and that determines the relative contributions of phytoplankton, periphyton, and macrophytes to total ecosystem production. In the riverine zone, high turbidity, high nutrients, often low temperature, and short water residence time lead to low overall productivity and reduced abundance of macrophytes and phytoplankton. Except where inflowing rivers carry high concentrations of suspended algae, periphyton may be the most significant primary producers in this zone. In the transitional zone, lower turbidity, high nutrient concentrations, and longer residence times create conditions that favor phytoplankton production. In some reservoirs this zone supports the highest phytoplankton productivity (e.g., Thornton et al. 1981). The most lake-like conditions are found in the lacustrine zone. Reduced turbidity (maximal light penetration), increased water residence time, lower nutrient concentrations, and warmer water temperatures favor macrophyte growth in littoral areas and phytoplankton growth (although algal productivity may become nutrient-limited) in the open water.

Groeger and Kimmel (1988) investigated the responses of the phytoplankton community to the development of a gradient in nitrogen availability along the longitudinal axis of Normandy Reservoir, a nitrogen-limited impoundment of the Duck River in south-central Tennessee. They documented the development of corresponding gradients (decreasing uplake to downlake) in phytoplankton biomass and productivity. Additionally, the development of longitudinal gradients in rates of phytoplankton lipid synthesis, in NH_4-enhanced uptake of carbon in the dark, and in effects of nutrient enrichment on photosynthetic carbon metabolism provided direct evidence of algal physiological responses to the uplake-to-downlake decline in nitrogen availability. These data clearly demonstrate the responsiveness of the reservoir phytoplankton to longitudinal gradients in physical and chemical factors influencing algal growth (e.g., see Fig. 9).

Primary productivity in reservoirs is dependent upon physical, chemical, and biological variables which relate to climate, the size, topography, geology, and land use of the watershed, the shape of the reservoir basin, inflow characteristics, water residence time, and operational practices. Water-level fluctuations and turbidity in reservoirs can eliminate or severely reduce the littoral zone communities (periphyton and aquatic macrophytes) that, in many natural lakes, are central to food webs. As a result, the food web in many southeastern impoundments is dominated by pelagic (open water) components (e.g., Ellis 1936, Isom 1971, Ryder 1978, Kimmel and Groeger 1984, Kimmel et al. 1990) and phytoplankton dominate ecosystem primary productivity. The phytoplankton discharged from southeastern reservoirs may contribute to the food webs of downstream river reaches and impoundments (Adams et al. 1983).

Consumers

The organic matter supplied to reservoir food webs by internal and external production supports a diverse array of consumer organisms. The structure of the consumer community in southeastern reservoirs is shaped by direct effects of physical and chemical factors on the consumer organisms themselves and indirectly by the effects of reservoir conditions on resource availability, predation, and competition. Reservoir consumer organisms can be categorized into three groups: (1) zooplankton (drifting and weakly swimming animals), (2) benthos (bottom dwellers), and (3) nekton (strongly swimming animals).

Zooplankton

The animal component of the plankton is composed primarily of protozoans, rotifers, cladocerans, and copepods. These animals are mostly microscopic and tend to be cosmopolitan in their distribution (Pennak 1953). Zooplankton species of southeastern reservoirs are generally typical of North American lakes, although variants in cladoceran species that are apparently unique to the southeast have been reported (see Chapter 2). In function, the zooplankton assemblage of some turbid reservoirs may differ slightly from that of most natural lakes in that these organisms may sustain themselves by feeding more on detritus and detrital aggregates than on phytoplankton (Goldman and Kimmel 1978, Marzolf and Arruda 1981, Marzolf 1981, 1984, Arruda et al. 1983). In addition, reservoir zooplankton must contend with changing patterns of stratification, horizontal transport, and predation.

The spatial distribution of reservoir zooplankton reflects the longitudinal zonation of physical and chemical characteristics within a reservoir, as described previously. In the riverine zone, rapid water renewal generally restricts zooplankton numbers. Rapidly growing forms (e.g., rotifers) typically show the highest relative abundance in the riverine zone. As longitudinal water movement slows, and the zooplankton food resource (primarily, phytoplankton) increases in the transition and upper lacustrine zones, the abundance and diversity of the zooplankton also increases (Meinecke et al. 1987, Applegate and Mullan 1967). Typically, the maximum abundance of zooplankton occurs downstream of the transition zone. It should also be noted that longitudinal patterns of zooplankton abundance and composition in southeastern reservoirs typically vary with time (e.g., Threlkeld 1983).

Direct effects of water level fluctuations on zooplankton are not likely to be significant, but the modification or elimination of littoral zone vegetation and habitat that results from water level changes may have important indirect effects on reservoir zooplankton. To our knowledge, however, this has never been carefully investigated.

According to Brook and Woodward (1956), a residence time of 18 days is required to develop a zooplankton assemblage, and, except for rapidly reproducing rotifers, rivers support a depauperate zooplankton. Thus, the zooplankton assemblage in rapidly flushed southeastern reservoirs (Fig. 11) is influenced strongly by flushing rate, as well as by longitudinal and vertical changes in water velocity (cf. Threlkeld 1982, Dirnberger and Threlkeld 1986).

Benthos

Benthic macroinvertebrate communities (benthos) of southeastern reservoirs and impounded rivers have been the subject of numerous studies because benthos are sensitive to environmental changes. Research has focused on the effect of impoundments on the relative abundance and species composition of benthos in below-dam reaches of southeastern rivers. Less is known about changes in species composition within the impoundments themselves. It is usually difficult to measure directly the food-web effects that result from changes in the benthic invertebrate fauna, and our knowledge of these effects in reservoirs is fragmentary and mostly inferred from correlations. There can be little doubt, however, that invertebrates provide an important pathway to higher consumers for organic matter derived from both the detrital–bacterial pool and the primary producers. Modeling studies by Ploskey and Jenkins (1982) and Adams et al. (1983) clearly demonstrate the pivotal role of these organisms in southeastern reservoirs. Ploskey and Jenkins (1982) estimated that, on average, about 36% of the total standing crop of fish in DeGray Reservoir (Arkansas) was supported by feeding on benthic organisms and that this fraction ranged as high as 45%. By comparison, they estimated that only about 10% of the fish standing crop was supported by feeding on zooplankton.

The abundance and species composition of the benthos is closely tied to three major factors: (1) characteristics of the substrate, (2) physical–chemical characteristics of the water, and (3) the food supply. As we have shown earlier, all of these factors can be strongly influenced by dam construction and reservoir operations. The most important effects of reservoirs on benthic organisms are related to (1) siltation and/or scouring, (2) anoxia, (3) water-level fluctuations, (4) altered flow, (5) changes in the temperature regimen, and (6) trapping and discharge or organic matter. In newly flooded impoundments, the presence and decomposition of terrestrial vegetation and soils can also be of major significance to the colonizing benthic community (Aggus 1971).

As with other components of the biota of new reservoirs, the benthic community changes dramatically following river impoundment. There is frequently a tremendous increase in the abundance of some forms (e.g., chironomids and oligochaetes) in the years immediately following inundation, while other groups (e.g., Ephemeroptera, Trichoptera, and Plecoptera) typically decrease (Baxter 1977). The effects of impoundments on the molluscan fauna of rivers in the Southeast have been particularly pronounced. For example, the closing of the Norris Dam on the Clinch River (Tennessee) in 1936 completely eliminated 45 species of pelecypods from the tailwater area in just four months (Isom 1971). Isom's (1971) review provides detailed information on the impacts of Tennessee Valley impoundments on the invertebrate fauna.

The benthic communities of impoundments continue to respond to the morphologic and operational characteristics of the system long after the initial effects of dam closure and inundation have faded. For example, standing timber provides habitat that persists for many decades, and seasonal anoxia in the hypolimnion appears to be a permanent feature of some impoundments. Likewise, fluctuating reservoir water levels continually affect the benthos by (1) the stranding and desiccation of benthic invertebrates in the littoral zone, (2) the burial of deep bottom-dwelling (profundal) organisms by erosion and redeposition of littoral sediments, and (3) the loss or lack of habitat and food sources due to changes in substrate particle size and absence of littoral macrophytes. Hale and Bayne (1980) documented the effects of a 3-m water-level fluctuation in West Point Reservoir (Alabama–Georgia) and indicated that recolonization of exposed areas by benthic organisms required 2–4 months of continuous inundation. Similar patterns have been reported elsewhere. Dendy (1946) found a general lack of benthic organisms in Norris Reservoir (Tennessee) where winter drawdowns often exceed 20 m. Grimas (1961) reported that littoral zone benthos declined by 70% and profundal organisms were reduced by 25% due to water-level changes in some Swedish reservoirs. Cowell and Hudson (1967) found that the negative impact of water-level fluctuations on benthic invertebrates can be ameliorated to some extent by reducing the amplitude of the water-level changes.

The spatial distribution of benthic organisms within a reservoir is closely tied to factors (e.g., water depth, substrate type, frequency and duration of anoxia) that are important to the abundance and composition of the benthos. Therefore, the benthic fauna of reservoirs should exhibit patterns of longitudinal zonation. Baxter (1977) reported considerable longitudinal variation in reservoir benthos, and that a common trend is for maximal abundance to occur near the river inflow. This may be explained by the relatively high concentration of organic matter (presumably, of allochthonous origin) in the riverine zone sediments of some impoundments (e.g., James et al. 1987).

The spatial patterns of abundance and species composition in the benthos of river–reservoir systems clearly show that impoundments must be viewed as integral parts of the larger river–reservoir ecosystems. Ward and Stanford (1983), in their "Serial Discontinuity Concept of Lotic Ecosystems," present a framework that treats impoundments as modifiers or "discontinuities" of the expected riverine pattern. This concept and the river continuum concept of Vannote et al. (1980), which it modifies, originated from observed patterns in species composition and abundance in river benthos.

Because the downstream impacts of impoundments on riverine benthos are well studied and excellent reviews are widely available (e.g., Hynes 1970, Ward and Stanford 1979, Baxter 1977), we will only summarize these effects here: (1) Hypolimnetic or surface release from an impoundment creates a thermal and flow regime that is unlike that of nearby unimpounded rivers, and the magnitude of temperature and flow fluctuations and their frequency and duration (both on a daily and seasonal scale) influence the suitability of downstream habitats. (2) The trapping of sediment within an impoundment, coupled with discharge of water from the dam, results in scouring of the downstream channel. (3) The trapping of inflowing particulate organic matter from upstream (terrestrial) sources and the discharge of autochthonous (planktonic) organic matter downstream changes the quality and quantity of the particulate food resource available to downstream benthos (Coutant 1968, Cowell 1970, Lind 1971, Stroud and Martin 1973). (4) Release of oxygen-depleted water that is relatively rich in nutrients and other (possibly toxic) dissolved material will affect food production and invertebrate survival in the downstream river. These various effects produce major shifts in the functional feeding-group composition of downstream and in-reservoir invertebrates, as well as the species composition of these functional groups. The abundance and species composition of higher consumers (e.g., fish) should also reflect these shifts.

Nekton

Fish are the most studied biotic component of southeastern reservoirs, and an extensive literature exists on reservoir fish populations and fisheries. Recent reviews and symposia cover the major aspects of reservoir fish ecology (e.g., Hall 1971, Hall and Van Den Avyle 1986). Our goal is to summarize major facets of reservoir fish ecology that relate to the ecological structure and functioning of southeastern impoundments.

The trophic relations of reservoir fish are complex (Fig. 14). Many of the abundant species are food generalists (Keast 1978) and their feeding behavior can typically span several trophic levels (e.g., from periphyton grazing to piscivory) and can vary with season and with fish age. For example, a top predator might progress from planktivore to insectivore to piscivore during its growth and development. To deal with this complexity, Ploskey and Jenkins (1982) and Adams et al. (1983), among others, have adopted a composite trophic-level index for various species and growth stages of reservoir fish. With this approach, the trophic relations and dynamics of reservoir fish communities can be analyzed at the ecosystem level and differing systems can be

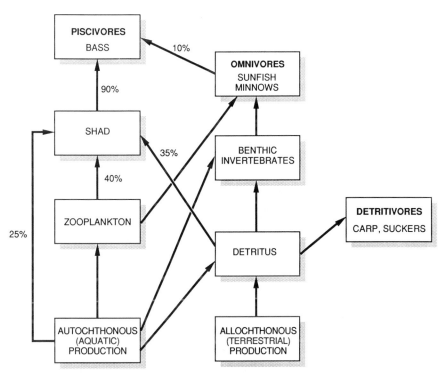

FIGURE 14. Generalized food-web diagram for the fish community of Watts Bar Reservoir, a main-stem storage impoundment of the Tennessee River, eastern Tennessee. The percentage values indicate the relative importance of various food resources to components of the fish community. Typically, in large impoundments in the southeastern United States, threadfin shad (*Dorosoma petenense*) and gizzard shad (*D. cepedianum*) populations occupy a critical food-web link between the lower trophic levels and the top piscivores such as striped bass (*Morone saxatilis*) and largemouth bass (*Micropterus salmoides*). (Diagram courtesy of S.M. Adams, Environmental Sciences Division, Oak Ridge National Laboratory, Oak Ridge, TN 37831–6036.)

compared. In a cross-sectional study using data from 17 reservoirs (12 from the Southeast), Adams et al. (1983) found that the general trophic structure across otherwise differing reservoirs appeared surprisingly constant. They suggest that this constancy may reflect the existence of complex feedback and compensatory mechanisms in reservoir food webs, a high degree of interconnection between grazer and detritus pathways, and the pivotal role of benthic detritivores. Adams et al. (1983) also showed a strong correlation between the simple morphoedaphic index (MEI = total dissolved solids/mean depth) of Ryder (1965) and fish production in southeastern reservoirs.

The standing crop and production of fish in southeastern reservoirs is dependent upon a host of environmental factors. Carline (1986) provides an excellent summary of single- and multivariable regression models that have been used to link fish standing crops and yields to abiotic and biotic variables in reservoirs. Most of the factors that proved significant in these models (e.g., Jenkins 1967) are closely tied to reservoir morphometry, water residence time, and nutrient concentrations. Other studies have shown that reservoir construction procedures and operational practices (i.e., standing timber and water-level fluctuations) that affect habitat and food resources are also important to fish standing crop and production.

Spatial heterogeneity in fish communities follows the pattern observed in abiotic components and in other biotic components of river–reservoir ecosystems. For example, Siler et al. (1986) showed that the abundance of threadfin shad and the harvest of sport fish in Lake Norman, North Carolina, were lowest in the near-dam area and increased steadily upstream into the riverine zone. A similar pattern for threadfin shad was found by Netsch et al. (1971) in Beaver Reservoir (Arkansas).

The effects of water-level fluctuations on reservoir fish populations and fisheries were reviewed by Ploskey (1986). In general, reservoir fisheries are most affected by water-level fluctuations that (1) are large (i.e., several meters), (2) last several months, (3) occur during the growing season, and (4) inundate or eliminate productive areas of littoral or terrestrial vegetation. Rapidly rising waters that inundate terrestrial areas can temporarily increase supplies of invertebrate foods; likewise, large drawdowns that concentrate prey fish for 2–3 months at temperatures above 13 °C can increase predator foraging success and growth (Aggus 1979). Loss of habitat by drawdown can negatively affect spawning success and egg mortality, but because year-class strength depends on factors beyond spawning success, drawdown may still have a net positive effect on recruitment in some species.

When a reservoir fills, most of the stream fish in the flooded area are ill-suited to lentic conditions and soon disappear. Natural colonization by lake species, however, can be extremely slow. To exploit the "trophic upsurge" that follows reservoir filling (see Kimmel and Groeger 1986) and to rapidly develop or maximize the sport fishery, reservoir managers have stocked a wide variety of game fish into southeastern impoundments (e.g., Kohler et al. 1986). In turn, forage species are also introduced to provide a food base for the "target" game fish (e.g., Kohler and Ney 1980, Kelso and Ney 1982).

The results in some cases have been the introduction of species that are fundamentally unsuited for the receiving waters and that cause a decline in native species. In the best cases, a rather spectacular sport fishery has resulted based on species (e.g., striped bass, rainbow trout, northern pike, yellow perch, and walleye) that require cool-water habitats that are rare in the southeast outside these deep reservoirs (e.g., Wilkins et al. 1967, Clugston et al. 1978). Reservoir aging (Kimmel and Groeger 1986) can change the reservoir fishery yield and species composition. To compensate for this phenomenon, fisheries managers must be adaptable enough to recognize changes in reservoir-use patterns, user expectations, and appropriate fish-stocking practices as the system ages.

The success of introduced species (particularly those requiring cool water) is closely tied to system morphometry and reservoir operational practices. The striped bass provides a good example. Because it is a prized sport fish, this species has been introduced to many of the major impoundments in the Southeast. However, larger specimens require cool water temperatures (<24 °C; Coutant and Carroll 1980) and thus often are forced into deeper water or spring-fed refuges in these impoundments during late summer. Hypolimnetic discharge may reduce the volume of cool-water habitat in a reservoir during late summer, but may improve the conditions in the receiving river or impoundment downstream. Further, the onset of summer anoxia may make the deeper water uninhabitable by these fish, forcing them into a squeeze between their preferred temperature range, dissolved oxygen concentrations required for survival, and the availability of prey (see Coutant 1985, 1986, 1987).

Reservoir tailwaters provide a valuable fisheries habitat that is heavily utilized in the Southeast. Hypolimnetic discharges from southeastern impoundments can provide cool-water fisheries (often "put-and-take"), if properly designed and managed (e.g., Hackney and Holbrook 1978, Barwick and Oliver 1982), but can also result in negative impacts on warm-water fishes. Middleton (1967) provides an example from the Narrows Reservoir on the Little Missouri River in south-central Arkansas. Without careful management of discharges, both warm-water and cool-water fishes are negatively affected, which results in a net overall reduction in fish abundance.

The management of reservoir fisheries is often attempted with very limited resources and incomplete information. The population dynamics of important predators and prey species in reservoirs are poorly understood (Noble 1986), although most manipulations have been directed at piscivorous game fishes and their prey. Reservoir fisheries management has relied heavily on stocking, yet the relationships between prey fishes and their food resources have received relatively little attention.

The trophic relations of reservoir fish are complex, but the standing crop and production of reservoir fish in the Southeast conform to patterns seen elsewhere and depend on reservoir morphometry, water residence time, nutrient levels, and operating characteristics. Deep impoundments with hypolimnetic discharges have created cool-water habitats in warm southeastern streams. In reservoirs and tailwaters that do not develop hypolimnetic oxygen depletion, fisheries based on cool-water species that are not indigenous to the Southeast have been developed (Pfitzer 1967). Although dam construction produces structural and thermal barriers to fish movement in the impounded rivers, in some cases, artificial stocking and removal of natural physical barriers (e.g., the Tombigbee project) have introduced new species into some southeastern drainage basins. Significant changes in the distribution of fish species in the Southeast have resulted.

Summary of Biotic Communities

The ecological structure and functioning of biotic communities in southeastern reservoirs are linked to water residence time, both because of the direct effects of water renewal on the plankton and other components of the biota and because water residence time correlates with other important limnological variables (e.g., nutrient loading, water depth, watershed size, turbidity, mixing regime). Usually reservoir productivity is high following initial basin inundation, but conditions stabilize at a lower level of productivity within 5–10 years (Ostrofsky and Duthie 1980, Kimmel and Groeger 1986).

The biotic communities of southeastern impoundments exhibit longitudinal zonation within reservoirs that coincides with the longitudinal zonation of physical and chemical variables. Reservoir primary productivity usually is dominated by phytoplankton and is often maximal in the zone of transition from riverine to lacustrine conditions. The irregular shorelines of southeastern impoundments provide extensive areas for littoral vegetation, but water-level fluctuations in many reservoirs impede macrophyte and periphyton growth.

Coves and embayments of southeastern reservoirs create tremendous spatial heterogeneity. Water exchange between the open lake and these shallow bays is often restricted, and nutrients, organic material, and planktonic biomass can accumulate in these areas. As a consequence, reservoir embayments are often more productive and have a more diverse biota than the main body of the reservoir. Unfortunately, reservoir cove studies have seldom included multiple coves on the same reservoir, and rarely have they examined in detail the physical, biological, and chemical connections between the cove and the adjacent open water. Thus, quantitative information on the variability among coves, and the contribution of coves to overall reservoir ecosystem functioning is still lacking.

The hypolimnetic and downstream releases of southeastern impoundments create conditions that differ markedly from unmodified waters in this region. Fisheries managers have taken advantage of these conditions to establish cool-water fish populations (e.g., rainbow trout) outside their usual geographic range, but warm-water fish populations have suffered (Pfitzer 1967). The biotic communities of reservoir tailwaters have received considerable attention and good reviews are available (Pfitzer 1967, Isom 1971, Baxter 1977).

REPRESENTATIVE AQUATIC COMMUNITIES

The biotic components of any river–reservoir system are shaped by site-specific conditions. Nonetheless, general patterns of reservoir structure and functioning are illustrated by examples from specific southeastern impoundments. In this section, we provide brief, case-study descriptions of two tributary storage reservoirs (Norris Reservoir, Tennessee, and DeGray Reservoir, Arkansas), a main-stem reservoir (West Point Reservoir, Georgia–Alabama), and a multiple-reservoir series (Norris, Melton Hill and Watts Bar reservoirs, Tennessee).

Tributary Reservoirs

Norris Reservoir, Tennessee Norris Reservoir is one of the oldest impoundments (closed in 1936), and perhaps the best studied, in the TVA system. Norris is located on the upper Clinch River, and its morphometry is typical of many southeastern tributary–storage reservoirs (Table 5). Norris Reservoir was constructed for hydropower generation and flood control. The impoundment is deep and elongate, stratifies strongly during summer, and discharges water from the hypolimnion. It is relatively unproductive (360 mg C/m^2 day^{-1} annual average, Taylor 1971). Norris Reservoir provides an excellent case study for longitudinal zonation, seasonal water-level fluctuations, hypolimnetic release for hydropower generation, downstream effects of hypolimnetic release, tailwater fisheries, and the ecology of reservoirs arranged in series.

Norris Reservoir has been the site of pioneering studies in reservoir limnology and fisheries ecology. Early work by Wiebe (1938, 1939, 1940, 1941) described the effects of density currents on the vertical profiles of temperature and dissolved oxygen with particular emphasis on the metalimnetic minimum in oxygen concentration produced by interflows. Tarzwell (1939) described the effect of the Norris Dam closure on the downstream benthic fauna, one of the earliest studies of this phenomenon. Eschmeyer and Smith (1943)

TABLE 5 Limnological Characteristics of Norris Reservoir, Tennessee

Surface area (km^2)	138
Volume (10^6 m^3)	2484
Maximum depth (m)	61
Mean depth (m)	18
Length (km)	208
Shoreline (km)	1290
Shoreline development ratio	31
Drainage area (km^2)	7542
Water residence time (days)	237
Mean annual water-level fluctuation (m)	18

documented the effects of cold-water releases from Norris Dam on the reproduction of warm-water fishes in the downstream river. More recently, Hickman and Hevel (1986) showed the effect of Norris releases on the growth and reproduction of largemouth bass in Melton Hill Reservoir, located downstream. Other studies have detailed the abundance and species composition of benthic communities downstream of the Norris dam site both before and after impoundment (Isom 1971).

DeGray Reservoir, Arkansas In recent years, intensive research (sponsored primarily by the U.S. Corps of Engineers and the U. S. Fish and Wildlife Service) has been conducted on DeGray Reservoir, a USCE tributary–storage impoundment in south-central Arkansas (Kennedy and Nix 1987). Data from fish studies in DeGray Reservoir were used in the trophic-relations analysis of Ploskey and Jenkins (1982), and a wide range of limnological and fisheries studies have been conducted on this reservoir.

Construction of DeGray Reservoir began in 1967, and power pool was reached in 1971. The reservoir was formed by impoundment of the Caddo River near Arkadelphia, Arkansas, approximately 12.7 km upstream from the confluence of the Caddo and Ouachita rivers. It provides flood control and hydropower generation. Inundation of the deeply cut Caddo River valley resulted in the formation of a deep (maximum depth = 60 m near the dam), long (32 km), and narrow impoundment with a complex shoreline (Table 6). Major features of the reservoir include numerous coves and islands, two major embayments (Brushy Creek and Big Hill Creek), a deeply cut submerged river channel, and a selective-withdrawal outlet structure capable of discharging from any of four gate depths. The Caddo River, which originates in the Ouachita Mountains to the west and north, enters the reservoir at its extreme western end. Riverine flows are frequently observed in the upper one-third of the reservoir, particularly during winter and spring high-flow seasons. Extensive stands of timber remain in this impoundment; it thus provides an excellent site to study the effects of standing timber on benthos and fisheries

TABLE 6 Limnological Characteristics of DeGray Reservoir, Arkansas

Surface area (km^2)	54.2
Volume (10^6 m^3)	808
Maximum depth (m)	60
Mean depth (m)	14.9
Length (km)	32
Shoreline length (km)	333
Shoreline development ratio	12.8
Drainage area (km^2)	1173
Water residence time (days)	511
Mean annual water-level fluctuation (m)	6

production. Hypolimnetic anoxia is common during summer months, but is restricted to shallow upstream areas and the area immediately upstream from the dam. This impoundment has shown the common pattern of reservoir aging, with a brief period of excellent fishing followed a by gradual decline in fishery production (Ploskey and Jenkins 1982).

Typical of tributary impoundments, DeGray Reservoir has a relatively long water residence time (1.4 years) and, because it is highly elongate, shows pronounced longitudinal zonation. Nutrient and chlorophyll concentrations are relatively low in this impoundment and exhibit marked longitudinal gradients from upstream to downstream (Thornton et al. 1982). Kennedy et al. (1986) and James et al. (1987) documented longitudinal variations in nutrient status and in suspended particle deposition rates in DeGray Reservoir. Groeger and Kimmel (1987) documented the longitudinal zonation of phytoplankton abundance and productivity (maximal uplake) and showed that phytoplankton assemblages were generally in poor physiological condition during the growing season, probably because of nitrogen deficiency. Meinecke et al. (1987) found longitudinal zonation (maximal abundances near the transition zone, but minimal at the riverine site) in zooplankton abundance. Moen and Dewey (1987) reported that the greatest fish biomass was also found in the uplake segments of this impoundment.

Main-Stem Reservoirs

Main-stem reservoirs in the southeast are extremely diverse. On the Tennessee River, for example, main-stem reservoirs are elongate and riverine (residence times less than 15 days). However, on the Savannah River, the main-stem impoundments formed by Hartwell, Richard B. Russell, and J. Strom Thurmond (formerly Clarks Hill) dams have residence times exceeding 3 months and, thus, are relatively lacustrine. West Point Reservoir, a main-stem impoundment on the Chattahoochie River (Georgia–Alabama), has a mean residence time of about 2 months (55 days) and is intermediate between the riverine reservoirs of the Tennessee River and the more lacustrine systems of the Savannah River.

West Point Reservoir was built by the Corps of Engineers (completed in 1975) and formed by impoundment of the Chattahoochee River near West Point, Georgia. The Chattahoochee River originates in the Blue Ridge Mountains of northern Georgia and flows southwesterly over metamorphic and igneous rock substrata. High soil erosion rates are common throughout the drainage basin and result in seasonally high turbidity in the impoundment.

West Point Reservoir provides flood control, hydroelectrical power, water supply, and recreation, and is a good case study for southeastern impoundments affected by urban development in the upstream watershed. The treated industrial and municipal effluents from over 50% of metropolitan Atlanta are discharged into the Chattahoochee River about 112 km upstream from the reservoir (Shelton and Davies 1981). The impoundment is elongate (Table 7) and exhibits sharp longitudinal zonation (Kennedy et al. 1982, 1984). The

TABLE 7   Limnological Characteristics of West Point Reservoir, Alabama–Georgia

Surface area (km²)	104.8
Volume (10⁶ m³)	733
Maximum depth (m)	25
Mean depth (m)	7.0
Length (km)	52
Shoreline length (km)	845
Shoreline development ratio	23
Drainage area (km²)	8908
Water residence time (days)	55
Mean annual water-level fluctuation (m)	3

transition zone is extensive and located well within the reservoir pool owing to the high flow of the nutrient- and sediment-laden river. Algal biomass is minimal at extreme upstream and downstream locations due to high levels of turbidity and reduced nutrient concentrations, respectively. Algal biomass is highest near midpool in an area coincident with the transition zone. Anoxic conditions develop during summer months in the deep areas of the reservoir downstream from the plunge line and transition zone.

Thermal stratification is well developed in the lacustrine zone of this impoundment. The direct influence of flow from the tributary stream is minimal. Processes that occur upstream in the riverine and transition zones modify both the quantity and quality of material loadings to the lacustrine zone. Autotrophic production in this system likely exceeds the supply of allochthonous organic matter. Algal production in the downstream (lacustrine) reaches of this impoundment is probably nutrient limited, because of nutrient deposition and retention in upstream areas and the transport of inflowing nutrients to the bottom layer by density flows.

As with other southeastern reservoirs, West Point is subjected to a seasonal drawdown (October–May) and has water-level fluctuations of about 3 m that result from flood-control operations. The seasonal lowering of the water level exposes about 2900 ha of littoral zone, and the effect of this exposure on benthic invertebrate populations has been examined (Hale and Bayne 1980).

The total standing stock of fish was estimated by cove rotenone samples from 1975 to 1979 (Shelton and Davies 1981). The fish fauna includes 59 species; however, the number of species collected from the reservoir has diminished each year as a result of the disappearance of riverine species unsuited for the more lacustrine, reservoir environment. The preimpoundment fish community of the Chattahoochee River consisted of 53 species, but the numbers declined to 48, 41, and 40 from 1975 to 1977 (Timmons et al. 1979), and further to 35 to 32 in 1978 and 1979. The loss of the original species has been greater than indicated by numbers alone because 6 species that did not occur in the streams prior to impoundment now occur in the reservoir.

Three years after West Point Reservoir was filled, striped bass × white bass hybrids were stocked in large numbers to (1) establish a fishery for a large, open-water game fish, and (2) exert predatory pressure on the larger, under-utilized gizzard shad population (Ott and Malvestuto 1981). Threadfin shad (*Dorosoma pretenense*) were found to be the major prey item in the diet of these hybrid fish, with bluegill, (*Lepomis macrochirus*), gizzard shad (*D. cepedianum*), and largemouth bass (*Micropterus salmoides*) also found in the stomachs of larger hybrids. The smaller hybrids preyed heavily on insect larvae (chironomidae). There seemed to be little or no relationship between prey size and the size of hybrid fish.

Reservoirs in Series

Nearly all the major streams of the Southeast have been impounded, and most of these impoundments are arranged in series. Thus, the rivers of the Southeast have been converted from lotic environments to stair-step chains of lentic or semilentic pools. The ecology of these interacting pools is not well understood, but progress is being made in recent and on-going studies.

Norris, Melton Hill, and Watts Bar reservoirs within the Tennessee River system form a three-reservoir subseries of relatively well-studied impoundments. The uppermost reservoir in this series, Norris, was previously described (Table 5). Melton Hill, the second reservoir in this series, is a highly elongate riverine impoundment on the Clinch River (Table 8). The third impoundment, Watts Bar is a typical main-stem reservoir (Table 8). Watts Bar receives inflow from the Clinch River and from the main-stem of the Tennessee River which has several impoundments located on its main-stem and its tributaries above Watts Bar Dam.

The temperature regime in this series of impoundments and its effect on the biota has received considerable attention (Eschmeyer and Smith 1943, Pfitzer 1967, Isom 1971, Hickman and Hevel 1986). The combination of

TABLE 8   Limnological Characteristics of Melton Hill and Watts Bar Reservoirs, Tennessee

Characteristic	Melton Hill	Watts Bar
Surface area (km²)	23	156
Volume (10^6 m³)	138	1 248
Maximum depth (m)	21	21
Mean depth (m)	6	8
Length (km)	71	154
Shoreline length (km)	238	1 239
Shoreline development ratio	14	28
Drainage area (km²)	8658	44 833
Water residence time (days)	15	18
Mean annual water-level fluctuation (m)	3	2

hypolimnetic discharges, hydropower generation, and the serial arrangement of these impoundments creates complex hydrodynamic conditions. In the summer months, the hypolimnetic discharge from Norris Reservoir enters the headwaters of Melton Hill Reservoir and may displace the warmer surface water several miles downstream (depending on the generation schedule at Norris Dam) before plunging and continuing downstream as an underflow. This underflow may then be released into Watts Bar Reservoir by the hypolimnetic discharge from Melton Hill Dam. Several hours after hydropower generation ceases at Norris, there is a rapid upstream flow of warm surface water (return flow) in Melton Hill, as the plunge line retreats upstream. A similar phenomenon occurs in Watts Bar Reservoir below the Melton Hill Dam. This longitudinal ebb and flow of water and the vertical shear created by daily pulsations in reservoir discharges can be expected to affect material fluxes and plankton ecology in these systems.

Elser and Kimmel (1985) investigated the availability of nitrogen (N) and phosphorus (P) for phytoplankton production in the Norris–Melton Hill–Watts Bar reservoir series and found that P was the more limiting nutrient for phytoplankton growth. Phytoplankton P deficiency, as indicated by chlorophyll-specific alkaline phosphatase activity (APA), peaked in late August and then decreased during the fall and winter. Late in the growing season, APA levels increased downstream within each reservoir, reflecting uplake-to-downlake decreases in P availability, but decreased downstream from reservoir to reservoir, indicating increased P availability down the impoundment series.

The hypolimnetic discharge of water and dissolved nutrients from upstream impoundments increased nutrient availability during the growing season and thus enhanced phytoplankton production in the Norris–Melton Hill–Watts Bar reservoir series. Similar patterns in nutrient availability and phytoplankton productivity have been observed in the Hartwell–Russell–Thurmond reservoir series on the Savannah River (Kimmel et al. 1988), and probably occur commonly in multiple-impoundment series throughout the Southeast.

RECOMMENDATIONS FOR FUTURE RESEARCH AND MANAGEMENT

Critical reservoir issues, like the reservoirs themselves, vary dramatically. Long-term siltation, eutrophication, and accumulation of toxic contaminants all constitute serious threats to the value of reservoirs as water resources. There is a need to deal with increasing point-source wastewater loadings from municipalities and industry, as well as with nonpoint-source loadings of nutrients and sediments from agricultural and lumbering activities. The need to understand the dynamics and effects of material transfers among reservoirs in series is of particular importance. These systems have the greatest potential for manipulation, and yet are perhaps the least understood ecologically.

Inundation of river-bottom habitat and the creation of physical–chemical barriers is a major ecological effect of reservoirs. River fishes, mussels, and terrestrial communities of river-bottom areas have all been negatively impacted. Impoundment of remaining segments of free-flowing river must be viewed with caution because such action may eliminate the last vestiges of now-critical, uncontrolled river habitat. In the future, there may be attempts to reintroduce species that were extirpated in the wake of reservoir construction. To improve the chances for success of these, and other mitigation efforts, there must be a more complete understanding of the ecology of these now-altered systems.

Dewatering of stream channels by hydropower diversions has been a major effect of southeastern impoundments, and there is a pressing need to understand the instream flow requirements of river biota and to coordinate these requirements with hydropower operations. In some cases, existing hydropower projects should be modified to ameliorate the negative effects of impoundments on downstream areas. However, these effects have been with us for some time now, and although mitigation is desirable, it may not be considered urgent on a regional scale.

Some reservoirs in the Appalachian region (and possibly the Ozarks) are potentially sensitive to acidic deposition, especially those at higher elevations with large watersheds. Acid mine drainage already affects some reservoirs in this region. However, the acid-neutralizing capacity in most southeastern reservoirs and their watersheds is high enough that acidification is unlikely to be a regional-scale problem in the immediate future.

There are unanswered questions about the ecological requirements of reservoir biota and their relationship to the conditions created by reservoir design and operations. The effectiveness of stocking programs, tailwater fisheries, coexistence among introduced and native species, and the creation or maintenance of wetland habitat all depend on these answers. Reservoir aging has dramatically affected the composition and yield of the fish communities in many southeastern impoundments, and in systems that support a strong recreational or commercial fishery, dealing with this phenomenon should be a high priority for future research.

The aquatic habitat and recreational use of many southeastern impoundments has been influenced by the exotic aquatic plants *Hydrilla verticulata* and *Myriophyllum spicatum*, which have been introduced into many, if not all, southeastern reservoirs and have grown to nuisance proportions in some systems. There is little hope for completely eliminating or even stopping the spread of these plants, but the full ecological impact of activities designed to limit their abundance (herbicide applications, water drawdown, harvesting, biological control) must also be understood.

River impoundments are now the dominant surface water feature of the southeastern United States, and growth in population and industrial activity is increasing the demand on southeastern reservoirs for water supply, hydropower, transportation, and recreational uses. As a result, the maintenance of reservoir water quality will become an increasingly important aspect of water resources management in this region. Reservoir construction in the southeastern United States is virtually complete (Fig. 1) and emphasis has shifted from building new impoundments to optimizing the uses of existing systems. Long-term degradation of southeastern reservoir resources due to siltation, reservoir aging, eutrophication, and contamination by pollutants and toxic wastes is likely unless countered by informed scientific management and regulation.

Solutions to resource problems in a multiple-user setting are as complex and interconnected as the ecosystems that are being managed. For example, a drawdown that results from one beneficial use can have negative effects on other users of both the impoundment and downstream systems. Some multiple uses are inherently incompatible. For example, maximum fish yields require mesotrophic or eutrophic conditions that may be undesirable for water supply and some recreational uses. In some cases, the longitudinal zonation of reservoirs may enable managers to optimize benefits by managing for different uses within different zones. The lacustrine zone is typically best for water supply and water-contact recreation, whereas irrigation withdrawals and fish and wildlife uses may best be located in the riverine and transition zones.

The management of reservoir systems at the river-basin scale can often present special institutional problems. However, there is a pressing need for a river-basin perspective in the management of reservoir resources. For example, the operation of headwater impoundments influences more than just the quantity of water flowing into downstream reservoirs. The physical–chemical characteristics of the discharge and its timing, on both short-term (daily) and long-term (seasonal) scales, is of direct ecological importance to receiving systems downstream.

Development of coordinated, technically sound, management plans and their implementation will be a long-term and essential requirement. In most decisions involving water resources, economic or legal constraints dictate the priorities, but a purely economic or legalistic approach to reservoir management without an adequate technical base is unlikely to optimize the utilization of these valuable resources in a multiple-user setting. To make intelligent choices, the nature of the ecological linkages within and between reservoir systems must be recognized by decision makers.

REFERENCES

Ackermann, W. C., G. F. White, and E. B. Worthington (eds.). 1973. *Man-Made Lakes: Their Problems and Environmental Effects*, Geophys. Monogr. No. 17. Washington, DC: American Geophysical Union.

Adams, S. M., B. L. Kimmel, and G. R. Ploskey. 1983. Sources of organic matter for reservoir fish production: a trophic-dynamic analysis. *Can. J. Fish. Aquat. Sci.* 40:1480–1495.

Aggus, L. R. 1971. Summer benthos in newly flooded areas of Beaver Reservoir during the second and third years of filling 1965–66. In G. E. Hall (ed.), *Reservoir*

Fisheries and Limnology, Spec. Publ. No. 8. Bethesda, MD: American Fisheries Society, pp. 139–152.

Aggus, L. R. 1979. Effects of weather on freshwater fish predator-prey dynamics. In R. H. Stroud and H. Clepper (eds.), *Black Bass Biology and Management*. Washington, DC: Sport Fishing Institute, pp. 47–56.

Applegate, R. L., and J. W. Mullan. 1967. Zooplankton standing crops in a new and an old Ozark reservoir. *Limnol. Oceanogr.* 12:592–599.

Arruda, J. A., G. R. Marzolf, and R. T. Faulk. 1983. The role of suspended sediments in the nutrition of zooplankton in turbid reservoirs. *Ecology* 64:1225–1235.

Barwick, D. H., and J. L. Oliver. 1982. Fish distribution and abundance below a southeastern hydropower dam. *Proc. Annu. Conf. Southeast. Assoc. Fish. Wild. Agencies* 36:135–145.

Baxter, R. M. 1977. Environmental effects of dams and impoundments. *Annu. Rev. Ecol. Syst.* 8:255–283.

Benson, N. G. 1973. Evaluating the effects of discharge rates, water levels, and peaking on fish populations in Missouri River mainstem impoundments. In W. C. Ackermann, G. F. White, and E. B. Worthington (eds.), *Man-Made Lakes: Their Problems and Environmental Effects*, Geophys. Monogr. No. 17. Washington, DC: American Geophysical Union, pp. 683–689.

Benson, N. G. 1982. Some observations on the ecology and fish management of reservoirs in the United States. *Can. J. Water Resour.* 7:2–25.

Benson, N. G., and B. Cowell. 1967. The environment and plankton density in Missouri River reservoirs. In *Reservoir Fishery Resources*. Bethesda, MD: Reservoir Committee, American Fisheries Society, pp. 358–373.

Briggs, J. C., and J. F. Ficke. 1977. *Quality of Rivers of the United States, 1975 Water Year—Based on the National Stream Quality Accounting Network (NASQUAN)*, Open-File Rep. 78–200. Reston, VA: U.S. Department of the Interior, Geological Survey.

Brook, A. J., and W. B. Woodward. 1956. Some observations on the effect of water inflow and outflow on the plankton of small lakes. *J. Anim. Ecol.* 25:22–35.

Carline, R. F. 1986. Indices as predictors of fish community traits. In G. E. Hall and M. J. Van Den Avyle (eds.), *Reservoir Fisheries Management: Strategies for the 80's*. Bethesda, MD: Reservoir Committee, American Fisheries Society, pp. 46–56.

Claflin, T. O. 1968. Reservoir aufwuchs on inundated tress. *Trans. Am. Microsc. Soc.* 87:97–104.

Clugston, J. P., J. L. Oliver, and R. Ruelle. 1978. Reproduction, growth, and standing crops of yellow perch in southern reservoirs. In R. L. Kendall (ed.), *Selected Coolwater Fishes of North America*, Spec. Publ. No. 11. Bethesda, MD: American Fisheries Society, pp. 89–99.

Coutant, C. C. 1963. Stream plankton above and below Green Lane Reservoir. *Proc. Pa. Acad. Sci.* 37:122–126.

Coutant, C. C. 1985. Striped bass, temperature, and dissolved oxygen: a speculative hypothesis for environmental risk. *Trans. Am. Fish. Soc.* 114:31–61.

Coutant, C. C. 1986. Thermal niches of striped bass. *Sci. Am.* 254:98–104.

Coutant, C. C. 1987. Thermal preference: when does an asset become a liability? *Environ. Biol. Fishes* 18:161–172.

Coutant, C. C., and D. S. Carroll. 1980. Temperatures occupied by ten ultrasonic-tagged striped bass in freshwater lakes. *Trans. Am. Fish. Soc.* 109:195–202.

Cowell, B. C. 1970. The influence of plankton discharges from an upstream reservoir on standing crops in a Missouri River reservoir. *Limnol. Oceanogr.* 15:427–441.

Cowell, B. C., and P. L. Hudson. 1967. Some environmental factors influencing benthic invertebrates in two Missouri River reservoirs. In *Reservoir Fishery Resources*. Bethesda, MD: Reservoir Committee, American Fisheries Society, pp. 541–555.

Cox, J. 1984. Evaluating reservoir trophic status: the TVA approach. In *Lake and Reservoir Management*, EPA 440/5/84-001. Washington, DC: U.S. Environmental Protection Agency, pp. 11–16.

Cummins, K. W. 1974. Structure and function of stream ecosystems. *BioScience* 24:631–641.

Cummins, K. W. 1979. The natural stream ecosystem. In J. V. Ward and J. A. Stanford (eds.), *The Ecology of Regulated Streams*. New York: Plenum, pp. 7–24.

Dendy, J. S. 1946. Food of several species of fish, Norris Reservoir, Tennessee. *J. Tenn. Acad. Sci.* 21:105–127.

Dirnberger, J. M., and S. T. Threlkeld. 1986. Advective effects of a reservoir flood on zooplankton abundance and dispersion. *Freshwater Biol.* 16:387–396.

Ellis, M. M. 1936. Erosion silt as a factor in aquatic environments. *Ecology* 17:29–42.

Elser, J. J., and B. L. Kimmel. 1985. Nutrient availability for phytoplankton production in a multiple-impoundment series. *Can. J. Fish. Aquat. Sci.* 42:1359–1370.

Eschmeyer, R. W., and C. G. Smith. 1943. Fish spawning below Norris Dam. *J. Tenn. Acad. Sci.* 24:41–51.

Evans, T. D., and L. J. Paulson. 1983. The influence of Lake Powell on the suspended sediment-phosphorus dynamics of the Colorado River inflow to Lake Mead. In V. D. Adams and V. A. Lamarra (eds.), *Aquatic Resources Management of the Colorado River Ecosystems*. Ann Arbor, MI: Ann Arbor Science, pp. 56–68.

Fair, G. M., J. C. Geyer, and D. A. Okun. 1966. *Water and Wastewater Engineering*, Vol. 1. *Water Supply and Wastewater Removal*. New York: Wiley.

Forbes, S. A. 1887. The lake as a microcosm. Reprinted 1925. *Bull.—Ill. Nat. Hist. Surv.* 15:537–550.

Gloss, S. P., R. C. Reynolds, L. M. Mayer, and D. E. Kidd. 1981. Reservoir influences on salinity and nutrient fluxes in the arid Colorado River basin. In H. G. Stefan (ed.), *Symposium on Surface Water Impoundments*. New York: American Society of Civil Engineers, pp. 1618–1629.

Goldman, C. R., and B. L. Kimmel. 1978. Biological processes associated with suspended sediment and detritus in lakes and reservoirs. In J. Cairns, E. F. Benfield, and J. R. Webster (eds.), *Current Perspectives on River-Reservoir Ecosystems*. Blacksburg, VA: North American Benthological Society, pp. 14–44.

Grimas, U. 1961. The bottom fauna of natural and impounded lakes in northern Sweden. *Inst. Freshwater Res. Drottingholm. Rep.* 42:183–237.

Groeger, A. W., and B. L. Kimmel. 1987. Spatial and seasonal patterns of photosynthetic carbon metabolism in DeGray Lake phytoplankton. In R. H. Kennedy and J. Nix (eds.), *Proceedings of the DeGray Lake Symposium*. Vicksburg, MS: U.S. Army Corps of Engineers, Waterways Experiment Station, Corps of Engineers, pp. 327–350.

Groeger, A. W., and B. L. Kimmel. 1988. Photosynthetic carbon metabolism by phytoplankton in a nitrogen-limited reservoir. *Can. J. Fish. Aquat. Sci.* 45:720–730.

Hackney, P. A., and J. A. Holbrook. 1978. Sauger, walleye, and yellow perch in the southeastern United States. In R. L. Kendall (ed.), *Selected Coolwater Fishes of North America*, Spec. Publ. No. 11. Bethesda, MD: American Fisheries Society, pp. 74–81.

Hale, M. M., and D. R. Bayne. 1980. Effects of water level fluctuations on the littoral macroinvertebrates of West Point Reservoir. *Proc. Annu. Conf. Southeast. Assoc. Fish. Wildl. Agencies* 34:175–180.

Hall, G. E. (ed.). 1971. *Reservoir Fisheries and Limnology*, Spec. Publ. No. 8. Bethesda, MD: American Fisheries Society.

Hall, G. E., and M. J. Van Den Avyle (eds.). 1986. *Reservoir Fisheries Management Strategies for the 80's.* Bethesda, MD: Reservoir Committee, American Fisheries Society.

Hammond, E. H. 1964. Analysis of properties in land form geography: an application to broad-scale land form mapping. *Ann. Assoc. Am. Geogr.* 54:11–23.

Henderson, H. G., R. A. Ryder, and A. W. Kudhongania. 1973. Assessing fishery potentials of lakes and reservoirs. *J. Fish. Res. Board Can.* 20:2000–2009.

Hickman, G. D., and K. W. Hevel. 1986. Effect of a hypolimnetic discharge on reproductive success and growth of warmwater fish in a downstream impoundment. In G. E. Hall and M. J. Van Den Avyle (eds.), *Reservoir Fisheries Management: Strategies for the 80's.* Bethesda, MD: Reservoir Committee, South. Division, American Fisheries Society, pp. 286–293.

Hoyer, M. V., and J. R. Jones. 1983. Factors affecting the relationship between phosphorus and chlorophyll-*a* in midwestern reservoirs. *Can. J. Fish. Aquat. Sci.* 40:192–199.

Hutchinson, G. E. 1957. *A Treatise on Limnology* Vol. 1, Part 1, *Geography and Physics of Lakes.* New York: Wiley.

Hynes, H. B. N. 1970. *The Ecology of Running Waters.* Toronto, Ontario, Canada: University of Toronto Press.

Hynes, H. B. N. 1975. The stream and its valley. *Verh.—Int. Ver. Theor. Angew. Limnol.* 19:1–15.

Isom, B. G. 1971. Effects of storage and mainstream reservoirs on benthic macroinvertebrates in the Tennessee Valley. In G. E. Hall (ed.), *Reservoir Fisheries and Limnology*, Spec. Publ. No. 8. Bethesda, MD: American Fisheries Society, pp. 179–191.

James, W. F., R. H. Kennedy, and R. H. Montgomery. 1987. Seasonal and longitudinal variations in apparent deposition rates within an Arkansas reservoir. *Limnol. Oceanogr.* 32:1169–1176.

Jenkins, R. M. 1967. The influence of some environmental factors on standing crop and harvest of fishes in U.S. reservoirs. In *Reservoir Fishery Resources*, Spec. Publ. Washington, DC: American Fisheries Society, pp. 298–321.

Jenkins, R. M. 1982. The morphoedaphic index and reservoir fish production. *Trans. Am. Fish. Soc.* 111:133–140.

Jones, J. R., and R. W. Bachmann. 1976. Predictions of phosphorus and chlorophyll levels in lakes. *J. Water Pollut. Control Fed.* 48:2176–2182.

Keast, A. 1978. Trophic and spatial interrelationships in the fish species of an Ontario temperate lake. *Environ. Biol. Fishes* 3:7–31.

Kelso, W. E., and J. J. Ney. 1982. Nocturnal foraging by alewives in reservoir coves. *Proc. Annu. Conf. Southeast. Assoc. Fish. Wildl. Agencies* 36:125–134.

Kennedy, R. H., and J. Nix (eds.). 1987. *Proceedings of the DeGray Lake Symposium,* Tech. Rep. E-87-4. Vicksburg, MS: Department of the Army, Waterways Experiment Station, Corps of Engineers.

Kennedy, R. H., K. W. Thornton, and R. C. Gunkel. 1982. The establishment of water quality gradients in reservoirs. *Can. Water Resour. J.* 7:71–87.

Kennedy, R. H., R. C. Gunkel, and J. M. Carlile. 1984. *Riverine Influence on the Water Quality Characteristics of West Point Lake*, Tech. Rep. E-84-1 Vicksburg, MS: U.S. Army Engineer Waterways Experiment Station.

Kennedy, R. H., K. W. Thornton, and D. E. Ford. 1985. Characterization of the reservoir ecosystem. In D. Gunnison (ed.), *Microbial Processes in Reservoirs.* Boston, MA: Junk, pp. 27–38.

Kennedy, R. H., W. F. James, R. H. Montgomery, and J. Nix. 1986. The influence of sediments on the nutrient status of DeGray Lake, Arkansas. In P. G. Sly (ed.), *Sediments and Water Interactions.* New York: Springer, pp. 53–62.

Kimmel, B. L., and A. W. Groeger, 1984. Factors controlling primary production in lakes and reservoirs: a perspective. In *Lake and Reservoir Management*, EPA 440/5/84–001. Washington, DC: U.S. Environmental Protection Agency, pp. 277–281.

Kimmel, B. L. , and A. W. Groeger. 1986. Limnological and ecological changes associated with reservoir aging. In G. E. Hall and M. J. Van Den Avyle (eds.), *Reservoir Fisheries Management: Strategies for the 80's.* Bethesda, MD: Reservoir Committee, American Fisheries Society, pp. 103–109.

Kimmel, B. L., D. M. Soballe, S. M. Adams, A. V. Palumbo, C. J. Ford, and M. S. Bevelhimer. 1988. Inter-reservoir interactions: effects of a new impoundment on organic matter production and processing in a multiple-impoundment series. *Verh. —Int. Ver. Theor. Angew. Limnol.* 23:985–994.

Kimmel, B. L., O. T. Lind, and L. J. Paulson. 1990. Reservoir primary production. In K. W. Thornton, B. L. Kimmel, and F. E. Payne (eds.), *Reservoir Limnology: Ecological Perspectives.* New York: Wiley, pp. 133–193.

Kohler, C. C., and J. J. Ney. 1980. Suitability of alewife as a pelagic forage fish for southeastern reservoirs. *Proc. Annu. Conf. Southeast. Assoc. Fish. Wildl. Agencies* 34:137–150.

Kohler, C. C., J. J. Ney, and W. E. Kelso. 1986. Filling the void: development of a pelagic fishery and its consequences to littoral fishes in a Virginia mainstem reservoir. In G. E. Hall and M. J. Van Den Avyle (eds.), *Reservoir Fisheries Management: Strategies for the 80's.* Bethesda, MD: Reservoir Committee, American Fisheries Society, pp. 166–177.

Leidy, G. R., and R. M. Jenkins. 1977. *The Development of Fishery Compartments and Population Rate Coefficients for Use in Reservoir Ecosystem Modeling*, Contract Rep. Y-77-1. Prepared by the National Reservoir Research Program, U. S. Fish and Wildlife Service, for the U.S. Corps of Engineers, Waterways Experiment Station, Vicksburg, MS.

Leopold, L. B., M. G. Wolman, and J. P. Miller. 1964. *Fluvial Processes in Geomorphology.* San Francisco, CA: Freeman.

Likens, G. E. 1984. Beyond the shoreline: a watershed-ecosystem approach. *Verh.— Int. Verein. Theor. Angew. Limnol.* 22:1–22.

Lind, O. T. 1971. The organic matter budget of a central Texas reservoir. In G. E. Hall (ed.) *Reservoir Fisheries and Limnology.* Washington, DC: American Fisheries Society, pp. 193–202.

Margalef, R. 1975. Typology of reservoirs. *Verh.—Int. Ver. Theor. Angew. Limnol.* 19:1841–1848.

Martin, R. G., and R. H. Stroud. 1973. *Influence of Reservoir Discharge Location on Water Quality, Biology and Sport Fisheries of Reservoirs and Tailwaters, 1968– 1971,* Tech. Rep. Vicksburg, MS: U.S. Corps of Engineers, Waterways Experiment Station.

Marzolf, G. R. 1981. Some aspects of zooplankton existence in surface water impoundments. In H. G. Stefan (ed.), *Proceedings of the Symposium on Surface Water Impoundments.* New York: American Society of Civil Engineers, pp. 1392– 1399.

Marzolf, G. R. 1984. Reservoirs in the Great Plains of North America. In F. B. Taub (ed.), *Ecosystems of the World,* Vol. 23. *Lakes and Reservoirs.* New York: Elsevier, pp. 291–302.

Marzolf, G. R., and J. A. Arruda. 1981. Roles of materials exported by rivers into reservoirs in the nutrition of cladoceran zooplankton. In *Restoration of Lake and Inland Waters,* EPA 440/5–81–010. Washington, DC: U.S. Environmental Protection Agency, pp. 53–55.

Meinecke, J. I., G. R. Ploskey, L. Aggus, S. M. Heinrichs, R. Roseberg, E. H. Schmitz, and M. D. Schram. 1987. Spatial and temporal distribution of zooplankton in DeGray Reservoir, Arkansas. In R. H. Kennedy and J. Nix (eds.), *Proceedings of the DeGray Lake Symposium,* Tech. Rep. E-87-4. Vicksburg, MS: U.S. Corps of Engineers, Waterways Experiment Station, pp. 351–367.

Middleton, J. B. 1967. Control of water discharged from a multipurpose reservoir. In *Reservoir Fishery Resources.* Bethesda, MD: Reservoir Comm., Southern Division, American Fisheries Society, pp. 37–46.

Minshall, G. W., R. C. Petersen, K. W. Cummins, T. L. Bott, J. R. Sedell, C. E. Cushing, and R. L. Vannote. 1983. Interbiome comparison of stream ecosystem dynamics. *Ecol. Monogr.* 53:1–25.

Moen, T. E., and M. R. Dewey. 1987. Population dynamics of fishes in DeGray Lake, Arkansas, during epilimnial and hypolimnial release. In R. H. Kennedy and J. Nix (eds.), *Proceedings of the DeGray Lake Symposium,* Tech. Rep. E-87-4. Vicksburg, MS: U.S. Army Corps of Engineers, Waterways Experiment Station, Corps of Engineers, pp. 397–436.

Neel, J. K. 1963. Impact of reservoirs. In D. G. Frey (ed.), *Limnology in North America.* Madison: Wisconsin WP, pp. 575–593.

Netsch, N. F., G. M. Kersch, Jr., A. Houser, and R. V. Kilambi. 1971. Distribution of young gizzard and threadfin shad in Beaver Reservoir. In G. E. Hall (ed.), *Reservoir Fisheries and Limnology.* Washington, DC: American Fisheries Society, pp. 95–106.

Noble, R. L. 1986. Predator-prey interactions in reservoir communities. In G. E. Hall and M. J. Van Den Avyle (eds.), *Reservoir Fisheries Management: Strategies for the 80's.* Bethesda, MD: Reservoir Committee, American Fisheries Society, pp. 137–143.

Ostrofsky, M. L., and H. C. Duthie. 1980. Trophic upsurge and the relationship between phytoplankton biomass and productivity in Smallwood Reservoir, Canada. *Can. J. Bot.* 58:1174–1180.

Ott, R. A., and S. P. Malvestuto. 1981. The striped bass x white bass hybrid in West Point reservoir. *Proc. Annu. Conf. Southeast. Assoc. Fish. Wildl. Agencies* 35:641–646.

Paulson, L. J., and J. R. Baker. 1981. Nutrient interactions among reservoirs on the Colorado River. In H. G. Stefan (ed.), *Proceedings of the Symposium on Surface Water Impoundments*. New York: American Society of Civil Engineers, pp. 1647–1658.

Pennak, R. W. 1953. *Freshwater Invertebrates of the United States*. New York: Ronald Press.

Pfitzer, D. W. 1967. Evaluation of tailwater fishery resources resulting from high dams. In *Reservoir Fishery Resources*. Bethesda, MD: Reservoir Committee, American Fisheries Society, pp. 477–488.

Placke, J. F. 1983. *Trophic Status Evaluation of TVA Reservoirs*, Rep. No. TVA/ONR/WR-83-7. Chattanooga: Office of Natural Resources, Division of Natural Resource Operations, Tennessee Valley Authority.

Ploskey, G. R. 1986. Effects of water-level changes on reservoir ecosystems, with implications for fisheries management. In G. E. Hall and M. J. Van Den Avyle (eds.), *Reservoir Fisheries Management: Strategy for the 80's*. Bethesda, MD: Reservoir Committee, American Fisheries Society, pp. 86–97.

Ploskey, G. R., and R. M. Jenkins. 1982. Biomass model of reservoir fish and fish-food interactions, with implications for management. *North Am. J. Fish. Manage.* 2:105–121.

Regier, H. A., and H. F. Henderson. 1973. Towards a broad ecological model of fish communities and fisheries. *Trans. Am. Fish. Soc.* 102:56–72.

Ryder, R. A. 1965. A method for estimating the potential fish production of north-temperate lakes. *Trans. Am. Fish. Soc.* 94:214–218.

Ryder, R. A. 1978. Ecological heterogeneity between north-temperate reservoirs and glacial lake systems due to differing succession rates and cultural uses. *Verh. — Int. Ver. Theor. Angew. Limnol.* 20:1568–1574.

Ryder, R. A., S. R. Kerr, K. H. Loftus, and H. A. Regier. 1974. The morphoedaphic index, a fish yield estimator: review and evaluation. *J. Fish. Res. Board Can.* 38:663–688.

Shelton, W. L., and W. D. Davies. 1981. West Point Reservoir—a recreation demonstration project. In H. G. Stefan (ed.), *Symposium on Surface Water Impoundments*. New York: American Society of Civil Engineers, pp. 1419–1431.

Siler, J. R., W. J. Foris, and M. C. McInerny. 1986. Spatial heterogeneity in fish parameters within a reservoir. In G. E. Hall and M. J. Van Den Avyle (eds.), *Reservoir Fisheries Management: Strategies for the 80's*. Bethesda, MD: Reservoir Committee, American Fisheries Society, pp. 122–136.

Soballe, D. M., and R. W. Bachmann. 1984. Removal of Des Moines River phytoplankton by reservoir transit. *Can. J. Fish. Aquat. Sci.* 41:1803–1813.

Soballe, D. M., and B. L. Kimmel. 1987. A large-scale comparison of factors influencing phytoplankton abundance in rivers, lakes, and impoundments. *Ecology* 68:1943–1954.

Soballe, D. M., and S. T. Threlkeld. 1985. Advection, phytoplankton biomass, and nutrient transformations in a rapidly flushed impoundment. *Arch. Hydrobiol.* 105: 187–203.

Stroud, R. H., and R. G. Martin. 1973. The influence of reservoir discharge location on the water quality, biology, and sport fisheries of reservoirs and tailwaters. In W. C. Ackermann, G. F. White, and E. B. Worthington (eds.), *Man-Made Lakes: Their Problems and Environmental Effects*, Geophys. Monogr. No. 17. Washington, DC: American Geophysical Union, pp. 540–558.

Tarzwell, C. M. 1939. Changing the Clinch River into a trout stream. *Trans. Am. Fish. Soc.* 68:228–233.

Taylor, M. P. 1971. Phytoplankton productivity response to nutrients correlated with certain environmental factors in six TVA reservoirs. In G. E. Hall (ed.), *Reservoir Fisheries and Limnology*. Washington, DC: American Fisheries Society, pp. 209–217.

Thornton, K. W. 1984. Regional comparisons of lakes and reservoirs: geology, climatology, and morphology. In *Lake and Reservoir Management*, EPA 440/5/84-001. Washington, DC: U.S. Environmental Protection Agency, pp. 261–265.

Thornton, K. W., R. H. Kennedy, J. H. Carrol, W. W. Walker, R. C. Gunkel, and S. Ashby. 1981. Reservoir sedimentation and water quality—an heuristic model. In H. G. Stefan (ed.), *Symposium on Surface Water Impoundments*. New York: American Society of Civil Engineers, pp. 654–661.

Thornton, K. W., R. H. Kennedy, A. D. Magoun, and G. E. Saul. 1982. Reservoir water quality sampling design. *Water Resour. Bull.* 18:471–480.

Threlkeld, S. T. 1982. Water renewal effects on reservoir zooplankton communities. *Can. Water Resour. J.* 7(1):151–167.

Threlkeld, S. T. 1983. Spatial and temporal variation in the summer zooplankton community of a riverine reservoir. *Hydrobiologia* 107:249–254.

Timmons, T. J., W. L. Sheldon, and W. D. Davies. 1979. Sampling reservoir fish populations with rotenone in littoral areas. *Proc. Annu. Conf. Southeast. Assoc. Fish Wildl. Agencies* 32:474–484.

U.S. Department of the Interior. 1964. *Generalized Map Showing Annual Runoff and Productive Aquifers in the Conterminous United States*, Hydrol. Invest. Atlas HA-194. Washington, DC: VSDI.

U.S. Environmental Protection Agency (USEPA). 1975. *A Compendium of Lake and Reservoir Data Collected by the National Eutrophication Survey in the Northeast and North-Central United States*, Working Paper No. 474. Corvallis, OR; Environmental Research Laboratory, USEPA.

U.S. Environmental Protection Agency (USEPA). 1978. *A Compendium of Lake and Reservoir Data Collected by the National Eutrophication Survey in the Eastern, North-Central and Southeastern United States*, Working Paper No. 475. Corvallis, OR: Environmental Research Laboratory, USEPA.

Vannote, R. L., G. W. Minshall, K. W. Cummins, J. R. Sedell, and C. E. Cushing. 1980. The river continuum concept. *Can. J. Fish. Aquat. Sci.* 37:130–137.

Walker, W. W., Jr. 1981. *Empirical Methods for Predicting Eutrophication in Impoundments*, Rep. 1, Phase I. *Data Base Development*, Tech. Rep. E-81–9. Vicksburg, MS: U.S. Army Engineer Waterways Experiment Station.

Walker, W. W., Jr. 1982. An empirical analysis of phosphorus, nitrogen, and turbidity effects on reservoir chlorophyll-*a* levels. *Can. Water Resour. J.* 7:88–108.

Walker, W. W., Jr. 1984. Empirical predication of chlorophyll in reservoirs. In *Lake and Reservoir Management*, EPA 440/5/84–001. Washington, DC: U.S. Environmental Protection Agency, pp. 292–297.

Walker, W. W., Jr. 1985. *Empirical Methods for Predicting Eutrophication in Impoundments*, Rep. 3, Phase II. *Model Refinements*, Tech. Rep. E-81–9. Vicksburg, MS. U.S. Army Engineer Waterways Experiment Station.

Ward, J. V. 1976. Comparative limnology of differentially regulated sections of a Colorado mountain river. *Arch. Hydrobiol.* 78:319–342.

Ward, J. V., and J. A. Stanford (eds.). 1979. *The Ecology of Regulated Streams*. New York: Plenum.

Ward, J. V., and J. A. Stanford. 1981. Tailwater biota: ecological response to environmental alterations. In H. G. Stefan (ed.), *Symposium on Surface Water Impoundments*. New York: American Society of Civil Engineers, pp. 1516–1525.

Ward, J. V., and J. A. Stanford. 1983. The serial discontinuity concept of lotic ecosystems. In T. D. Fontaine and S. M. Bartell (eds.), *Dynamics of Lotic Ecosystems*. Ann Arbor, MI: Ann Arbor Science, pp. 29–42.

Wiebe, A. H. 1938. Limnological observations on Norris Reservoir with especial reference to dissolved oxygen and temperature. *Trans. North Am. Wildl. Conf.* 3:440–457.

Wiebe, A. H. 1939. Density currents in Norris Reservoir. *Ecology* 20:446–450.

Wiebe, A. H. 1940. The effect of density currents upon the vertical distribution of temperature and dissolved oxygen in Norris Reservoir. *J. Tenn. Acad. Sci.* 15:301–308.

Wiebe, A. H. 1941. Density currents in impounded waters—their significance from the standpoint of fisheries management. *Trans. North Am. Wildl. Conf.* 6:256–264.

Wetzel, R. W. 1983. *Limnology*, 2d ed. New York: Saunders College.

Wilkins, P., L. Kirkland, and A. Hulsey. 1967. The management of trout fisheries in reservoirs having a self-sustaining warm water fishery. In *Reservoir Fishery Resources*. Bethesda, MD: Reservoir Committee, American Fisheries Society, pp. 444–452.

Wright, J. C. 1967. Effects of impoundments on productivity, water chemistry, and heat budgets of reservoirs. In *Reservoir Fishery Resources*. Bethesda, MD: Reservoir Committee, American Fisheries Society, pp. 188–199.

Wunderlich, W. O. 1971. The dynamics of density-stratified reservoirs. In G. E. Hall (ed.), *Reservoir Fisheries and Limnology*, Spec. Publ. No. 8. Bethesda, MD: American Fisheries Society, pp. 219–232.

Young, W. C., H. H. Hannan, and J. W. Tatum. 1972. The physicochemical limnology of a stretch of the Guadalupe River, Texas, with five mainstream impoundments. *Hydrobiologia* 40:297–319.

12 Natural Lakes of The Southeastern United States: Origin, Structure, and Function

THOMAS L. CRISMAN

Department of Environmental Engineering Sciences, University of Florida, Gainesville, FL 32611

This chapter is a review of the natural lakes of the southeastern United States, including their origin, ecology, and management problems. Particular emphasis has been placed on ecosystem structural and functional aspects including trophic-level interactions. Where appropriate, data will be discussed relative to the wide range of climate (temperate to tropical) covered by southeastern lakes.

Unlike north temperate systems, limnological data on most lake types in the Southeast are sparse and are usually presented in unpublished reports. The notable exceptions are subtropical and tropical Florida lakes, for which the refereed literature has been increasing progressively since the late 1960s. The emphasis on Florida lakes in the following discussion is a reflection of the paucity of data on other lake types rather than a regional bias of the author.

LAKE TYPES OF THE SOUTHEAST

The southeastern United States likely displays the greatest array of natural lake types of any unglaciated region in North America (Fig. 1). Examples have been documented of basins created by (1) fluviatile processes, (2) local subsidence associated with earthquakes, (3) landslides, (4) deflation, (5) epeirogenetic uplift of irregular marine surfaces, (6) erosional processes along marine coastlines, and (7) solution of carbonate bedrock.

Fluviatile Features

Although fluviatile lakes are found throughout the Southeast, those of the Mississippi River and its principal tributaries are greatest in both number and

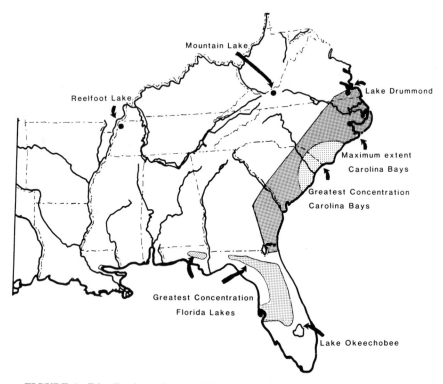

FIGURE 1. Distribution of natural lakes in the southeastern United States.

size (Fig. 1). Moore (1963) discussed four types of fluviatile lakes found in the Southeast. Lateral lakes form when sediments deposited by a primary stream block the flow of a tributary stream to produce a drowned valley (e.g., lakes occupying the side valleys of the Red River in western Louisiana). Lateral levee lakes, exemplified by Catahoula Lake, Louisiana, are positioned between a levee and a scarp defining the floodplain. Such systems display pronounced seasonal variations in both depth and extent and frequently disappear completely. Cane River Lake and Chaplin's Lake in Louisiana are the best-studied examples of lakes occupying abandoned river channels. Such lakes tend to be long and narrow and are quite similar in origin to oxbow lakes, which represent previous bends in a river that have become isolated as the watercourse meanders across its floodplain. Oxbows are by far the most numerous fluviatile lakes of the southeastern United States.

Although fluviatile lakes in general have been poorly investigated, limnological research has concentrated on oxbow lakes of the Mississippi River drainage, especially Lake Chicot, the largest natural lake of Arkansas (Nix and Schiebe 1984). Oxbow lakes are relatively short-lived geologically. Ages of 600–700 years have been calculated for two of the larger lakes, Hovey Lake of the Ohio River (Crisman and Whitehead 1975) and Lake Chicot of the Mississippi River (Nix and Schiebe 1984). Furthermore, sedimentation rates of 0.75–2.45 cm/year for Hovey lake, Indiana, based on palynological and archeological evidence (Crisman and Whitehead 1975), and 1.00–5.00 cm/year and 1.00–4.00 cm/year for Wolf Lake, Mississippi (McHenry et al. 1982), and Lake Chicot, Arkansas (McHenry et al. 1984), respectively, based on ^{137}Cs dating, are the greatest recorded for any lake type in the southeastern United States. With the exception of a few early surveys (summarized by Moore (1963)), the few biological investigations conducted on fluviatile lakes are for the most part hidden in nonrefereed reports and theses.

Subsidence Features

A series of shallow irregular lakes were created on the Mississippi River floodplain of Tennessee, Arkansas, and Missouri as a result of a crustal subsidence associated with the New Madrid earthquake of 1811 (Hutchinson 1957). A biological station was established in 1931 on the largest of these lakes, Reelfoot Lake, Tennessee (Fig. 1). Before the lab closed in 1969, numerous faunal and floral taxonomic works and macrophyte and fisheries investigations were published (partially summarized by Gerking 1963). Although less numerous, investigations have continued on taxonomy (McCullough et al. 1985), paleolimnology (Pardue et al. 1986), and macrophyte ecology (Guthrie 1989, Henson 1990a,b). A detailed synthesis of the functioning of this unique system remains to be done.

Landslide Features

Mountain Lake, Virginia (Fig. 1), was created 9180 years ago when a landslide initiated by solifluction dammed a valley (Marland 1967). A permanent lake did not become established until approximately 2000 years ago due to the leaky nature of the dam. With a maximum depth in excess of 30 m, Mountain Lake is the sole example of a dimictic natural lake in the Southeast. Beginning with the work of Hutchinson and Pickford (1932), limnological research at the lake has remained active, facilitated by the Mountain Lake Biological Station of the Virginia Polytechnic Institute and State University.

Deflation Features

Based on the presence of Cretaceous sands in the area, Watts (1980a) suggested that White Pond, a 19 100-year-old, 500-m-diameter pond on the innermost edge of the coastal plain of South Carolina, was created by deflation. It is likely that many of the small interdunal coastal ponds of the Southeast (Wharton 1978), as well as numerous lakes in peninsular Florida, are of similar origin.

Although the most plausible, deflation is one of 18 mechanisms proposed for the creation of Carolina Bays (Ross 1987). Carolina Bays are shallow, elliptical depressions usually oriented along a northwest–southeast linear axis and possessing a low sandy rim at the southeastern end (Prouty 1952). The term Carolina Bay was coined by early European settlers because of the characteristic presence of "bay" trees (*Gordonia, Magnolia, Persea*) along the margin (Sharitz and Gibbons 1982). Although the term has been applied to depressions dominated by either marsh/swamp or open-water lakes, this chapter considers only the latter category.

Carolina Bays are restricted to the coastal plain of the southeastern United States with a total geographic range of 23 310 km^2 (Sharitz and Gibbons 1982) extending from southeastern Virginia to northeastern Florida (Fig. 1). Approximately 80% of the estimated 500 000 bays (Prouty 1952) are located in South Carolina and southern North Carolina, with a majority being classified as marsh or swamp. Bay lakes all tend to be shallow and range in length between 50 m and 8 km (Lake Waccamaw, North Carolina). Most estimates of the age of the bay lakes fall in the range 10 000–100 000 years (Sharitz and Gibbons 1982), and Whitehead (1981) suggested that bay formation occurred in at least two distinct time periods. In addition to a lack of temporal synchroneity in formation, it has been suggested by Delcourt and Delcourt (1985) that the mechanism of formation varied according to latitude. They proposed a shift in the principal mechanism of origin from thermokarst depressions created during times of extreme periglacial conditions (north of 39°N latitude) to deflation basins that expanded in size during full-glacial times (33–39°N latitude) to basins resulting from a perched groundwater table as a result of humate-cemented sands during the late-Quaternary (south of 33°N latitude). No clear concensus has been reached regarding the complex issue of the origin of Carolina Bays and further research is warranted.

While a great deal of research has centered on the origin of bays and their value in palynological reconstruction of the regional history of terrestrial vegetation, comparatively little work has been devoted to the limnology of bay lakes themselves. Frey, in addition to doing classical investigations on regional vegetation history (1951a, 1953, among others) and basin formation and evolution (1954a,b), also collected some of the first detailed limnological data on Carolina Bay lakes (1948, 1949, 1951b). Limnological research since that time has concentrated on Lake Waccamaw, the largest bay lake, with particular interest being paid to its endemic fish and invertebrate fauna. A complete bibliography of all investigations on Carolina Bays has been compiled by Ross (1987). Although seemingly sparse, the limnological literature on Carolina Bays is the most comprehensive for any lake type in the southeastern United States except Florida solution lakes.

Lake Drummond is a large (7855-ha), shallow (2-m maximum) basin in the center of the Dismal Swamp (Fig. 1). Although it is near the northern geographical limit of Carolina Bays in southeastern Virginia, Whitehead and Oaks (1979) noted that because of the lack of characteristic bay geomorphic features and its relatively young age (4000 years), Lake Drummond could not be classified as a Carolina Bay. Instead, they suggested that the lake represented either an area of deep peat burn during a period of lowered water level or a solution-collapse feature. Lake Drummond again illustrates the

need for detailed investigations on the formation and ontogeny of southeastern coastal plain lakes. Although not extensive, the limnological database
on Lake Drummond has been summarized by Marshall (1979).

Uplift Features

Lake Okeechobee, Florida, is the second largest lake (1840 km^2) wholly within
the United States (Fig. 1). It has long been accepted that the lake represents
an irregular marine surface formed at the bottom of a Pliocene sea that became
a freshwater lake after epeirogenetic uplift (Parker and Cooke 1944, Hutchinson 1957). Brooks (1974) postulated that differential subsidence of an unusually thick Miocene clay sequence coupled with deposition of clastic sediments during Plio-Pleistocene eustatic high stands of sea level have been the
dominant formative processes for the Okeechobee basin. He noted that although there is evidence that a lake existed in the basin at least periodically
(12 000 and 6300 years ago) since formation, the present lake did not become
established until approximately 4000–4700 years ago. Subsequently, lake level
gradually increased, associated with the buildup of organic marsh sediments
along the southern shore. Brooks used old beach ridges to establish 1700
years ago as the time of maximum water level. Paleolimnological evidence
from sponge and diatom remains suggest that the lake has been moderately
eutrophic for at least the past 4700 years (Gleason and Stone 1975). Although
a reasonable limnological database exists for Lake Okeechobee, especially
for its water chemistry, most of the data appear only in nonrefereed reports
and university theses. Given that the lake sits at the subtropical–tropical
boundary, it deserves much more scientific attention than it has received in
the past.

Erosional Features

Several lakes are found in the central Florida panhandle immediately behind
the coastline of the Gulf of Mexico. These lakes generally are oriented with
their linear axis perpendicular to the coast, and moderately high beach dunes
separate them from the Gulf. It is likely that these lakes fall into three
categories, based on their mode of origin. The first category includes those
lakes that never had connection to the sea, but formed as blow-out depressions
behind the coastal dune complex. A second category includes coastline in-
dentations that were isolated from the ocean via the deposition of sand from
longshore currents. Many of the lakes in this category originated from blocked
stream mouths. Finally, it appears that several lakes were created when storm
surges breached the coastal dunes and promoted erosion of back dune areas.
Lakes were established in these erosional depressions once longshore currents
blocked their marine connection with sediments. Although both clear and
organically colored coastal lakes are found in Florida, even basic limnological
data on them appear to be totally lacking.

Solution Features

Solution lakes of the Southeast fall into three categories: sagponds of the southern Appalachians, limestone sinks of the Interior Low Plateaus regions between the Appalachian Mountains and the Ozark Plateau, and the sub-tropical lakes of Florida (Fig. 1). Sagponds result from subsidence of overlying noncalcareous Tertiary deposits into collapsed solution caverns in deep deposits of dolomite. These small (<1 ha), shallow (<1 m) ponds are common in the Ridge and Valley province from Pennsylvania southward to northern Georgia and Alabama (Watts 1979). As part of palynological investigations, the ages of two sagponds in Bartow County, Georgia, Quicksand and Bob Black, have been estimated at 20 000 and 22 900 years, respectively (Watts 1970). Because sagponds display a great deal of seasonal water level fluctuation and frequently dry completely, they have received little attention from limnologists. Sagponds in northwest Georgia are acidic (pH 4.59–5.77), with amphipods and isopods as the principal benthic invertebrates and mosquitofish (*Gambusia*), if present, as the only fish (Greear 1967, Wharton 1978).

Karst depressions formed from subterranean solution of limestone and dolomite dot the landscape between the Appalachian Mountains and the Ozark Plateau. Such features are not of uniform age and in several areas are still actively forming. Permanent lentic systems often become established in those depressions that develop a clay or shale seal. Delcourt and Delcourt (1985, among others) have documented records of 20 000–40 000 years of lacustrine sediment deposition in many of these basins between 33 and 40°N latitude that have proven extremely important in palynological reconstructions of the vegetational history for the Mid-South. As with sagponds, the generally small size (<1 ha) and extreme water level fluctuations (including periodic dessication) have precluded a great deal of attention from aquatic ecologists.

Florida solution lakes (7748) clearly comprise the dominant lake type in the Southeast (Shafer et al. 1986). A majority (71%) are closed basins 27–62 ha (42%) in size. Although lakes are found throughout the state and extreme southern Georgia, they are most numerous in a region from just north of Lake Okeechobee and east of Tampa Bay and extending up the center of the peninsula to near the Georgia border (Fig. 1). Two smaller lake regions are centered in the eastern (Leon County) and central (Washington County) panhandle.

Although numerous examples of single and complex doline lakes abound, many Florida solution lakes were formed by a variety of complex mechanisms. Long Lake (Levy County) is an example of a polje, a lake created when a series of sinks develop along a fault line resulting in a tectonokarstic depression (Vernon 1951). Additional lakes near Tallahassee developed in polje-like depressions created when sinks appeared in normal valleys (Hutchinson 1957). Such sinks drain both the upper and lower portions of the valley, thus increasing erosion into the sink that can eventually plug and form an elongate closed basin with a convoluted shoreline. Lake Jackson (Leon County) is an example of such a lake that experiences periodic complete desiccation when the sink reopens for a short time. Finally, it has been suggested that Lake

Tsala Apopka represents a high sea level estuary that has been greatly modified by a complicated pattern of solution depressions (Cooke 1939).

Florida lakes range in age from >44 000 years to those that are actively forming today. The three oldest Florida lakes discovered to date as a result of the extensive palynological investigations of Watts (1969, 1980b, among others) are Annie in Highlands County (44 300 years), Mud in Marion County (35 000 years), and Sheelar in Clay County (23 800 years). Watts (1980b) suggested that those lakes currently deeper than 20 m are likely to have remained as lentic environments during the full-glacial period when sea level was significantly lowered. The majority of Florida lakes are shallow and likely became permanent water bodies only following postglacial sea level rise.

PHYSICAL/CHEMICAL ENVIRONMENT

Thermal Regimes

Mountain Lake, Virginia, is the only example of a natural dimictic lake in the southeastern United States. The lake behaves as a typical north temperate system with summertime epilimnetic temperatures rarely exceeding 24 °C (Marland 1967). The remainder of the Southeast falls within the latitudinal range of warm monomictic lakes (Hutchinson and Loffler 1956), but only a few of the larger oxbow lakes of the Mississippi River and numerous Florida solution lakes are deep enough to display pronounced thermal stratification.

Beaver et al. (1981) divided the lakes of peninsular Florida into three latitudinal groups based on annual thermal regimes. Their zones coincided with major floral and faunal zones in the state. The northern-central zonal boundary was placed at 29.5°N, while that of the central-southern transition was at 28°N latitude.

All peninsular Florida lakes displayed summertime maxima of 30–32 °C, while differences in winter minima (10 °C north, 16 °C south) were responsible for statistically defining the latitudinal groupings. Latitudinal winter temperature differences are clearly a reflection of the progressively diminishing influence of continental air masses with increasing distance down the Florida peninsula. Although comparable thermal data are lacking for lakes of the panhandle, this area experiences both the greatest influence of continental air masses and consistently the lowest winter temperatures in Florida.

Lakes of northern and central Florida >5 m depth generally display thermal stratification from spring through early fall (March–October) and are holomictic during late fall–winter (October/November–February/March). The data of Beaver et al. (1981) suggest that, while all lakes in Florida are classified as warm monomictic, the length of the mixing period diminishes southward in the peninsular as a reflection of a progressive decrease in seasonal temperatures with decreasing latitude. While lakes in northern Florida can mix for up to 6 months, those of southern Florida may undergo mixis for a month or less.

Once established, thermal stratification in Florida lakes appears to be very stable. Although Yount (1961) demonstrated the destabilizing effect of hurricanes on shallow (<6 m) weakly stratified lakes, Nordlie (1972) found that the stratification of Lake Mize, a small, deep (25 m) sheltered forest lake in northern Florida, remained intact during two hurricanes that passed through the area. The only noticeable effect of the storms was a depression of temperature in the upper 3 m of the epilimnion by 2–6 °C as a result of the massive amounts of rainfall associated with the storms.

Water Chemistry

Canfield (1981) clearly demonstrated the wide range of water chemistry displayed by Florida lakes. A total of 165 lakes (approximately 2% of Florida lakes) representing all physiographic regions of the state were sampled three times during a single year. The lakes spanned a broad range of specific conductance (11–5600 μS/cm at 25 °C), organic color (0–416 mg/L as Pt), and total phosphorus (3–834 mg/m^3). Only 13% of the lakes were classified as oligotrophic, while a majority fell in the mesotrophic (42%) and eutrophic (35%) categories.

The linkage between water chemistry and geology is clearly demonstrated by Florida lakes (Shannon and Brezonik 1972). Phosphatic sands and clays of the Miocene Bone Valley and Hawthorn formations (Fig. 2) underlay an extensive area east of Tampa Bay and also form a mostly continuous 150-km band in northern Florida down the central axis of the peninsula (Puri and Vernon 1959, Vernon and Puri 1964). A majority of these deposits are located in the Polk Upland and Lake Upland physiographic regions east of Tampa Bay. Lakes of this region are mostly mesotrophic and eutrophic and display conductivity >100 μS (at 25 °C) and circumneutral pH (Table 1, Fig. 2). Moderately, organically colored lakes are common.

A contrasting geology (Fig. 2) is offered by both the mostly continuous band of relict beach ridges that extend two-thirds the length of peninsular Florida from the Georgia border to just north of Lake Okeechobee and the remnant portion of formerly extensive highlands located in the central Panhandle (Vernon and Puri 1964, White 1970). Two physiographic regions broadly representative of this geology, the Trail Ridge region of northern peninsular Florida and the New Hope Ridge of the central panhandle, are both dominated by highly leached sands of the Fort Preston formation (Vernon and Puri 1964). In contrast to those of the previously discussed phosphatic sand regions, lakes of these two areas are oligotrophic, clear water (<20 mg/L as Pt), acidic (<6.0 pH) systems displaying conductivity values <50 μS/cm (Fig. 2, Table 1). Although lakes influenced by cultural eutrophication can mask a geological control over water chemistry, the striking regional differences displayed by slightly to moderately impacted lakes situated in areas of contrasting geology clearly demonstrate the influence of geology over the baseline water chemistry of Florida lakes.

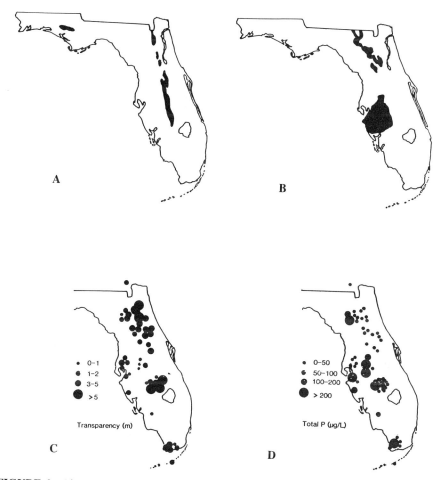

FIGURE 2. A) Trail Ridge region peninsular Florida and New Hope Ridge of the panhandle; B) phosphatic sands and clays of the Miocene Bone Valley and Hawthorn Formations; C) transparency of Florida lakes; and D) total phosphorus concentrations in Florida lakes.

The database on water chemistry for the remainder of lake types in the Southeast is rather sparse (Table 2). Fluviatile lakes of the Mississippi and Ohio Rivers tend to be highly turbid systems characterized by high conductivity (>100 μS/cm) and trophic state. In contrast, Carolina Bays display a broad range of trophic state and organic color and as a group are the most acidic lakes of the Southeast, with the exception of the Florida sandhill lakes. Finally, a recent study of dune ponds on the Outer Banks of North Carolina (Kling 1986) showed that the water chemistry of these shallow circumneutral pH ponds more closely approximated groundwater than seawater values.

TABLE 1 Physical and Chemical Parameters for Four Lake Regions in Florida

Parameter	Polk Upland	Lake Upland	Trail Ridge	New Hope Ridge
n	6	6	3, 12	5
Conductivity (μS)	115	97	33, 45	18
pH	6.5	5.8	5.7, 5.2	5.2
Total phosphorus (μg/L^3)	82	16	12, 14	4
Chlorophyll a (mg/m^3)	12.0	4.9	1.8, 2.1	0.8
Color (Pt units)	58	92	3, 18	3
Secchi (m)	1.1	1.8	4.3, 3.6	5.3

Source. Data from Hendry and Brezonik (1984) and Canfield (1981).

PLANT COMMUNITIES

Phytoplankton

As clearly demonstrated for temperate zone lakes, there is a strong positive relationship between estimates of algal biomass and concentrations of chlorophyll a ($r = .80$) in subtropical Florida lakes (Canfield et al. 1985a). Canfield (1983b) noted a positive linear relationship between chlorophyll a and total phosphorus concentrations in Florida lakes up to phosphorus values of 100 mg/m^3. Above this value, the relationship curves and weakens, and chlorophyll values approach a maximum of 277 mg/m^3. Lakes in excess of 100 mg/m^3 phosphorus tend to have total nitrogen : total phosphorus ratios <10 suggesting nitrogen rather than phosphorus limitation of phytoplankton production (Baker et al. 1981, Kratzer and Brezonik 1981, Canfield 1983b). Nitrogen-limited lakes comprise approximately 27% of all Florida lakes (Canfield 1983b).

While chlorophyll a values alone can be used broadly to define the trophic state of lakes (Wetzel 1983), it has become fashionable in recent years to use chlorophyll alone or in combination with other parameters to classify lakes according to a numerical scale (trophic state index). Carlson's (1977) trophic state index, perhaps the most popular of the indices for temperate lakes, is clearly not broadly applicable for Florida lakes because of its omission of nitrogen as an important trophic state variable. Developed specifically for Florida lakes, the trophic state index of Huber et al. (1982) is a modified Carlson index that incorporates nitrogen concentrations in model construction when the nitrogen : phosphorus ratio is <10. Huber et al. (1982) also present a detailed chronology of the decade-long history of construction and refinement of trophic state models specific to Florida lakes.

Strict reliance on trophic state indices to classify Florida lakes is somewhat restricted by factors associated with the generally shallow nature of such lakes. The photic zone often extends to the bottom of Florida lakes, thus favoring the development of an extensive vegetated littoral zone. Canfield et al. (1984b) noted that predictive models for chlorophyll based on nutrient concentrations in lakes are strongly influenced by the extent of aquatic macrophytes, especially once the percentage of lake volume occupied by macrophytes exceeds 50%. A new predictive model for chlorophyll concentrations in Florida lakes that factored in the percent plant volume infestation (PVI) was provided.

Relative to those in temperate lakes, chlorophyll concentrations in Florida lakes display little seasonal variation (Fig. 3). Midsummer chlorophyll values in eutrophic Florida lakes often are over 300% greater than values recorded for eutrophic natural lakes elsewhere in the Southeast (Table 3), with the latter values approximating winter levels in Florida. In spite of the persistence of abundant algal populations (estimated by chlorophyll) year round, Beaver et al. (1988) have demonstrated that there may be a pronounced seasonality in the availability of such a food resource to invertebrate grazers in Florida lakes. They found that much of the midsummer phytoplankton biomass in oligotrophic organically colored Florida lakes is contributed by zoochlorellae-

TABLE 2 Water Chemistry for Representative Lakes in the Southeastern United States

Lake Type	Lake	Trophic State[a]	Mean Depth (m)	pH	Alkalinity (mg/L as CaCO₃)	Conductivity (µS/cm at 25 °C)	Secchi (m)	Total P (mg/L)	Total N (mg/L)	Color (mg/L as Pt)	Reference
Fluviatile	Chicot, AR	E	2.7		85	189	0.4	0.162	1.200		USEPA 1978
	Grand, AR	E	2.1		95	160	0.5	0.101	1.220		USEPA 1978
	Hovey, IN	E	1.2		77	300	0.3	0.062	1.610		USEPA 1978
	Black, LA	E	2.6		10	125	1.2	0.077	0.620		USEPA 1978
	Bruin, LA	E	9.1		92	168	1.3	0.057	0.780		USEPA 1978
	Cocodrie (Con.), LA	E	4.9		56	141	0.5	0.090	0.995		USEPA 1978
	Cocodrie (Rap.), LA	E	0.5		82	162	0.5	0.106	0.820		USEPA 1978
	Concordia, LA	E	6.1		147	297	0.8	0.076	0.820		USEPA 1978
	False River, LA	E	7.1		136	235	1.5	0.082	0.860		USEPA 1978
	Saline LA	E	2.7		14	99	0.2	0.111	1.080		USEPA 1978
	Verret, LA	E	1.5		93	253	0.5	0.163	1.260		USEPA 1978
Earthquake	Reelfoot, TN	HE	1.4		91	200	0.6	0.233	2.170		USEPA 1978
Landslides	Mountain, VA	O				12					Marland 1967
Carolina Bays	Black, NC	E	<2.0	6.4	8	98	0.3	0.168	0.730	496	Weiss and Kuenzler 1976
	Jones, NC	O–M	<2.0	3.2	0	65	0.4	0.025	0.565	265	Weiss and Kuenzler 1976

Mattamuckett, NC	E	<2.0	6.7	4	2770	0.6	0.038		12	Weiss and Kuenzler 1976
Phelps, NC	M	<2.0	6.1	1	120	2.1	0.022	0.386	1	Weiss and Kuenzler 1976
Salters, NC	O-M	<2.0	4.1	0	59	0.6	0.016	0.374	65	Weiss and Kuenzler 1976
Singletary, NC	M	<3.0	3.3	0	55	0.6	0.049	0.414	171	Weiss and Kuenzler 1976
Waccamaw, NC	M	1.5	7.1	11	62	1.1	0.036	0.245	81	Weiss and Kuenzler 1976
White, NC	O-M	<2.0	4.6	11	63	1.1	0.018	0.630		USEPA 1978
Clear, SC	O	<2.0	6.4	0	65	>3.0	0.017	0.211	3	Weiss and Kuenzler 1976
Cypress, SC		0.5	4.3	1						Tilly 1973
Thunder, SC		0.6	4.3	3						Schalles 1979
Long Pond, SC		0.5	4.3	2						Schalles 1979
Coastal plain Drummond, VA	E	1.5	4.4			0.3	0.400			Marshall 1979
Moultrie, SC	E	5.7		17	72	1.1	0.026	0.625		USEPA 1978
Epeirogenetic uplift Okeechobee, FL	E	2.8		132	640	0.7	0.063	1.745		USEPA 1978
Marine coastline Pond A, NC		1.0	6.6	16	112	0.7				Kling 1986
Pond B, NC		2.4	7.1	73	381	0.5				Kling 1986
Pond C, NC		1.3	6.2	13	116	0.4				Kling 1986
Pond D, NC		0.5	7.2	81	205	0.4				Kling 1986
Pond E, NC		1.1	7.1	49	174	0.7				Kling 1986

[a]E, eutrophic; H, hypereutrophic; M, mesotrophic; O, oligotrophic.

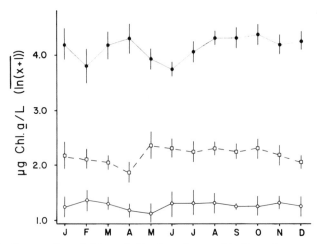

FIGURE 3. Seasonality of chlorophyll *a* for oligotrophic (lower), mesotrophic (middle), and eutrophic (upper) trophic state categories. (From Bays (1983).)

bearing ciliated protozoans, with peak contribution exceeding 90% of autotrophic biomass. The importance of myxotrophs as contributors to total autotrophic biomass generally decreased with increasing trophic state, with the mean annual myxotrophic contribution ranging from 19% (oligotrophic) to 3% (hypereutrophic). Such a symbiotic relationship could be especially important to algae during periods of nutrient limitation characterizing oligotrophic lakes during summer.

Few data are available on phytoplankton productivity in southeastern lakes outside Florida (Table 4). Florida has both the least productive as well as the most productive lakes of the Southeast. Maximum algal production in Florida lakes usually occurs during summer (Nordlie 1976, McDiffett 1980, McDiffett 1981), but an analysis of seasonal production data (>300 events) from over 75 lakes (Beaver and Crisman 1991) indicates that algal production in oligotrophic and mesotrophic Florida lakes displays no significant seasonality, whereas in eutrophic and hypereutrophic lakes summer production is significantly greater than winter levels. Fontaine and Ewel (1981) also found that phytoplankton metabolism remained relatively constant in mesotrophic Lake Conway, Florida, with respiration exceeding production on an annual basis.

Florida lakes, like most lakes of the Southeast, are generally shallow with extensive vegetated littoral zones covering most of the bottom. While phytoplankton metabolism remained constant throughout the year at Lake Conway, Fontaine and Ewel (1981) noted seasonal pulses in total system metabolism, which they attributed to the contribution of submergent macrophytes and their associated epiphytic algae. According to their calculations, littoral production dominated planktonic production throughout the year, in spite of some evidence for supression of the former during the summer period of elevated temperature. Although comparable data are unavailable for other lakes in the Southeast, Fontaine and Ewel (1981) and Ewel and Fontaine (1983) suggested that the Conway data supported the contention that production and respiration become more balanced with increasing latitude.

Estimates of algal abundance in southeastern lakes derived from direct cell counts display a great deal of variation even between lakes of comparable trophic state (Table 2). In addition, there appears to be little relationship between algal counts and measured chlorophyll levels in southeastern lakes. This is not surprising given the fact that such counts often are both based on arbitrarily defined algal "units" and not reported on the basis of algal size classes. When combined with biovolume estimates of individual taxa, direct cell counts can provide reasonable approximations of chlorophyll levels (Canfield et al. 1985a).

Except for a pulse of cryptophyte production during winter, Florida lakes fail to display the pronounced seasonal succession of major algal groups seen in comparable temperate lakes (Fig. 4). The classical diatom peak associated with springtime mixis in temperate lakes is absent in subtropical Florida lakes, and diatoms rarely comprise over 20% of phytoplankton abundance, regardless of trophic state. The extent of seasonal succession declines with increasing trophic state with the abundance of the most productive lakes being dominated by blue-green algae (>90%) all year. Unfortunately, detailed phytoplankton data for other southeastern lakes are sparse, so direct comparison with Florida lakes is not possible.

Unicellular ultraplanktonic blue-greens dominate the phytoplankton assemblages of acidic oligotrophic Florida lakes (Table 3). With increasing eutrophication, there is a progressive shift to net plankton dominance of the phytoplankton, especially taxa of blue-green algae. Green algae reach their maximum abundance in mesotrophic lakes. Such trends are consistent with those from temperate lakes (Wetzel 1983). Florida lakes differ from temperate lakes in the taxa expected to dominate eutrophic systems. While considered a major management problem in eutrophic temperate lakes, *Aphanizomenon* is only infrequently encountered in comparable Florida lakes. Instead, *Microcystis* and *Lyngbya* are most often cited as two of the principal problem-producing algae in Florida lakes. Natural lakes throughout the Southeast also fail to record *Aphanizomenon* as a dominant (Table 2). Unfortunately, the importance of latitude as a controlling variable in the distribution of *Aphanizomenon* has not been assessed.

Epiphytic Algae

With the exception of one study each for oligotrophic (Brown 1976), mesotrophic (Feerick 1977), and eutrophic (Hodgson et al. 1986) Florida lakes, very little attention has been paid to epiphytic algae in Southeast lakes. Although limited in number, these studies have proven valuable in delineating seasonal and trophic gradient trends for this normally overlooked autotrophic component.

TABLE 3 Trophic State and Phytoplankton Parameters for Representative Lakes in the Southeastern United States

Lake Type	Lake	Trophic State[a]	Mean Chlorophyll (mg/L)	Color (mg/L as Pt)	Phytoplankton Primary Productivity (mg C/m³ h⁻¹)								Reference
					Mean Summer		Maximum		Minimum		Mean Annual		
					Gross	Net	Gross	Net	Gross	Net	Gross	Net	
Fluviatile	Chicot, AR	E				206		500		0		35	Cooper et al. 1984
	Big Snooks, SC			78		11		28				4	Tilly 1973
Landslide	Mountain, VA	O									10^a		Marland 1967
Carolina Bays	Black, NC	E		496		35		56				4	Weiss and Kuenzler 1976
	Jones, NC	O–M		265		1		1				4	Weiss and Kuenzler 1976
	Mattamuskeet, NC	E		12		34		43				2	Weiss and Kuenzler 1976
	Phelps, NC	M		1		12		13				2	Weiss and Kuenzler 1976
	Salters, NC	O–M		65				5				1	Weiss and Kuenzler 1976
	Singletary, NC	M		171		17		22				4	Weiss and Kuenzler 1976
	Waccamaw, NC	M		81		16		22				3	Weiss and Kuenzler 1976
	White, NC	O–M		3				5				1	Weiss and Kuenzler 1976
	Clear Pond, SC	O		15		32		72				5	Tilly 1973

	Trophic state[a]												Reference
Epeirogenetic uplift													
Okeechobee, FL	E		15	130	123	298	274	12	12	113	108	12	Marshall 1979
Florida solution lakes													
Newnans, FL	E	65	163	1372	1143	1658	1472	166	0	824	701	18	Crisman et al. 1986
Wauberg, FL	E	95	45	2014	1403	3166	1900	505	291	1344	839	13	Crisman et al. 1986
Apopka, FL	E	33	73	383	292	1131	1500	23	0	208	140	35	Brezonik et al. 1978
Beauclair, FL	E	70	90	512	320	1028	769	89	24	242	138	11	Brezonik et al. 1978
Dora, FL	E	68	80	311	196	1232	777	67	20	124	95	25	Brezonik et al. 1978
Eustis, FL	E	23	43	118	42	178	106	59	5	88	33	13	Brezonik et al. 1978
Griffin, FL	E	42	61	348	222	758	426	54	40	170	114	19	Brezonik et al. 1978
Brooker, FL	E	54	28	96	46	143	66	69	27	83	40	6	Cowell and Dawes 1984
Weir, FL	M	6	8	0	18		30		5		12	6	Shannon 1970
Sante Fe, FL	M	5	59		17		30		1		9	6	Shannon 1970
Geneva, FL	O	1	10		5		5		1		3	6	Shannon 1970
Magnolia, FL	O	2	11		1		2		−1		1	6	Shannon 1970
Santa Rosa, FL	O	2	2		1		1		−1		1	6	Shannon 1970
Cowpen, FL	O	2	5		1		2		−1		2	6	Shannon 1970

[a]E, eutrophic; M, mesotrophic; O, oligotrophic.
[b]mg C/m^2.

491

TABLE 4 Phytoplankton Productivity for Representative Lakes in the Southeastern United States

Lake Type	Lake	Trophic State[a]	Limiting Nutrient[b]	Chlorophyll (mg/L)	Phytoplankton (No./mL)	Summer Algal Dominants: First Dominant	Second Dominant	Reference
Fluviatile	Chicot, AR	E	P	7.4	13.7	3 009 ... Merismopedia	Melosira	USEPA 1978
	Grand, AR	E	N	12.07	62.9	152 792 ... Dactylococcopsis	Stephanodiscus	USEPA 1978
	Hovey, IN	E	N	25.9	84.3	154 639 ... Oscillatoria	Pennate Diatom spp.	USEPA 1978
	Black, LA	E	N	8.05	12.7	18 903 ... Lyngbya	Dactylococcopsis	USEPA 1978
	Bruin, LA	E	N	13.68	16.4	52 369 ... Dactylococcopsis	Nitzschia	USEPA 1978
	Cocodrie (Con.), LA	E	P	11.05	35.3	17 161 ... Cyclotella	Stephanodiscus	USEPA 1978
	Cocodrie (Rap.), LA	E	N	7.7	33.4	10 149 ... Flagellate spp.	Dactylococcopsis	USEPA 1978
	Concordia, LA	E	N	10.78	33	16 929 ... Dactylococcopsis	Nitzschia	USEPA 1978
	False River, LA	E	N	10.48	24.6	37 764 ... Dactylococcopsis	Flagellata spp.	USEPA 1978
	Saline, LA	E	P	9.72	15.3	4 660 ... Flagellate spp.	Melosira	USEPA 1978
	Verret, LA	E	N	7.73	62	134 220 ... Oscillatoria	Dactylococcopsis	USEPA 1978
Earthquake	Reelfoot, TN	HE	N	9.3	81	54 366 ... Melosira	Merismopedia	USEPA 1978
Carolina Bays	Waccamaw, NC	M	P	35	3.6	2 618 ... Aphanocapsa	Anacystis (Microcystis)	USEPA 1978
Coastal plain	Drummond, VA					Asterionella		Marshall 1979
	Moultrie, SC	E	P	24	8.8	13 603 ... Melosira	Lyngbya	USEPA 1978
Epeirogenetic uplift	Okeechobee, FL	E	P-N	27.6	14.5	31 927 ... Lyngbya	Fragilaria	USEPA 1978
Florida solution lakes	Apopka, FL	HE	P-N	36.9	46.6	105 280 ... Lyngbya	Melosira	

Lake	Trophic	Limiting[b]				Dominant 1	Dominant 2	Reference
Effie, FL	HE	N	3.31	261.4	150 737	Scenedesmus	Cyclotella	USEPA 1978
Munson, FL	HE	N	2.71	140.3	11 975	Anabaena	Anacystic (Microcystis)	USEPA 1978
Eloise, FL	E	N	5.1	70.2	78 065	Lyngbya	Chroococcus	USEPA 1978
Glenada, FL	E	N	9.73	27.7	17 864	Anacystis (Microcystis)	Kirchneriella	USEPA 1978
Horseshoe, FL	E	N	27.35	12.1	6 314	Botryococcus	Flagellate spp.	USEPA 1978
Kissimmee, FL	E	P	44.55	24.1	47 740	Dactylococcopsis	Lyngbya	USEPA 1978
Monroe, FL	E	N	10.15	14.2	19 079	Melosira	Lyngbya	USEPA 1978
Seminole, FL	E	N	10.9	102	136 675	Dactylococcopsis	Oscillatoria	USEPA 1978
Yale, FL	E	N	47	25.4	48 308	Lyngbya	Synedra	USEPA 1978
E. Lk. Toho-pekaliga, FL	M	N	22.1	5.2	7 653	Lyngbya	Dactylococcopsis	USEPA 1978
Minnehaha, FL	M	N	15.52	8.7	33 784	Oscillatoria	Dactylococcopsis	USEPA 1978
Jumper, FL	M		12.7		15 419	Anaebaena	Flagellate sp.	Medley and T. L. Crisman (unpublished data)
Fore. FL	O			3.8	36 046	Oocystis	Ultraplankton blue green	Medley and T. L. Crisman (unpublished data)
Boyd, FL	O			3	80 224	Ultraplankton blue green	Ultraplankton blue green	Medley and T. L. Crisman (unpublished data)
Mill Dam. FL	O			1.7	32 363	Chroococcus	Chromulina	Medley and T. L. Crisman (unpublished data)
Sunrise. FL	O			1.4	54 030	Ultraplankton blue green	Ultraplankton blue green	Medley and T. L. Crisman (unpublished data)
Round. FL	O			1	52 270	Elakatothrix	Ultraplankton blue green	Medley and T. L. Crisman (unpublished data)
West Clear-water, FL	O			0.8	28 460	Aphanothece	Chromulina	Medley and T. L. Crisman (unpublished data)
Shoesole. FL	O			0.6	31 270	Chromulina	Ultraplankton blue green	Medley and T. L. Crisman (unpublished data)

[a]E, eutrophic; H, hypereutrophic; M, mesotrophic; O, oligotrophic.
[b]N, nitrogen; P, phosphorus.

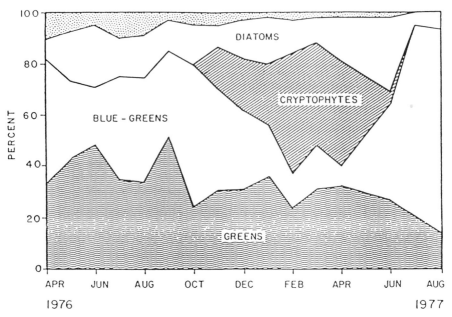

FIGURE 4. Seasonality of major algal groups in the phytoplankton of Lake Conway, Florida. (From Conley et al. (1979).)

Epiphyte productivity appears to be positively related to lake trophic state (Feerick 1977) with peak values recorded during summer. Abundance, however, is at a maximum during fall, suggesting poorer colonization of macrophytes during the active growing season (Fig. 5). Feerick (1977) attributed reduced midsummer abundance to inhibition from high light and temperature, but Hodgson et al. (1986) suggested that an antagonistic relationship exists between epiphytic algae and phytoplankton that is independent of light availability.

Regardless of trophic state, diatoms reach their maximum representation in Florida epiphytic assemblages during fall and winter, and green algae peak in the spring. Both diatoms and green algae can be important components throughout summer in oligotrophic lakes (Brown 1976), but they are progressively replaced by blue-green algae in mesotrophic and eutrophic lakes (Feerick 1977, Hodgson et al. 1986). Major differences are often noted between contemporaneous epiphytic algal and phytoplankton assemblages in Florida lakes. While rarely an important element in the plankton of any Florida lake, diatoms are abundant in epiphytic assemblages during the cooler months, regardless of trophic state.

Finally, although there may not be strong interspecific differences in epiphytic assemblages between macrophyte taxa (Feerick 1977), clear differences relative to water depth have been noted. Feerick (1977) found that epiphyte biomass in winter decreased with depth but displayed an opposite trend during

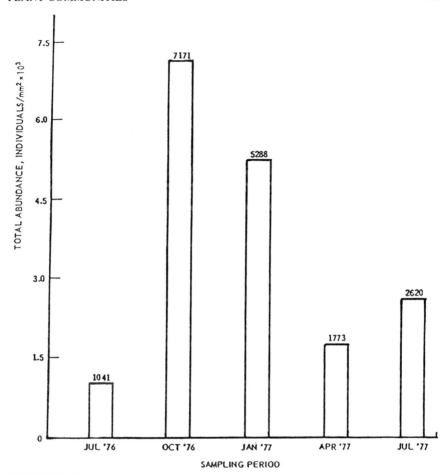

FIGURE 5. Seasonality of epiphytic algae on macrophytes in Lake Conway, Florida. (From Conley et al. (1979).)

summer, and Brown (1976) found that green algae were restricted to shallow depths, while chrysophytes had a broader depth distribution. Such observations were attributed to algal sloughing from wave action and light inhibition during summer.

Macrophytes

The submergent macrophyte community of Lake Waccamaw, a eutrophic Carolina Bay lake, is dominated by *Nitella* and *Najas*; *Panicum* characterizes the emergent vegetation zone (Stager and Cahoon 1987). The only detailed data on macrophytes in other Carolina Bay lakes are those of Schalles (1979), but the results may not be broadly applicable to large bay lakes because of the small size and shallow (<1.5 m) nature of the study ponds. Kling (1986) presented a detailed listing of macrophyte taxa characterizing coastal dune ponds on the Outer Banks, North Carolina, but did not present information on either the structure or function of the littoral zone.

With the exception of Reelfoot Lake (Guthrie 1989, Henson 1990a,b), macrophytes in oxbow lakes appear to have received little attention. Macrophyte communities are often restricted in oxbow lakes as a result of unstable substrate associated with both scouring and rapid deposition of inorganic sediments during frequent river flooding. Silt often remains suspended in the water column for prolonged periods and may further limit macrophyte development via light limitation. The recent replacement of *Zannichellia palustris*, *Najas guadalupensis*, and *Ranunculus flabellaris* as dominant submergents in Reelfoot Lake by *Potamogeton crispus* and *Ceratophyllum demersum*, and a shift in emergent vegetation from monotypic marshes (*Zizaniopsis miliacea*) to marsh–swamp vegetation appear to be related to maintenance of higher average water level during spring and early summer as part of the management plan for the Reelfoot National Wildlife Refuge (Henson 1990a,b).

Our understanding of macrophytes in natural lakes of the Southeast is most complete for subtropical Florida. Owing to a general lack of pronounced slopes and shallow mean depths, Florida lakes often are characterized by broad zones of emergent vegetation and richly developed submergent macrophytes covering most of the bottom. Over 28% of the lake surface area surveyed (520 848 ha) recently by the Florida Department of Natural Resources in 349 lakes was considered to have well-developed macrophyte communities (Schardt 1986). Although over 146 species were identified during the survey, 10 taxa were responsible for 59% of the observed macrophyte extent.

Most plant taxa are broadly distributed among lake types. The total number of species characterizing a lake does not change significantly relative to gradients of either trophic state or acidity (Garren 1982, Clarkson 1985), but pronounced changes in the structure of the littoral zone are often apparent. Free floating taxa are generally restricted to more productive lakes, largely because of a dependence on water column levels of nutrients for growth (Garren 1982). Although there are not major species replacements with increasing acidity, even in lakes as low as 4.2 pH (Clarkson 1985), submergent taxa displaying rosulate growth increase in importance, as often reported in comparable temperate lakes (Garren 1982).

Two general patterns of macrophyte community structure are exhibited by acidic (4–5.5 pH) peninsular Florida lakes (Crisman et al. 1986a). In many lakes of the Sandhill area, submergent macrophytes are totally absent, and the littoral zone is restricted to a narrow fringe of emergent grasses. Epiphytic algal biomass is great in the emergent zone. The second community structure type is exhibited by comparably acidic lakes located primarily in the Ocala National Forest. These lakes display no reduction in either species richness or overall extent of the submergent macrophyte community. Submergent vegetation in such lakes is frequently 2–3 m high and covered with extremely lush epiphytic algal "clouds" in excess of 2 m in diameter.

Overall, the response of Florida littoral zones to increasing acidity is markedly different from that observed in the temperate zone. Florida littoral zones fail to show either a major loss of species or increased importance of *Sphagnum*, and the aerial extent of submergent vascular vegetation may or may not be reduced. Interlake differences that do exist in the extent of submergent vegetation among acidic Florida lakes display no consistent pattern relative to pH and are likely attributable to basin morphometric parameters.

In spite of the lack of major taxonomic replacements with increasing acidity in Florida lakes, the uptake of both macronutrients and heavy metals by macrophytes appears to be strongly pH dependent (Clarkson 1985). Heavy metal uptake was greatest for submergent macrophytes and least for emergent plants with increasing acidity (6.4–4.2 pH) based on plant collections of 20 species common to 21 lakes. Within each habitat group, however, great interspecific differences often were noted, with presumably closely related species displaying opposite uptake patterns relative to the acidity gradient.

Nall and Schardt (1978) presented detailed data on the seasonality of submergent macrophyte biomass in Lake Conway, a mesotrophic central Florida lake (Fig. 6). Biomass peaked during early summer following a period of minimum values during midwinter, which is considered broadly representative of the peninsular lakes. As noted earlier, epiphytic algal abundance peaked coincidental with declining macrophyte biomass during fall. Thus, while it is likely that nutrients are leached from submergent macrophytes during fall as in temperate systems, Florida littoral zones differ from temperate systems most notably by the general absence of a pronounced winter dieback.

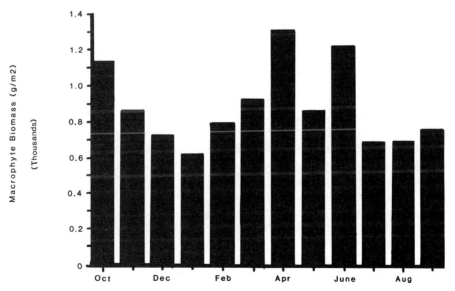

FIGURE 6. Seasonality of macrophyte biomass in the Middle Pool of Lake Conway, Florida, in 1977–78. (Data from Nall and Schardt (1978).)

Given the shallow nature of most lake basins and the general absence of pronounced seasonal variability in plant biomass, littoral zones in Florida may be expected to play an expanded role in system metabolism in the subtropics relative to the temperate zone. Canfield et al. (1984a) noted that predictive models for chlorophyll based on nutrient concentrations in Florida lakes are strongly influenced by the extent of aquatic macrophytes, especially once the percentage of lake volume occupied by plants exceeds 50%. Ewel and Fontaine (1983) found that the littoral zone was largely responsible for observed seasonal fluctuations in system metabolism in Lake Conway, and on an annual basis contributed approximately 56% of total autotrophic production in this mesotrophic lake. In addition, Crisman et al. (1986a) have suggested that total system production becomes increasingly dominated by the littoral zone in Florida lakes along a gradient of increasing acidity as phytoplankton biomass declines in direct response to decreasing dissolved nutrients. Total system productivity does not necessarily decline with increasing acidity, rather it simply shifts from a planktonic to a predominately benthic base.

The overall extent of the littoral zone can have a pronounced effect on both invertebrate and fish communities. Bryozoan abundance increased in Florida lakes with increasing littoral zone extent up to a point where macrophytes occupied approximately 50% of the water column (Crisman et al. 1986c). Above this point, abundance progressively declined, presumably in response to food limitation imposed by declining phytoplankton concentrations.

Chaoborid abundance in Florida lakes displayed the same distributional pattern as bryozoans, with maximum abundance at approximately 50% colonization of the water column by macrophytes (Crisman and Crisman 1988). It was suggested that predation intensity on chaoborids increased progressively with reduction in the extent of the littoral zone, while prey capture probability by chaoborids was seriously impaired with elevated macrophyte (>50% water column) extent. Finally, numerous studies have demonstrated the beneficial effect of vegetated littoral zones on fish populations in Florida lakes (Guillory et al. 1979, Williams et al. 1985), but the work of Colle and Shireman (1980) suggests that gamefish populations may be impaired if macrophyte colonization exceeds approximately 20% of the water column.

Any overview of macrophytes in the Southeast would be incomplete without mentioning floating macrophyte islands or mats. Although common throughout the lowland tropics (Hutchinson 1975), floating islands in the southeastern natural lakes appear to be largely restricted to Florida. Reid (1952) suggested that such features in Florida were formed by three possible processes: (1) water level fluctuations causing the leading edge of the littoral zone to break free, (2) buoyancy of macrophyte masses associated with lacunar spaces and aerenchyma in plant roots, and (3) ebullition of gases trapped in sediments during decomposition, thereby loosening sections of the littoral zone from the substrate. Such floating islands often support a diverse assemblage of macrophytes and often small trees. In many of the larger lakes, macrophyte islands are moved freely by the wind, their ultimate fate being either reincorporation into the littoral zone or disintegration and subsequent sinking.

Bacterioplankton

As with epiphytic algae, the database for bacterioplankton is extremely limited except for Florida lakes. Organically colored (>80 mg Pt/L) Florida lakes displayed peak bacterioplankton numbers during spring (March–May), regardless of trophic state (Crisman et al. 1984a, Crisman et al. 1984b). Bacteria numbers in all lakes increased sharply as lake temperature increased from 23 to 26 °C but declined progressively as water temperature increased further to the normal summer maximum of 31 °C. Crisman et al. (1988) also recorded a sharp decline in bacterial numbers at temperatures >25 °C in clear-water mesotrophic Lake Weir, but peak bacterioplankton abundance in this lake and noncolored eutrophic Lake Wauberg (Crisman et al. 1986b) occurred during fall (September–November).

Ongoing analysis of a bacterioplankton database of 425 events from 18 Florida lakes (T. L. Crisman, J. R. Beaver, and A. E. Keller, unpublished data) also noted significant differences between clear and colored lakes. Bacterioplankton abundance displayed a significant ($p < .05$) positive relationship with chlorophyll in clear lakes ($r^2 = .81$), but only weakly so ($r^2 = .38$) in colored lakes spanning the same range of trophic state. Within individual trophic state classes, bacteria concentrations were generally higher throughout the year in colored lakes. Annual mean bacterioplankton abundance for clear-water lakes spanning a gradient from oligotrophic to eutrophic ($1–8 \times 10^6$ cells/mL) were consistently lower than humically colored lakes of comparable trophic state ($5–9 \times 10^6$ cells/mL).

ANIMAL COMMUNITIES

Pelagic Zooplankton

Several researchers have addressed or are addressing questions regarding the structure and function of zooplankton communities in man-made lakes of the Southeast, but few studies have been conducted on zooplankton in natural lakes outside Florida. The limited number of studies available provide little more than species lists for individual systems.

Cladocerans dominated the zooplankton assemblage of Lake Drummond, Virginia (Marshall 1979) and displayed annual maximum and minimum abundance during fall and winter, respectively. Copepod and rotifer populations peaked during summer. *Bosmina longirostris* was clearly the dominant zooplanktor, and the cladoceran genus *Daphnia*, a common element of north temperate assemblages, was not reported.

Bosmina longirostris was also a common species in circumneutral dune ponds on the Outer Banks, North Carolina, and the genus *Daphnia* was represented only by small (*D. ambigua*) and intermediate (*D. laevis*) sized species (Kling 1986). The zooplankton assemblages of these coastal plain systems appear to be transitional between those of more temperate and sub-tropical lakes. Unlike more temperate systems, *Daphnia* appears to be of reduced importance and, if present, represented only by small bodied taxa. Florida lakes display a further reduction in *Daphnia*, with only one of the two smallest species in North America, *D. ambigua*, being present. In addition, *Eubosmina tubicen* is normally the only bosminid found, while *Bosmina longirostris* has rarely been reported (Crisman 1981).

Relation to Trophic State   Numerous papers have been published on structural and functional aspects of zooplankton communities in Florida lakes relative to gradients of trophic state, organic color, and acidity. While both the abundance and biomass of total zooplankton (including ciliated protozoans) display a strong positive regression with increasing trophic state in Florida lakes (Bays and Crisman 1983), not all zooplankton components contribute equally to this relationship.

The positive relationship between ciliate biomass and chlorophyll ($r^2 = .77$) is the strongest regression established between individual zooplankton components and trophic state in Florida lakes (Bays and Crisman 1983). Comparison with a similar suite of lakes from Canada (Beaver and Crisman 1989b) demonstrated that although Florida lakes tend to have greater ciliate biomass for a given level of chlorophyll, there is not a significant difference between the regression lines derived for the two lake regions (Fig. 7). Within the Florida data set, there were no differences in expected ciliate biomass between clear and organically colored lakes of the same trophic state (Beaver et al. 1988).

In addition to biomass, Florida ciliate assemblages display a number of structural changes relative to gradients of trophic state. While small-bodied taxa ($<30 \mu m$) dominate the ciliate assemblages of all lakes, larger-bodied forms ($40-50 \mu m$) are progressively replaced with increasing eutrophy (Beaver and Crisman 1982, 1989b). Size changes coincide with a major taxonomic replacement of predominately large-bodied Oligotrichida by small-bodied Scuticociliatida. Even within the Oligotrichida, the size replacement trend is also evident.

Such trends are even apparent within the oligotrophic category when lakes are ranked according to increasing acidity (7.0–4.7 pH). Beaver and Crisman (1981) reported that both abundance and biomass of ciliates were profoundly reduced in lakes <5.0 pH and dominance of the assemblages by large-bodied Oligotrichida (73%) was the greatest reported from Florida lakes. In addition, the largest-bodied ($150-200 \mu m$) euplanktonic ciliated protozoan found in Florida, *Stentor niger*, is restricted mainly to acidic lakes, where it often accounts for over 60% of total ciliate biomass (Beaver and Crisman 1989a).

Several authors have observed that rotifer abundance in Florida reaches a maximum in eutrophic lakes (Cowell et al. 1975, Elmore et al. 1984). Canfield and Watkins (1984) found that rotifers were the only zooplankton

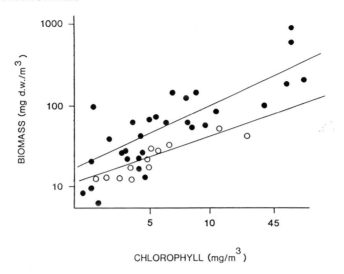

FIGURE 7. Comparison of annual means of ciliate protozoa and chlorophyll for a suite of Florida (●) and Quebec (○) lakes. Regression equations for Florida and Quebec are log 10 TB = 0.681 log 10 Chl + 1.237 and 10 TB = 0.486 log 10 Chl + 1.074, respectively. (From Beaver and Crisman (1989b).)

group (ciliates were omitted) whose midsummer abundance was positively related (r^2 = .72) to chlorophyll concentrations. Bays and Crisman (1983) noted that only ciliated protozoans displayed a stronger relationship than rotifers between mean annual biomass and chlorophyll (r^2 = .54).

Florida lakes are characterized by a depauperate planktonic cladoceran fauna (Crisman 1981, Wyngaard et al. 1982, Blancher 1984) relative to temperate lakes. *Bosmina longirostris*, a common temperate element, is rarely reported, and *Daphnia ambigua*, one of the two smallest North American species, is the only member of a common genus represented in temperate lakes by a broad array of body sizes (Crisman 1981). The latter species is found in all Florida lake types, and there is no evidence to suggest that *Daphnia* are eliminated from acidic Florida lakes (>4.7 pH), as noted for many comparable temperate systems (Brezonik et al. 1984). In addition to the lack of major species replacements within the fauna, Bays and Crisman (1983) reported that annual mean biomass displays only a weak (r^2 = .23) positive relationship with increasing trophic state.

Bays and Crisman (1983) also noted weak positive relationships between the annual mean biomass of both nauplii (cyclopoid + calanoid) and cyclopoid adults and chlorophyll. However, calanoid biomass appeared to be independent of lake trophic state. Closer examination of the data suggested that part of the apparent lack of a relationship for calanoids was a species replacement from *Diaptomus floridanus* and *D. mississippiensis* dominance in oligotrophic and mesotrophic lakes to *D. dorsalis* as the only representative of the genus normally found in eutrophic systems. Elmore (1983a,b) noted the same replacement series in central Florida lakes and suggested that *D. floridanus* and *D. mississippiensis* were not likely excluded from eutrophic lakes due to competition with *D. dorsalis*. The latter species, however, was likely excluded from oligotrophic and mesotrophic lakes because of competition. Fish predation by pump–filter-feeding gizzard shad (*Dorosoma cepedianum*) in eutrophic lakes, has been advanced as the likely reason for the exclusion of *D. floridanus* and *D. mississippiensis* from eutrophic lakes (Elmore et al. 1984).

Zooplankton are often grouped according to body size into microzooplankton (ciliates, rotifers, nauplii) and macrozooplankton (copepodite + adult copepods, cladocerans). Microzooplankton biomass displays a strong positive relationship to lake trophic state in Florida (Fig. 8), while no trend is evident for macrozooplankton (Bays and Crisman 1983). The microzooplankton trend is controlled principally by ciliated protozoans, with the regression of individual zooplankton components decreasing with increasing body size from rotifers to nauplii. Each of the microzooplankton groups displayed a stronger relationship to trophic state than was noted for any single macrozooplankton contributor. Beaver and Crisman (1990b) demonstrated that bactivorous microzooplankton are much more sensitive to small changes in lake trophic state than are macrozooplankton, and therefore can be used as an advanced warning of cultural eutrophication.

Finally, while Young (1979) noted that the abundance of total crustacean zooplankton declined with increasing eutrophy in central Florida lakes, Bays and Crisman (1983) found the opposite. Although weaker than that reported for a comparable suite of Canadian lakes, neither the slope or intercepts of the regression model were significantly different between subtropical and temperate lakes.

Seasonality   Several authors have considered the seasonality of various zooplankton groups in individual Florida lakes (Reid and Squibb 1971, Nordlie 1976, Cowell et al. 1975, Mallin 1978, Wyngaard et al. 1982, 1985, Elmore et al. 1984). Unfortunately, none of these studies investigated whether seasonal patterns varied according to trophic state, and, with the exception of Elmore et al. (1984), no data are available on the seasonality of ciliated protozoans. Recently, however, Beaver and Crisman (1990a) presented data on the seasonality of ciliates from 20 lakes spanning a broad trophic gradient. Total ciliate biomass displayed pronounced seasonal differences according to trophic state, with oligotrophic lakes peaking in fall and late winter, mesotrophic lakes in fall, and eutrophic lakes during summer. Abundance also displayed differences according to trophic state and did not necessarily peak coincident with biomass even within an individual trophic state category. The large-bodied Oligotrichida were largely responsible for abundance and biomass peaks in oligotrophic lakes, while the small-bodied Scuticociliatida were the major contributors to both at higher trophic state. Seasonal patterns of ciliates in Florida lakes have been attributed to pulses in primary production

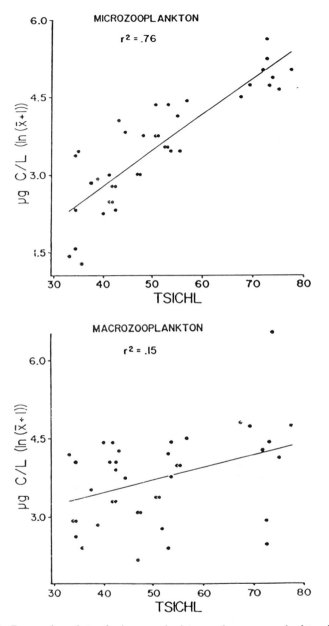

FIGURE 8. Regression plots of microzooplankton and macrozooplankton biomass in Florida versus Carlson's (1977) trophic state index (TSICHL). (From Beaver and Crisman (1982).)

as well as the influence of mixing regimes providing locally dense food resources within the water column (Beaver and Crisman 1989a, 1990a).

Beaver et al. (1988) provided evidence that the seasonality of ciliates differs between clear and organically colored Florida lakes. Unlike oligotrophic clear water lakes, which displayed peak biomass during fall and late winter, colored lakes of comparable trophic state peaked during summer. Ciliate seasonality in colored lakes of higher trophic state resembled that documented for similarly productive clear lakes.

One major difference between clear and colored lakes of all trophic states was the presence of large numbers of myxotrophic taxa during summer in colored lakes. *Strombidium* cf *oculatum*, a large-bodied ciliate containing zoochlorellae, was the principal myxotroph. Such autotrophic ciliates clearly dominated total ciliate biomass during summer and accounted for >50% of total annual biomass in several lakes. In addition, the proportion of total autotrophic biomass contributed by autotrophic ciliates often exceeded that of phytoplankton during summer, with one lake exhibiting a 96% contribution on one occasion. This pronounced trend toward myxotrophic dominance of autotrophic production during midsummer was most apparent in oligotrophic lakes, where it was considered a response to nutrient limitation.

Unlike ciliates, the two remaining contributors to the microzooplankton, rotifers and copepod nauplii, display seasonal patterns that appear independent of trophic state (Bays 1983). The biomass of both peaks during midsummer and remains high throughout fall.

Cladoceran biomass in all Florida lakes peaks during spring and declines to minimum levels during summer (Bays 1983). Cladocerans often disappear completely from lakes during midsummer, following the appearance of males in the population (Crisman 1981, 1986). This pattern is markedly different from that of the temperate zone where cladoceran biomass tends to parallel chlorophyll during warmer months and males appear during fall immediately prior to the cold weather crash of populations.

The ultimate cause for the midsummer crash in cladocerans observed in Florida lakes is not clear. Foran (1986b) suggested that temperature alone is not limiting to Florida cladocerans, and both Maslin (1969) and Bienert (1987) have suggested that while cladocerans are likely food limited during fall and winter, predation, especially by young bluegill, appears to be the dominant structuring force during spring and summer in acidic oligotrophic lakes. Experimental evidence suggests that enhanced summertime predation by gizzard shad is an important factor controlling the biomass of cladocerans in eutrophic lakes (Crisman and Kennedy 1982, Crisman 1986, Crisman and Beaver 1988, 1990a).

The seasonality of cyclopoid copepods in Florida lakes becomes more pronounced with increasing trophic state (Bays 1983). While the biomass of cyclopoids does not vary greatly throughout the year in oligotrophic lakes, mesotrophic and eutrophic lakes display peak biomass during late winter and early spring with minimum values occurring during early summer. Calanoid biomass displays less pronounced interlake differences with peak biomass during spring and minimum values during summer, regardless of trophic state.

Most research on copepod seasonality in Florida lakes has concentrated on the herbivorous calanoid *Diaptomus* and the carnivorous cyclopoid *Mesocyclops*. In addition to determining the overall species composition of *Diaptomus* according to trophic state, intralake temporal differences in the intensity of invertebrate (*Chaoborus*, *Mesocyclops*) and fish (gizzard shad, bluegill) predation are likely responsible for the observed seasonality of diaptomids in Florida lakes (Confer 1971, Elmore 1983a, Elmore et al. 1983). The seasonality of one of the principal invertebrate predators on *Diaptomus*, *Mesocyclops*, is also strongly controlled by temporal variability in the intensity of predation from chaoborids and gizzard shad (Wyngaard et al. 1982).

The overall seasonality of microzooplankton is comparable in all Florida lakes, regardless of trophic state (Bays 1983). Following minimum levels during late spring, biomass increases during early summer to maximum levels and remains relatively constant until fall. The seasonal pattern for macrozooplankton, however, is roughly the inverse of that for microzooplankton, with peak biomass during spring and minimum values during summer.

Vertical Migration   In addition to pronounced seasonal changes in zooplankton composition, Florida lakes also display diel variability in the distribution of taxa within the water column. Although some investigations failed to document rotifer migration (Shireman and Martin 1978, Comp 1979), Russell (1982) noted that not only did rotifers display pronounced vertical migration in Lake Mize, but the diel migration pattern was the opposite of that observed for cladocerans and chaoborids. Rotifers moved deeper in the water column at night and returned to the surface during the day. Both Shireman and Martin (1978) and Russell (1982) also documented cladoceran migration. The amplitude of change for *Eubosmina tubicen* in the water column of Lake Mize exceeded 4 m on some dates (Russell 1982).

It is generally agreed that cyclopoid copepods vertically migrate in Florida lakes (Reid and Blake 1969, Shireman and Martin 1978, Comp 1979). Although Shireman and Martin (1978) failed to document calanoid migration, *Diaptomus* was one of the principal migrating taxa in Lake Conway (Comp 1979). The diel migratory behavior of copepods in general displayed a clear seasonality in Lake Conway, with pronounced migration within the water column limited to the period of winter lake mixing. During the stratified summer period, copepods remained concentrated at the thermocline. Comp (1979) presented evidence that the lack of migration during summer was the result of a concentration of food at the thermocline in addition to possible avoidance of elevated temperature higher in the water column.

Factors Controlling Community Structure   Several investigations have assessed the importance of individual factors as controlling variables for observed changes in the structure of zooplankton communities relative to trophic state in Florida. The zooplankton communities of oligotrophic lakes are dominated by macrozooplankton, with increased importance of larger bodied taxa noted even for ciliated protozoan assemblages. Although less pronounced in more productive lakes, ciliate assemblages of organically colored oligotrophic lakes tend to be dominated during midsummer by large-bodied (150–200 μm) myxotrophic species.

Elmore (1983a) suggested that *Diaptomus dorsalis* is excluded from oligotrophic and mesotrophic lakes principally by food limitation. Both Maslin (1969) and Bienert (1987) stressed that while zooplankton communities in oligotrophic lakes are generally food limited throughout the year, predation can have a strong seasonal impact on zooplankton community structure especially during spring and summer. Crisman et al. (1981) found that while microzooplankton were important bacterial grazers, macrozooplankton were not and actually stimulated bacterial growth presumably from wastes generated during algal grazing. The contention of Beaver and Crisman (1982) that small-bodied (<30 μm) ciliates are of reduced importance in oligotrophic lakes principally because of their dependence on bacterial sized food particles is supported by Bienart (1987). He found that bacteria concentrations in an oligotrophic lake were always below the threshold needed to support ciliates and other microzooplankton dependent on a bacterial food base. The frequently observed dominance of ciliate assemblages in both clear (Bienert 1987) and organically colored (Beaver et al. 1988) lakes by large bodied myxotrophic taxa is considered a direct consequence of seasonally severe food limitation.

In addition to food limitation, the community structure of zooplankton communities in oligotrophic Florida lakes also appears to be influenced by both vertebrate and invertebrate predation. Maslin (1969) calculated that the planktivorous fish *Labidesthes sicculus* consumed approximately 7% of annual zooplankton production in an acidic oligotrophic lake in northern Florida, while predation by current year-class bluegill was responsible for the observed midsummer crash of cladocerans. Bienert (1987), in a detailed analysis of the zooplankton community from the same lake, also stressed the importance of fish predation as a controlling factor for cladoceran abundance especially during spring and summer.

The role of invertebrate predation has received less attention. Nordlie (1976) suggested that intense predation from *Chaoborus* in a highly colored meromictic lake was largely responsible for the scarcity of cladocerans. The other major invertebrate predator in oligotrophic Florida lakes, the copepod *Mesocyclops edax*, while apparently not having much impact on cladocerans, may be a controlling factor for *Diaptomus floridanus* abundance (Confer 1971). *Mesocyclops*, in turn, appears to be subject to fish predation (Wyngaard et al. 1982).

Microzooplankton replace macrozooplankton as the dominant size group in eutrophic lakes (Bays and Crisman 1983). Within the microzooplankton, the most pronounced increase is displayed by ciliates, especially those <30 μm, and secondarily by rotifers. Macrozooplankton biomass does not appear to change markedly with increasing eutrophy; rather, the dominance shift is largely the result of a disproportionate increase in the microzooplankton. The most pronounced taxonomic replacement noted within the macrozooplankton is for *Diaptomus*, where *D. dorsalis* replaces *D. floridanus* and *D. mississippiensis* in eutrophic lakes (Elmore et al. 1983).

Beaver and Crisman (1982) attributed the pronounced increase in small-bodied ciliates with increasing eutrophy to enhanced bacteria abundance. While the dampened response of macrozooplankton to elevated phytoplankton levels may in part be a reflection of an algal compositional shift to relatively unpalatable and unmanageable cyanophytes of large size, it is likely that concurrent variation in the intensity of fish predation plays an important role in observed zooplankton community structure.

Elmore et al. (1983) provided experimental evidence that predation from bluegill and especially gizzard shad (*Dorosoma cepedianum*) was likely largely responsible for the exclusion of *Diaptomus floridanus* and *D. mississippiensis* from eutrophic lakes. The contention of Bays and Crisman (1983) that gizzard shad predation is the primary factor structuring eutrophic macrozooplankton in general and cladocerans in particular was supported by field enclosure experiments (Crisman 1986, Crisman and Beaver 1988, 1990a). When freed from fish predation, cladoceran abundance in Lake Wauberg increased two orders of magnitude over ambient levels, while an experimental threefold elevation of shad biomass at shad-dominated Lake Apopka resulted in a further depression of cladoceran abundance from levels observed in the lake.

Although biomass for individual zooplankton components can be similar between comparable Florida and temperate lakes, subtropical macrozooplankton communites are characterized by a reduction in both the number of taxa and maximum body size. In particular, the genus *Daphnia* is rarely abundant in Florida lakes and is represented by only one species (*D. ambigua*), one of the smallest two species in North America. In addition, Florida cladoceran populations normally crash during midsummer, an event often preceded by an increase in the incidence of males (Crisman 1981, 1986).

Foran (1986a) suggested that large-bodied daphnids are not excluded from Florida by high temperature alone, but also by competition with small-bodied taxa at high temperature. Smaller *Daphnia* species may have a competitive advantage as a result of both their earlier age at first reproduction at higher temperatures and the relative temporal constancy of environmental conditions in Florida (Foran 1986b). Less pronounced seasonal variability in both water temperature and aquatic production allows daphnid populations to approach their carrying capacity, in effect reducing possibilities for niche separation and thus maximizing competition between ecologically equivalent taxa.

Similar differences in the reproductive biology of temperate and subtropical copepods have also been noted. Elmore (1983b), working with three species of *Diaptomus* found in Florida, found that egg development time was inversely related to the extent of the geographical range of individual species into the temperate zone, while Wyngaard et al. (1985) found that *Mesocyclops edax* produced seven more generations per year in Florida than Michigan. These data also suggest that broadly distributed taxa have adapted to the warmer subtropical temperature regime. While these studies are informative, additional detailed investigations are needed to fully delineate causal factors for observed differences between temperate and subtropical zooplankton assemblages.

Littoral Zooplankton (Meiofauna)

While several investigations have compared planktonic zooplankton abundance between vegetated littoral areas and open water (Shireman and Martin 1978, Crisman and Kooijman 1981, Watkins et al. 1983), surprisingly little attention has been paid to those zooplankton components that are closely associated with macrophytes and sediments and rarely are encountered in the plankton. Although all zooplankton groups have species that are largely restricted to littoral areas, the majority of the research in the Southeast on littoral zooplankton has concentrated on cladocerans of the family Chydoridae.

Chydorid biogeography in the Mississippi River Valley was examined by DeCosta (1964) based on a survey of assemblages from 45 lakes representing a latitudinal transect from Minnesota to Louisiana. Natural lakes of the Southeast that were investigated included Reelfoot Lake and 15 oxbow and fluviatile lakes in Arkansas, Mississippi, and Louisiana. Based on the distribution of 15 indicator species, three latitudinal species groupings (northern, southern, eurytopic) were statistically defined for the Mississippi Valley. North of 39.25°N and south of 35.80°N the northern and southern assemblages were dominant, respectively. The widest distribution was displayed by the eurytopic assemblage (*Alona rectangula, Camptocercus rectirostris, Chydorus globosus, Leydigia leydigi, Pleuroxus denticulatus*), which dominated the chydorid assemblages in the transition zone between the northern and southern faunas (35.80–39.25°N). DeCosta noted that the northern–southern faunal transition coincided with a region where lake thermal regimes alternated between dimixis and warm monomixis depending on the severity of winter, which led him to suggest that faunal assemblages were geographically limited by macroclimatic conditions and thermal regimes of lakes.

Using a survey of chydorid remains collected from the surficial sediments of 52 Florida lakes of known water chemistry and trophic state, Crisman (1980) was able to statistically separate hardwater and softwater species associations. In general, chydorid assemblages of soft water lakes were characterized by much greater species richness and diversity than hard water assemblages. While in close agreement with similar investigations relating chydorids to water chemistry in temperate lakes, the Florida data did not support many of the biogeographic conclusions that DeCosta (1964) derived for the Mississippi Valley. All five of the species of DeCosta's "northern" assemblage (*Acroperus harpae, Alona quadrangularis, Alonella excisa, Eurycercus lamellatus, Graptoleberis testudinaria*) in addition to 11 others are equally common in comparable Florida and temperate lakes. No species were eliminated from the "southern" category (*Alona karua, Alonella hamulata, Chydorus* spp., *Euryalona occidentalis, Leydigia acanthocercoides*) as a result of the Florida investigation; rather, it was supplemented by an additional four taxa. It was suggested that latitudinal changes in both dominant lake type from glacial to fluviatile origin and baseline water chemistry were at least partially responsible for the latitudinal species groupings defined by DeCosta.

Without a doubt, the extensive field surveys of D. G. Frey on a variety of water bodies (including natural lakes) throughout the Southeast represent the most detailed investigations on chydorid taxonomy (Frey 1978, 1982a–d) and ecology for the region. Chydorid populations display pronounced latitudinal variation in the timing and extent of gametogenesis (Frey 1982a). Whereas north temperate populations display intensive temporally coordinated gametogenesis during fall prior to winter dieback, gametogenesis becomes less coordinated and intense with decreasing latitude, and some parthenogenesis is observed even during winter. The latitudinal trend ends with fall gametogenesis being virtually absent from coastal regions of Louisiana, Mississippi, and Florida, and being replaced by spring gametogenesis.

The shift toward spring gametogenesis in northern Florida (Frey 1982a) is accompanied by a seasonal change in chydorid dominance from "northern" to "southern" taxa during summer. Northern species reach their maximum abundance during winter, spring, and early summer, undergo gametogenesis during spring, and die out during summer in response to elevated water temperatures. They are replaced by small-bodied taxa displaying principally southern geographical distributions that become gametogenic during autumn prior to drastic population reduction or disappearance during winter.

A clear replacement of a northern chydorid assemblage by a southern assemblage is evident from a paleolimnological reconstruction of the trophic history of 30 000 + -year-old Lake Annie in Florida (T. L. Crisman, unpublished data). The establishment of a southern chydorid assemblage comparable to the extant fauna of the lake is a relatively recent event (post 2500 B.C.) that coincided with the establishment of the current climate and terrestrial plant community (Watts 1975). While it is likely that climate played a direct role in the observed faunal change, the importance of covariance in lake level, the structure of the littoral, as well as general trophic state cannot be discounted.

Finally, unlike planktonic zooplankton, few data exist on the seasonality of littoral zooplankton abundance in lakes of the Southeast. Elmore et al. (1984) reported that the abundance of ostracods, harpacticoid copepods, and "benthic cladocerans" in Florida generally declines with increasing trophic state, but presented seasonal abundance data only for total meiobenthos (minus nematodes). The only other seasonal data for littoral zooplankton are those of Maslin (1970) who concentrated on chydorid cladocerans from two acidic oligotrophic Florida lakes. It was concluded that in spite of an apparent feeding preference for chydorids, predation by littoral dwelling fish was insufficient to account for the pronounced seasonal abundance shifts displayed by this zooplankton group. Rather, both community structure and abundance appeared to be closely linked with water temperature and overall lake productivity.

Benthic Macroinvertebrates

Except for Florida, very few studies of benthic macroinvertebrates exist for the Southeast. Most of the work has been taxonomic in nature, including a species list for Lake Drummond (Matta 1979) and a discussion of the endemic mollusk fauna of Lake Waccamaw (Fuller 1977). The notable exception is the year-long investigation of Schalles (1979) on the macroinvertebrate community of a Carolina Bay lake. Unfortunately, the shallow nature of the site (<1.5 m) hinders comparison with studies of Florida or temperate lake benthos.

The limited survey of Elmore et al. (1984) suggested a negative relationship between the abundance of both macrobenthos and meiobenthos (excluding nematodes) and the trophic state of Florida lakes. A more detailed investigation of 20 oligotrophic Florida lakes (Schulze 1980) suggested that both abundance and biomass of benthos displayed little relationship to acidity (4.7–6.6 pH) but that observed interlake variability was controlled by trophic state. Unlike abundance, diversity of benthic assemblages displays no relationship to trophic state (Schulze 1980, Osborne et al. 1976).

The importance of chironomids and oligochaetes in Florida lakes is assumed to be positively related to lake trophic state (Elmore et al. 1984), with the latter frequently exceeding 50% of total invertebrate abundance in highly eutrophic systems (Cowell et al. 1987a). The chironomid fauna of such lakes often is dominated by the Chironomini (*Chironomus, Glyptotendipes*), with the Tanypodinae as principal subdominants (Cowell et al. 1975). Although commonly observed in the temperate zone, the abundance of *Chaoborus* does not appear to be positively related to trophic state in Florida (Crisman and Crisman 1988, Crisman and Beaver 1990b).

While Elmore et al. (1984) suggested that amphipod abundance declined with increasing trophic state, Schulze (1980) found an opposite trend relative to increasing pH (4.7–6.6) and presumably trophic state. In addition, chironomids were found to decrease with increasing pH, while mollusks were absent below 5.6 pH.

Several studies have examined macroinvertebrate seasonality in Florida lakes. Elmore et al. (1984) suggested that total macrobenthos abundance peaks during either late fall or winter in Florida lakes, and is lowest during spring and summer. Seasonal patterns appear to be independent of trophic state and extent of the vegetated littoral zone (Scott and Osborne 1981, Leslie and Kobylinski 1985). The winter maximum in total macrobenthos has been attributed to enhanced water column oxygen levels during seasonal destratification in algal dominated eutrophic lakes (Cowell and Dawes 1984), and to a seasonal reduction of plant biomass in excessively vegetated lakes (Scott and Osborne 1981). Regarding the major invertebrate groups in lakes, oligochaetes maintain fairly constant population levels in eutrophic lakes (Cowell et al. 1975), while chironomids display a more temporally spiked pattern, with peak abundance occurring during late spring and summer (Cowell and Vodopich 1981). The latter paper also presents seasonal data for some of the dominant taxa in eutrophic Lake Thonatosassa, Florida.

The structure of macrobenthic assemblages in Florida lakes appears to be strongly influenced by water depth, dissolved oxygen, sediment composition, and the extent of aquatic macrophytes. Both the number of taxa and total macrobenthos abundance decline with increasing water depth in both productive (Cowell et al. 1975, Cowell and Vodopich 1981, Cowell and Dawes 1984) and unproductive (Elmore et al. 1984) lakes. Even in shallow nearshore areas (0.2–1.5 m), the abundance of some taxa is negatively correlated with water depth (Vodopich and Cowell 1984). Chaoborid abundance in Florida, however, is positively related to mean lake depth (Crisman and Beaver 1990b).

Both the species richness and abundance of macrobenthos and meiobenthos are positively related to dissolved oxygen concentrations at the sediment–water interface (Elmore et al. 1984). Depth-related reduction in oxygen seems to explain most of the apparent differences observed in macrobenthic assemblages between deep silt and shallow sand stations in eutrophic Lake Thonatosassa (Cowell and Vodopich 1981). This interpretation is supported by the observation of Cowell and Dawes (1984) that oligochaetes in eutrophic Lake Brooker are able to colonize deep water only when seasonal destratification eliminates anoxia during winter. Finally, Scott and Osborne (1981) noted that macrobenthos during summer were absent from sediments, while abundant on vegetation due to reduced oxygen levels at the sediment–water interface.

Nearshore sediments in Florida lakes are characterized by low organic, large-grained sediments. With increasing depth and distance from shore, grain size decreases and sediment organic content increases. Although both the number of taxa and abundance of macrobenthos (Cowell and Vodopich 1981) and meiobenthos (Elmore et al. 1984) are positively related to sediment grain size in eutrophic lakes, the relationship is also strongly controlled by covariance in oxygen concentrations with increasing depth.

The best systems for testing the influence of sediment type on invertebrate distributions are oligotrophic lakes that display little oxygen reduction with depth. Such an approach has recently been taken by T. L. Crisman (unpublished data), who assessed intralake differences in macrobenthic assemblages from sand and mud stations for 20 acidic oligotrophic Florida lakes. Species richness was greater at sand stations, as was the abundance of all taxonomic groups including chironomids. In addition, all major feeding types except predators reached maximum abundance at sand stations. This study demonstrates that sediment composition does play an important role in structuring macrobenthic communities in oligotrophic lakes, while this role is muted in more productive lakes by the importance of oxygen stress.

Iovino and Bradley (1969) demonstrated that, in addition to being influenced by sediment composition, the macrobenthos can directly control sediment composition through their feeding activities. Laboratory studies showed that *Chironomus* fed blue-green algae produced coherent fecal pellets, while the pellets of those fed green algae quickly disintegrated. In addition, the shape of pellets was age dependent, with early instars producing ovoid pellets and later instars producing longer cylindrical pellets. Close examination of the 1-m-thick organic sediment characterizing Mud Lake, Florida, revealed that it was made up almost entirely of chironomid fecal pellets, many of which still contained viable blue-green cells. They hypothesized that the Mud Lake sediment was a modern analog of the Eocene age Green River formation oil shale of the western United States.

Watkins et al. (1983) found that macrobenthos abundance was greater in macrophyte habitats of Orange Lake, Florida, with *Hydrilla* displaying significantly greater numbers than the other two plant taxa examined (*Nuphar* and *Panicum*). It was suggested that the morphology and growth habit of *Hydrilla* afforded the benthos protection from fish predation. In generally shallow lakes such as those found in Florida, a degree of plant cover is essential for the survival of a vertically migrating species like *Chaoborus punctipennis* (Leslie and Kobylinski 1985). Crisman and Crisman (1988) noted that while chaoborid abundance is positively related to progressively increasing macrophyte cover up to approximately a 50% filling of the water column, abundance is progressively reduced as lake dominance by plants increases further. It was suggested that moderate macrophyte cover provided a refuge from fish predation, while overly abundant plants hindered feeding activities. A similar pattern has been delineated by Crisman et al. (1986c) for bryozoans, with 50% plant cover serving as the inflection point. In this case, increasing plant abundance to moderate levels serves to increase bryozoan habitat, but food availability for these filter feeders is markedly reduced in richly developed vegetated littoral zones as macrophytes and phytoplankton compete for the nutrient pool.

Excessive macrophyte growth, especially from exotic taxa such as *Hydrilla*, can be a negative influence on the total macrobenthos community. Scott and Osborne (1981) found that both species richness and total abundance increased in a *Hydrilla*-infested central Florida lake during winter, which was associated with a seasonal reduction in plant biomass, while a dramatic reduction in macrophytes by grass carp in a panhandle lake led to increased benthos abundance, especially chironomids (Leslie and Kobylinski 1985). It was suggested in the latter study that accumulation of partially digested plant material from grass carp feeding stimulated epiphytic algal production via increased nutrient availability and that the positive benthos response was directly related to this enhanced food source.

Macroinvertebrates appear to be able to recolonize littoral areas fairly rapidly following moderate disturbance. The exotic cichlid *Tilapia aurea* has built up large populations in several central Florida lakes. Fuller and Cowell (1985) estimated that construction of shallow depression nests during their breeding season (February–September) disturbed approximately 12% of the littoral area at eutrophic Lake Thonatosassa. Recolonization of nesting sites by the benthos was rapid during the warm season, with 90% similarity between disturbed and undisturbed sites within 15 days. Winter rates were longer, with only 67% similarity after one month. Recolonization was fastest for early instars, especially those of planktonically dispersed taxa (Cowell 1984).

Florida macroinvertebrate communities differ from those of comparable temperate lakes in a number of respects. Unlike temperate lakes, there is no marked decline in the number of species or the abundance and biomass of total benthos with increasing acidity (Schulze 1980), although it has been suggested that chironomid species diversity is lower in Florida (Iovino and Bradley 1969). Chaoborid abundance in Florida does parallel increasing trophic state, as observed in temperate lakes (Crisman and Crisman 1988), and a chironomid subfamily especially important in oligotrophic and mesotrophic temperate lakes, the Orthocladiinae, is only a minor element in Florida lakes

(Crisman 1989). While many taxa are common to both temperate and Florida lakes, most display pronounced multivoltinism in Florida, with rapid generation times of 14–22 days at 27–31 °C (Cowell and Vodopich 1981). Finally, it has been suggested that observed invertebrate community differences between temperate and Florida systems may be reflective of differences in major controlling factors. Iovino and Bradley (1969) felt that the apparent lower diversity of chironomid assemblages in Florida was the result of more intense fish predation in such generally shallow lakes, and Cowell and Vodopich (1981) suggested that both temperature and food availability play less of a role in structuring macrobenthic communities in Florida than in temperate lakes. Such speculations hopefully will stimulate additional research on the structure and function of benthic communities.

Fish

Two fish families characteristic of cold temperate lakes in North America, Salmonidae and Coregonidae, are not native to the warm temperate and subtropical lakes of the southeastern United States (Scott and Crossman 1973). In addition, the Clupeidae are absent from both southern Florida and the Appalachian Mountains. While most species of Centrarchidae are widespread throughout the eastern United States, smallmouth bass (*Micropterus dolomieui*) does not naturally extend below approximately 33°N latitude. Werner et al. (1978), in a comparison of the fish communites of Michigan and Florida oligotrophic lakes, suggested that the species pool for Michigan is approximately three times richer than the native species pool of south Florida, based on an estimated number for the latter of approximately 40–50 species (Kushlan and Lodge 1974). A recent update of the fish fauna identified a native species pool for south Florida of 80 (Loftus and Kushlan 1987). Relative to cold temperate fauna, the Catastomidae, Cyprinidae, and Percidae are poorly represented in Florida. While the Cyprinidae were the dominant small-bodied fishes in Michigan, they were replaced by Cyprinodontiformes in Florida. In spite of some faunal differences, Werner et al. (1978) felt that fish faunas of these two regions displayed reasonable functional convergence.

While most fish species in natural lakes of the Southeast are widespread throughout eastern North America, some endemism, especially at Lake Waccamaw, North Carolina, has been reported. In spite of its suggested relatively young (15 000 years) age (Stager and Cahoon 1987), two [(Glassminnow (*Menidia extensa*), Waccamaw darter (*Etheostoma perlongum*)] of the 25 fish species found in the lake are considered endemic (Lee et al. 1980). Two additional species once considered endemic to the lake (Frey 1951b) have been reclassified.

In an investigation of the fish communities at two oligotrophic Florida lakes employing underwater surveys with scuba, Werner et al. (1978) noted that almost the entire fish community was restricted to the vegetated portions of the lake. In addition, Williams et al. (1985) suggested that while the biomass of all fish groups is greater in the littoral zone of Florida lakes than in open water, the relationship is most pronounced for sport fish (Fig. 9). Although both largemouth bass (*Micropterus salmoides*) and redear sunfish (*Lepomis microlophus*) prefer vegetated habitats year round, black crappie (*Pomoxis nigromaculatus*) utilizes the littoral zone only during winter spawning.

Even within the vegetated littoral zone, Werner et al. (1978) observed a definite preference of most fish for submergent over emergent vegetation zones (Fig. 10). Although such distribution patterns may be largely a reflection of foraging preferences and predator avoidance, several studies have clearly demonstrated a preference of several species for emergent and floating-leaved macrophyte taxa as breeding sites. Carr (1942) observed largemouth bass nesting most commonly at the base of emergent vegetation, while Bruno (1984), using radiotagging, demonstrated a preference of this important sport fish for hard organic substrates found within zones of emergent vegetation (*Polygonum* and *Panicum*). A secondary preference was noted for the rhizomes of water lilies (*Nuphar*). While a similar preference for emergent vegetation has been demonstrated for Florida gar (*Lepisosteus platyrhincus*), brook silversides (*Labidesthes sicculus*), and bluegill (*Lepomis macrochirus*), it has been suggested that the best habitat for the latter, in addition to largemouth bass and chain pickeral (*Esox niger*), is a littoral zone of dense emergent vegetation interspersed with patches of submergent and floating-leaved plants (Schramm et al. 1983).

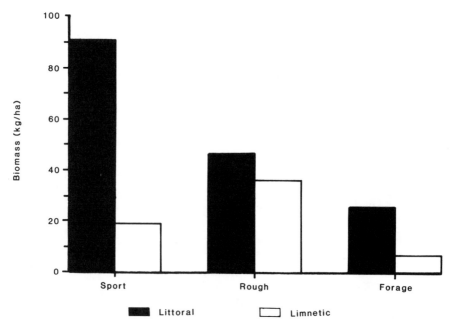

FIGURE 9. Biomass distribution of major fish groups in Florida for littoral and pelagic habitats. (Data from Williams et al. (1985).)

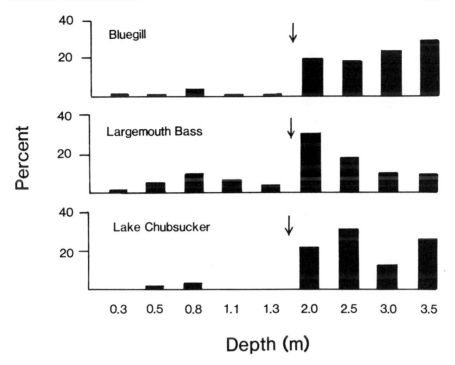

FIGURE 10. Partitioning of the abundance of three common fish species according to depth in Lake Annie, Florida. Arrows indicate the outer edge of emergent vegetation. (From Werner et al. (1978).)

As a means of assessing the impact of the exotic macrophyte *Hydrilla verticillata* on Florida ecosystems, several studies have examined the relationship between the percentage of the water column colonized by submergent vegetation (plant volume infestation) and the success of fish communities. While most sportfish, including largemouth bass, prefer shallow vegetated littoral zones (Williams et al. 1985), a great deal of interspecific variability has been noted in the response of individual fish taxa to a progressive dominance of the water column by submergent vegetation. The percentage of bluegill and redear sunfish populations considered harvestable decreases with increasing extent of hydrilla (Colle et al. 1987), but the populations are not considered to be adversely affected until plant volume infestation (PVI) exceeds 80% of the water column (Colle and Shireman 1980). At PVI values > 80% populations become food limited as the foraging gradient between macrophytes and open water is dramatically reduced. Although the harvestable portion of black crappie populations does not appear to be adversely impacted by the extent of hydrilla (Colle et al. 1987), it has been suggested that fish growth rates may be reduced at >50% aerial plant coverage because populations of their principal food, *Chaoborus*, are reduced by weed interference with diel migration patterns (Maceina and Shireman 1982). This interpretation is supported by the recent study of Crisman and Crisman (1988), which demonstrated an inverse relationship between chaoborid abundance and progressively increasing PVI > 50% in Florida lakes.

It appears that gamefish populations respond more to the percentage of the water column infested (PVI) by *Hydrilla* than to its total aerial coverage (Colle and Shireman 1980). Often, however, the relationship between fish and submergent vegetation is not clear. Largemouth bass biomass quadrupled and recruitment increased following complete removal of submergent macrophytes at Lake Baldwin, Florida (Shireman et al. 1984a). Colle and Shireman (1980) indicated that largemouth bass populations were adversely affected when *Hydrilla* coverage exceeded 30%, while Colle et al. (1987) could not demonstrate a reduction in the harvestable population of bass at *Hydrilla* coverage > 80%. Shireman et al. (1984b) concluded after a 20-pond macrophyte manipulation study that there was no clear predictable effect of macrophyte extent on fish populations, a sentiment echoed by Hoyer et al. (1985), who stressed that such an approach is overly simplistic and that additional factors, including food availability and lake trophic state, are likely exerting a major influence on the structure of fish communities.

Total fish biomass (Fig. 11) is positively related to the trophic state of Florida lakes (Kautz 1981, Bays and Crisman 1983). In spite of a great deal of interlake variability, the regression (r^2 = .42) remains significant (Bays and Crisman 1983). In marked contrast, fish abundance peaks in meso-eutrophic lakes and progressively declines with increasing trophic state (Kautz 1981), suggesting dominance of hypereutrophic fish communities by large-bodied individuals.

Sport fish biomass peaks in meso-eutrophic lakes, while maximum abundance occurs at the eutrophic–hypereutrophic transition (Kautz 1981). Both parameters decline sharply in the hypereutrophic range (Fig. 11). This trend is consistent with the observation of Kautz (as cited in Williams et al. 1985) that sport fish biomass increases linearly with increasing total nitrogen up to a value of 1.5 mg/L. However, the relationship totally breaks down above this concentration, reflecting a possible shift to phosphorus as the limiting nutrient for algal productivity. In spite of a trend of increasing biomass up through moderately eutrophic conditions, Bays and Crisman (1983) noted a progressive decrease (r^2 = .50) in the percent contribution of sport fish to total fish biomass in Florida lakes along a gradient of increasing trophic state.

While rough fish are nearly absent from oligotrophic Florida lakes, both their biomass (Fig. 11) and their percent contribution to total fish biomass increase linearly from mesotrophic through moderately hypereutrophic conditions (Kautz 1981, Bays and Crisman 1983). This relationship is dictated principally by the contribution of gizzard shad (*Dorosoma cepedianum*), which can constitute >80% of total fish biomass in hypereutrophic lakes (Bays and Crisman 1983).

The response of forage fish to trophic state is less clear than that of the three previous groups (Fig. 11). In spite of great interlake variability within a trophic state category, both total biomass of forage fish (Kautz 1981) and

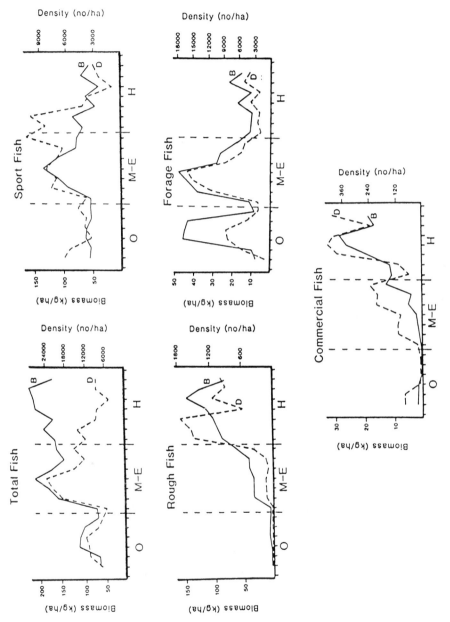

FIGURE 11. Density (D) and biomass (B) of major fish groups relative to the trophic state of Florida lakes. (Adapted from Kautz (1981).) O = oligotrophic; M–E, mesotrophic–eutrophic; H, hypereutrophic.

517

their percent contribution to total fish biomass (Bays and Crisman 1983) tend to decrease along a gradient of increasing eutrophy. The seemingly poor relationship for this predominantly littoral dwelling group is strongly influenced by the facts that trophic state is normally defined by pelagic parameters and there is a great deal of interlake variability in both the extent and structure of vegetated littoral zones.

A thriving commercial fishery exists for catfish throughout Florida, blue tilapia (*Tilapia aurea*) in central Florida, and bluegill, redear sunfish, and black crappie predominantly in Lake Okeechobee (Williams et al. 1985). Kautz (1981) noted that commercial fish (principally catfish) are found only nominally in oligotrophic and mesotrophic lakes, but that their biomass increases dramatically at progressively higher trophic state (Fig. 11). Blue tilapia was introduced into Florida in 1961 (Williams et al. 1985) and have become a major freshwater commercial fishery in central Florida, yielding 445 214 pounds in 1987 with a wholesale market value of $103 476 (Florida Game and Fresh Water Fish Commission, unpublished data). Similarly, during 1976–1981, commercial fishing removed 24% of the standing crop of bluegill + redear and 65% of the black crappie from Lake Okeechobee for a combined market value of $27 million (Schramm et al. 1985). The catch of crappie alone was estimated at $4.39 million wholesale and added approximately $8.38 million to the local economy (Williams et al. 1985). Schramm et al. (1985) suggested that the removal of bluegill and redear in Lake Okeechobee actually increased both the growth rate and condition factor of crappie by increasing the food base for the latter.

While it is clear that fish communities respond directly to structural aspects of their physical and biological environment, the direct relationship with water chemistry is less evident. The notable exception is anthropogenic lake acidification. Loss of biomass and individual species has been attributed to recent lake acidification in Europe (Wright and Snekvik 1978), Canada (Beamish and Harvey 1972), and the northeastern United States (Colquhoun et al. 1984). Several of these lakes are currently fishless.

Florida has not only the lakes of lowest pH but also the greatest number of lakes with pH less than 5.0 of any lake region in the eastern United States (Linthurst et al. 1986). While the number of fish species present in both temperate and Florida lakes declines with increasing acidity, the coincidental loss of individual taxa along this gradient observed in temperate lakes is absent from Florida (Keller and Crisman 1990). The controlling influence of water chemistry over fish composition in acidic Florida lakes is discounted by the absence of aluminum concentrations considered toxic to fish and commonly observed in acidic temperate lakes (Hendry and Brezonik 1984). Largemouth bass collected in acidic Florida lakes showed only slight reductions in both blood plasma osmotic and electrolytic concentrations (Canfield et al. 1985b). Canfield et al. found deformed otoliths in bass collected from a pH 3.7 lake (suggesting disturbance in calcium metabolism) and significantly lower growth and condition of fish > 305 mm, but they were unable to separate the direct influence of acidity from that of generally lowered lake productivity as controlling variables for their observations. Additional investigations by Lamia

(1987) and Romie (1989) indicated that total fish biomass and largemouth bass condition were related more to lake trophic state than acidity.

It is appropriate to end this review of southeastern lakes with a discussion of the role of fish in the structuring and functioning of lacustrine ecosystems. While most sport fish and forage fish prey on zooplankton to some degree (Maslin 1969, Wyngaard et al. 1982, Elmore et al. 1983), the ascendence of pump–filter-feeding gizzard shad as the dominant fish in eutrophic–hypereutrophic lakes is a major factor for observed differences along a trophic state gradient regarding nutrient cycling and the structure and productivity of bacterio-, phyto-, and zooplankton.

Although gizzard shad are known to consume sediments and macroinvertebrates, they feed primarily as pump–filterers of plankton (Berry 1955, McLane 1955, Crisman and Beaver 1988, 1990a). Excluded from oligotrophic lakes by a paucity of plankton, shad biomass increases progressively throughout the mesotrophic and eutrophic ranges to peak in hypereutrophic lakes. Biomass in the most hypereutrophic systems may actually decline (Kautz 1981) as the fish populations become self-limiting.

Gizzard shad can affect phytoplankton community structure both indirectly by enhancing orthophosphate concentrations (Crisman and Kennedy 1982, Crisman and Beaver 1988) and directly through differential digestion of algal taxa (Berry 1955, Crisman and Kennedy 1982). Digestion of blue-green algae is especially poor, with many species remaining viable upon gut passage (Crisman and Kennedy 1982). It has been suggested that once total plankton biomass reaches a level critical for maintenance of a self-sustaining shad population, the fish actually promotes system dominance by blue-green algae (Crisman 1981). Recent in situ enclosure experiments (Crisman and Beaver 1988) demonstrated that the poor digestion efficiency of shad also significantly stimulates phytoplankton productivity and system respiration. Reduction in shad biomass in the most hypereutrophic Florida lakes (Kautz 1981) likely results from extreme dominance of relatively inedible blue-green algae coupled with elevated nighttime respiratory demands.

Shad affect zooplankton community structure both indirectly through an enhancement of bacterioplankton levels that support microzooplankton, including ciliated protozoa (Crisman and Beaver 1988), and directly through predation on macrozooplankton, especially cladocerans (Crisman and Kennedy 1982, Crisman et al. 1986b, Crisman and Beaver 1988, 1990a). In spite of a 1–3 order of magnitude increase in cladoceran biomass when freed from shad predation, the associated enhancement of zooplankton grazing appears to be insufficient to reduce phytoplankton biomass significantly in hypereutrophic Florida lakes.

RESOURCE MANAGEMENT

The entire Southeast is experiencing increased development pressures from a rapidly expanding population. In Florida alone, approximately 900 new residents enter the state daily. Increasing pressure to meet the multiple-use needs of this expanding human population for agricultural, domestic, and recreational purposes has posed serious management problems for the natural lakes of the region. Such problems are related principally to eutrophication, acidification, land management practices, and the introduction of exotic species.

Eutrophication

Eutrophication may be defined as the enhancement of autotrophic production (algae and/or macrophytes) when supplied with elevated concentrations of some essential nutrient (phosphorus or nitrogen) considered limiting to photosynthesis. Whether algae or macrophytes become the dominant plant component depends on a combination of factors including the pre-impact partitioning of the autotrophs, basin configuration, nutrient loading rates, and water residence time. Shallow lakes that possess extensive vegetated littoral zones and are experiencing relatively slow rates of nutrient input tend to remain dominated by macrophytes during cultural eutrophication. Conversely, lakes whose pre-impact state was phytoplankton dominated or lakes undergoing high nutrient input usually shift to blue-green algal dominance. As evidenced by the example of Lake Apopka, Florida, it is often difficult to predict the final outcome of the eutrophication process (Crisman 1986). Prior to 1947, the vegetated littoral zone of the lake was able to absorb the progressive nutrient loading to the lake coming from agricultural and domestic sources. Catastrophic destruction of the macrophyte beds during the 1947 hurricane stimulated massive blooms of blue-green algae. Over 40 years later, the lake continues to be algal dominated and submergent macrophytes are essentially absent.

Most phytoplankton problems are associated with a buildup of total algal biomass and a shift to dominance by blue-green species. Management strategies have centered on ways to control algae, either indirectly by reducing nutrient availability or directly through biological control of phytoplankton biomass and composition. The State of Florida has employed lake drawdown at several sites both for compaction and oxidation of bottom sediments to reduce nutrient cycling and reestablishment of vegetated littoral zones (Williams et al. 1985). Total water column aeration has also been examined in Florida for reducing nutrient availability and the competitive advantage afforded several blue-green algae by an ability to control their position in the water column (Bateman and Laing 1977, Crisman et al. 1984b, Cowell et al. 1987b).

Mechanical harvesting of algal biomass is not currently cost effective, but a variety of chemicals, from copper sulfate to algicides and algistatics, can provide effective phytoplankton control in small basins (Crisman 1986). Biological control research has concentrated on both the use of cyanophyte-specific pathogens, especially cyanovirus and bacteria, and the effectiveness of grazing by fish and zooplankton for reducing algal biomass and promoting taxonomic compositional shifts (Crisman 1986).

The potential of biomanipulation for algal control in subtropical Florida lakes is hindered by the fact that macrozooplankton are small-bodied regardless of trophic state and seem unable to reduce algal biomass in hypereutrophic lakes significantly even when freed from fish predation (Crisman 1981, Crisman and Beaver 1988, 1990a). Phytoplankton biomass is sufficient year round to support phytophagous fish in productive Florida lakes, but the native fish species (gizzard shad) displays poor digestive efficiency for many of the dominant blue-green taxa in eutrophic lakes (Crisman and Kennedy 1982). If biomanipulation is to be an effective management tool for phytoplankton in eutrophic lakes, an exotic fish, such as *Tilapia*, that has a high digestive efficiency for blue-greens may have to be employed (Crisman and Beaver 1988). Finally, grazing experiments have demonstrated that the exotic filter-feeding clam *Corbicula fluminea* is found in densities sufficient in many Florida lakes to exert a major controlling influence over plankton structure and biomass (Beaver et al. 1991).

Most of the eutrophication control research in the Southeast has been directed toward problem aquatic macrophytes. Three exotic species have posed the most serious problems in the region. Water hyacinth (*Eichhornia crassipes*) is a free-floating plant that was introduced into the United States during the 1884 Cotton Exposition in New Orleans. This plant was already posing management problems on the St Johns River, Florida, by the 1890s, and it spread progressively in subsequent years, so that by 1975 approximately 80 000 ha of Florida surface water was considered densely covered (Crisman 1986).

Hydrilla verticillata is an African submergent plant likely introduced into Florida by the aquarium industry. Since first collected in 1960, it has spread rapidly throughout peninsular Florida to infest approximately 141 000 ha in 1986 (Schardt 1986). Although *Hydrilla* is present in many other areas of the Southeast, it is usually replaced by another exotic submergent, Eurasian watermilfoil (*Myriophyllum spicatum*) as the major problem plant in lakes of the Florida panhandle and northward. Watermilfoil was perhaps the first exotic macrophyte introduced into the United States (17th century), but it has caused widespread serious management problems only within the past 30 years (Crisman 1986).

A number of techniques are available for mechanically harvesting both free-floating and submergent macrophytes (Crisman 1986). Unfortunately, these usually do not provide a permanent solution to the problem, owing to the persistence of seeds and underground tubers and roots. Haller et al. (1980) demonstrated that one of the detrimental side effects of mechanical harvesting of submergents may be the associated harvest of sport fish, which in Orange Lake, Florida, was estimated to have a replacement cost of $6000/ha.

Chemical formulations are very effective at controlling excessive macrophyte growth and are applied as solutions, emulsions, suspensions, oil solutions, slow-release pellets, or granules (Crisman 1986). While development of formulations displaying great plant specificity has proliferated in recent years, chemical treatment, like mechanical treatment, may serve as only a short-term prophylaxis, which if performed too rapidly may promote algal blooms from nutrients released from decaying plant tissue.

Biological control of macrophytes has only recently been investigated. Three general areas have been considered: plant pathogens, insects, and herbivores. Research on the first two categories has concentrated on finding organisms whose effects on macrophytes are highly species specific, while research with herbivores is mostly concerned with a general reduction of plant biomass. A complete review of the topic is given in Crisman (1986).

Fungal pathogens have proven useful for control of water hyacinth in Florida, and a variety of lepidopterans (e.g., *Sameodes albiguttalis*) and coleopterans (e.g., *Agasicles hygrophila, Neochetina bruchi, N. eichhorniae*) are effective insect pests for water hyacinth and alligatorweed (*Alternanthera philoxeroides*). To date, such agents are effective only for control of free-floating and emergent macrophytes. While a variety of herbivores have been proposed (Crisman 1986), the grass carp (*Ctenopharyngodon idella*) currently is the best biotic control for submergent macrophytes. The grass carp has effectively removed submergent macrophytes in several Florida lakes (Shireman and Maceina 1981, Miller 1982, Crisman 1986), but the resulting effect on water quality depends on the pretreatment percentage of the lake volume infested with macrophytes and the overall macrophyte standing crop (Canfield et al. 1983). Too rapid release of nutrients tied up in macrophyte tissue can stimulate algal growth and shift the principal management problem to blue-green-dominated phytoplankton.

The best approach to management of aquatic macrophytes may involve an integration of several control techniques (Canfield 1983a, Crisman 1986). In dealing with problem free-floating macrophytes, an integrated approach might involve an initial reduction in plant biomass via chemical or mechanical techniques followed by the introduction of multiple biocontrol agents including pathogens and insect species. Such a control strategy would be less likely to promote rapid, drastic changes in water quality and would have a greater chance at long-term control. Unfortunately, a comparable integrated approach for algal management is not yet available.

Acidification

While acidic lakes in the Southeast are represented principally by Carolina Bays and Florida solution lakes, research on the potential impact of acidic deposition on ecosystem structure and function has been limited principally to the latter. From an EPA survey it was determined that Florida has not only the lakes of lowest pH but also the greatest number of lakes <5.0 pH of any lake region in the eastern United States (Linthurst et al. 1986). Approximately 30% of Florida's 7700 lakes are considered potentially sensitive to acidic deposition (Canfield 1983c, Hendry and Brezonik 1984). Florida is currently experiencing acidic precipitation, with annual average rainfall pH values <4.7 for the northern 75% of the peninsula (Brezonik et al. 1980). The available database indicates that both rainfall and lake pH values at several sensitive sites have declined in the past 25 years (Brezonik et al. 1980, Hendry and Brezonik 1984).

A number of surveys have provided data on most biotic components relative to a gradient of lake acidity (4.0 → 7.0 pH) in Florida. Bacterioplankton density appears little affected by acidity (Crisman and Bienert 1983), while reduced phytoplankton abundance and decreased representation of blue-green algae with increasing acidity are similar to findings in temperate lakes (Brezonik et al. 1984). Unlike comparable temperate systems, macrozooplankton biomass does not decline with increasing acidity and no major species replacements are noted. Within the microzooplankton, however, the abundance and biomass of ciliate protozoa are significantly lower in acidic lakes (Beaver and Crisman 1981), and a large-bodied myxotrophic species (*Stentor niger*) that is principally restricted to acidic systems can constitute >30% of total zooplankton biomass in such systems (Beaver and Crisman 1989a). Although biomass may be reduced, the Florida response of benthic invertebrates, fish, and macrophytes appears less pronounced than noted for temperate systems and is not accompanied by major species eliminations (Crisman and Bienert 1983, Crisman 1984, 1989).

The apparent differences in the biotic response to lake acidity displayed by temperate and Florida lakes has been attributed to significantly lower aluminum concentrations, the absence of seasonal pulses of acidity, and major faunal differences in the latter lakes (Crisman 1984). Results from recent in situ enclosure and laboratory experiments, however, suggest that major alterations in all autotrophic and heterotrophic communities will occur if the pH of Florida lakes drops below 3.5–3.8 and aluminum is mobilized from sediments (Crisman 1984).

Numerous authors have suggested that observed changes in all biotic components from plankton through fish relative to increasing acidity in Florida likely are attributable more to lake trophic state than to pH directly (Crisman and Bienert 1983, Canfield 1983c, Brezonik et al. 1978, Crisman 1984, Canfield et al. 1985b). In addition, Crisman et al. (1986a) have suggested that while littoral and benthic communities are major contributors to the autotrophic production of small lakes in general, it is likely that their share assumes greater importance in acidic lakes where phytoplankton are often severely nutrient limited. Primary production may not be reduced in acidic lakes, but merely reapportioned between plankton and benthic components.

Land Management Practices

Sharitz and Gibbons (1982) identified four major watershed practices adversely affecting Carolina Bays: forestry, agriculture, drainage systems, and peat mining. Throughout the coastal plain from southern Virginia (Marshall and Robinson 1979) to Georgia (Wharton 1978), Carolina Bays have undergone either extensive logging or total drainage for conversion to farming or commercial pine plantations. Similar conversion to planted pines covers vast areas of north Florida. For the most part, assessment of the impact of such practices on natural lakes of the Southeast is incomplete.

Many smaller Carolina Bays in Georgia have been totally drained for agricultural purposes (Wharton 1978). Elsewhere, utilization of nutrient-poor acidic soils surrounding Carolina Bays has necessitated substantial additions of nitrogen, phosphorus, potassium, and lime, which, if coupled with increased field drainage, can promote eutrophication of bay lakes (Sharitz and Gibbons 1982).

Extensive wetlands surrounding Lakes Apopka and Okeechobee in Florida have been drained for vegetable and sugar production. In both cases, excess water is backpumped from fields to the lakes. A great deal of research is currently ongoing to assess the historical significance of this agricultural practice as well as that of extensive dairy farms, in the case of Lake Okeechobee, on the progression of cultural eutrophication. Elsewhere in the state, agricultural discharge is of concern throughout the St. Johns River basin.

Drainage practices are of major concern throughout the Southeast because of their impact on hydrology and nutrient loading to lakes. Special concern has developed recently regarding extensive drainage of peatlands for agricultural or mining purposes (Sharitz and Gibbons 1982). Most research on the topic has centered on the impact on Carolina Bays. Peat drainage promotes increases in the volume, duration, and peak flow of discharge water, but actually reduces baseflow between storm events (Gregory et al. 1984). Drainage of peatlands in North Carolina has led to a slight elevation of mercury concentrations in surface waters receiving discharge, but these are within state standards (Evans et al. 1984).

Finally, control structures have been built at the outlets of most of the larger lakes in the Southeast to stabilize water levels for wildlife or recreational purposes. Reduction in the amplitude of water level fluctuation at Reelfoot Lake, Tennessee, following establishment of the Reelfoot National Wildlife Refuge in 1941 has been linked in part to major species replacements within the submerged macrophyte community and a shift in emergent vegetation from monotypic marshes (*Zizaniopsis miliacea*) to mixed marsh–swamp vegetation (Henson 1990a). Similar stabilization of water level at both Newnan's Lake, Florida (Gottgens and Crisman 1991), and Lake Weir, Florida (Crisman et al. 1990), resulted in increased annual accumulation of sediment, phosphorus, and nitrogen, resulting in an acceleration of cultural eutrophication.

Exotic Species

As with exotic macrophytes, the most serious problems with exotic fish and invertebrates occur in peninsular Florida. Of the 25 exotic fish species maintaining breeding populations in the United States in 1970, a majority were restricted to Florida (Lachner et al. 1970). While most of the exotic species in the rest of the Southeast are salmonids introduced into mountain streams for sport fishing, the common carp (*Cyprinus carpio*), introduced from Europe in 1877, is found in lakes throughout the Southeast with the exception of a majority of peninsular Florida (Lee et al. 1980).

Since the beginning of the Florida aquarium industry in 1929 (Courtenay and Robins 1973), the number of established exotic fish species in the state has grown steadily. For example, 14 species were established by 1979 (Shafland 1979) and the number increased to 17 by 1986 (Shafland 1986). The potential impact of the aquarium industry is illustrated by the fact that 80–95% of all ornamental fish grown in North America come from Florida (Shafland 1986).

Although most exotic fish species are limited to the Florida peninsula due to temperature constraints, exotic fish are of concern because they (1) alter energy flow through ecosystems, (2) respond unpredictably when introduced into new habitats, (3) may compete indirectly and directly with native fish species, and (4) harbor exotic diseases and parasites (Shafland 1986). While most attention has been paid to highly visible (walking catfish, *Clarias batrachus*) or pelagic commercially important taxa (blue tilapia, *Tilapia aurea*), a majority of the 17 species established in Florida are small predaceous cichlids mainly restricted to vegetated littoral zones of lakes and canals. Given the importance of the littoral zone as a native fish breeding and nursery ground, more attention should be paid to the impact of introduced predators on the structure and function of such areas.

A number of exotic invertebrates have established populations in the Southeast (Lachner et al. 1970). Two mollusks are of special concern. The ampullariid snail *Marisa cornuarietis* was introduced into Florida from South America by the aquarium trade. Concern over the establishment of this herbivorous snail is especially acute in south Florida where competition is feared with the apple snail (*Pomacea paludosa*), the principal food of the endangered Everglades kite (*Rostrhamus sociabilis*). Since its introduction into Washington State in 1938, the Asian clam *Corbicula fluminea* has spread throughout the United States and has caused serious management problems in rivers and reservoirs of the Southeast. This clam is well established in Lake Okeechobee, Florida, where it has attained some of the greatest densities ($2300/m^2$) recorded in North America (Florida Game and Fresh Water Fish Commission, unpublished data). Although the impact of such *Corbicula* "reefs" on the ecology of Lake Okeechobee has not been investigated, recent experimental evidence suggest that the clam is at sufficient densities in the lake to alter plankton biomass and structure significantly (Beaver et al. 1991).

RESEARCH NEEDS

A great deal has been learned about the structure and function of natural lakes in the Southeast in the 25 years since the first reviews of southeastern limnology were published (Gerking 1963, Moore 1963, Yount 1963). Given the diversity of lake types in the Southeast, it is indeed unfortunate that most of the recent literature has dealt primarily with Florida solution lakes.

A number of questions remain regarding the formation and ontogeny of lakes in the Southeast. Much of our current knowledge is conjectural and needs to be documented. With the exception of Florida, basic understanding of the structure and function of the individual biotic components of lakes is often lacking, as well as their response to trophic state and climatic gradients. While recent research has begun to put subtropical lakes in perspective, we lack a clear understanding of the transitional nature of warm temperate systems between cold temperate and subtropical lakes. The linkage between biogeography and lake structure has not been made.

The Southeast is experiencing rapid population growth and increased development pressure, both agricultural and residential, on natural lakes. It is most unfortunate that so little baseline information is available on individual lake types against which we can gauge the impact of anthropogenic disturbances. If there are no data on the expected range of pre-impact conditions for a given lake type, rational management or restoration plans cannot be developed.

It is thus imperative that representative lakes be protected for use as controls for assessing the impact of continued human disturbance. The Archbold Biological Station in southern Florida and the Katherine B. Ordway Preserve (University of Florida) in northern Florida protect several clear and organically colored soft water lakes. The value of these living laboratories is becoming increasing apparent every year as extensive development pressures encroach on the boundaries of these preservation areas. Much of our understanding of the response of Florida lakes to acidity is based in large part on lakes located in these preserves.

Elsewhere in the Southeast, many of the early limnological investigations on natural lakes were conducted at Reelfoot Lake Biological Station, Tennessee, and Mountain Lake Biological Station, Virginia. Although there is no organized biological station for Carolina Bay lakes, lakes under state protection as preserves or parks, especially in North Carolina, have provided most of our understanding of how these fascinating systems function. The lack of data on oxbow lakes is likely directly related to the absence of a biological research facility devoted specifically to this lake type. It is imperative that past oversights be corrected and that a strong system of lake preservation areas and biological stations be developed to serve as regional models for limnological research.

REFERENCES

Baker, L. A., P. L. Brezonik, and C. R. Kratzer. 1981. *Nutrient Loading—Trophic State Relationships in Florida Lakes*, OWRT Project No. A-038-FLA, Publ. No. 56. Gainesville: Water Resources Research Center, University of Florida.

Bateman, J. M., and R. L. Laing. 1977. Restoration of water quality in Lake Weston, Orlando, Florida. *J. Aquat. Plant Manage.* 15:69–73.

Bays, J. S. 1983. Zooplankton-trophic state relationships and seasonality in Florida lakes. Thesis, University of Florida, Gainesville.

Bays, J. S., and T. L. Crisman. 1983. Zooplankton trophic state relationships in Florida lakes. *Can. J. Fish. Aquat. Sci.* 40:1813–1819.

Beamish, R. J., and H. H. Harvey. 1972. Acidification of the La Cloche Mountain lakes, Ontario, and resulting fish mortalities. *J. Fish. Res. Board Can.* 29:1131–1143.

Beaver, J. R., and T. L. Crisman. 1981. Acid precipitation and the response of ciliated protozoans in Florida lakes. *Verh.—Int. Ver. Theor. Angew. Limnol.* 21:353–358.

Beaver, J. R., and T. L. Crisman. 1982. The trophic response of ciliated protozoans in freshwater lakes. *Limnol. Oceanogr.* 27:246–253.

Beaver, J. R., and T. L. Crisman. 1989a. The role of ciliated protozoa in pelagic freshwater ecosystems. *Microb. Ecol.* 17:111–136.

Beaver, J. R., and T. L. Crisman. 1989b. Analysis of the community structure of planktonic ciliated protozoa relative to trophic state in Florida lakes. *Hydrobiologia* 174:177–184.

Beaver, J. R., and T. L. Crisman. 1990a. Seasonality of planktonic ciliated protozoa in 20 subtropical Florida lakes of varying trophic state. *Hydrobiologia* 190:127–135.

Beaver, J. R., and T. L. Crisman. 1990b. Use of microzooplankton as an early indicator of advancing cultural eutrophication. *Verh.—Int. Ver. Theor. Angew. Limnol.* 24:532–537.

Beaver, J. R., and T. L. Crisman. 1991. Temporal variability in algal biomass and primary productivity in Florida lakes relative to latitudinal gradients, organic color and trophic state. *Hydrobiologia* 224:89–97.

Beaver, J. R., T. L. Crisman, and J. S. Bays. 1981. Thermal regimes of Florida lakes. *Hydrobiologia* 83:267–273.

Beaver, J. R., T. L. Crisman, and R. W. Bienert, Jr. 1988. Distribution of planktonic ciliates in highly coloured subtropical lakes: comparison with clearwater ciliate communities and the contribution of myxotrophic taxa to total autotrophic biomass. *Freshwater Biol.* 20:51–60.

Beaver, J. R., T. L. Crisman, and R. J. Brock. 1991. Grazing effects of an exotic bivalve (*Corbicula fluminea*) on hypereutrophic lake water. *Lake Reservoir Manage.* 7:45–51.

Berry, F. H. 1955. Age, growth, and food of the gizzard shad, *Dorosoma cepedianum* in Lake Newnan, Florida. Thesis, University of Florida, Gainesville.

Bienert, R. W., Jr. 1987. Zooplankton dynamics in an acidic subtropical lake. Thesis, University of Florida, Gainesville.

Blancher, E. C., II. 1984. Zooplankton–trophic state relationships in some north and central Florida lakes. *Hydrobiolgia* 109:251–263.

Brezonik, P. L., C. D. Pollman, T. L. Crisman, J. N. Allinson, and J. L. Fox. 1978. *Limnological Studies on Lake Apopka and the Oklawaha Chain of Lakes. I. Water Quality in 1977*, Rep. ENV-07-78-01. Gainesville: Dept. Environ. Eng. Sci., University of Florida.

Brezonik, P. L., E. S. Edgerton, and C. D. Hendry. 1980. Acid precipitation and sulfate deposition in Florida. *Science* 208:1027–1029.

Brezonik, P. L., T. L. Crisman, and R. L. Schulze. 1984. Planktonic communities in Florida softwater lakes of varying pH. *Can. J. Fish. Aquat. Sci.* 41:46–56.

Brooks, H. K. 1974. Lake Okeechobee. In P. J. Gleason (ed.), *Environments of South Florida: Present and Past*, Mem. 2. Miami, FL: Miami Geological Survey, pp. 256–286.

Brown, H. D. 1976. A comparison of the attached algal communities of a natural and an artificial substrate. *J. Phycol.* 12:301–306.

Bruno, N. A. 1984. Nest site selection by and spawning season vegetation associations of Florida largemouth bass, *Micropterus salmoides floridanus*, in Orange Lake, Florida. Thesis, University of Florida, Gainesville.

Canfield, D. E., Jr. 1981. *Chemical and Trophic State Characteristics of Florida Lakes in Relation to Regional Geology*, Final Rep. Gainesville: Florida Cooperative Fish and Wildlife Unit., U.S. Fish and Wildlife Service.

Canfield, D. E., Jr. 1983a. Impact of integrated aquatic weed management on water quality in a citrus grove. *J. Aquat. Plant Manage.* 21:69–73.

Canfield, D. E., Jr. 1983b. Prediction of chlorophyll-a concentrations in Florida lakes: the importance of phosphorus and nitrogen. *Water Resour. Res.* 19:255–262.

Canfield, D. E., Jr. 1983c. Sensitivity of Florida lakes to acidic precipitation. *Water Resour. Res.* 19:833–839.

Canfield, D. E., Jr., and C. E. Watkins, II. 1984. Relationships between zooplankton abundance and chlorophyll a concentrations in Florida lakes. *J. Freshwater Ecol.* 2:335–344.

Canfield, D. E., Jr., K. A. Langeland, M. J. Maceina, W. T. Haller, J. V. Shiremann, and J. R. Jones. 1983. Trophic state classification of lakes with aquatic macrophytes. *Can. J. Fish. Aquat. Sci.* 40:1713–1718.

Canfield, D. E., Jr., S. B. Linda, and L. M. Hodgson. 1984a. Relations between color and some limnological characteristics of Florida lakes. *Water Resour. Bull.* 20:323–329.

Canfield, D. E., Jr., J. V. Shireman, D. E. Colle, W. T. Haller, C. E. Watkins, II, and M. J. Maceina. 1984b. Prediction of chlorophyll a concentrations in Florida lakes: importance of aquatic macrophytes. *Can. J. Fish. Aquat. Sci.* 41:497–501.

Canfield, D. E., Jr., S. B. Linda, and L. M. Hodgson. 1985a. Chlorophyll-biomass-nutrient relationships for natural assemblages of Florida phytoplankton. *Water Resour. Bull.* 21:381–391.

Canfield, D. E., Jr., M. J. Maceina, F. G. Nordlie, and J. V. Shireman. 1985b. Plasma osmotic and electrolyte concentrations of largemouth bass from some acidic Florida lakes. *Trans. Am. Fish. Soc.* 114:423–429.

Carlson, R. E. 1977. A trophic state index for lakes. *Limnol. Oceanogr.* 22:361–369.

Carr, M. H. 1942. The breeding habits, embryology, and larval development of the large-mouthed black bass in Florida with notes on the feeding habitats of the fry. Thesis, University of Florida, Gainesville.

Clarkson, C. L. 1985. Macrophyte species distribution and tissue chemical composition in acid Florida Lakes. Thesis, University of Florida, Gainesville.

Colle, D. E., and J. V. Shireman. 1980. Coefficients of condition for largemouth bass, bluegill and redear sunfish in hydrilla-infested lakes. *Trans. Am. Fish. Soc.* 109:521–531.

Colle, D. E., J. V. Shireman, W. T. Haller, J. C. Joyce, and D. E. Canfield, Jr. 1987. Influence of hydrilla on angler utilization and monetary expenditure at Orange Lake, Florida. *North Am. J. Fish. Manage.* 7:410–417.

Colquhoun, J., W. Kretser, and M. Pfeiffer. 1984. *Acidity Status Update of Lakes and Streams in New York State*. Albany: New York Dept. Environ. Conserv.

Comp, G. S. 1979. Diel and seasonal patterns in the vertical distribution of zooplankton in Lake Conway, Florida. Thesis, University of Florida, Gainesville.

Confer, J. L. 1971. Intrazooplankton predation by *Mesocyclops edax* at natural prey densities. *Limnol. Oceanogr.* 16:663–666.

Conley, R., E. C. Blancher, F. Kooijman, C. Ferrick, J. L. Fox, and T. L. Crisman. 1979. *Large-Scale Operations Management Test of Use of the White Amur for Control of Problem Aquatic Plants*, Rep. 1, Vol. III, Techn. Rep. A-78-2. Vicksburg, MS: U.S. Army Corps of Engineers, Waterways Experiment Station.

Cooke, C. W. 1939. Scenery of Florida. *Fl., Geol. Surv., Geol. Bull.* 17:1–118.

Cooper, C. M., E. J. Bacon, and J. C. Ritchie. 1984. Biological cycles in Lake Chicot, Arkansas. In J. F. Nix and F. R. Schiebe (eds.), *Limnological Studies of Lake Chicot, Arkansas*. Arkadelphia, AR: Ouachita Baptist University, pp. 48–61.

Courtenay, W. R., Jr., and C. R. Robins. 1973. Exotic aquatic organisms in Florida with emphasis on fishes: a review and recommendations. *Trans. Am. Fish. Soc.* 102:1–12.

Cowell, B. C. 1984. Benthic invertebrate recolonization of small-scale disturbances in the littoral zone of a subtropical Florida lake. *Hydrobiologia* 109:193–205.

Cowell, B. C., and C. J. Dawes. 1984. *Algal Studies of Eutrophic Florida Lakes: The Influence of Aeration on the Limnology of a Central Florida Lake and Its Potential as a Lake Restoration Technique*, Final Rep. Tallahassee: Bureau of Aquatic Plant Control and Research, Florida Dept. Nat. Resour.

Cowell, B. C., and D. S. Vodopich. 1981. Distribution and seasonal abundance of benthic macroinvertebrates in a subtropical Florida lake. *Hydrobiologia* 78:97–105.

Cowell, B. C., C. W. Dye, and R. C. Adams. 1975. A synoptic study of the limnology of Lake Thonotosassa, Florida. Part I. Effects of primary treated sewage and citrus wastes. *Hydrobiologia* 16:301–345.

Cowell, B. C., H. C. Hull, Jr., and A. Fuller. 1987a. Recolonization of small-scale disturbances by benthic invertebrates in Florida freshwater ecosystems. *Fla. Entomol.* 70:1–14.

Cowell, B. C., C. J. Dawes, W. E. Gardiner, and S. M. Scheda. 1987b. The influence of whole lake aeration on the limnology of a hypereutrophic lake in central Florida. *Hydrobiologia* 148:3–24.

Crisman, T. L. 1980. Chydorid cladoceran assemblages from subtropical Florida. In W. C. Kerfoot (ed.), *Evolution and Ecology of Zooplankton Communities*. Biddeford, ME: University Press of New England, pp. 657–668.

Crisman, T. L. 1981. Algal control through trophic-level interactions: a subtropical perspective. In *Proceedings of Workshop on Algal Management and Control*. Vicksburg, MS: U.S. Army Engineers, Waterways Experiment Station, pp. 131–145.

Crisman, T. L. 1984. Temperate and subtropical biotic responses to lake acidity: a preliminary assessment of the importance of direct and indirect pH effects. In M.

E. Vittes and H. Kennedy (eds.), *Acid Rain: Its Impact on Florida and the Southeast.* Orlando: University of Central Florida, pp. 139–157.

Crisman, T. L. 1986. Eutrophication control with emphasis on macrophytes and algae. In N. Polunin (ed.), *Ecosystem Theory and Application.* New York: Wiley, pp. 200–239.

Crisman, T. L. 1989. A preliminary assessment of chironomid distributions in acidic Florida lakes. In D. F. Charles and D. R. Whitehead (eds.), *Paleoecological Investigation of Recent Lake Acidification (PIRLA): 1983–1985,* EPRI EN-6526. Palo Alto, CA: Electric Power Research Institute, pp. 14-1–14-10.

Crisman, T. L., and J. R. Beaver. 1988. *Lake Apopka Trophic Structure Manipulation,* Final Rep. Palatka, FL: St. Johns River Water Management District.

Crisman, T. L., and J. R. Beaver. 1990a. Applicability of planktonic biomanipulation for managing eutrophication in the subtropics. *Hydrobiologia* 200/201:177–185.

Crisman, T. L., and J. R. Beaver. 1990b. A latitudinal assessment of distribution patterns in chaoborid abundance for eastern North American lakes. *Verh.—Int. Ver. Theor. Angew. Limnol.* 24:547–553.

Crisman, T. L., and R. W. Bienert, Jr. 1983. Perspectives on biotic responses to acidification in Florida lakes. In A. E. S. Green and W. H. Smith (eds.), *Acid Deposition: Causes and Effects.* Government Institutes, Inc., pp. 307–315.

Crisman, T. L., and U. A. M. Crisman. 1988. Subtropical chaoborid populations and their representation in the paleolimnological record. *Verh.—Int. Ver. Theor. Angew. Limnol.* 23:2157–2164.

Crisman, T. L., and H. M. Kennedy. 1982. *The Role of Gizzard Shad (Dorosoma cepedianum) in Eutrophic Florida Lakes,* Publ. No. 64. Gainesville: Water Resources Research Center, University of Florida.

Crisman, T. L., and F. M. Kooijman. 1981. *The Plankton and Benthos of Lake Conway, Florida,* Rep. 2, Vol. II. Tech. Rep. A-78-2. Jacksonville, FL: U.S. Army Engineer District.

Crisman, T. L., and D. R. Whitehead. 1975. Environmental history of Hovey Lake, southwestern Indiana. *Am. Midl. Nat.* 93:198–205.

Crisman, T. L., J. R. Beaver, and J. S. Bays. 1981. Examination of the relative impact of microzooplankton and macrozooplankton on bacteria in Florida lakes. *Verh.—Int. Ver. Theor. Angew. Limnol.* 21:359–362.

Crisman, T. L., P. Scheuerman, R. W. Bienert, Jr., J. R. Beaver, and J. S. Bays. 1984a. A preliminary characterization of bacterioplankton seasonality in subtropical Florida lakes. *Verh.—Int. Ver. Theor. Angew. Limnol.* 22:620–626.

Crisman, T. L., P. R. Scheuerman, A. E. Keller, U. A. M. Crisman, D. J. Medina, J. S. Bays, J. R. Beaver, and M. W. Binford. 1984b. *Algal Management Through Lake Aeration,* Final Project Rep. Tallahassee: Bureau Aquatic Plant Research and Control, Florida Dept. Nat. Resour.

Crisman, T. L., C. L. Clarkson, A. E. Keller, R. A. Garren, and R. W. Bienert, Jr. 1986a. A preliminary assessment of the importance of littoral and benthic autotrophic communities in acidic lakes. In B. G. Isom, S. D. Dennis, and J. M. Bates (eds.), *Impact of Acid Rain and Deposition on Aquatic Biological Systems,* Spec. Tech. Publ. No. 928. Philadelphia, PA: Am. Soc. Test. Mater. (ASTM), pp. 17–27.

Crisman, T. L., J. A. Foran, J. R. Beaver, A. E. Keller, P. D. Sacco, R. W. Bienert, Jr., R. W. Ruble, and J. S. Bays. 1986b. *Algal Control Through Trophic-Level*

Interactions: Investigations at Lakes Wauburg and Newnans, Florida, Final Rep. Tallahassee: Bureau Aquatic Plant Research and Control. Florida Dept. Nat. Resour.

Crisman, T. L., U. A. M. Crisman, and M. W. Binford. 1986c. Interpretation of bryozoan microfossils in lacustrine sediment cores. *Hydrobiologia* 143:113–118.

Crisman, T. L., A. E. Keller, J. K. Jones, H. Meier, and J. R. Beaver. 1988. *Lake Weir Eutrophication Study*, Final Report of Phases I and II. Palatka, FL: St. Johns River Water Management District.

Crisman, T. L., J. R. Beaver, J. K. Jones, A. G. Neugaard, and V. Nilakantan. 1990. *Historical Assessment of Cultural Eutrophication in Lake Weir, Florida*. Palatka, FL: St. Johns River Water Management District.

DeCosta, J. J. 1964. Latitudinal distribution of chydorid cladocera in the Mississippi Valley, based on their remains in surficial lake sediments. *Invest. Indiana Lakes Streams* 6:65–101.

Delcourt, H. R., and P. A. Delcourt. 1985. Quaternary palynology and vegetational history of the southeastern United States. In V. M. Bryant, Jr. and R. G. Holloway (eds.), *Pollen Records of Late-Quaternary North American Sediments*. Am. Assoc. Stratigraphic Palynol. Found., pp. 1–37.

Elmore, J. L. 1983a. Factors influencing *Diaptomus* distributions: an experimental study in subtropical Florida. *Limnol. Oceanogr.* 28:522–532.

Elmore, J. L. 1983b. The influence of temperature on egg development times of three species of *Diaptomus* from subtropical Florida. *Am. Midl. Nat.* 109:300–308.

Elmore, J. L., D. S. Vodopich, and J. J. Hoover. 1983. Selective predation by bluegill sunfish (*Lepomis macrochirus*) on three species of *Diaptomus* (Copepoda) from subtropical Florida. *J. Freshwater Ecol.* 2:183–192.

Elmore, J. L., B. C. Cowell, and D. S. Vodopich. 1984. Biological communities of three subtropical Florida lakes of different trophic character. *Arch. Hydrobiol.* 100:455–478.

Evans, D. W., R. T. Digiulio, and E. A. Ryan. 1984. *Mercury in Peat and Its Drainage Waters in Eastern North Carolina*, Rep. UNC-WRRI-84-218. Chapel Hill: Water Resources Research Institute, University of North Carolina.

Ewel, K. C., and T. D. Fontaine, III. 1983. Structure and function of a warm monomictic lake. *Ecol. Modell.* 19:139–161.

Feerick, C. P., Jr. 1977. Studies of the periphyton community of Lake Conway, Florida. Thesis, University of Florida, Gainesville.

Fontaine, T. D., III, and K. C. Ewel. 1981. Metabolism of a Florida lake ecosystem. *Limnol. Oceanogr.* 26:754–763.

Foran, J. A. 1986a. A comparison of the life history of a temperate and a subtropical *Daphnia* species. *Oikos* 46:185–193.

Foran, J. A. 1986b. The relationship between temperature, competition and the potential for colonization of a subtropical pond by *Daphnia magna*. *Hydrobiologia* 134:102–112.

Frey, D. G. 1948. A biological survey of Lake Waccamaw. *Wildl. N. C.* 12:4–6, 17–19.

Frey, D. G. 1949. Morphometry and hydrography of some natural lakes of the North Carolina coastal plain: the bay lakes as a morphometric type. *J. Elisha Mitchell Sci. Soc.* 65:1–37.

Frey, D. G. 1951a. Pollen succession in the sediments of Singletary Lake, North Carolina. *Ecology* 32:518–533.

Frey, D. G. 1951b. The fishes of North Carolina's Bay lakes and their intraspecific variation. *J. Elisha Mitchell Sci. Soc.* 67:1–44.

Frey, D. G. 1953. Regional aspects of the Late-Glacial and Post-Glacial pollen succession of southeastern North Carolina. *Ecol. Monogr.* 23:289–313.

Frey, D. G. 1954a. Evidence of recent enlargement of the "Bay" lakes of North Carolina. *Ecology* 35:78–88.

Frey, D. G. 1954b. Stages in the ontogeny of the Carolina Bays. *Verh.—Int. Ver. Theor. Angew. Limnol.* 12:660–668.

Frey, D. G. 1978. A new species of *Eurycercus* (Cladocera, Chydoridae) from the southern United States. *Tulane Stud. Zool. Bot.* 20:1–25.

Frey, D. G. 1982a. Contrasting strategies of gameogenesis in northern and southern populations of cladocera. *Ecology* 62:223–241.

Frey, D. G. 1982b. Relocation of *Chydorus barroisi* and related species (Cladocera, Chydoridae) to a new genus and description of two new species. *Hydrobiologia* 86:231–239.

Frey, D. G. 1982c. The honeycombed species of *Chydorus* (Cladocera, Chydoridae): comparison of *C. bicornutus* and *C. bicollaris* n.sp. with some preliminary comments on *C. faviformis*. *Can. J. Zool.* 60:1892–1916.

Frey, D. G. 1982d. The reticulated species of *Chydorus* (Cladocera, Chydoridae): two new species with suggestions of convergence. *Hydrobiologia* 93:255–279.

Fuller, A., and B. C. Cowell. 1985. Seasonal variation in benthic invertebrate recolonization of small-scale disturbances in a subtropical Florida lake. *Hydrobiologia* 124:211–221.

Fuller, S. L. H. 1977. Freshwater and terrestrial molluscs. In J. E. Cooper, S. S. Robinson, and J. B. Funderburg (eds.), *Endangered and Threatened Plants and Animals of North Carolina*. Raleigh: North Carolina State Museum of Natural History, pp. 143–194.

Garren, R. A. 1982. Macrophyte species composition trophic-state relationships in fourteen north and north central Florida lakes. Thesis, University of Florida, Gainesville.

Gerking, S. D. 1963. Central states. In D. G. Frey (ed.), *Limnology in North America*. Madison: University of Wisconsin Press, pp. 239–268.

Gleason, P. J., and P. A. Stone. 1975. *Prehistoric Trophic Level Status and Possible Cultural Influences on the Enrichment of Lake Okeechobee*. West Palm Beach: South Florida Water Management District.

Gottgens, J. F. and T. L. Crisman. 1991. Newnan's Lake, Florida: removal of particulate organic matter and nutrients using a short-term partial drawdown. *Lake and Reservoir Management* 7:53–60.

Greear, P. F. 1967. Composition, diversity, and structure of the vegetation of some natural ponds in northwest Georgia. Thesis, University of Georgia, Athens.

Gregory, J. D., R. W. Skaggs, R. G. Broadhead, R. H. Culbreath, J. R. Bailey, and T. L. Foutz. 1984. *Hydrologic and Water Quality Impacts of Peat Mining in North Carolina*, Rep. UNC-WRRI-83-214. Chapel Hill: Water Resources Research Institute, University of North Carolina.

Guillory, V., M. D. Jones, and M. Rebel. 1979. A comparison of fish communities in vegetated and beach habitats. *Fla. Sci.* 42:113–122.

Guthrie, M. 1989. A floristic and vegetational overview of Reelfoot Lake. *J. Tenn. Acad. Sci.* 64:113–116.

Haller, W. T., J. V. Shireman, and D. F. DuRant. 1980. Fish harvest resulting from mechanical control of hydrilla. *Trans. Am. Fish. Soc.* 109:517–520.

Hendry, C. D., and P. L. Brezonik. 1984. Chemical composition of softwater Florida lakes and their sensitivity to acid precipitation. *Water Resour. Bull.* 20:75–86.

Henson, J. W. 1990a. Aquatic and certain wetland vascular vegetation of Reelfoot Lake, 1920's–1980's. I. Floristic survey. *J. Tenn. Acad. Sci.* 65:63–68.

Henson, J. W. 1990b. Aquatic and certain wetland vascular vegetation of Reelfoot Lake, 1920's–1980's. II. Persistent marshes and marsh–swamp transitions. *J. Tenn. Acad. Sci.* 65:69–74.

Hodgson, L. M., S. B. Linda, and D. E. Canfield, Jr. 1986. Periphytic algal growth in a hypereutrophic Florida lake following a winter decline in phytoplankton. *Fla. Sci.* 49:234–241.

Hoyer, M. V., D. E. Canfield, Jr., J. V. Shireman, and D. E. Colle. 1985. Relationship between abundance of largemouth bass and submerged vegetation in Texas reservoirs: a critique. *North Am. J. Fish. Manage.* 5:613–616.

Huber, W. C., P. L. Brezonik, J. P. Heaney, R. E. Dickinson, S. D. Preston, D. S. Dwornik, and M. A. DeMaio. 1982. *A Classification of Florida Lakes*, Publ. No. 72. Gainesville: Water Resources Research Center, University of Florida.

Hutchinson, G. E. 1957. *A Treatise on Limnology*, Vol. 1. *Geography, Physics, and Chemistry*. New York: Wiley.

Hutchinson, G. E. 1975. *A Treatise on Limnology*, Vol. III. *Limnological Botany*. New York: Wiley.

Hutchinson, G. E., and H. Loffler. 1956. The thermal classification of lakes. *Proc. Natl. Acad. Sci. U.S.A.* 42:84–86.

Hutchinson, G. E., and G. E. Pickford. 1932. Limnological observations on Mountain Lake, Virginia. *Int. Rev. Hydrobiol. Hydrogr.* 27:252–264.

Iovino, A. J., and W. H. Bradley. 1969. The role of larval Chironomidae in the production of lacustrine copropel in Mud lake, Marion County, Florida. *Limnol. Oceanogr.* 14:898–905.

Kautz, R. S. 1981. Effects of eutrophication on the fish communites of Florida lakes. *Proc. Annu. Conf. Southeast. Assoc. Fish Wildl. Agencies* 34:67–80.

Keller, A. E., and T. L. Crisman. 1990. Factors influencing fish assemblages and species richness in subtropical Florida lakes and a comparison with temperate lakes. *Can. J. Fish. Aquat. Sci.* 47:2137–2146.

Kling, G. W. 1986. The physicochemistry of some dune ponds on the Outer Banks, North Carolina. *Hydrobiologia* 134:3–10.

Kratzer, C. R., and P. L. Brezonik. 1981. A Carlson-type trophic state index for nitrogen in Florida lakes. *Water Resour. Bull.* 17:713–715.

Kushlan, J. A., and T. E. Lodge. 1974. Ecological and distributional notes on the freshwater fish of southern Florida. *Fla. Sci.* 37:110–128.

Lachner, E. A., C. R. Robins, and W. R. Courtenay, Jr. 1970. Exotic fishes and other aquatic organisms introduced into North America. *Smithson. Contrib. Zool.* 59:1–29.

Lamia, J. A. 1987. The limnological and biological characteristics of Cue Lake, an acidic lake in north Florida. Thesis, University of Florida, Gainesville.

Lee, D. S., C. R. Gilbert, C. H. Hocutt, R. E. Jenkins, D. E. McAllister, and J. R. Stauffer, Jr. 1980. *Atlas of North American Freshwater Fishes*. North Carolina State Museum, Raleigh, N.C.

Leslie, A. J., Jr., and G. J. Kobylinski. 1985. Benthic macroinvertebrate response to aquatic vegetation removal by grass carp on a North-Florida reservoir. *Fla. Sci.* 48:220–231.

Linthurst, R. A., D. H. Landers, J. M. Eilers, D. F. Brakke, W. S. Overton, E. P. Meier, and R. E. Crowe. 1986. *Characteristics of Lakes in the Eastern United States*, Vol. I. *Population Descriptions and Physico-Chemical Relationships*, EPA/600/4-86/007a. Washington, DC: U.S. Environ. Prot. Agency.

Loftus, W. F., and J. A. Kushlan. 1987. Freshwater fishes of southern Florida. *Bull. Fla. State Mus. Biol. Sci.* 31:147–344.

Maceina, M. J., and J. V. Shireman. 1982. Influence of dense hydrilla infestation on black crappie growth. *Proc. Annu. Conf. Southeast Assoc. Fish Wildl. Agencies* 36:394–402.

Mallin, M. A. 1978. Zooplankton population dynamics in the Oklawaha lake chain. Thesis, University of Florida, Gainesville.

Marland, F. C. 1967. The history of Mountain Lake, Giles County, Virginia: an interpretation based on paleolimnology. Thesis, Virginia Polytechnic Institute and State University, Blacksburg.

Marshall, H. G. 1979. Lake Drummond: with a discussion regarding its plankton composition. In P. W. Kirk (ed.), *The Great Dismal Swamp*. Charlottesville: University Press of Virginia, pp. 169–182.

Marshall, H. G., and W. W. Robinson. 1979. Notes on the overlying substrate and bottom contours of Lake Drummond. In P. W. Kirk (ed.), *The Great Dismal Swamp*. Charlottesville: Virginia UP, pp. 183–187.

Maslin, K. R. 1970. The interaction of littoral zooplankton and their fish predators. Thesis, University of Florida, Gainesville.

Maslin, P. E. 1969. Population dynamics and productivity of zooplankton in two sandhills lakes. Thesis, University of Florida, Gainesville.

Matta, J. F. 1979. Aquatic insects of the Dismal Swamp. In P. W. Kirk (ed.), *The Great Dismal Swamp*. Charlottesville: Virginia UP, pp. 200–221.

McCullough, J. D., T. E. Venneman, A. A. Hartung, and D. M. Wedding. 1985. Occurrence of the rotifer *Trochosphaera solstitialis* in Reelfoot Lake, Tennessee. *J. Tenn. Acad. Sci.* 60:20–22.

McDiffett, W. F. 1980. Limnological characteristics of several lakes on the Lake Wales Ridge, south-central Florida. *Hydrobiologia* 71:137–145.

McDiffett, W. F. 1981. Limnological characteristics of eutrophic lake Istokpoga, Florida. *Fla. Sci.* 44:172–181.

McHenry, J. R., C. M. Cooper, and J. C. Ritchie. 1982. Sedimentation in Wolf Lake, Lower Yazoo River, Mississippi. *J. Freshwater Ecol.* 1:547–558.

McHenry, J. R., F. R. Shiebe, J. C. Ritchie, and C. M. Cooper. 1984. Deposited sediments of Lake Chicot. In J. F. Nix and F. R. Shiebe (eds.), *Limnological Studies of Lake Chicot, Arkansas*. Arkadelphia, AR: Ouchita Baptist University, pp. 18–47.

McLane, W. Mc. 1955. The fishes of the St. Johns River system. Thesis, University of Florida, Gainesville.

Miller, H. D. 1982. *The Water and Sediment Quality of Lake Conway, Florida*, Rep. 2. Tech. Rep. A-78-2. Jacksonville, FL: U.S. Army Engineer District.

Moore, W. G. 1963. Central gulf states and the Mississippi embayment. In D. G. Frey (ed.), *Limnology in North America*. Madison: Wisconsin UP, pp. 287–300.

Nall, L. E., and J. D. Schardt. 1978. *The Aquatic Macrophytes of Lake Conway, Florida*, Tech. Rep. A-78-2. Vicksburg, MS: U.S. Army Engineer Waterways Experiment Station.

Nix, J. F., and F. R. Schiebe (eds.). 1984. *Limnological Studies of Lake Chicot, Arkansas*. Arkadelphia, AR: Ouachita Baptist University.

Nordlie, F. G. 1972. Thermal stratification and annual heat budget of a Florida sinkhole lake. *Hydrobiologia* 40:183–200.

Nordlie, F. G. 1976. Plankton communities of three central Florida lakes. *Hydrobiologia* 48:65–78.

Osborne, J. A., M. P. Wanielista, and Y. A. Yousef. 1976. Benthic fauna species diversity in six central Florida lakes in summer. *Hydrobiologia* 48:125–129.

Pardue, J., D. H. Kesler, J. Dabezies, and L. Prufert. 1986. Phosphorus and chlorophyll degradation product profiles in sediment from Reelfoot Lake, Tennessee. *J. Tenn. Acad. Sci.* 61:46–49.

Parker, G. G., and C. W. Cooke. 1944. Late Cenozoic geology of southern Florida, with a discussion of the ground water. *Fla., Geol. Surv., Geol. Bull.* 27:1–119.

Prouty, W. F. 1952. Carolina Bays and their origin. *Geol. Soc. Am. Bull.* 63:167–224.

Puri, H. S., and R. O. Vernon. 1959. *Summary of the Geology of Florida and a Guidebook to the Classic Exposures*, Special Publ. No. 5. Tallahassee: Florida Geological Survey.

Reid, G. K., Jr. 1952. Some considerations and problems in the ecology of floating islands. *Q. J. Fla. Acad. Sci.* 15:63–66.

Reid, G. K., Jr., and N. J. Blake. 1969. Diurnal zooplankton ecology in a phosphate pit lake. *Q. J. Fla. Acad. Sci.* 32:275–284.

Reid, G. K., Jr., and S. D. Squibb. 1971. Limnological cycles in a Phosphatic Limestone Mine Lake. *Q. J. Fla. Acad. Sci.* 34:17–47.

Romie, K. F. 1989. Trophic status and fish populations in five lakes of low pH in northern Florida. Thesis, University of Florida, Gainesville.

Ross, T. E. 1987. A comprehensive bibliography of the Carolina Bays literature. *J. Elisha Mitchell Sci. Soc.* 103:28–42.

Russell, S. A. 1982. Diel vertical migration in a subtropical doline Lake. Thesis, University of Florida, Gainesville.

Schalles, J. F. 1979. Comparative limnology and ecosystems analysis of Carolina Bay ponds on the upper coastal plain of South Carolina. Thesis, Emory University, Atlanta, GA.

Schardt, J. D. 1986. *1986 Florida Aquatic Plant Survey*. Tallahassee: Bureau of Aquatic Plant Management, Florida Dept. Nat. Resour.

Schramm, H. L., Jr., M. V. Hoyer, and K. J. Jirka. 1983. *Relative Ecological Value of Common Aquatic Plants*, Final Rep. Tallahassee: Bureau of Aquatic Plant Research and Control, Florida Dept. Nat. Resour.

Schramm, H. L., Jr., J. V. Shireman, D. E. Hammond, and D. M. Powell. 1985. Effect of commercial harvest of sport fish on the black crappie populations in Lake Okeechobee, Florida. *North Am. J. Fish. Manage.* 5:217–226.

Schulze, R. L. 1980. The biotic response to acid precipitation in Florida lakes. Thesis, University of Florida, Gainesville.

Scott, S. L., and J. A. Osborne. 1981. Benthic macroinvertebrates of a hydrilla infested central Florida lake. *J. Freshwater Ecol.* 1:41–49.

Scott, W. B., and E. J. Crossman. 1973. Freshwater fishes of Canada. *Bull. Fish. Res. Board Can.* 184.

Shafer, M. D., R. E. Dickinson, J. P. Heaney, and W. C. Huber. 1986. *Gazetteer of Florida Lakes*, Publ. No. 96. Gainesville: Water Resources Research Center, University of Florida.

Shafland, P. L. 1979. Non-native fish introductions with special reference to Florida. *Fisheries* 4:18–24.

Shafland, P. L. 1986. A review of Florida's efforts to regulate, assess and manage exotic fishes. *Fisheries* 11:20–25.

Shannon, E. E. 1970. Eutrophication–trophic state relationships in north and central Florida lakes. Thesis, University of Florida, Gainesville.

Shannon, E. E., and P. L. Brezonik. 1972. Limnological characteristics of north and central Florida lakes. *Limnol. Oceanogr.* 17:97–110.

Sharitz, R. R., and J. W. Gibbons. 1982. *The Ecology of Evergreen Shrub Bogs, Pocosins and Carolina Bays of the Southeast: A Community Profile*, FWS/OBS-82/04. Washington, DC: U.S. Fish Wildl. Serv.

Shireman, J. V., and M. J. Maceina. 1981. The utilization of grass carp, *Ctenopharyngodon idella* Val., for hydrilla control in Lake Baldwin, Florida. *J. Fish Biol.* 19:629–636.

Shireman, J. V., and R. G. Martin. 1978. Seasonal and diurnal zooplankton investigations of a south-central Florida lake. *Fla. Sci.* 41:193–201.

Shireman, J. V., D. E. Canfield, D. E. Colle, D. F. DuRant, and W. T. Haller. 1984a. *Evaluation of Biological, Chemical, and Mechanical Aquatic Vegetation Control upon Fish Populations in 0.2 ha Research Ponds*, Final Rep. USDA-ARS No. 58-7B30-0-177. Gainesville: University of Florida.

Shireman, J. V., M. V. Hoyer, M. J. Maceina, and D. E. Canfield, Jr. 1984b. The water quality and fishery of Lake Baldwin, Florida: 4 years after macrophyte removal by grass carp. *Proc. 4th Annu. Conf. North Am. Lake Manage. Soc.*, pp. 201–206.

Stager, J. C., and L. B. Cahoon. 1987. The age and trophic history of Lake Waccamaw, North Carolina. *J. Elisha Mitchell Sci. Soc.* 103:1–13.

Tilly, L. J. 1973. Comparative productivity of four Carolina lakes. *Am. Midl. Nat.* 90:356–365.

U. S. Environmental Protection Agency. 1978. *A Compendium of Lake and Reservoir Data Collected by the National Eutrophication Survey in Eastern, North-Central, and Southeastern United States*, Working Paper No. 475. Corvallis, OR: Corvallis Environ. Res. Lab., U.S. Environ. Prot. Agency.

Vernon, R. O. 1951. Geology of Citrus and Levy Counties, Florida. *Fla., Geol. Surv., Geol. Bull.* 33:1–256.

Vernon, R. O., and H. S. Puri. 1964. *Geologic Map of Florida. Map Series.* 3. Tallahassee: Florida Department of Natural Resources, Bureau of Geology.

Vodopich, D. S., and B. C. Cowell. 1984. Interaction of factors governing the distribution of a predatory aquatic insect. *Ecology* 65:39–54.

Watkins, C. E., II, J. V. Shireman, and W. T. Haller. 1983. The influence of aquatic vegetation upon zooplankton and benthic macroinvertebrates in Orange lake, Florida. *J. Aquat. Plant Manage.* 21:78–83.

Watts, W. A. 1969. A pollen diagram from Mud Lake, Marion County, north-central Florida. *Geol. Soc. Am. Bull.* 80:631–642.

Watts, W. A. 1970. The full-glacial vegetation of northwestern Goergia. *Ecology* 51:17–33.

Watts, W. A. 1975. A late Quaternary record of vegetation from Lake Annie, south-central Florida. *Geology* 3:344–346.

Watts, W. A. 1979. Late Quaternary vegetation of central Appalachia and the New Jersey coastal plain. *Ecol. Monogr.* 49:427–469.

Watts, W. A. 1980a. Late-Quaternary vegetation history at White Pond on the inner coastal plain of South Carolina. *Quat. Res. (N.Y.)* 13:187–199.

Watts, W. A. 1980b. The late Quaternary vegetation history of the southeastern United States. *Annu. Rev. Ecol. Syst.* 11:387–409.

Weiss, C. M., and E. J. Kuenzler. 1976. *The Trophic State of North Carolina Lakes*, Rep. No. 119. Raleigh: Water Resources Research Institute, North Carolina State University.

Werner, E. E., D. J. Hall, and M. D. Werner. 1978. Littoral zone fish communities of two Florida lakes and a comparison with Michigan lakes. *Environ. Biol. Fishes* 3:163–172.

Wetzel, R. G. 1983. *Limnology*. Philadelphia, PA: Saunders.

Wharton, C. H. 1978. *The Natural Environments of Georgia*. Atlanta: Georgia Department of Natural Resources.

White, W. A. 1970. *The Geomorphology of the Florida Peninsula*, Bull. No. 51. Tallahassee: Florida Bureau of Geology.

Whitehead, D. R. 1981. Late Pleistocene vegetational changes in northeastern North Carolina. *Ecol. Monogr.* 51:451–471.

Whitehead, D. R., and R. Q. Oaks, Jr. 1979. Developmental history of the Dismal Swamp. In P. W. Kirk, Jr. (ed.), *The Great Dismal Swamp*. Charlottesville: Virginia UP, pp. 25–43.

Williams, V. P., D. E. Canfield, Jr., M. M. Hale, W. E. Johnson, R. S. Kautz, J. T. Krummrich, F. H. Langford, K. Langland, S. P. McKinney, D. M. Powell, and P. L. Shafland. 1985. Lake habitat and fishery resources of Florida. In W. Seaman, Jr. (ed.), *Florida Aquatic Habitat and Fishery Resources*. Gainesville, FL: Am. Fish. Soc., pp. 43–120.

Wright, R. F., and E. Snekvik. 1978. Acid precipitation: chemistry and fish populations in 700 lakes in southernmost Norway. *Verh. — Int. Ver. Theor. Angew. Limnol.* 20:765–775.

Wyngaard, G. A., J. L. Elmore, and B. C. Cowell. 1982. Dynamics of a subtropical plankton community, with emphasis on the copepod *Mesocyclops edax*. *Hydrobiologia* 89:39–48.

Wyngaard, G. A., E. Russek, and J. D. Allen. 1985. Life history in north temperate and subtropical populations of *Mesocyclops edax* (Crustacea: Copepoda). *Verh. — Int. Ver. Theor. Angew. Limnol.* 22:3149–3153.

Young. S. N. 1979. Relationship between abundance of crustacean zooplankton and trophic state in fourteen central Florida lakes. Thesis, University of South Florida, Tampa.

Yount, J. L. 1961. A note on stability in central Florida lakes, with discussion of the effect of hurricanes. *Limnol. Oceanogr.* 6:322–325.

Yount, J. L. 1963. South Atlantic states. In D. G. Frey (ed.), *Limnology in North America*. Madison: Wisconsin UP, pp. 269–286.

PART IV
Tidal Systems

13 Low-Salinity Backbays and Lagoons

RICHARD H. MOORE

Department of Biology and Center for Marine and Wetland Studies,
University of South Carolina, Coastal Carolina College, Conway, SC
29526, and Belle W. Baruch Institute for Marine Biology and Coastal
Research, University of South Carolina.

PHYSICAL CHARACTERISTICS

Low-salinity backbays and lagoons are varieties of estuarine habitats, which
are characterized by very low (if any) salinity and which contrast with typical
estuaries (Fig. 1) in their hydrography and sedimentology as well as salinity
regime (Emery and Stevenson 1957). Their greater proximity to surrounding
terrestrial environments means that backbays are influenced to a greater
extent by events that occur in the neighboring terrestrial or upstream freshwater habitats than are more typical estuarine environments where oceanic
tides and currents play a dominant role in establishing hydrologic and sedimentary regimes (Odum 1980).

According to the classification of Cowardin et al. (1979), freshwater backbays and lagoons are habitats that are either riverine/tidal or palustrine/tidal
and are distinguished as lying between the upstream limit of tidal variation
and the downstream point where the concentration of ocean-derived salts
exceeds 0.5 ppt. Lagoons are typically shallow and because of their shallow-
ness, lagoon floors are subject to reworking by wind-driven waves, while
hydrologic and sedimentary forces in typical estuaries are usually tidally or
riverine generated. In actuality, the distinctions are not exclusive and there
are few "pure" estuaries or lagoons; most coastal systems are compound,
consisting of estuaries with adjoining lagoons and backbays. Such compound
systems are apparently the result of alternating periods of coastal submergence
and emergence, such as when offshore barrier islands are created through

Contribution No. 696 of the Belle W. Baruch Institute for Marine Biology and Coastal Research,
University of South Carolina.

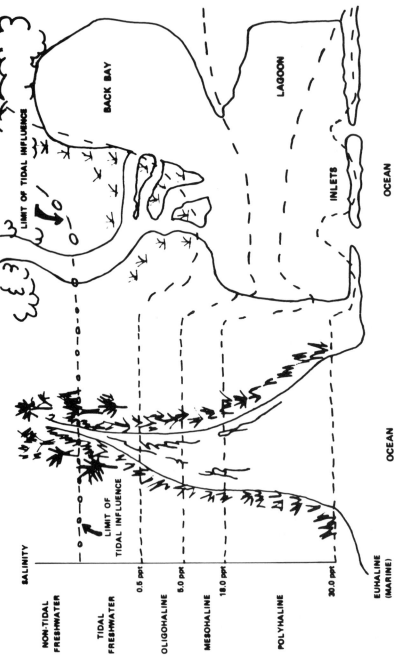

FIGURE 1. Comparison of physical features in a river-valley estuary with those of a lagoonal system.

TABLE 1 Major Freshwater and Oligohaline Estuarine Systems of the Southeastern United States

Feature	Estuarine System	State	Area (ha)	Discharge (m³/year)	References Physical	References Biotic
Trinity Bay	Galveston Bay	Texas	36 000	6×10^9	4, 8	2, 4, 8, 19, 25–27, 34
Lake Pontchartrain	Lake Pontchartrain	Louisiana	164 500	6×10^9	1, 35, 36, 38, 45	1, 12–15, 18, 32, 35, 39, 40, 44, 48
Mobile River Delta	Mobile Bay	Alabama	8 225	5×10^{10}	11, 27, 47	3, 5, 7, 11, 46
East Bay	Apalachicola Bay	Florida	3 981	2×10^{10}	29, 31	24, 29, 31
Indian River Lagoon	Indian River Lagoon	Florida	150 000	$<1 \times 10^9$	41–43, 50, 51	20, 23, 28, 33
Pamlico River Estuary	Pamlico River	North Carolina	30 500	5×10^9	10, 22, 29, 30, 49	6, 9, 10, 16, 21

References. 1, Bahr et al. 1980; 2, Bechtel and Copeland 1970; 3, Beshears 1959; 4, Blackwater 1977; 5, Boschung and Davis 1976; 7, Christmas 1973; 8, Copeland and Fruh 1970; 9, Copeland et al. 1974; 10, Copeland et al. 1984; 11, Crance 1971; 12, Darnell 1958; 13, Darnell 1961; 14, Darnell 1962; 15, Darnell 1979; 16, Davis and Brinson 1976; 17, Dean and Ballis 1975; 18, Deselle et al. 1978; 19, Diener 1975; 20, Eiseman & Benz 1975; 21, Epperly 1984; 22, Giese et al. 1979; 23, Gore at al. 1981; 24, Gorsline 1963; 25, Holland et al. 1973; 26, Holt and Strawn 1983; 27, Johnson 1974; 28, Kerschner et al. 1985; 29, Kuenzler et al. 1979; 30, Kulczycki et al. 1981; 31, Livingston 1984; 32, Meyers and Iverson 1981; 33, Montz 1978; 34, Mulligan and Snelson 1983; 35, Parker 1970; 36, Perret et al. 1971; 37, Rounsefell 1964; 38, Schroeder 1977; 39, Sikora and Kjerfve 1985; 40, Sikora and Sikora 1982; 41, Sikora et al. 1981; 42, Smith 1983a; 43, Smith, 1983b; 44, Smith 1986; 45, Suttkus et al. 1954; 46, Swenson 1981; 47, Swingle 1971; 48, Swingle and Bland 1974; 49, Turner et al. 1980; 50, Twilley et al. 1985; 51, Virnstein and Carbonara 1985; 52, Young and Young 1978.

deposition by longshore drift across the mouths of drowned river valleys (Collier and Hedgpeth 1950). Table 1 summarizes information on the major freshwater/oligohaline estuarine systems of the southeastern United States and Table 2 compares structural and functional features of low-salinity estuarine regions with higher salinity (mesohaline) estuaries.

Backbays   While a search of the literature revealed no formal definition, backbays may be satisfactorily defined by their common name. They are bays, that is, semi-enclosed or detached estuarine areas, that are "back" from the estuary's center or connection(s) with the sea. Backbays may represent re-

TABLE 2   Comparison of Tidal Freshwater and Mesohaline Ecosystems

System Characteristic	Tidal Freshwater Ecosystem	Mesohaline/ Estuarine Ecosystem
Location	Head of estuary	Mid and lower estuary
Salinity	Average below 0.5 ppt	Average above 8 ppt but below 35 ppt
Hydrology	Riverine influence, tidal influence may be present or minor	Strong (largely) tidal influence
Sediments	Silt–clay, high organic content, moderately reducing	Sandy, low organic content, strongly reducing (due to sulphur reduction)
Dissolved oxygen	Very low (summer)	Low (summer)
Vegetation	Freshwater species	Marine and estuarine species
Plant diversity	High	Low
Plant zonation	Present, but not always distinct	Pronounced
Seasonal succession of plant species	Pronounced	Absent or minor
Aboveground primary production	Very high (?) or comparable	High
Ratio of decomposition of intertidal plants	Extremely rapid	Moderate
Food quality of plant detritus	Very nutritious	Moderately nutritious
Primary consumers	Insects, oligochaetes and amphipods	Estuarine mollusks, crustaceans, and polychaetes
Fish community	Freshwater species, anadromous larvae and juveniles	Estuarine species, euryhaline marine juveniles
Waterfowl	High usage	Medium to low usage
Reptiles and amphibians	High diversity	Low diversity
Furbearers	High population densities	Medium to low population densities
Nutrient cycles	Spring uptake and high "leakage" in fall and winter	Even processing and slower release

Source. Modified from Odum and Smith (1981) and Odum et al. (1984).

juvenations of formal coastal lagoons resulting from the reemergence of an old coastline (Collier and Hedgpeth 1950) or they may form by sedimentary processes along the shorelines of large open-water estuarine bays. As befitting their lagoon-like origins, backbays resemble lagoons in their hydrology and sedimentary patterns.

In regions with moist climates, backbays can receive considerable freshwater input from rivers, bayous, swamps, or marshes and thus remain essentially freshwater habitats throughout most of the year. During periods of low rainfall, however, salinities in backbays may rise to marine or higher levels (Collier and Hedgpeth 1950).

Other backbays experience regular tidally induced salinity variations, with riverine conditions prevailing near low tide and higher salinities (usually in the oligohaline range of 0.5–5 ppt) near high tide. Trinity Bay, Texas (Copeland and Fruh 1970), Biloxi Back Bay, Mississippi, East Bay of the Apalachicola Bay system, Florida (Livingston et al. 1975), various tributaries of Albemarle and Pamlico sounds in North Carolina (Copeland et al. 1983, 1984), and the upper reaches of most east coast estuaries contain habitat that falls into this general category (Figs. 3–7).

Lagoons While the categorization of backbays may be hampered by the lack of a formal definition, coastal lagoons are even more difficult to delimit because of the many definitions in the literature, which are often contradictory or based on widely different criteria (Mee 1978, Nixon 1982). Various definitions of lagoons have been provided by Pritchard (1967), Lankford (1977), Barnes (1980), and R. A. Davis (1983), but perhaps Nixon (1982) adopted the wisest strategy: He avoided restrictive definitions and included any open body of water separated from the sea by a spit, bar, or island. Kjerfve (1986) has similarly proposed a broadly inclusive definition of a lagoon, which he suggests should replace "bar-built estuary" (Pritchard 1967) as one of the three major types of estuary in the world.

Along the South Atlantic and Gulf coasts of the United States there are many small freshwater to oligohaline embayments, as well as a few large open-water areas that fit the standard definitions of "lagoon." Emery and Stevenson (1957) define lagoons as "bodies of water, separated in most cases from the ocean by offshore bars or islands, of marine origin and [which] are usually parallel to the coastline." Included in their definition would be such diverse coastal systems as the Laguna Madre of Texas; East Lagoon on Galveston Island, Texas; Lake Pontchartrain, Louisiana; Mississippi Sound, Mississippi; Appalachicola Bay and the Indian River Lagoon, Florida; and most of North Carolina's extensive bay and sound complex lying behind the Outer Banks (Figs. 3–7).

Not all of these systems contain fresh or low-salinity regions. The Laguna Madre is typically hypersaline (Simmons 1957), while the Indian River Lagoon and the outer North Carolina sounds are poly- to euhaline. Despite these differences in salinity, all lagoons share certain similarities in their hydrography, sediments, nutrients, and biota that permit generalizations to be drawn about all these systems.

Hydrography

Excluding river mouths with deltas extending out into the sea, such as the Mississippi, Colorado, and Santee rivers, southeastern estuaries primarily are lagoonal embayments, characterized by their particular combinations of hydrography, meteorological forcing, tidal effects, freshwater inflow, and density currents (Ward 1980).

In microtidal lagoons circulation patterns are often wind driven. In the absence of tidal currents or regular strong riverine inflows, wind-driven currents promote horizontal homogeneity throughout the lagoon's area and provide a powerful mixing force that prevents stratification and so ensures greater vertical homogeneity. Normal patterns of circulation are established and maintained by normal wind, offshore winds of greater than normal strength can quickly increase the lagoon's volume and area through the addition of water from offshore, and a shore breeze can just as quickly drive waters seaward, exposing considerable stretches of bottom along the lagoon's landward margin (Copeland et al. 1968), a phenomenon often referred to as "tilting" (Ward 1980).

Sediments

Shepard and Moore (1960) recognized four estuarine sedimentary facies: (1) deep central bays with muddy sediments; (2) river mouths, like central bays but with more terrigenous sand; (3) oyster reefs; and (4) channels and narrows, where tidal scouring leaves predominately sand. All four facies may be found in low salinity backbays and lagoons; however, oyster (*Crassostrea virginiana*) shells are replaced by the valves of species more tolerant of low salinities, such as the brackish-water clam (*Rangia cuneata*). The fine-grained sediments are an important reservoir for nutrients, acting as both sinks and sources, depending on water chemistry (Pritchard and Schubel 1981). Seasonal variations in river discharge, which occur in all southeastern estuaries, affect sedimentary patterns. Materials moving seaward with freshwater discharge are deposited where the effects of tidal currents are first encountered (Biggs and Cronin 1981) and are redistributed by tidal or wind currents during periods of low discharge (Dyer 1986). The final depth of a lagoon represents an equilibrium of opposing forces of deposition of river-derived sediments and the removal of sediments by wind, wave, and current action (Price 1947).

Organisms can also contribute significantly to sediments and sedimentation through their skeletal debris (plankton tests and shelled benthos) and feeding/excretory activities producing mud/fecal pellets (R. A. Davis 1983). Rapid sedimentation as a result of the low velocity of freshwater discharges and low tidal energy characteristic of most lagoons means low water turbidity and high transparency and, consequently, greater development of photosynthetic bottom communities in contrast to plankton dominated ecosystems found in more turbid, more turbulent estuaries with greater freshwater input and/or tidal amplitudes. This benthic vegetation can also affect sedimentation. Montgomery et al. (1983) found sediments in areas of the Indian River Lagoon vegetated with *Halodule wrightii* to contain only 56% particles in their smallest size fraction (250 μm), while sediments from adjacent unvegetated areas consisted of up to 71% particles in this size fraction. In addition, seagrass vegetated areas have up to three times the population of benthic invertebrates (Virnstein et al. 1983), which have been shown to play an active role in restructuring the sediment (Aller 1978).

Nutrients

The water chemistry of freshwater lagoons and backbays is dominated by infusions of freshwater. Like other marine and estuarine systems the major dissolved inorganics found in these low-salinity systems are sodium, potassium, calcium, chloride, magnesium, and sulfates; biologically active compounds include nitrates, ammonium, phosphates, and silicates (Carpenter et al. 1969).

Nutrients transported into the lagoon from terrestrial or marine sources may remain in the water column or become incorporated into sediments (Bowden 1984). The high bottom area to water volume ratio found in most shallow lagoons means that benthic-pelagic coupling must play a very important role in nutrient dynamics and benthic community metabolism, including heterotrophic remineralization activity, the production of organic matter, and associated nutrient uptake by benthic autotrophs (Nowicki and Nixon 1985). Incorporation of nutrients into sediments may be enhanced by two processes: adsorption of nutrients by suspended particles, which may flocculate and settle rapidly in low energy environments (Kramer et al. 1972), and agglomeration of inorganic particles and phytoplankton by filter-feeding zooplankton (Pritchard and Schubel 1981).

Salinity

Fresh water is essential to estuaries. A standard definition of an estuary includes the requirement that seawater be "measurably diluted with freshwater derived from land drainage" (Pritchard 1967). However, fresh water also assumes a functional importance by providing a means of transport for nutrients and organic materials into the estuary (Copeland 1966, Duke and Rice 1967) as well as playing an important role in regulating chemical reactions due to the dramatic changes in thermodynamic equilibria of many dissolved materials that occur at salinities near 1 ppt (Morris et al. 1978) where the flocculation and settlement of river-borne colloidal sediment particles is promoted by the higher electrolyte content of brackish waters. Freshwater input into estuaries is also recognized for its biological importance. Gunter et al. (1964) correlated commercial shrimp landings (an indication of production) with rainfall and riverine input in states bordering the Western Gulf of Mexico. Freshwater input to estuaries is also important because it is believed to exclude parasites, disease organisms, and many predators (Chapman 1959, Galtsoff

1964, Gunter et al. 1973, Hackney 1978, Gauthier et al. 1990) from the estuarine environment.

Freshwater input into estuaries therefore requires the existence of various fresh to very-low-salinity habitats bordering the estuary and forming a buffer between the typical estuarine habitats and the adjacent upstream or terrestrial habitats. These freshwater habitats include open water areas known as back-bays or lagoons as well as fringing tidal marshes, swamps, and inflowing systems.

Along coastlines characterized by low diurnal tidal amplitude, such as the Gulf coast of the southeastern United States, freshwater marshes extend far inland, replacing the salt marsh of the Eastern seaboard as the most extensive fringing wetland. Along the northern Gulf coast freshwater tidal marshes found at the mouths of rivers and bordering the upper estuaries merge with extensive floodplain hardwood swamps or surround open waters of coastal lakes. These marshes and swamps drain via bayous into backbays such as those that occur along the Mississippi and northwest Florida (panhandle) coasts. Similar floodplain swamps and freshwater marshes are also found along the major river systems of the south Atlantic coast where blackwater streams and ponds may be tidally influenced far inland and share a great deal in common with more coastal habitats (Doumlele et al. 1985). In southern Florida mangals or mangrove swamps replace the marsh/swamp complexes of more temperate latitudes (Reimold 1977). Like their temperate counterparts, mangals are contiguous with open water backbays, as in the Ten Thousand Islands area (Caloosahatchee River/Big Cypress watershed) along the south-west Florida coast or with lagoons, as in the Indian River Lagoon of Florida. On the more arid Texas coast extensive freshwater and brackish marshes grade into upland coastal grassland prairies.

Barnes (1980) notes that coastal lakes are formed from lagoons that per-manently loose their connections to the sea; however, coastal lakes may retain their marine connections provided that sufficient quantities of freshwater are available year round, as demonstrated by Lake Pontchartrain in Louisiana.

Although freshwater backbays and lagoons may be treated as distinct hab-itats, they are generally situated along a continuum of habitats intergrading without clear distinction from freshwater riparian to mesohaline estuarine regimes. Not only will the salinity decline as one moves away from the sources of salt water, but there will also be marked changes in the plant and animal communities between those found in pure fresh water and those typical of brackish/estuarine habitats. Freshwater backbays and lagoons are typically inhabited by communities of organisms drawn from surrounding biotopes rather than possessing a unique definable community of their own. Diversity will typically be lower (Fig. 2) than in either pure freshwater or marine habitats, because the brackish water community will consist only of those species that tolerate intermediate salinities (Remane and Schlieper 1971, Hedgpeth 1983, Barnes 1980). These habitats fall geographically between the "classic" freshwater and marine (or estuarine) environments. Perhaps because of their isolation or because they also fall between or overlap the limits of

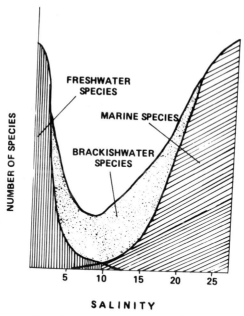

FIGURE 2. Salinity vs. diversity in estuarine ecosystems. (Modified from Remane and Schlieper (1971).)

authority of major funding agencies they also fall outside of the habitats traditionally studied by freshwater and marine ecologists and consequently they are among our least well-studied coastal environments.

PLANT COMMUNITIES

Plant communities associated with backbays and lagoons may be divided into three principal types: planktonic, submerged benthic, and fringing emergent marshes. In all cases, community structure and function may vary from system to system, depending on physicochemical variables such as salinity, tidal range and duration, exposure, substrate elevation, temperature, and latitude (photoperiod, growing season).

Phytoplankton

Although commonly perceived as the base for most aquatic food chains, the importance of phytoplankton varies tremendously in different estuarine systems. Estuaries may not develop autochthonous phytoplankton populations unless water residence times are greater than a few weeks. If flushing rates are higher, as in most river-dominated systems, the phytoplankton consist of a mixture of allochthonous riverine and oceanic species that can survive, but that exhibit lower productivity rates in low-salinity habitats (Kennish 1986). In shallow lagoons the high surface to volume ratio will favor benthic vegetation over plankton if water clarity permits.

Most studies of phytoplankton in fresh and low-salinity estuarine open waters and marshes indicate the predominance of diatoms and green, and blue-green algae, roughly in that order and including temporarily suspended benthic forms (Barnes 1980, Gilmartin and Revelante 1978, Livingston 1984, Odum et al. 1984). However, flagellated algae, especially the dinoflagellates and chlorophytes may be quite abundant or even dominant at certain seasons or in certain systems (Copeland et al. 1974, Odum et al. 1984).

In contrast to the dominance of diatoms in most southeastern freshwater tidal marshes, dinoflagellates of the genus *Peridinium* were the dominant phytoplankton group in the James River of Virginia (Odum et al. 1984). Dinoflagellates (primarily *Heterocapsa* = *Peridinium*) were also the most abundant phytoplankters in the oligohaline upper Pamlico River estuary (Copeland et al. 1984), while downstream, diatoms "characteristic of more mesohaline east coast estuaries" (Kuentzler et al. 1979) were more abundant. Not all North Carolina estuaries resemble the Pamlico Sound. Despite the numerical dominance of dinoflagellates, diatoms contributed nearly half the total number of species in Pamlico Sound tributaries (Stanley and Daniel 1985). Carpenter (1971) also reported that diatoms were the most diverse group of phytoplankters in the Cape Fear River estuary, comprising 134 of the 203 total species, while dinoflagellates were represented by only 15 species. Goldstein and Manzi (1976), Jacobs (1971), and Manzi and Zingmark (1978) reported that diatom species were second in abundance only to chlorophytes in South Carolina coastal freshwaters.

In backbay and lagoonal ecosystems of the Gulf of Mexico, diatoms also appear to constitute the predominant phytoplanktonic type in freshwater to moderate salinity areas. In Wiggins Bay and in adjacent Chokoloskee Bay (Caloosahatchee System) diatoms were predominant phytoplankters, comprising 68–70% of the cell numbers. In both bays dinoflagellates were very minor components (Drew and Schomer 1984). Frazer and Wilcox (1981) reported diatoms as the major phytoplankton group in subtropical Charlotte Harbor, where dinoflagellates and golden-brown algae were occasionally dominant at isolated stations.

Estabrook (1973, cited in Livingston 1984) states that diatoms were dominant in phytoplankton taken from the Apalachicola estuary throughout the year. In oligohaline East Bay, *Melosira granulata* was the dominant species versus *Chaetoceros lorenzianus* in the outer bay system. Gosselink (1984) states that benthic and epibenthic diatoms are important producers in Louisiana tidal marshes. They are "bite-sized" and hence important to consumers and are probably "metabolically more important than previously considered" (Gosselink 1984). In oligohaline Trinity Bay, diatom species of *Leptocylindricus*, *Nitzschia seriata*, *Skeletonema costatus* and *Thalassionema nitzchoides* and blue-green algae alternated as co-dominants (Copeland and Fruh 1970).

Benthic and Epiphytic Algae

Diatoms and blue-greens form a nearly continuous cover on the salt marsh surface, with lower salinity marshes having a greater diversity than higher salinity marshes (Sullivan 1978). Blue-green algae such as species of *Anacystis*, *Chromulina*, and *Schizothrix*, green algae such as *Clostreium* and *Cosmarium*, and diatoms such as the genera *Coscinodiscus*, *Cyclotella*, *Navicula*, and *Stephanodiscus* are also the predominate forms of benthic and epiphytic algae found in low-salinity and freshwater estuarine habitats on both the Atlantic and Gulf coasts. The marsh microflora of Atlantic and Gulf coast estuaries has been documented by Kurz and Wagner (1957), Maples (1982), Maples and Watson (1979), Sage and Sullivan (1978) Sullivan (1975, 1977, 1978, 1981, 1982a,b), and Otte and Bellis (1985). One hundred of the 111 diatom taxa reported by Sullivan (1978) from Mississippi also occur in Delaware (Sullivan 1975) and New Jersey (Sullivan 1977). Otte and Bellis (1985) however, noted that the North Carolina edaphic diatom flora differed appreciably from those of the mid-Atlantic and Gulf coasts. Many of these same species are also reported from open waters, suggesting that suspension of benthic and epiphytic diatoms plays an important role in maintaining the planktonic diatom community structure.

In addition to benthic unicellular algae, multicellular filamentous green algae such as *Enteromorpha* and *Compsopogon* are especially common in the cooler months (Stowe 1972, Davis and Brinson 1976). These large algae have a functional significance in the plant community similar to that of the other large submerged plants: the seagrasses and submerged macrophytes.

Seagrasses and Submerged Macrophytes

The shallow, calm waters of backbays and lagoons afford ideal habitat for seagrasses and other aquatic spermatophytes. As the name implies, most seagrasses typically occur at higher salinities and so are absent from freshwater estuarine systems (Phillips 1960). Freshwater marsh plants such as *Najas guadalupensis* and *Ceratophyllum demersum* that can tolerate, at least for short periods, salinities up to about 6–10 ppt (Haller et al. 1974) and other aquatic macrophytes such as wild celery or tapegrass (*Vallisneria americana*), sago pondweed (*Potamogeton perfoliatus*), the exotic Eurasian waterfoil (*Myriophyllum spicatum*), water hyacinth (*Eichornia crassipes*), and widgeongrass (*Ruppia maritima*) as well as other species such as *Zannichellia palustris*, *Najas marina*, *Heteranthera dubia*, and *Nitella* sp are characteristic of low-salinity estuarine areas (Bourne 1934) from the Chesapeake Bay region (Anderson 1972), North Carolina (Davis and Brinson 1976), Florida (Phillips 1960, Drew and Schomer 1984), Alabama (Swingle 1971), Mississippi (Christmas 1973), Louisiana (Penfound and Hathaway 1938) to Texas (Diener 1975). Although considered by Humm (1973) as a freshwater species, *Ruppia mar-*

itima commonly occurs at salinities between 10 and 25 ppt (Stevenson and Confer 1978), especially when true seagrasses are absent (Barnes 1980). Davis et al. (1985b) reported that *Ruppia maritima* replaces *Zannichellia palustris* as the most abundant macrophyte in Pamlico Sound tributary creeks during the summer.

Marsh Vegetation

Marshes found in association with freshwater backbays and lagoons contrast sharply with salt marshes dominated by *Spartina alterniflora* that surround higher salinity estuarine systems. Species diversity is much greater (Table 3) and species richness increases with decreasing salinities from salt, to brackish, to intermediate, to freshwater marshes. According to Chabreck (1972), the plant species richness in Louisiana marshes was salt marsh, 17 species; brackish marsh, 40; intermediate marsh, 54; and freshwater marsh, 93. Although different geographic regions differ appreciably in species composition, similar data from east coast marshes show strikingly similar species richnesses: 14, 45, 55, and 72 species, respectively (Anderson et al. 1968). Furthermore, the differences are not as great as might appear at first because Gulf coast species are often replaced by ecologically similar cogeners in Atlantic coast marshes. East coast freshwater marshes include a mixed community of aquatic annuals and perennials. Pickerelweed (*Pontederia cordata*) and arrow arum (*Peltandra virginica*) form predominant elements of the flora of southeast Atlantic marshes, while in the Gulf of Mexico arrowheads (*Sagittaria* spp.) replace these two species as the dominant emergent species and maiden cane (*Panicum hemitomom*) and sedges (*Carex* spp. and *Cyperus* spp.) predominate at higher elevations (Mitsch and Gosselink 1986). Other dominant species in Gulf of Mexico fresh and brackish marshes include *Panicum hemitomon*, *Cladium jamaciense*, *Sagittaria falcata*, *Eleocharis* sp., *Phragmites communis*, and *Alternanthera philoxeroides* (Swingle 1971, Chabreck 1972) and the same species occupy similarly dominant positions in east coast tidal freshwater marshes (Dahlberg 1972, Sandifer et al. 1980). While freshwater species are quickly excluded by increased salinities from brackish or saline marshes, the opposite is not nearly as true. Typical brackish marsh species such as *Spartina patens*, *Spartina cynosuroides*, and especially *Juncus roemerianus* may frequently be found mixed with more typical freshwater vegetation in very low and intermediate salinity marshes (Philipp and Brown 1965). In fact, most salt marsh species exhibit reduced growth and fertility with increasing salinity (Adams 1963, Phleger 1971), making the lower salinity habitats more favorable for these species in the absence of competition from freshwater species. Along the Gulf coast, extensive low-salinity (but not fresh water) marshes surrounding river mouths, bays, and lagoons account for a majority of the United States's *Juncus* marshes (Eleuterius 1976, de la Cruz 1981). At the upper limits of tidal variation, these brackish marshes grade into *Cladium*-dominated freshwater marshes and then into hardwood forest swamps (Eleuterius and Eleuterius 1979).

ANIMAL COMMUNITIES

The animal communities of freshwater backbays and lagoons, like the plant communities, are composed of a mixture of species drawn from the surrounding fresh water and brackish water habitats, which exhibit the ability to tolerate the salinity variations that typify many of these habitats. Hedgpeth (in Emery and Stevenson 1957) noted that euryhalinity is a phylogenetic characteristic. Most marine and freshwater organisms are stenohaline, while true euryhaline species are rare and often occur in genera or families including other euryhaline species. Euryhaline species that exist in both pure fresh and marine waters were classified by Myers (1938) as diadromous species if this salinity transition is an obligate part of their life cycle. Among diadromous species, catadromous species spawn in salt water, but spend an often lengthy, mandatory residence period in fresh water. Anadromous species have the reverse life cycle: they spawn in freshwater, but grow and mature in marine or estuarine waters. The American eel (*Anguilla rostrata*) (Helfman et al. 1984) is an example of a catadromous species while shad (*Alosa* spp.) (Adams 1970), sturgeon (*Acipenser* spp.) (Van Den Avyle 1984), and striped bass (*Morone saxatilis*) are common east coast examples of anadromous fishes. Some organisms, such as striped mullet (*Mugil cephalus*) (Thompson 1966, Moore 1974) and blue crab (*Callinectes sapidus*) (Tagatz 1971) appear to be able to move freely between fresh and marine waters without the massive, and often irreversible, physiological changes that characterize the migrations of diadromous species such as eels and salmon. Even species regarded as primary freshwater fish can usually tolerate salinities in the 12- to 15-ppt range for a few hours to days (Renfro 1959, Hoese 1963, Schwartz 1964).

Invertebrates

Zooplankton   The zooplankton community is typically dominated by the copepods *Acartia tonsa* and *Eurytemora affinis*, with other estuarine species, such as *Pseudodiaptomus coronatus*, *Paracalanus crassirostris*, and *Oithona* spp., all typically classified as brackish water forms (Cuzon du Rest 1963), occurring in much lower densities in the lowest salinity areas than in contiguous brackish water habitats (Darnell 1961, Stickney and Knowles 1975, Show 1980). *Acartia tonsa* is the dominant estuarine copepod from the Chesapeake Bay (Davis 1944) to the Laguna Madre of Texas (Simmons 1975), wherever considerable land drainage occurs. Darnell (1961) found that the abundance of this species in Lake Pontchartrain was greatest in nearshore regions characterized by moving waters. Cuzon du Rest (1963) reported that *A. tonsa* dominated zooplankton populations of Louisiana brackish and salt marshes in all seasons and salinities. Despite its overwhelming abundance, Conover (1956) states that *A. tonsa* appears to be important only when salinity restricts the occurrence of other species of copepods.

Eurytemora affinis (or *E. hirundoides*), the other common estuarine copepod, appears to tolerate ranges of salinity and temperature at least as great as *A. tonsa* (Deevey 1960) and has a similar geographically widespread and ecologically ubiquitous estuarine distribution (Cuzon du Rest 1963). The life history and energy requirements of this species have been well documented in Chesapeake Bay waters (Heinle and Flemmer 1975) and might be expected to be similar in other geographic locations. However, appreciable anatomical and physiological differences have been demonstrated in geographically distinct populations of this as well as other common estuarine "species" such as the isopod *Cyathura polita* (Miller and Burbank 1961, Frankenberg and Burbank 1963), the polychaete *Capitella capitata* (Grassle and Grassle 1976) and the mummichog *Fundulus heteroclitus* (Fritz and Garside 1975, Morin and Able 1984, Able and Felley 1986), to the degree that it has been suggested that the populations may represent different species.

Euryhaline and brackish-water mysids such as *Neomysis americana* and *Mysidopsis bigelowi* are typical members of the zooplankton community on the east coast (Odum et al. 1984). In south Florida and along the Gulf of Mexico, *Mysidopsis almyra*, *M. bahia*, and *Taphromysis bowmani* are the dominant estuarine species, commonly occurring in salinities between 0 and 15 ppt (Bascescu 1961, Odum and Heald 1972, Beck 1977, Stuck et al. 1979), and *T. louisianae* is abundant in freshwater and up to about 2 ppt (Stuck et al. 1979, Conte 1972).

Freshwater zooplankters such as rotifers, ostracods, cladocerans, and cyclopoid copepods may also be common, especially during periods of higher river flow when these typically freshwater forms will be washed into estuarine habitats (Cuzon du Rest 1963, Bahr et al. 1983, Livingston 1984, Copeland et al. 1984).

Macrocrustaceans   Larger crustaceans form a highly visible and important part of the biota of freshwater backbays and lagoons, supporting a number of valuable commercial and recreational fisheries as well as filling vital ecological roles in the ecosystem. Because many of the shrimps and crabs have well-developed osmoregulatory abilities, the low and often variable salinities that occur in these habitats do not pose such a stress as they do for smaller organisms.

Penaeid shrimp are among the most common larger organisms in these habitats along the Gulf of Mexico. The white shrimp (*Penaeus setiferus*), common in freshwater habitats along the Gulf of Mexico and in many parts of the southern United States, is known as the river or lake shrimp in deference to its habitat (Spaulding 1908) (this common name is also applied to various species of the genus *Macrobrachium*, especially *M. ohione*; Huner 1978). Although occurring in low-salinity environments on the Atlantic coast as well, it apparently does not penetrate fresh waters there as readily as on the Gulf coast (Odum et al. 1984). Both brown shrimp (*P. aztecus*) and pink shrimp (*P. duorarum*) also occur in low-salinity habitats, but neither appears to be as common as the white shrimp in these habitats or to occur in as low as a salinity (Hackney and de la Cruz 1981).

Perhaps the most abundant macrocrustaceans are the grass shrimps (genus *Palaemonetes*), which occur from nontidal fresh waters to full-strength seawater. Five species of *Palaemonetes* are known from fresh and brackish waters of the southeastern United States (Strenth 1976). *Palaemonetes paludosus* is restricted to freshwaters of rivers in the Atlantic and Gulf coastal plains, where it occurs in tidal freshwater marshes adjacent to the mouths of rivers (Meehean 1956, Turner et al. 1975). *Palaemonetes kadiakensis* is another freshwater species found in Gulf coastal drainages (McGuire 1961, Heard 1982). The most abundant grass shrimp in low-salinity estuarine areas is *Palaemonetes pugio* (Wood 1967, Knowlton and Williams 1970, Welsh 1975, Sikora 1977). *P. pugio* may be replaced in Florida estuaries at low salinities by *P. intermedius*, which may increase rapidly in abundance during periods of unusually low salinity in the normally higher salinity habitats (Dugan and Livingston 1982). The fifth species, *P. vulgaris*, does not usually occur in waters below 10 ppt (Knowlton and Williams 1970); 3 ppt appears to represent the natural limit of salinity tolerance (Thorp and Hoss 1975), although it can survive in much lower salinities in the laboratory (McFarland and Pickens 1965). *Macrobrachium ohione* and *M. acanthurus* are two other palaemonid shrimps that occur in freshwater and low-salinity estuarine regions of the southeastern United States (Hedgpeth 1949, Hobbs 1952, Reimer et al. 1974, Huner 1978).

Freshwater crawfish (*Procambarus* spp.) have been reported in most surveys of low-salinity estuarine areas (Livingston 1984, Rozas and Hackney 1984); however, these have rarely been identified to species. Huner (1981) notes that the red swamp crawfish (*Procambarus clarkii*) lives and reproduces in estuarine areas with salinities ranging up to 10 ppt. This species is native to fresh waters draining into the Gulf of Mexico west of Mobile Bay, but has been widely introduced in North America and elsewhere in the world as an aquacultural product. Salinity tolerances of other North America crawfishes are generally unknown.

One of the largest organisms in individual size as well as biomass that occurs in southern marshes and low-salinity estuaries is the blue crab (*Callinectes sapidus*). The life history of this important commercial species is well understood and includes considerable utilization of low-salinity or even freshwater habitat (More 1969, Tagatz and Hall 1971). Fully grown blue crabs (especially males) are not uncommon far upstream in coastal rivers. However, it is the juveniles that characteristically occur at the lowest salinities (Darnell 1959, Tagatz 1968). As with the white shrimp, blue crabs appear to enter fresh water more readily along the Gulf than on the Atlantic coast (Williams 1966).

Fiddler crabs constitute a prominent part of the intertidal biota of marshes. Teal (1958), Miller (1965), Kerwin (1971), and Aspey (1978) discuss the distribution of temperate Atlantic fiddler crabs and show soil moisture and salinity to be important features effecting their distribution. Powers and Cole (1976) found that density of cover, ambient temperature, and substrate type distinguished fiddler crab habitats in a nontidal environment.

For whatever reasons, the red-jointed fiddler (*Uca minax*), the largest North American species, is the common fiddler crab in fresh water/low-salinity marshes of the southeastern United States, occurring in nontidal as well as tidal habitats from Texas to Massachusetts (Dahlberg 1972, Montague 1980). The species is also more able to tolerate desiccation than the other Gulf species of *Uca* (Heard 1982) and is frequently found well away from any water. Ringold (1979) found that *U. minax* burrows only in soil with thick root mats and stated that its burrows may be preempted by other members of the genus that do not (or cannot?) dig their own burrows through such root mats.

Although neither of the other Atlantic coast species, *U. pugnax* or *U. pugilator*, typically occur in freshwater habitats, the low-salinity marshes of the Gulf coast also provide suitable habitat for *U. spinicarpa* (*U. speciosa spinicarpa* according to Crane 1975). Other crabs that are common in low-salinity areas include the wharf crab (*Sesarma cinereum*), which shares the terrestrial habits of *U. minax*, and the mud crab (*Rithropanopeus harrissi*), which inhabits muddy, shelly, and grassy habitats from tidal freshwaters to 20 ppt (Heard 1982, Williams 1966, May 1974).

Benthos Because most benthic organisms are relatively immobile, benthic communities exhibit greater (and more consistent) differences along natural salinity gradients. Boesch et al. (1976) observed low benthic diversity in the oligohaline reaches of the James River, Virginia, which gradually increased both up and downstream, following the classical pattern described by Remane and Schlieper (1971). However, as noted by Ellison and Nichols (1976), the transition from fresh to brackish water may be as sharply marked as the transition in dominant microbenthic organisms from freshwater shelled amoebae (Thecoamoebinids) to oligohaline foraminiferans.

Oligochaetes, which are not generally considered a component of the typical estuarine fauna because of their freshwater affinities and origins (Shirley and Loden 1982), are far more common in low-salinity habitats than in more typical estuarine salinities (Heard 1982, Diaz 1980). The oligochaete *Limnodriloides* is one of the most common and widespread benthic organisms in many east coast freshwater tidal marshes (Odum et al. 1984). Insect larvae, especially chironomids like *Chironomus tanypus*, are also an important part of the freshwater and low-salinity benthic community on both the Gulf and Atlantic coasts (Harrel et al. 1976, Odum et al. 1984).

The occurrences of insect larvae and oligochaetes may be characteristic of freshwater estuarine benthos, but by no means is the community limited to these organisms. While there may be differences in community composition and species abundance between the Atlantic and Gulf coasts, the majority of reports indicate a basically similar benthic community in freshwater and brackish water areas from the Chesapeake Bay to Texas, including such widely distributed forms such as *Cyathura polita, Laeoneris culveri, Streblospio benedicti*, and *Gammarus fasciatus* on the Atlantic coast and several closely allied species from the Gulf coast (Thomas 1976) and *Capitella capitata*. McBee and Brehm (1979, 1982) showed that the benthic community in St. Louis Bay, Mississippi, contained several species that were also common in Chesapeake Bay. The faunal similarity with Tampa Bay, Florida, was much greater, but greatest similarity was with Trinity Bay, Texas.

One of the dominant macrobenthic organisms is the brackish water clam (*Rangia cuneata*), which is common in low-salinity environments along the South Atlantic and Gulf coasts (Hoese 1973). Odum (1967) noted that densities of this clam could be as great as 12 individuals/ft^2 (or approximately 128/m^2). *Rangia cuneata* is capable of tolerating salinities between 0 and 25 ppt (Castagna and Chanley 1973) and can survive at 0.3 ppt for at least 7 months (Hopkins and Andrews 1970). According to Bedford and Anderson (1972), members of the species are "unique among marine/estuarine bivalves in their ability to osmoregulate," although Fingerman and Fairbanks (1956) demonstrated limited osmoregulatory capacity in the oyster (*Crassostrea virginica*). Otto and Pierce (1981) showed that the osmoregulatory abilities of *Rangia* are limited to salinities below 10 ppt and that at higher salinities it is an osmoconformer, thus combining physiological responses that are normally limited to either freshwater or marine bivales. In this sense it may be unique. Floods and other higher discharge events may wash *Rangia* from fresh upstream regions into freshened downstream areas where they may survive until normal river discharges and normal estuarine salinity regimes are reestablished (Boesch et al. 1976). Tenore et al. (1968) demonstrated that *Rangia* derive much of their nutrition from organic matter brought into the system by river input. Another common brackish water bivalve is *Polymesoda caroliniana* (Bosc), which is abundant in brackish marshes near mouths of rivers from Virginia to Texas (Duobinis-Gray and Hackney 1982). *P. caroliniana* appears to tolerate exposure better than *Rangia*, which is more abundant in subtidal as well as very low-salinity habitats. A third bivalve of increasing abundance is the introduced Asiatic clam (*Corbicula fluminea*, often reported as *C. maniliensis*) (Sandifer et al. 1980, Diaz and Boesch 1977).

Among the few studies that have specifically concentrated on benthic populations in low-salinity marshes, Bishop and Hackney (1987) showed significant differences in benthic mollusk populations, not only between salt and oligohaline marshes, but also between different types of brackish water marshes. They found that, in comparison to salt marshes, brackish water marshes (1) have fewer species, (2) have unique species combinations, (3) exhibit different patterns of seasonal abundance (greatest biomass in summer), and (4) exhibit potentially lower productivities. Further, in the Mississippi habitats that they investigated, frequently flooded *Juncus* marsh communities were dominated by the bivalves *Polymesoda caroliniana, Cyrenoida floridana*, and *Geukensia demissa granosissima*, while higher, better drained *Spartina cynosuroides* marshes primarily contained gastropods, such as *Detracia floridana, Littoridinops palustris*, *Onobops jacksoni*, *Melampus bidentatus*, *Succinea ovalis*, *Vertigo ovata*, and *Deroceras laeve*. The first three species are typical low-salinity forms; *Melampus*, a euryhaline estuarine species, and the last three species are primarily terrestrial (*Deroceras* is a slug), demonstrating the greater upland influence in the higher marsh as well as the unique species assemblages that typify each type of oligohaline marsh.

Insects   A major biotic difference between fresh and saline estuarine marshes is seen in the extent and composition of their insect communities. Estuarine and marine aquatic communities typically contain no insects and salt marsh communities contain relatively few resident insects, notably grasshoppers (Smalley 1960) and salt marsh dipterans (Davis and Gray 1966). Although many transient species may be found in salt marshes (Foster and Treherne 1976), they and their ecological functions have been ignored by most authors working in east coast salt marshes (Cameron 1972). In contrast with salt marshes, insects are abundant and characteristic of freshwater marsh habitats (Heard 1982), and insects, especially larvae, represent an important component in the benthic and occasionally the planktonic communities of freshwater backbays and lagoons (Boesch et al. 1976, Menzie 1980). Gulf of Mexico *Juncus* marshes, with their generally lower salinities (de la Cruz 1981), bear greater resemblances to east coast freshwater marshes (Simpson et al. 1979) than to east coast salt marshes (Davis and Gray 1966, Parsons 1978). LaSalle and Bishop (1987) reported 19 species, representing 7 families of aquatic diptera alone in oligohaline *Juncus* marshes in Mississippi.

The importance of insects in very low-salinity estuarine ecosystems is documented not only by their abundance in surveys but also by their occurrence in the diets of larger organisms. Sheridan (1979) noted the abundance of insects in fish stomachs collected from backbay and river mouth stations in East Bay (Apalachicola Bay System, Florida), where Livingston (1984) reported the occurrence of 30 species of aquatic insects, 22 of them dipterans. Insects belonging to eight families occurred in the diets of three out of four species of sunfish (*Lepomis* spp.) studied in Lake Pontchartrain by Deselle et al. (1978). The importance of mosquito larvae in the diets of many freshwater and brackish water marsh fishes was also demonstrated by Harrington and Harrington (1961).

Vertebrates

Fishes   Freshwater and oligohaline estuarine areas support a large, but not necessarily speciose fish community (Gunter 1961) consisting of the more euryhaline members of both the freshwater and estuarine/marine fish faunas, plus a few truly diadromous species.

Numerous studies have demonstrated the composition of estuarine fish communities from Texas to Virginia; however, relatively few studies have concentrated on those low-salinity habitats (Hastings et al. 1988, Rozas and Odum 1987, McIvor and Odum 1988, Felley 1989). These studies have indicated that despite the distance involved, these communities are quite uniform in composition. This feature is undoubtedly due to the high similiarity seen in both freshwater (Swift et al. 1986) and marine fish faunas (Hoese and Moore 1977) from which the brackish water community is drawn. Two important characteristics of the community are the predominance of young individuals (Gunter 1957, Talbot and Able 1984, Rogers and Herke 1985) and the presence of many migratory species such as the eel (*Anguilla rostrata*) (Hansen and Eversole 1984, Helfmann et al. 1984) or anadromous shads and herring (*Alosa* spp.) (Davis and Cheek 1966, Adams 1970) and moronids (perichthyids) (Pierce and Davis 1972). Generally, the same explanations are cited for juvenile fish abundance in fresh low-salinity habitats as in salt marsh creeks: the abundance of food and a lack of predators (Boesch and Turner 1984). In contrast to many higher salinity communities that are divided between resident and migratory components (Dahlberg and Odum 1970, Haedrich and Haedrich 1974), there are few permanent resident species. Instead, several migratory patterns overlap. First, predominately freshwater and predominantly marine communities may alternately occupy the same physical locality as long-term salinity fluctuations permit (Felley 1987, Rogers et al 1984). Second, diadromous species will occupy the habitat at appropriate times in their life cycles (Godwin and Adams 1969, Tyus 1974, Bozeman et al. 1985). Finally, opportunistic migrants will utilize low-salinity areas for nursery or feeding habitats (Weinstein 1979, Epperly 1984). Diadromous species together with several opportunistic migratory species such as menhaden (*Brevoortia* spp.) and anchovies (*Anchoa* spp.) contribute a disproportionate amount of biomass to the oligohaline fish community (Reintjes and Pacheco 1966, Rogers and Herke 1985). Their seasonal migrations are also important features of energy flow in these ecosystems.

Dahlberg (1972) reported that in Georgia estuaries, fish species declined in number up-estuary until pure freshwater was reached. He reported 39 species from the lowest salinity range; over half were widespread coastal plain species and a few migratory species. Hastings et al. (1988) collected 67 species of fish from the oligohaline Lake Maurepas, Louisiana. Included in this number were 37 species (55%) of freshwater fishes, 27 species (40%) of marine fishes, and 3 species (4%) of diadromous fishes. The study occurred during a period when salinities rarely rose above 2.5 ppt. Hastings et al. noted that earlier studies, conducted during times of slightly higher salinities (up to 10 ppt), had encountered fewer species; however, most of their "new" species, including several marine forms, were rare species, whose inclusion was probably due more to increased sampling effort and efficiency than to environmental conditions per se.

Keup and Bayless (1964) found fish diversity high in freshwater reaches of the Neuse River, North Carolina, declining between approximately 5 and 8 ppt, and high again in mesohaline Pamlico Sound (12.25 ppt). The composition of the fish fauna nearly completely changed from a freshwater fauna dominated by pirate perch (*Aphredoderus sayanus*), centrarchids, cyprinids, and catostomids, to an estuarine fauna dominated by *Anchoa* spp. and sciaenids. Renfro (1960) and Merriner et al. (1976) studied fish distributions in relation to salinity in the Aransas River, Texas, and the Piankatank River, Virginia. Although these rivers are extralimital to the geographic scope of this review, their fish faunas are essentially identical with those in areas of East Texas and Virginia that are included. Both studies indicated that migratory marine species (chiefly menhaden) were common near the mouths of rivers and were replaced by centrachid-dominated, freshwater faunas upstream. The highest overall diversity was observed at the freshwater station in the summer. The middle reaches again had the lowest diversity (Merriner et al. 1976). In the St. Andrews Bay system of Florida, highest standing crops but lowest number of species and lowest number of individuals occurred in brackish areas (Naughton and Saloman, 1978).

Cyprinodontids, such as the sheepshead minnow (*Cyprinodon variegatus*), numerous species of killifish of the genus *Fundulus*, and live-bearers, such as the mosquitofish (*Gambusia affinis*) and the sailfin molly (*Poecilia latipinna*), are small fishes that occur along the entire eastern coast of the United States (Parenti 1981). Of all fishes, perhaps no group is as ubiquitous or characteristic of low-salinity habitats as these. Most cyprinodontids live at or near the water's surface in shallow, vegetated, often intertidal habitats, and exhibit morphological and physiological adaptations to this habitat and the environmental extremes that often characterize it (Lewis 1970, Mitton and Koehn 1976). The basic ecology and life histories of many of the species have been documented by Simpson and Gunter (1956), Kilby (1955), Hastings and Yerger (1971), Griffith (1974), Byrne (1978), Weisberg and Lotrich (1982), and Kneib (1984). Unlike many other estuarine species, the cyprinodontids are permanent inhabitants, which complete their life cycle in ponds, backbays, lagoons, and marshes. Their tolerance of broad ranges of temperature, salinity, and dissolved oxygen levels allow cyprinodontids to exploit habitats that many other species would find intolerable or suboptimal. Surveys of fishes living in coastal ponds, pools, and backwater areas throughout the Gulf of Mexico and along the Eastern Seaboard all attest to the dominance of the cyprinodontids in such marginal habitats (Kilby 1955, Franks 1970, Martin 1972, Swingle and Bland 1974). Many species are effective predators on insect larvae (Harrington and Harrington 1972), which makes them desirable species and an important part of mosquito abatement programs in many regions of the United States (Gilmore et al. 1982, Talbot et al. 1986).

Odum et al. (1984) contrasted fish communities of freshwater and low-salinity tidal marshes of the mid-Atlantic and south Atlantic coasts of the United States. Many of their observations are also true for the Gulf coast. While communities in both regions were numerically dominated by the cyprinodontoid and atherinid fishes, Odum et al. noted that juvenile sciaenids were more typical of mid-Atlantic tidal freshwater marshes, but were much less common in southeastern and Gulf marshes. However, these southern marshes were characterized by the more frequent occurrence of other marine species, such as flounder (*Paralichthys* spp.), striped mullet (*Mugil cephalus*), and mojarra (*Eucinostomus* spp.), as well as more freshwater species such as sunfish and black bass, family Centrarchidae, and catfishes, family Ictaluridae (Perry 1968, Heard 1975). These families, together with the cyprinodontoids, typically contribute much of the biomass in low-salinity fish communities.

Centrarchids have a well-known tolerance for salt water (Bailey et al. 1954) and have been widely reported from low-salinity open water areas on the Gulf and south Atlantic coasts (Brockmann 1974, Swift et al. 1977, Schwartz et al. 1982), although Dahlberg (1972) found none in Georgia coastal waters at salinities greater than 0.5 ppt. Hackney and de la Cruz (1981) found centrarchids to be the most important component of the Bay St. Louis, Mississippi marsh community during the summer months, when typical estuarine sciaenids were rare. Livingston and Duncan (1979) found the East Bay (Apalachicola Bay system) fish community was primarily estuarine, consisting of anchovies and sciaenids, but also including significant "freshwater" representatives such as white catfish, channel catfish, gar, bluegill, and blue-spotted sunfish. A similar situation was reported by Naughton and Salomon (1978) for the St. Andrews Bay system of Florida where fish communities of low-salinity, backbay areas (East and West Bays) were dominated by cyprinodontids *Gambusia affinis* and *Menidia beryllina*, while the fish community of a brackish/freshwater interface region (North Bay) was dominated by these same species plus mojarras (*Eucinostomus* spp.) and freshwater species, including various centrarchids and the coastal shiner (*Notropis petersoni*) (Davis and Louder 1971, Cowell and Resico 1975). Kilby (1955) collected largemouth bass (*Micropterus salmoides floridanus*), spotted sunfish (*Lepomis punctatus*), redear sunfish (*L. microlophus*), bluegill (*L. macrochirus*), and warmouth (*L. gulosus*, reported as *Chaenobryttus coronarius*) in his study of the fish fauna of marshes in the Bayport, Florida, area. All species except the warmouth, which was restricted to freshwater, occurred in waters with salinities up to 11 ppt, although all were most common below 5 ppt. Largemouth bass feed in brackish water on a variety of typical marine and estuarine species, including pinfish, mojarras, needlefish, snapper, blue crab, and penaeid shrimp (Kilby 1955, Rozas and Hackney 1984).

In addition to these common species, oligohaline portions of Atlantic coast estuaries provide important habitat for the shortnosed sturgeon (*Acipenser brevirostrum*) (Heidt and Gilbert 1981), one of the few species of marine fishes that is classified as endangered by the U.S. Fish and Wildlife Service (Reiger 1977). Frequent confusion of this species with the young of the Atlantic sturgeon (*Acipenser oxyrhynchus*) (Leland 1968), and its inclusion in the by-catch of both the Atlantic sturgeon and shad fisheries have provided most southeastern states with the incentive to further study the biology of this species and to enforce management restrictions on commercial fisheries (Van Den Avyle 1984).

Characteristic changes in salinity can restrict growth and reproduction in species that can otherwise survive normally in these habitats. Peters and Boyd (1972) have shown that feeding and growth are regulated in the hogchoker (*Trinectes maculatus*) by a combination of salinity and temperature; however, the migratory habits of this species appear to consistently place it in less-than-optimal habitats. It is widely known that many species of mullet (*Mugil cephalus*, as well as other members of the mullet family) will not reproduce in freshwater (Thompson 1966, Eckstein 1975). Tebo and McCoy (1964) noted that many freshwater fish can survive but cannot reproduce in salt water. For example, largemouth bass and bluegill can reproduce in salinities up to only 4.5 ppt. However, fingerlings of both species survive well up to 10 ppt and adults may be caught occasionally at even higher salinities. Salt water has also been shown to effect the rate of digestion in several euryhaline freshwater species (Hunt 1960). Such adaptiveness can vary from population to population and has been shown to have a genetic basis. Hallerman et al. (1986) detected electrophoretic differences between populations of largemouth bass living in and immediately upstream of the Mobile River delta. They concluded that a genetically distinct breeding population of bass was confined to the meso- and oligohaline reaches of the river. Conversely, freshwater intrusion into typically low-salinity estuarine areas can limit their role as nurseries for some estuarine species (Rogers et al. 1984, Weinstein et al. 1980), especially those that actively avoid areas with low or unstable salinities (Tabb 1966, Perez 1969).

Amphibians and Reptiles   Amphibians are the only class of vertebrates that have not adapted to saline waters. Consequently, if they occur irregularly in estuarine environments, their presence usually goes unnoticed (Neill 1958). Reptiles, while more common than amphibians, are also relatively scarce in brackish water habitats. However, the two most common varieties, alligators and turtles, often exist in sizable populations with significant biomass (Chabreck 1971, Iverson 1982).

Very low-salinity and freshwater habitats that are contiguous with brackish water areas support a much larger community of amphibians and reptiles than brackish habitats. The only such habitat that might be expected to have a depauperate herpetofauna are freshwater areas on coastal barrier islands where intervening saltwater barriers or the lack of a permanent freshwater supply, such as provided by a water table, prevent the establishment and maintenance of herptile populations (Gibbons and Coker 1978).

The species of reptiles and amphibians that inhabit tidal freshwater marshes and impoundments do not differ significantly from those in adjoining nontidal freshwater habitats. Emergent vegetation creates greater habitat diversity ("edge effect") in marsh environments that allows for an increased species diversity relative to both riverine or terrestrial habitats (Odum 1971). Aquatic salamanders such as sirens (*Amphiuma*) and dwarf mudpuppies live in subtidal waters, while emergent plants provide habitat for hylid frogs. Ranid frogs are often a conspicuous part of the community due to their larger sizes. The southern leopard frog (*Rana utricularia*) occurs in marshes with salinities up to 21 ppt (Pearse 1936), where they eat *Uca minax* and other marsh organisms, including small alligators (Springer 1938). Salinities greater than 5 ppt are lethal to eggs of the closely related northern leopard frog (*R. pipiens*), and so limit the reproductive habitat of this species (Ruibal 1959). Toads (*Bufo*), spadefoot toads (*Scaphiopus*), and narrow-mouthed toads (*Gastrophyrne*) inhabit drier, more terrestrial habitats within and surrounding freshwater marshes.

Snakes, especially the highly aquatic rainbow and eastern mud snakes, may be abundant, but their secretive, nocturnal habits makes these species hard to find and they usually go unreported in surveys. Rainbow and mud snakes feed on eels and eel-like salamanders, respectively (Martof et al. 1980). However, in brackish marshes the mud snake also eats eels and other fish due to scarcity of salamanders in saline habitats (Neill 1958). Other aquatic snakes, such as the water snakes (*Nerodia*, formerly *Natrix*, spp.), and garter and ribbon snakes (*Thamnophis* spp.) are common, within the ranges of each species. The cottonmouth (*Agkistrodon piscivorous*) is the only poisonous snake regularly found in coastal aquatic environments; however, several species of rattlesnake (*Crotalus* spp.) commonly occur in nearby maritime forest and coastal plain habitats. Lizards are scarce in aquatic habitats in general; however, the green anole (*Anolis carolinensis*), may be found in coastal marshes, and skinks, especially the arboreal broadheaded skink (*Eumeces laticeps*), inhabits tidal swamps and neighboring maritime forests. Glass lizards (*Ophiosaurus* spp.) are legless fossorial lizards that occur in moist habitats, including fresh and brackish swamps (McConkey 1954). The eastern glass lizard (*O. ventralis*) has been found in the burrows of *Uca minax* (Neill 1958). I have often observed this species (usually near the upland border) in *Spartina alterniflora* and *Juncus roemerianus* marshes, and individuals are occasionally found on the forebeach and even in the surf.

The only reptile truly characteristic of subtidal estuarine habitats is the diamondback terrapin (*Malaclemys terrapin*), several subspecies of which are found in coastal salt marshes from the Chesapeake Bay to Texas (Cagle 1952). Although normally found in salt marshes, *M. terrapin* can survive for extended periods in freshwater and regularly occurs in coastal streams and marshes above the limit of tidal influence. Other species of turtles are more common in these freshwater habitats. Virtually any of the many species and subspecies of coastal plain turtles can be expected to occur in tidally influenced freshwaters.

The American alligator (*Alligator mississippiensis*) is also a regular inhabitant of fresh and brackish marshes from North Carolina to Texas, although as Chabreck (1971) noted, the brackish habitat appears to be a less favorable one. Alligator hatchlings were smaller in brackish water and experienced weight loss when compared with similar aged animals in freshwater habitats (Chabreck 1971). In Dutchman's Creek, North Carolina, alligator nesting occurred only in highest reaches characterized by freshwater marsh vegetation (Birkhead and Bennett 1981). Louisiana alligators nested in more saline *Spartina patens* marsh but are most abudant in fresh, brackish, and intermediate marshes (Joanen 1969). Alligators are among the largest inhabitants of the marsh and function as the top carnivore in the tidal marsh food web. Alligators are rivaled in size only by alligator gar (*Lepisoteus spatula*) and alligator and common snapping turtles (*Macroclemys temmincki* and *Chelydra serpentina*) (Gunter 1979). Only the latter occurs with the alligator in east coast marshes.

Declining alligator populations in the mid 1960s led to the species being placed on the Federal list of endangered species in 1967. This protection afforded the alligator the opportunity to recover in numbers to the extent that its special status was changed in 1977 to threatened. Previous restrictions were removed and a limited hunting season was reestablished in portions of the animal's range.

Birds Freshwater backbays and lagoons, together with their surrounding swamps and marshes provide habitats for an abundance of birdlife in the southeastern United States. Many birds are described as being typical of a particular habitat, as the [long-billed] marsh wren is supposed to occur in freshwater marshes and the sedge [short-billed marsh] wren is supposed to occur in salt marshes. However, there is a lot of overlap in distribution of such species (Kale 1965). This realization and the fact that birds, especially when migrating, often occur in very "untypical" habitats mean that virtually any of the over 300 species of birds reported from the eastern United States could be expected to be seen in such coastal habitats.

Three major categories of birds contribute many of the species, and probably most of the individuals and biomass found in coastal freshwater and low-salinity habitats: waterfowl (ducks, geese, and swans), gruiform birds (cranes, rails, gallinules, and coots), and wading birds (herons, egrets, ibises, and their allies).

Dabbling (puddle) ducks and geese are the most abundant waterfowl in freshwater marshes (Shaw and Ferdine 1956, Palmisano 1972). These include mallards, American black ducks, mottled ducks, blue-wing teal, green-wing teal, pintail, gadwall, American widgeon, and shovelers. As their common names imply, these ducks typically feed in shallow waters (including puddles) by dabbling or tippng forward without submerging their entire body. Shallow artificial impoundments afford ideal habitat for those ducks if salinity regimes are properly managed (Chabreck 1961). Diets are chiefly characterized as herbivorous/granivorous (Kerwin and Webb 1972), but animal food may also be taken. Although dabbling ducks frequently occur together in mixed flocks, Chabreck (1979b) was able to determine subtle physical, chemical, and biological differences in the feeding habitats. Species of geese that are common include the Canada and greater snow geese of the Atlantic coast and the white-fronted and lesser snow geese on the Gulf coast. Geese feed on a variety of aquatic plants, but also commonly eat terrestrial grasses and sedges from fields (including agricultural) and exposed tidal marshes (Bellrose 1976). Large populations of geese and whistling swans winter at the head of the Chesapeake Bay and south to the Pamlico Sound area of North Carolina. Smaller populations of geese and occasional groups of swans may be farther south in North and South Carolina (Bellrose 1976). The marshes of Louisiana and east Texas also afford excellent goose feeding habitat (Lowery 1974a).

Diving ducks or polchards (the ring-necked, canvasback, redhead, lesser and greater scaup) form large flocks that feed primarily on aquatic vegetation in deeper water of open bays. While most of these species are normally found in oligohaline waters (Stewart 1962), the ring-neck duck more commonly occurs in freshwater habitats (Shanholtzer 1974) and the canvasback also occurs there more frequently than do the other diving ducks (Potter et al. 1980). Sea ducks are relatively rare in freshwater habitats, but the hooded merganser lives and nests in southeastern freshwater marshes (Lowery 1974a). Most of the above are migratory waterfowl that use southeastern marshes for winter feeding grounds. Only the wood duck, the mottled duck on the Gulf coast, the American black duck on the Atlantic coast, and occasional mallards and blue-winged teal and fulvous whistling ducks breed in southeastern freshwater marshes (Stotts and Davis 1960, Lowery 1974a, Odum et al. 1978).

Along the Gulf coast, sandhill cranes may occasionally be found in coastal freshwaters, although the species is more common on inland prairie wetlands. Rails, gallinules, and coots, however, are commonly found in freshwater bays and marshes. King rail are found in freshwater *Typha* marshes, while the clapper rail is more common in salt, *Spartina alterniflora*, marshes. Sora, yellow, and Virginia rails are also more abundant in low-salinity marshes. The American coot, purple gallinule, and common moorhen (gallinule) are permanent inhabitants of fresh and brackish marshes.

Wading birds, such as the herons, egrets, storks, and ibises, constitute what are perhaps the most conspicuous part of the bird community. One of the more spectacular aspects of the biology of these birds is their habit of nesting in colonies that contain hundreds or even thousands of individuals. These colonies may occur anywhere vegetation is present for nesting. Preferred nesting sites are high in trees or tall shrubs, open on at least one side, and surrounded by water or otherwise isolated, as on small islands in swamps or bays. These locations presumably afford some protection from predators while allowing the birds room in which to fly. While natural lakes and swamps are certainly suitable nesting localities, in South Carolina the abandoned ricefield freshwater storage areas known as "reserves," often hundreds of acres in size, provide the ideal combination of appropriate vegetation, plentiful and stable water regime, and abundant food. Such reserves are usually adjacent to old rice fields, which have typically been allowed to revert to tidal marshes.

While the most common species of herons and egrets are seasonal migrants north of South Carolina, the same species are generally year-round residents in the southeastern United States. Sandifer et al. (1980) list 12 species of birds, 11 of them year-round permanent residents, that regularly nest in rookeries of the southeastern United States. Four species (white ibis, cattle egret, snowy egret, and tricolored [Louisiana] heron) are listed as dominant, four other species (greater egret, little blue heron, glossy ibis, and black-crowned night heron) are of moderate importance or abundance, and four additional species (great blue heron, yellow-crowned night heron, green heron, and anhinga [nonwader]) are of minor importance. Nine of the same species, all except the anhinga, cattle egret, and yellow-crowned night heron, were also common in more saline habitats, although the relative abundances of the species differed between habitats. Wading birds may roost or nest in one habitat but use an entirely different habitat for feeding. Custer and Osburn

(1978) studied patterns of migration between roosting and nesting sites in nine species of colonial nesting shorebirds in the Beaufort, North Carolina, area. Kushlan (1977) and Maxwell and Kale (1977) noted that white ibises may feed at a considerable distance from their rookery, especially after their young have hatched. Although the same species of herons and egrets are commonly found over the entire southeastern United States, relative abundances may also differ between different geographic areas. Tricolor herons were apparently the most common heron throughout the South Carolina–Georgia Sea Island area; however, greater egrets and yellow-crowned night herons had their greatest concentration in lower salinity habitats (Sandifer et al. 1980). Custer and Osburn (1978) found that most herons and egrets in the Beaufort, North Carolina, area fed in salt marshes. Of the individuals that did feed in freshwater habitats, greater egrets and white ibis fed there most frequently. Greater egrets showed the most diversity in their choice of feeding habitats, so their occurrence in freshwater habitats may not indicate a preference. In Florida, Bent (1926) observed that little blue herons and tricolored herons (Louisiana herons) feed most frequently in freshwater sites, and Jenni (1969) noted that three species of Florida herons that nest together, utilized different feeding areas. Portnoy (1977) reported that greater egrets and snowy egrets were the most common wading birds in Gulf coast rookeries.

Herons and egrets are fairly opportunistic in their feeding, eating whatever is most abundant that they can swallow. Fish, various macroinvertebrates, amphibians, reptiles, small mammals, and even birds, especially nestlings of other herons, are common food items (Jenni 1969, Maxwell and Kale 1977).

Gulls and terns, which are common and typical members of the avifauna in salt and brackish marshes, are far less common in freshwater marshes, where their typically aerially searching, piscivorous, or scavenging feeding niches may be occupied by belted kingfishers (*Ceryle alcyon*), southern bald eagles (*Halieiaeetus leucocephalus*), ospreys (*Pandion haliaetus*), American crows (*Corvus brachyrhynchos*), and fish crows (*C. ossifragus*) (Odum et al. 1984). On the southeast coast, three species of gull, the laughing gull (*Larus atricilla*), the ring-bill gull (*L. delawarensis*), and the herring gull (*L. argenteus*), and four terns, the common tern (*Sterna hirundo*), Forster's tern (*S. forsteri*), the Caspian tern (*S. caspia*), and the gull-billed tern (*Gelochelidon nilotica*), occur regularly. All seven of these species were recorded as occurring in salt marshes and more-saline estuarine habitats (Sandifer et al. 1980) and four were included as dominant species; however, no species of gulls or terns were among 78 species of birds reported from coastal freshwater marshes where kingfishers were included among the dominant species.

Forsythe (1973) reported that ring-bill gulls are regularly found in coastal freshwater habitats near Charleston, South Carolina, during winter. The same species also occurs regularly in south Florida and Louisiana tidal freshwater habitats in the winter (Drew and Schomer 1984, Lowery 1974a). Although they may be the most common gull in these habitats, they are never regarded as abundant. Similarly, laughing gulls, which occur primarily in saline habitats during their summer breeding season, may occur more frequently in inland

and coastal freshwater habitats in the winter (Forsythe 1973). Despite these occurrences, most winter records of gulls are from nontidal waters further inland (Lowery 1974a).

Terns are even less frequently encountered than gulls in freshwater habitats, although on the Gulf coast, Caspian, Foster's, and gull-billed terns, as well as black terns (*Childonias niger*), occur regularly in fresh and intermediate salinity marshes (Gosselink 1984). The first two species are year-round inhabitants, the gull-billed tern is primarily a winter resident, and the black tern is a summer but nonbreeding resident of these marshes.

Although waterfowl, rails, wading birds, and seabirds constitute a large part of the coastal bird community, both in numbers and biomass, they are certainly not the only varieties of birds to occur in such habitats. A variety of small birds, including phoebes, wrens (chiefly long-billed marsh wrens), swallows, and martins, feed on the abundant insects that live in and above coastal freshwater habitats. Savannah and sharp-tailed sparrows and, on the south Atlantic coast, bobolinks feed on the abundant crop of seeds produced by marsh grasses in the fall. Blackbirds and common grackles are abundant omnivores, feeding on wild rice and other seeds and grain in autumn when these are available, and on insects during the spring and summer when this prey is most abundant. Such shifts in dietary preference may be opportunistic or related to greater energy needs associated with raising young (Meanley 1972).

Mammals   Freshwater coastal ecosystems differ from analogous saline ecosystems in their greater mammalian biota. Gosselink et al. (1979) reported 14 species of mammals from Louisiana and East Texas freshwater marshes versus only eight species in salt marshes. Odum et al. (1984) list a total of 45 species as occurring in tidal freshwater marshes of the Atlantic coast.

In addition to terrestrial mammal species inhabitating marshlands, Odum et al. (1984) include one species of porpoise, the common dolphin (*Delphinus delphis*). Along the southern coastlines of the United States another species of porpoise, the bottle-nosed dolphin (*Tursiops truncatus*) is known from estuarine waters and occasionally enters freshwater (Caldwell and Caldwell 1973). Both species of porpoise must be considered rare, at best, in low-salinity habitats. A third species of marine mammal, the manatee (*Trichechus manatus*), is widespread throughout fresh and low-salinity coastal rivers, bays, and waterways in Florida and occurs less frequently along the northern Gulf Coast and south to Mexico (Powell and Rathbum 1984). This species is believed to be a poor osmoregulator and hence may be prevented from penetrating higher salinity areas in which abundances of suitable foods, such as the submerged aquatic grasses *Syringodium*, *Thalassia*, and *Halodule*, occur (Powell and Rathbun 1984).

The most abundant mammals in southeastern coastal marsh habitats are the marsh rabbit (*Sylvilagus palustris*) on the Atlantic and Gulf coast west to Mobile Bay, the swamp rabbit (*S. aquaticus*), which occurs from Mobile Bay west to the central Texas coast, and small rodents, including meadow voles

(Harris 1953), rice rats (Negus et al. 1961, Abernethy et al. 1985), shrews, and mice (Golley 1962, Abernethy et al. 1985). Larger mammals that occur regularly in Atlantic and Gulf coast marshes include the eastern racoon, river otter, mink, white-tailed deer, muskrat, and nutria.

The last two species are perhaps the most characteristic and important mammals in the marshes of Louisiana, where they are commercially valuable fur-bearing species, as well as ecologically significant components of the marsh fauna (Chabreck 1979a). Muskrats (*Odonata zibethica*) have been implicated as controlling the observed 10- to 14-year cycles in marsh growth and collapse. As muskrat populations increase, the marshes are subject to "eat outs," the almost total destruction of marsh vegetation by overgrazing. The marsh surface may be so disturbed by muskrat digging and burrowing that revegetation is slow. Only the collapse of the muskrat population will allow the vegetation to recover, which will initiate another cycle of muskrat population increase (Lowery 1974b).

Nutria (*Myocastor coypus*) were introduced in the southern United States in the 1930s as another commercial fur-bearing species (Evans 1970, Willner et al. 1979). In Louisiana they appear to favor freshwater marshes and have successfully displaced the more valuable muskrat into less-favored brackish and salt marsh habitats (Wilson 1968). Nutria are also common on the Atlantic coast in northeastern North Carolina and around the Chesapeake (Evans 1970), where they coexist with muskrats in nontidal freshwater, tidal freshwater, and salt marshes. At this northern limit of its range, the nutria suffers from hard winter freezes (Willner et al. 1979), which stress or limit population and give the more cold-tolerant muskrat an advantage or parity (Willner et al. 1975).

An interesting contrast between Gulf and south Atlantic coastal marshes is the almost total absence of muskrats and nutria from these habitats in South Carolina and Georgia (Sandifer et al. 1980), although the muskrat at least occurs inland in both states. It has been suggested that the greater tidal amplitude on the southeastern coast prevents these animals from exploiting the coastal marsh habitat (Sandifer et al. 1980).

The Florida water rat (round-tailed muskrat, *Neofiber alleni*) is a diminutive version of the muskrat that occurs south of the Indian River, Florida, north on the Florida west coast to the Apalachicola River, and in the Okeefenokee Swamp of Georgia (Burt and Grossenheider 1976). It too, would appear to be adapted, or adaptable, to the coastal marshes of Georgia and South Carolina and its absence is a part of the larger mystery surrounding the absence of all three large semiaquatic rodents from these habitats.

Although no predatory mammals presently are permanent residents of coastal freshwater marsh habitats, omnivores and predators from adjacent upland areas make frequent forays into the marsh, especially at low tide, to feed. Raccoon, foxes (gray and red), bobcat, skunks (spotted and striped), mink, weasels, coyote, and black bear may be found in these habitats. The river otter (*Lutra canadensis*) is found in coastal rivers and swamps, never far from water where it feeds on fishes and crustaceans. Formerly, the red wolf (*Canis rufus*) was a common inhabitant of coastal forests, prairies, and marshes. Hunting and hybridization with other canid species has decimated this species and it may now be extinct in the wild, although captive breeding populations have been established and the species has recently been reintroduced in the Carolinas.

ENERGY SOURCES AND TROPHIC RELATIONSHIPS

Low-salinity backbays and lagoons exhibit some fundamental differences when contrasted with *Spartina*-dominated higher salinity systems (Odum 1988). Furthermore, within different salinity regimes, shallow-water habitats with emergent or submerged vegetation may differ from adjacent deeper open-water habitats in their patterns of energy and nutrient flow. These differences may be attributable to both biologic and physicochemical features of each type of habitat.

Like salt marshes, freshwater marshes derive energy from tidal subsidy (Odum 1980), but their closer proximity to land means that terrestrial energy sources will play a greater role. Odum et al. (1984) present a conceptual scheme of energy flow in tidal freshwater marshes. They hypothesize three major sources of organic carbon: vascular marsh plants, terrestrial runoff, and phytoplankton. The relative contributions of these three sources appear to vary considerably in different seasons (Simpson et al. 1983, Caffrey and Kemp 1990).

Primary production in fresh and brackish marshes is comparable to or higher than that in *Spartina alterniflora* dominated salt marshes at the same latitude (Simpson et al. 1983, Mitsch and Gosselink 1986). Productivity maxima exhibit less seasonality due to contributions of many species throughout the annual growing season (Mendelssohn and McKee 1982), with varying contributions (especially in the winter) from other plants and benthic algae (Roman et al. 1990). Lower salinity per se may also affect higher productivity (Linthurst and Seneca 1980) and photosynthesis (Pearcy and Ustin 1984), as in *Spartina alterniflora*, although Roman et al. (1984) present evidence that this species may be excluded from low-salinity habitats in which it has a high potential productivity by competition from other species. Perhaps this is due to the burden of synthesizing compounds that aid with salt and sulfate exclusion or elimination (Cavalieri and Huang 1979, Mendelssohn and Burdick 1987), which requires additional energy not demanded from freshwater marsh plants.

Fresh and salt marshes differ in how their productivity is processed and made available to consumers. Herbivory may be more intense in low-salinity marshes (Vince et al. 1981, Cahoon and Stevenson 1986) and decomposition rates may vary from patterns and rates observed for salt marshes (Odum et al. 1984). While both fresh and brackish marshes are generally described as detritus-based systems (Odum et al. 1984, Findlay et al. 1990) detritus from macrophyte litter tends to both accumulate and decompose faster in freshwater marshes (Simpson et al. 1983) and so can serve as a source of carbon and nutrients both to the marsh and to adjacent waters with greater continuity than in higher salinity marshes (Findlay et al. 1990). According to Jordan et al. (1989), the litter incorporated in the upper layers of freshwater marsh sediments alternately takes up and releases nutrients, acting as a buffer to sedimentary nutrient pools. As in salt marshes, carbon export, if and when it occurs, is primarily in the form of dissolved rather than particulate matter (Weigert et al. 1975, Axelrad et al. 1976). Decomposition processes also differ because freshwater sediments present a very different type of reducing environment from that found in salt marsh sediments. Sulfur and sulfates are much less common in freshwater and sulfate reduction does not normally occur in freshwater marsh soils (Odum 1988). The appearance of H_2S gas in freshwater marshes is often used as an indication of saltwater intrusion. Carbon may be lost from the marshes via exchange with the neighboring open waters or by incorporation in the sediments. Significant amounts of carbon may be lost through reduction to CH_4 in anaerobic sediments (Smith et al. 1982). In this sense, despite the differences in species composition, seasonal patterns of productivity, and decomposition rates, low-salinity marshes appear to function much the same as salt marshes.

Fresh and brackish marshes also support a more diverse and highly productive assemblage of consumer organisms. Although the aquatic fauna in brackish marshes may be somewhat less speciose than in either fresh or more saline waters, the lower tidal amplitude and greater plant species diversity create a greater variety of microhabitats, which support a greater diversity of insects. Different palatabilities of plants combined with feeding/digestion adaptations of insects produce the variable, sometimes high rates of insect grazing observed in low-salinity marshes, which may vary from 0 up to 25% of the plant biomass (Cahoon and Stevenson 1986) compared to no more than 7% biomass consumption in a *Spartina* marsh (Smalley 1959). Low nutrient content of live *Spartina* (Vince et al. 1981) and the relative unpalatability of other halophile species, such as *Juncus roemerianus* (Eleuterius 1980), apparently keeps herbivory low in salt marshes, although Pfeiffer and Weigert (1981) have indicated that insect grazing may be more substantial than had been believed earlier (Cameron 1972, Smalley 1959). As previously noted, freshwater marshes also support a large diversity of bird and mammal species, which feed on plants, insects, and each other, creating a broad and complex food web. Herbivory is also common among these larger consumers. Many bird species feed on the fruits and seeds of marsh plants, while mammals feed on fruits, seeds, shoots, and roots. Intense grazing by muskrats or geese ("eat outs") may completely denude large stretches of marsh vegetation in Louisiana (Gosselink 1984).

In contrast to the similarities in function seen in marshes, low (freshwater to oligohaline) and higher (meso- to euhaline) salinity open water habitats differ appreciably. Because the latter have been more extensively studied, they are often regarded as typical of the estuary as a whole and many investigators have not recognized that differences exist within estuarine complexes.

Odum and Smith (1981) noted that due to paucity of information specific to these low-salinity systems, ecologists have been forced to make assumptions and extrapolations from inland lakes and more saline estuarine systems. These result in hypotheses that may be "slightly incorrect to dead wrong." Furthermore, comparisions between estuaries are made difficult by the lack of uniform methods (Kemp and Boynton 1980, Fisher et al. 1981) and presentation of results.

Typical open-water estuarine habitats are assumed to be autotrophic, with either phytoplankton or benthic plant production as the predominant source of carbon. This generalization appears reasonable based on the investigations reviewed by Mann (1982), as well as by the logical extension of the river continuum concept of limnologists (Minshall et al. 1985), which states that low-order (headwater) streams tend to be heterotrophic with detritus based foodchains, while higher order rivers are autotrophic and plankton based. As Odum and Smith (1981) predict, these generalizations break down at the freshwater–saltwater interface, where physicochemical processes drastically alter biological community structure and function and produce an environment in which external influences and carbon sources dominate.

Variations in estuarine primary production have been linked to freshwater input in a number of studies. However, the results are not uniform. In some cases fresh water appears to enhance productivity; in others fresh water appears to limit it. Seasonal and interannual variations in light, temperature, nutrients, physical transport, and herbivory were shown to have effects on phytoplankton species composition and productivity. Among the mechanisms presented to account for these changes were changing nutrient sources from terrestrial watershed to estuary, changing the rates of dilution or advection of phytoplankton cells out of the estuary, and changing the light availability through stratification, gravitational circulation, and longitudinal position of turbidity maxima (Boynton et al. 1982). Large inputs of fresh water, as from hurricanes or other storms, often result in increased productivity in subsequent years (Boynton et al. 1982). Kemp and Boynton (1984) suggest that this is due to the remineralization and recycling of nutrients that had been trapped in estuarine sediments. Bowden and Hobbie (1977) found that higher river flows following Hurricane Ginger in the fall of 1970 washed algae out of Albemarle Sound and no spring bloom occurred in 1971; however, conditions had normalized by the following year.

Boynton et al. (1982) present a conceptual model of estuarine primary production derived by a discriminant analysis of 63 estuarine systems. River-dominated estuaries show a wide variation. In the Chesapeake Bay, spring runoff introduces nutrients, which results in high summer productivity, an interesting contrast to many southern estuaries, which also have winter–spring peaks in phytoplankton production. Different seasonal patterns of turbidity producing rainfall and warmer southern temperatures probably account for these differences.

Fisher et al (1988) have also presented a similar model of estuarine production developed from studies of other east coast estuaries. Based on their observations that estuaries received continuous inputs of nutrients from fresh water, the existence of turbidity maxima coinciding with the mixing of saline and fresh water in the oligohaline zone and nutrient depletion at the highest salinities, they hypothesize that net heterotrophy will occur in oligohaline reaches in contrast to net autotrophy downstream of the turbidity maximum. Light limitation in the zone of mixing keeps primary production low, while further downstream clearer water permits more photosynthesis and phytoplankton population growth, resulting in nutrient depletion. A zooplankton maximum downstream of the phytoplankton maximum is also predicted. This model was developed for linear, drowned river valley estuaries typical of the Atlantic coast. Lagoons and bar-built estuaries, with their different morphologies and water residence times, should support different patterns of phyto- and zooplankton abundances. In larger open bays of the Gulf coast, phytoplankton may account for as much as 75%–99% of the system's productivity (Armstrong and Hinson 1973, Ward et al. 1980, Armstrong 1987, Lewis and Estavez 1988).

Two of the most extensively studied systems that possess gradients from freshwater to essentially marine conditions are the Barataria Basin of Louisiana (Conner and Day 1987) and the complex of riverine estuaries that surround Pamlico Sound, North Carolina (Fisher et al. 1982).

Comparisons of three of the North Carolina estuarine rivers showed that freshwater input had a positive influence on two measures of biological activity, carbon primary production and nitrogen demand, while tidal range had a negative influence, acting to disperse nutrients.

The combination of high freshwater input and low tidal amplitude in the Neuse River, a tributary to Pamlico Sound, yielded the highest productivity values. Low freshwater input and high tidal flushing in the Newport River (which exchanges directly with the Atlantic Ocean) yielded the lowest. The South River, a tributary of the Neuse with low runoff and little tidal flushing, yielded intermediate values. Davis et al. (1985a) found highly significant positive correlations between ammonium and phosphorous and turbidity, suggesting that watershed runoff was the major source of these nutrients in three other Pamlico tributary creeks. They also showed a positive correlation between nutrient concentrations and freshwater runoff after a suitable (6-day) time lag was included. The input of nutrients, including organic carbon from river flow, may result in greater numbers of heterotrophic organisms as well as autotrophs, which are responsible for primary production. Unfortunately, neither study presents community respiration values, which are necessary to determine the extent to which each system is autotrophic or heterotrophic.

Data from many studies in the Barataria Basin, reviewed by Conner et al. (1987), show a similar pattern for primary production, but the inclusion of respiration data allows a more complete analysis of energy flow. Upper basin waterways exhibit no measurable salinity, high productivity, pronounced seasonality, and net heterotrophy, while more saline areas have lower productivity and no consistent seasonal trends, and tend to be autotrophic. Lac des

Allemands is a fresh, turbid lake in the upper basin. It is highly eutrophic due to nutrient loading from adjacent wetland and upland drainage and exhibits pronounced seasonality in primary production. A large community of benthic invertebrates and fishes is supported by the influx of organic carbon from adjacent swamps and bayous and by a detritus/microbial-based food chain within the benthos. Three lakes in the central basin (Lake Cataouatche, Lake Salvador, and Little Lake) show different productivity patterns according to the extent of upland runoff entering each lake. All three lakes are slightly brackish and are effected by tidal currents originating in the Gulf of Mexico, which increase flushing and mixing and combine with seasonal river flows to establish a cycle of varying salinities. All three are also heterotrophic; only Little Lake has any significant benthic algae production. Diversity and biomass values for consumer organisms were intermediate in these middle reaches. The more saline lakes in the lower basin have production values slightly higher than Little Lake and also lack any strong seasonal patterns. Secondary consumers were more diverse and abundant than in the middle lakes, but were less diverse and abundant than in the freshwater and oligohaline regions of the system. Tidal fluxes provide an important link between lower basin marshes and the open water of bays.

Such contrasts are not restricted to southeastern and mid-Atlantic estuaries. Data for two estuaries in the Netherlands indicate that detritus from land runoff and phytoplankton production vary inversely in the Grevelingen (Wolff 1977) and Ems (VanEs 1977) estuaries. During periods of high runoff, detritus can contribute as much as 41% of the total food available in these estuaries. During the same periods, phytoplankton contribute as little as 1%. However, when runoff is low these figures are reversed. Allochthonous sources also supply a major part of the nutrients and particulate organic carbon in the Morlaix River estuary of France. Here anthropogenic rather than natural sources provide most of the nutrients, and nitrogen loading in this heavily polluted estuary is offset by the high tidal flushing, which dilutes and purges nutrients from the estuary despite low freshwater runoff (Wafer et al. 1989).

Secondary production may be fueled by allochthonous carbon and other nutrients brought into low-salinity backbays and lagoons and by in situ primary production. The dividing line between in situ and imported carbon may be vague, especially in large estuarine systems with numerous secondary bays and backbays. Sheridan and Livingston (1979) noted that fish and invertebrate abundance increased throughout the river-dominated Appalachicola Bay system each year following the peak periods in freshwater input. Within the system, biological and physical activities by higher tropic-level organisms seem to contribute to nutrient remineralization and so help increase the potential for primary production if it is not limited by other factors (Smith 1978, Vargo 1979). Nixon (1982) noted that high fishery yields from many lagoonal systems were due to relatively immobile resident forms, largely shellfish. Their high production was supported by "auxiliary energy" (Nixon 1982), a subsidy from tidal, riverine, or wind-driven currents bringing nutrients and organic matter into the lagoon from sources covering a much greater area than the lagoon itself. Zooplankton and benthic deposit feeders could feed directly on organic matter swept into the bay, while the infusion of nutrients would stimulate phytoplankton production, which could also serve as a first link in a food chain leading to fish and larger invertebrates.

Within a single system, peaks in primary and secondary production may occur at different times or locations but be functionally linked by currents. Fisher et al. (1988) cite evidence of a peak in zooplankton abundance downstream from the phytoplankton maximum. They hypothesize that the zooplankton peak occurs here rather than concurrent with the phytoplankton maximum because the greater width of the estuary downstream provides a longer residence time, which then allows zooplankton populations with their longer generation times to become established. The zooplankton, however, are dependent on the advection of phytoplankton from upstream as their chief source of food.

REPRESENTATIVE SYSTEMS

The previous sections reviewed the biota of freshwater and low-salinity estuarine areas of the southeastern United States. Many similarities have been noted in the flora and fauna of these systems throughout the region, with the most noticeable differences occurring between congeneric species in the Atlantic and Gulf regions. Important physical, chemical, and biological differences between Gulf and Atlantic coastal marshes have been reviewed by Kurz and Wagner (1957), Linton (1968), and de la Cruz (1981) and are summarized in Table 3.

Despite their overwhelming similarities, each system is unique, especially when relationships with surrounding systems, impacts of anthropogenic developments, or resource management problems are considered. Although the following five systems are not the only freshwater backbays and lagoonal systems in the southeastern United States, they are the largest and generally the best documented, and so serve as examples of the varieties of systems and their problems.

Lake Pontchartrain, Louisiana

Lake Pontchartrain (Fig. 3) is the largest single oligohaline ecosystem in the southeastern United States and one of the most heavily impacted by human activities. Mean salinity ranges from 1.2 ppt in the west to 5.4 ppt in the east nearest the Gulf of Mexico (Sikora and Kjerfve 1985). The most important factor governing salinity distribution is freshwater input, both direct from the Tchefuncte River and occasionally the Pearl and Mississippi Rivers, and indirect input, which enters the lake through the surrounding marshes, swamps, and bayous. Pearl River discharge does not normally enter the lake, but rather flows into the adjoining Lake Bourne in such a manner as to restrict the entry of more saline Gulf of Mexico waters into Lake Pontchartrain (Bahr

TABLE 3 Comparison of South Atlantic and Gulf Coastal Marshes

Feature	South Atlantic	Gulf Coast
Salinity range	10–30 ppt	0–15 ppt
Tidal cycle	Semidiurnal	Diurnal
Tidal amplitude	2 m	0.3 m
Freshwater input		
Major river system	Savannah River	Mississippi River
Annual precipitation	120 cm	140 cm
Primary production	1000–2000 g/m^2 year^{-1}	1500–3000 g/m^2 year^{-1}
Secondary production		
Penaeid shrimp	5964 tonnes	55 193 tonnes
	(13.2 kg/ha)	(33.9 kg/ha)
Total fish yield	144 245 tonnes	790 625 tonnes
	(320.2 kg/ha)	(485.4 kg/ha)
Size of intertidal area	450 500 ha	1 628 900 ha
Plants		
Dominant vascular plant	*Spartina alterniflora*	*Juncus roemerianus*
High marsh plant association	*Salicornia–Distichlis– Juncus*	*Spartina patens–Distich- lis–Scirpus*
Inland marsh plant association	*Spartina cynosuroides– Typha–Phragmites communis*	*Spartina cynosuroides– Cladium*
Animals		
Dominant insects	*Orchelimum fidicinium Prokelisia marginata*	*Orchelimum concinnum Conocephalus* spp.
Benthic bivalves	*Modiolus demissus*	*Polymesoda caroliana Modiolus demissus*
Snails	*Littorina irrorata*	*Melampus bidentatus Littorina irrorata*
Crabs	*Uca* spp. and *Sesarma*	*Uca* spp. and *Sesarma*
Mammals	*Procyon lotor*	*Ondatra zibethica Myocastor coypus*

Source. Modified from de la Cruz (1981).

et al. 1980). Mississippi River water also does not normally enter the lake; however, the Bonnet Carré spillway at the west end of the lake was constructed to allow the diversion of Mississippi River flood waters away from the city of New Orleans and into Lake Pontchartrain. Such diversions in 1973 and 1975 had immediate and long-term impacts on the lake's salinity regime as well as on the benthic plant and animal communities (Poirrier and Mulino 1975, 1977), although, in time, both physical and biological parameters returned to their preflood conditions.

Extensive ecological surveys of the lake, conducted at intervals during the past 35 years, provide a valuable comparative database from which to access the impacts of human activities and natural perturbations. Surveys conducted

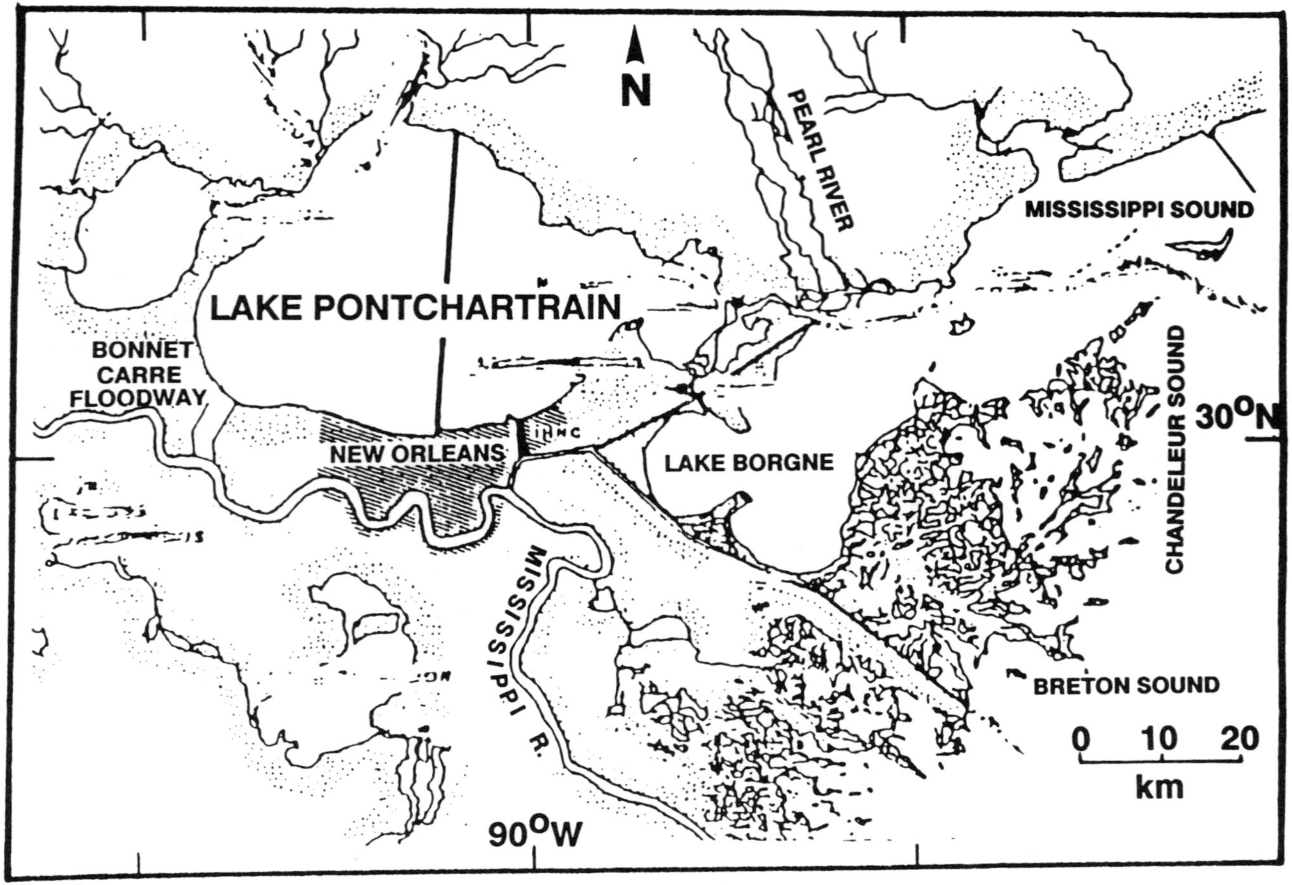

FIGURE 3. Map of Lake Pontchartrain, Louisiana.

in the 1950s showed an ecosystem that contained a sizable and diverse biota that was typical of similar southeastern habitats with respect to fishes and macrocrustaceans (Suttkus et al. 1954, Darnell 1958, 1961), zooplankton (Cuzon du Rest 1963), benthic invertebrates (Cory 1967, Dean and Ballis 1975, Wurtz and Roback 1955), and plants (Turner et al. 1980). However, by the late 1960s survey data showed considerable changes in the size and diversity of the community. Plant diversity and areal coverage decreased (Hayes 1968, Perret et al. 1971, Montz 1978), apparently due to eutrophication and the addition of herbicides, pesticides, and industrial and domestic chemicals (Perret et al. 1971), while zooplankton communities, typically dominated by *Acartia tonsa*, became even more monospecific and more restricted in distribution. *Rangia cuneata*, which thrived in 1953, was practically absent in the later surveys (Darnell 1979, Bahr et al. 1980). Sikora and Sikora (1982) controversially attributed much of the loss of benthic community diversity and numbers to the widespread hydraulic dredging (or mining) of clam shells, which produces considerable tax revenues for the State of Louisiana as well as considerable quantities of materials necessary for highway construction.

Indian River Lagoon, Florida

The Indian River Lagoon (Fig. 4) is a narrow (1–8 km in width) longitudinal complex of lagoonal embayments: the Mosquito Lagoon, the Indian River, and the Banana River, 220 km in length and covering approximately 150 000 ha in area (Mulligan and Snelson 1983). Although it is connected with the Atlantic Ocean by only four small inlets, there is no consistent pattern of freshwater input, and the effects of freshwater inflow are localized, so that salinities in most of the lagoon are near seawater strength. Despite this significant difference in salinity regime, the lagoon shares a great many physicochemical and biotic characteristics with freshwater lagoons. The lagoon is shallow (average depth 1.5 m) and currents are primarily wind driven. Lunar tides have little effect on water level or flow except near the inlet mouths (Smith 1983a). Approximately 2% of the bottom area (2776 ha) is vegetated (Thompson 1978) by typical seagrasses, *Syringodium filiforme*, *Halodule wrightii*, and *Thalassia testundinum*, *Ruppia maritima*, and macroalgae (Eiseman and Benz 1975, Gilbert and Clark 1981, Virnstein and Carbonara 1985). The grass flats and algal habitats support most of the Indian River Lagoon's invertebrate and fish populations (Kulczycki et al. 1981). Grass flats and open bottom areas near the inlets support a typical generalized estuarine infaunal community: *Crepidula*, bivalves, calanoid and harpacticoid copepods, caprellid and gammarid amphipods, and polychaetes (Young and Young 1978, Kerschner et al. 1985). Macrocrustaceans are represented by 38 species from 28 genera and 17 families, including caridean shrimp and brachyuran crabs. *Libinia dubia*, *Penaeus duorarum*, and *Palaemonetes intermedius* were the numerically dominant macrocrustaceans associated with seagrasses and drift algae (Gore et al. 1981). Food is so abundant that there is little possibility of competition between species for food (Zimmerman et al. 1979) and the

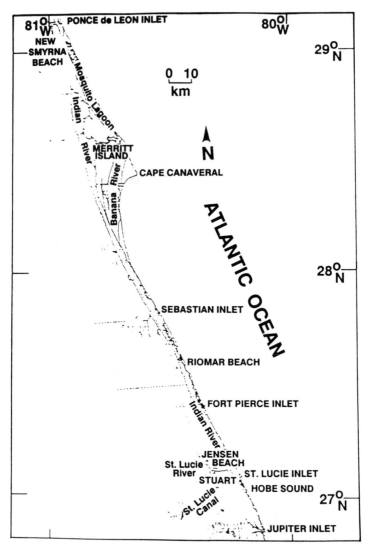

FIGURE 4. Map of the Indian River Lagoon system, Florida.

various species apparently successfully coexist by partitioning resources, although amphipod diversity may be inexplicably lower than in other seagrass-dominated subtropical estuaries at the same latitude (Virnstein et al. 1984). The fouling invertebrate community of the Loxahatchee River, a freshwater-dominated estuary near the Indian River Lagoon, consisted of species typical of other similar east coast habitats: oyster spat, *Balanus subalbidus*, *Neritina clenchii*, *Neritina reclivata*, and several species of serpulid worms (McPherson et al. 1984). Only *N. reclivata* and *B. subalbidus* regularly occurred at salinities less than 1 ppt (Poirrier and Partridge 1979).

The fish community of the Indian River Lagoon consists chiefly of the same euryhaline species found in other subtropical estuaries of the southeastern United States, although it is richer than most in seagrass-associated species and those with tropical affinities (Gilmore 1977, Kulczycki et al. 1981, Mulligan and Snelson 1983). While such habitat is not extensive, the Indian River Lagoon complex does contain several essentially freshwater habitats, chiefly river mouths and canals, the shores of which are lined with *Panicum* and *Typha*, with dense submerged populations of *Elodea* and *Hydrilla verticillata*, and with surface cover of *Eichornia crassipes*, *Pistice stratioites*, or *Pontederia lanceolata* (Gilmore 1977). Fishes in these freshwater tributaries included *Esox americanus*, *Aphredoderus sayanus*, *Centropomus parallelus*, *Cichlasoma octofasciatum*, *Xiphopherus helleri*, *Tilapia mariae*, *T. melanotheron*, and *Coryphopterus punctipectophorus* (Gilmore et al. 1983). Kushlan and Lodge (1974) noted that very few primary freshwater fishes occurred in Florida canals. Among fish collected in Indian River Lagoon tributaries they found *Amia calva*, *Notemigonus crysoleucas*, *Notropis petersoni*, and *N. maculatus*, *Erimyzon* sp., *Ictalurus catus*, *I. natalis*, *I. nebulosus*, *I. punctatus*, *Noturus gyrinus*, *Clarius batrachus*, *Elassoma evergladei*, *Enneacanthus gloriosus*, and *E. obesus*, *Pomoxis*, several species of *Lepomis*, *Micropterus salmoides*, but no percids. Gilmore and Hastings (1983) noted that peninsular Florida has no primary freshwater fishes representative of the tropical North American region, rather freshwater fishes have been recruited from the temperate fauna via existing land connections and so consist of the "typical" North American fish families: Centrarchidae, Percidae, Catostomidae, and Cyprinidae. While some tropical freshwater fishes have become established through introduction, the only fishes living in freshwater with tropical affinities are euryhaline species that occur seasonally in the Indian River Lagoon: *Oostethus lineatus* (= *O. brachyurus*), *Awaous tajasica*, *Gobionellus pseudifasciatus*, and *Gobiomorus dormitator*. The latter species also occurs in subtropical south Texas (Hoese and Moore 1977) and *Oostethus lineatus* occurs in fresh and salt marshes from Texas to North Carolina (Dawson 1970).

The shores of the Indian River were formerly bordered by *Spartina alterniflora* marshes. However, these have been mostly eliminated by impoundments for mosquito control (Gilmore et al. 1982). Mangroves (*Laguncularia racemosa* and *Avicennia germinans*) border most of the undeveloped shores of the lagoon. Such developmental pressures on the shorelines and the increasing demands upon the lagoon's limited freshwater supplies are the greatest threats to its ecology in the future.

Pamlico River Estuary, North Carolina

The Pamlico Estuary (Fig. 5) was highlighted by Copeland et al. (1974) as the typical "Oligohaline Regime" for the United States. It typifies a type of low-salinity estuarine habitat, which, while not unique to, is perhaps best represented on the Atlantic coast: a "classical" riverine estuary formed in a drowned river valley that demonstrates a longitudinal change in salinity re-

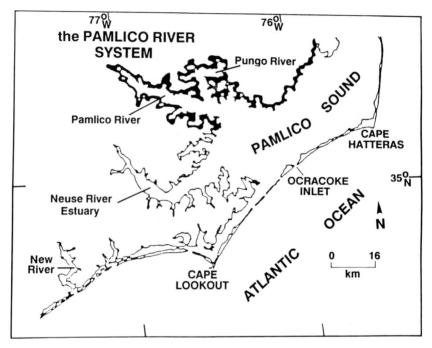

FIGURE 5. Map of Pamlico River Estuary, North Carolina.

gimes from freshwater to nearly marine conditions. These regimes do not occupy set areas, but fluctuate up and downstream daily and seasonally with fluctuations in tidal currents and riverine flow. Thus, riverine estuaries show little of the long-term stability that characterizes lagoonal habitats such as Lake Pontchartrain or the Indian River. The biota of such riverine systems is also characterized by changes—a transitory, often migratory, biota that experiences great fluctuations in community composition and population size, with concurrent variations in controlling factors.

The unique physicochemical environment that characterizes the freshwater–brackish water interface (Morris et al. 1978) certainly contributes to phytoplankton growth and production and may be responsible for the noted abundance of diatoms in freshwater reaches, in contrast to dinoflagellates commonly reported from the slightly higher salinity, oligohaline reaches. In this regard, the Pamlico River appears to be similar to other east coast river-dominated estuaries including the James, the Albemarle, the Cape Fear, and the Altamaha. Anderson (1986) found that silica, necessary for the growth of diatoms, is more readily available at the freshwater–brackish water interface, and hence diatom growth in this zone is enhanced relative to growth in either adjacent zone. Large spring–early summer and fall blooms are common in low-salinity waters (Stanley and Hobbie 1977, Witherspoon et al. 1979, Filardo and Dunstan 1985). Some investigators have assumed that nutrient input from rivers has been responsible for high phytoplankton growth rates (Thayer 1971). In Charlotte Harbor, Florida, highest cell counts were correlated with maximum river flow (Frazer and Wilcox 1981). Peaks in Chl-*a* remained in the riverine portion during low flow periods and moved into the harbor during high flow. Riverloads of silica varied seasonally and annually. Concentrations in the harbor remained relatively constant, although sharp decreases occurred with increases in chlorinity, which Frazer and Wilcox (1981) attribute to silica uptake by diatom populations in the harbor. Such silica-enhanced blooms must be ephemeral and transitory, moving up and down the estuary within a narrow "proper" salinity regime. However, many diatom species are stenohaline and cannot tolerate either fresh or high-salinity estuarine waters. Williams (1964) showed that lowered salinities reduced the growth and rate of division of marine diatoms, while increased salinities had analogous effects on freshwater species. In the James River phytoplankton maxima were inversely correlated with river discharge (Filardo and Dunstan 1985). During low discharge periods more than 50% of the chlorophyll *a* biomass that had been measured at 0 ppt disappeared as salinities increased to 2 ppt. Cell enumeration showed that freshwater species were the ones that disappeared. Filardo and Dunstan (1985) hypothesized that blooms of freshwater phytoplankton were advected into oligohaline waters where salinity stresses created mass mortalities. Analogous changes in distributions of freshwater/brackish invertebrates have also been noted following alterations in river discharge or salinity regime from natural or artificial causes (Harrel et al. 1976, Boesch et al. 1976). High flows in the Albemarle River of North Carolina following Hurricane Ginger in the fall of 1970 washed algae out of Albemarle Sound and no spring bloom occurred in 1971. However, conditions had normalized by the following year (Bowden and Hobbie 1977). During periods of low flow, saltwater encroachment may depress blue-green algae production in the Albemarle (Copeland et al. 1983) and Chowan rivers (Stanley and Hobbie 1977, Witherspoon et al. 1979).

The estuary is utilized by most animal species (macrocrustaceans, zooplankton, fish, and waterfowl) for only part of the life cycle, most often as either a spawning, nursery, or feeding ground. These seasonal cycles in species occurrence, abundance, and diversity are correlated with freshwater inflow patterns to permit the species to take advantage of different aspects of the oligohaline salinity regime, such as increased protection due to a depauperate predatory fauna or a food production maximum. East coast freshwater and oligohaline reaches are important spawning habitat for many anadromous species of fishes including the sturgeons (*Acipenser*) and shad (*Alosa*), which spawn near the freshwater/brackish water interface (Tyus 1974, Loesch and Lund 1977).

East Bay, Apalachicola Bay System, Florida

East Bay is a 3941-ha oligohaline backbay in the Apalachicola Bay system of Florida's west coast (Fig. 6). East Bay receives freshwater from the Apa-

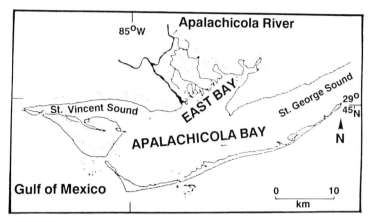

FIGURE 6. Map of the East Bay/Apalachicola Bay system, Florida.

lachicola River (see Chapter 9), which has the largest flow rate and broadest flood plain of any Florida river (Leitman et al. 1982) and whose discharge amounts to 35% of total runoff for Florida's west coast, as well as runoff from Tates Hell Swamp (Livingston 1984). Studies of the marshes above East Bay (Coultas 1980, Coultas and Gross 1975) indicate that the predominant vegetation is sawgrass (*Cladium jamaicensis*) in the lower salinity areas and black needlerush (*Juncus roemerianus*) in the higher salinity areas (Livingston and Joyce 1977).

Sediments in East Bay are silty sands and shelly sands (Livingston et al. 1977) and beds of submerged macrophytes cover 10% of the bottom area in upper East Bay where freshwater and brackish water species *Ruppia maritima*, *Vallisneria americana*, *Potamogeton pusillus*, *Cladophora*, and the exotic *Myriophyllum spicatum* predominate. The ecology of these grass beds, which have an annual production of 320–350 g C/m^2 year^{-1} has been studied by Livingston and Duncan (1979), Purcell (1977), Sheridan (1978, 1979), and Sheridan and Livingston (1979, 1983). These grass beds serve as habitat for many species of invertebrates. Among epibenthic forms *Neritina reclivata*, *Gammarus macromucronatus*, *Taphromysis bowmani*, and *Odostoma* spp. have been reported as common, while the infaunal community is dominated by polychaetes, such as *Loandalia americana* and *Mediomastus ambiseta*, the amphipod *Grandidierella bonnieroides*, and chironomids *Dicrontendipes* sp. The zooplankton mean standing crop is not high (4.0 mg/m^3) and shows the greatest amount of seasonal variation of any part of the bay system (Edmisten 1979). East Bay also supports the major oyster beds of the bay system.

East Bay is the most terrestrially dominated area of the Apalachicola Bay system (Livingston and Duncan 1979). The fish community is primarily estuarine, consisting of anchovies, sciaenids, and so on, but with significant "freshwater" influences, such as white catfish, channel catfish, gar, several species of cyprinodontids, *Menidia beryllina*, *Microgobius gulosus*, bluegill, and bluespotted sunfish (Purcell 1977, Livingston et al. 1978, Livingston and Duncan 1979). Sheridan (1979) noted that landward stations in East Bay, especially 5A (backbay) and 6 (river mouth), showed greater numbers of insects and freshwater crustaceans, *Daphnia*, etc., than open bay stations. Increased terrestrial/freshwater influences were also seen in the food habits of fishes collected at these stations. The Apalachicola River is the only Florida river that supports a *Morone saxatilis* fishery. However, this has declined drastically since the construction in 1955 of the Jim Woodruff Dam above the bay at the confluence of the Chattahoochee and Flint rivers (Livingston 1984).

Trinity Bay, Galveston Bay System, Texas

Trinity Bay (Fig. 7) is a secondary or backbay in the Galveston Bay System, Texas, which receives its freshwater inflow directly from the Trinity River and from surrounding marshlands (Diener 1975). Its area (36 000 ha) makes it one of the largest oligohaline systems on the Gulf coast.

Although the biota and ecology of Trinity Bay have received considerable attention in the past 20 years, a majority of scientific information on Trinity Bay still exists as contractual reports to governmental agencies or private corporations. Most of these sources are cited in Bechtel and Copeland (1970), Parker (1970), Holland et al. (1973), and Holt and Strawn (1983).

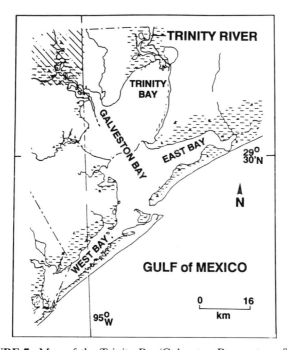

FIGURE 7. Map of the Trinity Bay/Galveston Bay system, Texas.

With respect to its biota, Trinity Bay appears to be typical of most other freshwater and oligohaline habitats in the southeastern United States, bearing the strongest resemblance to East Bay (Apalachicola Bay system, Florida). Both systems contain extensive open bottom habitat as well as oyster reefs and submerged macrophyte beds. Except where developed, their borders are lined with fresh or brackish marshes and floodplain swamps near the mouths of the major tributary rivers. Both bays receive freshwater input from major rivers that have upstream impoundments, diversions, and other significant demands of water for agricultural, industrial, and domestic purposes. These demands, coupled with natural fluctuations in river inflow caused by hurricanes, droughts, and so on, affect the critical salinity regimes that regulate the communities of plants and animals that sustain important commercial and recreational harvests in these estuarine systems. Reduced freshwater input associated with the filling of Lake Livingston on the Trinity River in 1971 affected macrozooplankton community composition (Holt and Strawn 1983). The construction of the Wallisville Dam in 1973 near the mouth of the Trinity River not only reduced Trinity River inflow but also flooded 4000 ha of brackish marsh estimated by Blackwater (1977) to annually produce 7 million tons of commercial fish and shellfish. While cause and effect were not firmly established, Parker (1970) confirmed a strong correlation between shrimp distribution, abundance, and migrations and salinity variations in the system. Both Bechtel and Copeland (1970) and Holland et al. (1973) noted that Trinity Bay communities were less diverse than those of lower Galveston Bay, except at stations nearest the river mouth, a condition they attributed to the "natural stress" of low and more variable salinities.

Atypical Systems

Although the five systems noted above may be taken as typical freshwater/oligohaline estuarine systems, they are by no means the only such systems, nor are they representative of all varieties of such systems in the southeastern United States. To try and note specifics of each low-salinity habitat would require more space and time than is available. There is, however, one different and atypical habitat that needs to be mentioned: the rivers of east Florida in which high concentrations of dissolved nonmarine salts permit the invasion by and establishment of populations of marine organisms in inland "fresh" waters. The St. John's River and Homosassa Spring of Florida's Atlantic coast and the Crystal River of the Gulf coast are unique in the extent of marine-derived organisms that penetrate into their freshwaters. The occurrence of marine fauna in these rivers has been explained by Odum (1953), Hulet et al. (1967), and Carrier and Evans (1976) as being due to the high calcium concentration and the presence of salt springs in these rivers. Indeed, Homossassa Spring, which has the largest number of marine species has the highest calcium concentration (Herald and Strickland 1949), and Crystal River, with the second largest number of marine species, also has the second greatest calcium concentration (Swift et al. 1986). The occurrence of these marine species in fresh water is not necessarily a year-round phenomenon, as Parrish and Yerger (1974) noted. In the Ochlockonee River, Florida, more estuarine species occurred during warmer months than in winter. The occurrence of a marine/estuarine fauna in Florida's rivers is not limited to fishes. The estuarine mysids *Mysidopsis almyra* (Price and Vodopich 1979) and *Taphromysis bowmani* (Beck 1977) as well as the blue crab (*Callinectes sapidus*) (Tagatz 1968) also occur in Floridian freshwaters.

Outside Florida the occurrence of resident populations of marine fish in freshwater is rarer. The pipefish *Syngnathus scovelli* has reproductive populations in the Gulf of Mexico (Whatley 1962, 1969) and Atlantic coasts (Target 1984) and may be more common in freshwater than has been previously supposed. Although species of *Fundulus* are quite euryhaline, *F. heteroclitus* does not normally inhabit fresh water. However, a breeding population has been reported from Pennsylvania (Denoncourt et al. 1978).

RESOURCE USE AND MANAGEMENT EFFECTS

Resource Potential and Past Impacts

Freshwater estuarine ecosystems act as buffers between upstream, freshwater riparian habitats and the downstream, typical estuarine zone. They store and diffuse freshwaters (especially storm runoff), minimize industrial and domestic pollution effects, increase the variety of food and cover available to fish and wildlife, and serve as a source for recharging coastal aquifers. Wetlands vegetation stabilizes pulsed movements of water into the system, modifies the hydrologic regime, and minimizes the natural instability of the system (Livingston and Duncan 1979). Freshwater estuarine areas are ecotones and, by definition, are variable, dynamic systems. Despite their natural variability, or perhaps because of it, these ecosystems are quite sensitive to natural and anthropogenic perturbations. Large systems, such as Lake Pontchartrain, exhibit considerable environmental stability, but this might be better thought of as environmental inertia, for when such systems are stressed by rapid environmental changes, their response may be dramatic and their recovery slow unless conditions return rapidly to normal.

Environmental impacts, whether induced by human or natural causes, can have profound influences on the biota of freshwater backbays and lagoons. The effects of natural perturbations, such as hurricanes and floods have been well documented (Chabreck and Palmisano 1973, Poirrier and Mulino 1975, 1977, Boesch et al. 1976, Schroeder 1977) and illustrate both the responsiveness and resilience of these communities. Habitats that provide reasonably stable seasonal environments can host different communities at different times and thus appear more stable than they actually are. The biota in such habitats is very dependent on the predictability of variations to which seasonal migrations may be cued (Benson 1981). Freshwater estuarine habitats typically contain a depauperate resident community that is highly resistant to change.

Reduction of species richness and abundance was characteristic of the upper parts of East and Trinity bays (Livingston and Duncan 1979, Holland et al. 1973). These natural tendencies were aggravated by upstream river modifications that affected freshwater inflow to these estuaries. However, these ecosystems also provide habitat for a large number of seasonal migrants, especially fish, invertebrates, and waterfowl, many of which utilize low-salinity marshes for breeding, winter feeding grounds, or nurseries for the young.

Within the marsh habitat many factors aside from salinity, such as soil characteristics, nutrient content, and tidal elevation, contribute to the distribution of each species. The variability of marsh habitats and the higher species diversity in freshwater marshes causes pronounced differences in community structure between even geographically neighboring marshes (Odum et al. 1984).

While true brackish water organisms have evolved adaptive mechanisms that permit their survival in variable salinities (Kinne 1964), the variations that normally occur in low-salinity habitats often exceed the compensatory abilities and result in massive mortalities and displacement of a significant part of the community. Many of these organisms could be considered "r-selected," that is, capable of rapid reproduction and colonization of an area when the environment becomes suitable. *Ruppia maritima*, although slow to reproduce vegetatively, possesses a very large seed bank, which gives this species the capability for rapid colonization (McMillan and Moseley 1967). Other freshwater marsh plants also possess large seed banks (Leck and Graveline 1979). Within the animal community, a similar capacity for rapid colonization and expansion has been noted in *Palaemonetes intermedius* when this species quickly appeared in Apalachee Bay, Florida, seagrass meadows after salinities decreased (Dugan and Livingston 1982). *P. pugio* has also been noted to favor areas with low salinities, while *P. vulgaris*, which favors higher salinities, was slow to recolonize North Carolina estuarine areas after the salinity had returned above 3 ppt (Thorp and Hoss 1975).

Resource Management Problems and Concerns

Coastal salt marsh productivity has been recorded as among the highest gross productivities found in natural ecosystems. Because of their greater diversity, freshwater marshes do not show the pronounced seasonal patterns of productivity found in salt marshes, and Whigham et al. (1978) have suggested that freshwater marshes might be at least as productive as salt marshes at the same latitude. Freshwater backbays and lagoons, including their constituent marshes, may be managed to enhance their natural functions, to mitigate the effects of natural or human-induced stresses, or to provide resources or services that are not normally associated with these ecosystems, but that are of value to society.

Coastal marshes have been managed to attract and provide feeding habitat for waterfowl (Chabreck 1961). Freshwater impoundments attract the most ducks, while brackish impoundments were utilized similar to natural marshes. Another potential use for managed freshwater coastal ecosystems is in aquaculture. Perry et al. (1970) have shown that similar food plants and water management schemes permit rearing crawfish in impoundments being managed for waterfowl. Whether this managed productivity is greater than, or of greater value to society than, the natural productivity of these systems remains an open question.

Like coastal floodplain swamps, which absorb nutrients and reduce levels of bacteria in river water (Kitchens et al. 1975), coastal freshwater marshes have been proposed to be used for tertiary sewage treatment. Seasonal patterns in nutrient fluxes as well as the marshes' natural limit for absorbing nutrients may limit this use of coastal freshwater marshes. Whigham and Simpson (1976) demonstrated that marshes take up nutrients from secondary treated effluents; however, Whigham et al. (1980) showed that this assimilation occurred only during the growing season in the spring and summer, and that the same marshes would release nutrients in fall and winter.

Often the value of freshwater estuarine ecosystems can be assessed through the real or potential fisheries supported by these habitats. Utilization of low-salinity habitats may be measured and changes in the population sizes of any organisms that have commercial or recreational value can be assessed in the time following any environmental manipulation. For example, freshwater fish were more common before the installation of weirs in Louisiana marshes (Fontenot and Rogillo 1970). Species such as *Scaphirhynchus platorhynchus*, *Esox niger*, *Morone mississippiensis*, *Lepomis* spp., *Micropterus salmoides*, and *Stizostedion canadense* disappeared as salinity increased, while gars, cyprinodonts, catfish, and drum (*Aplodinotus*) did not. The latter species are generally considered as less desirable from the angler's point of view than are most of the species that disappeared.

Natural and Human-Induced Stresses

Darnell (1976) divided the major activities that affect estuarine ecosystems into three categories: channelization, dredge and spoil, and impoundment. To these could be added the effects of pollution and biological contamination and the diversion of water for agricultural, industrial, and municipal uses. The estuarine environment is a complex one, and stresses due to single factors such as these cannot usually be studied without considering synergistic combinatory effects (Livingston 1979), which makes it difficult to discuss or assess the effects of any single modification or class of modifications.

Channelization refers to the creation of new channels or the deepening of existing channels to alter hydrological patterns or to permit or "improve" navigational uses. Channelization reduces the length of naturally meandering streams and channels, which in turn increases the velocity of currents, which can increase scouring and sediment transport. Channelization of upstream areas usually increases the severity of flooding in downstream areas by restricting flood waters to the river channel. Nutrient transport is reduced when waters are prevented from spreading over floodplains and floodplain swamps, important sources for river-borne nutrients (Day et al. 1980). Channelization of coastal streams or marshes not only effects the drainage pattern and dynamics of these systems, but also permits greater saltwater encroachment and often results in the need for construction of saltwater barriers as corrective measures (Stone et al. 1978). While such barriers may effectively exclude salt water (Weaver and Holloway 1974, Harrel et al. 1976), they also exclude those organisms that periodically migrate between salt water and the newly "protected" freshwater habitats. Management schemes that permit the passage of organisms may partially mitigate these effects (Davidson and Chabreck 1983), but are usually expensive and relatively ineffective (Herke 1971). Channelization anywhere leads to loss of habitat. In estuarine regions channelization may be used to help control or regulate salinities by creating additional channels or passes between an estuary and the sea (Reid 1957). The enlargement of Sykes' Cut to provide an additional connection between Apalachicola Bay and the Gulf of Mexico led to an increase in bay salinities and the reduction in nursery habitat and productivity (Livingston 1984). Grid ditching and channelization of brackish marshes has been widely practiced along the east coast from New Jersey (Talbot et al. 1986) to Florida (Gilmore et al. 1982) as a means of controlling the salt marsh mosquito (*Aedes solicitans*), which breeds in stagnant water pools in high marshes with restricted circulation. However, Daiber et al. (1976) note that biting insect problems are minimal in regularly flooded marshes without ditches or impoundments. Channelization usually involves dredging and so its effects cannot be entirely separated from those of dredge and spoil.

Dredge and fill operations create land in the form of spoil islands or "developed" marshlands, but reduce the size of open water habitat and inhibit natural circulation patterns, increasing turbidity and reducing productivity (Mee 1978). Adverse effects of sand, shell and gravel dredging practices remain controversial (Taylor and Salomon 1968, Sikora et al. 1981).

The construction of upstream dams and impoundments has been generally regarded as deleterious to normal estuarine functioning. Dam construction and subsequent lake filling on the Apalachicola River did not appear to be related to long-term cyclical patterns, but weekly cycles in river flow were established soon after dam operations began. Bay productivity, as measured by commercial fishery harvests, was not statistically related to river flow. However, standing crops of blue crab, oysters, and shrimp in the Apalachicola Bay system were correlated (Meeter et al. 1979).

Alterations in freshwater inflow include both reductions and increases, and either form of alteration can be harmful to the normal functioning of the estuarine environment (Armstrong 1980). Increased runoff and freshwater input carry larger quantities of pathogenic bacteria, such as coliforms (Skaggs et al. 1980), the presence of which are the commonest reason for oysters becoming unfit for human consumption (Bell and Canterbery 1974), and *Aeromonas hydrophila*, which causes red-spot disease in freshwater and estuarine fishes (Esch and Hazen 1980) and has been identified as the cause of serious infections and even deaths in humans (Hansen et al. 1977).

Decreased runoff can degrade the value of low- and medium-salinity habitats as nursery areas for fish and important crustacean species by leading to increases in salinity and decreases in important river-borne nutrients (Pate and Jones 1981). Although riverine flow is commonly the source of pollutants and contaminants, reduced freshwater inflow does not necessarily reduce pollutant inflow; rather, it reduces the dilution of pollutants (Funicelli and Rogers 1981).

FUTURE RESEARCH NEEDS

Freshwater backbays and lagoons lie between the habitats traditionally studied by freshwater and estuarine/marine ecologists. They also fall between or outside areas in which studies have traditionally been supported by major funding agencies. As a consequence, freshwater backbays and lagoons are among our least well-known estuarine habitats. While it is certainly true than virtually any study conducted in these habitats will contribute new information to our knowledge base, attention to certain questions will provide more immediately useful information.

Life history studies of valuable organisms (value in this case being determined from aesthetic and ecologic as well as commercial viewpoints) as well as studies at the population and community level are necessary to assess the importance to and utilization of low-salinity habitats by these species.

At the ecosystem level, comparative studies and those that investigate linkages between these systems and the contiguous true freshwater and mesohaline estuarine systems are necessary to properly assess the role of these low-salinity environments in the functioning of coastal ecosystems. As freshwater and low-salinity estuarine habitats become increasingly threatened by development, regulatory and resource management agencies have begun to apply results of studies from better-known higher salinity habitats (Nixon 1980) to predict the effects of modifications to low-salinity habitats. But are these results truly applicable? Until this question is answered, our best course of action would be to admit that we do not know what the effects will be and to disturb these environments as little as possible (Odum et al. 1984).

ACKNOWLEDGMENTS

I thank the staff of the Kimbel Library, Coastal Carolina College, especially J. Shuster, who assisted with bibliographic searches and acquisition of literature; J. Turner of Media Services, who prepared the figures; M. Brinson and C. Cordes, who provided constructive reviews of the manuscript; and my wife, R. Morris, who tolerated my monopolization of the study, and sometimes the dining room table, for the duration of this project.

REFERENCES

Abernethy, R. K., G. W. Petersen, and J. D. Gosselink. 1985. A small mammal trapping study of the floating freshwater marshes surrounding Lake Boeuf, Louisiana. *Northeast Gulf Sci.* 7:177–180.

Able, K. W., and J. D. Felley. 1986. Geographical variation in *Fundulus heteroclitus*: tests for concordance between egg and adult morphologies. *Am. Zool.* 26:145–157.

Adams, D. A. 1963. Factors influencing vascular plant zonation in North Carolina salt marshes. *Ecology* 44:445–455.

Adams, J. G. 1970. *Cluepeids of the Altamaha River, Georgia*, Contrib. Ser. No. 20. Brunswick: Coastal Fisheries Division, Georgia Fish and Game Commission.

Aller, R. C. 1978. Experimental studies of changes produced by deposit feeders on pore water, sediment, and overlying water chemistry. *Am. J. Sci.* 278:1185–1234.

Anderson, G. F. 1986. Silica, diatoms, and freshwater productivity maximum in Atlantic coastal plain estuaries, Chesapeake bay. *Estuarine, Coastal Shelf Sci.* 22:183–198.

Anderson, R. R. 1972. Submerged vascular plants of the Chesapeake Bay and tributaries. *Chesapeake Sci.* 13(Suppl.):S87–S89.

Anderson, R. R., R. G. Brown, and R. D. Rappleye. 1968. Water quality and plant distribution along the upper Patuxent River, Maryland. *Chesapeake Sci.* 9:145–156.

Armstrong, N. E. 1980. Effects of altered freshwater inflow on estuarine systems. In P. L. Fore and R. D. Peterson (eds.), *Proceedings of the Gulf of Mexico Coastal Ecosystem Workshop*, FWS/OBS 80-/30. Washington, DC: U.S. Fish Wildl. Serv., pp. 17–31.

Armstrong, N. E. 1987. The ecology of open bay bottoms of Texas: a community profile. *U.S. Fish Wildl. Serv., Biol. Rep.* 85(7.12):1–104.

Armstrong, N. E., and M. O. Hinson, Jr. 1973. Galveston Bay ecosystem freshwater requirements and phytoplankton productivity. In B. J. Copeland (ed.), *Toxicity Studies of Galveston Bay Project*. Austin: University of Texas, Marine Science Institute, pp. II-1–II-97.

Aspey, W. P. 1978. Fiddler crab behavioral ecology: burrow density in *Uca pugnax* (Smith) and *Uca pugilator* (Bosc) (Decapoda, Brachyura). *Crustaceana* 34:235–244.

Axelrad, D. M., K. A. Moore, and M. E. Bender. 1976. Nitrogen, phosphorus and carbon flux in Chesapeake Bay marshes. *Bull.—Va. Water Resour. Cent.* 79:1–182.

Bahr, L. M., J. P. Sikora, and W. B. Sikora. 1980. Macrobenthic survey of Lake Pontchartrain, Louisiana, 1978. pp. 659–696 In J. H. Stone (ed.), *Environmental Analysis of Lake Pontchartrain, Louisiana, Its Surrounding Wetlands and Selected Land Uses*, Report to U.S. Army Engineer District, New Orleans, Contract No. DACW29-77-C-0253. Baton Rouge: Coastal Ecology Laboratory, Center for Wetland Resources, Louisiana State University, Baton Rouge, Louisiana.

Bahr, L. M., R. Costanza, J. W. Day, Jr., S. E. Bayley, C. Neill, S. G. Leibowitz, and J. Fruci. 1983. Ecological characterization of the Mississippi deltaic plain region: a narrative with management recommendations. United States Fish and Wildlife Service, *Off. Biol. Serv.* [*Tech. Rep.*] FWS/OBS-82/69:1–189.

Bailey, R. M., H. E. Winn, and C. L. Smith. 1954. Fishes of the Escambia River, Alabama and Florida, with ecologic and taxonomic notes. *Proc. Acad. Nat. Sci. Philad.* 106:109–164.

Barnes, R. S. K. 1980. *Coastal Lagoons, the Natural History of a Neglected Habitat,* Cambridge Stud. Mod. Biol. 1. Cambridge, UK: Cambridge UP.

Bascescu, M. 1961. *Taphromysis bowmani* n. sp a new brackish water mysid from Florida. *Bull. Mar. Sci. Gulf Caribb.* 11:517–524.

Bechtel T. J., and B. J. Copeland. 1970. Fish species diversity indices as indicators of pollution in Galveston Bay, Texas. *Contrib. Mar. Sci.* 15:103–132.

Beck, J. T. 1977. Reproduction of the estuarine mysid *Taphromysis bowmani* (Crustacea: Malacostracea) in freshwater. *Mar. Biol. (Berlin)* 42:253–257.

Bedford, W. B., and J. W. Anderson. 1972. The physiological response of the estuarine clam *Rangia cuneata* (Gray) to salinity. I. Osmoregulation. *Physiol. Zool.* 45:255–260.

Bell, F. W., and E. R. Canterbery. 1974. *Estimated Historical Impact of Deteriorated Water Quality on United States Coastal Fishery Resources,* Contract WQ5AC008. Washington, DC: National Commission on Water Resources.

Bellrose, F. C. 1976. *Ducks, Geese and Swans of North America,* 2d ed. Harrisburg PA: Stackpole Books.

Benson, N. G. 1981. The freshwater-inflow-to estuaries issue. *Fisheries* 6:5–7.

Bent, A. C. 1926. Life histories of North American marsh birds. *Bull. — U.S. Natl. Mus.* 135:1–392.

Beshears, W. W. 1959. Aquatic plant studies in the Mobile Delta. *Ala. Conserv.* 31:14–17.

Biggs, R. B., and L. E. Cronin. 1981. Special characteristics of estuaries. In B. J. Neilson, and L. E. Cronin (eds.), *Estuaries and Nutrients.* Clifton, NJ: Humana, pp. 3–24.

Birkhead, W. S., and C. R. Bennett. 1981. Observations of a small population of estuarine-inhabiting alligators near Southport, North Carolina. *Brimleyana* 6:111–117.

Bishop, T. D., and C. T. Hackney. 1987. A comparative study of the mollusc communities of two oligohaline intertidal marshes: spatial and temporal distribution of abundance and biomass. *Estuaries* 10:141–152.

Blackwater, B. 1977. Dams, impoundments, reservoirs. In J. R. Clark (ed.), *Coastal Ecosystems: Ecological Considerations for Management of the Coastal Zone.* Washington, DC: Conservation Foundation, pp. 600–604.

Boesch, D. F., and R. E. Turner. 1984. Dependence of fishery species on salt marshes: the role of food and refuge. *Estuaries* 7:460–468.

Boesch, D. F., R. J. Diaz, and R. W. Virnstein. 1976. Effects of tropical storm Agnes on soft-bottom macrobenthic communities of the James and York estuaries and the lower Chesapeake Bay. *Chesapeake Sci.* 17:246–259.

Boshung, H. T., and A. F. Hemphill. 1960. Marine fishes collected from inland streams of Alabama. *Copeia* 1960:73.

Bourne, W. S. 1934. Sea-water tolerance of *Vallisneria spiralis* L. and *Potamogeton foliosus. Contrib. Boyce Thompson Inst.* 6:303–308.

Bowden, W. B. 1984. Nitrogen and phosphorous in the sediments of a tidal freshwater marsh in Massachusetts. *Estuaries* 7:108–118.

Bowden, W. B., and J. E. Hobbie. 1977. Nutrients in Albemarle Sound, North Carolina. *Univ. N. C. Sea Grant Coll. Program, Raleigh* 75-25:1–187.

Boynton, W. R., W. M. Kemp, and C. W. Keefe. 1982. A comparative analysis of nutrients and other factors influencing estuarine phytoplankton production. In V. S. Kennedy (ed.), *Estuarine Comparisons*. New York: Academic, pp. 69–90.

Bozeman, E. L., G. S. Helfman, and T. Richardson. 1985. Population size and home range of American eels in a Georgia tidal creek. *Trans. Am. Fish. Soc.* 114:821–825.

Brinson, M. M., and G. J. Davis. 1976. *Primary Productivity and Mineral Cycling in Aquatic Macrophyte Communities of the Chowan River, North Carolina*, Rep. No. 120. Raleigh: University of North Carolina, Water Resources Research Institute.

Brockmann, F. H. 1974. Seasonality of fishes in a south Florida brackish canal. *Fla. Sci.* 37:65–70.

Burt, W. H., and R. P. Grossenheider. 1976. *A Field Guide to the Mammals*, 3rd ed. Boston, MA: Houghton Mifflin.

Byrne, D. M. 1978. Life history of the spotfin killifish, *Fundulus luciae* (Pisces: Cyprinodontidae) in Fox Creek Marsh, Virginia. *Estuaries* 1:211–227.

Caffrey, J. M., and W. M. Kemp. 1990. Nitrogen cycling in sediment with estuarine populations of *Potamogeton perfoliatus* and *Zostera marina. Mar. Ecol.: Prog. Ser.* 66:147–160.

Cagle, F. R. 1952. A Louisiana terrapin population (*Malaclemys*). *Copeia* 1952:74–76.

Cahoon, D. R., and J. C. Stevenson. 1986. Production, predation, and decomposition in a low-salinity *Hibiscus* marsh. *Ecology* 67:1341–1350.

Caldwell, D. K., and M. C. Caldwell. 1973. Marine Mammals of the Eastern Gulf of Mexico. In J. I. Jones, R. E. Ring, M. O. Rinkel, and R. E. Smith (eds.), *A Summary of Knowledge of the Eastern Gulf of Mexico, 1973*. St. Petersburg: State University of Florida, Institute of Oceanography, pp. III-I-1 to III-I-24.

Cameron, G. N. 1972. Analysis of insect tropic diversity in two salt marsh communities. *Ecology* 53:58–73.

Carpenter, E. J. 1971. Annual phytoplankton cycle in the Cape Fear River estuary, North Carolina. *Chesapeake Sci.* 12:95–104.

Carpenter, J. H., D. W. Pritchard, and R. C. Whaley. 1969. Observations on eutrophication: causes, consequences and corrections. In *Eutrophication*. Washington, DC: National Academy of Sciences, pp. 210–221.

Carrier, J. C., and D. H. Evans. 1976. The role of environmental calcium in freshwater survival of the marine teleost, *Lagodon rhomboides. J. Exp. Zool.* 65:529–538.

Castagna, M., and D. Chanley. 1973. Salinity tolerance of some marine bivalves from inshore and estuarine environments in Virginia waters on the western mid-Atlantic coast. *Malacologia* 12:47–96.

Cavalieri, A. J., and A. H. C. Huang. 1979. Evaluation of proline accumulation in the adaptation of diverse species of marsh halophytes to the saline environment. *Am. J. Bot.* 66:307–312.

Chabreck, R. H. 1961. Coastal marsh impoundments for ducks in Louisiana. *Proc. Annu. Conf. Southeast. Assoc. Fish Game Comm.* 14:24–29.

Chabreck, R. H. 1971. The foods and feeding habits of alligators from fresh and saline environments in Louisiana. *Proc. Annu. Conf. Southeast. Assoc. Game Fish Comm.* 25:117–124.

Chabreck, R. H. 1972. Vegetation, water, and soil characteristics of the Louisiana Coastal region. *La., Agric. Exp. Stn., Bull.* 664.

Chabreck, R. H. 1979. Wildlife harvest in wetlands of the United States. In Greeson, J. R. Clark, and J. C. Clarke (eds.), *Wetland Functions and Values: The State of Our Understanding.* American Water Resources Association, Minneapolis, MN pp. 618–631.

Chabreck, R. H. 1979. Winter habitat of dabbling ducks—physical, chemical and biological aspects. *Waterfowl and Wetlands Symposium*, Madison, WI.

Chabreck, R. H., and A. W. Palmisano. 1973. The effects of hurricane Camile on the marshes of the Mississippi delta. *Ecology* 54:1118–1123.

Chapman, C. R. 1959. Oyster drill (*Thais haemostoma*) predation in Mississippi Sound. *Proc. Nat. Shellfish. Assoc.* 49:87–97.

Christmas, J. Y. 1973. *Cooperative Gulf of Mexico Estuarine Inventory and Study, Mississippi.* Ocean Springs: State of Mississippi Gulf Coast Research Laboratory.

Collier, A. C., and J. W. Hedgpeth. 1950. Introduction to the hydrography of Texas waters. *Publ. Inst. Mar. Sci., Univ. Tex.* 1:123–194.

Conner, W. H., and J. W. Day, Jr. (eds.). 1987. *The Ecology of Barataria Basin, Louisiana, an Estuarine Profile*, Biol. Rep. No. 85(7.13). Washington, DC: U.S. Fish & Wildlife Service.

Conner, W. H., J. W. Day, Jr., J. G. Gosselink, C. S. Hopkinson, Jr., and W. C. Stowe. 1987. Vegetation: composition and production. In W. H. Conner and J. W. Day, Jr. (eds.), *The Ecology of Barataria Basin, Louisiana, an Estuarine Profile*, Biol. Rep. No. 85(7.13). pp. 51–57. Washington, DC: U.S. Fish Wildl. Serv.

Conover, R. J. 1956. Oceanography of Long Island Sound 1952–1954. IV. Biology of *Acartia clausi* and *A. tonsa. Bull. Bingham Oceanogr. Collect.* 15:156–233.

Conte, F. S. 1972. Occurrence of *Taphromysis louisianae* Banner (Crustacea, Mysidacea) in marsh embayments on the Texas coast. *Southwest. Nat.* 17:203–204.

Copeland, B. J. 1966. Effects of decreased river flow on estuarine ecology. *J. Water Pollut. Control Fed.* 38:1831–1839.

Copeland, B. J., and E. G. Fruh. 1970. *Ecological Studies of Galveston Bay*, Final Rep. IAC (68-69)-408. Galveston Bay: Texas Water Quality Board.

Copeland, B. J., J. Thompson, and W. Ogletree. 1968. Effects of wind on water levels in the Texas Laguna Madre. *Tex. J. Sci.* 20:196–197.

Copeland, B. J., K. R. Tenore, and D. B. Horton. 1974. Oligohaline regime. In H. T. Odum, B. J. Copeland, and E. A. McMahon (eds.), *Coastal Ecological Systems of the United States*, Vol. 2, Washington, DC: Conservation Foundation, pp. 315–357.

Copeland, B. J., R. G. Hodson, S. R. Riggs, and J. E. Easley, Jr. 1983. The ecology of Albemarle Sound, North Carolina: an estuarine profile. *U. S. Fish Wildl. Serv., Off. Biol. Serv.* [*Tech. Rep.*] FWS/OBS-83/01:1–68.

Copeland, B. J., R. G. Hodson, and S. R. Riggs. 1984. The ecology of the Pamlico River, North Carolina: an estuarine profile. *U. S. Fish Wildl. Serv., Off. Biol. Serv.* [*Tech. Rep.*] FWS/OBS-79/31:1–103.

Cory, R. L. 1967. Epifauna of the Patuxent River estuary, Maryland for 1963 and 1964. *Chesapeake Sci.* 8:71–89.

Coultas, C. L. 1980. Soils of the marshes in the Apalachicola, Florida, estuary. *Soil Sci. Soc. Am. J.* 44:348–353.

Coultas, C. L., and E. R. Gross. 1975. Distribution and properties of some tidal marsh soils of Apalachee Bay, Florida. *Soil Sci. Soc. Am. Proc.* 39:914–919.

Cowardin, L. M., V. Carter, F. C. Golet, and E. T. LaRoe. 1979. Classification of wetlands and deepwater habitats of the United States. *U. S. Fish Wildl. Serv., Biol. Sci. Program* FWS/OBS-79/31:1–103.

Cowell B. C., and C. H. Resico, Jr. 1975. Life history patterns in the Coastal Shiner, *Notropis petersoni*, Fowler. *Fla. Sci.* 38:113–121.

Crance, J. H. 1971. Description of Alabama Estuarine Areas. Cooperative Gulf of Mexico Estuarine Inventory. *Ala. Mar. Resour. Bull.* 6:1–85.

Crane, J. 1975. *Fiddler Crabs of the World*. Princeton, NJ: Princeton UP.

Custer, T. W., and R. G. Osburn. 1978. Feeding habitat use by colonially breeding herons, egrets, and ibises in North Carolina. *Auk* 95:733–743.

Cuzon du Rest, R. P. 1963. Distribution of the zooplankton in the salt marshes of southeastern Louisiana. *Publ. Inst. Mar. Sci., Univ. Tex.* 9:132–155.

Dahlberg, M.D. 1972. An ecological study of Georgia coastal fishes. *Fish. Bull.* 70:323–354.

Dahlberg, M. D., and E. P. Odum. 1970. Annual cycles of species occurrence, abundance, and diversity in Georgia estuarine fish populations. *Am. Midl. Nat.* 83:382–392.

Daiber, F. C., L. L. Thornton, K. A. Bolster, T. G. Campbell, O. W. Crichton, G. L. Esposito, D. R. Jones, and J. M. Tyrawski. 1976. *An Atlas of Delaware's Wetlands and Estuarine Resources*, Tech. Rep. No. 2. Delaware Coastal Management Program, Wilmington, DE.

Darnell, R. M. 1958. Food habits of fishes and invertebrates of Lake Pontchartrain, Louisiana, an estuarine community. *Publ. Inst. Mar. Sci., Univ. Tex.* 5:353–416.

Darnell, R. M. 1959. Studies on the life history of the blue crab (*Callinectes sapidus* Rathbun) in Louisiana waters. *Trans. Am. Fish. Soc.* 88:294–304.

Darnell, R. M. 1961. Trophic spectrum of an estuarine community based on studies of Lake Pontchartrain, Louisiana. *Ecology* 42:553–568.

Darnell, R. M. 1962. Ecological history of Lake Pontchartrain, an estuarine community. *Am. Midl. Nat.* 68:434–445.

Darnell, R. M. 1976. Impacts of construction activities on wetlands of the United States. *Ecol. Res. Ser., U. S. Environ. Prot. Agency* 600/3-76-045, 1–392.

Darnell, R. M. 1979. *Hydrology of Lake Pontchartrain, Louisiana, during 1953–1955*. Coastal Ecology Laboratory, Center for Wetland Resources, Louisiana State University, Baton Rouge (unpublished manuscript).

Davidson, R. B., and R. H. Chabreck. 1983. Fish, wildlife and recreational values of brackish marsh impoundments. *Proc. Water Qual. Wetland Manage. Conf.*, New Orleans, LA.

Davis, C. C. 1944. On four species of copepods new to Chesapeake Bay with a description of a new variety of *P. crassirostris* Dahl. *Chesapeake Biol. Lab. Publ., Dep. Resour. Educ., Md.* 61:1–11.

Davis, G. J., and M. M. Brinson. 1976. *The Submerged Macrophytes of the Pamlico River Estuary, North Carolina*, Rep. No. 112. Raleigh: University of North Carolina, Water Resources Research Institute.

Davis, G. J., H. D. Bradshaw, M. M. Brinson, and C. M. Lekson. 1985a. Salinity and nutrient dynamics in Jacks, Jacob's and South Creeks in North Carolina, October 1981–November 1982. *J. Elisha Mitchell Sci. Soc.* 101:37–51.

Davis, G. J., H. D. Bradshaw, and S. M. Harlan. 1985b. Submerged macrophytes in Jacks and Jacobs Creeks, September 1981–February 1983. *J. Elisha Mitchell Sci. Soc.* 101:125–129.

Davis, J. R., and R. P. Cheek. 1966. Distribution, food habits and growth of young clupeids, Cape Fear River system, North Carolina. *Proc. Annu. Conf. Southeast. Assoc. Game Fish Comm.* 20:250–260.

Davis, J. R., and D. E. Louder. 1971. Life history and ecology of the cyprinid fish *Notropis petersoni* in North Carolina waters. *Trans. Am. Fish. Soc.* 100:726–733.

Davis, L. V., and I. E. Gray. 1966. Zonal and seasonal distribution of insects in North Carolina salt marshes. *Ecol. Monogr.* 36:275–295.

Davis, R. A., Jr. 1983. *Depositional Systems.* New York: Prentice-Hall.

Dawson, C. E. 1970. A Mississippi population of the opossum pipefish *Oostethus lineatus* (Syngnathidae). *Copeia* 1970:772–773.

Day, J. W., Jr., W. H. Conner, and G. P. Kemp. 1980. Contribution of wooded swamps and bottomland forests to estuarine productivity. In P. L. Fore and R. D. Peterson (eds.), *Proceedings of the Gulf of Mexico Coastal Ecosystem Workshop*, FWS/OBS 80-/30. Washington, DC: U. S. Fish Wildl. Serv., pp. 33–50.

Dean, T. A., and V. J. Ballis. 1975. Seasonal and spatial distribution of epifauna in the Pamlico River estuary, North Carolina. *J. Elisha Mitchell Sci. Soc.* 91:1–12.

Deevey, G. B. 1960. The zooplankton of the surface waters of the Delaware Bay region. *Bull. Bingham Oceanogr. Collect.* 17:5–53.

de la Cruz, A. 1981. Differences between South Atlantic and Gulf coastal marshes. In R. C. Carey, P. S. Markovits, and J. B. Kirkwood (eds.), *Proceedings of the Workshop on Coastal Ecosystems of the Southeastern United States*, FWS/OBS 80/59. Washington, DC: U. S. Fish Wildl. Serv., pp. 10–20.

Denoncourt, R. F., J. C. Fisher, and K. M. Rapp. 1978. A freshwater population of the mummichog, *Fundulus heteroclitus*, from the Susquehanna River drainage in Pennsylvania. *Estuaries* 1:269–272.

Deselle, W. J., M. A. Poirrier, J. S. Rogers, and R. C. Cashner. 1978. A discriminant functions analysis of sunfish (*Lepomis*) food habits and feeding niche segregation in the Lake Pontchartrain, Louisiana, estuary. *Trans. Am. Fish. Soc.* 107:713–719.

Diaz, R. J. 1980. Ecology of tidal freshwater and estuarine Tubificidae (Oligochaeta). In R. O. Brinkhurst and D. G. Cook (eds.), *Aquatic Oligochaete Biology.* New York: Plenum, pp. 319–330.

Diaz, R. J., and D. F. Boesch. 1977. Habitat development field investigations Windmill Point marsh development site, James River, Virginia. Appendix C: development with dredged material: acute impacts on the macrobenthic community. *U. S. Army Waterways Exp. Stn., Tech. Rep.* D-77-23.

Diener, R. A. 1975. Cooperative Gulf of Mexico Estuarine Inventory and Study— Texas: area description. *NOAA Tech. Rep., NMFS Circ* 393.

Doumlele, D. G., B. K. Fowler, and G. M. Silverhorn. 1985. Vegetative community structure of a tidal fresh-water swamp in Virginia. *Wetlands* 4:129–145.

Drew, R. D., and N. S. Schomer. 1984. An ecological characterization of the Caloosahatchee River/Big Cypress watershed. U. S. Fish Wildl. Serv., FWS/OBS-82/58.2: 1–225.

Dugan, P. J., and R. J. Livingston. 1982. Long-term variation of macroinvertebrate assemblages in Apalachee Bay, Florida. *Estuarine, Coastal Shelf Sci.* 14:391–403.

Duke, T. W., and T. R. Rice. 1967. Cycling of nutrients in estuaries. *Proc. Annu. Gulf Caribb. Fish. Inst.* 19:59–67.

Duobinis-Gray, E. M., and C. T. Hackney. 1982. Seasonal and spatial distribution of the carolina marsh clam *Polymesoda carolina* (Bosc) in a Mississippi tidal marsh. *Estuaries* 5:102–109.

Dyer, K. R. 1986. *Coastal and Estuarine Sediment Dynamics.* New York: Wiley (Interscience).

Eckstein, B. 1975. Possible reasons for the infertility of grey mullets confined to fresh waters. *Aquaculture* 5:41–51.

Edmisten, H. L. 1979. The zooplankton of the Apalachicola Bay System. Thesis, Florida State University, Tallahassee.

Eiseman, N. J., and M. C. Benz. 1975. Studies of the benthic plants of the Indian River region. In *Indian River Coastal Zone Study, 1974–1975*, Annu. Rep., Vol. 1. Ft. Pierce, FL: Harbor Branch Consortium, pp. 89–103.

Eleuterius, L. N. 1976. The distribution of *Juncus romerianus* in the salt marshes of North America. *Chesapeake Sci.* 17:289–292.

Eleuterius, L. N. 1980. Amino acids of the salt marsh rush *Juncus roemerianus. Arch. Hydrobiol.* 87:112–117.

Eleuterius, L. N., and C. K. Eleuterius. 1979. Tide levels and salt marsh zonation. *Bull. Mar. Sci.* 29:394–400.

Ellison, R. L., and M. M. Nichols. 1976. Modern and holocene foraminifera in the Chesapeake Bay region. *Mar. Sediments, Spec. Publ.* 1:131–151.

Emery, K. O., and R. E. Stevenson. 1957. Estuaries and Lagoons. In J. W. Hedgpeth (ed.), *Treatise on Marine Ecology and Paleoecology*, Mem. No. 67, Vol. 1. Boulder, CO: Geol. Soc. Am., pp. 673–750.

Epperly, S. 1984. Fishes of the Pamlico-Albemarle Peninsula, N. C.: area utilization and potential impacts. *N. C. Dep. Nat. Resour. Commun. Dev., Div. Mar. Fish., Spec. Sci. Rep.* 42/CEIP Rep. 23:1–129.

Esch, G. W., and T. C. Hazen. 1980. *The Ecology of Aeromonas hydrophila in Albemarle Sound, North Carolina*, Rep. No. 153. Raleigh: University of North Carolina, Water Resources Research Institute.

Estabrook, R. H. 1973. Phytoplankton ecology and hydrography of Apalachicola Bay. MS Thesis, Florida State University, Tallahassee.

Evans, J. 1970. About nutria and their control. *U. S. Fish Wildl. Serv., Resour. Publ.* 86:1–65.

Felley, J. D. 1987. Nekton assemblages of three tributaries to the Calcasieu estuary, Louisiana. *Estuaries* 10:321–329.

Felley, J. D. 1989. Nekton assemblages of the Calcasieu estuary. *Contrib. Mar. Sci.* 31:95–117.

Filardo, M. J., and W. M. Dunstan. 1985. Hydrodynamic control of phytoplankton in low salinity waters of the James River estuary, Virginia. *Estuarine, Coastal Shelf Sci.* 21:653–668.

Findlay, S., K. Howe, and H. K. Austin. 1990. Comparison of detritus dynamics in two tidal freshwater wetlands. *Ecology* 71:288–295.

Fingerman, M. F., and L. D. Fairbanks. 1956. Osmotic behavior and bleeding of the oyster *Crassostrea virginiana*. *Tulane Stud. Zool.* 3:151–168.

Fisher, T. R., P. R. Carlson, and R. J. Barber. 1981. Some problems in the interpretation of ammonium kinetics. *Mar. Biol. Lett.* 2:33–44.

Fisher, T. R., P. R. Carlson, and R. T. Barber. 1982. Carbon and nitrogen primary production in three North Carolina estuaries. *Estuarine, Coastal, Shelf Sci.* 15:621–644.

Fisher, T. R., L. W. Harding, D. W. Stanley, and L. G. Ward. 1988. Phytoplankton, nutrients and turbidity in the Chesapeake, Delaware and Hudson estuaries. *Estuarine, Coastal Shelf Sci.* 27:61–93.

Fontenot, B. J., Jr., and H. E. Rogillio. 1970. *A Study of the Estuarine Sportsfishes in the Biloxi Marsh Complex, Louisiana*, F-8 Completion Rep., Baton Rouge: Louisiana Wildlife and Fisheries Commission.

Forsythe, D. M. 1973. Gull populations at Charleston S. C., June 1971 to June 1972. *Chat* 37:57–62.

Foster, W. A., and J. E. Treherne. 1976. Insects of marine salt marshes: problems and adaptations. In L. Sheng, (ed.), *Marine Insects*. Amsterdam: Elsevier, pp. 5–42.

Frankenberg, D., and W. D. Burbank, 1963. A comparison of the physiology and ecology of an estuarine isopod *Cyathura polita* in Massachusetts and Georgia. *Biol. Bull. (Woods Hole Mass.)* 125:81–95.

Franks, J. S. 1970. An investigation of the fish population within the inland waters of Horn Island, Mississippi, a barrier island in the northern Gulf of Mexico. *Gulf Res. Rep.* 3:3–104.

Frazer, T. H., and W. H. Wilcox. 1981. Enrichment of a subtropical estuary with nitrogen, phosphorus and silica. In B. J. Neilson and L. E. Cronin (eds.), *Estuaries and Nutrients*. Clifton, NJ: Humana, pp. 481–498.

Fritz, E. S., and E. T. Garside. 1975. Comparison of age composition, growth and fecundity between two populations of *Fundulus heteroclitus* and *Fundulus diaphanus* (Pisces: Cyprinodontidae). *Can. J. Zool.* 53:361–369.

Funicelli, N., and H. M. Rogers. 1981. Reduced freshwater inflow impacts on estuaries. In R. C. Carey, P. S. Markovits, and J. B. Kirkwood (eds.), *Proceedings of the Workshop on Coastal Ecosystems of the Southeastern United States*, FWS/OBS-80/59. Washington, DC: U. S. Fish Wildl. Serv., pp. 214–219.

Galtsoff, P. S. 1964. The American oyster *Crassostrea virginiana*, Gmelin. *Fish. Bull.* 64:1–480.

Gauthier, J. D., T. M. Soniat, and J. S. Rogers. 1990. A parasitological survey of oysters along salinity gradients in coastal Louisiana. *J. World Aquacult. Soc.* 21:105–115.

Gibbons, J. W., and J. W. Coker. 1978. Herpetofaunal colonization patterns of Atlantic coast barrier islands. *Am. Midl. Nat.* 99:219–233.

Giese, G. L., H. B. Wilder, and G. G. Parker. 1979. Hydrology of major estuaries and sounds of North Carolina. *Water Resour. Invest. (U.S. Geol. Surv.)* 79–46:1–145.

Gilbert, S. S., and K. B. Clark. 1981. Seasonal variation in standing crop of the seagrass *Syringodium filiforme* and associated macrophytes in the northern Indian River, Florida. *Estuaries* 4:223–225.

Gilmartin, J., and N. Revelante. 1978. The phytoplankton characteristics of the barrier island lagoons of the Gulf of California. *Estuarine Coastal Mar. Sci.* 7:29–47.

Gilmore, R. G. 1977. Fishes of the Indian River Lagoon and adjacent waters, Florida. *Bull. Fla. State Mus. Biol. Sci.* 22:101–148.

Gilmore, R. G., and P. A. Hastings. 1983. Observations on the ecology and distribution of certain tropical peripheral fishes in Florida. *Fla. Sci.* 46:31–51.

Gilmore, R. G., D. W. Cooke, and C. J. Donohoe. 1982. A comparison of the fish populations and habitat in open and closed salt marsh impoundments in east-central Florida. *Northeast Gulf Sci.* 5:25–37.

Gilmore, R. G., P. A. Hastings, and D. J. Herrema. 1983. Ichthyofaunal additions to the Indian River Lagoon and adjacent waters, east-central Florida. *Fla. Sci.* 46:22–30.

Godwin, W. F., and J. G. Adams. 1969. Young clupeids of the Altamaha River, Georgia. *Ga. Game Fish Comm., Contrib. Ser.* 15:1–30.

Goldstein, A. K., and J. J. Manzi. 1976. Additions to the freshwater algae of South Carolina. *J. Elisha Mitchell Sci. Soc.* 92:9–13.

Golley, F. B. 1962. *Mammals of Georgia*. Athens: Georgia UP.

Gore, R. H., E. E. Gallaher, L. E. Scotto, and K. A. Wilson. 1981. Studies on decapod crustacea from the Indian River region of Florida IX. Community composition, structure, biomass and species-areal relationships of seagrass and drift algae associated macrocrustaceans. *Estuarine, Coastal Shelf Sci.* 12:485–508.

Gorsline, D. S. 1963. Oceanography of Apalachicola Bay, Florida. In *Essays in Honor of K. O. Emery*. Los Angeles: University of Southern California, pp. 146–176.

Gosselink, J. G. 1984. The ecology of delta marshes of coastal Louisiana: a community profile. *U. S. Fish Wildl. Serv.* FWS/OBS-84/09:1–134.

Gosselink, J. G., C. L. Cordes, and J. W. Parsons. 1979. *An Ecological Characterization Study of the Chenier Plain Coastal Ecosystem of Louisiana and Texas*, 3 vols., FWS/OBS-79/9 through 78/11. Washington, DC: U. S. Fish Wildl. Serv., Biol. Sci. Program.

Grassle J. P., and J. F. Grassle. 1976. Sibling species in the marine pollution indicator *Capitella* (Polychaeta). *Science* 192:567–569.

Griffith, R. W. 1974. Environment and salinity tolerance in the genus *Fundulus*. *Copeia* 1974:319–331.

Gunter, G. 1957. The predominance of the young among marine fishes found in fresh waters. *Copeia* 1957:13–16.

Gunter, G. 1961. Some relations of estuarine organisms to salinity. *Limnol. Oceanogr.* 6:182–190.

Gunter, G. 1979. Observations on territoriality in *Alligator mississippiensis*, the American alligator, and other points concerning its habits and conservation. *Gulf Res. Rep.* 6:79–81.

Gunter, G., H. Hildebrand, and R. Killebrew. 1964. Some relations of salinity to population distribution of motile estuarine organisms with specific reference to penaeid shrimp. *Ecology* 45:181–185.

Gunter, G., B. S. Ballard, and A. Venkataramiah. 1973. A review of salinity problems of organisms in United States coastal areas subject to the effects of engineering works. *Gulf Res. Rep.* 4:380–475.

Hackney, C. T. 1978. Summary of information: relationships of freshwater inflow to estuarine productivity along the Texas coast. *U. S. Fish Wildl. Serv.* FWS/OBS:78–73:1–16.

Hackney, C. T., and A. A. de la Cruz. 1981. Some notes on the macrofauna of an oligohaline tidal creek in Mississippi. *Bull. Mar. Sci.* 31:658–661.

Haedrich, R. L., and S. O. Haedrich. 1974. A seasonal survey of the fishes of the Mystic River, a polluted estuary in downtown Boston, Massachusetts. *Estuarine Coastal Mar. Sci.* 2:59–73.

Haller, W. T., D. I. Sutton, and W. C. Barlowe. 1974. Effects of salinity on growth of several aquatic macrophytes. *Ecology* 55:891–894.

Hallerman, E. M., R. O. Smitherman, P. B. Reed, W. H. Tucker, and R. A. Dunham. 1986. Biochemical generatics of largemouth bass in mesosaline and freshwater areas of the Alabama River system. *Trans. Am. Fish. Soc.* 115:15–20.

Hansen, A. G., J. Standridge, F. Jarrett, and D. Maki. 1977. Freshwater wound infection due to *Aeromonas hydrophila. JAMA, J. Am. Med. Assoc.* 238:1053–1054.

Hansen, R. G., and A. G. Eversole. 1984. Age, growth and sex ratio of American eels in brackish water portions of a South Carolina river. *Trans. Am. Fish. Soc.* 113:744–749.

Harrel, R. C., J. Ashcroft, R. Howard, and L. Patterson. 1976. Stress and community structure of macrobenthos in a Gulf coast riverine estuary. *Contrib. Mar. Sci.* 20:69–81.

Harrington, R. W., Jr., and E. S. Harrington. 1961. Food selection among fishes invading a high subtropical salt marsh from onset of flooding through the progress of a mosquisto brood. *Ecology* 42:646–666.

Harrington, R. W., Jr., and E. S. Harrington. 1972. Food of female marsh killifish *Fundulus confluentus* Goode and Bean in Florida. *Am. Mid. Nat.* 87:492–502.

Harris, V. T. 1953. Ecological relationships of meadow voles and rice rats in tidal marshes. *J. Mammal.* 34:479–487.

Hastings, R. W., and R. W. Yerger. 1971. Ecology and life history of the Diamond Killifish *Adenia xenica* (Jordan and Gilbert). *Am. Midl. Nat.* 86:276–291.

Hastings, R. W., D. A. Turner, and R. G. Thomas. 1988. The fish fauna of Lake Maurepas an oligohaline part of the Lake Pontchartrain estuary. *Northeast. Gulf Sci.* 9:89–98.

Hayes, R. R. 1968. *Potamogeton* in Louisiana. *Proc. La. Acad. Sci.* 31:82–90.

Heard, R. W. 1975. Feeding habits of white catfish from a Georgia estuary. *Fla. Sci.* 38:20–28.

Heard, R. W. 1982. *Guide to Common Tidal Marsh Invertebrates of the Northeastern Gulf of Mexico,* Mississippi–Alabama Sea Grant Consortium MASGP-79-004.

Hedgpeth, J. W. 1949. The North American species of *Macrobrachium. Tex. J. Sci.* 1:28–38.

Hedgpeth, J. W. 1983. Brackish waters, estuaries, and lagoons. In O. Kinne (ed.), *Marine Ecology,* Vol. 5, Pt. 2, Sect. 3.1. Chichester: Wiley, pp. 739–757.

Heidt, R. R., and R. J. Gilbert. 1981. *Seasonal Distribution and Daily Movements of Shortnosed Sturgeon in the Altamaha River, Georgia.* Final Report to the National Marine Fisheries Service Contract No. 03- 7-143-39165.

Heinle, D. R., and D. A. Flemmer. 1975. Carbon requirements of a population of estuarine copepods *Eurytemora afinis. Mar. Biol. (Berlin)* 31:235–247.

Helfman, G. S., E. L. Bozeman, and E. B. Brothers. 1984. Size, sex and age of American eels in a Georgia River. *Trans. Am. Fish. Soc.* 113:132–141.

Herald, E. S., and R. R. Strickland. 1949. An annotated list of the fishes of Homasassa Springs, Florida. *Q. J. Fla. Acad. Sci.* 11:99–109.

Herke, W. H. 1971. Use of natural, and semi-impounded, Louisiana tidal marshes as nurseries for fishes and crustaceans. Thesis, Lousiana State University, Baton Rouge.

Hobbs, H. H., Jr. 1952. The river shrimp *Macrobrachium ohione* (Smith) in Virginia. *Va. J. Sci.* 3:206–207.

Hoese, H. D. 1963. Salt tolerance of the eastern mudminnow, *Umbra pygmaea. Copeia* 1963:165–166.

Hoese, H. D. 1973. Abundance of the low salinity clam *Rangia cuneata* in southwestern Louisiana. *Proc. Natl. Shellfish Assoc.* 63:99–106.

Hoese, H. D., and R. H. Moore. 1977. *Fishes of the Gulf of Mexico, Texas, Louisiana, and Adjacent Waters.* College Station: Texas A & M UP.

Holland, J. S., N. J. Maciolek, and C. H. Oppenheimer. 1973. Benthic community structure as an indicator of water quality. *Contrib. Mar. Sci.* 17:169–188.

Holt, J., and K. Strawn. 1983. Community structure of macrozooplankton in Trinity and upper Galveston bays. *Estuaries* 6:66–75.

Hopkins S. H., and J. D. Andrews. 1970. *Rangia cuneata* on the east coast: thousand mile range extension or resurgence? *Science* 167:868.

Hulet, W. H., S. J. Mosel, L. H. Jodney, and R. C. Wehr. 1967. The role of Calcium in the survival of marine teleosts in dilute sea water. *Bull. Mar. Sci.* 17:677–688.

Humm, H. H. 1973. Seagrasses, In J. I. Jones, R. E. Ring, M. O. Rinkel, and R. E. Smith (eds.), *A Summary of Knowledge of the Eastern Gulf of Mexico, 1973.* St. Petersburg, FL: State University System Institute of Oceanography, pp. III-C-1 to III-C-10.

Huner, J. V. 1978. Observations on the biology of the river shrimp from a commercial bait fishery near Port Allen, Louisiana. *Proc. Annu. Conf. Southeast. Fish Wildl. Agencies* 31:380–386.

Huner, J. V. 1981. Information about the biology and culture of the red crawfish *Procambarus clarki* (Girard, 1852) (Decapoda, Cambaridae) for fisheries managers in Latin America. *An. Inst. Cien. Mar Limnol. Univ. Nat. Auton. Mex.* 8:43–50.

Hunt, B. P. 1960. Digestion rate and food consumption of Florida gars, warmouth, and largemouth bass. *Trans. Am. Fish. Soc.* 89:206–221.

Iverson, J. B. 1982. Biomass in turtle populations: a neglected subject. *Oecologia* 55:69–76.

Jacobs, J. E. 1971. A preliminary taxonomic survey of the freshwater algae of the Belle W. Baruch Plantation in Georgetown County, South Carolina. *J. Elisha Mitchell Sci. Soc.* 87:26–30.

Jenni, D. A. 1969. A study of the ecology of four species of herons during the breeding season at Lake Alice, Alachua County, Florida. *Ecol. Monogr.* 39:245–270.

Joanen, T. 1969. Nesting ecology of alligators in Louisiana. *Proc. Annu. Conf. Southeast. Assoc. Fish Game Comm.* 23:141–151.

Johnson, R. B., Jr. 1974. Ecological changes associated with the industrialization of Cedar Bayou and Trinity Bay, Texas. *Tex. Parks Wildl. Dep., Tech. Ser.* 16:1–79.

Jordan, T. E., D. F. Whigham, and D. L. Correll. 1989. The role of litter in nutrient cycling in a brackish tidal marsh. *Ecology* 70:1906–1915.

Kale, W. H., II. 1965. Ecology and bioenergetics of the long-billed marsh wren *Termatodytes palustris griseus* in a salt marsh ecosystem. *Nuttall Ornithol. Club Publ.* 5:1–142.

Kemp, W. M., and W. R. Boynton. 1980. Influence of biological and physical processes on dissolved oxygen dynamics in an estuarine system: implications for measuring community metabolism. *Estuarine Coastal Mar. Sci.* 11:407–431.

Kemp, W. M., and W. R. Boynton. 1984. Spatial and temporal coupling of nutrient inputs to estuarine primary production: the role of particulate transport and decomposition. *Bull. Mar. Sci.* 35:522–535.

Kennish, M. J. 1986. *Ecology of Estuaries*, Vol. 1. *Physical and Chemical Aspects.* Boca Ration, FL: CRC Press.

Kerschner, B. A., M. S. Peterson, and R. G. Gilmore, Jr. 1985. Ecotopic and ontogenetic trophic variation in mojarras (Pisces: Gerreidae). *Estuaries* 8:311–322.

Kerwin, J. A. 1971. Distribution of the fiddler crab *Uca minax* in relation to marsh plants within a Virginia estuary. *Chesapeake Sci.* 12:180–183.

Kerwin, J. A., and L. G. Webb. 1972. Foods of ducks wintering in South Carolina, 1965–1967. *Proc. Annu. Conf. Southeast. Game Fish Comm.* 25:223–245.

Keup, L., and J. Bayless. 1964. Fish distribution at varying salinities in Neuse River Basin, North Carolina. *Chesapeake Sci.* 5:119–123.

Kilby, J. D. 1955. The fishes of two Gulf coastal marsh areas of Florida. *Tulane Stud. Zool.* 2:175–247.

Kinne, O. 1964. The effects of temperature and salinity on marine and brackish water animals. II. Salinity and temperature-salinity combinations. *Oceanogr. Mar. Biol.* 2:281–339.

Kitchens, W. M., Jr., J. M. Dean, L. H. Stevenson, and J. H. Cooper. 1975. The Santee swamp as a nutrient sink. In F. G. Howell, J. B. Gentry, and M. H. Smith (eds.), *Mineral Cycling in Southeastern Ecosystems.* Washington, DC: U. S. Energy Research and Development Administration, Technical Information Center, pp. 349–366.

Kjerfve, B. 1986. Comparative oceanography of coastal lagoons. In D. A. Wolfe (ed.), *Estuarine Variability.* New York: Academic, pp. 63–81.

Kneib, R. T. 1984. Patterns in the utilization of the intertidal salt marsh by larvae and juveniles of *Fundulus heteroclitus* (Linnaeus) and *Fundulus luciae* (Baird). *J. Exp. Mar. Biol. Ecol.* 83:41–51.

Knowlton, R. E., and A. B. Williams. 1970. The life history of *Palaemonetes vulgaris* (Say) and *P. pugio* Holthuis in coastal North Carolina. *J. Elisha Mitchell Sci. Soc.* 86:185.

Kramer, J. R., S. E. Herbes, and H. E. Allen. 1972. Phosphorous, analyses of water, biomass and sediment. In *Nutrients in Natural Waters.* H. E. Allen and J. R. Kramer (eds.) New York: Wiley (Interscience), pp. 51–100.

Kuenzler, E. J., D. W. Stanley, and J. P. Koenings. 1979. *Nutrient Kinetics of Phytoplankton in the Pamlico River, N. C.*, Rep. No. 139. Raleigh: University of North Carolina, Water Resources Research Institute.

Kulczycki, G. R., W. G. Nelson, and R. W. Virnstein. 1981. The relationship between fish abundance and algal biomass in a seagrass-drift algae community. *Estuarine, Coastal Shelf Sci.* 12:341–347.

Kurz, H., and D. Wagner. 1957. Tidal marshes of the Gulf and Atlantic coasts of northern Florida and Charleston, South Carolina. *Fla. State Univ. Stud.* 24:1–165.

Kushlan, J. A. 1977. Population energetics of the white ibis. *Condor* 81:376–389.

Kushlan, J. A., and T. E. Lodge. 1974. Ecological and distributional notes on the freshwater fish of southern Florida. *Fla. Sci.* 37:110–128.

Lankford, R. R. 1977. Coastal lagoons of Mexico- their origins and classification. In M. L. Wiley (ed.), *Estuarine Processes*, Vol. 2, *Circulation, Sediments and Transport of Materials in the Estuary*. New York: Academic, pp. 182–216.

LaSalle, M. W., and T. D. Bishop. 1987. Seasonal abundance of aquatic diptera in two oligohaline tidal marshes in Mississippi. *Estuaries* 10:303–315.

Leck, M. A., and K. J. Graveline. 1979. The seed bank of a freshwater tidal marsh. *Am. J. Bot.* 66:1006–1015.

Leitman, H. M., J. E. Sohm, and M. A. Franklin. 1982. *Wetland Hydrology and Tree Distribution of the Apalachicola River Flood Plain, Florida*, Rep. No. 82: Washington, DC: U.S. Geol. Surv.

Leland, J. G. II. 1968. A survey of the sturgeon fishery of South Carolina. *Contrib. Bear's Bluff Lab.* 47:1–27.

Lewis, R. R., III, and E. D. Estevez. 1988. The ecology of Tampa Bay, Florida; an estuarine profile. *U.S., Fish Wildl. Serv., Biol. Rep.* 85(7.18):1–132.

Lewis, W. M., Jr. 1970. Morphological adaptations of cyprinodontoids for inhabiting oxygen deficient waters. *Copeia* 1970:319–326.

Linthurst, R. A., and E. D. Seneca. 1980. Aeration, nitrogen and salinity as determinants of *Spartina alterniflora* Loisel growth. *Estuaries* 4:53–63.

Linton, T. L. 1968. A description of the South Atlantic and Gulf Coast marshes and estuaries. In J. D. Newsom (ed.), *Proceedings of the Marsh Estuarine Management Symposium*. Baton Rouge: Louisiana State University, pp. 1–25.

Livingston, R. J. (ed.). 1979. *Ecological Processes in Coastal and Marine Systems*. New York: Plenum.

Livingston, R. J. 1984. The ecology of the Apalachicola Bay System: an estuarine profile. *U.S. Fish Wildl. Serv.* FWS/OBS 82/05:1–148.

Livingston, R. J., and J. L. Duncan. 1979. Climatological control of a North Florida coastal system and impact due to upland forestry management. In R. J. Livingston (ed.), *Ecological Processes in Coastal and Marine Systems*. New York: Plenum, pp. 339–381.

Livingston, R. J., and E. A. Joyce. 1977. *Proceedings of the Conference on the Apalachicola Drainage System*, Contrib. No. 26. Tallahassee, FL: Florida Marine Resources.

Livingston, R. J., R. L. Iverson, R. H. Estabrook, V. E. Keys, and J. Taylor. 1975. Major features of the Apalachicola Bay System: physiography, biota, and resource management. *Fla. Sci.* 37:245–271.

Livingston, R. J., P. S. Sheridan, B. G. McLane, F. G. Lewis, III, and G. G. Kobylinski. 1977. The biota of Apalachicola Bay System: functional relationships. *Fla. Mar. Resour. Pub.* 26:75–100.

Livingston, R. J., N. Thompson, and D. Meeter. 1978. Long-term variation of organochloride residues and assemblages of epibenthic organisms in a shallow north Florida (USA) estuary. *Mar. Biol. (Berlin)* 46:355–372.

Loesch, J., and W. A. Lund Jr. 1977. A contribution to the life history of the blueback herring *Alosa aestivalis. Trans. Am. Fish. Soc.* 106:583–589.

Lowery, G. H. 1974a. *Louisiana Birds*, 3d ed. Baton Rouge: Louisiana State UP.

Lowery, G. H. 1974b. *The Mammals of Louisiana and its Adjacent Waters.* Baton Rouge: Louisiana State UP.

Mann, K. 1982. *Ecology of Coastal Waters: A Systems Approach.* Berkeley: California UP.

Manzi, J. J., and R. G. Zingmark. 1978. Phytoplankton. In R. G. Zingmark (ed.), *An Annotated Checklist of the Biota of the Coastal Zone of South Carolina.* Columbia: South Carolina UP, pp. 2–36.

Maples, R. S. 1982. Bluegreen algae of a coastal salt panne and surrounding angiosperm zones in a Louisiana salt marsh. *Northeast Gulf Sci.* 6:39–43.

Maples, R. S., and J. C. Watson. 1979. Occurrence of microalgae in southwestern Louisiana salt marsh. *Gulf Res. Rep.* 6:301–303.

Martin, F. D. 1972. Factors influencing the local distribution of *Cyprinodon variegatus* (Pisces: Cyprinodontidae). *Trans. Am. Fish. Soc.* 101:89–93.

Martof, B. S., W. M. Palmer, J. R. Bailey, and J. H. Harrison, III. 1980. *Amphibians and Reptiles of the Carolinas and Virginia.* Chapel Hill: North Carolina UP.

Maxwell, G. R., III, and W. H. Kale, II. 1977. Breeding biology of five species of herons in coastal Florida. *Auk* 94:689–700.

May, E. B. 1974. Distribution of mud crabs (Xanthidae) in Alabama estuaries. *Proc. Natl. Shellfish. Assoc.* 64:33–37.

McBee, J. T., and W. J. Brehm. 1979. Macrobenthos of Simmons Bayou and an adjoining residential canal. *Gulf Res. Rep.* 6:211–216.

McBee, J. T., and W. T. Brehm. 1982. Spatial and temporal patterns in the macrobenthos of St. Louis Bay, Mississippi. *Gulf Res. Rep.* 7:115–124.

McConkey, E. H. 1954. A systematic study of the North American lizards of the genus *Ophiosaurus. Am. Midl. Nat.* 51:133–171.

McFarland, W. N., and P. E. Pickens. 1965. The effect of season, temperature, and salinity on standard and active oxygen consumption of the grass shrimp *Palaemonetes vulgaris* (Say). *Can. J. Zool.* 43:571–585.

McGuire, E. J. 1961. The influence of habitat NaCl concentrations on the distribution of two species of *Palaemonetes. Proc. La. Acad. Sci.* 24:71–75.

McIvor, C. C., and W. E. Odum. 1988. Food, predation risk, and microhabitat selection in a marsh fish assemblage. *Ecology* 69:1341–1351.

McMillan, C., and F. N. Moseley. 1967. Salinity tolerances of five marine spermatophytes of Redfish Bay, Texas. *Ecology* 48:503–506.

McPherson, B. F., W. H. Sonntag, and M. Sabanskas. 1984. Fouling community of the Loxahatchee River estuary, Florida, 1980–1981. *Estuaries* 7:49–157.

Meanley, B. 1972. *Swamps, River Bottoms, and Canebreaks*. Barre, MA: Barre.

Mee, L. E. 1978. Coastal lagoons. In J. P. Riley and R. Chester (eds.), *Chemical Oceanography*, 2d ed., Vol. 7. London: Academic, pp. 441–490.

Meehean, O. L. 1956. Notes on the freshwater shrimp *Palaemonetes paludosa* (Gibbes). *Trans. Am. Microsc. Soc.* 55:433–441.

Meeter, D. A., R. J. Livingston, and G. Woodsum. 1979. Short and long-term hydrological cycles of the Apalachicola drainage system with application to Gulf coastal systems. In R. J. Livingston (ed.), *Ecological Processes in Coastal and Marine Systems*. New York: Plenum, pp. 318–338.

Mendelssohn, I. A., and D. M. Burdick. 1987. The relationship of soil parameters and root metabolism to primary productivity in periodically innundated soils. In D. D. Hook et al. (eds.), *Ecology and Management of Wetlands*. London: Croom Helm, pp. 398–428.

Mendelssohn, I. A., and K. L. McKee. 1982. Sublethal stress controlling *Spartina alterniflora* productivity. In B. Gopal et al. (eds.), *Wetlands: Ecology and Management*. Jarpur, India: International Scientific, pp. 223–242.

Menzie, C. A. 1980. The chironomid (Insecta: Diptera) and other fauna of a *Myriophyllum spicatum* L. plant bed in the Lower Hudson River. *Estuaries* 3:38–54.

Merriner, J. V., W. H. Kriete, and G. C. Grant. 1976. Seasonality, abundance and diversity of fishes in the Piankantank River, Virginia (1970–1971). *Chesapeake Sci.* 17:238–245.

Meyers, V. B., and R. I. Iverson. 1981. Phosphorous and nitrogen limited phytoplankton production in northeastern Gulf of Mexico coastal estuaries. In B. J. Neilson and L. E. Cronin (eds.), *Estuaries and Nutrients*. Clifton, NJ: Humana, pp. 569–584.

Miller, D. C. 1965. Studies on systematics, ecology and geographical distribution of certain fiddler crabs. Thesis, Duke University, Durham, NC.

Miller, M. A., and W. D. Burbank. 1961. Systematics and distribution of an estuarine isopod crustacean, *Cyathura polita* (Stimpson, 1885) new comb. on the Gulf and Atlantic seaboard of the United States. *Biol. Bull. (Woods Hole, Mass.)* 120:62–84.

Minshall, G. W., K. W. Cummings, R. C. Petersen, C. E. Cushing, D. A. Burns, J. R. Sedell, and R. L. Vannote. 1985. Developments in stream ecosystem theory. *Can. J. Fish. Aquat. Sci.* 42:1045–1055.

Mitsch, W. J., and J. G. Gosselink. 1986. *Wetlands*. New York: Van Nostrand-Reinhold.

Mitton, J. B., and R. K. Koehn. 1976. Morphological adaptation to thermal stress in a marine fish, *Fundulus heteroclitus. Biol. Bull. (Woods Hole, Mass.)* 151:548–559.

Montague, C. L. 1980. A natural history of temperate western Atlantic fiddler crabs (genus *Uca*) with reference to their impact on the salt marsh. *Contrib. Mar. Sci.* 23:25–55.

Montgomery, J. R., C. F. Zimmermann, G. Petersen, and M. Price. 1983. Diel variations of dissolved ammonia and phosphate in estuarine sediments. *Fla. Sci.* 45:535–542.

Montz, G. N. 1978. The submerged vegetation of Lake Pontchartrain, Louisiana. *Castanea* 43:115–128.

Moore, R. H. 1974. General ecology, distribution, and abundance of the mullets *Mugil cephalus* and *Mugil curema* on the south Texas coast. *Contrib. Mar. Sci.* 18:242–255.

More, W. R. 1969. A contribution to the biology of the blue crab (*Callinectes sapidus* Rathbun). *Tex. Parks Wildl. Dep., Tech. Ser.* 1:1–31.

Morin, R. P., and K. W. Able. 1984. Patterns of geographic variation in the egg morphology of the fundulid fish, *Fundulus heteroclitus. Copeia* 1983:726–740.

Morris, A. W., R. F. C. Mantoura, A. J. Bale, and R. J. M. Howland. 1978. Very low salinity regions of estuaries: important sites for chemical and biological reactions. *Nature (London)* 274:678–680.

Mulligan, T. J., and F. N. Snelson, Jr. 1983. Summer-season populations of epibenthic marine fishes in the Indian River Lagoon System, Florida. *Fla. Sci.* 46:250–276.

Myers, G. S. 1938. Fresh-water fish and West Indian zoogeography. *Smithson. Inst. Annu. Rep.* (1937):339–364.

Naughton, S. P., and C. H. Saloman. 1978. Fishes of the nearshore zone of St Andrew Bay, Florida and adjacent coast. *Northeast Gulf Sci.* 2:43–55.

Negus, N. C., E. Gould, and R. K. Chipman. 1961. Ecology of the rice rat, *Oryzomys palustris* (Harlan) on Breton Island. Gulf of Mexico, with a critique of the social stress theory. *Tulane Stud. Zool.* 8:93–123.

Neill, W. T. 1958. The occurrence of amphibians and reptiles in salt-water areas, and a bibliography. *Bull. Mar. Sci. Gulf Caribb.* 8:1–97.

Nixon, S. W. 1980. Between coastal marshes and coastal waters—a review of twenty years of speculation and research on the role of salt marshes in estuarine productivity and water chemistry. In P. Hamilton and K. B. MacDonald (eds.), *Estuarine and Wetlands Resources*. New York: Plenum, pp. 437–525.

Nixon, S. W. 1982. Nutrient dynamics, primary production and fisheries yields of lagoons. Oceanologica Acta 1982. *Proc. Int. Symp. Coastal Lagoons SCOR/IABO/ UNESCO*, Bordeaux 8–14 Sept *1981*; pp. 357–371.

Nowicki, B. L., and S. W. Nixon. 1985. Benthic nutrient remineralization in a coastal lagoon ecosystem. *Estuaries* 8:182–190.

Odum, E. P. 1971. *Fundamentals of Ecology*. New York: Saunders.

Odum, E. P. 1980. The status of three ecosystem-level hypotheses: tidal subsidy, outwelling, and detritus-based food chains. In V. S. Kennedy (ed.), *Estuarine Perspectives* New York: Academic Press, pp. 485–495.

Odum, H. T. 1953. Factors controlling marine invasion into Florida fresh water. *Bull. Mar. Sci. Gulf Caribb.* 3:134–156.

Odum, H. T. 1967. Biological circuits and marine systems of Texas. In T. A. Olson and F. J. Burgess (eds.), *Pollution and Marine Ecology*. New York: Wiley (Interscience), pp. 99–157.

Odum, W. E. 1988. Comparative ecology of tidal freshwater and salt marshes. *Annu. Rev. Ecol. Syst.* 19:147–176.

Odum, W. E., and E. J. Heald. 1972. Trophic analysis of an estuarine mangrove community. *Bull. Mar. Sci.* 22:671–738.

Odum, W. E., and T. J. Smith, III. 1981. Ecology of tidal, low salinity ecosystems. In R. C. Carey, P. S. Markovits, and J. B. Kirkwood (eds.), *Proceedings of the Workshop on Coastal Ecosystems of the Southeastern United States*, FWS/OBS 80/59. Washington, DC: U. S. Fish Wildl. Serv., pp. 36–44.

Odum, W. E., M. L. Dunn, and T. J. Smith III. 1978. Habitat value of tidal freshwater wetlands. In P. E. Greeson, J. R. Clark, and J. E. Clark (eds.), *Wetland Functions and Values: The State of Our Understanding*, Proc. Nat. Symp. Wetlands. Minneapolis, MN: Am. Water Resour. Assoc., pp. 248–255.

Odum, W. E., T. J. Smith, III, J. K. Hoover, and C. C. McIvor. 1984. The ecology of tidal freshwater marshes of the United States east coast: a community profile. *U. S., Fish Wildl. Serv.* FWS/OBS-83/17:1–177.

Otte, A. M., and V. J. Bellis. 1985. Edaphic diatoms of a low salinity estuarine marsh system in North Carolina—a comparative floristic study. *J. Elisha Mitchell Sci. Soc.* 101:116–124.

Otto, J., and S. K. Pierce. 1981. Water balance systems of *Rangia cuneata*: ionic and amino acid regulation in changing salinities. *Mar. Biol. (Berlin)* 61:185–192.

Palmisano, A. W. 1972. Habitat preference of waterfowl and fur animals in the northern Gulf coast marshes. In R. H. Chabreck (ed.), *Proceedings of the Marsh and Estuary Management Symposium*. Baton Rouge: Louisiana State University Press, pp. 163–177.

Parenti, L. 1981. A phylogenetic and biogeographic analysis of cyprinodontiform fishes (Teleostei, Atherinomorpha). *Bull. Am. Mus. Nat. Hist.* 168:355–357.

Parker, J. C. 1970. Distribution of juvenile brown shrimp (*Penaeus aztecus* Ives) in Galveston Bay Texas, as related to certain hydrographic features and salinity. *Contrib. Mar. Sci.* 15:1–12.

Parrish, R. P., and R. W. Yerger. 1974. Ochlockonee River fishes: salinity-temperature effects. *Fla. Sci.* 36:339–364.

Parsons, K. A. 1978. Insect population diversity and the effects of grasshopper (Othroptera: Tetigoniidae) grazing on the energy flow of a Mississippi *Juncus roemerianus* marsh. Thesis, Mississippi State University, Mississippi State.

Pate, P. P., Jr. and R. Jones. 1981. Effects of upland drainage on estuarine nursery areas of Pamlico Sound, North Carolina. In R. D. Cross and D. L. Williams, eds. *Proc. Natl. Symp. Freshwater Inflow Estuaries* Vol. II. U.S. Fish and Wildl. Serv. FWS/OBS—81/04:402–418.

Pearcy, R. W., and S. L. Ustin. 1984. Effects of salinity on growth and photosynthesis of three California tidal marsh species. *Oecologia* 62:68–73.

Pearse, A. S. 1936. Estuarine animals at Beaufort, North Carolina. *J. Elisha Mitchell Sci. Soc.* 32:174–222.

Penfound, W. T., and E. S. Hathaway. 1938. Plant communities of the marshlands of southeastern Louisiana. *Ecol. Monogr.* 8:1–56.

Perez, K. 1969. An orthokinetic response to rate of salinity change in two estuarine fishes. *Ecology* 50:454–457.

Perret, W. S., B. Barrett, W. Latapie, J. Polland, W. Mock, G. Adkins, W. Gaidry, and C. White. 1971. *Cooperative Gulf of Mexico Estuarine Inventory and Study, Louisiana*, Phase I Area Description. New Orleans: Louisiana Wildlife and Fish Commission.

Perry, G. W. 1968. Distribution and relative abundance of the blue catfish, *Ictalurus furcatus* and channel catfish *Ictalurus punctatus* with relation to salinity. *Proc. Annu. Conf. Southeast. Assoc. Game Fish Comm.* 21:438–444.

Perry, W. G., Jr., T. Joanen, and L. McNease. 1970. Crawfish- waterfowl, a multiple use concept for impounded marshes. *Proc. Annu. Conf. Southeast. Assoc. Game Fish Comm.* 23:178–188.

Peters, D. S., and M. T. Boyd. 1972. The effect of temperature, salinity, and availability of food on the feeding and growth of the hogchoaker, *Trinectes maculatus* (Bloch & Schneider). *J. Exp. Mar. Biol. Ecol.* 7:201–207.

Pfeiffer, W. J., and R. G. Weigert. 1981. Grazers on *Spartina* and their predators. In L. R. Pomeroy and R. G. Weigert (eds.), *The Ecology of a Salt Marsh*. New York: Springer, pp. 87–112.

Phillip, C. C., and R. G. Brown. 1965. Ecological studies of transition-zone vascular plants in South River, Maryland. *Chesapeake Sci.* 6:73–81.

Phillips, R. C. 1960. Observations on the ecology and distribution of the Florida seagrasses. *Fla. State Board Conserv., Prof. Pap. Ser.* (2):i–iv, 1–72.

Phleger, C. F. 1971. Effect of salinity on growth of a salt marsh grass. *Ecology* 52:908–911.

Pierce, R. A., and J. Davis. 1972. Age, growth and mortality of white perch, *Morone americana*, in the James and York Rivers, Virginia. *Chesapeake Sci.* 13:272–281.

Poirrier, M. A., and M. M. Mulino. 1975. The effects of the 1973 opening of the Bonnet Carrè Spillway upon epifaunal invertebrates in southern Lake Pontchartrain. *Proc. La. Acad. Sci.* 38:36–40.

Poirrier, M. A., and M. M. Mulino. 1977. The impact of the 1975 Bonnet Carrè Spillway opening on epifaunal invertebrates in southern Lake Pontchartrain. *J. Elisha Mitchell Sci. Soc.* 93:11–18.

Poirrier, M. A., and M. R. Partridge. 1979. The barnacle *Balanus subalbidus* as a salinity indicator in the oligohaline estuarine zone. *Estuaries* 2:204–206.

Portnoy, J. W. 1977. Nesting colonies of seabirds and wading birds- coastal Louisiana, Mississippi, and Alabama. *U. S. Fish Wildl. Serv.* FWS/OBS-77/07:1–126.

Potter, E. F., J. F. Parnell, and R. P. Teulings. 1980. *Birds of the Carolinas*. Chapel Hill: North Carolina UP.

Powell, J. A., and G. B. Rathbun. 1984. Distribution and abundance of manatees along the northern coast of the Gulf of Mexico. *Northeast. Gulf Sci.* 7:1–28.

Powers, L. W., and J. F. Cole. 1976. Temperature variations in fiddler crab microhabitats. *J. Exp. Mar. Biol. Ecol.* 21:141–157.

Price, W. A. 1947. Equilibrium of form and forces in tidal basins on the coast of Texas and Louisiana. *Am. Assoc. Pet. Geol. Bull.* 31:1619–1663.

Price, W. W., and D. S. Vodopich. 1979. Occurrence of *Mysidopsis almyra* (Mysidacea, Mysidae) on the east coast of Florida. U.S.A. *Crustaceana* 36:194–196.

Pritchard, D. W. 1967. What is an estuary: physical viewpoint. In G. H. Lauff (ed.), *Estuaries*. Spec. Publ. No. 3. Washington, DC: Am. Assoc. Adv. Sci., pp. 3–5.

Pritchard, D. W., and J. R. Schubel. 1981. Physical and ecological processes controlling nutrient levels in estuaries. In B. J. Neilson and L. E. Cronin (eds.), *Estuaries and Nutrients*. Clifton, NJ: Humana, pp. 47–70.

Purcell, B. H. 1977. The ecology of epibenthic fauna associated with *Valisneria americana* beds in a north Florida estuary. Thesis, Department of Oceanography, Florida State University, Tallahassee.

Reid, G. K. 1957. Biologic and hydrographic adjustment in a disturbed Gulf Coast estuary. *Limnol. Oceanogr.* 2:198–212.

Reiger, G. 1977. Native fish in troubled waters. *Audubon* 79:18–41.

Reimer, R. D., K. Strawn, and A. Dixon. 1974. Notes on the river shrimp *Macrobrachium ohione* (Smith). *Trans. Am. Fish. Soc.* 103:120–126.

Reimold, R. J. 1977. Mangals and salt marshes of eastern United States. In V. J. Chapman (ed.), *Wet Coastal Ecosystems*, Vol. 1. New York: Elsevier, pp. 157–166.

Reintjes, J. W., and A. L. Pacheco. 1966. The relation of menhaden to estuaries. In R. F. Smith, A. H. Swartz, and W. H. Massman (eds.), *Symposium on Estuarine Fisheries*, Spec. Publ. No. 3. Washington, DC: Am. Fish. Soc., pp. 50–58.

Remane, A., and C. Schlieper. 1971. *The Biology of Brackish Water*, 2d ed. New York: Wiley (Interscience).

Renfro, W. C. 1959. Survival and migration of fresh-water fishes in salt water. *Tex. J. Sci.* 11:172–180.

Renfro, W. C. 1960. Salinity relations of some fishes in the Aransas River, Texas. *Tulane Stud. Zool.* 8:83–91.

Ringold, P. 1979. Burrowing, root mat density, and the distribution of fiddler crabs in the eastern United States. *J. Exp. Mar. Biol. Ecol.* 36:11–21.

Rogers, B. D., and W. H. Herke. 1985. *Temporal Patterns and Size Characteristics of Migrating Juvenile Fishes and Crustaceans in a Louisiana Marsh*, Res. Rep. No. 5. Baton Rouge: Louisiana Agricultural Experimental Station.

Rogers, S. G., T. E. Targett, and S. B. Van Sant. 1984. Fish-nursery use in Georgia salt-marsh estuaries: the influence of springtime freshwater conditions. *Trans. Am. Fish. Soc.* 113:595–606.

Roman, C. T., W. A. Niering, and R. S. Warren. 1984. Salt marsh vegetation change in response to tidal restriction. *Environ. Manage.* 8:141–150.

Roman, C. T., K. W. Able, M. A. Lazzari, and K. L. Heck. 1990. Primary productivity of angiosperm and macroalgae dominated habitats in a New England salt marsh: a comparative analysis. *Estuarine, Coastal Shelf Sci.* 30:35–46.

Rounsefell, G. A. 1964. Preconstruction study of the fisheries of the estuarine areas traversed by the Mississippi River-Gulf Outlet project. *Fish. Bull.* 63:373–393.

Rozas, L. P., and C. T. Hackney. 1984. Use of oligohaline marshes by fishes and macrofaunal crustaceans in North Carolina. *Estuaries* 7:213–224.

Rozas, L. P., and W. E. Odum. 1987. Use of freshwater marshes by fishes and macrofaunal crustaceans along a marsh stream-order gradient. *Estuaries* 10:36–43.

Ruibal, R. 1959. The ecology of a brackishwater population of *Rana pipiens. Copeia* 1959:315–322.

Sage, W. W., and M. J. Sullivan. 1978. Distribution of bluegreen algae in a Mississippi gulf coast salt marsh. *J. Phycol.* 14:333–337.

Sandifer, P. A., J. V. Miglarese, D. R. Calder, J. J. Manzi, and L. A. Barclay. Ecological characterization of the Sea Island coastal region of South Carolina and Georgia. Vol. III. Biological features of the characterization area. *U. S. Fish Wildl. Serv.* FWS/OBS-79/42, 1–620.

Schroeder, W. W. 1977. The impact of the 1973 flooding of the Mobile River system on the hydrography of Mobile Bay and East Mississippi Sound. *Northeast. Gulf Sci.* 1:68–76.

Schwartz, F. J. 1964. Natural salinity tolerances of some freshwater fishes. *Underwater Nat.* 2:13–15.

Schwartz, F. J., W. T. Hogarth, and M. P. Weinstein. 1982. Marine and freshwater fishes of the Cape Fear Estuary, North Carolina and their distribution in relation to environmental factors. *Brimleyana* 7:17–37.

Shanholtzer, G. F. 1974. Relationships of vertebrates to salt marsh plants. In R. J. Reimold and W. H. Queen (eds.), *Ecology of Halophytes.* New York: Academic Press, pp. 463–474.

Shaw, S. P., and C. G. Ferdine. 1956. Wetlands of the United States. *U. S., Fish Wildl. Serv., Circ.* 39:1–69.

Shepard, F. J., and D. G. Moore. 1960. Bays of the Central Texas Coast. In F. J. Shepard, F. B. Phleger, and T. J. VanAndel (eds.), *Recent Sediments of the Northwestern Gulf of Mexico.* Tulsa OK: Am. Assoc. Pet. Geol., pp. 117–152.

Sheridan, P. F. 1978. Food habits of the Bay Anchovy, *Anchoa mitchilli*, in Apalachicola Bay, Florida. *Northeast. Gulf Sci.* 2:126–132.

Sheridan, P. F. 1979. Trophic resource utilization by three species of sciaenid fishes in a northwest Florida estuary. *Northeast. Gulf Sci.* 3:1–14.

Sheridan, P. F., and R. J. Livingston. 1979. Cyclic tropic relationships of fishes in an unpolluted river-dominated estuary in North Florida. In R. J. Livingston (ed.), *Ecological Processes in Coastal and Marine Systems.* New York: Plenum, pp. 143–161.

Sheridan, P. F., and R. J. Livingston. 1983. Abundance and seasonality of infauna and epifauna inhabiting a *Halodule wrighti* meadow in Apalachicola Bay, Florida. *Estuaries* 6:407–419.

Shirley, T. C., and M. S. Loden. 1982. Tubificidae (Annelida: Oligochaeta) of a Louisiana estuary: ecology and systematics with the description of a new species. *Estuaries* 5:47–56.

Show, I. T., Jr. 1980. The movements of a marine copepod in a tidal lagoon. In P. Hamilton and K. B. MacDonald (eds.), *Estuarine and Wetland Processes with Emphasis on Modeling.* New York: Plenum, pp. 561–602.

Sikora, W. B. 1977. The ecology of *Paleomonetes pugio* in a southeastern salt marsh ecosystem with particular emphasis on the production and trophic relationships. Thesis, University of South Carolina, Columbia.

Sikora, W. B., and B. Kjerfve. 1985. Factors influencing the salinity regime of Lake Pontchartrain, a shallow coastal lagoon: analysis of a long-term data set. *Estuaries* 8:170–180.

Sikora, W. B., and J. P. Sikora. 1982. *Ecological Characterization of the Benthic Community of Lake Pontchartrain, Louisiana*, Contract No. DACW29-79-C-0099. Prepared for the U. S. Army Engineer District, New Orleans, LA.

Sikora, W. B., J. P. Sikora, and A. M. Prior. 1981. *Environmental Effects of Hydraulic Dredging for Clam Shells in Lake Pontchartrain, Louisiana*, LSU-CEL-81-18. Baton Rouge: Center for Wetland Resources, Louisiana State University.

Simmons, E. G. 1957. An ecological survey of the upper Laguna Madre of Texas. *Publ. Inst. Mar. Sci. Univ. Tex.* 4:156–200.

Simpson, D. G., and G. G. Gunter. 1956. Notes on habitats, systematic characters, and life histories of Texas salt water cyprinodontes. *Tulane Stud. Zool.* 4:115–134.

Simpson, R. L., D. F. Whigham, and K. Branigan. 1979. The midsummer insect communities of freshwater tidal wetland macrophytes. *Bull. N. J. Acad. Sci.* 24:22–28.

Simpson, R. L., R. E. Good, M. A. Leck, and D. F. Whigham. 1983. The ecology of freshwater tidal wetlands. *BioScience* 33:255–259.

Skaggs, R. N., J. W. Gilliam, T. J. Sheets, and J. S. Barnes. 1980. *Effects of Agricultural Land Development on Drainage Water on the North Carolina Tidewater Region*, Rep. No. 159. Raleigh: University of North Carolina, Water Resources Research Institute.

Smalley, A. E. 1959. The growth cycle of *Spartina* and its relation to the insect populations of the marsh. In *Proceedings of the Salt Marsh Conference, Sapelo Island, Georgia*. Athens: University of Georgia Marine Institute, pp. 96–100.

Smalley, A. E. 1960. Energy flow of a salt marsh grasshopper. *Ecology* 41:785–790.

Smith, C. J., R. D. DeLaune, and W. H. Patrick, Jr. 1982. Carbon and nitrogen cycling in a *Spartina alterniflora* salt marsh. In J. R. Franey and I. E. Galvally (eds.), *Cycling of Carbon, Nitrogen, Sulphur, and Phosphorous in Terrestrial and Aquatic Ecosystems*. New York: Springer-Verlag, pp. 97–104.

Smith, N. P. 1983a. Tidal and low frequency net displacement in a coastal lagoon. *Estuaries* 6:180–189.

Smith, N. P. 1983b. A comparison of winter and summer temperature variations in a shallow bar-built estuary. *Estuaries* 6:2–9.

Smith, N. P. 1986. The rise and fall of the estuarine intertidal zone. *Estuaries* 9:95–101.

Smith, S. L. 1978. The role of zooplankton in the nitrogen dynamics of a shallow estuary. *Estuarine Coastal Mar. Sci.* 7:555–565.

Spaulding, M. H. 1908. *Preliminary Report on the Life History and Habits of the "Lake Shrimp"* (*Pennaeus* [sic] *setiferus*), Bull. No. 11. Cameron, LA: Gulf Biological Laboratory.

Springer, S. 1938. On the size of *Rana sphenocephala. Copeia* 1938:49.

Stanley, D. W., and D. A. Daniel. 1985. Seasonal phytoplankton density and biomass changes in South Creek, North Carolina. *J. Elisha Mitchell Sci. Soc.* 101:130–141.

Stanley, D. W., and J. E. Hobbie. 1977. *Nitrogen Recycling in the Chowan River, N. C.*, Rep. No. 121. Raleigh: University of North Carolina, Water Resources Research Institute.

Stevenson, J. C., and N. M. Confer. 1978. Summary of available information on Chesapeake Bay submerged vegetation. *U. S. Fish Wildl. Serv., Biol. Serv. Program* USFWS/BA 76-66.

Stewart, R. E. 1962. Waterfowl populations in the upper Chesapeake region. *U. S. Fish Wildl. Serv., Spec. Sci. Rep., Wildl.* 65:1–208.

Stickney, R. B., and S. C. Knowles. 1975. Summer zooplankton distribution in a Georgia estuary. *Mar. Biol. (Berlin)* 33:147–154.

Stone, J. H., L. H. Bahr, Jr., and J. W. Day, Jr. 1978. Effects of canals on freshwater marshes in coastal Louisiana and implications for management. In R. E. Good, D.F. Whigham, and R. L. Simpson (eds.), *Freshwater Wetlands, Ecological Processes and Management Potential*. New York: Academic, pp. 299–320.

Stotts, V. D., and D. E. Davis. 1960. The black duck in the Chesapeake Bay of Maryland: breeding, behavior and biology. *Chesapeake Sci.* 1:127–154.

Stowe, W. C. 1972. Community structure and productivity of the epibenthic algae in the Barataria Bay area of Louisiana. Thesis, Louisiana State University, Baton Rouge.

Strenth, N. E. 1976. A review of the systematics and zoogeography of the freshwater species of *Palaemonetes* Heller (Crustacea: Decapoda) of North America. *Smithson. Contrib. Zool.* 228.

Stuck, K. C., H. M. Perry, and R. W. Heard. 1979. An annotated key to the Mysidacea of the north central Gulf of Mexico. *Gulf Res. Rep.* 6:225–238.

Sullivan, M. J. 1975. Diatom communities from a Delaware salt marsh. *J. Phycol.* 11:384–390.

Sullivan, M. J. 1977. Edaphic diatom communities associated with *Spartina alterniflora* and *S. patens* in New Jersey. *Hydrobiologia* 52:207–211.

Sullivan, M. J. 1978. Diatom community structure: taxonomic and statistical analysis of a Mississippi marsh. *J. Phycol.* 14:468–475.

Sullivan, M. J. 1981. Effects of canopy removal and nitrogen enrichment on a *Distichlis spicata*- edaphic diatom complex. *Estuarine, Coastal Shelf Sci.* 13:119–129.

Sullivan, M. J. 1982a. Similarity of an epiphytic and edaphic diatom community associated with *Spartina alterniflora. Trans. Am. Microsc. Soc.* 101:84–90.

Sullivan, M. J. 1982b. Distribution of edaphic diatoms in a Mississippi salt marsh: canonical correlation analysis. *J. Phycol.* 19:130–133.

Suttkus, R. D., R. M. Darnell, and J. Darnell. 1954. *Biological Study of Lake Pontchartrain—Annual Report (1953–1954).* New Orleans, LA: Tulane University.

Swenson, E. M. 1981. Physical effects of the 1979 opening of the Bonnet Carre spillway. *Proc. La. Acad. Sci.* 44:121–131.

Swift, C. C., R. W. Yerger, and P. R. Parrish. 1977. Distribution and natural history of the fresh and brackish water fishes of the Ochlockonee River, Florida and Georgia. *Bull. Tall Timbers Res. Stn.* 20:1–111.

Swift, C. C., C. R., Gilbert, S. A. Bartone, G. H. Burgess, and R. W. Yerger. 1986. Zoogeography of freshwater fishes of the Southeastern United States: Savannah River to Lake Pontchartrain. In C. H. Hocutt and E. O. Wiley (eds.), *Zoogeography of North American Freshwater Fishes.* New York: Wiley (Interscience), pp. 213–265.

Swingle, H. A. 1971. Biology of Albama estuarine areas—Cooperative Gulf of Mexico estuarine inventory. *Ala. Mar. Resour. Bull.* 5:123.

Swingle, H. A., and D. G. Bland. 1974. A study of the fishes of the coastal watercourses of Alabama. *Ala. Mar. Resour. Bull.* 10:17–102.

Tabb, D. C. 1966. The estuary as a habitat for spotted seatrout *Cynoscion nebulosus.* In R. F. Smith, A. H. Swatz, and W. H. Massman (eds.), *Symposium on Estuarine Fisheries*, Spec. Pub. No. 3. Washington, DC: Am. Fish. Soc., pp. 59–67.

Tagatz, M. E. 1968. Biology of the blue crab, *Callinectes sapidus* Rathbun, in the St. John's River, Florida. *Fish. Bull.* 67:17–33.

Tagatz, M. E. 1971. Osmoregulatory ability of blue crabs in different temperature-salinity combinations. *Chesapeake Sci.* 12:14–17.

Tagatz, M. E., and A. B. Hall. 1971. Annotated bibliography on the fishing industry and biology of the blue crab *Callinectes sapidus. NOAA Tech. Rep., NMFS SSRF* NMFS SSRF-640:1–94.

Talbot, C. W., and K. W. Able. 1984. Composition and distribution of larval fishes in New Jersey high marshes. *Estuaries* 7:434–443.

Talbot, C. W., K. W. Able, and S. K. Shisler. 1986. Fish species composition in New Jersey salt marshes: effects of marsh alterations for mosquito control. *Trans. Am. Fish. Soc.* 115:269–278.

Targett, T. E. 1984. A breeding population of Gulf pipefish (*Syngnathus scovelli*) in a Georgia estuary, with discussion on the ecology of the species. *Contrib. Mar. Sci.*, 27:169–174.

Taylor, J. L., and C. H. Saloman. 1968. Some effects of hydraulic dredging on coastal development in Boca Ciega Bay, Florida. *Fish. Bull.* 67:213–241.

Teal, J. M. 1958. Distribution of fiddler crabs in Georgia salt marshes. *Ecology* 39:185–193.

Tebo, L. B., Jr. and E. G. McCoy. 1964. Effects of seawater concentrations on the reproduction and survival of largemouth bass and bluegills. *Prog. Fish-Cult.* 26:99–106.

Tenore, K. R., D. B. Horton, and T. W. Duke. 1968. Effects of bottom substrate on the brackish water bivalve *Rangia cuneata*. *Chesapeake Sci.* 3:238–248.

Thayer, G. W. 1971. Phytoplankton production and the distribution of nutrients in a shallow unstratified estuarine system near Beaufort, North Carolina. *Chesapeake Sci.* 12:240–253.

Thomas, J. D. 1976. Survey of gammarid amphipods of the Barataria Bay, Louisiana, region. *Contrib. Mar. Sci.* 20:87–100.

Thompson, J. M. 1966. The grey mullets. *Oceanogr. Mar. Biol.* 4:301–355.

Thompson, M. J. 1978. Species composition and distribution of seagrasses in the Indian River Lagoon, Florida. *Fla. Sci.* 41:90–96.

Thorp, J. H., and D. E. Hoss. 1975. Effects of salinity and cyclic temperature on survival of two sympatric species of grass shrimp (*Palaemonetes*) and their relationship to natural distributions. *J. Exp. Mar. Biol. Ecol.* 18:19–28.

Turner, R. E., R. M. Darnell, and J. Bond. 1980. Changes in the submerged macrophytes of Lake Pontchartrain (Louisiana): 1954–1973. *Northeast Gulf Sci.* 4:44–49.

Turner, R. L., E. F. Lowe, and J. M. Lawrence. 1975. Isosmotic intracellular regulation in the freshwater palaemonid shrimp *Palaemonetes paludosus* (Crustacea: Decapoda). *Physiol. Zool.* 48:235–241.

Twilley, R. R., L. R. Blanton, M. M. Brinson, and G. J. Davis. 1985. Biomass production and nutrient cycling in aquatic macrophyte communities of the Chowan River, North Carolina. *Aquat. Bot.* 22:231–252.

Tyus, H. M. 1974. Movements and spawning of anadromous alewives at Lake Mattamuskeet, North Carolina. *Trans. Am. Fish. Soc.* 103:392–396.

Van Den Avyle, M. J. 1984. Species profile: life histories and environmental requirements of coastal fishes and invertebrates (South Atlantic)—Atlantic sturgeon. *U. S. Fish Wildl. Serv.* FWS/OBS-82/11.25:1–17.

VanEs, F. 1977. A preliminary carbon budget for a part of the Ems estuary: the Dollard. *Helgol. Wiss. Meersunters.* 30:283–294.

Vargo, G. 1979. The contribution of ammonia excreted by zooplankton to phytoplankton production in Narragansett Bay. *J. Plankton Res.* 1:75–84.

Vince, S. W., I. Valiela, and J. Teal. 1981. An experimental study of the structure of herbivorous insect communities in a salt marsh. *Ecology* 62:1662–1668.

Virnstein, R. W., and P. A. Carbonara. 1985. Seasonal abundance and distribution of drift algae and seagrasses in the mid-Indian River Lagoon, Florida. *Aquat. Bot.* 23:67–82.

Virnstein, R. W., P. S. Mikkelsen, K. D. Cairns, and M. A. Capone. 1983. Seagrass beds versus sand bottoms: the trophic importance of their associated benthic invertebrates. *Fla. Sci.* 46:363–381.

Virnstein, R. W., W. G. Nelson, F. G. Lewis, III, and R. K. Howard. 1984. Latitudinal patterns in seagrass epifauna: do patterns exist and can they be explained? *Estuaries* 7:310–330.

Wafer, M. V. M., P. LeCorre, and J. L. Borrien. 1989. Transport of carbon, nitrogen and phosphorous in a Brittany River, France. *Estuarine Coastal Shelf Sci.* 29:489–500.

Ward, G. H. 1980. Hydrography and circulation processes of Gulf estuaries. In P. Hamilton and K. MacDonald (eds.), *Estuarine and Wetland Processes*. New York: Plenum, pp. 183–215.

Ward, G. H., N. E. Armstrong, and the Matagorda Bay Project Teams. 1980. Matagorda Bay, Texas: its hydrology, ecology and fishery resources. *U.S. Fish Wildl. Serv., Biol. Serv. Program* FWS/OBS-81/52.

Weaver, J. E., and L. F. Holloway. 1974. Community structure of the fishes and macrocrustaceans in ponds of a Louisiana tidal march [sic] influenced by weirs. *Contrib. Mar. Sci.*, 18:57–70.

Weigert, R. G., R. R. Christian, J. L. Gallagher, J. R. Hall, R. D. H. Jones, and R. L. Wetzel. 1975. A preliminary ecosystem model of coastal Georgia *Spartina* marsh. In L. E. Cronin (ed.), *Estuarine Research*, Vol. I. New York: Academic, pp. 583–601.

Weinstein, M. P. 1979. Shallow marsh habitats as primary nurseries for fishes and shellfish, Cape Fear River, North Carolina. *Fish. Bull.* 77:339–357.

Weinstein, M. P., S. L. Weiss, R. G. Hodson, and L. R. Gerry. 1980. Retention of three taxa of postlarval fishes in an intensively flushed tidal estuary, Cape Fear River, North Carolina. *Fish. Bull.* 78:419–436.

Weisburg, S. B., and V. A. Lotrich, 1982. The importance of infrequently flooded intertidal marsh surface as an energy source for the mummichog *Fundulus heteroclitus*: an experimental approach. *Mar. Biol. (Berlin)* 66:307–310.

Welsh, B. L. 1975. The role of grass shrimp *Palaemonetes pugio* in a tidal marsh ecosystem. *Ecology* 56:513–530.

Whatley, E. C. 1962. Occurrence of breeding Gulf pipefish, *Syngnathus scovelli*, in the inland freshwaters of Louisiana. *Copeia* 1962:220.

Whatley, E. C. 1969. A study of *Syngnathus scovelli* in fresh waters of Louisiana and salt waters of Mississippi. *Gulf Res. Rep.* 2:437–474.

Whigham, D. F., and R. L. Simpson. 1976. The potential use of freshwater tidal marshes in the management of water quality in the Delaware River. In J. Tourbier and R. W. Pierson, Jr. (eds.) *Biological Control of Water Pollution*. Philadelphia: Pennsylvania UP, pp. 173–186.

Whigham, D. F., J. McCormick, R. E. Good, and R. L. Simpson. 1978. Biomass and primary production in freshwater tidal wetlands of the Middle Atlantic coast. In R. E. Good, D. F. Whigham, and R. L. Simpson (eds.), *Freshwater Wetlands: Ecological Processes and Management Potential*. New York: Academic, pp. 3–20.

Whigham, D. F., R. L. Simpson, and K. Lee. 1980. *The Effect of Sewage Effluent on the Structure and Function of a Freshwater Tidal Marsh Ecosystem*, Tech. Compliance Rep., Tech. Project B-60-NJ. Washington, DC: U. S. Dept. of the Interior, Water Resources.

Williams, A. B. 1966. Marine decapod crustaceans of the Carolinas. *Fish. Bull.* 65:1–298.

Williams, R. B. 1964. Division rates of salt marsh diatoms in relation to salinity and cell size. *Ecology* 45:877–880.

Willner, G. R., J. A. Chapman, and J. R. Goldsmith. 1975. A study and review of muskrat food habits with special reference to Maryland. *Md. Wildl. Admin. Publ. Wildl. Ecol.* 1:1–25.

Willner, G. R., J. A. Chapman, and D. Pursley. 1979. Reproduction, physiological responses, food habits, and abundance of nutria in Maryland marshes. *Wildl. Monogr.* 65:1–43.

Wilson, K. A. 1968. Fur production in southeastern coastal marshes. In J. D. Newsom (ed.), *Proceedings of the Marsh and Estuary Management Symposium*. Baton Rouge: Louisiana State UP, pp. 149–162.

Witherspoon, A. M., C. Baldwin, O. C. Boody, and J. Overton. 1979. *Response of Phytoplankton to Water Quality in the Chowan River System*, Rep. No. 129. Raleigh: University of North Carolina, Water Resources Research Institute.

Wolff, W. 1977. A benthic food budget for the Grevelinger Estuary, the Netherlands, and a comparison of the mechanisms causing high benthic secondary production in estuaries. In B. Coull (ed.), *Ecology of Marine Benthos*. Columbia: South Carolina UP, pp. 267–280.

Wood, C. E. 1967. The physioecology of the grass shrimp *Palaemonetes pugio* in the Galveston Bay estuarine system. *Contrib. Mar. Sci.* 12:54–79.

Wurtz, C. B., and S. S. Roback. 1955. The invertebrate fauna of some Gulf Coast rivers. *Proc. Acad. Nat. Sci. Philadelphia* 107:167–206.

Young, D. K., and M. W. Young. 1978. Regulation of species-densities of seagrass associated macrobenthos: evidence from field experiments in the Indian River Lagoon. *J. Mar. Res.* 36:560–593.

Zimmerman, R., R. Gibson, and J. Harrington. 1979. Herbivory and detritivory among gammaridean amphipods from a Florida seagrass community. *Mar. Biol. (Berlin)* 54:41–47.

14 Estuaries

MICHAEL R. DARDEAU*

Marine Environmental Sciences Consortium, Dauphin Island, AL 36528

RICHARD F. MODLIN

Department of Biological Sciences, University of Alabama, Huntsville, AL 35899

WILLIAM W. SCHROEDER

Marine Science Program, The University of Alabama, Dauphin Island, AL 36528

JUDY P. STOUT

University of South Alabama, Marine Environmental Sciences Consortium, Dauphin Island, AL 36528

Estuaries or estuarine-like environments are transition zones between freshwater and marine aquatic systems. Numerous definitions for the term estuary have been proposed (see Lauff 1967, Kennish 1986). Generally, the most widely accepted is that set forth by Pritchard (1967), who defined an estuary as "a semi-enclosed coastal body of water which has a free connection with the open sea and within which seawater is measurably diluted with fresh water from land drainage." However, the broader, more ecologically oriented definition presented in Cowardin et al. (1979) is more suited for this chapter: "deep-water tidal habitats and adjacent tidal wetlands which are usually semi-enclosed by land, but have open, partially obstructed, or sporadic access to the open ocean and in which ocean water is at least occasionally diluted by freshwater runoff from land."

ABIOTIC ENVIRONMENT

Drainage and Watershed Characteristics

Atlantic Estuaries   Thirty-four estuaries or estuarine-like environments (hereafter all are referred to as estuaries), encompassing 30 605 km², represent approximately 90% of the total estuarine area in the Southeastern United

*Authorship is listed in alphabetical order.

States (Fig. 1). Of these estuaries 18 occur along the SE Atlantic Coast from Albemarle Sound in North Carolina and Virginia to Biscayne Bay, Florida, and have a combined area of 11 029 km² or approximately 32% of the estuarine area in the Southeastern United States (Table 1). Generally, two types of shoreline physiography dominate this coastal region. The first is characterized by sounds or lagoons landward of very extensive outerbank and/or barrier island systems (e.g., northern North Carolina and southern Florida coasts). The barrier islands tend to be long and narrow with relatively few tidal inlets. The lagoon–marsh systems are narrow, shallow, and densely vegetated (Cleary et al. 1979). The second type is a less protected, low-lying, marshy shoreline with numerous sea islands (Terrell 1979) and a dendritic pattern of tributaries flowing into the sea (e.g., South Carolina and Georgia coasts). Barrier islands along this coastal region are short, with wide central sections and narrow ends broken by numerous tidal inlets (Bahr and Lanier 1981). One unique exception is the St. Johns River, a large tidally influenced drowned riverine system with restricted access to the sea.

Estuaries along the SE Atlantic coast have estuarine and fluvial drainage areas totaling 125 522 and 202 482 km², respectively, and the total average freshwater inflow into these estuaries is 4228 m³/s (Table 1). The highest average inflows occur in Albemarle Sound, followed closely by Pamlico Sound and Winyah Bay. The lowest average inflows occur in the New River (23 m³/s), the St. Catherine, Sapelo, and Doboy sounds complex, and the Broad River. The highest freshwater inflows occur from January through May, except in SE Florida where the highest inflows occur from June through September (U.S. Department of Commerce 1985a). Approximately 1.4 m/year of precipitation falls in the northern region, the central region receives between 1.2 and 1.3 m/year, while up to 1.5 m/year falls in south Florida (U.S. Department of Commerce 1983).

Tides along the SE Atlantic coast are semidiurnal, with two nearly equal high and low water periods each tidal day (Gerlack 1970). Average tidal amplitudes for estuaries in the northern and southern regions are 0.1–2.0 m and 0.1–1.4 m, respectively (Table 1), and fall into the microtidal range (Davis 1964). Estuaries in the central region have larger average amplitudes of from 1.2 to 2.3 m (Table 1) that are generally classified in the mesotidal range (Davis 1964).

Gulf of Mexico Estuaries   Sixteen estuaries occur along the northern coast of the Gulf of Mexico from the Ten Thousand Islands area of SW Florida to Sabine Lake on the Texas–Louisiana border (Fig. 1). They have a combined area of 19 576 km², or approximately 58% of the total estuarine area in the Southeastern United States (Table 2). This entire region is characterized by flat coastal plains with shoreline physiography that includes (1) small mangrove island–tidal channel complexes and extensive marshy and swampy areas (SW Florida); (2) rocky, drowned karst and wide shallows with extensive seagrass beds and marshes (central W Florida); (3) barrier island systems with sand beaches and dunes and mud-bottom bays (NW Florida, Alabama, and Mississippi); (4) the Mississippi–Atchafalaya delta complex; and (5) strand

FIGURE 1. Major estuaries or estuarine-like areas along the southeastern Atlantic and Gulf coasts of the United States. Major geological features and water characteristics of each are presented in Tables 1 and 2. Numbers denote the approximate location of each area identified as follows: 1, Albermarle Sound; 2, Pamilco Sound; 3, Bogue Sound; 4, New River; 5, Cape Fear River; 6, Winyaw Bay; 7, North and South Santee Rivers; 8, Charleston Harbor; 9, St. Helena Sound; 10, Broad Sound; 11, Savannah River; 12, Ossabaw Sound; 13, St. Catherine–Sapelo–Doboy sounds; 14, Altamaha River; 15, St. Simons and St. Andrews sounds; 16, St. Johns River; 17, Indian River; 18, Biscayne Bay; 19, Ten Thousand Islands; 20, Charlotte Harbor; 21, Tampa Bay; 22, Suwannee River; 23, Apalachee Bay; 24, Apalachicola Bay; 25, St. Andrews Bay; 26, Choctawhatchee Bay; 27, Pensacola Bay; 28, Perdido Bay; 29, Mobile Bay; 30, Mississippi Sound; 31, Mississippi Delta region; 32, Vermillion and Atchafalaya bays; 33, Calcasieu Lake; 34, Sabine Lake.

plain–chenier plain systems with extensive marshlands (W Louisiana) (Terrell 1979).

Estuaries in this region have the largest drainage areas in the nation, receiving runoff from nearly two-thirds of the contiguous United States (U.S. Department of Commerce 1985a). Estuarine and fluvial drainage areas total 169 686 and 3 443 966 km^2, respectively (Table 2). The combined fluvial drainage of the Mississippi River and the Atchafalaya and Vermilion bays

TABLE 1 General Characteristics of Southeastern Estuaries Along the Atlantic Coast

	Location		Estuarine Zones[c] (km²)		
Estuary	Fluvial Drainage[a]	Basic Physiographic Type[b]	Tidal Fresh	Mixing Zone	Seawater
(1) Albemarle Sound, NC and VA[d]	NC, VA	A, B	1880	508	0
(2) Pamlico Sound, NC[e]	NC	A, B	249	4898	104
(3) Bogue Sound, NC	—	A	3	54	207
(4) New River, NC	—	A, B	0	78	5
(5) Cape Fear River	NC	B, A	13	70	16
(6) Winyah Bay, SC	SC, NC	B, A	23	54	0
(7) North and South Santee rivers, SC	SC, NC	E	10	13	0
(8) Charleston Harbor, SC	SC, NC	B, C	16	80	0
(9) St. Helena Sound, SC	SC	C, B	8	106	106
(10) Broad River, SC	—	C, B	10	145	104
(11) Savannah River, SC and GA	SC, GA	C, B	8	60	18
(12) Ossabaw Sound, GA	GA	C, B	5	73	8
(13) St. Catherines, Sapelo, and Doboy sounds, GA	—	C, B	5	179	10
(14) Altamaha River and Sound, GA	GA	C, B	10	21	8
(15) St. Simons and St. Andrew sounds, GA	GA	C, B	3	176	8
(16) St. Johns River, FL	FL	B	308	246	114
(17) Indian River, FL	—	A	0	28	0
(18) Biscayne Bay, FL	—	A	0	16	268

	Tidal Prism[f] (m³)	Flow Ratios[g]		
		Low Flow Period	Average Annual	High Flow Period
(1)	1.45×10^8	0.163	0.216	0.280
(2)	8.10×10^8	0.024	0.033	0.038
(3)	1.35×10^8	0.008	0.012	0.015
(4)	1.72×10^8	0.040	0.058	0.080
(5)	1.00×10^8	0.072	0.128	0.224
(6)	0.86×10^8	0.148	0.301	0.498
(7)	0.25×10^8	0.050	0.138	0.306
(8)	1.35×10^8	0.112	0.150	0.201
(9)	3.94×10^8	0.010	0.015	0.024
(10)	5.41×10^8	0.001	0.002	0.004
(11)	1.75×10^8	0.065	0.092	0.137
(12)	1.76×10^8	0.013	0.022	0.059
(13)	4.16×10^8	0.001	0.003	0.004
(14)	0.67×10^8	0.117	0.283	0.557
(15)	3.88×10^8	0.005	0.008	0.014
(16)	0.53×10^8	0.125	0.185	0.252
(17)	0.65×10^8	0.013	0.027	0.044
(18)	3.00×10^8	0.007	0.013	0.021

Average Depth (m)	Drainage Area Estuarine[h] (km²)	Drainage Area Fluvial[a] (km²)	Average Freshwater Inflow (m³/s)	Prevailing Tide[j]	Phase Range of the tide[k] (m)	Stratification Classification[l] 3-month High Freshwater Inflow	Stratification Classification[l] 3-month Low Freshwater Inflow
(1) 3.3	14 193	32 204	702	SD	No data	VH	VH
(2) 3.0	14 996	14 364	600	SD	0.1–0.6	VH[m]	VH
(3) 0.9	1 761	0	37	SD	0.2–0.9	VH	VH
(4) 0.6	1 217	0	23	SD	0.9	VH	VH
(5) 4.0	11 241	12 302	268	SD	0.5–1.3	HS	MS
(6) 2.7	24 633	22 217	578	SD	0.2–1.2	MS	MS
(7) 2.3	1 860	37 767	7 613	SD	1.2–1.4	MS	VH
(8) 3.7	3 113	37 767	45 613	SD	1.5–1.7	MS	MS
(9) 4.9	3 981	8 397	130	SD	1.7–1.8	VH	VH
(10) 5.9	2 590	0	25	SD	2.0–2.1	VH	VH
(11) 3.5	2 372	24 563	362	SD	1.1–2.2	MS	MS
(12) 3.4	3 859	8 392	85	SD	2.0–2.3	VH	VH
(13) 4.2	2 499	0	23	SD	2.1–2.3	VH	VH
(14) 2.7	3 911	32 867	422	SD	1.4–2.0	MS	MS
(15) 4.9	8 443	2 002	71	SD	1.8–2.3	VH	VH
(16) 4.3	16 835	7 407	221	SD	0.1–1.4	VH	VH
(17) 1.8	3 227	0	40	SD	0.1–0.4	VH	VH
(18) 2.3	4 791	0	91	SD	0.2–0.6	VH	VH

[a]The land and water components of an entire watershed upstream of the estuarine drainage.

[b]A, bar-built or barrier island; B, drowned river valley or coastal plain; C, low-lying, marshy shoreline; D, low-lying, mangrove shoreline; E, delta front.

[c]Approximate salinity ranges (ppt): tidal fresh, 0.0–0.5; mixing zone, 0.5–25.0; seawater, 25.0.

[d]Includes Currituck Sound.

[e]Includes the Pamlico, Pungo, and Neuse rivers.

[f]The volume of water entering a coastal system during flood tide.

[g]Flow ratios are the proportion of the volume of freshwater entering a coastal system during a tidal cycle to the volume of the tidal prism.

[h]The land and water components of an entire watershed that most directly affects an estuary.

[i]Flows do not account for current significant rediversion within the basin.

[j]SD, semidiurnal; D, diurnal.

[k]The differences in tidal elevation at a particular location relative to the occurrence of flood and ebb tide at a tidal reference station.

[l]VH, vertically homogenous (surface salinity equals bottom salinity); MS, moderately stratified (surface salinity less than bottom salinity); HS, highly stratified (salt wedge-type estuary).

[m]MS in the Neuse River.

Source. From the *National Estuarine Inventory*, Vol. 1 (U.S. Department of Commerce 1985a).

TABLE 2 General Characteristics of Southeastern Estuaries Along the Gulf of Mexico Coast

Estuary	Fluvial Drainage[a]	Basic Physiographic Type[b]	Estuarine Zones[c] (km²) Tidal Fresh	Estuarine Zones[c] (km²) Mixing Zone	Estuarine Zones[c] (km²) Seawater
(1) Ten Thousand Islands, FL	—	D	21	168	179
(2) Charlotte Harbor, FL[d]	—	A, B	16	378	412
(3) Tampa Bay, FL	—	B, A	8	300	588
(4) Swannee River, FL	FL, GA	C	5	62	41
(5) Apalachee Bay, FL	FL, GA	C	3	88	321
(6) Apalachicola Bay, FL	FL, GA, AL	A	21	417	117
(7) St. Andrew Bay, FL	—	B, A	3	142	98
(8) Choctawhatchee Bay, FL	FL, AL	A, B	3	305	26
(9) Pensacola Bay, FL	FL, AL	B, A	3	313	54
(10) Perdido Bay, FL and AL	—	B, A	3	122	5
(11) Mobile Bay, AL	AL, MS, GA, TN	B, A	75	847	137
(12) Mississippi Sound, AL, MS, and LA[e]	AL, MS, LA	A	23	4527	241
(13) Mississippi River Delta Region, MS and LA[f]	g	E	316	4942	1924
(14) Atchafalaya and Vermilion Bays, LA	LA, AK, OK, TX	E	355	1466	0
(15) Calcasieu Lake, LA	LA	B	3	254	0
(16) Sabine Lake, LA and TX	LA, TX	B	3	241	0

Average Depth (m)	Drainage Area (km²)		Average Freshwater Inflow (m³/s)	Prevailing Tide[i]	Phase Range of the tide[j] (m)
	Estuarine[h]	Fluvial[a]			
(1) 1.4	10 878	0	51	M	0.1–1.0
(2) 3.4	13 028	0	136	M	0.1–0.7
(3) 4.7	6 729	0	68	M	0.5–0.7
(4) 2.3	4 817	21 600	317	M	0.7
(5) 4.0	9 635	2 396	150	M	0.9–1.1
(6) 2.6	7 690	45 405	824	M	0.2–0.6
(7) 3.8	2 927	0	127	D	0.3–0.4
(8) 4.8	5 851	8 055	241	D	0.2
(9) 4.2	9 013	9 091	328	D	0.3–0.4
(10) 2.4	3 121	0	62	D	No data
(11) 2.9	12 626	102 887	2 246	D	0.2–0.4
(12) 3.0	31 235	38 435	1 235	D	0.2–0.5
(13) 2.3	21 600	2 926 168	14 563	D	0.2–0.5
(14) 1.8	15 281	239 833	6 337	M	0.2–0.6
(15) 1.9	2 797	8 423	178	M	No data
(16) 1.8	12 458	41 673	487	M	0.4–0.6

TABLE 2 Continued

Stratification Classification[k]		Tidal Prism[l] (m³)	Flow Ratios[m]		
3 Month High Freshwater Inflow	3 Month Low Freshwater Inflow		Low Flow Period	Average Annual	High Flow Period
(1) VH	VH	1.55×10^8	0.006	0.015	0.029
(2) VH	VH	0.39×10^8	0.007	0.016	0.029
(3) VH	VH	5.24×10^8	0.003	0.006	0.011
(4) MS	VH	0.80×10^8	0.137	0.177	0.241
(5) MS	MS	4.25×10^8	0.010	0.016	0.022
(6) MS	MS	2.76×10^8	0.079	0.133	0.219
(7) MS	MS	0.96×10^8	0.113	0.120	0.125
(8) MS	VH	0.51×10^8	0.265	0.424	0.668
(9) MS	MS	1.47×10^8	0.111	0.200	0.335
(10) MS	MS	0.20×10^8	0.220	0.285	0.397
(11) HS	MS	4.47×10^8	0.173	0.449	0.870
(12) VH	VH	12.23×10^8	0.042	0.090	0.166
(13) VH[n]	VH[n]	24.84×10^8	2.57[o]	4.46[o]	6.60[o]
(14) MS	VH	6.68×10^8	0.437	0.848	1.296
(15) VH	VH	0.39×10^8	0.151	0.408	0.666
(16) VH	VH	0.41×10^8	0.487	1.067	1.564

[a] The land and water components of an entire watershed upstream of the estuarine drainage.
[b] A, bar-built or barrier island; B, drowned river valley or coastal plain; C, low-lying, marshy shoreline; D, low-lying, mangrove shoreline; E, delta front.
[c] Appropriate salinity ranges (ppt): tidal fresh, 0.0–0.05; mixing zone, 0.05–25.0; seawater, 25.0.
[d] Includes Pine Island Sound and the Caloosahatchee River.
[e] Includes Lakes Borgne, Pontchartrain, and Maurepas.
[f] Includes Chandeleur and Breton sounds and Barataria. Timbalier, Terrebonne, and Cailou bays.
[g] Central section of the United States.
[h] The land and water components of an entire watershed that most directly affects an estuary.
[i] M, mixed; D, diurnal.
[j] The difference in tidal elevation at a particular location relative to the occurence of flood and ebb tide at a tidal reference station.
[k] VH, vertically homogeneous (surface salinity equals bottom salinity); MS, moderately stratified (surface salinity less than bottom salinity); HS, highly stratified (salt wedge-type estuary).
[l] The volume of water entering a coastal system during flood tide.
[m] Flow ratios are the proportion of the volume of freshwater entering a coastal system during a tidal cycle to the volume of the tidal prism.
[n] HS in the Mississippi River.
[o] Represents only the Mississippi River proper.
Source. Data from the *National Estuarine Inventory*, Vol. 1 (U.S. Department of Commerce 1985b).

account for 3 166 001 km² or 91.9% of the total fluvial drainage. However, the remaining Gulf of Mexico estuaries still have a total fluvial drainage of 277 965 km² or 1.4 times that of the SE Atlantic estuaries.

Average freshwater inflow into these estuaries is 27 350 m³/s (Table 2). The combined inflows of the Mississippi and the Atchafalaya rivers account for 20 900 m³/s or 76.4% of the total inflow. The next highest inflows occur in Mobile Bay, followed by Mississippi Sound and Apalachicola Bay. The lowest inflows occur in Ten Thousand Islands, Perdido Bay, and Tampa Bay. The inflow from all sources other than the Mississippi and Atchafalaya rivers totals 6450 m³/s or 1.5 times that of the total inflow to SE Atlantic estuaries. Periods of high freshwater inflows vary from June through October in central and SW Florida to December through May for the remaining region (U.S. Department of Commerce 1985a). Coastal precipitation is in excess of 1.6 m/year for eastern Louisiana, Mississippi, Alabama, and NW Florida, less than 1.2 m/year for central Florida, and between 1.3 and 1.4 m/year in the other regions (U.S. Department of Commerce 1983, 1985b).

Tides along the northern Gulf of Mexico are of two types: mixed tides (two unequal high waters and/or two unequal low waters each tidal day) and diurnal tides (one high water and one low water each tidal day) (Gerlack 1970). Mixed tides occur from Ten Thousand Islands to Apalachicola Bay and from Atchafalaya Bay to Sabine Lake; diurnal tides occur from St. Andrew Bay through the Mississippi Delta region (Marmer 1954). The entire coastal system is classified as microtidal (Davis 1964). Average tidal amplitudes for estuaries along the west coast of Florida range from 0.1 to 1.1 m; for the remaining estuaries to the west the average tidal amplitudes range from 0.2 to 0.6 m (Table 2).

Currently, mean sea level along both the Atlantic and Gulf coasts is generally rising. Sea level measurements at Charleston, South Carolina, Savannah River entrance, Georgia, and Fernandina Beach (St. Mary's River), Florida, indicate a net rise of from 1.5 to 3.0 mm/year, since monitoring began in the early 1920s (Hicks 1973). Along the entire SW Atlantic coast Aubrey and Emery (1983) calculate a rise of 1.6–2.6 mm/year. In the Gulf, Provosts (1973) estimates a rise of 2.0–2.4 mm/year for the W Florida coast, while Aubrey and Emery (1983) calculate a rise of 1.5–1.8 mm/year for the NE Gulf coastal region.

Hydrology, Stratification, and Circulation

Of all aquatic systems, estuaries offer the greatest diversity in water composition and hydrology. Mixing of fresh and salt waters results in the creation of numerous physical and chemical subenvironments, which in turn support different communities of organisms particularly suited to a specific combination of water characteristics. Greater numbers of subenvironments support a greater variety of life forms (Vernberg and Vernberg 1981). The distribution and stability of these subenvironments depends principally on the basin morphology and temperature, salinity, and circulation regimes. These three water properties continually interact within a basin to determine the physical and partially the chemical nature of waters within an estuary at any given location or time (Kjerfve et al. 1982, Lewis and Platt 1982). For a general review of estuarine ecosystems see Dyer (1973), Officer (1976), Biggs and Cronin (1981), Cowardin et al. (1979), Kennish (1986), and Knox (1986).

Stratification and circulation patterns are primarily a function of the relative influences of freshwater input, tidal action, and winds, and produce different stratification classifications during high and low freshwater inflows (Tables 1 and 2). An individual estuary may change from one classification to another over seasonal or annual cycles of freshwater inflow or wind conditions (Kennish 1986). Estuaries with large riverine sources generally show a well-defined vertical salinity stratification during high inflow periods (e.g., Mobile Bay (Schroeder and Wiseman 1986)) or during flooding events (e.g., Pamlico River estuary (Hobbie 1970)). Fresh or brackish water overriding higher density coastal or open ocean water establishes, through the process of entrainment, a two-layer gravitational circulation system that results in a net outflow in the surface layer and a net inflow in the bottom layer (Pritchard 1955). Partially mixed estuaries occur when the tidal flow is sufficiently strong to prevent the riverine influence from dominating (e.g., Altamaha Sound (Bahr and Lanier 1981) and Pensacola Bay (Olinger et al. 1975)). A two-layered flow still exists in this type of estuary (Bowden 1980). Salinity gradients occur both longitudinally, increasing down estuary, and laterally, with more freshwater on the right side and more saline water on the left side (looking down estuary) in the Northern Hemisphere (Kennish 1986).

Estuaries that have a vertically homogeneous water column are either dominated by tidal mixing (e.g., Sapelo Sound (Finley 1975) and North Inlet (Kjerfve and Prochl 1979)), are not fed by major freshwater sources (e.g., lagoon-marsh complexes in southern North Carolina (Cleary et al. 1979)), or are shallow, wind–wave-dominated systems (e.g., Albemarle Sound (Copeland et al. 1983) and Mississippi Sound (Eleuterius 1978)). Persistently strong winds, occurring over large shallow expanses of water, have two consequences. First is the generation of intense surface waves which vertically mix the water column as described above. The second effect is "wind tides," the rise (setup) or fall (setdown) of the water surface. The specific response of the water surface is a function of the orientation of the water body relative to wind direction. Wind tides often exceed astronomical tides, particularly along microtidal coastlines. During periods of setdown, exposed basin bottoms are not true intertidal flats because they are exposed only at irregular intervals (Peterson and Peterson 1979). When conditions result in a setup, inundation of irregularly flooded marshes occurs (Stout 1984) and the high marsh meadow remains covered for extended periods of time (Provosts 1973, Gosselink 1984). Water level changes on time scales of weeks to seasons are described by Kjerfve et al. (1978) and Smith (1986) and have been examined in light of the role they play in the ecology of delta marshes (Gosselink 1984) and in annual material fluxes in a tidal creek (Kjerfve and McKellar 1980,

Kjerfve et al. 1982). The impact of aperiodic major storms (e.g., tropical storms or hurricanes) on marshes has been reported by Kuenzler and Marshall (1973) and Hackney and Bishop (1981).

Because estuaries are shallow features, their capacity to store heat over time is relatively small. As a result, water temperature fluctuates considerably, ranging from near freezing to above 30 °C in Pamlico Sound (Williams et al. 1973) and Calcasieu Lake (Barrett 1971) to between 16 and 31 °C in Biscayne Bay (Jaap 1984). The seasonal thermal cycle and the vertical profile structure, when present, are fairly predictable. Winter (December–February) water temperatures are the coldest of the year and are generally nearly constant from surface to bottom. During spring (March–May) and summer (June–August), surface waters heat up due to increased insolation. When these warmer waters overlay colder bottom waters, thermal stratification occurs and strong thermoclines can form. Often, by early to mid-summer, sufficient turbulence can occur to begin to break down this layering and by summer's end the water column can become homogeneous. In the fall (September–November), surface waters cool, increase in density, and sink toward the bottom. A very unstable water column can exist over this period.

Geology

Geologically, estuaries are ephemeral features having life spans over only thousands to a few tens of thousands of years. Once formed, they are rapidly destroyed by sediment filling (Schubel and Hirschberg 1978). Emery and Uchupi (1972) estimate that if all the suspended sediment carried by rivers, other than the Mississippi, into estuaries and lagoons along the U.S. Atlantic and Gulf coasts were deposited in these basins they would be filled in 9500 years on the average, assuming no sea level changes.

Sediment deposits are spatially confined to topographic lows within estuarine basins. Regardless of their origin, these basin depressions can receive sediments from fluvial, marine, eolian, or biological sources (Rusnak 1967). Fine sediments (silts and clays) indicate environments of low physical energy (weak turbulence), whereas coarser sediments (sands and even highly abraded shell fragments and pebbles) dominate high-energy (strong turbulence) environments (Sanders 1958, Warme 1971). Wind and wave processes are the principal forces dictating coastal morphology in microtidal coastal systems (Hayes 1975), while currents tend to be the dominant force along mesotidal and macrotidal coasts (Niedoroda et al. 1985). Thus, sediment type provides an integrated estimate of physical conditions in an estuary. For additional information on the geology of Atlantic Coast estuaries see Tanner (1960), Folger (1972), Lee (1973), Wanless (1976), Giese et al. (1979), Peterson and Peterson (1979), and Mathews et al. (1980). For Gulf coast estuaries see Kofoed and Gorsline (1963), Shier (1969), Barrett (1971), Huang and Goodell (1977), Doyle (1985), and Isphording et al. (1985).

Chemistry

Estuarine waters are chemically complex, consisting of varying kinds and amounts of dissolved inorganic and organic substances (Goldberg 1971, Phillips 1972, Carpenter et al. 1975, Burton 1976). The composition of more saline waters is chemically similar to seawater (Liss 1976), while the composition of freshwater inputs varies with the rates of river discharges and/or land runoff from drainage basins and with the geological/geochemical composition of each basin (Livingstone 1963, Gibbs 1970). In the absence of biochemical processes within an estuary and/or exchanges with the atmosphere, the concentration of a particular element will be a linear function of its concentration in the freshwater and seawater sources (Liss 1976, Biggs and Cronin 1981). Salinity is the common reference or conservative property upon which variations of other substances are compared.

In most estuaries oxygen is added by photosynthesis of submerged vegetation in shallow areas or near the surface of the water column by phytoplankton and to a lesser extent by diffusion from the atmosphere and by the aerating action of the wind. Therefore, surface waters normally remain at or near saturation during most of the year. Since deeper parts of estuaries are dependent on vertical mixing to transport oxygen-rich surface waters downward, concentrations there can range from saturated to anoxic. Hypoxic to anoxic conditions often occur in spring and summer during periods of vertical stratification when little or no mixing is occurring and when animal and bacteria respiration rates are high.

Nutrients to fuel coastal ecosystems (estuaries and adjacent communities) come primarily from terrestrial ecosystems via runoff and freshwater fluvial drainage and to a lesser extent from the adjacent marine ecosystem. Nutrients arrive as dissolved organic matter (DOM) or carbon (DOC), particulate organic matter (POM) or carbon (POC), and/or in the form of inorganic ions or elements (nitrites, nitrates, ammonia, phosphates, silicon, silicates, etc.). Estuaries act as a filter and these nutrient species are quickly, and almost totally, incorporated into the biota. Consequently, estuaries function as temporary or, in some cases, permanent nutrient sinks (Peterson and Peterson 1979).

Nutrient input to coastal ecosystems in the Southeastern United States is considerable, since the majority of the estuaries in this region are fed by complex river systems. Based on the magnitude of fluvial drainage (U.S. Department of Commerce 1985a), coastal ecosystems in the Southeast receive more input of terrestrial and upland resources than other coastal systems in the United States (Gunter 1967, Stickney 1984). The greatest proportion of nutrients impact coastal communities of the Gulf of Mexico. For example, the annual input of organic carbon by the Mississippi River is estimated to range from 3.0×10^9 (Happ et al. 1977) to 1.6×10^{10} kg organic C/year (Ho and Barrett 1975, as seen in Darnell and Soniat 1979). Such input coupled with an increased growing season (for the most part, growth continues throughout the year) and warmer water makes coastal ecosystems of the Southeast some of the most productive in the world (Table 3).

The magnitude of nutrient input to coastal ecosystems varies seasonally. In the Southeast input is strongly dependent on river discharge, with the largest nutrient load entering from January through May (Odum and Heald 1972, Odum et al. 1979, Livingston 1984c). During summer and autumn nutrient input from terrestrial systems is lowest. In summer and autumn estuarine production is more dependent on nutrients translocated from estuarine sediments, salt marshes, and marine water (Livingston 1984c). The seasonality observed in species composition, abundance, life cycles and/or behavior in coastal biota is strongly tied to this recurrent, periodic change in freshwater flow dynamics, that is, magnitude of nutrient input. Winter cold reduces estuarine production, species composition and abundance in northern estuaries, while the magnitude of freshwater and nutrient input similarly affects the ecological dynamics of SE estuaries.

Certain nutrients essential to production of all living matter, for example, phosphorus, nitrogen, and silicon, are considered the major limiting nutrients in aquatic systems (Stickney 1984). Although production limitations due to the absence of these nutrients do exist at some levels of ecosystem organization or in microhabitats, when viewed holistically such limitations have been suggested to be unimportant (Whitney et al. 1981). Nutrients are constantly produced, transformed, and degraded, resulting in a constant flux of essential nutrients. However, the rates at which these processes occur fluctuate predictably in rhythmic seasonal and/or tidal patterns and nonpredictably due to irregular changes in local climatic or other events.

Biogeochemical Cycles

Phosphate Although by no means completely understood, the phosphorous cycle is less complex than other nutrient cycles (Aston 1980, Webb 1981, Kennish 1986). In the estuarine environment phosphate is used by autotrophs (Taft and Taylor 1976) and heterotrophic microbes (Correll et al. 1975). In the available state, ortho-phosphate, residence time ranges from 0.05 to 200 h (Pomeroy 1960) before being taken up by bacteria and plants and/or adsorbed on suspended particles, sediment muds, and colloids. For reviews of the topic see Kennish (1986) and Knox (1986).

Phosphate input to coastal ecosystems is via freshwater inflow, surface runoff, saltwater intrusion, and pollution. The annual input to coastal waters from the Mississippi River drainage is estimated at about 5.7×10^6 kg/month (Darnell and Soniat 1979). Annual phosphorus inputs to Pamlico River Estuary are extremely high (331–843 metric tons/year), with most from a phosphate mine (843 metric tons/year) (Copeland et al. 1984). The greatest proportion of total phosphate enters bound to living or detrital organic matter, with minor amounts originating as inorganic phosphate adsorbed on sediment particles and as ortho-phosphate.

TABLE 3 Whole-System Productivity Estimates in the Southeast and Other Regions

| Estuary | Productivity (g C/m^2 year^{-1}) | | | |
| | Phytoplankton | | Benthic | |
	Estimate	References	Estimate	References
Pamlico Estuary, NC	200–500	Davis et al. 1978, Kuenzler et al. 1979	620	Fisher et al. 1982
Georgia Bight, GA	120–620		876	Hopkinson and Wetzel 1982
Tampa Bay, FL	360	Johansson et al. 1985		
Apalachicola Bay, FL	165	Livingston 1984c		
Barataria Bay, LA[a]	175	Day et al. 1973b	195	Day et al. 1973b
Corpus Christi Bay, TX	445	Flint and Kalke 1985	509	Flint and Kalke 1985
Mid Chesapeake Bay, VA	210	Boynton et al. 1982		
Patuxent Estuary, MD	307	Stross and Stottlemyer 1965		
Delaware Bay, DL	483	Pennock and Sharp 1986		
Lower New York Bay, NY	205	O'Reilly et al. 1976		
Mid Long Island Sound, NY		Riley 1956		
Mid Narragansett Bay, RI	310	Furnas et al. 1976	307	Hale 1975, Furnas et al. 1976
Buzzards Bay, MA				
San Francisco Bay, CA		Cole 1982	55	Rowe et al. 1975
Suisun Bay	95			
San Pablo Bay	100–130			
South Bay	150			
Kaneohe Bay, HA	165	Smith 1981		

[a]Happ et al. (1977), estimated that from 43 to 280 × 10^6 kg organic C/year is being exported from Barataria Bay to continental shelf waters.
Source. After Nixon (1983).

Studies undertaken in Georgia salt marshes show that phosphate is rarely limiting (see Whitney et al. 1981 for review). In the Apalachicola Bay system the annual level, output minus input, is about 1652 metric tons/year (Table 4). Although phosphates become limiting in the water column during summer/fall (Livingston 1984a), the change in seasonal concentrations are not of the same magnitude as observed for other nutrients (Table 5). Productivity at this time is dependent on nutrient transfer from sediments (Livingston 1984a).

Salt marshes were once considered the major supplier of phosphates via estuaries to coastal waters (Reimold and Daiber 1970, Gardner 1975). For some SE estuaries data do not support this contention. For example, at Sapelo Island, although phosphates are flushed from surrounding salt marshes into Duplin River and ultimately to the Altamaha River Estuary during local rain events, the whole phosphate concentrations in the Altamaha River are lower than those of coastal waters (Pomeroy et al. 1972). Consequently, coastal water, by its intrusion into the Altamaha River estuarine system, is considered the primary supplier of phosphates to the Sapelo Island salt marshes and estuarine system during periods of low river flow (Whitney et al. 1981). In contrast, phosphorous shows a net export to coastal waters at North Inlet, South Carolina (Dame et al. 1986).

Nitrogen The nitrogen cycle in coastal ecosystems is complex. Aston (1980) presents a very complete nitrogen cycle for the estuarine environment. Knox (1986) provides a detailed examination of nitrogen cycling at estuarine interfaces based primarily on the work of Kemp et al. (1982).

A comprehensive whole-system study of the nitrogen cycle in a SE coastal ecosystem was done for the Georgia Bight (Whitney et al. 1981). Nitrogen information on Barataria Bay, Louisiana (Hopkinson and Day 1977), Albemarle and Pamlico sounds, North Carolina, (Copeland et al. 1983, 1984), North Inlet, South Carolina (Dame et al. 1986), Florida seagrass communities (Zieman 1982), and Apalachicola Bay (Livingston 1984a) show that all systems function similarly. River input of total nitrogen is greatest in winter and least in summer (Table 5). Remineralization and exchange processes are most active during summer when biological demand is highest. Spatial distribution of inorganic nitrogen varies. In Apalachicola Bay concentrations are highest in rivers and lowest in the upper bay, with a general increase in concentration seaward (Livingston 1984a). A general seaward decrease in inorganic nitrogen concentration occurs in Albemarle and Pamlico sounds (Copeland et al. 1983, 1984). Spatial variations within an ecosystem occur throughout the year as a result of different degrees of biological activity. Concentrations in the Georgia Bight are lower than in other SE estuaries. This system receives little terrestrial drainage, but has high phytoplankton production year-round (Whitney et al. 1981).

A key transformation process in the nitrogen cycle is denitrification. Although denitrification results in the degradation of organic material, its impact is considerably less than that of sulfate reduction (Howarth and Teal 1980). However, because of the extensive areas of anaerobic soils in salt marshes,

TABLE 4 Nutrient Yields for Various Drainage Areas in the Apalachicola–Chattahoochee–Flint River System

Drainage Basin	Area (km²)	Annual Output Minus Input (metric tons)			Area Yield (g/m² year⁻¹)		
		Carbon	Nitrogen	Phosphorus	Carbon	Nitrogen	Phosphorus
Apalachicola–Chattahoochee–Flint	50 800	213 800	21 480	1652	4	0.4	0.03
Chattahoochee–Flint	44 500	142 700	17 860	1340	3	0.4	0.03
Apalachicola–Chipola	6 200	71 100	3 620	312	12	0.6	0.05
Apalachicola	3 100	41 500	1 060	237	13	0.3	0.08
Chipola	3 100	29 600	2 560	75	10	0.8	0.02
Apalachicola flood plain	393	34 300	674	206	87	1.7	0.62

Note. Data are presented on an areal basis.

Source. Adapted from Mattraw and Elder (1982) and Livingston (1984c).

TABLE 5   Nutrient Levels (Winter and Summer) for Stations in the Apalachicola Estuary (means ± 1 SD of 5 stations) and River (Station 2)

Nutrient	Site	Nutrient Values (μg/L)	
		17 February 1973	12 July 1973
NO₃	Bay T	179.53 ± 13.11	2.25 ± 2.84
	B	185.79 ± 19.48	4.24 ± 2.25
	River	232.90	219.54
NH₄	Bay T	25.13 ± 18.53	8.05 ± 3.30
	B	38.15 ± 30.61	14.26 ± 4.40
	River	7.81	7.57
PO₄	Bay T	6.92 ± 1.17	4.03 ± .76
	B	6.93 ± 1.29	5.78 ± 1.69
	River	12.53	9.53
Silicate (SiO₄)	Bay T	2531.90 ± 57.59	1939.66 ± 413.15
	B	2534.08 ± 62.88	1216.67 ± 802.98
	River	2532.55	3109.12

Source. From Livingston (1984c).

estuaries, and shelves, denitrification exceeds N-fixation (Whitney et al. 1981). This, coupled with the removal of biologically active nitrogen, results in nitrogen depletion. The magnitude of this impact in SE estuaries is not known.

Silica   The silicon cycle in the estuarine environment remains a subject of considerable debate (Kennish 1986). An extensive look at silicon behavior in marine systems can be found in Aston (1983). The major source of silica is terrestrial, entering the coastal ecosystem via river flow. The average input of silica from the Mississippi River to the Gulf of Mexico is approximately 1.8×10^8 kg/month (Ho and Barrett 1975). A small, but significant, quantity enters with intrusions of salt water during low river flow (Darnell and Soniat 1979).

Silica in SE estuaries is rarely limiting (Pomeroy and Imberger 1981). During high river flow a gradient of silica concentration is generally evident from the upper estuary to its mouth. Darnell and Soniat (1979) estimated silica transport to the ocean from estuaries to be on the order of 30–60 μg/L. Livingston (1984c) estimated temporal and spatial nutrient values for silica and showed little spatial variation within Apalachicola Bay during peak river flow in February, but a marked difference at this time was seen in silica concentration between the bay and the inflowing river. Concentration difference between upper and lower Apalachicola Bay was approximately 2.3 μg/L, while between the river and the bay it was 99 μg/L. In July during low river flow river concentrations remained high, while bay concentrations dropped considerably lower than what was observed in February. This suggests that relatively little silica is being transported seaward from the Apalachicola system during the summer (Livingston 1984c).

Silica concentration, exchange, and transport within and between the various components of the SE coastal ecosystem (e.g., saltmarsh, mangrove, seagrass, and water column) are poorly known. Most silicates produced in salt marshes around Sapelo Island, Georgia, were lost to the surrounding estuary, Doboy Sound, according to Pomeroy and Imberger (1981). They estimated a loss of approximately 90 mg/m^2 day^{-1}.

Sulfur   Sulfur is extremely active under anoxic conditions in estuaries. Sulfate in seawater is rapidly converted to hydrogen sulfide by sulfate-reducing bacteria, which react with soluble forms of iron to produce iron sulfides (Howarth and Teal 1980). Other nutrient cycles are impacted by these reactions, especially in anoxic sediments. Of particular importance is the potential mobilization of sediment phosphate, which can diffuse upward into the overlying oxic sediment or water column (Webb 1981). Energy from plant biomass can be uncoupled from carbon-based compounds by these bacteria in marshes and sediments and exported to nearshore waters through various sulfate compounds.

Carbon   Generally, sources of carbon input to coastal marine communities are from (1) primary production within the system, (2) dissolved and particulate carbon input via flooding tidal waters, (3) dissolved inorganic carbon in rainwater, and (4) dissolved organics and inorganics input with groundwater seepage. In addition, considerable amounts of terrestrially generated natural and anthropogenic particulate carbon enters estuaries and other near coastal systems by freshwater inflow and wind. A slight amount of terrigenous inorganic carbon is naturally input, but this contribution has rarely been considered in carbon flux models. Export from the coastal communities includes (1) dissolved and particulate carbon removed with the ebbing tide, (2) permanent burial in the bottom substrates, and (3) release of CO_2 and methane to the atmosphere. In general, the most significant source and sink of carbon in estuaries is primary production of phytoplankton and salt marsh plants and burial in the sediments, respectively (Odum et al. 1979, 1984).

Relatively few data exist that describe the carbon flux in any SE estuary. Davis et al. (1978) showed that inorganic carbon and dissolved organic carbon comprise about 95% of the total carbon budget in Pamlico River Estuary, while about 65% of this originates autochthonously (Copeland et al. 1984). Peterson and Peterson (1979) found that primary production from *Spartina* marshes, seagrasses, blue-green algal mats, diatoms and other pelagic phytoplanktors mediate carbon flux in the Newport River Estuary of North Carolina. In the Apalachicola estuarine system, phytoplankton production almost totally regulates carbon flux (Livingston 1984c). Zieman (1982) compared productivity of three dominant seagrasses, turtle grass (*Thalassia testudinum*), shoal grass (*Halodule wrightii*), and manatee grass (*Syringodium filiforme*), from different locations in the western Atlantic and Caribbean Sea. The Florida population of turtle grass fixed 0.9–16.0 g C/m^2 day^{-1}.

Another significant but poorly studied source of particulate and dissolved organic carbon to coastal waters is of anthropogenic origins. Oppenheimer et al. (1975, as cited in Armstrong 1987) calculated that carbon from sewage treatment plants and industrial complexes contributes about 1.1×10^6 kg C/year to estuarine waters. Naturally generated allochthonous organic material is also a significant source of carbon for coastal water. The Mississippi River alone supplies an estimated range of 3.0×10^9 (Happ et al. 1977) to 1.6×10^{10} kg organic C/year (Ho and Barrett 1975, as cited in Darnell and Soniat 1979), while Corpus Christi Bay receives about 8.2×10^6 kg C/year (Armstrong 1987).

Nutrient Ratios   If the relative concentrations of major nutrients (phosphorus, nitrogen, and carbon) in aquatic ecosystems reflect the ratios in typical plants, then productivity in the ecosystem is low. Usually the limiting nutrient is phosphorus, on occasion nitrogen, but rarely carbon. P:N ratios in estuaries range from 1:10 to 1:30 (Lackey 1967).

When adequate amounts of P and N are present, primary production can be limited by a minor nutrient. For example, the unavailability of silicates will inhibit diatom production even though N and P are available (Whitney et al. 1981). Whitney et al. (1981) calculated a P:N ratio of 1:2 for the Georgia and Massachusetts estuaries (from the data of Valiela and Teal 1979a,b), which suggested nitrogen depletion. This depletion was attributed primarily to denitrification, which occurs in the extensive anaerobic muds of salt marshes, estuaries, and coastal waters.

Calculation of P:N:C ratios for those estuarine systems in the Southeast where data were available shows P:N ratios from 1:0.5 to 1:1800; carbon data was available for the Apalachicola system (Livingston 1984c), where a P:N:C ratio from 1:3:166 to 1:1800:14 300 was found. These ratios were calculated using only those N and P species directly available to plants, for example, *ortho*-phosphate, nitrates, nitrites, and ammonia. Most estuaries, during periods of normal productivity, fell within the expected P:N range of 1:10 to 1:30. Albemarle Sound had a P:N of 1:26, based the on total annual concentration (Copeland et al. 1983). In contrast, ratios in Pamlico River Estuary and its watershed were 1:3 and 1:6, respectively, and suggest nitrogen depletion. Actually, nitrogen levels were normal, but effluent from phsophate mining made phosphate levels extremely high (Copeland et al. 1984). In contrast, the unusually high nitrogen level in the Chattahoochee–Flint River system of Apalachicola Bay estuary produced a P:N ratio of 1:1800 a condition attributed to phosphate trapping in reservoirs upstream. Removal of phosphates through sedimentation and production in these reservoirs resulted in an unusually low annual input of this nutrient (1.34 metric tons/year) to this portion of the bay (Livingston 1984c).

BIOTIC ENVIRONMENT

Complex physical and chemical processes are characteristic of the sharp gradients featured in SE estuaries. Carriker (1967) argues that the estuarine biotope is not a simple overlapping of marine and terrestrial conditions but a unique state, structured by materials and forces contributed by the surrounding environments. Distributional classes of estuarine organisms have been developed by Boesch (1977), which progress from stenohaline marine species found in tidal passes, through three classes of estuarine dominants, to freshwater species located at the head of the estuary (Fig. 2).

Southeastern estuaries are located along a latitudinal gradient from temperate to tropical waters. The estuarine biota of the upper gulf coast and lower east coast are similar and are best characterized as warm temperate (Hedgpeth 1953). On the other hand, estuaries of the Florida peninsula south of Cape Canaveral on the east and Cape Romano on the west (Fig. 1) support tropical species, including coral and mangrove associates (Hedgpeth 1953, McCoy and Bell 1985). Most estuarine organisms, however, are broadly ranging, eurythermic species that can be found from North Carolina to the southern Caribbean (Lyons and Collard 1974).

In conjunction with salinity and temperature, highly variable factors such as tidal amplitude, frequency and duration of flooding, dissolved oxygen, turbidity, depth, wave action, and pollutants determine the distribution and abundance of organisms in the estuary (Ketchum 1983). Substrate and vegetation, however, structure animal communities into recognizable assem-

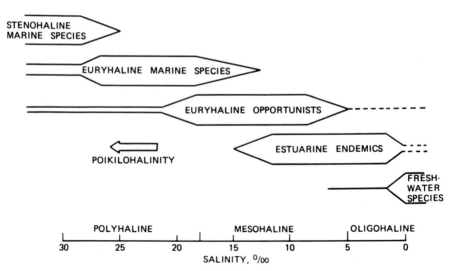

FIGURE 2. Schematic of the floral and faunal classes in a homeohaline estuary. (After Boesch (1977).)

blages. Accordingly, the estuarine biotope can be subdivided into six habitats: (1) intertidal emergent wetlands, (2) seagrass meadows, (3) unvegetated soft bottom, (4) hard substrate, (5) water column, and (6) aerial habitat. Each habitat, in turn, can be characterized by one or more biotic assemblages.

Intertidal Emergent Wetlands

Estuarine intertidal emergent wetlands are the transition zone between terrestrial and aquatic systems (Cowardin et al. 1979). In coastal estuaries of the Southeastern United States, the most common emergent wetlands are tidal marshes, dominated by black needlerush (*Juncus roemerianus*) and smooth cordgrass (*Spartina alterniflora*). Tidal marshes are replaced by mangrove forests south of Tampa Bay and Indian River in peninsular Florida.

Intertidal marshes are characterized by erect, rooted, herbaceous hydrophytes (Cowardin et al. 1979). The marsh ecosystem includes both emergent, grassy land zones and tidal creeks that exchange tidal water. Tidal marshes in the Southeast are the most extensive in the United States, comprising 78% of total U.S. coastal wetlands (Gosselink and Baumann 1980) and covering approximately 2.33 million ha (Table 6) (Fig. 3).

Marshes are typically areas of high environmental stress and low species diversity. The dominant vascular plant species are similar throughout the area, with regional dominance patterns reflecting different physical and chemical factors. Initiation and maintenance of tidal marsh habitats are controlled by conditions including (1) sufficient shelter from wave energy to ensure sedimentation and prevent excessive erosion; (2) sediment source for accretion;

TABLE 6   Distribution of Intertidal Salt Marshes in the Southeastern United States

State	Acreage		Reference
	Total	*Juncus* Dominated	
North Carolina	206 506	100 525	Critcher 1967
South Carolina	369 500	91 068	Alexander et al. 1986, South Carolina Water Resources Commission 1970
Georgia	374 500	74 906	Alexander et al. 1986, Eleuterius 1976
Florida			
Atlantic	95 900	69 887	Eleuterius 1976
Gulf	431 300	150 113	Alexander et al. 1986, Eleuterius 1976
Alabama	34 641	18 014	Crance 1971
Mississippi	66 981	61 444	Eleuterius 1972
Louisiana	4 203 171	157 820	Chabreck 1972
Totals	5 782 499	723 777	

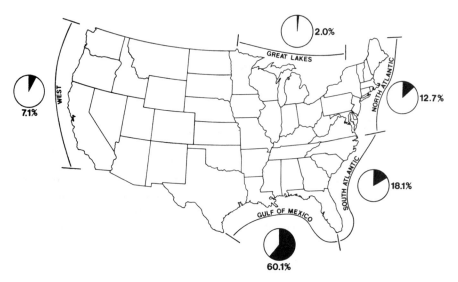

FIGURE 3. Percent distribution of coastal marshes in the U.S. (From Gosselink and Baumann (1980).)

(3) gentle, rather than steep, shoreline slope; and (4) tidal input. Tidal inundation itself influences physiographic, chemical, and biological processes, including sediment deposition and scouring, nutrient and organic matter dynamics, removal of toxins, and alterations of sediment redox potential (E. P. Odum 1980). Variation over the Southeast in tidal characteristics include microtidal versus mesotidal ranges in amplitude and diurnal versus equal semidiurnal patterns (see previous discussions and Tables 1 and 2 for details).

Plant species composition, abundance, and productivity are influenced by a number of these environmental factors. Earlier workers discussed obvious habitat differences that may be implicated as causes in plant species distribution. Among the parameters examined were patterns of tidal inundation, salinity, substrate type, nutrients, and aeration, both singly and in combination (Johnson and York 1915, Wells 1928, Chapman 1938, Penfound 1952, Hinde 1954, Bordeau and Adams 1956, Kurz and Wagner 1957, Kerwin 1966, Stalter 1968, Johnson et al. 1974, Baden et al. 1975). However, it has remained for more recent work with *Spartina alterniflora* to explain some of the direct relationships between environment and plant zonation.

On the Atlantic coast regularly flooded marshes (flooded on every tide) or zones within marshes, are covered by smooth cordgrass, the species best adapted to frequent and prolonged inundation, salt stress, and anaerobic conditions (Teal and Kanwisher 1966). Distinctive zonation can be seen within extensive stands of smooth cordgrass with decreasing height of *Spartina* from the tidal creek. Anderson and Treshow (1980), in a review of the subject, conclude that the differences are both genetic, with different ecotypes evolved, and environmental, with dominant factors being sediment anoxia and salinity stress.

In South Carolina, two smooth cordgrass zones have been described in the low, regularly flooded marsh: low low marsh with tall smooth cordgrass, and high low marsh with dwarfed smooth cordgrass (Stalter 1968, 1974, Stalter and Batson 1969). Greater tidal amplitude in Georgia results in more extensive development of natural levees in the low marsh; consequently, three cordgrass height zones can be delineated. Tall *Spartina* edge marsh occurs along the banks of creeks where smooth cordgrass reaches its maximum height (up to 3 m). Medium *Spartina* levee marsh is found atop natural levees along creek banks where smooth cordgrass averages 1 m in height. Short *Spartina* low marsh occurs between drainage creeks where smooth cordgrass ranges in height between 10 and 50 cm (Teal 1958, Bozeman 1975) (Fig. 4). The ultimate factor in variation in morphology and productivity appears to be the rate of interstitial exchange of water, more rapid on creek banks and slower inland (Valiela et al. 1978, Wiegert et al. 1983). Slower drainage rates may result in "stagnation" of interstitial waters with consequent elevated salinities, anoxia, toxins accumulation, and depleted nutrients. The gradient of these interrelated factors directly and indirectly affect growth and physiology of the plants (Dame and Kenny 1986).

As elevation increases toward the uplands, or wherever salinity of interstitial water is lowered, smooth cordgrass gives way to black needlerush. In areas of the marsh where groundwater does not intrude from the land, or where there is a large topographically isolated expanse of marsh, evaporation

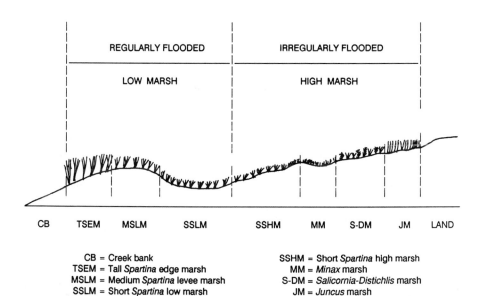

CB = Creek bank
TSEM = Tall *Spartina* edge marsh
MSLM = Medium *Spartina* levee marsh
SSLM = Short *Spartina* low marsh

SSHM = Short *Spartina* high marsh
MM = *Minax* marsh
S-DM = *Salicornia-Distichlis* marsh
JM = *Juncus* marsh

FIGURE 4. Diagram of South Atlantic saltmarsh vegetation zones. (Based on Teal (1958).)

leads to salinities high enough to exclude smooth cordgrass, leaving relatively bare expanses with a thin population of glassworts (*Salicornia* spp.) and salt grass (*Distichlis spicata*). A high marsh meadow of saltmeadow cordgrass (*Spartina patens*) with incidental upland forbs and grasses serves as a transition to the terrestrial ecosystem (Teal 1958, Tiner 1977).

North Carolina has two major types of salt marshes (Wilson 1962). Regularly flooded marshes, like those in Georgia and South Carolina, occur along the southern coast up to Morehead City and near inlets along the outer banks. The substratum is a gray, soft silt, and a complex system of tidal creeks is present. Irregularly flooded marshes (not flooded on every tide) are most extensively developed along the outer banks from Beaufort to Currituck County and on the inner fringes of sounds. Flooding in these marshes is primarily due to the effects of winds and storms. Salinity of the water is usually much below sea strength. The substratum is sandy, and tidal creeks are short and simple (Cooper and Waits 1973). Irregularly flooded marshes of the SE Atlantic are dominated by black needlerush if low enough in elevation to allow frequent flooding (Marshall 1974). Along the edge of heads of creeks, salt shrub (*Baccharis halimifolia*) and sea oxeye (*Borrichia frutescens*) occur. Along lower salinity creek edges giant cordgrass (*Spartina cynosuroides*) may occur. Saltmeadow cordgrass forms extensive stands on higher elevation marshes which occupy a brownish peat (Wilson 1967). Salt grass, three-cornered sedge (*Scirpus robustus*), and camphorweed (*Pluchea purpurascens*) are also scattered in this community.

Wind-dominated tides and tidal amplitudes of less than 1 m lead to infrequent flooding in marshes of the NE Gulf of Mexico (West Florida, Alabama, and Mississippi). This three-state arc has a greater proportion (52.9%) of its marshlands dominated by black needlerush than any other area of the United States except northern areas of North Carolina. Various authors have described and named zones of vegetation within these marshes (Uhler and Hotchkiss 1968, Eleuterius 1972, Subrahmanyam and Drake 1975, Stout 1979). A generalized NE Gulf marsh diagram is provided by Stout (1984) (Fig. 5).

Smooth cordgrass along the NE Gulf is restricted in its distribution to the regularly flooded intertidal zone of gulf marshes, quite frequently only as a narrow band. Needlerush dominates the majority of the marsh, occurring almost monospecifically at lower elevations and mixed with saltmeadow cordgrass near the upland transition. Needlerush may also exhibit zonation by height. High-salinity salt barrens or flats are also frequent small features of these marshes. Giant cordgrass may form large stands in low salinity marshes, especially in river delta marshes of Alabama and Mississippi.

Marshes of the Louisiana delta and Chenier Plain represent one of the largest contiguous salt/brackish marsh zones in the world and comprise over 72% of the U.S. coastal wetlands. Bahr et al. (1983) divide the marshes of the Mississippi River Deltaic Plain into salt marsh (182 000 ha) dominated by smooth cordgrass (61% of cover) and brackish marshes (404 000 ha) dominated by saltmeadow cordgrass (54% of cover). Brackish marshes include those classified as intermediate by Chabreck (1972). Delta marshes are de-

(A) PROTECTED LOW ENERGY SHORELINE

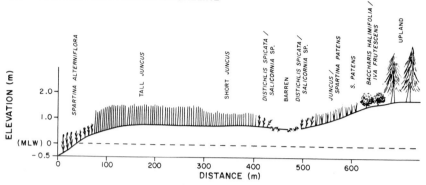

(B) OPEN MODERATE ENERGY SHORELINE

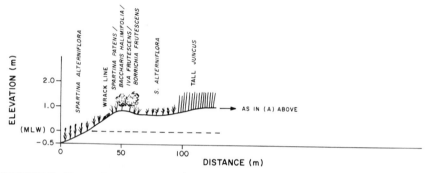

FIGURE 5. Generalized diagrams of Gulf coast salt marshes on protected low-energy shorelines and open moderate energy shorelines. (From Stout (1984).)

scribed in detail by Gosselink (1984). An additional 17 155 ha of salt marsh and 185 688 ha of brackish marsh occur in the Chenier Plain (Gosselink et al. 1979). Associated species are similar to other marshes of the Southeast.

The microalgal marsh flora of SE marshes is dominated by diatoms and myxophycean (blue-green algae) associations. Several hundred species of pennate diatoms have been identified in salt marshes, forming a continuous benthic marsh cover both in areas with and without a vascular plant canopy (Hustedt 1955, Williams 1962, Sullivan 1978). This association apparently represents a single, basic edaphic diatom community indigenous to Atlantic and Gulf coast salt marshes. Most abundant taxa include *Navicula* spp., *Nitzschia* spp., *Cylindrotheca* spp., and *Gyrosigma* spp. Differences in assemblage species composition may be seen under different spermatophyte canopies.

Sage and Sullivan (1978) and Ralph (1977) describe a single nearly homogeneous community of blue-green algae throughout Atlantic and Gulf marsh zones also suggesting an endemic temperate North American marsh algal association. Dominant taxa included are *Schizothrix arenaria*, *Schizothrix calcicola*, *Anacystis* spp., *Anabaena oscillaroides*, and *Entophysalis con-*

ferta. Light intensity was shown to be an important factor influencing the distribution of benthic blue-green algae in the marshes with open areas (without a canopy), such as salt flats, supporting development of a blue-green algae mat consisting primarily of *Microcoleus lyngbyaceous*, *S. calcicola*, *S. arenaria*, and *Oscillatoria* spp. (Sage and Sullivan 1978, Maples 1982).

Macroscopic algae are poorly represented in the marsh. Small species of red algae (*Caloglossa leprieurii* and *Bostrychia radicans*) may be found on standing dead *Spartina* culms during summer months (Chapman 1971, Blackwelder 1972) in Georgia and South Carolina. *Ectocarpus confervoides* can develop on the stems of streamside *Spartina* in mid-winter (Pomeroy et al. 1972), while a few other species from genera such as *Rhizoclonium*, *Ulva*, *Enteromorpha*, and *Vaucheria* are found only infrequently.

The paucity of macroscopic algae is likely caused by the high turbidity of estuarine waters and rapid sedimentation that occurs in the lower areas of the marsh (Williams 1962), and the extremes of temperature and desiccation (Pomeroy and Weigert 1981). The motility of the pennate diatoms, filamentous blue-green algae, and some euglenoids enables them to remain in or near the photic zone at the surface of shifting marsh sediments.

Teal (1962) described three categories of salt marsh animals: (1) those of terrestrial origin, (2) aquatic species with centers of abundance in the estuary proper, and (3) true marsh species derived from aquatic ancestors (Fig. 6). Few species in the salt marsh are endemic; most are estuarine aquatic species

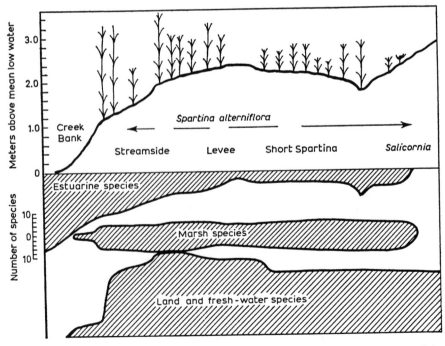

FIGURE 6. Representative section of a Georgia salt marsh. (From Teal (1962).)

able to tolerate environmental extremes. Consequently, animal zonation in salt marshes is more complex and less precise than vegetational zonation (Dörjes 1972, Subrahmanyam et al. 1976, Coull et al. 1979, Daiber 1982).

Despite differences in dominant vegetation and tidal regimes, faunal assemblages of Atlantic and Gulf marshes are similar (de la Cruz 1981). Niche substitutions have occurred in only a few cases, most notably in insects and vertebrates, even though populations were essentially isolated by the most recent emergence of peninsular Florida 3.75 million years ago.

The herbaceous strata of salt marshes are dominated by herbivorous insects (Davis and Gray 1966, McCoy and Rey 1981, 1987, Rey and McCoy 1982, 1986) and carnivorous spiders (Pfeiffer and Wiegert 1981, Rey and McCoy 1983, LaSalle and de la Cruz 1985). Seaside sparrows (*Ammodramus maritimus*, *A. miriablis*, and *A. nigrescens*), sharptail sparrows (*A. caudacuta*) and marsh wrens (*Cistothorus palustris*) also nest (Montagna 1942, Woolfenden 1956, Bent et al. 1968) and forage (Kale 1965, Bent et al. 1968, Pfeiffer and Wiegert 1981) in marsh vegetation. Marsh hawks (*Circus cyaneus*) glide low over coastal marshes while hunting small mammals (Sprunt 1955), but do not nest in the Southeastern United States. Other organisms, like the marsh periwinkle, *Littorina irrorata*, climb marsh vegetation to evade predation (Hamilton 1976, Warren 1985).

Edaphic and epiphytic algae and diatoms are consumed by epifaunal gastropods, especially the coffee bean snail, *Melampus bidentatus* (Hausman 1936a;b), and the olive nerite, *Neritina usnea* (Lehman and Hamilton 1980). Grazing of vascular marsh plants is generally limited to insects (E. P. Odum and Smalley 1959, Parsons and de la Cruz 1980), muskrat (*Ondatra ziebethica*) (O'Neil 1949), nutria (*Myocaster coypus*) (Evans 1970), and crabs of the genus *Sesarma* (Crichton 1974), and the entire crop is rarely grazed (Valiela 1984).

Macrophyte detritus is converted to microbial biomass on the floor of the marsh, where bacteria and fungi function as primary decomposers (Fenchel and Jørgensen 1977, Christian and Wetzel 1978). Bacteria are most abundant in the upper centimeter of sediment (Rublee 1982), where they are preyed upon by protozoans (Johannes 1965, Muller and Lee 1969) and meiofauna (Coull 1973, Coull and Bell 1979). Many infaunal and epifaunal salt marsh organisms, including the marsh periwinkle (Odum and Smalley 1959, Alexander 1979), mud snails (*Nassarius obsoletus*) (Pace et al. 1979), fiddler crabs (*Uca* spp.) (Teal 1958, Montague 1980), the mud crab (*Rhithropanopeus harrisii*) (Odum and Heald 1972), tanaids (Ogle et al. 1982), isopods (Frankenberg and Burbanck 1963), oligochaetes (Giere 1975), and the polychaetes, *Capitella capitata* (Grassle and Grassle 1974), *Neanthes succinea* (Cammen 1980), and *Streblospio benedicti* (Dauer et al. 1981), obtain all or part of their nutrition by deposit feeding, ingesting sediment and detritus along with associated bacteria, protozoa, and meiofauna (Cammen 1979, Tenore et al. 1982). Cammen (1979) notes that grazing by infauna on microbial populations and benthic algae communities may actually stimulate both detrital decomposition and primary production.

The ribbed mussel (*Geukensia demissa*) and the Carolina marsh clam (*Polymesoda caroliniana*), which occur in clumped distributions on the marsh floor, are filter feeders (Kuenzler 1961, Duobinis-Gray and Hackney 1982, West and Williams 1986). Another bivalve filter feeder, often present in great numbers but much less conspicuous because of its small size, is the Florida marsh clam (*Cyrenoida floridana*) (Heard 1982a, Bishop and Hackney 1987). Isotope studies have shown that ribbed mussels within the marsh consume detritus from *Spartina* and those at the edge consume mostly plankton (Peterson et al. 1985). Salt marsh bivalves filter very efficiently, removing substantial amounts of bacterioplankton (Olsen 1976, Wright et al. 1982, Chrzanowski et al. 1986).

Tabanid larvae and some dipteran larvae are among the few infaunal predators found in salt marshes (Meany et al. 1976, LaSalle and Bishop 1987). Predators that forage on the surface of the marsh include raccoons (*Procyon lotor*), clapper rails (*Rallus longirostris*), rice rats (*Oryzomys palustris*), and xanthid crabs (*Eurythium limosum*) (O'Neil 1949, Sharp 1967, Heard 1982a;b). The only reptiles that have adapted to coastal marshes are the alligator (*Alligator mississippiensis*), several subspecies of the diamondback terrapin (*Malaclemys terrapin*), and the Gulf salt marsh water snake (*Nerodia fasciata*) (Neill 1958).

The intricate pattern of tidal creeks characteristic of most salt marshes forms the connection between the marsh and the open bay or sound through which sediment, nutrients, phytoflora, and fauna are exchanged (Dame et al. 1977, Moore and Reis 1983, Spurrier and Kjerfve 1988, Wolaver et al. 1988). Bacteria in tidal creeks are either free living in the water column or associated with suspended particulate materials. Wilson and Stevenson (1980) found that bacteria associated with seston varied tidally and seasonally, while planktonic bacteria populations varied only seasonally. Bacterial densities were highest in late summer and lowest during winter months. Weiland et al. (1979) suggest that levels of total microbial biomass in salt marsh creeks are not influenced by changes in salinity. Natant macrofauna forage during flood tide in the vegetated intertidal zone and retreat to tidal pools and creeks as the tide ebbs. Blue crabs, grass shrimp (*Palaemonetes pugio*), penaeid shrimp (*Penaeus aztecus*, *P. setiferus*) sciaenids (*Micropogonias undulatus*, *Leiostomus xanthurus*), silversides (*Menidia* spp.), and killifishes (*Fundulus* spp.) are significant predators on benthic components of the salt marsh community (Lucas 1982, Smith et al. 1984, Kneib 1985, 1987a;b, West and Williams, 1986, Hunter and Feller 1987, O'Neil and Weinstein 1987, Smith and Coull 1987). Predation and disturbance control densities and structure the composition of both marsh surface and tidal creek infauna (Bell and Coull 1978, Kneib and Stiven 1982, Wiltse et al. 1984, West 1985). Tidal creeks support staggered pulses of recruitment of fishes and macroepifaunal invertebrates from early spring to late fall (Subrahmanyam and Drake 1975, Hackney and Burbanck 1976, Hackney et al. 1976, Subrahmanyam and Coultas 1980, Horlick and Subrahmanyam 1983, Rogers and Herke 1985) and serve as an important winter nursery ground for spot (*Leiostomus xanthurus*), pinfish (*Lagodon rhomboides*), and menhaden (*Brevoortia* spp.) (Weinstein 1979, Bozeman and Dean 1980, Rogers and Herke 1985). Tidal creeks also serve as feeding areas for wading birds, including the great white heron (*Ardea herodias occidentalis*), great blue heron (*A. herodias*), green-backed heron (*Butorides virescens*), little blue heron (*Egretta caerulea*), great egret (*Casmerodius albus*), snowy egret (*Egretta thula*), tricolor heron (*E. tricolor*), and black crowned night heron (*Nycticorax nycticorax*) (Custer and Osborn 1978a,b, Daiber 1982, Bildstein et al. 1982).

Animal communities in salt marshes are composed of large numbers of relatively few species (Teal 1962, Kneib 1984). Distribution and abundance are affected by vegetation, microtopography, tidal flooding, and salinity, as well as biological factors such as predation, competition, and density-dependent processes (Fig. 7) (Daiber 1982, Kneib 1984). Interactions between dominant macrophytes and other marsh organisms are well documented (Kraeuter and Wolf 1974, Shanholtzer 1974, Bell et al. 1978, Osenga and Coull 1983, Rader 1984). Peak infaunal densities in Southeastern marshes occur during spring and, in some cases, autumn, while lowest densities are reported for summer (Teal 1962, Day et al. 1973b, Subrahmanyam et al. 1976, Cammen 1979, Kneib 1984). Cammen (1979) suggests that predation by juvenile fishes, combined with mortality of short-lived invertebrate cohorts spawned in the spring, is responsible for summer declines in abundance. He also notes that standing stock (excluding mussels) in North Carolina *Spartina* marshes is twice that of Georgia and an order of magnitude greater than

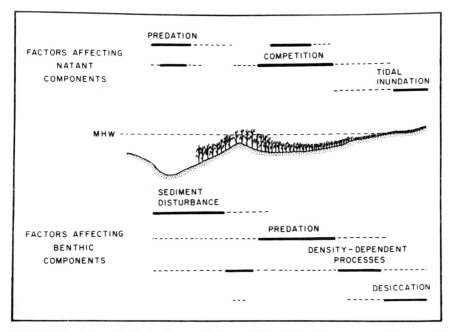

FIGURE 7. A hypothetical scheme of site-specific relative importance of factors affecting abundance of natant and benthic community components of an intertidal salt marsh (Sapelo Island, Georgia). (From Kneib (1984).)

Louisiana. Comparison of densities on a species by species basis, however, indicates that a Florida *Juncus* marsh has comparable numbers of marsh periwinkles, coffee bean snails, and isopods (*Cyathura polita*), more tanaeids and orbiniid polychaetes (*Leitoscoloplos fragilis*), and fewer nereid polychaetes (*Neanthes succinea, Laeonereis culveri*) and ribbed mussels than *Spartina* marshes (Subrahmanyam et al. 1976).

Salt marshes provide a nursery for commercially valuable fish and wildlife species through high vascular plant detritus production and reduced predation (de la Cruz 1980, Odum and Smith 1981). However, energy flow studies (Teal 1962, Day et al. 1973a, Cammen et al. 1982) and stable isotope ratio studies (Haines 1979, Haines and Montague 1979, Hackney and Haines 1980) suggest that net fluxes of organic matter and nutrients vary in magnitude and direction among marshes and that generalizations are not appropriate (Nixon 1980). Current thinking recognizes that export of marsh plant production is in the form of both animal carbon and particulate detritus (Boesch and Turner 1984), thus emphasizing the importance of the refuge role of salt marshes (Zimmerman and Minello 1984). Reviews of the structure and function of salt marsh fauna can be found in Ranwell (1972), Cooper (1974), Chapman (1977), Pomeroy and Wiegert (1981), Heard (1982a), Daiber (1982), Stout (1984), Durako et al. (1985), and Mitsch and Gosselink (1986).

Seagrass Meadows

Seagrass meadows constitute one of the most conspicuous and common subtidal estuarine habitat types, often providing significant contributions to estuarine primary productivity. Roots and rhizomes transfer from sediments important nutrients that are subsequently released through the leaves into the water column (McRoy and Barsdate 1970, McRoy et al. 1972). Leaves of submerged seagrasses provide substrate for other epiphytic primary producers as well as invertebrates (Nagle 1968, Kikuchi and Perès 1977, Harlin 1980, Zieman 1982). The complex vertical and horizontal structure of grassbeds is important in predator–prey relationships (Coen et al. 1981, Heck and Thoman 1981, Peterson 1982, Orth et al. 1984). Reduced current velocities across grassbeds, due to drag forces on the leaves (Fonseca et al. 1982), promote sedimentation and reduce turbulence and scouring. Roots and rhizomes enhance sediment stability and aeration.

As rooted plants, seagrasses need a soft bottom to colonize. The depth range of meadows on suitable substrate is dependent upon species resistance to desiccation and high insolation near the low tide mark and water clarity at the deeper limits. Salinity tolerance and adaptability to rapid changes in salinity vary between seagrass species and affect local community structure. Reduced wave energy required for grassbed survival may be provided by barrier islands and bars, offshore reefs, and bottom configurations.

Seven species of seagrasses may be found in the tropical to subtropical estuaries of the U.S. Atlantic and Gulf coasts (Eiseman 1980): turtle grass (*Thalassia testudinum*), manatee grass (*Syringodium filiforme*), shoal grass

(*Halodule wrightii* syn. *Diplanthera wrightii*), widgeon grass (*Ruppia maritima*), and *Halophila engelmanni*, *Halophila decipens*, and *Halophila johnsonii*.

Subtropical seagrass meadows are generally dominated by turtle grass; in areas of lower salinity, such as mouths of rivers and in intertidal zones, shoal grass frequently dominates. In the upper, brackish end of estuaries widgeon grass dominates. Manatee grass typically occurs as a successional stage in the development of a turtle grass community. Reviews of seagrass distribution in the Southeast include Humm (1956), Phillips (1960), Moore (1963), McNulty et al. (1972), Earle (1972), Phillips et al. (1974), Continental Shelf Associates and Martel Laboratories (CSAML) (1985), and Iverson and Bittaker (1986).

Turtle grass-dominated meadows occur as extensive beds from Biscayne Bay, Florida, southward and westward to the Dry Tortugas, providing almost complete cover from the mainland to the keys and offshore to the outer reef where they cease in the coral zone (Voss and Voss 1955). Seagrass meadows cover approximately 5500 km^2 in Florida Bay (Iverson and Bittaker 1986). North of Biscayne Bay, on the Florida Atlantic coast, winter storms off the Atlantic Ocean produce such high energy onshore that seagrasses are limited to only a few inlets in which turbidity is minimal and salinities are high.

On the Gulf coast from Cape Romano north to Tampa Bay, wave action and high turbidity caused by discharge from the Everglades–Big Cypress region, similarly reduce the occurrence of seagrasses. Seagrasses are again abundant in the region from Tampa Bay north to Apalachee Bay (Florida Big Bend). This region supports meadows over an area of about 3000 km^2, including the shallow, nearshore shelf (Iverson and Bittaker 1986). Beds may extend up to 18 miles offshore and as patches to depths of 20 m (CSAML 1985).

In the northern Gulf of Mexico, seagrass meadows are locally abundant in St. Joseph Bay, St. Andrews Bay, Santa Rosa Sound, Perdido Bay, Mississippi Sound, and the Chandeleur Islands of Louisiana. High turbidities and low salinity in Apalachicola Bay, Mobile Bay, and most of the Louisiana coast limit and often exclude seagrass species (Moore 1963).

Within seagrass meadows, zonation patterns, relative to tidal exposure, reflect physiological differences between species (Fig. 8). Shoal grass is tolerant of higher water temperatures and longer air exposure than other species (Humm 1956), and is, therefore, the most abundant species between neap high and neap low tide lines (Phillips 1960). Widgeon grass is commonly mixed with shoal grass in low salinities. In estuaries with salinities of 25 ppt or greater, turtle grass does not become abundant above the spring low tide line, but may be scattered among shoalgrass and widgeon grass at higher elevations. In lower salinity areas, manatee grass may replace turtle grass from the spring low tide line down.

Of all the seagrasses, widgeon grass tolerates the broadest range of salinity, occurring in freshwater and in areas with salinities in excess of 35 ppt (Fig. 9). Turtle grass is relatively stenohaline (Moore 1963), usually restricted to areas with salinity over 20 ppt (Phillips 1960). Shoalgrass and manatee grass exhibit maximum growth in moderately brackish water (Phillips 1960). *Hal-*

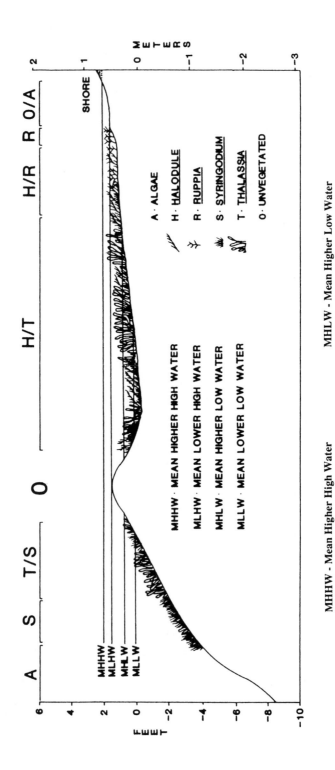

FIGURE 8. Tropical seagrass species zonation. (From Lewis et al. (1982).)

MHHW - Mean Higher High Water   MHLW - Mean Higher Low Water
MLHW - Mean Lower High Water   MLLW - Mean Lower Low Water

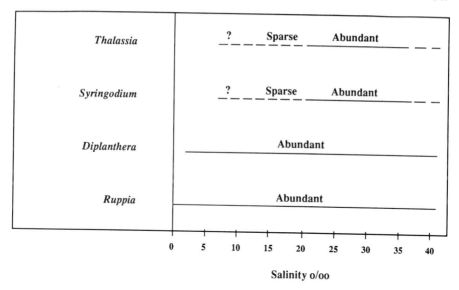

FIGURE 9. Salinity preferences and tolerances of seagrasses. (Modified from Phillips 1960 and Moore (1963).)

ophila is of relatively low abundance, possibly due to a requirement for high salinity (Taylor 1974).

Turtle grass generally has a dense rhizome/root system, often penetrating the sediment to 25 cm (Zieman 1982). Anaerobic nitrogen fixation in reducing sediments seems to be the source of nitrogen for this species (Patriquin 1972). Shoal grass occurs on the same muddy sand substrates as turtle grass, as well as on extremely coarse muddy sands (Phillips 1960). However, shoalgrass is more prevalent on oxidized substrates, with roots and rhizomes only a few centimeters deep. Though manatee grass is also shallowly rooted, it occurs in both oxidized and reduced sediments (Patriquin and Knowles 1972). Widgeon grass is found on predominantly mud and silt substrates containing finer textured sand than that associated with other species (Phillips 1960).

Estuaries of lower salinities and the upper end of estuaries subjected to fluctuating river flow and salinities may have submerged grassbeds comprised of typically freshwater species tolerant of periodic elevated salinity. Most commonly encountered species are tapegrass (*Vallisneria americana*) and pondweed (*Potamogeton* spp.)

Areas inhabited by seagrasses do not offer the hard substrate for attachment required by benthic algae. Mat-forming algae and rhizophytic chlorophyta may use seagrass sediments as substrate. *Halimeda* and *Penicillus* are the most abundant, but *Caulerpa*, *Udotea*, *Rhizocephalus*, and *Avrainvillea* are also common (Zieman 1982). All except *Caulerpa* produce a skeleton of calcium carbonate. A major function of these algae in the early successional stage of grassbeds is the contribution of sedimentary particles (S. L. Williams 1981). In addition to the rhizomatous, calcareous forms, several algae are present in grass beds as large clumps of drift algae, the most abundant being *Laurencia.*

An additional diverse community of microalgae utilizes seagrasses themselves as a substrate. Several forms of attachment have been described (van Montfrans et al. 1984): *epiphytes*, microalgae and macroalgae colonizing seagrass blades; *periphyton*, microalgae and algae sporelings, which are found in a mucus-like layer coating the seagrass blades; and *encrusting algae*, calcareous coralline species growing on seagrasses. The availability of a substrate appears to be the most important role played by the seagrasses. Harlin (1975) listed factors important to the epiphyte–seagrass relationship, including physical substrate, access to photic zone, exposure to moving water, nutrient exchange with host, and organic carbon source.

Seagrass leaves are more heavily epiphytized near the older tips than the young bases. Several authors have suggested that moisture retention by epiphytes may assist the seagrass in desiccation resistance (Penhale and Smith 1977, Richardson 1980). Trocine et al. (1981) demonstrated that shading by epiphytes may allow seagrasses to inhabit shallow areas where photoinhibition would preclude them.

Heavy epiphyte loads may threaten seagrass communities with local extinction (Borum and Wirem-Anderson 1980, Sand-Jenson and Borum 1983). Potential mechanisms include competition for similar wavelengths of light (Sand-Jenson 1977), shading (Taylor and Lewis 1970, Larkum 1976, Sand-Jenson and Borum 1983), and suppression of carbon (HCO_3^-) (Sand-Jenson 1977) and phosphorus (PO_4) assimilation (Johnstone 1979). Light attenuation by epiphytes may cause leaf senescence and affect both vegetative and sexual reproductive capabilities (Richardson 1980, Rice et al. 1983). For a summary of epiphyte effects on seagrasses see Table 1 in van Montfrans et al. (1984).

In temperate climates, eelgrass (*Zostera marina*) is the most common seagrass species. Estuaries of the Carolinas fall within the range of this species on the Atlantic coast and it is particularly abundant in the sounds and bays landward of the outer banks. South of approximately 33°N latitude, eelgrass is replaced by dominants of more tropical affinities. Eelgrass inhabits areas that have sediments ranging from soft mud to coarse sand, average salinities of 10–30 ppt, and a water temperature range from less than 0 °C to greater than 30 °C. The annual mean temperature at the southern end of its range is 17.6 °C (Thayer et al. 1984a). Light availability appears to be the primary factor limiting both depth and up-estuary penetration of eelgrass. The depth distribution of eelgrass on the Atlantic coast also has a range proportional to tidal ranges characteristic of the geographic region.

Shoalgrass, a pantropical species, grows over a tidal range similar to eelgrass, but shoalgrass beds may extend into the upper intertidal zone and are frequently exposed at low tide. In temperate North America, shoalgrass occurs from the Atlantic coast of Florida to North Carolina. Eelgrass and shoalgrass co-occur in North Carolina estuaries, with shoalgrass dominant in late summer and early fall and eelgrass dominant in winter to early summer (Kenworthy 1981). Because of this bimodal seasonal abundance pattern, the coexistence of both species in a mixed stand provides a continuous cover and source of estuarine primary production throughout most of the year.

A third species, widgeon grass, may co-occur with shoalgrass and eelgrass in temperate estuaries. This species is both eurythermal and euryhaline (Phillips 1974), but may be limited in depth distribution by available light (Congdon and McComb 1979). Widgeon grass grows almost exclusively in brackish water and frequently in low-salinity pools in salt marshes. Eelgrass, however, dominates the mid- to high-salinity ranges within the estuary.

Seagrass fauna have been grouped on the basis of their mode of existence and microhabitat structure (Kikuchi and Pérès 1977): (1) species living on the leaves, including crawling or swimming epifauna found on the blades; (2) species attached to the stem and rhizomes; (3) nekton utilizing the leaf canopy; and (4) infaunal species. Because of sampling constraints, however, most studies have concentrated on only a single component (Zieman 1982). Discussions of animal communities in seagrass meadows must be tempered with the awareness of their enormous diversity and variability (Zieman 1982). With hundreds of species in a small area, dramatic changes in composition and density commonly occur on minute temporal and spatial scales.

Seagrasses dramatically increase bottom surface area and provide a substrate for attachment of epiphytic plants and animals. Hydroids, bryozoans, sponges, mollusks, barnacles, and tube-building polychaetes attach to seagrasses (Nagle 1968, Harlin 1980). Bacteria, protozoans, nematodes, oligochaetes, and other micro- and meiofauna live among the attached algae and animals (Lewis and Hollingworth 1982, Bell et al. 1984). These organisms are extensively grazed by macroepifauna (Kitting 1984, Orth and van Montfrans 1984, van Montfrans et al. 1984).

The macroepifaunal grazers of leaves are predominantly gastropods (*Bittium varium*, *Cerithium* spp., *Anachis avara*, *Modulus modulus*, *Mitrella lunata*) and crustaceans, especially amphipods (*Cymadusa compta*, *Melita nitida*, *Grandidierella bonnieroides*, *Gammarus mucronatus*) and shrimps (*Penaeus* spp., *Palaemonetes* spp., *Perclimenes* spp., *Hippolyte* spp., *Thor* spp., *Tozeuma carolinense*) (Brook 1977, 1978, Zimmerman et al. 1979, Morgan 1980, Morgan and Kitting 1984). Large epibenthic organisms include the herbivorous queen conch (*Strombus gigas*) and sea urchins (*Lytechinus variegatus*, *Tripneustes ventricosus*), which feed largely on the seagrasses and algae, and the carnivorous whelks (*Busycon* spp.), tulip shells (*Fasciolaria tulipa*), and horse conchs (*Pleuroploca gigantea*) (Paine 1963, Randall 1964, Thayer et al. 1984b). Many mobile epifauna appear to enter the canopy from the underlying sediment at night (Greening and Livingston 1982, Howard 1987). Several studies have demonstrated that abundance of epifaunal invertebrates are increased when drift algae are present in seagrass meadows (Hooks et al. 1976, Thorhaug and Rossler 1977, Gore et al. 1981, Lewis 1987, Virnstein and Howard 1987).

Macroinfaunal components of seagrass meadows are often an extension of the community on adjacent unvegetated bottom (Thayer et al. 1975a, Virnstein et al. 1983), but usually have more species and greater abundances (Santos and Simon 1974, Stoner 1980a, Lewis and Stoner 1983, Lewis 1984). Sediment-dwelling meiofauna exhibited the opposite pattern. Barren sand harbored greater densities of most taxa than did seagrass sediments (Decho et al. 1985). Densities of macroinfaunal organisms of SE seagrass meadows vary widely, ranging from 328 to nearly 40 000 m^{-2} (Table 7). Because comparisons are affected by composition and density of seagrass species, sieve mesh size, and the duration and frequency of sampling, few infaunal dominants are common to geographically separated seagrass meadows (Tables 8 and 9). The most abundant groups are polychaetes and amphipods.

Fulton (1985) found higher abundance of an epibenthic littoral copepod (*Pseudodiaptomus coronatus*) in an eelgrass bed than in an unvegetated channel. On the other hand, abundances of pelagic copepods were reduced in grassbeds relative to the unvegetated bottom, suggesting intense predation by postlarval fishes. Laboratory experiments substantiated the refuge role of eelgrass for the epibenthic copepod and selective predation on pelagic copepods (*Centropages* spp.) by estuarine planktivores.

Most fishes found in grassbeds are seasonal residents, present only during their spawning season or, more commonly, during their juvenile or subadult stages (Adams 1976, Brook 1977, Livingston 1975, 1982). Many commercially and recreationally important species use grass beds as a nursery ground (Table 10). Permanent residents are usually small, cryptic species such as the emerald clingfish (*Acyrtops beryllina*), pipefishes (*Syngnathus* spp.), seahorses (*Hippocampus* spp.), gobies (*Gobisoma robustum*), and blennies (Zieman 1982, Huh and Kitting 1985, Sogard et al. 1987).

The rich, abundant fauna characteristic of seagrass meadows are probably controlled by a combination of factors, including food availability (Thayer ad Ustach 1981), competition (Coen et al.1981), predation (Orth et al. 1984), and physical factors (Livingston 1984a). Structural complexity affects predator–prey interactions (Heck and Orth 1980, Heck and Thoman, 1981, Nelson 1981) and influences habitat selection directly (Stoner 1980b, Leber 1985). One index of habitat complexity, seagrass biomass, is a good indicator of fish and crustacean abundance within a single meadow (Heck and Wetstone 1977, Stoner 1983a, b). Blade density, however, proved more useful in predicting fish and crustacean abundances across several monotypic beds of different seagrass species (Stoner 1983a, b). In the case of infauna, the depth at which they live relative to the root–rhizome system, rather than canopy architecture, determines their vulnerability to predation (Virnstein 1979, Peterson 1982). Infaunal amphipod abundance, for example, was inversely related to seagrass biomass (Stoner 1980a).

TABLE 7 Comparison of Average Densities of Infaunal Communities Associated with Seagrass Meadows in Southeastern Estuaries

Location	Seagrass	Average Density (per m²)	Seive Size (mm)	References
Back Sound, NC	Z. marina and H. wrightii	3 223	1.2	Summerson and Peterson 1984
Newport River	Zostera marina	8 500	0.5	Homziak et al. 1982
	Zostera marina	328	1.2	Thayer et al. 1975a
Indian River	Halodule wrightii	8 291	1.0	Young and Young 1977
	T. testudinum and H. wrightii	17 479	0.5	Virnstein et al. 1983
Biscayne Bay	Halodule wrightii	5 000	1.6	Moore et al. 1968
	Halodule wrightii	1 113	3.0	O'Gowar and Wacasey 1967
Tampa Bay	Thalassia testudinum	33 485[a]	0.5	Santos and Simon 1974
	Halodule wrightii	13 313[a]	0.5	Santos and Simon 1974
Apalachee Bay	Syringodium filiforme and T. testudinum	3 185	0.5	Stoner 1980b
Apalachicola Bay	Halodule wrightii	38 780	0.5	Sheridan and Livingston 1983
	Halodule wrightii	17 815[a]	0.5	Sheridan and Livingston 1983
Biloxi Bay	Ruppia maritima	10 366	0.52	McBee and Brehm 1979

[a] Polychaetes only.

Source. From Sheridan and Livingston (1983).

TABLE 8 Comparison of Seagrass Meadow Benthic Dominants (ranked in decreasing order of abundance) in Several Southeastern Estuaries Along the Gulf Coast

	Apalachicola Bay	Apalachee Bay	Cedar Key
	Hargeria rapax	*Aricidea taylori*	*Laeonereis culveri*
	Heteromastis filiformis	*Elasmopus levis*	*Onuphis simoni*
	Ampelisca vadorum	*Lembos* sp. A	*Aricidea philbinae*
	Oligochaetes	*Cymadusa* sp. A	Oligochaetes
	Aricidea fragilis	*Lysianopsis alba*	*Ampelisca holmesii*
	Streblospio benedicti	*Platynereis dumerilii*	*Capitella capitata*
	Gammarus mucronatus	*Syllis* sp. B	*Streblospio benedicti*
	Fabricia sp.	*Exogone dispar*	*Clymenella mucosa*
	Haploscoloplos fragilis	*Mediomastus californiensis*	*Prionospio heterobranchia*
	Hobsonia florida	*Dorvillea sociabilis*	*Scoloplos foliosus*
	Halodule wrightii	*Thalassia testudinum* and	*Halodule wrightii*
		Syringodrum filiforme	
Vegetation			
Gear	Hand core	Hand core	Hand core
Sieve size	0.5 mm	0.5 mm	0.5 mm
Stations (replicates)	1(10)	3(12)	1(25)
Frequency	Monthly for 1 year	Monthly for 1 year	Monthly for 25 months
Source	Sheridan and Livingston 1983	Stoner 1980b	Bloom 1983

TABLE 9 Comparison of Seagrass Meadow Benthic Dominants (ranked in decreasing order of abundance) in Several Southeastern Estuaries Along the Atlantic Coast

	Biscayne Bay	Biscayne Bay	Indian River	Indian River	Newport River
	Phascolion sp. A *Chione cancellata* *Anodontia alba*	*Codakia orbiculais* *Chione cancellata* *Semiodera roberti* *Amphioplus pulchella* *Prunum apicinum* *Loimia medusa* *Panopeus occidentalis* *Notomastus luridus* *Terebellides stroemi*	*Clymenella mucosa* *Polydora ligni* *Phascolion* sp. *Exogone dispar* Paratanaidae *Cymadusa* sp. *Streblospio benedicti* Nemertines *Cerithium muscarum* *Erichsonella filiformis*	*Diastoma varium* *Spiochaetopterus costarum* *Cymadusa compta* Capitellidae *Crepidula convexa* *Amphithoe longimana* *Tharyx annulosus* *Phascolion cryptus* *Astyris lunata* *Branchioasychis americana*	*Nereis pelagica* *Tellina versicolor* *Solemya velum* *Abra aequalis* *Clymenella torquata*
Vegetation	*Halodule wrightii*	*Thalassia testudinum*	*Halodule wrightii*	*Thalassia testudinum* and *Halodule wrightii*	*Zostera marina*
Gear	Self-closing sampler	Self-closing sampler	Hand core	Post hole-type sampler	Post hole digger
Sieve Size	3.0 mm	3.0 mm	1.0 mm	0.5 mm	1.2 mm
Stations (replicates)	1(50)	1(50)	3(5)	1(2)	1(110)
Frequency	Summer	Summer	4 times between Oct and Feb	Spring	Monthly for 15 months
	O'Gower and Wacasey 1967	O'Gower and Wacasey 1967	Young and Young 1977	Virnstein et al. 1983	Thayer et al. 1975a

TABLE 10 Representative Species of Commercial and Recreational Organisms Using Seagrass Beds, the Major Geographic Habitat, and Life History Stage

Common Name	Scientific Name	Range[a]	Life Stage[b]
Spotted seatrout	*Cynoscion nebulosus*	A, G	J
Mullet	*Mugil cephalus*	A, G	J
Sea bream	*Archosargus rhomboides*	T	J
Spot	*Leiostomus xanthurus*	A, G	A, J
Pinfish	*Lagodon rhomboides*	A, G	A, J
Pigfish	*Orthopristis chrysopterus*	A, G	J
Gag grouper	*Mycteroperca microlepis*	T	J
Gray snapper	*Lutjanus griseus*	T	A, J
Sheepshead	*Archosargus probatocephalus*	T	A, J
Holbrooks porgy	*Diplodus holbrooki*	A, G	J
Halfbeak	*Hyporhamphus unifasciatus*	T	J
Red drum	*Sciaenops ocellata*	T	L, J
Thread herring	*Opisthonema oglinum*	A	J
Permit (pompano)	*Trachinotus falcatus*	A, G	J
White grunt	*Haemulon plumieri*	A	J
Silver perch	*Bairdiella chrysura*	T	J, A
Mojarra	*Gerres cinereus*	G	J
Green sea turtle	*Chelonia mydas*	T	A
Queen conch	*Strombus gigas*	T	A
Bay scallop	*Argopecten irradians*	A, G	A, J, L
Pink shrimp	*Penaeus duorarum*	A, G	A, J
Brown shrimp	*Penaeus aztecus*	G	A, J
Blue crab	*Callinectes sapidus*	A, G	A
Brant	*Branta bernicla*	A	M
Scaup	*Aythya marilataffinis*	A, G	M
Canada geese	*Branta canadensis*	A, G	M
Redhead duck	*Aythya americana*	A, G	M

[a] A, Atlantic; G, gulf; T, tropical Florida.
[b] A, adult; J, juvenile; L, larvae; E, eggs; M, migratory.
Source. From Thayer et al. (1979).

On a geographic scale, species richness of seagrass-associated fishes, amphipods, and decapods increases with decreasing latitude, but only amphipods clearly demonstrate this relationship within the relatively narrow latitudinal range of SE estuaries (Virnstein et al. 1984).

As in the intertidal vegetated habitats, organic material is made available to consumers in seagrass meadows through macrophyte production, epiphyte production, and production by benthic algae, as well as from allochthonous sources (Kikuchi 1980). Only a few species of animals graze seagrass directly, notably green turtles (*Chelonia mydas*), echinoderms, and, in temperate areas, waterfowl (Thayer et al. 1984b). Most organic matter decomposes within the grass bed (Zieman 1975, Fenchel 1977) or is transported out of the system (Zieman et al. 1979). The breakdown of plant material is complex, involving the loss of plant compounds (Knauer and Ayers 1977, Thayer et al. 1977) and the synthesis of microbial products (Lee 1980). Microorganisms serve not only as processors but as links to higher trophic levels. Kitting et al. (1984) and van Montfrans et al. (1984) emphasized the role of seagrass epiphytes in detrital food webs. Secondary consumers, principally shrimps and fishes, exhibit ontogenetic transitions in food habits in conjunction with seasonal or nocturnal migrations into seagrass habitats in order to optimize available food resources (Livingston 1980, 1982, 1984a;b, Leber 1985).

The significance of subtidal vegetated habitat to faunal communities was underscored by catastrophic declines in populations of two disparate species, brant (Cottam 1934) and scallops (Thayer and Stuart 1974), following drastic reductions in the areal extent of eelgrass meadows of the east coast in the early 1930s due to wasting disease.

Reviews of seagrass animal communities can be found in H. T. Odum (1974), Phillips (1974), Kikuchi and Pérès (1977), Kikuchi (1980), Ogden (1980), Zieman (1982), Thayer et al. (1984a), Gilmore (1987), and Virnstein (1987).

Unvegetated Soft Bottom

In terms of area, unvegetated soft bottom represents the most extensive submerged habitat in SE estuaries. Sediment type varies with the energy level of the physical environment. Silts and clays predominate in low-energy environments, whereas sands are found in areas of relatively high water turbulence and rapid velocities. An entire food chain is associated with unvegetated soft bottom habitat, including benthic microalgae, bacteria, benthic invertebrates, shorebirds, waders, and demersal fishes.

Organisms most closely associated with unvegetated soft bottom are the benthic infauna. Infauna are frequently subdivided into microfauna, meiofauna and macrofauna, size classes that correspond roughly with major taxonomic and functional groupings. As in salt marsh sediments, microfauna and meiofauna tend to be decomposers, trophic intermediates between detritus and detrital consumers (Tenore 1977). Both groups also function in remineralization, converting dead organic matter to inorganic nutrients (Hobbie and Lee 1980, Coull 1973). In addition, bacteria offer first-level trophic support by concentrating dissolved organic and inorganic compounds into consumable biomass. Estuarine microbial populations are grazed by protozoans (Hamilton 1973), meiofauna (Coull 1973, Montagna 1984), and macrofauna (Levinton 1979, Federle et al. 1983).

Sediment type may determine the relative importance of each functional role. Silts and clays support populations of bacteria as much as two orders of magnitude greater than sand (Montagna et al. 1983), perhaps because of their greater surface area (Zobell 1938). Meiofauna, particularly burrowing nematodes, are also more abundant in fine sediments than in sand substrates, which support interstitial harpacticoid copepods and gastrotrichs as well as nematodes (Findlay 1981, Coull 1985, Coull and Dudley 1985, Eskin and

Coull 1987). Although meiofauna in sand may be distributed to depths of 10–15 cm, in mud the animals are often restricted to the upper centimeter by anoxic sediment (Coull and Bell 1979). Coull and Bell (1979) suggest that meiofaunal communities of sand bottoms do not provide as much trophic support to higher levels as those of mud bottoms because the organisms are not as available to indiscriminate browsers. Gill raker morphology of bottom-feeding fishes may also play a role in limiting feeding in sandy substrates (Smith and Coull 1987). On the other hand, the remineralization role of meiofauna becomes predominant as the sand content of sediments increases.

Further information on the function and interactions of estuarine microbial and meiofaunal communities may be found in Lackey (1967), Fenchel and Jørgensen (1977), Fenchel 1978, Stevenson and Colwell (1973), and Tietjen (1980).

Macroinfauna, by virtue of their sedentary nature, are susceptible to the complex interplay of physical and biological forces found in SE estuaries. Dominant species tolerate variable environmental conditions and have short life cycles and high reproductive rates, resulting in populations that superimpose seasonal dynamics on wide fluctuations in abundance from one year to the next. Average densities of infauna may be high, particularly in sand substrates (Table 11). Sand bottoms in temperate estuaries support an abundance of small bivalves, haustorid amphipods, and polychaetes, whereas echinoderms predominate in the carbonate sands of tropical Biscayne Bay (Tables 12 and 13). The mixture of clays, silts, fine sands, and organic matter covering the bottom of most SE estuaries supports communities dominated by polychaetes along the Gulf and southeast Atlantic coasts, and bivalves, particularly *Macoma balthica* and *Mulinia lateralis*, in the estuaries of North Carolina (Tables 14 and 15). It should be noted, however, that Tables 12–15 are ranked

TABLE 11   Comparison of Average Densities of Estuarine Soft-Bottom Infauna Found Along the Southeast Coast of the United States

Location	Average Density (per m²)		Sieve Size (mm)	Source
	Mud	Sand		
Santa River Delta	597		0.5	Calder et al. 1977
Ogeechee Estuary	475		0.9	Dörjes and Howard 1975
Indian River		5 844	0.5	Virnstein et al. 1983
Biscayne Bay		166	1.0	Singletary and Moore 1974
Tampa Bay		510	1.0	Bloom et al. 1972
East Bay	2655[a]		0.5	Mahoney 1982
Escambia Bay	41	486	1.0	USEPA 1975
Mississippi Sound	3915	12 970	0.5	U.S. Army Corps of Engineers 1982
Galveston Bay	828[b]		1.5	Holland et al. 1973

[a]Vegetated and short-term stations excluded.
[b]Barnacles excluded.

TABLE 12 Comparison of Infaunal Dominants (ranked in decreasing order of abundance) from Sand Substrate in Several Southeastern Estuaries Along the Gulf Coast

	Mississippi Sound	Escambia Bay	Tampa Bay	Tampa Bay
	Gemma gemma	*Grandidierella bonnieroides*	*Ampelisca abdita*	*Bittium varium*
	Paraonis cf. fulgens	*Mulinia lateralis*	*Mysella planulata*	*Tagelus divisus*
	Acanthohaustorius sp. A	*Laonereis culveri*	*Streblospio benedicti*	*Onuphis eremita*
	Lepidactylus sp. A	*Odostomia* sp.	*Mulinia lateralis*	*Nassarius vibex*
		Tagelus plebeius	*Mediomastus californiensis*	*Acanthohaustorius* sp.
Gear	Box core	Ponar	Hand core	Corer
Sieve size	0.5 mm	1.0 mm	0.5 mm	1.0 mm
Stations (replicates)	1(8)	2(12)	1(10)	9(4)
Frequency	Fall and spring	Winter and summer	Monthly for 42 months	Seasonally for 1 year
Source	U.S. Army Corps of Engineers 1982	USEPA 1975	Santos and Simon 1980b	Bloom et al. 1972

TABLE 13 Comparison of Infaunal Dominants (ranked in decreasing order of abundance) from Sand Substrate in Several Southeastern Estuaries Along the Atlantic Coast

	Biscayne Bay	Indian River	North Inlet
	Ophionephthys limicola	*Cylinchnella canaliculata*	*Lepidactylus dysticus*
	Moira atropos	*Lysilla* sp.	*Acanthohaustorius millsi*
	Amphioplus coniortodes	Capitellidae	*Haploscoloplos fragilis*
	Tellina alternata	*Axiothella mucosa*	*Heteromastis filiformis*
	Eucratopsis crassimanus	*Spiochaetopterus costarum*	*Protohaustorius* cf. *deichmannae*
Gear	Van Veen grab	Post hole-type sampler	Shovel and 0.25 m^2 frame
Sieve size	1.0 mm	0.5 mm	1.0 mm
Stations (replicates)	2(20)	1(2)	36(1)
Frequency	Eight times over 1 year	Spring	Midsummer
Source	McNulty et al. 1962	Virnstein et al. 1983	Holland and Dean 1977

TABLE 14 Comparison of Infaunal Dominants (ranked in decreasing order of abundance) from Mud Substrate in Several Southeastern Estuaries Along the Gulf Coast

	Galveston Bay[a]	Mississippi Sound	Escambia Bay	Apalachicola Bay[b]
	Nereis succinea	*Myriochele oculata*	*Sigambra bassi*	*Mediomastus ambiseta*
	Streblospio benedicti	*Owenia fusiformis*	*Paraprionospio pinnata*	*Streblospio benedicti*
	Pista spp.	*Balanoglossus* cf. *auranticus*	*Odostomia* sp.	*Hobsonia florida*
	Eupomatus dianthus	*Linopherus-Paramphinome*	*Cerebratulus lacteus*	*Grandidierella bonnieroides*
	Lyonsia hyalina floridana	*Mediomastus* spp.	*Parandalia fauveli*	
Gear	Jackson sampler	Box core	Van Veen grab	Hand core
Sieve size	1.5 mm	0.5 mm	1.0 mm	0.5 mm
Stations (replicates)	5(4)	24(8)	2(5)	3(10)
Frequency	Seasonally for 1 year	Fall and spring	Winter and summer	Monthly for 5 years
Source	Holland et al. 1973	U.S. Army Corps of Engineers 1982	USEPA 1975	Mahoney 1982

[a]Barnacles excluded.
[b]Vegetated and short-term station excluded

TABLE 15 Comparison of Infaunal Dominants (ranked in decreasing order of abundance) from Mud Substrates in Several Southeastern Estuaries Along the Atlantic Coast

	Ogeechee Estuary	Santee River Delta	Pamlico River Estuary	Newport River Estuary
	Spiophanes bombyx	*Streblospio benedicti*	*Macoma balthica*	*Mulinia lateralis*
	Scolecolepides viridis	*Peloscolex heterochaetus*	*Nereis succinea*	*Nereis falsa*
	Oxyurostylis smithi	Oligochaeta (unidentified)	*Rangia cuneata*	*Ilyanassa obsoleta*
	Solen viridis	*Brachidontes exustus*	*Mulinia lateralis*	*Clymenella torquata*
	Pinnixa cf. *chaetopterana*	*Scolecolepides viridis*	*Macoma phenax*	*Macoma balthica*
Gear	Box core	Peterson grab	Van Veen grab	Knudson sampler
Sieve Size	0.8 mm²	0.5 mm	1.0 mm	1.2 mm
Stations (replicates)	63(2)	6(3)	36(3)	34(?)
Frequency	Spring	Seasonally for 1 year	Seasonally for 1 year	Seasonally for 2 years
Source	Dörjes and Howard 1975	Calder et al. 1977	Tenore 1972	Chester et al. 1983

on total abundances of organisms throughout the estuary over the entire sampling period. The bias toward irruptive species, combined with differences in sampling gear and processing techniques, limits utility of the tables to comparisons of general patterns.

Salinity sets broad limits on the distribution of infauna along the halocline within estuaries (Fig. 2; Gunter et al. 1974). The oligohaline portions of most SE estuaries are characterized by high densities of relatively few species (See Chapter 13), with species richness increasing along the salinity gradient (Holland et al. 1973, Chester et al. 1983, Livingston 1984c). Substrate parameters such as particle size distribution, organic content, and the depth of redox potential discontinuity have also been shown to influence the spatial distribution of infaunal trophic groups (Carriker 1967). Additional abiotic factors known to influence spatial and seasonal distribution of infauna include temperature, dissolved oxygen, turbidity, pollution, current velocity, and depth (Kendall 1983).

McNulty et al. (1962) and Bloom et al. (1972) found that in SE estuaries, as in the northeastern infaunal communities studied by Saunders (1958), surface deposit feeders are generally associated with fine sediments, while filter feeders are more abundant in sand substrates. Rhoads and Young (1970) argue that deposit feeders create an environment in which sediment is easily resuspended, preventing colonization by suspension feeding larvae and clogging the filtering apparatus of adult and juvenile suspension feeders, a theory known as trophic group amensalism. Tube builders may create exceptions to trophic group amensalism by stabilizing sediments and allowing suspension feeders to coexist with deposit feeders (Young and Rhoads 1971, Woodin 1976, Eckman et al. 1981).

Infaunal populations also vary seasonally. Numerical abundance and species richness usually peak sometime between fall and spring and decline in the summer (Tenore 1972, Livingston 1984c). This pattern often results from the interplay of summer anoxia, recruitment dynamics of infauna, and predation by young of the year fishes entering the estuary in the spring (Tenore 1972, Santos and Simon 1980b, Livingston 1984c, Holland 1985).

Although not all paradigms of soft bottom ecology are applicable to SE estuaries, disturbance or predation by large mobile predators such as shore birds, fish, and crustaceans, by large-scale sediment processors like enteropnests and holothuroids, or even by meiofauna seems to be important in structuring infaunal communities, particularly in the meso- and polyhaline portions of the estuary (Peterson 1979, Virnstein 1979, Woodin 1981, Livingston 1984c, Luckenbach 1984, Watzin 1985, Williams et al. 1986, Peckol and Baxter 1986). The resulting interactions, including adult–larval relationships, offer an attractive explanation for dominance patterns commonly seen in SE infaunal communities, micro-scale patches that fluctuate over time (Frankenberg 1971, Boesch et al. 1976, Frankenberg and Leiper 1977, Flint and Younk 1983, Livingston 1987a). On a larger scale, when natural disturbances like dinoflagellate blooms result in mass mortality in areas of the Tampa Bay estuary, infaunal populations exhibit high resilience, returning to their previous state within three months (Dauer and Simon 1976, Santos and Simon 1980a, Dauer 1984), but these are cyclic defaunations, which hold the community in the predictable initial successional stages. The end points of community succession in SE estuaries are less predictable.

From a functional standpoint, macrofaunal biomass generally exceeds meiofaunal biomass, augmenting the transfer of carbon to secondary consumer levels begun by the meiofauna (Wolff 1977, Coull and Bell 1979, Arntz 1980). Macrofauna also efficiently burrow and rework sediments. The resulting bioturbation exposes nutrient reservoirs in the sediment to overlying waters, a process known as nutrient regeneration (Zeitzschel 1980, Flint and Kalke 1983, 1986, Flint and Kamykowski 1984).

Skates, rays, flatfishes, and other demersal fishes are intimately associated with subtidal bottoms and often forage on intertidal flats at high tide. Because of the frequent exposure, however, fishes are not permanent residents of intertidal flats (Peterson and Peterson 1979).

Avian predators obtain trophic support from both subtidal and intertidal soft bottom and many species of birds use intertidal flats for roosting. Four guilds are represented: (1) floating and diving water birds, (2) aerial-searching birds, (3) waders, and (4) probing shorebirds. These species are rarely preyed upon and often represent end points of estuarine food webs (Peterson and Peterson 1979).

Many diving ducks, including scaup, goldeneye, bufflehead, scoters, and ruddy duck, winter in SE estuaries where they feed predominately on benthic clams (Clapp et al. 1982b). Aerial searching birds like gulls and terns, do not directly depend on unvegetated bottom for food but do spend a significant portion of their time loafing and roosting on intertidal flats. Herons, egrets, ibises, and yellowlegs stalk or stand in shallow water and feed on small fishes and crustaceans (Kushlan 1978). Custer and Osborn (1978a) found productivity and water clarity to be important determinants in the selection of feeding sites by wading birds. Shorebirds that probe sediment with their bills are the most abundant and diverse guild found on intertidal flats. The many species of sandpipers, plovers, and dowitchers that comprise this group feed primarily on small invertebrates and are heavily dependent on intertidal habitat for trophic support (Clapp et al. 1983). Recher (1966) has suggested that the increased area made available by lower tides during spring months allows these migratory birds more feeding opportunities, thereby compressing that migratory period relative to the fall.

Further information on the structure and function of soft bottom estuarine communities can be found in Carriker (1967), Coull (1977), Peterson and Peterson (1979), Tenore and Coull (1980), Kendall (1983), and Armstrong (1987).

Hard Substrates

Only three types of hard substrates are encountered in SE estuaries. Oyster reefs are common in temperate estuaries and are supplemented in the tropical estuaries of South Florida by live bottom communities of coral reef species. All SE estuaries contain increasing areas of artificial hard substrates in the form of jetties, seawalls, and pilings.

Oyster reefs begin when spat of the American oyster (*Crassostrea virginica*) aggregate on a suitable substrate, for example, rocks, wood, shells, or other types of hard debris. Reefs often produce fringes and clumps in shallow water adjacent to salt marshes. In open waters, particularly along the northern Gulf coast, reefs become more extensive, up to 40 km long (Price 1954). Subtidal oyster reefs grow above substrates to form shoals with the surface remaining in the photic zone, enabling oysters to feed on phytoplankton (Galtsoff 1964).

Structurally, a typical Georgia oyster reef is composed of 61% living oysters, 21% dead oyster shell, and 18% silt, clay, and nonoyster macrofauna (Bahr and Lanier 1981). Vertical zonation of the nonoyster macrofauna of intertidal reefs may be determined by the organisms' tolerance to desiccation (Grant and McDonald 1979) or by biological interactions (Menendez 1987). Oysters attached to pilings are limited to a zone 1.5 m above MLW, the height corresponding to maximum vertical extent of adjacent oyster shoals above bottom sediments (Fig. 10). The upper intertidal zone of both pilings and oyster reefs is inhabitated by the barnacle *Chthamalus fragilis* (McDougall 1943). Below, in the lower intertidal and upper subtidal zones, other barnacles (*Balanus* spp.), the hooked mussel (*Ischadium recurvum*), and the ribbed mussel (*Guekensia demissa*) are attached (Gunter and Geyer 1955).

Since oyster reefs usually provide the only solid substrate in an otherwise soft bottom habitat, most sessile organisms characteristic of a given geographic area are associated with reefs in that area. Species diversity is usually high, with the number of species ranging from 22 to 303 and densities from 3300 to 38 000 m² (Wells 1961, Dame 1979, Bahr and Lanier 1981). Decapod omnivores common to oyster reefs of the Southeast include the mud crabs *Panopeus obesus*, *P. simpsoni*, and *Eurypanopeus depressus* (Ryan 1956, May 1974). They feed on the epiphytes encrusting the growing margins of the reef and on young oysters and small crustaceans (McDonald 1982, Reames and Williams 1984). The hooked mussel is ten times more abundant on oyster reefs than is the ribbed mussel (Dame 1979, Bahr and Lanier 1981). *Neanthes succinea* is the most abundant polychaete on reefs along the coast of Georgia, but *Polydora websteri*, *Heteromastus filiformis*, and *Streblospio benedicti* can be equally abundant on reefs off other SE states (Owen 1957, Dame 1979, Bahr and Lanier 1981). Fishes commonly found on oyster reefs include gobies (*Gobiosoma* spp.), blennies (*Chasmodes* spp., *Hypleurochilus* spp., *Hypsoblennius* spp.), skilletfish (*Gobiesox strumosus*), and toadfish (*Opsanus* spp.) (Gunter 1967, Crabtree and Middaugh 1982).

Oysters tolerate salinities ranging from to 5 to 30 ppt and grow best in high-salinity waters. (Loosanoff 1965). Predator pressure and disease determine the distribution of oyster reefs. A major oyster predator, the oyster drill (*Thais haemastoma*), does not enter portions of estuaries where salinities are lower than 18 ppt, so reefs are often most extensive in the mesohaline zone (Burkenroad 1931, Gunter 1979). A protozoan pathogen, *Perkinsus*

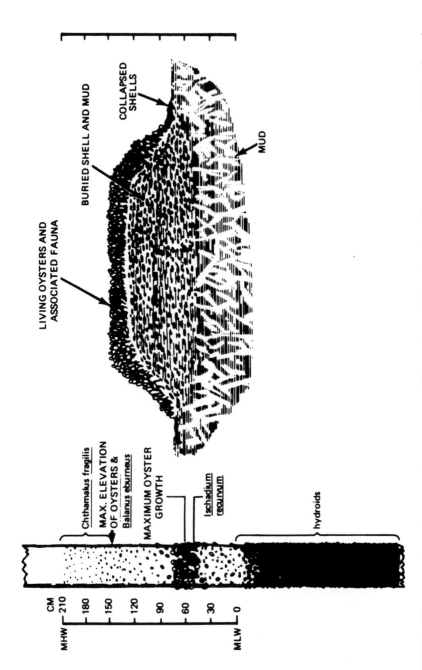

FIGURE 10. Diagrammatic section through an oyster reef illustrating relative elevation with respect to mean tidal levels and corresponding fouling pattern on piling. (From Bahr and Lanier (1981).)

marinus, has also caused high mortalities during periods of high salinity and high water temperature (Overstreet 1978, Soniat 1985). Other predators include the crown conch (*Melongena corona*), lightning whelk (*Busycon contrarium*), flatworms (*Stylochus ellipticus*), crabs (*Menippe adina, Callinectes sapiens*), American oystercatchers, (*Haematopus bachmani*), and black drum (*Pogonias cromis*) (Tomkins 1947, Gunter and Menzel 1957, Menzel and Nichy 1958, May 1968, Overstreet 1978).

In addition to the increased habitat diversity provided by the reef and the food web role of the inhabitants, the oyster reef community also plays a major role in mineralizing organic carbon and releasing nitrogen and phosphorus in forms that can be used by other organisms (Bahr and Lanier, 1981). Bahr and Lanier (1981) also note that the enormous surface area of the reef supports large numbers of aerobic bacteria and macrofauna, which, along with the oysters themselves, contribute to a very high community metabolic rate. Bahr (1976) characterizes oyster reefs in SE estuaries as heterotrophic "hot spots."

The presence of shell material in Indian middens indicates that oysters have been harvested from SE estuaries for food for over 10 000 years (Goodyear and Warren 1972). Landings from the south Atlantic and the Gulf of Mexico averaged 1383 and 8015 metric tons of meat, respectively, for the years 1950–1979 (Sellers and Stanley 1984). Periodic cultch planting is frequently necessary to sustain reefs where oyster harvesting occurs.

Summaries of the biology of oysters and oyster communities may be found in Hedgpeth (1953, 1954, 1957), Butler (1954), Galtsoff (1964), Wells (1961), Chestnut (1974), Dame (1979), and Bahr and Lanier (1981).

In some of the more tropical estuaries along the SE coast, sponges, octocorals, anemones, and hard corals colonize exposed fossil coral reef formations, limestone, and other naturally occurring hard substrates. Relief of the live bottom community is generally less than 1.0 m and these structures do not actively accrete (Jaap 1984). The faunal elements are hardy, tolerant species. Gorgonians, often *Leptogorgia virgulata*, visually dominate the habitat (Derrenbacker and Lewis 1985, Gotelli 1988). Sheepshead (*Archosargus probatocephalus*), grunt (*Haemulon* sp.), and pinfish (*Lagodon rhomboides*) are common on live bottoms in Tampa Bay (Derrenbacker and Lewis 1985). Size of the habitat may vary from ten to hundreds of square meters (Jaap 1984). Voss (1976) noted that some hardbottom areas in Biscayne Bay have been covered by silt.

The greatest area of hard substrate in estuaries is often manmade. Jetties, seawalls, and pilings occur in nearly all SE estuaries. Major estuarine algal assemblages utilize these substrates for attachment. Humm (1969) recognized seven distributional groups of species along the north Atlantic coast (Fig. 11). Associations of the Southeastern U.S. coasts include elements of a tropical flora and a cooler North Atlantic flora, but are dominated by elements that are more eurythermal than either of these two zones of origin. Dominant species include *Polysiphonia* spp., *Ulva curvata, Ulva lactuca, Enteromorpha* spp., *Lyngbya* spp., *Gracilaria* spp., *Dasya baillouviana, Gelidium crivale*, and *Padina vickersiae* (Wiseman 1978, Kapraun and Zechman 1982, Richardson 1986, 1987).

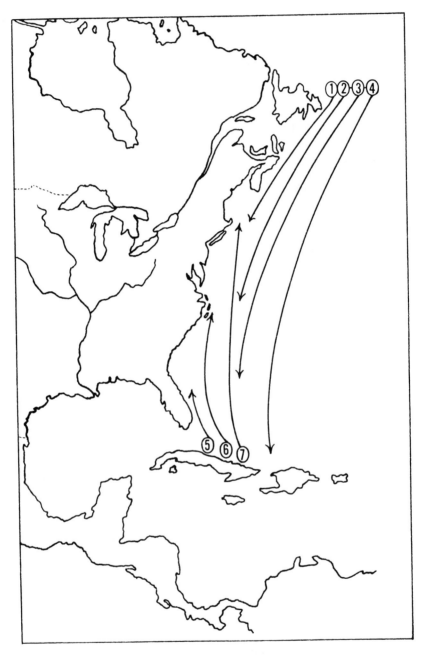

FIGURE 11. Diagrammatic representation of the distribution of the inshore marine algae of the Atlantic coast of North America. (From Humm (1969).)

Kapraun and Zechman (1982) described a complex vertical gradient of overlapping algal species ranges instead of discrete species bands on a North Carolina jetty. Zonation patterns were dynamic, rather than static, with the vertical ranges of species alternately expanding and retreating seasonally. Sedimentation appeared to limit space competition from sessile invertebrate dominants and predation limited the extent of space domination on rock faces by polychaetes (Richardson 1988). Thus, space was available for filamentous and leafy algal coverage.

The northern and NW Gulf of Mexico inshore algal flora (Tampa Bay, Florida, to Port Aransas, Texas) is related to flora of the Atlantic coasts between Cape Canaveral and Cape Hatteras. There are a number of species in the Gulf that are not continuous in distribution around the Florida peninsula, extending on the east coast southward only to Cape Canaveral. Conversely, a number of tropical species are found in the Gulf that do not extend north of Cape Canaveral (Humm and Caylor 1957, Humm and Darnell 1959, Taylor 1965).

Kapraun (1974) found slight algal vertical zonation on Louisiana rock jetties with conspicuous seasonal fluctuations. Two distinct suites of seasonally dominant species were described for summer and winter/spring. Summer dominants included *Caloglossa lepricurrii*, *Chaetomorpha limim*, and *Polysiphonia subtilissima* with tropical affinities. Cool temperate and warm temperate species dominate the winter/spring assemblages and include *Enteromorpha* spp., *Ectocarpus sliliculosus*, *Streblonema oligosporum*, *Ulothrix flacca*, and *Blidingia* spp. Sessile faunal communities are similarly arranged in distinct zones composed of cirripeds (*Chthamalus* sp., *Balanus* spp.), a limpet (*Siphonaria naufragum*), snails (*Littorina* spp.), a hermit crab (*Clibanarius vittatus*), an isopod (*Ligia exotica*), and the hooked mussel (*Ischadium recurvum*) as well as ascidians, hydroids, and bryozoans (McDougall 1943, Whitten et al. 1950, Hedgpeth 1953, 1954). Because hard substrate is unusual in temperate estuaries, tropical species are often attracted to jetties, especially during late summer and early fall (Hastings 1979). Artificial substrates support communities similar in diversity and number of species to those of natural hard substrates (Abele 1974, Courtney 1975). Because of the ease with which hard substrate communities may be manipulated, they are a favorite subject of experimental studies (McDougall 1943, Sutherland and Karlson 1977, Mook 1980). Hay and Sutherland (1988) provide an introduction to the ecology of communities on artificial hard substrates in the South Atlantic Bight.

Water Column

The water column transports sediment, organic matter, nutrients, and organisms throughout the estuary, providing the connection between each of the previous habitats. Two major groups of organisms, the weakly mobile plankton and the strongly motile nekton, inhabit the water column.

Phytoplankton in estuaries consist of four principal microalgal groups: phytomicroflagellates (7 or more classes), diatoms, dinoflagellates, and blue-green algae. Trends in numbers of species and abundance usually reveal an inverse horizontal relationship with increasing salinity from the estuary head to its mouth (Hulbert 1965, Kinne 1967). Many species are cosmopolitan and endemic populations are rare in estuaries (Wood 1965, Lackey 1967, Steidinger 1973), but brackish assemblages may be distinct spatially and seasonally. Most species are "neritic" and represent an estuarine/neritic assemblage due to freshwater discharge or oceanic intrusions, respectively (Wood 1965). Environmental factors limiting phytoplankton occurrence, diversity, and abundance include light (water clarity), temperature (metabolic processes, division rates), salinity (osmoregulation), micro- and macronutrients (growth factors), and circulation patterns (spatial distribution).

Boynton et al. (1982) reviewed data on phytoplankton production, chlorophyll *a*, and associated physical and chemical variables from 63 estuarine systems. Algal production and biomass were consistently high in warm periods of the year. Ratios of available nitrogen to phosphorus were low during periods of high production except in highly eutrophic systems.

Phytoplankton community structure is similar to that of the zooplankton in that a single species, the diatom *Skeletonema costatus*, is usually dominant. Steidenger (1973) listed assemblages that characterize eastern Gulf of Mexico estuarine conditions, including the diatoms *Skeletonema costatus*, *Chaetoceros* spp., and *Thalassiosira* spp., and the dinoflagellates *Ceratium hircus*, *Gymnodinium splendens*, *Gyrodinium* spp., *Polykrikos* spp., *Peridinium* spp., and *Gonyaulax* spp. Species composition of the phytoplankton communities in the estuaries of Georgia and North and South Carolina are similar (Hustedt 1955, Zingmark 1978). Pelagic diatoms, such as *Skeletonema costatum*, *Rhizosolenia* spp., *Asterionella* spp., and *Coscinodescus* spp., are dominant. Several species of dinoflagellates and several green flagellates are present and may at times dominate the community (Whitney et al., in ms., reported in Pomeroy et al. 1981).

Blooms of photosynthetic planktonic microalgae may occur in estuarine settings. Cell densities above background levels, usually greater than 50 000 cells/L, are considered a bloom. Blooms may be either mixed species and then become monospecific by competitive inhibition or exclusion, or remain mixed. Most blooms are autochthonous and originate in the estuarine system either from a benthic or planktonic inoculum. Two exceptions are *Ptychodiscus brevis* and *Oscillatoria erythreae*, which are oceanic/coastal "invaders" and of allochthonous origin. Commonly occurring bloom species include *Gonyaulax* spp., *Gymnodinium* spp., *Peridinium* spp., *Ceratium hircus*, *Prorocentrum micans*, *Amphidinium* spp., and *Noctiluca* (Steidinger 1985). Whether the species produces a toxin or not, blooms can lead to oxygen depletion and animal mortalities, especially in the early morning following high respiration rates at night.

Water column bacterial population densities declined along a transect from high marsh to stations located about 1 km offshore North Carolina (Wilson and Stevenson 1980). A similar decline from the upper estuary to the mouth of the system was reported by Palumbo and Ferguson (1978) for the Newport River Estuary.

Zooplankton may be holoplanktonic, spending their entire life cycle in the plankton, or meroplanktonic, remaining in the plankton for only a few days or weeks during their earliest life stages. Nearly all estuarine organisms, including commercially important penaeid shrimps, blue crabs, oysters, and most fishes, spend time in the water column as larvae or postlarvae before assuming their adult form (Williams 1969, 1971, Williams and Deubler 1968, Williams and Porter 1971, Stancyk and Feller 1986). Although it has long been assumed that larvae actively control their spatial distributions and ultimately their recruitment to various habitats (Carriker 1967, Weinstein et al. 1980, Cronin and Forward 1982, Woodin 1986), passive entrainment has been demonstrated in several taxa (Eckman 1983, Hannan 1984). After metamorphizing and settling out, many adult benthic invertebrates periodically enter the plankton by vertically migrating, which in a two-layer system leads to transport, either flushing or retention (Williams 1972, Williams and Bynum 1972, Dauer et al. 1982).

Numerous studies have examined the structure of SE zooplankton communities (Tables 16 and 17). Although subject to all the cautions concerning differences in mesh size, sampling technique, location, timing, and taxonomic calibration that applied to earlier comparisons of infauna, several broad patterns are evident. The copepod *Acartia tonsa* is the most abundant plankter in the majority of SE estuaries and appears among the five most abundant species in every estuary examined. Because of its large size relative to the next most frequently occurring dominants, *Olithona* spp. and *Paracalanus* sp., *A. tonsa* nearly always dominates in terms of biomass. The most frequently occuring noncopepod holoplankton were appendicularians (*Oikopleura dioica*) and cladocerans (*Penilia avirostris*). Dominant species of holoplankton of the tropical estuaries of south Florida are similar to those of temperate SE estuaries. However, cirriped larvae are the most abundant meroplankters in temperate estuaries, while mollusk veligers dominate in tropical estuaries. Standing stock is highest in estuaries of Florida and the northern Gulf coast and lowest in those of North Carolina and Texas (Tables 16 and 17). Although not included in Tables 16 and 17, several species of jellyfish are common in SE estuaries (Kraeuter and Setzler 1975).

Many SE estuaries show similar seasonal patterns of high abundances of copepods during summer months regardless of latitude (Copeland et al. 1984, Fulton 1984, Livingston 1984c, Armstrong 1987, Conner and Day 1987). Holt and Strawn (1983) and Livingston (1984c) have noted that temperature has a significant effect on species composition of estuarine plankton communities, particularly the meroplanktonic component. In the copepod community, a seasonal succession of dominance by *Centropages* in winter, *A. tonsa* in spring, and *Paracalanus* and *Oithona* in summer has been described by Fulton (1984).

In a review of latitudinal patterns of calanoid and cyclopoid copepod diversity in estuarine waters of eastern North America, Turner (1981) showed that although numbers of copepod species increased along a north–south gradient to about 40°N, species richness did not continue to increase. Numbers of cyclopoid species actually did increase, but calanoid diversity was reduced because in the warmer waters of SE estuaries, calanoid species persist all year

TABLE 16 Dominant Zooplankton (listed in descending order of abundance) from Several Southeastern Estuaries Along the Gulf Coast

	East Lagoon, TX	Vermilion/ Atchafalaya Bay	Biloxi Bay/ Mississippi Sound	St. Andrews Bay	Apalachicola Bay	Tampa Bay
	Acartia tonsa	*Acartia* spp.	*Acartia tonsa*	*Paracalanus crassirostris*	*Acartia tonsa*	*Oithona colcarva*
	Cirripedia nauplii	Decapod larvae	*Penilia avirostris*	Copepod nauplii	Cirripedia larvae	*Acartia tonsa*
	Oithona spp.	*Brachyura megalops*	*Labidocera aestiva*	*Acartia tonsa*	*Paracalanus crassirostris*	*Paracalanus crassirostris*
	Paracalanus crassirostris	*Brachyura zoea*	*Sagitta* spp.	*Oithona colcarva*	*Pseudo-diaptomus coronatus*	*Oikopleura dioica*
	Copepod nauplii	Crustacea larvae	*Brachyura zoea*	*Oikopleura dioica*	Cladocerans	
Mean biomass (mg DW/m^3)	14.5		32.8	42.8	32.1	39.6
Mesh size (μm)	158	366	366	153	202	74
Sampling periodicity	Weekly for 18 months	Weekly for 2 years	Monthly for 1 year	Monthly for 13 months	Monthly for 13 months	Seasonally for 1 year
Source	Fleminger 1959[a]	Juneau 1975	Perry and Christmas 1973	Hopkins 1966	Livingston 1984c[b]	Hopkins 1977

[a]Plankton pump.
[b]Some oblique tows.

TABLE 17 Dominant Zooplankton (listed in descending order of abundance) from Several Southeastern Estuaries Along the Atlantic Coast

	Biscayne Bay	Biscayne Bay	Indian River	North Inlet, SC	Newport/North River Estuary, NC	Newport/North River Estuary, NC
	Gastropod veligers	Copepod nauplii	Copepods	Copopodids	*Acartia tonsa*	*Paracalanus crassirostris*
	Acartia tonsa	Gastropod veligers	*A. tonsa*	*Paracalanus crassirostris*	*Oithona* spp.	*Oithona* spp.
	Paracalanus quasimodo	Bivalve veligers	Veliger larvae	Copepod nauplii	*Corycaeus* spp.	*Acartia tonsa*
	Decapod zoea	*Oithona nana*	Cirriped larvae	Cirripedia nauplii	*Centropages* spp.	*Euterpina acutifrons*
		Acartia tonsa	Caridean larvae	*Acartia tonsa*		*Centropages* spp.
			Larvaceans			
Mean biomass (mg DW/m³)	53.4			16.1	17.5	
Mesh size (μm)	239	300/35	202	153	153	75
Sampling periodicity	Weekly for 14 months	Biweekly for 13 months	Weekly for 9 months	Biweekly for 20 months	Biweekly for 28 months	Biweekly for 30 months
Source	Woodmansee 1958	Reeve 1970	Youngbluth and Gamble 1976[a]	Lonsdale and Coull 1977[a]	Thayer et al. 1974	Fulton 1984[b]

[a]Some oblique tows.
[b]Pump samples 5 cm off bottom.

and do not exhibit seasonal alternations of congeners characteristic of northern estuaries.

Fulton (1984) also examined vertical distribution of estuarine zooplankton and concluded that three distinct groupings occur: (1) those that aggregate on the bottom during the day, (2) those that avoid the surface but are not necessarily most abundant at the bottom, and (3) those that are uniformly distributed in the water column, presumably in response to predation. Fulton (1984) concluded that traditional daytime surface tows with 150-μm nets underestimate plankton populations.

Predators of zooplankton populations include larval and adult fishes, ctenophores (*Mnemiopsis mccradyi*), chaetognaths (*Sagitta hispida*), mysids, and the larger copepods (Miller 1974, Reeve 1975, Kjelson et al. 1975, Fulton 1982; 1984). Trophodynamics within plankton communities are not well defined; many of the smaller copepods seem to be omnivores (Paffenhöfer and Knowles 1980), while the jellyfish are predatory (Philips et al. 1969)

The nekton represent most of the secondary consumer biomass found in estuaries and are the group of organisms most responsible for the reputation of SE estuaries as nursery areas. Because of their mobility, they are able to take advantage of seasonal opportunities for food supply, refuge, and other favorable conditions. Fishes, mobile epibenthic invertebrates, birds, marine mammals, and marine turtles are a few of the many nektonic organisms that may be found in the water column.

McHugh (1967) classifies estuarine fishes into six categories according to their distribution within the estuary: (1) freshwater fishes that occasionally enter brackish waters; (2) truly estuarine species, which spend their entire lives in the estuary; (3) anadromous and catadromous species that pass through the estuary; (4) marine species that pay regular seasonal visits to the estuary, usually as adults; (5) marine species that use the estuary as a nursery ground, usually spawning and spending much of their adult life at sea, but often returning seasonally to the estuary; and (6) adventitious visitors, which appear irregularly and have no apparent estuarine requirements. The most important patterns of use in SE estuaries are shown in Fig. 12. Spatial distributions of nekton are often determined by freshwater discharge to the estuary and its effects on salinity (Gunter 1961, Livingston 1984c), with the number of species in an estuary declining with decreasing salinity (Comp 1985).

Typically, a few species dominate both abundance and biomass of estuarine fish communities (Hester and Copeland 1975, Comp 1985). These are usually young of the year, often sciaenids, which enter the estuary as eggs or postlarvae in the spring and emigrate back offshore by fall. A small group of species reverse this pattern, spawning in the fall and overwintering in the estuary (Miller et al. 1984). The strong seasonal replacement pattern effectively partitions available resources (Weinstein 1982). Spatial distribution within the estuary may likewise depend on life stage (Weinstein and Brooks 1983, Deegan and Day 1984).

The six most abundant species, as well as the total number of species collected in trawl studies of SE estuaries are shown in Tables 18 and 19.

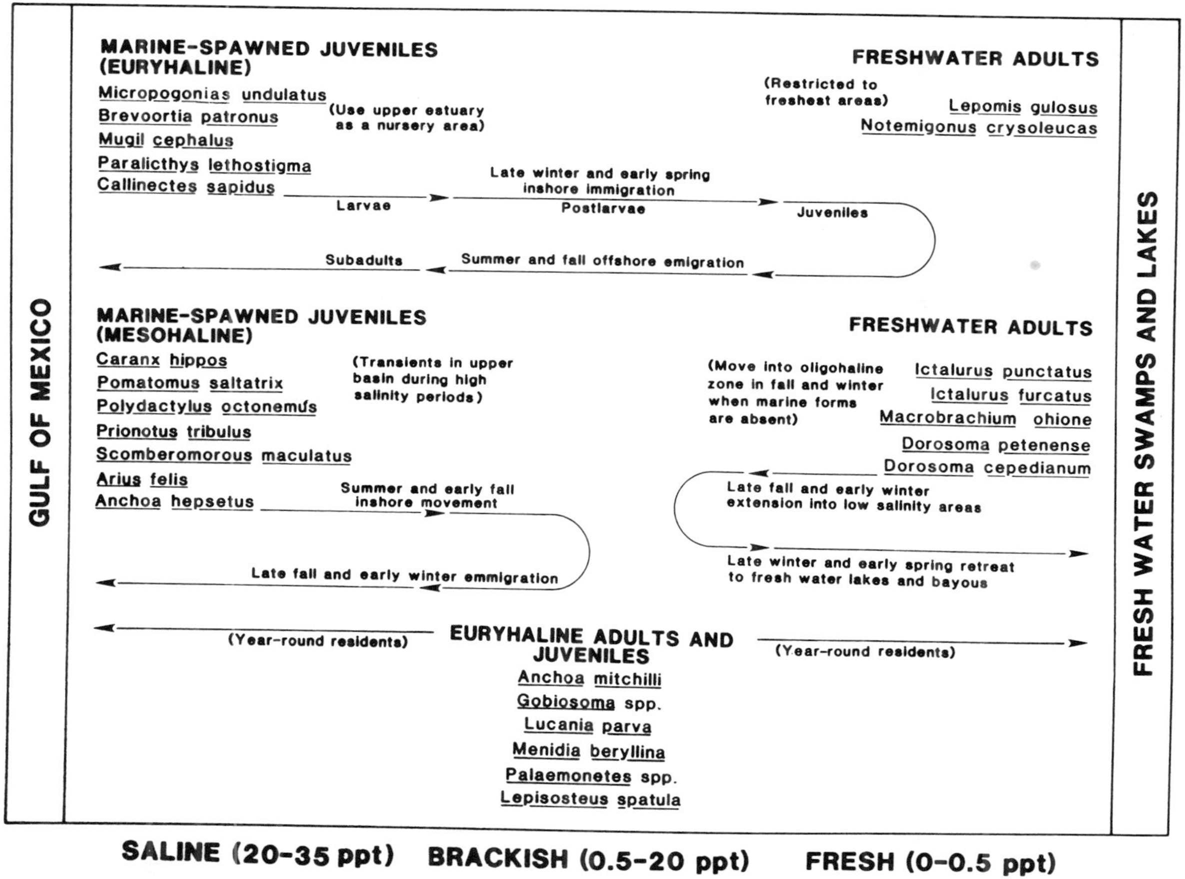

FIGURE 12. Patterns of estuarine use by nektonic animals. (From Day et al. (1982).)

Figure 13 shows the number of times a species occurred among the six most abundant fishes in 18 SE estuaries. The bay anchovy (*Anchoa mitchilli*) was a dominant in 16 of the 18 estuaries examined and was most abundant in eight. The next most frequently occurring dominants were the sciaenids, croaker (*Micropogonias undulatus*), spot (*Leiostomus xanthurus*), and sea trout (*Cynoscion* spp.), which appeared among the dominants in 14, 11, and 8 estuaries, respectively. Of the 19 species to appear as dominants in only a single estuary, 10 were from the Florida estuaries. All six dominants of one Florida estuary (Biscayne Bay) did not occur as dominants at any other estuary examined. With the exception of the tropical Florida estuaries, however, a marked similarity exists among estuarine fish communities of the Southeast.

Many mobile epibenthic invertebrates follow a pattern of estuarine use similar to that of fishes, entering the estuary seasonally as postlarvae and utilizing resources of the estuary during early life stages. Chief among invertebrates exibiting this pattern are commercially harvested shrimps (*Penaeus* spp.) and blue crab (*Callinectes sapidus*) (Williams 1955, Anderson 1970, Livingston et al. 1976). A nektonic mollusk, the brief squid (*Lolliguncula brevis*), can also be common in lower reaches of estuaries (Hoese 1973, Laughlin and Livingston 1982).

Adult estuarine fishes may be classified as planktivores (e.g., anchovies and menhaden), herbivores (e.g., pinfish), benthic omnivores (e.g., croaker and spot), detritivores (e.g., mullet), or epibenthic carnivores (e.g., seatrout, silver perch), (W. E. Odum 1970b, Sheridan and Livingston 1979). Crabs and penaeid shrimps are generally omnivores (Williams 1955, Leber 1985, Hunter and Feller 1987). Sheridan and Livingston (1979) and Livingston (1984c) argue that the seasonal progressions of nektonic young of the year and their ontogenetic changes in feeding habits contribute to a highly structured estuarine trophic organization involving competition, predation, and trophic resource availability, which are in turn dependent on river flow, detrital input, and plankton production. Currin et al. (1984) compared secondary production values and population parameters of spot and croaker in several SE estuaries and found that, despite obvious differences in hydrography and detrital input, P/B ratios differed only slightly, indicating a basic similarity in the trophic function of different systems.

Smith et al. (1966), McHugh (1967), Haedrich (1983), and Comp and Seaman (1985) provide syntheses of the structure and function of estuarine nekton and the role of estuaries in fisheries production. Nektonic estuarine food webs have been discussed in detail by Darnell (1958, 1961), de Sylva (1975), Livingston (1980, 1982, 1984b), and Miller and Dunn (1980).

The extent to which avian species have adapted to estuarine conditions is nowhere more apparent than in the floating and diving species of birds like cormorants, loons, ducks, and grebes. Using webbed or lobed feet for locomotion, they swim on the surface and dive, sometimes to depths of 20 m, to capture fish (Clapp et al. 1982a). Many species appear on SE coasts only as migrants or winter residents, breeding at either inland or more northern locations (Clapp et al. 1982a,b).

TABLE 18 Dominant Fish Species (listed in descending order of abundance) Trawled from Several Southeastern Estuaries Along the Gulf Coast

	Galveston Bay	Vermillion– Atchafalaya Bay	Mississippi Sound	Mobile Bay
	Micropogonias undulatus	*Anchoa mitchilli*	*Anchoa mitchilli*	*Anchoa mitchilli*
	Anchoa mitchilli	*Micropogonias undulatus*	*Micropogonias undulatus*	*Micropogonias undulatus*
	Arius felis	*Ictalurus furcatus*	*Peprilus burti*	*Leiostomus xanthurus*
	Cynoscion arenarius	*Stellifer lanceolatus*	*Cynoscion arenarius*	*Arius felis*
	Leiostomus xanthurus	*Cynoscion arenarius*	*Leiostomus xanthurus*	*Cynoscion arenarius*
	Brevoortia patronus	*Polydactylus octonemus*	*Anchoa hepsetus*	*Etropus crossotus*
Sampling periodicity	Seasonally for 4 seasons	Biweekly for 2 years	Monthly for 1 year	Monthly for 15 months
Number of species	66	56	110	65
Source	Bechtel and Copeland (1970)	Juneau (1975)	Christmas and Waller (1973)	Swingle (1971)

TABLE 18—*Continued*

	Escambia Bay	St. Andrew's Bay	Apalachicola Bay	Tampa Bay	Caloosahatchee Estuary
	Leiostomus xanthurus	*Polydactylus octonemus*	*Anchoa mitchilli*	*Anchoa mitchilli*	*Anchoa mitchilli*
	Lagodon rhomboides	*Micropogonias undulatus*	*Micropogonias undulatus*	*Lagodon rhomboides*	*Arius felis*
	Micropogonias undulatus	*Leiostomus xanthurus*	*Cynoscion arenarius*	*Orthopristis chrysoptera*	*Leiostomus xanthurus*
	Brevoortia patronus	*Symphurus plagiusa*	*Leiostomus xanthurus*	*Eucinostomus gula*	*Trinectes maculatus*
	Anchoa hepsetus	*Lagodon rhomboides*	*Polydactylus octonemus*	*Bairdiella chrysoura*	*Bairdiella chrysoura*
	Anchoa mitchilli	*Anchoa mitchilli*	*Arius felis*	*Eucinostomus argenteus*	*Micropogonias undulatus*
Sampling periodicity	Seasonally for 5 seasons	Biweekly for 1 year	Monthly for 2 years	Monthly for 1 year	Seasonally for 3 years
Number of species	92	128	79	40	39
Source	Cooley 1978	Ogren and Brusher 1977	Livingston et al. 1977	McNulty et al. 1974	Gunter and Hall 1965

TABLE 19 Dominant Fish Species (listed in descending order of abundance) Trawled from Several Southeastern Estuaries Along the Atlantic Coast

	Biscayne Bay	St. Lucie Estuary	Indian River	St. Johns River	Doboy Sound
	Sparisoma rubripinne	*Micropogonias undulatus*	*Anchoa mitchilli*	*Micropogonias undulatus*	*Stellifer lanceolatus*
	Haemulon plumieri	*Arius felis*	*Bairdiella chrysoura*	*Anchoa mitchilli*	*Anchoa mitchilli*
	Monacanthus hispidus	*Anchoa mitchilli*	*Gobiosoma robustum*	*Leiostomus xanthurus*	*Micropogonias undulatus*
	Monacanthus ciliatus	*Pomoxis nigromaculatus*	*Syngnathus scovelli*	*Trinectes maculatus*	*Symphurus plagiusa*
	Syngnathus floridae	*Ictalurus catus*	*Leiostomus xanthurus*	*Gobiosoma bosci*	*Cynoscion regalis*
	Opsanus beta	*Eucinostomus gula*	*Opisthonema oglinum*	*Menidia menidia*	*Leiostomus xanthurus*
Sampling periodicity	10 samples between Apr. and Aug.	10 samples over 2 years	Monthly June–Sep. for 2 years	Irregularly for 44 months	Monthly for 1 year
Number of species	57	—	57	100	—
Source	Roessler 1965	Gunter and Hall 1963	Mulligan and Snelson 1983	Tagatz 1967	Hoese 1973

677

TABLE 19—*Continued*

	Sapelo/ St. Catherines Sound	Santee River Delta	Neuse River	Albemarle Sound
	Stellifer lanceolatus	*Micropogonias undulatus*	*Micropogonias undulatus*	*Anchoa mitchilli*
	Cynoscion regalis	*Anchoa mitchilli*	*Leiostomus xanthurus*	*Micropogonias undulatus*
	Symphurus plagiusa	*Trinectes maculatus*	*Anchoa mitchilli*	*Morone americana*
	Arius felis	*Bairdiella chrysoura*	*Trinectes maculatus*	*Alosa aestivalis*
	Menticirrhus americanus	*Ictalurus catus*	*Brevoortia tyrannus*	*Ictalurus catus*
	Bairdiella chrysoura	*Stellifer lanceolatus*	*Cynoscion regalis*	*Trinectes maculatus*
Sampling periodicity	3-week intervals for 14 months	Monthly for 2 years	Monthly for 13 months	Monthly for 13 months
Number of species	70	89	24	25
Source	Dahlberg and Odum 1970	Wenner et al. 1982	Hester and Copeland 1975	Hester and Copeland 1975

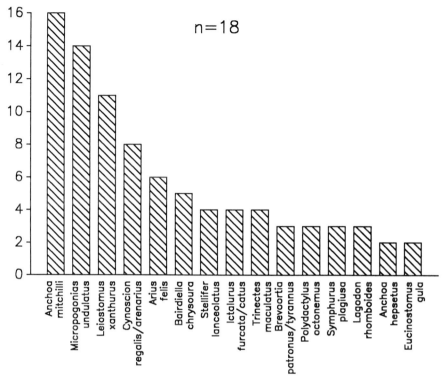

FIGURE 13. Number of times in which each of 15 species of fishes were among the six most abundant species. Eighteen estuaries were examined (see Tables 18 and 19).

Several mammalian species regularly appear in SE estuaries. The bottlenose dolphin (*Tursiops truncatus*) enters estuaries from nearshore coastal waters and may actually be more abundant within the estuary (Hoese 1971, Odell 1975, Wells et al. 1980, Fritts et al. 1983). Manatees (*Trichechus manatus*), a federally listed endangered species, frequent both fresh and estuarine waters in SW Florida (Caldwell and Caldwell 1973, Fritts et al. 1983) and harbor seals (*Phoca vitulina*) occasionally enter estuaries along the Atlantic coast (Schmidley 1980).

The most seriously endangered species of marine turtle, the Kemp's Ridley (*Lepidochelys kempi*) utilizes the inshore waters of the northern Gulf coast as feeding grounds, often entering estuaries in search of its preferred prey, portunid crabs (Hildebrand 1981). Loggerheads (*Caretta caretta*) and green turtles (*Chelonia mydas*) occasionally enter estuaries (Mendonca and Ehrhart 1982), and may hibernate in the deeper channels leading to bays during winter (Ogren and McVea 1981).

Aerial Habitat

Although inhabitants of aerial habitat are, of necessity, transient, they are often the most conspicuous members of the estuarine community. Gulls, terns, osprey, and brown pelicans dive from flight to capture living or dead fish at the waters surface (Palmer 1962, Schreiber et al. 1975). The gulls and terns nest colonially during summer on barrier island and mainland beaches (Parnell and Soots 1979). Populations of the brown pelican are showing signs of recovering, following precipitous declines in the 1950s and 1960s due to reproductive failure associated with organochlorine pollutants (Schreiber and Risebrough 1972, Blus et al. 1979a,b, Clapp et al. 1982a, Fritts et al. 1983). Clapp et al. (1982a,b; 1983) have provided detailed synopsis of the life histories and environmental requirements of many SE coastal birds.

INTERSYSTEM COUPLING

Coastal ecosystems are some of the most productive systems in nature. Each community comprising a coastal ecosystem contributes significantly to its overall productivity, but processes coupling the communities into an ecosystem are complex and poorly known (Pomeroy et al.1981). Suggestions that coastal communities are detritus-based systems (Teal 1962, Odum and de la Cruz 1967, Wiegert et al. 1975) and function primarily in exchanging detrital material with contiguous communities have been questioned (Haines 1977, 1979, Dame 1982, Dame and Stilwell 1984). Studies of coastal areas in Georgia suggest that detrital materials like plant and animal parts (POM) largely remain within the community where they were produced, or are normally transported to landward communities from open estuarine waters and shelf waters, rather than exported seaward from the community where they were produced. Studies of the North Inlet Estuary, South Carolina, indicate a net export of macrodetritus, but equivalent only to 1% of new marsh primary productivity (Dame 1982, Dame and Stilwell 1984). POM accumulates in the sediments and is consumed by flora and fauna and remineralized. Degraded and remineralized materials are exported to adjacent communities in the form of inorganic nutrients (IN) (Howarth and Teal 1980), dissolved organic matter (DOM), and dissolved organic carbon (DOC) (Haines 1979).

Primary Production

In estuaries, the most significant contributors to primary productivity are bacteria and benthic algae. Bacteria reduce POM through aerobic and anaerobic processes, to DOM, DOC and IN, metabolites of degradation that are quickly used by benthic microalgae covering the surface of coastal muds. Less significant contributions come from allochthonous sources, as well as salt marsh, seagrass, mangrove, and phytoplankton communities in the form of POM and POC (Ragotzkie 1959, Nixon et al. 1976, Howarth and Teal 1980, Haines 1979). The magnitude of POM and POC input to adjacent communities and near-shelf waters from a particular estuary depends largely on its geomorphometry and hydrology (Odum et al. 1979).

The contribution of allochthonous sources to the productivity of the estuary was shown by Livingston (1984c), who found annual flooding supplied organic detritus to Apalachicola Bay, Florida. The amount of organic detritus input from the forested floodplain during spring floods alone (March and April) was estimated to be 35 000 metric tons; annually this input amounted to 214 000 metric tons. Annual input of detrital carbon was estimated at 30 000 metric tons, about 53% of the total organic carbon and about 60% of the POC annually produced on the floodplain.

Pomeroy et al. (1981), using the gas-exchange dynamics of CO_2 to estimate net primary production of several major macrophytes, found primary production in Georgia coastal marshes to vary within and between species and within and between locations in the marsh (Table 20). Seasonal variation as a function of light intensity and temperature and because of intrinsic plant characteristics was also noted. Additionally, annual belowground production was also found to vary between species (Pomeroy et al. 1981). Net primary production of smooth cordgrass and needlerush is shown in Table 21. Smooth cordgrass was the major producer in salt marshes of Georgia where black needlerush occupied only 6.0% of the total area of marshes. The contribution of such common species as saltmarsh cordgrass, saltgrass, and coastal dropseed (*Sporobolus virginicus*) varied spatially but was considered negligible (Pomeroy et al. 1981). On the Gulf coast where black needlerush dominates, productivity in northwest Florida marshes was 850 g C/m^2 year^{-1} in the low marsh, 570 g C/m^2 year^{-1} in the upper intermediate, and 180 g C/m^2 year^{-1} in the high marsh (Kruczynski et al. 1978a,b); values comparable to salt marshes around Apalachicola Bay and other areas along the Atlantic Coast (Livingston 1984c).

Production of a Georgia salt marsh epibenthic algal community was estimated at 190 g C/m^2 year^{-1}, most of which occurred during ebb tide. Estimates along the east coast from Delaware to Massachusetts range from about 80 to 105 g C/m^2 year^{-1}. In these salt marshes, as well as in those in Georgia, epibenthic algal primary productivity amounts to about 25–33% of the aboveground macrophyte productivity (Pomeroy et al. 1981).

TABLE 20   A Comparison of Respiration Rates in the Dark and Light Relative to Photosynthetic Rates in Salt Marsh Macrophytes

Species	Respiration Rate (mg CO_2/m^2 h^{-1})			Night DR/Day Ps %	PR/Ps %
	Ps	PR	DR		
S. alterniflora					
Tall	21.3	2.4	1.4	11	11
Short	14.4	6.5	1.9	26	45
J. roemerianus	16.9	9.1	2.1	14	54
B. frutescens	20.2	14.8	2.2	16	65
B. maritima	14.3	13.6	3.6	17	95
S. virginica	45.5	38.7	4.0	29	85

Note. Ps, photosynthesis; PR, photorespiration; DR, dark respiration.

Source. From Pomeroy et al. (1981)

TABLE 21 Net Aboveground Primary Productivity of *Juncus roemerianus* and *Spartina alterniflora* Along the South Atlantic and Gulf of Mexico

	Location	Net 1° Productivity (g/m² year⁻¹)	References
Juncus roemerianus			
	NC	560	Foster 1968
		754	Williams and Murdoch 1968
		796	Stroud and Cooper 1968
		895	Waits 1967
		870–1900	Kuenzler and Marshall 1973
	GA	1538	Gallagher et al. 1972
		1300	Reimold et al. 1975
	FL	849	Heald 1969
		390–1140	Kruczynski et al. 1978a
	MS	2000	Eleuterius 1972
		1300	de la Cruz 1974
	LA	3295	Hopkinson et al. 1980
Spartina alterniflora			
	NC	329–1296	Stroud and Cooper 1968
		610–1300	Marshall 1970
	SC	724–2188	Dame and Kenny 1986
	GA	985	Smalley 1959
		1158	Teal 1962
		2000	Schelske and Odum 1961
		2883	Odum and Fanning 1973
	FL	550	Young 1974
		1201–1281	Turner and Gosselink 1975
		130–700	Kruczynski et al. 1978a
	MS	1473	Turner 1976
	LA	750–2600	Kirby and Gosselink 1976
		1005–1410	Day et al. 1973b
		1381	Hopkinson et al. 1980

Source. From Durako et al. (1985)

Both salinity and turbidity affect production in seagrass beds. Salinity affects species composition, while turbidity affects productivity. Primary productivity of eelgrass in North Carolina is low, with aerial production ranging from 0.59 to 1.23 g C/m² day⁻¹ and underground production ranging from 0.15 to 0.28 g C/m² day⁻¹ (Penhale 1977, Thayer et al. 1984b). The biomass of tapegrass (*Vallisneria americana*), the numerically dominant species in the Pamlico River, ranged from 0.0 g in February to 18 g ash-free dry wt/m² in August (Copeland et al. 1984). Biomass of pond weed (*Potamogeton* spp.) and widgeon grass (*Ruppia maritima*), the numerical subdominants, ranged respectively, from 2.0 in August to 14 g ash-free dry wt/m² in October, and from 0.5 in August to 2.0 g ash-free dry wt/m² in April.

Estuaries farther to the south and along the eastern Gulf of Mexico are dominated by turtle grass (*Thalassia testudinum*), shoal grass (*Halodule wrightii*), and manatee grass (*Syringodium filiforme*). The biomass and productivity of each is shown in Tables 22 and 23. In addition to the above species, tapegrass and widgeon grass beds also occur in more oligohaline areas of estuaries in the northern Gulf of Mexico. The annual net productivity of tapegrass in the East Bay portion of Apalachicola Bay, Florida, ranged from 320 to 350 g C/m^2 year^{-1}; net primary production of turtle grass beds in the more saline portions of Apalachicola Bay was estimated to be about 500 g C/m^2 year^{-1} (Livingston 1984c).

Phytoplankters, primarily diatoms, are important in estuarine primary production and as basic elements in food webs of coastal ecosystems. In SE estuaries the phytoplankton community is rich in species, but usually has a low biomass (Pomeroy et al. 1981, Livingston 1984c). Although phytoplankton production continues throughout the year, increased nutrient input due to increases in river discharge results in major production peaks in spring and secondary peaks in autumn, with lows in winter and summer. Species composition of the various peaks appears to be temperature dependent (Livingston 1984c).

Annual phytoplankton primary production in SE estuaries is shown in Table 24. Throughout the year, however, production varied from 63 to 1694 mg C/m^2 day^{-1} for Apalachicola Bay, Florida (Livingston 1984c). Chlorophyll *a* content in Albermarle Sound ranged from 3 to 40 µg/L (Copeland et al. 1983) and from 5 to 100 µg/L in the Pamlico River estuary (Copeland et al. 1984). Production in these two systems was greatest in midestuary. In the Duplin River estuary and Doboy Sound, Georgia, phytoplankton production was very similar to that of other estuaries (Table 19), but constituted only 6.0% of the system's production when compared on an areal basis (Whitney et al. 1981). Greatest peaks of production occurred in waters covering salt marshes during spring tides (Pomeroy et al. 1981). However, the magnitude of this was considered negligible when compared to the whole system. Phytoplankton primary production in Albermarle Sound and the Pamlico River was dominated by dinoflagellates (Copeland et al. 1983, 1984). Diatoms were the primary photosynthesizers in the Apalachicola Bay system (Livingston 1984c).

Seven hypotheses concerning the relative importance of energy sources within mangrove communities, proposed by Odum et al. (1982), can also be applied to other communities comprising the estuarine biotope. The relative importance of each source can vary seasonally from one location to the next (Table 25). Phytoplankton sources are most important in locations associated with large bodies of relatively deep clear water. Seagrasses and benthic algae are important sources of particulates to communities fringing large expanses of open water. Epiphytes growing on salt marsh sediments, seagrasses, and mangrove prop roots are a significant source of carbon within the community as well as to adjacent communities. Leaf litter generated in marshes, seagrass beds, and mangrove forests is an important source of energy to secondary

TABLE 22 Biomass Comparison of Three Seagrass Species from Three Sites in Pine Channel, Florida Bay, Florida, June 1980

Species	Component	Central		Portion		Northern End	
		g/m²	%	g/m²	%	g/m²	%
Thalassia	Leaves	206	11	58	15	267	10
	Roots and rhizomes	1669	89	321	85	2346	90
	Total biomass	1875		379		2613	
Syringodium	Leaves	58	24	102	16	28	47
	Roots and rhizomes	182	76	521	84	31	53
	Total biomass	240		623		59	
Halodule	Leaves	54	21	15	11	5	33
	Roots and rhizomes	200	79	120	89	10	67
	Total biomass	254		135		15	
All species	Total biomass	2369		1137		2687	

Source. From Zieman (1982).

TABLE 23 Representative Seagrass Productivities

Species	Location	Productivity ($g\ C/m^2\ day^{-1}$)	References
Halodule wrightii	North Carolina	0.5–2.0	Dillon 1971
Syringodium filiforme	Florida	0.8–3.0	J. C. Zieman, Jr., un-published data
	Texas	0.6–9.0	Odum and Hoskins 1958, McRoy 1974
Thalassia testudinum	Florida (east coast)	0.9–16.0	E. P. Odum 1963, H. T. Odum 1957, Jones 1968, Zieman 1975
	Cuba	0.6–7.2	Buesa 1972, 1974
	Puerto Rico	2.5–4.5	Odum et al. 1960
	Jamaica	1.9–3.0	Greenway 1974
	Barbados	0.5–3.0	Patriquin 1972

Source. From Zieman (1982).

TABLE 24 Annual Phytoplankton Primary Production in Estuaries Along the Southeastern Coast of the United States and Eastern Gulf of Mexico

Location	Production (g C/m²)	References
Duplin River, GA	248[a]	Ragotzkie 1959
Beaufort Channel, NC	225[a]	Williams and Murdoch 1966
Bogue Sound, Newport River, North River, and Core Sound, NC	100[a]	Williams 1966
	67	Thayer 1971
North Inlet, SC	346	Sellner and Zingmark 1976
Doboy Sound-Duplin River, GA	375	Whitney et al. 1981
Apalachicola Bay	371	Livingston 1980

[a] Gross production values.

Source. Modified from Pomeroy et al. (1981).

consumers. Herbivorous invertebrates and vertebrates have evolved to feed directly on the aerial portions of marsh and mangrove vegetation and, finally, anaerobic decomposition in all coastal sediments supports an extensive bacterial-based food web that involves methanogenesis and the processing of reduced sulfur compounds. Anaerobically produced products are translocated via deposit-feeding consumers, for example, mullet and grass shrimp, adapted to graze on the sulfur-oxidizing bacterial community in the surface sediments (Lackey 1967, Howarth and Teal 1980). The most important pathway proceeds from leaf detritus substrate to the microbe detritus consumer to the higher consumers, with the critical links mediated by bacteria and fungi (W. E. Odum, 1970a,b, Fell 1975, Odum and Heald 1972, Odum et al. 1982).

Organic Transport

In riverine-influenced estuaries, seasonal timing and magnitude of flooding determine levels of export to the estuary from river wetlands and deltas (Hackney 1978, Livingston 1984c, Armstrong 1987). The geomorphology of the basin and the relative magnitude of the tidal range determine the degree of organic flux within estuarine communities and between coastal ecosystems (Odum et al. 1979).

Harriss et al. (1980) summarized tidally related characteristics of suspended aggregates as follows: (1) The concentrations of suspended particulates in ebb-tide waters are typically greater than concentrations measured in flood-tide waters. (2) The concentration of suspended particulates is commonly higher in daytime ebb-tide waters than in nighttime ebb-tide waters. (3) The rate of change in suspended particulate concentrations reaches values up to approximately 9 mg/L h^{-1} during the transition from flood to ebb or vice versa when other factors (e.g., wind, benthic animal activity) are at a maximum.

TABLE 25 Total Annual Net Productivity and Net Input to the Apalachicola Estuary and Bay System.

Vegetation	Apalachicola estuary (metric tons C/year)		Apalachicola Bay System (metric tons C/year)		
	Net in Situ Productivity	Net Input	Net in Situ Productivity	Net Input	Season of Maximum Input
Freshwater wetlands	360 000	30 000	360 000	30 000	Winter/spring
Coastal marshes	37 714	37 714(?)	46 905	46 905(?)	Late summer, fall(?)
Phytoplankton	103 080	103 080	233 284(?)	233 284(?)	Spring and fall
Seagrass beds	8 953	8 953	27 213	27 213	Summer–fall

Note. Productivity includes organic carbon (metric tons) produced by the Apalachicola River wetlands, coastal marshes, phytoplankton and seagrasses.
Source. From Livingston (1984c).

Seasonally, concentrations of detrital aggregates are lower in the winter months and increase through the spring and early summer to maximum values in August and September. The scale of the seasonal variability in suspended particulate concentrations was similar to the variability observed on an hourly to daily basis during intensive sampling periods. Suspended particulate concentrations in flood-tide water typically did not vary by more than a factor or two on a daily or seasonal basis. The suspended particulate concentrations in ebb-tide waters vary by an order of magnitude on a seasonal basis. Qualitatively these data suggest that the *Juncus* marsh probably exports particulates to estuarine waters during approximately 9 months of the year in northwest Florida (Harriss et al. 1980).

In *Spartina* marshes of the South Carolina coast, Dame and Stilwell (1984) found processes within the marsh (high water and biomass peaks) were the most important determinants of ebb flux and that oceanic processes controlled flood flux. High-energy events are necessary to move organic material from high marshes and may account for substantial portions of the total annual export (Hackney and Bishop 1981, Dame 1982, Stout 1984). The relatively great variation in abiotic factors among SE estuaries results in net fluxes that vary in magnitude and direction (Nixon 1980). Additional transport is in the form of animal carbon (Boesch and Turner 1984).

Secondary Production

Although high levels of secondary production are assumed to occur within estuaries because of the high yields of commercial species, few quantitative estimates are available (Wolff 1977, W. E. Odum 1984). Over 90% of the fishery landings on the SE and Gulf coasts involve species that spend some portion of their lives in the estuary (Gunter et al. 1974, Nixon 1980), but annual fishery production is difficult to measure because many estuarine nekton move between inshore and offshore waters (McHugh 1967). The nursery role of subtidal and intertidal vegetation, the long growing season and the variable, but frequently large, riverine input common in SE estuaries are often cited as reasons for high secondary production (Gunter 1967, Lindhall and Saloman 1977, Hackney 1978, Turner 1979, Nixon 1980, Day et al. 1982).

The organisms and processes generating primary production are linked to consumers by feeding relationships. Direct grazing by herbivores of macrophytes is often less intense than cropping of phytoplankton resulting in conspicuous amounts of dead plant material entering the estuary as detritus (Valiela 1984, Mann 1988). The large inputs of detrital material characteristic of SE estuaries, combined with the increased significance of benthic–pelagic coupling in shallow water, underlies the pivotal role of decomposers in estuarine productivity (Fig. 14). On the other hand, Hughes and Sherr (1983) note that grazing of phytoplankton and benthic algae is responsible for no less than 56% of body carbon in subtidal estuarine populations (Fig. 15). In estuaries with large populations of intertidal bivalves, significant quantities of suspended particulate matter are filtered from the water column and re-

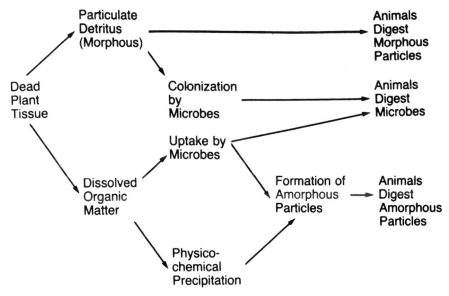

FIGURE 14. Transfer processes by which detritus is converted to secondary production. (From Mann (1988).)

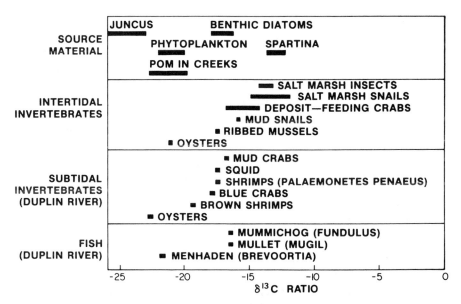

FIGURE 15. Stable carbon isotope ratios of detritus source materials, intertidal and subtidal invertebrates, and fish from a Georgia salt marsh. (From Mann (1988).)

packaged as animal biomass, feces, and pseudofeces (Dame et al. 1980). Feeding strategies of estuarine infauna are dominated by detritivory and deposit feeding. For example, in the Apalachicola Bay system, of the ten dominant infaunal species, five are detrital feeders, four are deposit feeders, and one is a filter feeder (Livingston 1984c). In the nekton, top carnivores are not as conspicuous as relatively unspecialized omnivores. In general, food chains of SE estuaries are short and feeding strategies are flexible. Sites of secondary production within estuaries may vary because many estuarine species are highly mobile and common to several habitats. Livingston (1984c) has developed a model that links physical forcing functions to the various trophic levels on a seasonal basis in a river influenced SE estuary (Fig. 16).

Specific production values for selected invertebrates from SE estuaries are listed in Table 26. Not surprisingly, the easily accessible intertidal species are the best studied. Production rates vary greatly but turnover rates, where available, do not, which indicates a basic functional similarity.

RESOURCE ISSUES AND ENVIRONMENTAL CONCERNS

The ability of estuaries to sustain the myriad of activities and conflicting uses common to the coastal zone is becoming increasingly impaired as coastal population growth continues to accelerate. As estuarine environments become degraded, fewer natural resources are available for use by increasing numbers of people. In no other environment are multiple uses and demands so great or increasing at the rate seen in estuaries.

Population in U.S. coastal areas increased from 44.4 million to 75.2 million between 1950 and 1980 (an increase of 69%) and is still growing rapidly. Florida's coast is being settled at the rate of 3000–4000 people a week. The U.S. Census Bureau predicted that by 1990, 75% of the U.S. population would be living within 50 miles of a coastline (President's Council on Environmental Quality 1984).

Table 27 lists 23 land and water activities impacting SE estuaries. The cumulative result of all these activities is that estuarine environments are suffering threats to public health, the health and abundance of living resources, the coastal economy, and the enjoyment of coastal areas and resources. The U.S. Environmental Protection Agency (1986) has categorized impacts into five major environmental problem areas (Table 28).

Toxics Contamination

The introduction of toxic materials (pesticides, herbicides, heavy metals, dioxin, petroleum compounds, radionuclides, and polychlorinated biphenyls, among others) has become an increasingly serious problem in SE estuaries. Humans contact toxics during recreational use of coastal waters and by consuming contaminated fish and shellfish. Accumulated toxic chemicals may impose physiological and ecological stress on the system, for example, disease

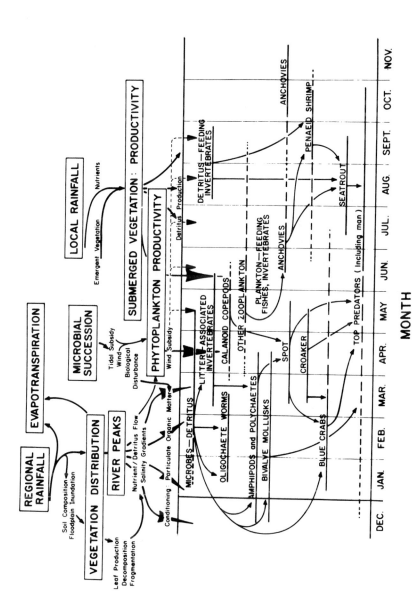

FIGURE 16. Model that associates the population distributions and seasonal relationships of dominant macroinvertebrates and fishes with seasonal changes in key physical parameters, productivity, and predator–prey relationships in the Apalachicola Bay Estuary, Florida. (From Livingston (1984c).)

TABLE 26 Production Estimates for Invertebrate Species from SE Estuaries

Taxa	Location	Production (kcal/m^2 year^{-1})	P/B Ratio	References
HERBIVORES				
Littorina irrorata	Georgia	41	0.7	Odum and Smalley 1959
Littorina irrorata	North Carolina	8		Cammen et al. 1980
Marsh snails (grouped)	Louisiana	40[a]		Day et al. 1973b
Orchelimum fidicinium	Georgia	11		Smalley 1960
SUSPENSION FEEDERS				
Geukensia demissa	Georgia	16.7	0.38	Kuenzler 1961
Crassostrea virginica	South Carolina	3828	1.87	Dame 1976
DEPOSIT FEEDERS				
Panopeus herbstii	South Carolina	87	0.38	Dame and Vernberg 1982
Uca pugnax	Georgia	15	0.35	Shanholtzer 1973
Uca pugnax	North Carolina	51		Cammen et al. 1980
Uca minax	North Carolina	13		Cammen et al. 1980
Marsh crabs (grouped)	Georgia	35		Teal 1962
Marsh crabs (grouped)	Louisiana	*13		Day et al. 1973b
Macroinfauna (grouped)	North Carolina	*30		Cammen 1979

[a]Converted from grams organic matter.

in fish and invertebrates, increased population mortality rates, and shifts in species composition within the biological community. Higher rates of cancer mortality have been statistically linked to drinking water from the Mississippi River system and to heavy concentrations of the petroleum industry (Hoover et al. 1975, Page et al. 1975, Bolt et al. 1977). Respiratory cancer has also been correlated with wetlands residency in coastal Louisiana (Voors et al. 1978).

Eutrophication and Hypoxia

Eutrophication is the natural or artificial addition of nutrients to water bodies and the effects of these added nutrients. Eutrophic waters, characterized by overproduction of algae, increased biological oxygen demand, and oxygen depletion, often lead to fish and shellfish mortality.

Many Gulf coast estuaries are seriously hypoxic every summer, and the extent and duration of hypoxia and periodic anoxia are believed to be increasing (Copeland et al. 1984, Livingston 1984c, Whitledge 1985) (Fig. 17). Major sources of nutrients causing eutrophication are agricultural and urban nonpoint and point source runoff (Hopkinson and Day 1980, Gates et al. 1985). Significant natural sources of nutrient input include precipitation, feces of wildlife and waterfowl, and organic mineralization (Bahr et al. 1983).

Pathogen Contamination

The major cause for the closure of commercial shellfish areas is bacterial pollution at sublethal contamination levels. The primary criterion used to limit shellfish harvest is contamination of the water by coliform bacteria associated with human fecal material. Of 11 coastal states shown in Fig. 18, ten show a loss of productive shellfish beds, with the greatest decline in Florida. Closings of Louisiana's shellfish beds went from 2388 ha in 1965 to 80459 ha in 1971, a 3200% increase of closure (Livingston 1984c). Septic tank effluents, sewage waters, and municipal and industrial runoff account for most of these problems, though natural pollution from pathogens may occur. Sewage contamination of estuarine waters has led to numerous outbreaks of hepatitis A, Norwalk illness, and nonspecific viral gastroenteritis among shellfish consumers (Richards 1985).

Habitat Loss and Alteration

Since the 1700s, approximately 50% of U.S. coastal wetlands have been destroyed, with the highest loss rates occurring over the last three decades. Losses averaged 8100 ha per year over the past 25 years (Alexander et al. 1986). Estuarine wetland loss has been greatest in five states: Louisiana, Florida, California, New Jersey, and Texas (Tiner 1984).

Louisiana, which contains 25% of U.S. coastal wetlands, has lost 40% of its wetlands due to natural causes (erosion, subsidence) and man-induced

TABLE 27 Agents of Impact from Activities Affecting Southeastern U.S. Estuaries

Agents of Impact

Activities	Toxics		Sediments	Nutrients	Pathogens	Oil	Organic Material	Physical Modification	Salinity	Heat	Radio-nuclides
	Organic	Inorganic									
Industrial discharge	●	●	●	○		○	◐	○	○	◐	○
Municipal discharge	●	●	◐	●	●	○	●	○	○		
Suburban/tourist development	◐	○	◐	◐	●	◐		●			
Urban construction and redevelopment	○	○	◐	○		●		●			
Dredging	●	●	●	○		○		●	◐		
Railroads/airports/highways	○	○	○					○	○		
Shipping and ports	◐	◐	●			●	○	●	◐		
Agriculture/livestock	●	○	●	●	●		○	●			
Forestry/timber harvest			○				○	●			
Oil and gas operations	●	◐	◐			●		●	◐		
Mining		●	◐					○			

Marinas/
recreational
boating

Flood control/
diversions

Fishing

Wetlands drain/fill

Military oper./facil.

Barge waste
disposal

Atmospheric
deposition

Dredge mtrl.
disposal

Hazardous waste
facil.

Groundwater
discharge

Shoreline and
estuary erosion
control

Aquaculture

Note. Range of impact: ●, major; ⊖, moderate; ○, minor.

Source. Modified from U.S. Environmental Protection Agency (1986).

TABLE 28 Contribution of Land and Water Activities to Environmental Problems in Southeastern U.S. Estuaries

Activities	Major Environmental Problem				
	Toxic Contamination	Eutrophication	Pathogens Contamination	Habitat Loss/Alteration	Changes in Living Resources
Industrial discharge	●	◐		○	●
Municipal discharge	●	●	●	◐	●
Suburban/tourist development	◐	●	●	●	●
Urban construction and redevelopment	○	○	○	○	○
Dredging	◐		○	●	◐
Railroads/airports/highways	○			◐	
Shipping and ports	◐		○	◐	●
Agriculture/livestock	●	●	○	●	◐
Forestry/timber harvest				●	○
Oil and gas operations	○			●	◐

Activity					
Mining	○			○	⊖
Marinas/recreational boating	○	○		○	○
Flood control/diversions			●	●	⊖
Fishing				○	
Wetlands drain/fill				●	●
Military oper./facil.	⊖			⊖	●
Barge waste disposal	○			○	
Atmospheric deposition					
Dredge mtrl. disposal	⊖		○	●	○
Hazardous waste facil.	●		⊖	●	●
Groundwater discharge	⊖			⊖	○
Shoreline and estuary erosion control				⊖	
Aquaculture				⊖	

Note. Range of contribution: ●, major; ⊖, moderate; ○, minor.

Source. Modified from U.S. Environmental Protection Agency (1986).

PRIORITY HYPOXIC AREAS
POTENTIAL HYPOXIC AREAS
POTENTIAL HYPOXIC AREAS BUT LACK DATA

FIGURE 17. Areas of concern for hypoxia and eutrophication in the U.S. (From U.S. Environmental Protection Agency, (1986).)

losses from channelization, pipelines, impoundment, dredging, spoil disposal, and levying of the Mississippi River. From a loss rate of 41.4 km²/year in the early 1970s, the loss in Louisiana has increased to 129.5 km²/year or 85% of the total national loss rate (Boesch et al. 1983). Forty-four percent of emergent wetlands, mangrove forests, and tidal marshes in Florida have been lost (Dial and Deis 1986). Dredged canals and marinas contributed to over 50% of coastal wetlands loss, especially in South Carolina, North Carolina, Texas, Louisiana, and Florida. Urbanization and residential development are also major factors contributing to losses in South Carolina, Florida, and Mississippi. Other contributors to alteration or loss of wetlands include fill from dredged material or beach creation and man-induced or natural transition to freshwater and agricultural conversions (Office of Technology Association 1984). In South Carolina presently, 14–16% of coastal marshes are former rice field impoundments in various stages of management or planning for waterfowl habitat, aquaculture, and waste treatment (Miglarese and Sandifer 1982).

Additional estuarine habitat is being lost through disappearance and alteration of estuarine submerged grassbeds. In Florida, seagrass meadows have been virtually eliminated in portions of Pensacola Bay and Tampa Bay systems and significant losses have been noted over the last 20–40 years in other estuaries. Various activities have been implicated in seagrass decline, including thermal pollution (Zieman 1970, Burton et al. 1976), turbidity/sedimen-

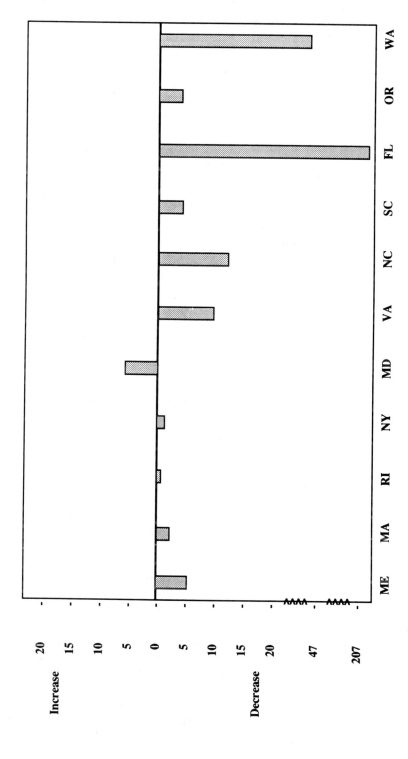

FIGURE 18. Changes in approved shellfish waters, 1980–1985 (thousands of acres). (From U.S. Environmental Protection Agency, (1986).)

tation, and toxic agents (McRoy and Helfferich 1980). Thayer et al. (1975b) have listed, in addition, upland runoff, oil spills, industrial discharges, and commercial fishing. However, few detailed studies exist that document the causes of observed long-term decline of estuarine seagrass systems. The association of historic declines with increased industrial development has been established, although detailed mechanisms of such changes remain largely undefined (Livingston 1987b). Studies in Apalachee Bay, Florida (Zimmerman and Livingston 1976), correlated severe damage to seagrass beds with pollution from kraft pulp-mill effluents. However, following pollution abatement measures and water quality improvements only slow recovery of macrophytes was seen, indicating long-term impacts (Livingston 1984a).

Coastal population growth and development continue to diminish the quality and area of estuaries. Cumulative and synergistic environmental impacts threaten continued production of fishery resources through destruction and alteration of aquatic and estuarine habitats. As an example of the types and extent of loss, Table 29 summarizes projects reviewed by the National Marine Fisheries Service (NMFS) in the Southeastern United States from 1981 to 1985.

Changes In Living Resources

Both commercial and recreational fishing industries of the Southeast are dependent on estuaries. National statistics indicate for 1976 that a volume of 800 000 metric tons of fish and shellfish were landed in northern Gulf of Mexico estuaries, producing a value of nearly $390 million. Of these totals about 89% of the volume and 92% of the value consisted of species considered to be dependent on estuaries. Additionally, 70% of the volume entering the sport fishery was similarly dependent (Lindall and Saloman 1977).

In the last several decades, catches have increased in quantity, primarily due to expansion of fishing efforts. Consequently, increased pressure on fishery stocks is necessary to maintain current catch levels. Increases in fishing effort coupled with pollution, eutrophication, and habitat loss and alteration threaten specific fishery species as well as total species catch per unit effort.

Livingston (1984c) has projected that habitat alterations in the Apalachicola Estuary will cause changes in fish species composition, shifting from commercial and game species to rough and forage species. Trends (1967–1980) in Albemarle Sound, North Carolina, indicate declines in commercial landings from a high of 10 500 metric tons in 1970 to a low of 4500 metric tons in 1980 (Street 1982) (Fig. 19). This decline has primarily been attributed to a decline in water quality (Johnson 1982). Tampa Bay commercial finfish landings peaked in 1964 when a total of more than 8000 metric tons were landed (Fig. 20) (Lombardo and Lewis 1985). Since then, there has been a steady decline with current harvest levels at about 5000 metric tons per year. Several of the sought-after species, such as spotted seatrout (*Cynoscion nebulosus*) and red drum (*Sciaenops ocellatus*), show signs of much lower availability to commercial fishermen than landing data indicate for Charlotte Har-

TABLE 29 Number of Proposed Projects and Acres of Habitat by State Involved in NMFS Habitat Conservation Efforts from 1981 through 1985

State	No. of Permit Applications	Acreage Proposed by Applicants				Acreage NMFS Accepted or Did Not Object To				Potential Acreage Conserved				Mitigation Recommended by NFMS	
		Dredge	Fill	Drain	Impound	Dredge	Fill	Drain	Impound	Dredge	Fill	Drain	Impound	Restore Acreage	Generate Acreage
LA	1229	70657.3	36513.5	5139.9	37563.8	24558.2	7939.3	160.5	6274.3	46099.1	28574.2	4979.2	31289.5	92471.8	10914.3
TX	684	3541.9	3916.3	0.1	9186.0	2017.8	688.9	0	987.4	1524.1	3227.4	0.1	8198.6	3904.8	557.2
MS	94	120.3	386.6	5.0	6.3	61.4	245.7	0	0	58.9	140.9	5.0	6.3	6.2	37.4
AL	206	423.5	533.9	0	1.9	226.6	53.0	0	0	196.9	480.9	0	1.9	40.1	7.1
FL	1806	2733.1	2870.8	0.5	275.1	1914.3	926.1	0.2	5.8	818.8	1944.7	0.3	269.3	475.4	765.1
GA	194	625.6	411.5	0	69.2	94.4	109.9	0	0.2	531.2	301.6	0	69.0	34.1	213.0
SC	576	844.5	419.3	0	4345.9	333.6	103.4	0	12.5	510.9	315.9	0	4333.4	68.0	41.1
NC	547	1133.6	612.0	700.0	673.9	869.0	435.1	300.0	68.4	264.6	176.9	400.0	605.5	491.4	85.0
PR	42	66.7	180.2	0	100.0	18.1	15.3	0	0	48.6	164.9	0	100.0	147.7	11.5
VI	7	80.4	48.7	0	0	45.4	35.5	0	0	35.0	13.2	0	0	0	134.5
Total	5385	80226.9	45892.8	5845.5	52222.1	30138.8	10552.2	460.7	7348.6	50088.1	35340.6	5384.8	44873.5	97639.5	12766.2

Source. From Mager and Thayer (1986).

701

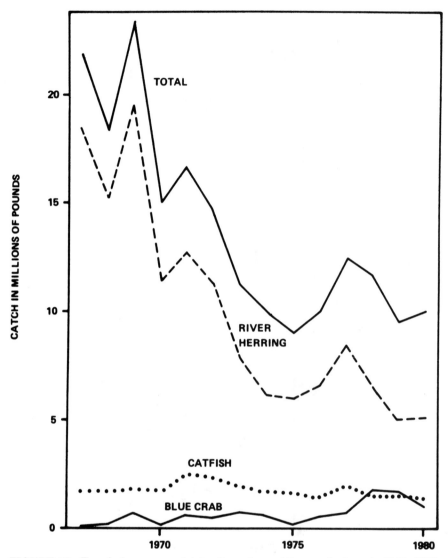

FIGURE 19. Trends in commercial landings from Albermarle Sound, 1967–1980. (From Street (1982).)

bor, a smaller but healthier estuary south of Tampa Bay (Lewis and Estevez 1988).

In the Barataria Basin, Louisiana, salinity intrusion (increased incidence of oyster drill predation) coupled with eutrophication (lowered water quality) has effectively reduced healthy nursery grounds for oysters. Rising salinity and eutrophication can also be expected to seriously affect commercial fisheries for other species such as crabs, shrimp, and fish (Craig and Day 1977).

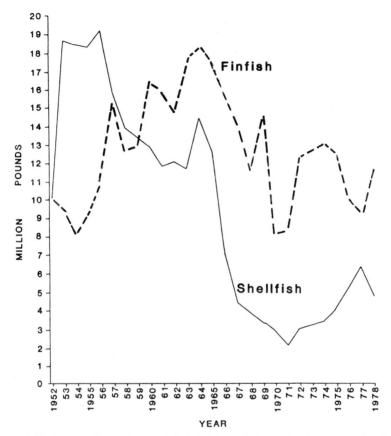

FIGURE 20. Tampa Bay commercial shellfish and finfish landings. (Modified from Lombardo and Lewis (1985).)

Concurrent with the expansion of commercial fishing, a marked increase in recreational fishing has taken place, increasing the already heavy pressure on species common to both fisheries. User conflicts and management crises over commercial and sport fishing regulations are common in many SE states. Many game species are exhibiting declines in yield. Loss of habitats and seagrass food sources have also caused declines in waterfowl and manatees using estuaries. For more indepth reviews of environmental problems in estuaries consult the publications in Table 30.

RESEARCH AND MANAGEMENT NEEDS

Though the intensity of uses and specific data gaps differ from estuary to estuary, Table 31 indicates areas in which information is lacking or poorly understood. The list is not exhaustive and includes only the more important questions for SE estuaries.

TABLE 30   Uses and Impacts on Estuaries: Literature Reviews and Compilations

Topic	References
Toxic contamination	Gilbert et al. 1981
	Fritts and McGehee 1982
	Olsen 1984
	Broutman and Leonard 1988
Oil and gas	Longley et al. 1978
	Longley et al. 1981
Navigation	Allen and Hardy 1980
	Dial and Deis 1986
Shoreline development	Mulvihill et al. 1980
Habitat loss and alteration	Georgia Department of Natural Resources 1976
	Darnell et al. 1976
	Davis and Brinson 1980
	Boesch 1982
	Wolf et al. 1986
	Newton, no date
	Livingston 1987b
Fisheries resources	Lindall and Saloman 1977
	Seaman 1985
General	Fore and Peterson 1980
	Carey and Markovits 1981
	Flint 1984

TABLE 31   Research and Management Needs in Southeastern Estuaries

I.   Living Resources
    A.   Recruitment and early life history
        Environmental requirements and ecological factors affecting recruitment
        Partitioning of resources
        Role of different nursery areas and carrying capacity of each
        Period of residency in each habitat and exchange between habitats
        Relationship to shelf and nearshore waters
    B.   Adult populations
        Standing stocks
        Habitat use, community relationships and relationship to productivity
    C.   Management
        Impact of commercial and recreational fishing
        Multispecies management plans
        Economics of conflicting uses
        Population enhancement methods: resource protection, habitat
          enhancement and protection
        Control of freshwater inflows and salinity fluctuations
II.   Nutrient loading and eutrophication
        Nutrient sources and control technology
        Role in stimulating primary producers: water column and benthic

TABLE 31 (*II Nutrient loading and eutrophication continued*)

       Coupling of primary production and secondary production
       Exchange rates and recycling of nutrients and organic material:
          geochemical and biological
       Carrying capacity

III. Anoxia–hypoxia
       Geochemical processes
       Benthic boundary layer
       Turbidity effect
       Behavior and impact on fauna exposed
       Impact on environmental carrying capacity
       Interaction with toxics

IV. Toxics
       Sources and control
       Biological impact: exposure and effects
       Relationship of sediment dynamics to removal, mobilization and/or
          transport of toxics
       Human health effects

V. Pathogens
       Indicator species: methodology and critical levels
       Biological and environmental interactions
       Identification of nonpoint sources
       Techniques to distinguish point and non-point sources
       Transport and fate
       Develop more effective treatment systems

VI. Loss and alteration of habitats
       Methods to predict and minimize changes due to accelerated sea level
          rise
       Better management strategies for canals, use of spoil, flood diversion
       Relationships between habitats
       Effects of cumulative modifications and losses
       Causes of aquatic vegetation declines
       Restoration and enhancement methodologies

VII. Assimilative capacity and cumulative impacts
       Temporal and spatial boundaries, variability and processes
       Synergisms
       Incremental and decremental impacts: rates and relationships
       Ecological and economic evaluation of system components
       Legal aspects
       Methodologies for assessment and management

The sheer size and inherent variability of estuarine ecosystems create problems for resource management specialists. Data gaps exist at every biological level from the individual to the population to the community and especially at the ecosystem level. Current management practice should recognize the need for additional research, for minimizing additional impact until consequences are known, and for determining acceptable levels of resource utilization.

ACKNOWLEDGMENTS

This paper is Contribution No. 158 from the Marine Environmental Sciences Consortium, Dauphin Island, Alabama, and Contribution No. 116 from the Aquatic Biology Program, University of Alabama. We thank Courtney T. Hackney and two anonymous reviewers for substantially helping us improve the manuscript. The excellent secretarial assistance of Lynn Bryant, Rita George, and Carolyn Wood is appreciated.

REFERENCES

Abele, L. G. 1974. Species diversity of decapod crustaceans in marine habitats. *Ecology* 55:156–161.

Adams, S. A. 1976. The ecology of eelgrass. *Zostera marina* (L.) fish communities, 1: structural analysis. *J. Exp. Mar. Biol. Ecol.* 22:269–291.

Alexander, C. E., M. A. Brontman, and D. W. Field. 1986. *An Inventory of Coastal Wetlands of the USA*. Washington, DC: NOAA, National Ocean Survey, U.S. Department of Commerce.

Alexander, S. K. 1979. Diet of the periwinkle *Littorina irrorata* in a Louisiana salt marsh. *Gulf Res. Rep.* 6:293–295.

Allen, K. O., and J. W. Hardy. 1980. Impacts of navigation dredging on fish and wildlife: a literature review. *U.S., Fish Wildl. Serv., Off. Biol. Serv.* [*Tech. Rep.*] *FWS/OBS* FWS/OBS/80-07:1–82.

Anderson, C. M., and M. Treshow. 1980. A review of environmental and genetic factors that affect height in *Spartina alterniflora*, Loisel. *Estuaries* 3:168–176.

Anderson, W. W. 1970. Contributions to the life histories of several penaeid shrimps (Penaeidae) along the South Atlantic Coast of the United States. *U.S., Fish Wildl. Serv., Spec. Sci. Rep.—Fish.* 605:1–24.

Armstrong, N. E. 1987. The ecology of open-bay bottoms of Texas: a community profile. *U.S., Fish Wildl. Serv., Biol. Rep.* 85(7.12):1–104.

Arntz, W. E. 1980. Predation by demersal fish and its impact on the dynamics of macrobenthos. In K. R. Tenore and B. C. Coull (eds.), *Marine Benthic Dynamics*. Columbia: South Carolina UP, pp. 121–150.

Aston, S. R. 1980. Nutrients, dissolved gases and general biogeochemistry in estuaries. In E. Olausson and I. Cato (eds.), *Chemistry and Biogeochemistry of Estuaries*. Chichester: Wiley, pp. 233–260.

Aston, S. R. 1983. *Silicon Geochemistry and Biogeochemistry*. London: Academic.

Aubrey, D. G., and K. O. Emery. 1983. Eigenanalysis of recent United States sea levels. *Cont. Shelf Res.* 2:21–33.

Baden, J., W. T. Batson, and R. Stalter. 1975. Factors affecting the distribution of vegetation of abandoned rice fields in Georgetown County, South Carolina. *Castanea* 40:171–184.

Bahr, L. M., Jr. 1976. Energetic aspects of the intertidal oyster reef community at Sapelo Island, Georgia (U.S.A.). *Ecology* 57:121–131.

Bahr, L. M., Jr., and W. P. Lanier. 1981. The ecology of intertidal oyster reefs of the south Atlantic coast: a community profile. *U.S., Fish Wildl. Serv., Off. Biol. Serv.* [*Tech. Rep.*] *FWS/OBS* FWS/OBS/81-15:1–105.

Bahr, L. M., Jr., R. Costanza, J. W. Day, Jr., S. E. Bayley, C. Niell, S. G. Leibowitz, and J. Fruci. 1983. Ecological characterization of the Mississippi Deltaic Plain region: a narrative with management recommendations. *U.S., Fish Wildl. Serv., Off. Biol. Serv.* [*Tech. Rep.*] *FWS/OBS* FWS/OBS/82-69:1–189.

Barrett, B. B. 1971. *Cooperative Gulf of Mexico Estuarine Inventory and Study, Louisiana*, Phase II. *Hydrology*; Phase III. *Sedimentology*. New Orleans: Louisiana Wildlife and Fisheries Commission.

Bechtel, T. J., and B. J. Copeland. 1970. Fish species diversity indices as indicators of pollution in Galveston Bay, Texas. *Contrib. Mar. Sci.* 15:103–132.

Bell, S. S., and B. C. Coull. 1978. Field evidence that shrimp predation regulates meiofauna. *Oecologia* 35:141–148.

Bell, S. S., M. C. Watzin, and B. C. Coull. 1978. Biogenic structure and its effect on the spatial heterogeneity of meiofauna in a salt marsh. *J. Exp. Mar. Biol. Ecol.* 35:99–107.

Bell, S. S., K. W. Walters, and J. C. Kern. 1984. Meiofauna from seagrass habitats: a review and prospectus for future research. *Estuaries* 7:331–338.

Bent, A. C., and collaborators. Compiled and edited by O. L. Austin, Jr. 1968. Life histories of North American cardinals, grosbeaks, buntings, towheads, finches, sparrows and allies. *Bull.—U.S. Natl. Mus.* 237:1–1889.

Biggs, R. B., and L. E. Cronin. 1981. Special characteristics of estuaries. In B. J. Neilson and L. E. Cronin (eds.), *Estuaries and Nutrients*. Clifton, NJ: Humana, pp. 3–24.

Bildstein, K. L., R. Christy, and P. DeCoursey. 1982. The role of avian consumers in saltmarsh ecosystems. *Bird Behav.* 4:50–52.

Bishop, T. D., and C. T. Hackney. 1987. A comparative study of the mollusc communities of two oligohaline intertidal marshes: spatial and temporal distribution of abundance and biomass. *Estuaries* 10:141–152.

Blackwelder, B. C. 1972. Algae of the marshlands. In *Port Royal Sound Environmental Study*. Columbia: South Carolina Water Resources Commission, pp. 265–269.

Bloom, S. A. 1983. Seasonality and structure of a macrobenthic seagrass community on the Florida gulf coast. *Int. Rev. Gesamten Hydrobiol.* 68:539–564.

Bloom, S. A., J. L. Simon, and V. D. Hunter. 1972. Animal sediment relations and community analysis of a Florida estuary. *Mar. Biol. (Berlin)* 13:43–56.

Blus, L. J., T. G. Lamont, and B. S. Nealy, Jr. 1979a. Effects of organochlorine residues on eggshell thickness, reproduction and population status of brown pelicans (*Pelecanus occidentalis*) in South Carolina and Florida, 1969–76. *Pestic. Monit. J.* 12:172–184.

Blus, L. J., E. Cromartie, L. McNease, and T. Joanen. 1979b. Brown Pelican: population status, reproductive success and organochlorine residues in Louisiana, 1971–1976. *Bull. Environ. Contam. Toxicol.* 22:128–135.

Boesch, D. F. 1977. A new look at the zonation of benthos along the estuarine gradient. In B. C. Coull (ed.), *Ecology of Marine Benthos*. Columbia: South Carolina UP, pp. 245–266.

Boesch, D. F. (ed.). 1982. Proceedings of the conference on coastal erosion and wetland modification in Louisiana: causes, consequences and options. *U.S., Fish Wildl. Serv., Off. Biol. Serv.* [*Tech. Rep.*] *FWS/OBS* FWS/OBS/82-59:1–257.

Boesch, D. F., and R. E. Turner. 1984. Dependence of fishery species on salt marshes: roles of food and refuge. *Estuaries* 7:460–468.

Boesch, D. F., M. L. Wass, and R. W. Virnstein. 1976. The dynamics of estuarine benthic communities. In M. Wiley (ed.), *Estuarine Processes*, Vol. 1. New York: Academic, pp. 177–196.

Boesch, D. F., D. Levin, D. Nummedal, and K. Bowles. 1983. Subsidence in coastal Louisiana: causes, rates and effects on wetlands. *U.S., Fish Wildl. Serv., Off. Biol. Serv.* [*Tech. Rep.*] *FWS/OBS* FWS/OBS/83-26:1–30.

Bolt, W. J., L. A. Brinston, J. F. Fraumeni, and B. J. Stone. 1977. Cancer mortality in U.S. counties with petroleum industries. *Science* 198:51–53.

Bordeau, P. F., and D. A. Adams. 1956. Factors in vegetational zonation of salt marshes near Southport, N.C. *Bull. Ecol. Soc. Am.* 37:68.

Borum, J., and S. Wirem-Anderson. 1980. Biomass and production of epiphytes on eelgrass (*Zostera marina* L.) in Oresund, Denmark. *Ophelia, Suppl.* 1:57–64.

Bowden, K. F. 1980. Physical factors: salinity, temperature, circulation and mixing processes. In E. Olausson and I. Cato (eds.), *Chemistry and Biogeochemistry of Estuaries*. Chichester: Wiley, pp. 37–51.

Boynton, W. R., W. M. Kemp, and C. W. Keefe. 1982. A comparative analysis of nutrients and other factors influencing estuarine phytoplankton production. In V. S. Kennedy (ed.), *Estuarine Comparisons*. New York: Academic, pp. 69–90.

Bozeman, E. L., Jr., and J. M. Dean. 1980. The abundance of estuarine larval and juvenile fish in a South Carolina intertidal creek. *Estuaries* 3:89–97.

Bozeman, J. R. 1975. Vegetation. In H. O. Hillstad, J. R. Bozeman, A. S. Johnson, C. W. Berisford, and J. I. Richardson (eds.), *The Ecology of the Cumberland Island National Seashore, Camden County, Georgia*, Tech. Rep. Ser. No. 75-5. Skidaway Island: Georgia Marine Science Center, pp. 63–117.

Brook, I. M. 1977. Trophic relationships in a seagrass community (*Thalassia testudinum*), in Card Sound, Florida: fish diets in relation to macrobenthic and cryptic faunal abundance. *Trans. Am. Fish. Soc.* 106:219–229.

Brook, I. M. 1978. Comparative macrofaunal abundance in turtlegrass (*Thalassia testudinum*) communities in south Florida characterized by high blade density. *Bull. Mar. Sci.* 28:213–217.

Broutman, M. A., and D. L. Leonard. 1988. *National Estuarine Inventory: The Quality of Shellfish Growing Waters in the Gulf of Mexico*. Washington, DC: NOAA, Ocean Assessments Division.

Buesa, R. J. 1972. Producion primaria de las praderas de *Thalassia testudinum* de la plataforma norroccidental de Cuba. *I. N. P. Cont. Inv. Pesq. Renuni. Bal. Trab.* 3:101–143.

Buesa, R. J. 1974. Population and biological data on turtle grass (*Thalassium testudinum* Konig 1805) on the northwestern Cuban shelf. *Aquaculture* 4:207–266.

Burkenroad, M. D. 1931. Notes on the Louisiana conch, *Thais haemastoma* Linn., in its relation to the oyster *Ostrea virginica*. *Ecology* 12:656–664.

Burton, D. T., L. B. Richardson, S. L. Margrey, and P. R. Able. 1976. Effects of low powerplant temperatures on estuarine invertebrates. *J. Water Pollut. Control Fed.* 48:2259–2272.

Burton, J. D. 1976. Basic properties and processes in estuarine chemistry. In J. D. Burton and P. S. Liss (eds.), *Estuarine Chemistry*. New York: Academic, pp. 1–36.

Butler, P. A. 1954. Summary of our knowledge of the oyster in the Gulf of Mexico. *Fish. Bull.* 55:479–490.

Calder, D. R., B. B. Boothe, Jr., and M. S. Maclin. 1977. *A Preliminary Report on the Estuarine Macrobenthos of the Edisto and Santee River Systems, South Carolina*, Tech. Rep. No. 22. South Carolina Marine Resources Center.

Caldwell, D. K., and M. C. Caldwell. 1973. Marine mammals of the Eastern Gulf of Mexico. In. J. Jones et al. (eds.), *Summary of Knowledge of the Eastern Gulf of Mexico*. Gainesville: Florida State University System, Inst. Oceanogr., pp. III-I-1–23.

Cammen, L. M. 1979. The macro-infauna of a North Carolina salt marsh. *Am. Midl. Nat.* 102:244–253.

Cammen, L. M. 1980. The significance of microbial carbon in the nutrition of the deposit feeding polychaete, *Nereis succinea. Mar. Biol. (Berlin)* 61:9–20.

Cammen, L. M., E. D. Seneca, and L. M. Stroud. 1980. Energy flow through the fiddler crabs *Uca pugnax* and *U. minax* and the marsh periwinkle *Littorina irrorata* in a North Carolina salt marsh. *Am. Midl. Nat.* 103:238–150.

Cammen, L. M., U. Blum, E. D. Seneca, and C. M. Stroud. 1982. Energy flow in a North Carolina salt marsh: a synthesis of experimental and published data. *ASB Bull.* 29:111–134.

Carey, R. C., and P. S. Markovits. 1981. Proc. U.S. Fish and Wildlife Service Workshop on Coastal ecosystems of the southeastern United States. *U.S., Fish Wildl. Serv., Off. Biol. Serv. [Tech. Pap.] FWS/OBS* FWS/OBS/80-59:1–257.

Carpenter, J. H., W. L. Bradford, and J. Grant. 1975. Processes affecting the composition of estuarine waters. In L. E. Cronin (ed.), *Estuarine Research I*. New York: Academic Press, pp. 188–214.

Carriker, M. R. 1967. Ecology of estuarine and benthic invertebrates: A perspective. In G. H. Lauff (ed.), *Estuaries*, Publ. No. 83. Washington, DC: Am. Assoc. Adv. Sci., pp. 442–487.

Chabreck, R. H. 1972. *Vegetation, Water and Soil Characteristics of the Louisiana Coastal Region*, Bull. No. 664. Baton Rouge: Louisiana Agric. Exp. Stn.

Chapman, R. L. 1971. The macroscopic marine algae of Sapelo Island and other sites on the Georgia coast. *Bull. Ga. Acad. Sci.* 29:77–89.

Chapman, V. J. 1938. Studies in salt marsh ecology. Sections I–III. *J. Ecol.* 26:144–179.

Chapman, V. J. 1976. *Mangrove Vegetation*. Germany: J. Cramer.

Chapman, V. J. 1977. *Wet Coastal Ecosystems*. Amsterdam: Elsevier.

Chester, A. J., R. L. Ferguson, and G. W. Thayer. 1983. Environmental gradients and benthic macroinvertebrates in a shallow North Carolina estuary. *Bull. Mar. Sci.* 33:282–295.

Chestnut, A. F. 1974. Oyster reefs. In H. T. Odum, B. J. Copeland, and E. A. McMahan (eds.), *Coastal Ecological Systems*, Vol. 2. Washington, DC: Conservation Foundation, pp. 171–230.

Christian, R. R., and R. L. Wetzell. 1978. Interactions between substrate microbes and consumers of *Spartina* "detritus" in estuaries. In M. Wiley (ed.), *Estuarine Interactions*. New York: Academic Press, pp. 93–114.

Christmas, J. T., and R. S. Waller. 1973. Estuarine vertebrates, Mississippi. In J. Y. Christmas (ed.), *Cooperative Gulf of Mexico Estuarine Inventory and Study, Mississippi*. Ocean Springs, MS: Gulf Coast Research Laboratory, pp. 320–434.

Chrzanowski, T. H., J. D. Spurrier, R. F. Dame, and R. G. Zingmark. 1986. Processing of microbial biomass by an intertidal reef community. *Mar. Ecol.: Prog. Ser.* 30:181–189.

Clapp, R. B., R. C. Banks, D. Morgan-Jacobs, and W. A. Hoffman. 1982a. Marine birds of the southeastern United States and Gulf of Mexico. Part I. Gaviiformes through Pelecaniformes. *U.S., Fish Wildl. Serv., Off. Biol. Serv.* [*Tech. Rep.*] *FWS/OBS* FWS/OBS/82-01:1–637.

Clapp, R. B., D. Morgan-Jacobs, and R. C. Banks. 1982b. Marine birds of the southeastern United States and Gulf of Mexico, 2: Anseriformes. *U.S., Fish Wildl. Serv., Off. Biol. Serv.* [*Tech. Rep.*] *FWS/OBS* FWS/OBS/82-20:1–492.

Clapp, R. B., D. Morgan-Jacobs, and R. C. Banks. 1983. Marine birds of the southeastern United States and Gulf of Mexico, 3: Charadriiformes. *U.S., Fish Wildl. Serv., Off. Biol. Serv.* [*Tech. Rep.*] *FWS/OBS* FWS/OBS/83-30:1–853.

Cleary, W. J., P. E. Hosier, and G. R. Wells. 1979. Genesis and significance of marsh islands within southeastern North Carolina lagoons. *J. Sediment. Petrol.* 49:703–710.

Coen, L. D., K. L. Heck, and L. G. Abele. 1981. Experiments on competition and predation among shrimps of seagrass meadows. *Ecology* 62:1484–1493.

Cole, B. E. 1982. Size-fractionation of phytoplankton production in San Francisco Bay: 1980. *Abstracts for the Symposium on Factors Controlling Biological Production in the Sacramento-San Joaquin Estuary*. Concord, CA: Contra Costa County Water District Center.

Comp, G. S. 1985. A survey of the distribution and migration of the fishes in Tampa Bay. In S. F. Treat, J. L. Simon, R. R. Lewis, III, and R. L. Whitman, Jr. (eds.), *Proceedings of the Tampa Bay Area Scientific Information Symposium*, Rep. No. 65. Bellwether Press, Florida Sea Grant College, pp. 393–425.

Comp, G. S., and W. Seaman, Jr. 1985. Estuarine habitat and fishery resources of Florida. In W. Seasman, Jr. (ed.), *Florida Aquatic Habitat and Fishery Resources*. Fla. Chptr., Am. Fish. Soc., Eustis, FL pp. 337–435.

Congdon, R. A., and A. J. McComb. 1979. Productivity of *Ruppia*: seasonal changes and dependence on light in an Australian estuary. *Aquat. Bot.* 6:121–132.

Conner, W. H., and J. W. Day, Jr. (eds.). 1987. The ecology of Barataria Basin, Louisiana: an estuarine profile. *U.S., Fish Wildl. Serv., Biol. Rep.* 85(7.13):1–165.

Continental Shelf Associates and Martel Laboratories (CSAML). 1985. *Florida Big Bend Seagrass Habitat Study Narrative Report*. Final report, Contract No. 14-12-0001-30188. Metairie, LA: U.S Minerals Management Service.

Cooley, N. R. 1978. An inventory of the estuarine fauna in the vicinity of Pensacola, Florida. *Fla. Mar. Res. Publ.* 31:1–119.

Cooper, A. W. 1974. Salt marshes. In H. T. Odum, B. J. Copeland, and E. A. McMahan (eds.), *Coastal Ecological Systems of the United States*, Vol. 2. Washington, DC: Conservation Foundation, pp. 55–98.

Cooper, A. W., and E. D. Waits. 1973. Vegetation types in an irregularly flooded salt marsh on the North Carolina outer banks. *J. Elisha Mitchell Sci. Soc.* 89:78–91.

Copeland, B. J., R. G. Hodson, S. R. Riggs, and J. E. Easley, Jr. 1983. The ecology of Albemarle Sound, North Carolina: an estuarine profile. *U.S., Fish Wildl. Serv., Off. Biol. Serv.* [*Tech. Rep.*] *FWS/OBS* FWS/OBS/83-01:1–68.

Copeland, B. J., R. G. Hodson, and S. R. Riggs. 1984. The ecology of the Pamlico River, North Carolina: an estuarine profile. *U.S., Fish Wildl. Serv., Off. Biol. Serv.* [*Tech. Rep.*] *FWS/OBS* FWS/OBS/82-06:1–83.

Correll, D. L., M. A. Faust, and D. J. Severn. 1975. Phosphorus flux in estuaries. In L. E. Cronin (ed.), *Estuarine Research*, Vol. 1. New York: Academic, pp. 108–136.

Cottam, C. 1934. The eelgrass shortage in relation to waterfowl. *Am. Game Conf. Trans.* 20:272–279.

Coull, B. C. 1973. Estuarine meiofauna: a review: trophic relationships and microbial interactions. In L. H. Stevenson and R. R. Caldwell (eds.), *Estuarine Microbial Ecology*. Columbia: South Carolina UP, pp. 499–512.

Coull, B. C. (ed.). 1977. *Ecology of Marine Benthos*. Columbia: University of South Carolina.

Coull, B. C. 1985. Long-term variability of estuarine meiobenthos: an 11 year study. *Mar. Ecol.: Prog. Ser.* 24:205–218.

Coull, B. C., and S. S. Bell. 1979. Perspectives of meiofaunal ecology. In R. J. Livingston (ed.), *Ecological Processes in Coastal and Marine Systems*. New York: Plenum, pp. 189–216.

Coull, B. C., and B. W. Dudley. 1985. Dynamics of meiobenthic copepod populations: a long-term study (1973–1983). *Mar. Ecol.: Prog. Ser.* 24:219–229.

Coull, B. C., S. S. Bell, A. M. Savory, and B. W. Dudley. 1979. Zonation of meiobenthic copepods in a southeastern United States salt marsh. *Estuarine Coastal Mar. Sci.* 9:181–188.

Courtney, W. R. 1975. Mangrove and seawall oyster communities at Marco Island, Florida. *Bull. Am. Malacol. Union* 41:29–32.

Cowardin, L. M., V. Carter, F. C. Golet, and E. T. LaRoe. 1979. Classification of wetlands and deepwater habitats of the United States. *U.S., Fish Wildl. Serv., Off. Biol. Serv.* [*Tech. Rep.*] *FWS/OBS* FWS/OBS/79-31:1–103.

Crabtree, R. E., and D. P. Middaugh. 1982. Oyster shell size and the selection of spawning sites by *Chasmodes bosquianus*, *Hypleurochilus germinatus*, *Hypsoblennius ionthas* (Pisces, Blenniidae) and *Gobiosoma bosci* (Pisces, Gobiidae) in two South Carolina estuaries. *Estuaries* 5:150–155.

Craig, N. J., and J. W. Day, Jr. 1977. *Cumulative Impact Studies in the Louisiana Coastal Zone: Eutrophication and Land Loss*. Final Rep. to St. Plng. Off. Baton Rouge: Louisiana State University, Center for Wetland Research.

Crance, J. H. 1971. *Description of Alabama Estuarine Areas—Cooperative Gulf of Mexico Estuarine Inventory*, Ala. Mar. Resour. Bull. No. 6. Dauphin Island: Alabama Marine Resources Lab.

Crichton, O. W. 1974. Caloric studies of *Spartina* and the marsh crab *Sesarma reticulatum* (Say). Pp. 142–144 in C. Daiber, Tidal marshes of Delaware. In H. T. Odum, B. J. Copeland, and E. A. McMahan (eds.), *Coastal Ecological Systems of the United States*, Vol. 2. Washington, DC: Coastal Conservation Foundation, pp. 99–149.

Critcher, T. S. 1967. *The Wildlife Values of North Carolina Estuarine Lands and Waters*. North Carolina Wildlife Resources Commission.

Cronin, T. W., and R. B. Forward. 1982. Tidally timed behavior: effects on larval distributions in estuaries. In V. S. Kennedy (ed.), *Estuarine Comparisons*. New York: Academic Press, pp. 505–520.

Currin, B. M., J. P. Reed, and J. M. Miller. 1984. Growth, production, food consumption and mortality of juvenile spot and croaker: a comparison of tidal and nontidal nursery areas. *Estuaries* 7:451–459.

Custer, T. W., and R. G. Osborn. 1978a. Feeding-site description of three heron species near Beaufort, North Carolina. In A. Sprunt, IV, J. C. Ogden, and S. Winckler (eds.), *Wading Birds*, Res. Rep. No. 7. New York: National Audubon Society, pp. 355–360.

Custer, T. W., and R. G. Osborn. 1978b. Feeding habitat use by colonially-breeding herons, egrets and ibises in North Carolina. *Auk* 95:733–743.

Dahlberg, M. D., and E. P. Odum. 1970. Annual cycles of species occurrence, abundance and diversity in Georgia estuarine fish populations. *Am. Midl. Nat.* 83:382–392.

Daiber, F. C. 1982. *Animals of the Tidal Marsh*. New York: Van Nostrand-Reinhold.

Dame, R. F. 1976. Energy flow in an intertidal oyster population. *Estuarine Coastal Mar. Sci.* 4:243–283.

Dame, R. F. 1979. The abundance, diversity and biomass of macrobenthos on North Inlet, South Carolina, intertidal oyster reefs. *Proc. Natl. Shellfish. Assoc.* 69:6–10.

Dame, R. F. 1982. The flux of floating macrodetritus in the North Inlet estuarine ecosystem. *Estuarine, Coastal Shelf Sci.* 15:337–344.

Dame, R. F., and P. D. Kenny. 1986. Variability of *Spartina alterniflora* primary production in the euhaline North Inlet estuary. *Mar. Ecol.: Prog. Ser.* 32:71–80.

Dame, R. F., and D. Stilwell. 1984. Environmental factors influencing macrodetritus flux in North Inlet Estuary. *Estuarine, Coastal Shelf Sci.* 18:721–726.

Dame, R. F., and F. T. Vernberg. 1982. Energetics of a population of the mud crab *Panopeus herbstii* (Milne Edwards) in the North Inlet Estuary, South Carolina. *J. Exp. Mar. Biol. Ecol.* 63:183–193.

Dame, R. F., F. J. Vernberg, R. Bonnell, and W. Kitchens. 1977. The North Inlet marsh-estuarine ecosystem: a conceptual approach. *Helgol. Wiss. Meeresunters.* 30:343–356.

Dame, R. F., R. Zingmark, H. Stevenson, and D. Nelson. 1980. Filter feeder coupling between the estuarine water column and benthic subsystems. In V. S. Kennedy (ed.), *Estuarine Perspectives*. New York: Academic, pp. 521–526.

Dame, R. F., T. Chrzanowshi, K. Bildstein, B. Kjerfue, H. McKellar, D. Nelson, J. Spurrier, S. Stancyk, H. Stevenson, T. Vernberg, and R. Zingmark. 1986. The outwelling hypothesis and North Inlet, South Carolina. *Mar. Ecol.: Prog. Ser.* 33:217–229.

Darnell, R. M. 1958. Food habits of fishes and larger invertebrates of Lake Ponchartrain, Louisiana, an estuarine community. *Publ. Inst. Mar. Sci.* 5:353–416.

Darnell, R. M. 1961. Trophic spectrum of an estuarine community based on studies of Lake Pontchartrain, La. *Ecology* 42:553–568.

Darnell, R. M., and T. M. Soniat. 1979., The Estuary/Continental Shelf as an interactive system. In R. J. Livingston (ed.), *Ecological Processes in Coastal and Marine Systems*. Ecol. Study Ser., No. 10. New York: Plenum, pp. 487–525.

Darnell, R. M., W. E. Pequegnat, B. M. James, F. J. Benson, and R. A. Defenbach. 1976. *Impacts of Construction Activities in Wetlands of the United States*, EPA-600/3-76-045. Washington, DC: U.S. Environ. Prot. Agency.

Dauer, D. M. 1984. High resilience to disturbance of an estuarine polychaete community. *Bull. Mar. Sci.* 34:170–174.

Dauer, D. M., and J. L. Simon. 1976. Repopulation of the polychaete fauna of an intertidal habitat following defaunation: species equilibrium. *Oecologia* 22:99–117.

Dauer, D. M., C. A. Maybury, and R. M. Ewing. 1981. Feeding behavior and general ecology of several spionid polychaetes from the Chesapeake Bay. *J. Exp. Mar. Biol. Ecol.* 54:21–38.

Dauer, D. M., R. M. Ewing, J. W. Sourbeer, W. T. Harlan, and T. L. Stokes, Jr. 1982. Nocturnal movements of the macrobenthos of the Lafayette River, Virginia. *Int. Rev. Gesamter Hydrobiol.* 67:761–775.

Davis, G. J., and M. M. Brinson. 1976. The submerged macrophytes of the Pamlico River Estuary, North Carolina. *Rep.—Water Resour. Res. Inst., Univ. N. C.* 112:1–202.

Davis, G. J., and M. M. Brinson. 1980. Responses of submerged vascular plant communities to environmental change: summary. *U.S., Fish Wildl. Serv., Off. Biol. Serv. [Tech. Rep.] FWS/OBS* FWS/OBS/80-42:1–16.

Davis, G. J., M. M. Brinson, and W. A. Burke. 1978. Organic carbon and deoxygenation in the Pamlico River Estuary. *Rep.—Water Resour. Res. Inst., Univ. N. C.* 131:1–123.

Davis, J. L. 1964. A morphogenetic approach to world shorelines. *Z. Geomorphol.* 8:127–142.

Davis, L. V., and I. E. Gray. 1966. Zonal and seasonal distribution of insects in North Carolina salt marshes. *Ecol. Monogr.* 36:275–295.

Day, J. W., Jr., W. G. Smith, and C. S. Hopkinson, Jr. 1973a. Some tropic relationships of marsh and estuarine areas. In R. H. Chabreck (ed.), *Proceedings of the Coastal Marsh and Estuary Management Symposium*. Baton Rouge: Louisiana State University, pp. 115–135.

Day, J. W., Jr., W. G. Smith, P. R. Wagner, and W. C. Stowe. 1973b. *Community Structure and Carbon Budget of a Salt Marsh and Shallow Bay Estuarine System in Louisiana*, Publ. LSU-SG-72-04. Baton Rouge: Louisiana State University, Center for Wetland Resources.

Day, J. W., Jr., C. S. Hopkinson, and W. H. Conner. 1982. An analysis of environmental factors regulating community metabolism and fisheries production in a Louisiana estuary. In V. S. Kennedy (ed.), *Estuarine Comparisons*. New York: Academic Press, pp. 121–136.

Decho, A. W., W. D. Hummon, and J. W. Fleeger. 1985. Meiofauna-sediment interactions around subtropical seagrass sediments using factor analysis. *J. Mar. Res.* 43:237–255.

Deegan, L. A., and J. W. Day, Jr. 1984. Estuarine fishery habitat requirements. In *Research for Managing the Nation's Estuaries: Proceedings of a Conference in Raleigh, North Carolina*, Publ. UNC-SG-84-08. Raleigh: University of North Carolina, Sea Grant College.

de la Cruz, A. A. 1974. Primary productivity of coastal marshes in Mississippi. *Gulf Res. Rep.* 4:351–356.

de la Cruz, A. A. 1980. Recent advances in our understanding of salt marsh ecology. In P. L. Fore and R. D. Peterson (eds.), *Proceedings of the Gulf of Mexico Coastal*

Ecosystems Workshop, FWS/OBS-80/30. Albuquerque, NM: U.S. Fish Wildl. Serv., pp. 51–65.

de la Cruz, A. A. 1981. Differences between south Atlantic and gulf coast marshes. In R. C. Corey, P. S. Markovits, and J. B. Kirkwood (eds.), *Proceedings of the U.S. Fish and Wildlife Service Workshop on Coastal Ecosystems of the Southeastern United States*, FWS/OBS-80-59. Washington, DC: U.S. Fish Wildl. Serv., pp. 10–20.

Derrenbacker, J. A., Jr., and R. R. Lewis, III. 1985. Live bottom communities of Tampa Bay. In S. F. Treat, J. L. Simon, R. R. Lewis III, and R. L. Whitman, Jr. (eds.), *Proceedings of the Tampa Bay Area Scientific Information Symposium*, Rep. No. 65. Bellwether Press, Florida Sea Grant College, pp. 385–392.

de Sylva, D. P. 1975. Nektonic food webs in estuaries. In L. E. Cronin (ed.), *Estuarine Research*, Vol. 1. New York: Academic, pp. 420–447.

Dial, R. J., and D. R. Deis. 1986. Mitigation options for fish and wildlife resources affected by port and other water-dependent developments in Tampa Bay, Florida. *U.S., Fish Wildl. Serv., Biol. Rep.* 86(6):1–150.

Dillon, C. R. 1971. A comparative study of the primary productivity of estuarine phytoplankton and macrobenthic plants. Thesis, University of North Carolina, Chapel Hill.

Dörjes, J. 1972. Georgia coastal region, Sapelo Island, U.S.A.: sedimentology and biology, 7: distribution and zonation of macrobenthic animals: *Senckenbergiana Marit.* 4:183–216.

Dörjes, J., and J. D. Howard. 1975. Estuaries of the Georgia coast, U.S.A. sedimentology and biology, 4: fluvial–marine transition indicators in an estuarine environment, Ogeechee–Ossabaw Sound. *Senckenbergiana Marit.* 7:137–179.

Doyle, L. J. 1985. A short summary of the geology of Tampa Bay, May 1982. In S. F. Treat, J. L. Simon, R. R. Lewis, III, and R. L. Whitman, Jr. (eds.), *Proceedings of the Tampa Bay Area Scientific Information Symposium*, Rep. No. 65. Bellwether Press, Florida Sea Grant College, Gainesville, FL. pp 27–30.

Duobinis-Gray, E. M., and C. T. Hackney. 1982. Seasonal and spatial distribution of the Carolina marsh clam *Polymesoda caroliniana* (Bosc) in a Mississippi tidal marsh. *Estuaries* 5:102–109.

Durako, M. J., J. A. Browder, W. L. Kruczynski, C. B. Subrahmanyam, and R. E. Turner. 1985. Salt marsh habitat and fishery resources of Florida. In W. Seaman, Jr. (ed.), *Florida Aquatic Habitat and Fishery Resources*. Fla. Chptr., Am. Fish. Soc., pp. 189–280.

Dyer, K. R. 1973. *Estuaries: A Physical Introduction*. New York: Wiley.

Earle, S. A. 1972. Benthic algae and seagrasses. In C. V. Bushnell (ed.), *Chemistry, Primary Productivity and Benthic Algae of the Gulf of Mexico*, Serial Atlas of the Marine Environment, Folio 22. Washington, DC: Am. Geogr. Soc., pp. 15–18.

Eckman, J. E. 1983. Hydrodynamic processes affecting benthic recruitment. *Limnol. Oceanogr.* 28:241–257.

Eckman, J. E., A. R. Nowell, and P. A. Jumars. 1981. Sediment destabilization by animal tubes. *J. Mar. Res.* 39:361–374.

Eiseman, N. J. 1980. *An Illustrated Guide to the Seagrasses of the Indian River Region of Florida*, Tech. Rep. No. 31. Harbor Branch Foundation, Inc., Ft. Pierce, FL.

Eleuterius, C. K. 1978. Classification of Mississippi Sound as to estuary hydrological type. *Gulf Res. Rep.* 6:185–187.

Eleuterius, L. N. 1972. The marshes of Mississippi. *Castanea* 37:153–168.

Eleuterius, L. N. 1976. The distribution of *Juncus roemerianus* in the salt marshes of North America. *Chesapeake Sci.* 17:289–292.

Emery, K. O., and E. Uchupi. 1972. Western North Atlantic Ocean: topography, rocks, structure, water, life and sediments. *Mem.—Am. Assoc. Pet. Geol.* 17:1–532.

Eskin, R. A., and B. C. Coull. 1987. Seasonal and three-year variability of meiobenthic nematode populations at two estuarine sites. *Mar. Ecol.: Prog. Ser.* 41:295–303.

Evans, J. 1970. *About Nutria and Their Control*, Resour. Publ. No. 86. Fish Wildl. Serv., U.S. Dept. Int.

Federle, T. W., J. R. Livingston, D. A. Meeter, and D. C. White. 1983. Modifications of estuarine sedimentary microbiota by exclusion of epibenthic predators. *J. Exp. Mar. Biol. Ecol.* 73:81–94.

Fell, J. W. 1975. Phycomycetes associated with degrading mangrove leaves. *Can. J. Bot.* 53:2908–2922.

Fenchel, T. M. 1977. Aspects of the decomposition of seagrasses. In C. P. McRoy and C. Helfferich (eds.), *Seagrass Ecosystems: A Scientific Perspective*. New York: Dekker, pp. 123–145.

Fenchel, T. M. 1978. The ecology of micro- and meiobenthos. *Annu. Rev. Ecol. Syst.* 9:99–121.

Fenchel, T. M., and B. B. Jørgensen. 1977. Detritus food chains of aquatic ecosystems: the role of bacteria. *Adv. Microb. Ecol.* 1:1–58.

Findlay, S. E. G. 1981. Small-scale spatial distribution of meiofauna on a mud- and sandflat. *Estuarine, Coastal Shelf Sci.* 12:471–484.

Finley, R. T. 1975. Hydro-dynamics and tidal deltas of North Inlet, South Carolina. In L. E. Cronin (ed.), *Estuarine Research*, Vol. 2. New York: Academic, pp. 277–291.

Fisher, T. R., P. R. Carlson, and R. T. Barber. 1982. Sediment nutrient regeneration in three North Carolina estuaries. *Estuarine, Coastal Shelf Sci.* 14:101–116.

Fleminger, A. 1959. East lagoon zooplankton. *Fish Wildl. Serv., Bull. Comm. Fish. Circ.* 62:114–118.

Flint, R. W. 1984. Estuarine management—the integrated picture. In T. N. Vezirglu (ed.), *The Biosphere: Problems and Solutions*. Amsterdam: Elsevier, pp. 387–406.

Flint, R. W., and R. D. Kalke. 1983. Environmental disturbances and estuarine benthos functioning. *Bull. Environ. Contam. Toxicol.* 31:501–511.

Flint, R. W., and R. D. Kalke. 1985. Benthos structure and function in a south Texas estuary. *Contrib. Mar. Sci.* 28:33–53.

Flint, R. W., and R. D. Kalke. 1986. Biological enhancement of estuarine benthic community structure. *Mar. Ecol.: Prog. Ser.* 31:23–33.

Flint, R. W., and D. Kamykowski. 1984. Benthic nutrient regeneration in south Texas coastal waters. *Estuarine, Coastal Shelf Sci.* 18:221–230.

Flint, R. W., and J. A. Younk. 1983. Estuarine benthos: long-term community structure variations, Corpus Christi Bay, Texas. *Estuaries* 6:126–144.

Folger, D. W. 1972. Characteristics of estuarine sediments of the United States. *Geol. Surv. Prof. Pap. (U.S.)* 742:1–94.

Fonseca, M. S., J. J. Fisher, J. C. Zieman, and G. W. Thayer. 1982. Influence of the seagrass, *Zostera marina* L., on current flow. *Estuarine, Coastal Shelf Sci.* 15:351–364.

Fore, P. L., and R. D. Peterson (eds.). 1980. *Proceedings of the Gulf of Mexico Coastal Ecosystems Workshop*, FWS/OBS-80/30. Washington, DC: U.S., Fish Wildl. Serv.

Foster, W. A. 1968. Studies on the distribution and growth of *Juncus roemerianus* in southeastern Brunswick Co., North Carolina. Thesis, North Carolina State University, Raleigh.

Frankenberg, D. 1971. The dynamics of benthic communities off Georgia, USA. *Thalassia Jugosl.* 7:49–55.

Frankenberg, D., and W. D. Burbanck. 1963. A comparison of physiology and ecology of the estuarine isopod *Cyathura polita* in Massachusetts and Georgia. *Biol. Bull. (Woods Hole, Mass.)* 125:81–95.

Frankenberg, D., and A. S. Leiper. 1977. Seasonal cycles in benthic communities of the Georgia continental shelf. In B. C. Coull (ed.), *Ecology of Marine Benthos*. Columbia: South Carolina UP, pp. 383–397.

Fritts, T. H., and M. A. McGehee. 1982. Effects of petroleum on the development and survival of marine turtle embryos. *U.S., Fish Wildl. Serv., Off. Biol. Serv.* [*Tech. Rep.*] *FWS/OBS* FWS/OBS/82-37:1–41.

Fritts, T. H., A. B. Irvine, R. D. Jennings, L. A. Collum, W. Hoffman, and M. A. McGehee. 1983. Turtles, birds and mammals in the northern Gulf of Mexico and nearby Atlantic waters. *U.S., Fish Wildl. Serv., Off. Biol. Serv.* [*Tech. Rep.*] *FWS/OBS* FWS/OBS/82-65:1–455.

Fulton, R. S., III. 1982. Predatory feeding of two marine mysids. *Mar. Biol. (Berlin)* 72:183–191.

Fulton, R. S., III. 1984. Distribution and community structure of estuarine copepods. *Estuaries* 7:38–50.

Fulton, R. S., III. 1985. Predator-prey relationships in an estuarine littoral copepod community. *Ecology* 66:21–29.

Furnas, M. J., G. O. Hitchcock, and T. J. Smayda. 1976. Nutrient-phytoplankton relationships in Narrangansett Bay during the 1974 summer bloom. In M. L. Wiley (ed.), *Estuarine Processes*. New York: Academic, pp. 118–133.

Gallagher, J. L., R. J. Reimold, and D. E. Thompson. 1972. Remote sensing and salt marsh productivity. In W. J. Kosco (Chairman), *Proceedings of the 38th Annual Meeting*, Am. Soc. Photogram. Washington, DC: pp. 338–348.

Galtsoff, P. S. 1964. The American oyster, *Crassostrea virginica* (Gmelin). *Fish. Bull.* 64:1–480.

Gardner, L. R. 1975. Runoff for an intertidal marsh during tidal exposure: regression curves and chemical characteristics. *Limnol. Oceanogr.* 20:81–89.

Gates, K. W., B. E. Perkins, J. G. EuDaly, A. S. Harrison, and W. A. Bough. 1985. The impact of discharges from seafood processing on southeastern estuaries. *Estuaries* 8:244–251.

Georgia Department of Natural Resources. 1976. *The Environmental Impact of Freshwater Wetland Alterations on Coastal Estuaries*, Conf. Rep. Atlanta; Georgia Dept. of Natural Resources.

Gerlack, A. C. (ed.). 1970. *National Atlas of the U.S.A.* Washington, DC: U.S. Dept. of the Interior, U.S. Geol. Surv.

Gibbs, R. J. 1970. Mechanisms controlling world water chemistry. *Science* 170:1088–1090.

Giere, O. 1975. Population structure, food relations and ecological role of marine oligochaetes, with special reference to meiobenthic species. *Mar. Biol. (Berlin)* 31:139–156.

Giese, G. L., H. B. Wilder, and G. G. Parker, Jr. 1979. Hydrology of major estuaries and sounds of North Carolina. *Water Resour. Invest. (U.S. Geol. Surv.)* 79-46:1–175.

Gilbert, T., T. King, and B. Barnett. 1981. An assessment of wetland habitat establishment at a central Florida phosphate mine site. *U.S., Fish Wildl. Serv., Off. Biol. Serv. [Tech. Rep.] FWS/OBS* FWS/OBS/81-38:1–96.

Gilmore, R. G. 1987. Subtropical-tropical seagrass communities of the southeastern United States: fishes and fish communities. *Fla. Mar. Res. Publ.* 42:117–137.

Goldberg, E. D. 1971. River-ocean interactions. In J. D. Costlow, Jr. (ed.), *Fertility of the Sea*, Vol. 1. New York: Gordon & Breach, pp. 143–156.

Goodyear, A. C., and L. O. Warren. 1972. Further observations on the submarine oyster shell deposits of Tampa Bay, Fl. *Anthropologist* 25:52–66.

Gore, R. H., E. E. Gallagher, L. E. Scotto, and K. A. Wilson. 1981. Studies on decapod crustacea from the Indian River region of Florida, 11: community composition, structure, biomass and species-areal relationships of seagrass and drift algae-associated macrocrustaceans. *Estuarine, Coastal Shelf Sci.* 12:485–500.

Gosselink, J. G. 1984. The ecology of delta marshes of coastal Louisiana: a community profile. *U.S., Fish Wildl. Serv., Off. Biol. Serv. [Tech. Rep.] FWS/OBS* FWS/OBS/84-09:1–134.

Gosselink, J. G., and R. H. Baumann. 1980. Wetland inventories: wetland loss along the United States coast. *Geomorphology, Suppl.* [N.S.] 34:173–187.

Gosselink, J. G., C. L. Cordes, and J. W. Parsons. 1979. An ecological characterization study of the Chenier Plain coastal ecosystem of Louisiana and Texas. Vol. I. Narrative report. *U.S., Fish Wildl. Serv., Off. Biol. Serv. [Tech. Rep.] FWS/OBS* FWS/OBS/78-9:1–302.

Gotelli, M. J. 1988. Determinants of recruitment, juvenile growth and spatial distribution of a shallow-water gorgonian. *Ecology* 69:157–166.

Grant, J., and J. McDonald. 1979. Desiccation tolerance of *Eurypanopeus depressus* (Smith)(Decapoda: Xanthidae) and the exploitation of microhabitat. *Estuaries* 2:172–177.

Grassle, J. F., and J. P. Grassle. 1974. Opportunistic life histories and genetic systems in marine benthic polychaetes. *J. Mar. Res.* 32:253–284.

Greening, H. S., and R. J. Livingston. 1982. Diel variation in the structure of seagrass-associated epibenthic macroinvertebrate communities. *Mar. Ecol.: Prog. Ser.* 7:147–157.

Greenway, M. 1974. The effect of cropping on growth of *Thalassia testudinum* (Konig) in Jamaica. *Aquaculture* 4:199–206.

Gunter, G. 1961. Some relations of estuarine organisms to salinity. *Limnol. Oceanogr.* 6:182–190.

Gunter, G. 1967. Some relationships of estuaries to the fisheries of the Gulf of Mexico. In G. H. Lauff (ed.), *Estuaries*, Publ. No. 83. Washington, DC: Am. Assoc. Adv. Sci., pp. 621–638.

Gunter, G. 1979. Studies of the southern oyster borer, *Thais haemastoma. Gulf Res. Rep.* 6:249–260.

Gunter, G., and R. A. Geyer. 1955. Studies on fouling organisms of the northwest Gulf of Mexico. *Publ. Inst. Mar. Sci., Univ. Tex.* 4:37–67.

Gunter, G., and G. E. Hall. 1963. Biological investigations of the St. Lucie estuary (Florida) in connection with Lake Okeechobee discharges through the St. Lucie canal. *Gulf Res. Rep.* 1:189–207.

Gunter, G., and G. E. Hall. 1965. A biological investigation of the Caloosahatchee Estuary of Florida. *Gulf Res. Rep.* 2:1–72.

Gunter, G., and R. W. Menzel. 1957. The crown conch, *Melongena corona*, as a predator upon the Virginia oyster. *Nautilus* 70:84–87.

Gunter, G., B. S. Ballard, and A. Venkataramiah. 1974. A review of salinity problems of organisms in United States coastal areas subject to the effects of engineering works. *Gulf Res. Rep.* 4:380–475.

Hackney, C. T. 1978. Summary of information: Relationship of freshwater inflow to estuarine productivity along the Texas coast. *U.S., Fish Wildl. Serv., Off. Biol. Serv.* [*Tech. Rep.*] *FWS/OBS* FWS/OBS/78-73:1–25.

Hackney, C. T., and T. D. Bishop. 1981. A note on relocation of marsh debris during a storm surge. *Estuarine Coastal Mar. Sci.* 12:621–624.

Hackney, C. T., and W. D. Burbanck. 1976. Some observations on the movement and location of juvenile shrimp in coastal waters of Georgia. *Bull. Ga. Acad. Sci.* 34:129–136.

Hackney, C. T., and E. B. Haines. 1980. Stable carbon isotope composition of fauna and organic matter collected in a Mississippi estuary. *Estuarine Coastal Mar. Sci.* 10:703–708.

Hackney, C. T., W. D. Burbanck, and O. P. Hackney. 1976. Biological and physical dynamics of a Georgia tidal creek. *Chesapeake Sci.* 17:271–280.

Haedrich, R. L. 1983. Estuarine fishes. In B. H. Ketchum (ed.), *Estuaries and Enclosed Seas*, Ecosystems of the World 26. New York: Elsevier, pp. 183–207.

Haines, E. B. 1977. The origins of detritus in Georgia salt marsh estuaries. *Oikos* 29:254–260.

Haines, E. B. 1979. Interactions between Georgia salt marshes and coastal waters: a changing paradigm. In R. J. Livingston (ed.), *Ecological Processes in Coastal and Marine Systems*. New York: Plenum, pp. 35–46.

Haines, E. B., and C. L. Montague. 1979. Food sources of estuarine invertebrates analyzed using 13C/12C ratios. *Ecology* 60:48–56.

Hale, S. S. 1975. The role of benthic communities in the nitrogen and phosphorus cycles of an estuary. *Recent Adv. Estuarine Res.* 1:291–308.

Hamilton, P. V. 1976. Predation of *Littorina irrorata* by *Callinectes sapidus*. *Bull. Mar. Sci.* 26:403–409.

Hamilton, R. D. 1973. Interrelationships between bacteria and protozoa. In L. H. Stevenson and R. R. Colwell (eds.), *Estuarine Microbial Ecology*. Columbia: University of South Carolina, pp. 491–497.

Hannan, C. A. 1984. Planktonic larvae may act like passive particles in turbulent near-bottom flows. *Limnol. Oceanogr.* 29:1108–1116.

Happ, G., J. G. Gosselink, and J. W. Day, Jr. 1977. The seasonal distribution of organic carbon in a Louisiana estuary. *Estuarine Coastal Mar. Sci.* 5:695–705.

Harlin, M. M. 1975. Epiphyte-host relations in seagrass communities. *Aquat. Bot.* 1:125–131.

Harlin, M. M. 1980. Seagrass epiphytes. In R. C. Phillips and C. P. McRoy (eds.), *Handbook of Seagrass Biology: An Ecosystem Perspective*. New York: Garland STPM, pp. 117–151.

Harriss, R. C., B. W. Ribeun, and C. Dreyer. 1980. Sources and variability of suspended particulates and organic carbon in a salt marsh estuary. In P. Hamilton and K. B. MacDonald (eds.), *Estuarine and Wetland Processes: With Emphasis on Modeling*. New York: Plenum, pp. 371–384.

Hastings, R. W. 1979. The origin and seasonality of the fish fauna on a new jetty in the northeastern Gulf of Mexico. *Bull. Fla. State Mus., Biol. Sci.* 24:1–22.

Hausman, S. A. 1936a. A contribution to the ecology of the salt marsh snail, *Melampus bidentatus*. Say. *Am. Nat.* 66:541–545.

Hausman, S. A. 1936b. Food and feeding activities of the salt marsh snail (*Melampus bidentatus*). *Anat. Rec.* 67:127.

Hay, M. E., and J. P. Sutherland. 1988. The ecology of rubble structures of the south Atlantic Bight: a community profile. *U.S., Fish Wildl. Serv., Biol. Rep.* 85(7.20):1–67.

Hayes, M. D. 1975. Morphology of sand accumulation in estuaries: an introduction to the symposium. In L. E. Cronin (ed.), *Estuarine Research*, Vol. 2. New York: Academic, pp. 2–22.

Heald, E. J. 1969. The production of organic detritus in a south Florida estuary. Thesis, University of Miami, Coral Gables, FL.

Heard, R. W. 1982a. *Guide to Common Tidal Marsh Invertebrates of the Northeastern Gulf of Mexico*, Miss.-Ala. Sea Grant Consortium Publ. No. MASGP-79-004. 82. Ocean Springs, MS: Gulf Coast Research Lab.

Heard, R. W. 1982b. Observations on the food and food habits of clapper rails (*Rallus longirostris* Boddaert) from tidal marshes along the east and gulf coasts of the United States. *Gulf Res. Rep.* 7:125–135.

Heck, K. L., and R. J. Orth. 1980. Seagrass habitats: the roles of habitat complexity, competition and predation in structuring associated fish and motile macroinvertebrate assemblages. In V. S. Kennedy (ed.), *Estuarine Perspectives*. New York: Academic, pp. 449–464.

Heck, K. L., and T. A. Thoman. 1981. Experiments on predator-prey interactions in vegetated aquatic habitats. *J. Exp. Mar. Biol. Ecol.* 53:125–134.

Heck, K. L., and G. S. Wetstone. 1977. Habitat complexity and invertebrate species richness and abundance in tropical seagrass meadows. *J. Biogeogr.* 4:135–143.

Hedgpeth, J. W. 1953. An introduction to the zoogeography of the northwestern Gulf of Mexico with reference to the invertebrate fauna. *Publ. Inst. Mar. Sci., Univ. Tex.* 3:110–224.

Hedgpeth, J. W. 1954. Bottom communities of the Gulf of Mexico. Gulf of Mexico, its origin, waters and marine life. *Fish. Bull.* 55:203–214.

Hedgpeth, J. W. 1957. Biological aspects. *Mem. — Geol. Soc. Am.* 67:693–749.

Hester, J. M., Jr., and B. J. Copeland. 1975. *Nekton Population Dynamics in the Albemarle Sound and Neuse River Estuaries*, Raleigh: University of North Carolina, Sea Grant College Program.

Hicks, S. D. 1973. *Trends and Variability of Yearly Mean Sea Level, 1893–1971*, Tech. Memo. 12. Rockville, MD: NOAA, National Ocean Survey.

Hildebrand, H. H. 1981. A historical review of the status of sea turtle populations in the western Gulf of Mexico. In K. A. Bjorndal (ed.), *Biology and Conservation of Sea Turtles*. Washington, DC: Smithsonian Institution Press, pp. 447–453.

Hinde, H. P. 1954. The vertical distribution of salt marsh phanerogams in relation to tide levels. *Ecol. Monogr.* 24:209–225.

Ho, C. L., and B. B. Barrett. 1975. *Distribution of Nutrients in Louisiana's Coastal Waters Influenced by the Mississippi River*, Tech. Bull. No. 17. Baton Rouge: Louisiana Wildlife and Fisheries Commission.

Hobbie, J. E. 1970. Hydrography of the Pamlico River, N.C. *Rep.—Water Resour. Res. Inst. Univ. N.C.* 39:1–69.

Hobbie, J. E., and C. Lee. 1980. Microbial production of extracellular material: importance in benthic ecology. In K. R. Tenore and B. C. Coull (eds.), *Marine Benthic Dynamics*. Columbia: South Carolina UP, pp. 341–346.

Hoese, H. D. 1971. Dolphins out of water in a salt marsh. *J. Mammal.* 52:222–223.

Hoese, H. D. 1973. A trawl study of near-shore fishes and invertebrates of the Georgia coast. *Contrib. Mar. Sci.* 17:63–98.

Holland, A. F. 1985. Long-term variation of macrobenthos in a mesohaline region of Chesapeake Bay. *Estuaries* 8:93–113.

Holland, A. F., and J. M. Dean. 1977. The community biology of intertidal macrofauna inhabiting sand bars in the North Inlet area, South Carolina. In B. C. Coull (ed.), *Ecology of Marine Benthos*. Columbia: South Carolina UP, pp. 423–438.

Holland, J. S., N. J. Maciolek, and C. H. Oppenheimer. 1973. Galveston Bay benthic community structure as an indicator of water quality. *Contrib. Mar. Sci.* 17:169–188.

Holt, J., and K. Strawn. 1983. Community structure of macrozooplankton in Trinity and upper Galveston Bays. *Estuaries* 6:66–75.

Homziak, J., M. S. Fonseca, and W. J. Kenworthy. 1982. Macrobenthic community structure in a transplanted eelgrass (*Zostera marina*) meadow. *Mar. Ecol.: Prog. Ser.* 9:211–221.

Hooks, T. A., K. L. Heck, and R. J. Livingston. 1976. An inshore marine invertebrate community: structure and habitat associations in the northeastern Gulf of Mexico. *Bull. Mar. Sci.* 26:99–109.

Hoover, R., T. Mason, F. McKay, and J. Frawmeni, Jr. 1975. Cancer by country: new resources of etiologic clues. *Science* 189:1005–1007.

Hopkins, T. L. 1966. The plankton of the St. Andrew Bay system, Florida. *Publ. Inst. Mar. Sci., Univ. Tex.* 11:12–64.

Hopkins, T. L. 1977. Zooplankton distribution in surface waters of Tampa Bay, Florida. *Bull. Mar. Sci.* 27:467–478.

Hopkinson, C. S., and J. W. Day, Jr. 1977. A model of the Barataria Bay salt marsh ecosystem. In C. A. S. Hall and J. W. Day (eds.), *Ecosystem Modeling in Theory and Practice*. New York: Wiley (Interscience), pp. 235–266.

Hopkinson, C. S., and J. W. Day, Jr. 1980. Modeling the relationships between development and storm water and nutrient runoff. *Environ. Manage.* 4:315–324.

Hopkinson, C. S., and R. L. Wetzel. 1982. In situ measurements of nutrient and oxygen fluxes in a coastal marine benthic community. *Mar. Ecol.: Prog. Ser.* 10:29–35.

Hopkinson, C. S., J. G. Gosselink, and R. T. Parrondo. 1980. Production of coastal Louisiana marsh plants calculated from phenometric techniques. *Ecology* 61:1091–1098.

Horlick, R. G., and C. B. Subrahmanyam. 1983. Macroinvertebrate fauna of a salt marsh tidal creek. *Northeast Gulf Sci.* 6:79–90.

Howard, R. K. 1987. Diel variation in the abundance of epifauna associated with seagrasses of the Indian River, Florida, USA. *Mar. Biol. (Berlin)* 96:137–142.

Howarth, R. W., and J. M. Teal. 1980. Energy flow in a salt marsh ecosystem: the role of reduced inorganic sulfur compounds. *Am. Nat.* 116:862–872.

Huang, T. C., and H. G. Goodell. 1977. Sediments of Charlotte Harbour, southwestern Florida. *J. Sediment. Petrol.* 37:449–474.

Hughes, E. H., and E. B. Sherr. 1983. Subtidal food webs in a Georgia estuary: $^{13}C/^{12}C$ analysis. *J. Exp. Mar. Biol. Ecol.* 67:227–242.

Huh, S. H., and C. L. Kitting. 1985. Trophic relationships among concentrated populations of small fishes in seagrass meadows. *J. Exp. Mar. Biol. Ecol.* 92:29–43.

Hulbert, E. M. 1965. Flagellates from brackish waters in the vicinity of Woods Hole, Massachusetts. *J. Phycol.* 1:87–94.

Humm, H. J. 1956. Seagrasses on the northern gulf coast. *Bull. Mar. Sci. Gulf Caribb.* 6:305–308.

Humm, H. J. 1969. Distribution of marine algae along the Atlantic coast of North America. *Phycologia* 7:43–53.

Humm, H. J., and R. L. Caylor. 1957. The summer marine flora of Mississippi Sound. *Publ. Inst. Mar. Sci., Univ. Tex.* 4:228–264.

Humm, H. J., and R. M. Darnell. 1959. A collection of marine algae from the Chandeleur Islands. *Publ. Inst. Mar. Sci., Univ. Tex.* 6:265–276.

Hunter, J., and R. J. Feller. 1987. Immunological dietary analysis of two penaeid shrimp species from a South Carolina tidal creek. *J. Exp. Mar. Biol. Ecol.* 107:61–70.

Hustedt, F. 1955. *Marine Littoral Diatoms of Beaufort, North Carolina*, Mar. Stn. Bull. No. 6. Durham, NC: Duke University.

Isphording, W. C., J. A. Stringfellow, and G. C. Flowers. 1985. Sedimentary and geochemical systems in transitional marine sediments in the northeastern Gulf of Mexico. *Trans.—Gulf Coast Assoc. Geol. Soc.* 35:397–408.

Iverson, R. L., and H. F. Bittaker. 1986. Seagrass distribution and abundance in eastern Gulf of Mexico coastal waters. *Estuarine, Coastal Shelf Sci.* 22:577–602.

Jaap, W. C. 1984. The ecology of the south Florida coral reefs: a community profile. *U.S.. Fish Wildl. Serv., Off. Biol. Serv. [Tech. Rep.] FWS/OBS* FWS/OBS/82-08:1–138.

Johannes, R. E. 1965. Influence of marine protozoa on nutrient regeneration. *Limnol. Oceanogr.* 10:434–442.

Johansson, J. O. R., K. A. Steidinger, and D. C. Carpenter. 1985. Primary production in Tampa Bay: a review. In S. F. Treat, J. L. Simon, R. R. Lewis III, and R. L. Whitman, Jr. (eds.), *Proceedings of the Tampa Bay Area Scientific Information Symposium*, Rep. No. 65. Bellwether Press, Gainesville, FL, Florida Sea Grant College, pp. 279–298.

Johnson, A. S., H. O. Hillestad, S. F. Shanholzer, and G. G. Shanholzer. 1974. *An Ecological Survey of the Coastal Region of Georgia*, Sci. Monogr. Ser. No. 3, Washington, DC: National Park Service.

Johnson, D. S., and H. H. York. 1915. The relation of plants to tide levels. *Carnegie Inst. Washington Publ.* 206:1–162.

Johnson, H. 1982. Fisheries production in Albemarle Sound. In *Albemarle Sound Trends and Management*, 82-02. Raleigh: University of North Carolina, Sea Grant College Program, p. 55.

Johnstone, I. M. 1979. Papua New Guinea seagrasses and aspects of the biology and growth of *Enhalus acoroides* (L.f.) Royle. *Aquat. Bot.* 7:197–208.

Jones, J. A. 1968. Primary productivity by the tropical marine turtle grass *Thalassia testudinum*, Konig. and its epiphytes. Thesis, University of Miami, Coral Gables, FL.

Juneau, C. L., Jr. 1975. An inventory and study of the Vermilion Bay—Atchafalaya Bay complex. *La., Dep. Wildl. Fish, Tech. Bull.* 13:1–153.

Kale, H. W., II. 1965. Ecology and bioenergetics of the long-billed marsh wren *Telmatodytes palustris griseus* Brewster in Georgia salt marshes. *Publ. Nuttall Ornithol. Club* 5:589–591.

Kapraun, D. F. 1974. Seasonal periodicity and spatial distribution of benthic marine algae in Louisiana. *Contrib. Mar. Sci.* 18:139–167.

Kapraun, D. F., and F. W. Zechman. 1982. Seasonality and vertical zonation of benthic marine algae on a North Carolina coastal jetty. *Bull. Mar. Sci.* 32:702–714.

Kemp, W. M., R. L., Wetzel, W. R. Boynton, C. F. D'Elia, and J. C. Stephenson. 1982. Nitrogen cycling and estuarine interfaces: some current concepts and research direction. In V. S. Kennedy (ed.), *Estuarine Interactions*. New York: Academic, pp. 209–230.

Kendall, D. R. 1983. *The Role of Physical-Chemical Factors in Structuring Subtidal Marine and Estuarine Benthos*, Tech. Rep. EL-83-2. Vicksburg, MS: U. S. Army Engineer Waterways Exp. Stn.

Kennish, M. J. 1986. *Ecology of Estuaries*, Vol. I. *Physical and Chemical Aspects*. Boca Raton, FL: CRC Press.

Kenworthy, W. J. 1981. The interrelationship between seagrasses, *Zostera marina* and *Halodule wrightii* and the physical and chemical properties of sediments in a mid-Atlantic coastal plain estuary near Beaufort, North Carolina (USA). Thesis, University of Virginia, Charlottesville.

Kerwin, J. A. 1966. Classification and structure of the tidal marshes of the Poropotank River, Virginia. *ASB Bull.* 13:40 (abstr.).

Ketchum, B. H. 1983. Estuarine characteristics. In B. H. Ketchum (ed.), *Ecosystems of the World*, Vol. 26. *Estuaries and Enclosed Seas*. Amsterdam: Elsevier Scientific, pp. 1–14.

Kikuchi, T. 1980. Faunal relationships in temperate seagrass beds. In R. C. Phillips and C. P. McRoy (eds.), *Handbook of Seagrass Biology: An Ecosystem Perspective*. New York: Garland STMP, pp. 153–172.

Kikuchi, T., and J. M. Pérès. 1977. Consumer ecology of seagrass. In C. P. McRoy and Helfferich (eds.), *Seagrass Ecosystems: A Scientific Perspective*. New York: Dekker, pp. 147–193.

Kinne, O. 1967. Physiology of estuarine organisms with special reference to salinity and temperature: general aspects. In G. H. Lauff (ed.), *Estuaries*, Publ. No. 83. Washington, DC: Am. Assoc. Adv. Sci., pp. 525–540.

Kirby, C. J., and J. G. Gosselink. 1976. Primary production in a Louisiana Gulf coast *Spartina alterniflora* marsh. *Ecology* 57:1052–1059.

Kitting, C. L. 1984. Selectivity by dense populations of small invertebrates foraging among seagrass blade surfaces. *Estuaries* 7:276–288.

Kitting, C. L., B. D. Fry, and M. D. Morgan. 1984. The base of seagrass meadow food webs: inconspicuous algae, not seagrass detritus. *Oecologia* 62:145–149.

Kjelson, M. A., D. S. Peters, G. W. Thayer, and G. N. Johnson. 1975. The general feeding ecology of postlarval fishes in Newport River estuary. *Fish. Bull.* 73:137–144.

Kjerfve, B., and H. N. McKellar, Jr. 1980. Time series measurements of estuarine fluxes. In V. S. Kennedy (ed.), *Estuarine Perspectives.* New York: Academic, pp. 341–357.

Kjerfve, B., and J. A. Prochl. 1979. Velocity variability in a cross-section of a well mixed estuary. *J. Mar. Res.* 37:409–418.

Kjerfve, B., J. E. Greer, and R. L. Crout. 1978. Low-frequency response for estuarine sea level to non-local forcing. In M. L. Wildy (ed.), *Estuarine Interactions.* New York: Academic, pp. 497–513.

Kjerfve, B., J. A. Proehl, F. B. Schwing, H. E. Seim, and M. Marozas. 1982. Temporal and spatial considerations in measuring estuarine water fluxes. In V. S. Kennedy (ed.), *Estuarine Comparisons.* New York: Academic, pp. 37–52.

Knauer, G. A., and A. V. Ayers. 1977. Changes in carbon, nitrogen, adenosine triphosphate and chlorophyll in decomposing *Thalassia testudinum* leaves. *Limnol. Oceanogr.* 22:408–414.

Kneib, R. T. 1984. Patterns of invertebrate distribution and abundance in the intertidal salt marsh: causes and questions. *Estuaries* 7:392–412.

Kneib, R. T. 1985. Predation and disturbance by grass shrimp, *Palaemonetes pugio* Holthius, in soft-substrate benthic invertebrate assemblages. *J. Exp. Mar. Biol. Ecol.* 93:91–102.

Kneib, R. T. 1987a. Seasonal abundance, distribution and growth of postlarval and juvenile grass shrimp (*Palaemonetes pugio*) in a Georgia, USA, salt marsh. *Mar. Biol. (Berlin)* 96:215–223.

Kneib, R. T. 1987b. Predation risk and use of intertidal habitats by young fishes and shrimp. *Ecology* 68:379–386.

Kneib, R. T., and A. E. Stiven. 1982. Benthic invertebrate responses to size and density manipulations of the common mummichog, *Fundulus heteroclitus*, in an intertidal salt marsh. *Ecology* 63:1518–1532.

Knox, G. A. 1986. *Estuarine Ecosystems: A System Approach.* Vol. 1. Boca Raton, FL. CRC Press.

Kofoed, J. S., and D. S. Gorsline. 1963. Sediment environments in Apalachicola Bay and vicinity, Florida. *J. Sediment. Petrol.* 33:205–223.

Kraeuter, J. N., and E. M. Setzler. 1975. The seasonal cycle of *Scyphozoa* and *Cubozoa* in Georgia estuaries. *Bull. Mar. Sci.* 25:66–74.

Kraeuter, J. N., and P. L. Wolf. 1974. The relationship of marine macro-invertebrates to salt marsh plants. In R. J. Reimold and W. H. Queen (eds.), *Ecology of Halophytes.* New York: Academic, pp. 449–462.

Kruczynski, W. L., C. B. Subrahmanyam, and S. H. Drake. 1978a. Studies on the plant community of a north Florida salt marsh. Part I. Primary production. *Bull. Mar. Sci.* 28:316–334.

Kruczynski, W. L., C. B. Subrahmanyam, and S. H. Drake. 1978b. Studies on the plant community of a north Florida salt marsh, Part 2: nutritive value and decomposition. *Bull. Mar. Sci.* 28:707–715.

Kuenzler, E. J. 1961. Structure and energy flow of a mussel population in a Georgia salt marsh. *Limnol. Oceanogr.* 6:191–204.

Kuenzler, E. J., and H. L. Marshall. 1973. *Effects of Mosquito Control Ditching on Estuarine Ecosystems*, Project B-026-NC. Raleigh: University of North Carolina, Water Resources Research Institute.

Kuenzler, E. J., D. W. Stanley, and J. P. Koenings. 1979. Nutrient kinetics of phytoplankton in the Pamlico River, N.C. *Rep.—Water Resour. Res. Inst., Univ. N.C.* 139:1–163.

Kurz, H., and K. Wagner. 1957. *Tidal Marshes of the Gulf and Atlantic Coasts of Florida and Charleston, South Carolina*, Stud. No. 24. Tallahassee: Florida State University.

Kushlan, J. A. 1978. Feeding ecology of wading birds. In A. Sprunt, IV, J. C. Ogden, and S. A. Winckler (eds.), *Wading Birds*. New York: National Audubon Society, pp. 249–296.

Lackey, J. B. 1967. The microbiota of estuaries and their roles. In G. Lauff (ed.), *Estuaries*, Publ. No. 83. Washington, DC: Am. Assoc. Adv. Sci., pp. 291–302.

Larkum, A. W. D. 1976. Ecology of Botany Bay. I. Growth of *Posidonia australis* (Brown) Hook. f. in Botany Bay and other bays of the Sydney basin. *Aust. J. Mar. Freshwater Res.* 27:117–127.

LaSalle, M. W., and T. D. Bishop. 1987. Seasonal abundance of aquatic diptera in two oligohaline tidal marshes in Mississippi. *Estuaries* 10:303–315.

LaSalle, M. W., and A. A. de la Cruz. 1985. Seasonal abundance and diversity of spiders in two intertidal marsh plant communities. *Estuaries* 8:381–393.

Lauff, G. H. (ed.). 1967. *Estuaries*, Publ. No. 83. Washington, DC: Am. Assoc. Adv. Sci.

Laughlin, R. A., and R. J. Livingston. 1982. Environmental and trophic determinants of the spatial/temporal distribution of the brief squid (*Lolliguncula brevis*) in the Apalachicola estuary (North Florida, U.S.A.). *Bull. Mar. Sci.* 32:489–497.

Leber, K. M. 1985. The influence of predatory decapods, refuge and microhabitat selection on seagrass communities. *Ecology* 66:1951–1964.

Lee, C. W. 1973. The recent lithofacies and biofacies of North Santee Bay and lower reaches of North Santee River, South Carolina, U.S.A. Thesis, University of South Carolina, Columbus.

Lee, J. E. 1980. A conceptual model of marine detrital decomposition and organisms associated with the process. *Adv. Microb. Ecol.* 2:257–291.

Lehman, H. K., and P. V. Hamilton. 1980. Some factors influencing the distribution of the snail *Neritina reclivata*. *Northeast Gulf Sci.* 4:67–72.

Levinton, J. S. 1979. Deposit feeders, their resoures and the study of resource limitation. In R. J. Livingston (ed.), *Ecological Processes in Coastal and Marine Systems*. New York: Plenum, pp. 117–141.

Lewis, F. G., III. 1984. The distribution of macrobenthic crustaceans associated with *Thalassia, Halodule* and bare sand substrata. *Mar. Ecol.: Prog. Ser.* 19:101–113.

Lewis, F. G., III. 1987. Crustacean epifauna of seagrass and macroalgae in Apalachee Bay, Florida, USA. *Mar. Biol. (Berlin)* 94:219–229.

Lewis, F. G., III, and A. W. Stoner. 1983. Distribution of macrofauna within seagrass beds: an explanation for patterns of abundance. *Bull. Mar. Sci.* 33:296–304.

Lewis, J. B., and C. E. Hollingworth. 1982. Leaf epifauna of the seagrass *Thalassia testudinum*. *Mar. Biol. (Berlin)* 71:41–49.

Lewis, M. R., and T. Platt. 1982. Scales of variability in estuarine ecosystems. In V. S. Kennedy (ed.), *Estuarine Comparisons*. New York: Academic, pp. 3–20.

Lewis, R. R., III, and E. D. Estevez. 1988. The ecology of Tampa Bay, Florida: an estuarine profile. *U.S., Fish Wildl. Serv., Biol. Rep.* 85(7.18):1–132.

Lewis, R. R., III, M. J. Durako, M. D. Moffler, and R. C. Phillips. 1982. Seagrass meadows of Tampa Bay—a review. In S. F. Treat, J. L. Simon, R. R. Lewis III, and R. L. Whitman, Jr. (eds.), *Proceedings of the Tampa Bay Area Scientific Information Symposium*, Bellwether Press, Gainesville, FL, Florida Sea Grant College, pp. 210–246. Rep. No. 65.

Lindall, E. N., Jr., and C. H. Saloman. 1977. Alteration and destruction of estuaries effecting fishery resources. *Mar. Fish. Rev.* 39:1–7.

Liss, P. S. 1976. Conservative and non-conservative behavior of dissolved constituents during estuarine mixing. In J. D. Burton and P. S. Liss (eds.), *Estuarine Chemistry*. New York: Academic, pp. 93–130.

Livingston, R. J. 1975. Impact of Kraft pulp-mill effuents on estuarine and coastal fishes in Apalachee Bay, Florida, USA. *Mar. Biol. (Berlin)* 32:19–48.

Livingston, R. J. 1980. Ontogenetic trophic relationships and stress in a coastal seagrass system in Florida. In V. S. Kennedy (ed.), *Estuarine Perspectives*. New York: Academic, pp. 423–435.

Livingston, R. J. 1982. Trophic organization of fishes in a coastal seagrass system. *Mar. Ecol.: Prog. Ser.* 7:1–12.

Livingston, R. J. 1984a. The relationship of physical factors and biological response in coastal seagrass meadows. *Estuaries* 7:377–391.

Livingston, R. J. 1984b. Trophic response of fishes to habitat variability in coastal seagrass systems. *Ecology* 65:1258–1275.

Livingston, R. J. 1984c. The ecology of the Apalachicola Bay system: an estuarine profile. *U.S., Fish Wildl. Serv., Off. Biol. Serv. [Tech. Rep.] FWS/OBS* FWS/OBS/82-05:1–148.

Livingston, R. J. 1987a. Field sampling in estuaries: the relationship of scale to variability. *Estuaries* 10:194–207.

Livingston, R. J. 1987b. Historic trends of human impacts on seagrass meadows in Florida. *Fla. Mar. Res. Publ.* 42:139–151.

Livingston, R. J., G. J. Kobylinski, F. G. Lewis III, and P. F. Sheridan. 1976. Long-term fluctuations of epibenthic fish and invertebrate populations in Apalachicola Bay, Florida. *U.S. Fish. Wildl. Serv., Fish Bull.* 74:311–321.

Livingston, R. J., P. S. Sheridan, B. G. McLane, F. G. Lewis III, and G. J. Kobylinski. 1977. The biota of the Apalachicola Bay system: functional relationships. *Fla. Mar. Res. Publ.* 26:75–100.

Livingston, D. A. 1963. Chemical composition of rivers and lakes. *Geol. Surv. Prof. Pap. (U.S.)* 440-G:1–64.

Lombardo, R., and R. R. Lewis III. 1985. A review of commercial fisheries data. In S. F. Treat, J. L. Simon, R. R. Lewis III, and R. L. Whitman, Jr. (eds.), *Proceedings of the Tampa Bay Area Scientific Information Symposium*, Rep. No. 65. Bellwether Press, Gainesville, FL Florida Sea Grant College, pp. 614–634.

Longley, W. L., R. Jackson, and B. Snyder. 1978. Managing oil and gas activities in coastal environments. *U.S., Fish Wildl. Serv., Off. Biol. Serv.* [*Tech. Rep.*] *FWS/OBS* FWS/OBS/78-54:1–66.

Longley, W. L., R. Jackson, and B. Snyder. 1981. Managing oil and gas activities in coastal environments: refuge manual. *U.S., Fish Wildl. Serv., Off. Biol. Serv.* [*Tech. Rep.*] *FWS/OBS* FWS/OBS/81-22:1–452.

Lonsdale, D. L. J., and B. C. Coull. 1977. Composition and seasonality of zooplankton of North Inlet, South Carolina. *Chesapeake Sci.* 18:272–283.

Loosanoff, V. L. 1965. The American or eastern oyster. *Fish Wildl. Serv. Bur. Comm. Fish. Circ.* 205:1–36.

Lucas, J. R. 1982. Feeding ecology of the Gulf silverside, *Menidia peninsulae*, near Crystal River, Florida, with notes on its life history. *Estuaries* 5:138–144.

Luckenbach, M. W. 1984. Settlement and early post-settlement survival in the recruitment of *Mulinia lateralis* (Bivalva). *Mar. Ecol.: Prog. Ser.* 17:245–250.

Lyons, W. G., and S. B. Collard. 1974. Benthic invertebrate communities of the eastern Gulf of Mexico. In R. E. Smith (ed.), *Proceedings of the Marine Environmental Implications of Offshore Drilling in the Eastern Gulf of Mexico: 1974.* St. Petersburg: State University System, Florida Institute of Oceanography, pp. 157–165.

Mager, A., Jr., and G. W. Thayer. 1986. National Marine Fisheries Service habitat conservation efforts in the southeast region of the United States from 1981 through 1985. *Mar. Fish. Rev.* 48:1–8.

Mahoney, B. M. S. 1982. Seasonal fluctuations of benthic macrofauna in the Apalachicola estuary, Florida. The role of predation and larval availability. Thesis, Florida State University, Tallahassee.

Mann, K. H. 1988. Production and use of detritus in various freshwater, estuarine and coastal marine systems. *Limnol. Oceanogr.* 33:910–930.

Maples, R. S. 1982. Blue-green algae of a coastal salt panne and surrounding angiosperm zones in a Louisiana salt marsh. *Northeast Gulf Sci.* 5:39–43.

Marmer, H. A. 1954. Tides and sea level in the Gulf of Mexico. *Fish. Bull.* 55:101–118.

Marshall, D. E. 1970. Characteristics of *Spartina* marsh receiving treated municipal wastes. In H. T. Odum and A. F. Chestnut (eds.), *Studies of Marine Estuarine Ecosystems Development with Treated Sewage Wastes*, Annu. Rep. 1969–1970. Chapel Hill: University of North Carolina, Inst. Mar. Sci., pp. 317–359.

Marshall, H. L. 1974. Irregularly flooded marsh. In H. T. Odum, B. J. Copeland, and E. A. McMahan (eds.), *Coastal Ecological Systems of the United States*, Vol. II. Washington, DC: Conservation Foundation, pp. 150–169.

Mathews, T. D., F. W. Stapor, Jr., C. R. Richter, et al. (eds.). 1980. *Ecological Characterization of the Sea Island Coastal Region of South Carolina and Georgia*, Vol. I: *Physical Features of the Characterization Area*, Washington, DC: FWS/OBS/79-40. U.S. Fish Wildl. Serv., Off. Biol. Serv.

Mattraw, H. C., and J. F. Elder. 1982. Nutrient and detritus transport in the Apalachicola River, Florida. *Geol. Surv. Water-Supply Pap. United States* 2196-C:1–64.

May, E. B. 1968. Summer oyster mortalities in Alabama. *Prog. Fish. Cult.* 30:99.

May, E. B. 1974. The distribution of mud crabs (*Xanthidae*) in Alabama estuaries. *Proc. Natl. Shellfish. Assoc.* 64:33–37.

McBee, J. T., and W. T. Brehm. 1979. Macrobenthos of Simmons Bayou and an adjoining residential canal. *Gulf Res. Rep.* 6:211–216.

McCoy, E. D., and S. S. Bell. 1985. Tampa Bay: the end of the line? In S. F. Treat, J. L. Simon, R. R. Lewis, III, and R. L. Whitman (eds.), *Proceedings of the Tampa Bay Area Information Symposium*, Rep. No. 65. Bellwether Press, Gainesville, FL Florida Sea Grant College, pp. 460–474.

McCoy, E. D., and J. R. Rey. 1981. Terrestrial arthropods of northwest Florida salt marshes: coleoptera. *Fla. Entomol.* 64:405–411.

McCoy, E. D., and J. R. Rey. 1987. Terrestrial arthropods of northwest Florida salt marshes: hymenoptera (Insecta). *Fla. Entomol.* 70:90–97.

McDonald, J. 1982. Divergent life history patterns in the co-occurring intertidal crabs *Panopeus herbstii* and *Eurypanopeus depressus* (Crustacea: Brachyura: Xanthidae). *Mar. Ecol.: Prog. Ser.* 8:173–180.

McDougall, K. D. 1943. Sessile marine invertebrates at Beaufort, North Carolina. *Ecol. Monogr.* 13:321–374.

McHugh, J. L. 1967. Estuarine nekton. In G. H. Lauff (ed.), *Estuaries*, Publ. No. 83. Washington, DC: Am. Assoc. Adv. Sci., pp. 581–620.

McNulty, J. K., R. C. Work, and H. B. Moore. 1962. Some relationships between the infauna of the level bottom and the sediment in south Florida. *Bull. Mar. Sci. Gulf Caribb.* 12:322–332.

McNulty, J. K., W. N. Lindall, Jr., and J. E. Sykes. 1972. Cooperative Gulf of Mexico estuarine inventory and study, Florida. Phase I. Area description. *NOAA Tech. Rep. NMFS Circ.* 368:1–126.

McNulty, J. K., W. N. Lindall, Jr., and E. A. Anthony. 1974. Data of the Biology phase, Florida portion, cooperative Gulf of Mexico estuarine inventory. *Natl. Mar. Fish. Serv., Data Rep.* 95:1–229.

McRoy, C. P. 1974. Seagrass productivity: carbon uptake experiments in eelgrass, *Zostera marina. Aquaculture* 4:131–137.

McRoy, C. P., and R. J. Barsdate. 1970. Phosphate absorption in eelgrass. *Limnol. Oceanogr.* 15:14–20.

McRoy, C. P., and C. Helfferich. 1980. Applied aspects of seagrasses. In R. C. Phillips and C. P. McRoy (eds.), *Handbook of Seagrass Biology: An Ecosystem Approach.* New York: Garland, pp. 297–342.

McRoy, C. P., R. J. Barsdate, and M. Nebert. 1972. Phosphorus cycling in an eelgrass (*Zostera marina* L.) ecosystem. *Limnol. Oceanogr.* 17:58–67.

Meany, R. A., I. Valiela, and J. M. Teal. 1976. Growth, abundance and distribution of larval tabanids in experimentally fertilized plots on a Massachusetts salt marsh. *J. Appl. Ecol.* 13:323–332.

Mendonca, M. T., and L. M. Ehrhart. 1982. Activity, population size and structure of immature *Chelonia mydas* and *Caretta caretta* in Mosquito Lagoon, Florida. *Copeia* 1982:161–167.

Menendez, R. J. 1987. Vertical zonation of the xanthid mud crabs *Panopeus obesus* and *Panopeus simpsoni* on oyster reefs. *Bull. Mar. Sci.* 40:73–77.

Menzel, R. W., and F. E. Nichy. 1958. Studies of the distribution and feeding habits of some oyster predators in Alligator Harbor, Florida. *Bull. Mar. Sci. Gulf Caribb.* 8:125–145.

Miglarese, J. V., and P. A. Sandifer (eds.). 1982. *An Ecological Characterization of South Carolina Wetland Impoundments*, Tech. Rep. No. 51. Columbia: South Carolina University, Mar. Res. Cent.

Miller, J. M., and M. L. Dunn. 1980. Feeding strategies and patterns of movement in juvenile estuarine fishes. In V. S. Kennedy (ed.), *Estuarine Perspectives*. New York: Academic, pp. 437–448.

Miller, J. M., J. P. Reed, and L. J. Pietrafesa. 1984. Patterns, mechanisms and approaches to the study of migrations of estuarine-dependent fish larvae and juveniles. In McCleave, J. D., G. P. Arnold, J. J. Dodson, and W. H. Neill (eds.), *Mechanisms of Migration in Fishes*. New York: Plenum, pp. 209–226.

Miller, R. J. 1974. Distribution and biomass of an estuarine ctenophore population, *Mnemiopsis leidyi* (A. Agassiz). *Chesapeake Sci.* 15:1–8.

Mitsch, W. J., and J. G. Gosselink. 1986. *Wetlands*. New York: Van Nostrand-Reinhold.

Montagna, P. A. 1984. *In situ* measurement of meiobenthic grazing rates on sediment bacteria and edaphic diatoms. *Mar. Ecol.: Prog. Ser.* 18:119–130.

Montagna, P. A., B. C. Coull, T. L. Herring, and B. W. Dudley. 1983. The relationship between abundances of meiofauna and their suspected microbial food (diatoms and bacteria). *Estuarine, Coastal Shelf Sci.* 17:381–394.

Montagna, W. 1942. The sharp-tailed sparrows of the Atlantic coast. *Wilson Bull.* 54:107–120.

Montague, C. L. 1980. A natural history of temperate western Atlantic fiddler crabs (genus *Uca*) with reference to their impact on the salt marsh. *Contrib. Mar. Sci.* 23:25–55.

Mook, D. 1980. Seasonal variation in species composition of recently settled fouling communities along an environmental gradient in the Indian River Lagoon, Florida. *Estuarine Coastal Mar. Sci.* 11:572–581.

Moore, D. R. 1963. Distribution of the seagrass, *Thalassia* in the United States. *Bull. Mar. Sci.* 13:329–342.

Moore, H. B., L. T. Davies, T. H. Fraser, R. H. Gore, and N. R. Lopez. 1968. Some biomass figures from a tidal flat in Biscayne Bay, Florida. *Bull. Mar. Sci.* 18:261–279.

Moore, R. H., and R. R. Reis. 1983. Analysis of spatial and temporal variations in biomass and community structure of motile organisms in Town Creek, a South Carolina tidal pass. *Contrib. Mar. Sci.* 26:111–125.

Morgan, M. D. 1980. Grazing and predation of the grass shrimp *Palaemonetes pugio*. *Limnol. Oceanogr.* 25:896–902.

Morgan, M. D., and C. L. Kitting. 1984. Productivity and utilization of the seagrass *Halodule wrightii* and its attached epiphytes. *Limnol. Oceanogr.* 29:1066–1076.

Muller, W. A., and J. J. Lee. 1969. Apparent indispensability of bacteria in foraminifera nutrition. *J. Protozool.* 16:471–478.

Mulligan, T. J., and F. F. Snelson, Jr. 1983. Summer-season populations of epibenthic marine fishes in the Indian River Lagoon system, Florida. *Fla. Sci.* 46:250–276.

Mulvihill, E. L., C. A. Francisco, J. B. Glad, K. B. Kaster, and R. E. Wilson. 1980. *Biological Impacts of Minor Shoreline Structures on the Coastal Environment: State of the Art Review*, 2 vols, FWS/OBS-77/51. Washington, DC: U.S. Fish Wildl. Serv.

Nagle, J. S. 1968. Distribution of the epibiota of macroepibenthic plants. *Contrib. Mar. Sci.* 13:105–114.

Neill, W. T. 1958. The occurrence of amphibians and reptiles in saltwater areas, and a bibliography. *Bull. Mar. Sci. Gulf Caribb.* 8:1–97.

Nelson, W. G. 1981. The role of predation by decapod crustaceans in seagrass ecosystems. *Kiel. Meeresforsch., Sonderh.* 5:529–536.

Newton, R. B. no date. *Wetlands: Uses and Misuses*, Natl. Wetl. Inventory Working Pap. Washington, DC: U.S. Fish Wildl. Serv., Off. Biol. Serv.

Niedoroda, A. W., D. J. P. Swift, and T. S. Hopkins. 1985. The shoreface. In R. A. Davis, Jr. (ed.), *Coastal Sedimentary Environments*, 2d ed. New York: Springer, pp. 533–624.

Nixon, S. W. 1980. Between coastal marshes and coastal waters: a review of twenty years of speculation and research on the role of salt marshes in estuarine productivity and water chemistry. In P. Hamilton and K. B. MacDonald (eds.), *Estuarine and Wetland Processes*. New York: Plenum, pp. 437–525.

Nixon, S. W. 1983. Estuarine ecology—a comparative and experimental analysis using 14 estuaries and the MERL microcosms. (unpublished report). Washington, DC: U.S. Environ. Prot. Agency.

Nixon, S. W., C. A. Oviatt, J. Gerber, and V. Lee. 1976. Diel metabolism and nutrient dynamics in a salt marsh embayment. *Ecology* 57:740–750.

Odell, D. K. 1975. Status and aspects of the life history of the bottlenose dolphin, *Tursiops truncatus*, in Florida. *J. Fish. Res. Board Can.* 32:1055–1058.

Odum, E. P. 1963. Primary and secondary energy flow in relation to ecosystem structure. *Proc. 16th Int. Cong. Zool.* 4:336–338.

Odum, E. P. 1980. The status of three ecosystem-level hypotheses regarding salt marsh estuaries: tidal subsidy, outwelling and detritus-based food chains. In V. S. Kennedy (ed.), *Estuarine Perspectives*. New York: Academic, pp. 485–495.

Odum, E. P., and A. A. de la Cruz. 1967. Particulate organic detritus in a Georgia salt marsh-estuarine ecosystem. In G. H. Lauff (ed.), *Estuaries*, Publ. No. 83. Washington, DC: Am. Assoc. Adv. Sci., pp. 383–388.

Odum, E. P., and M. E. Fanning. 1973. Comparison of the productivity of *Spartina alterniflora* and *S. cynosuroides* in Georgia coastal marshes. *Bull. Ga. Acad. Sci.* 31:1–12.

Odum, E. P., and A. E. Smalley. 1959. Comparison of population energy flow of a herbivorous and a deposit-feeding invertebrate in a salt-marsh ecosystem. *Proc. Natl. Acad. Sci. U. S. A.* 45:617–622.

Odum, H. T. 1957. Primary production of eleven Florida springs and marine turtle grass community. *Limnol. Oceanogr.* 2:85–97.

Odum, H. T. 1974. Tropical marine meadows. In H. T. Odum, B. J. Copeland, and E. A. McMahan (eds.), *Coastal Ecological Systems of the United States*, Vol. 1. Washington, DC: Conservation Foundation, pp. 442–497.

Odum, H. T., and C. M. Hoskins. 1958. Comparative studies on the metabolism of marine waters. *Publ. Inst. Mar. Sci., Univ. Tex.* 5:16–46.

Odum, H. T., R. R. Burkholder, and J. Rivero. 1960. Measurements of productivity of turtle grass flats, reefs and the Baha Fosferlescente of southern Puerto Rico. *Publ. Inst. Mar. Sci., Univ. Tex.* 6:159–170.

Odum, W. E. 1970a. Pathways of energy flow in a south Florida estuary. Thesis, University of Miami, Coral Gables, FL.

Odum, W. E. 1970b. Utilization of the direct grazing and plant detritus food chains by the stripped mullet *Mugil cephalus*. In J. Steele (ed.), *Marine Food Chains, a Symposium*. Berkeley: California UP, pp. 222–240.

Odum, W. E. 1984. Estuarine productivity: unresolved questions concerning the coupling of primary and secondary production. In B. J. Copeland, K. Hart, N. Davis, and S. Friday (eds.), *Research for Managing the Nation's Estuaries: Proceedings of a Conference in Raleigh, North Carolina*, UNC-SG-84-08. Raleigh: University of North Carolina, Sea Grant College Program, pp. 231–253.

Odum, W. E., and E. J. Heald. 1972. Trophic analyses of an estuarine mangrove community. *Bull. Mar. Sci.* 22:671–738.

Odum, W. E., and T. G. Smith. 1981. Habitat value of coastal wetlands. In R. C. Carey, P. S. Markovits, and J. B. Kirkwood (eds.), *Proceedings of the U.S. Fish and Wildlife Service Workshop on Coastal Ecosystems of the Southeastern United States*, FWS/OBS-80/59. Washington, DC: U.S. Fish Wildl. Serv., pp. 30–35.

Odum, W. E., J. S. Fisher, and J. Pickral. 1979. Factors controlling the flux of particulate organic carbon from estuarine wetlands. In R. J. Livingston (ed.), *Ecological Processes in Coastal and Marine Systems*. Ecological Study Ser., No. 10. New York: Plenum, pp. 69–80.

Odum, W. E., C. C. McIvor, and T. J. Smith, III. 1982. The ecology of the mangroves of south Florida: A community profile. *U.S., Fish Wildl. Serv., Off. Biol. Serv.* [*Tech. Rep.*] *FWS/OBS* FWS/OBS 81/24:1–144.

Odum, W. E., T. J. Smith, III, J. K. Hoover, and C. C. McIvor. 1984. The ecology of tidal freshwater marshes of the United States east coast: a community profile. *U.S., Fish Wildl. Serv., Off. Biol. Serv.* [*Tech. Rep.*] *FWS/OBS* FWS/OBS-83/17:1–177.

Office of Technology Assessment. 1984. *Wetlands: Their Use and Regulation*, OTA-0-206. Washington, DC: U.S. Congress, Off. Technol. Assess.

Officer, C. B. 1976. *Physical Oceanography of Estuaries (and Associated Coastal Waters)*. New York: Wiley.

Ogden, J. C. 1980. Faunal relations in Carribean seagrass beds. In R. C. Phillips and C. P. McRoy (eds.), *Handbook of Seagrass Biology: An Ecosystem Perspective*. New York: Garland STMP, pp. 173–198.

Ogle, J. T., R. W. Heard, and J. Sieg. 1982. Tanaidacea (Crustacea: Peracarida) of the Gulf of Mexico, 1: introduction and an annotated bibliography of Tanaidacea previously reported from the Gulf of Mexico. *Gulf Res. Rep.* 7:101–104.

O'Gower, A. K., and J. W. Wacasey. 1967. Animal communities associated with *Thalassia, Diplanthera*, and sand beds in Biscayne Bay, 1: analysis of communities in relation to water movements. *Bull. Mar. Sci.* 17:175–210.

Ogren, L. H., and C. McVea, Jr. 1981. Apparent hibernation by sea turtles in North American waters. In K. A. Bjorndal (ed.), *Biology and Conservation of Sea Turtles*. Washington, DC: Smithsonian Institution Press, pp. 127–132.

Ogren, L. H., and H. A. Brusher. 1977. The distribution and abundance of fishes caught with a trawl in the St. Andrew Bay system, Florida. *Northeast Gulf Sci.* 1:83–105.

Olinger, L. W., R. G. Rogers, P. L. Fore, R. L. Todd, B. L. Mullins, F. T. Bisterfeld, and L. A. Wise, II. 1975. *Environmental and Recovery Studies of Escambia Bay and Pensacola Bay System, Florida*, EPA 904/9-76-016. Atlanta, GA: U.S. Environ. Prot. Agency.

Olsen, L. A. 1976. Ingested material in two species of estuarine bivalves: *Rangia cuneata* (Gray) and *Polymesoda caroliniana* (Bosc). *Proc. Natl. Shellfish. Assoc.* 66:103–104.

Olsen, L. A. 1984. Effects of contaminated sediment on fish and wildlife: review and annotated bibliography. *U.S., Fish Wildl. Serv., Off. Biol. Serv. [Tech. Rep.] FWS/OBS* FWS/OBS/82-66:1–104.

O'Neil, S. P., and M. P. Weinstein. 1987. Feeding habitats of spot, *Leiostomus xanthurus*, in polyhaline versus meso-oligohaline tidal creeks and shoals. *Fish. Bull.* 85:785–796.

O'Neil, T. 1949. *The Muskrat in the Louisiana Coastal Marshes.* Baton Rouge: Louisiana Dep. Wildl. Fish.

Oppenheimer, C. H., T. Isensee, W. B. Brogden, and D. Bowman. 1975. *Establishment of Operational Guidelines for Texas Coastal Zone Management. Biological Uses Criteria.* Final report to National Science Foundation (Grant GI-34870x) and Office of the Governor of Texas IAC (74-75)-0685 from the University of Texas Marine Sciences Institute, Austin.

O'Reilly, J. E., J. P. Thomas, and C. Evans. 1976. Annual primary production (nannoplankton, net plankton, dissolved organic matter) in the Lower New York Bay. In W. H. McKeon and G. H. Lauer (eds.), *Fourth Symposium on Hudson River Ecology*, Pap. No. 19. New York. Hudson River Environmental Society.

Orth, R. J., and J. van Montfrans. 1984. Epiphyte–seagrass relationships with an emphasis on the role of micrograzing: a review. *Aquat. Bot.* 18:43–69.

Orth, R. J., K. L. Heck, Jr., and J. van Montfrans. 1984. Faunal communities in seagrass beds: a review of the influence of plant structure and prey characteristics on predator-prey relationships. *Estuaries* 7:339–350.

Osenga, G. A., and B. C. Coull. 1983. *Spartina alterniflora* Loisel root structure and meiofaunal abundance. *J. Exp. Mar. Biol. Ecol.* 67:221–225.

Overstreet, R. M. 1978. *Marine Maladies: Worms, Germs and Other Symbionts from the Northern Gulf of Mexico*, Miss.-Ala. Sea Grant Consortium Publ. No. MASGP-78-021. Ocean Springs, MS: Gulf Coast Res. Lab.

Owen, H. M. 1957. Etiological studies on oyster mortality. II. *Polydora websteri* Hartman (Polycheaeta: Spionidae). *Bull. Mar. Sci. Gulf Caribb.* 7:35–46.

Pace, M. L., S. Shimmel, and W. M. Darley. 1979. The effect of grazing by a gastropod, *Nassarius obsoletus*, on the benthic microbial community of a salt marsh mudflat. *Estuarine Coastal Mar. Sci.* 9:121–134.

Paffenhöfer, G. A., and S. C. Knowles. 1980. Omnivorousness in marine planktonic copepods. *J. Plankton Res.* 2:355–365.

Page, T., R. Harris, and S. S. Epstine. 1975. Drinking water and cancer mortality in Louisiana. *Science* 193:55–57.

Paine, R. T. 1963. Trophic relationships of eight sympatric predatory gastropods. *Ecology* 44:63–73.

Palmer, R. S. (ed.). 1962. *Handbook of North America Birds*, Vol. 1, *Loons Through Flamingos.* New Haven, CT: Yale UP. London, England. 567 pp.

Palumbo, A. V., and R. Ferguson. 1978. Distribution of suspended bacteria in the Newport River Estuary, North Carolina. *Estuarine Coastal Mar. Sci.* 7:521–529.

Parnell, J. F., and R. F. Soots, Jr. 1979. *Atlas of Colonial Waterbirds of North Carolina Estuaries*, UNC-SG-78-10. Raleigh: University of North Carolina, Sea Grant College Program.

Parsons, K. A., and A. A. de la Cruz. 1980. Energy flow and grazing behavior of conocephaline grasshoppers in a *Juncus roemerianus* marsh. *Ecology* 61:1045–1050.

Patriquin, D. G. 1972. The origin of nitrogen and phosphorus for growth of the marine angiosperm, *Thalassia testudinum. Mar. Biol. (Berlin)* 15:35–46.

Patriquin, D. G., and R. Knowles. 1972. Nitrogen fixation in the rhizosphere of marine angiosperms. *Mar. Biol. (Berlin)* 16:49–58.

Peckol, P., and D. Baxter. 1986. Population dynamics of the onuphid polychaete *Diopatra cuprea* (Bosc) along a tidal exposure gradient. *Estuarine Coastal Shelf Sci.* 22:371–377.

Penfound, W. T. 1952. Southern swamps and marshes. *Bot. Rev.* 18:413–446.

Penhale, P. A. 1977. Macrophyte-epiphyte biomass and productivity in an eelgrass (*Zostera marina* L.) community. *J. Exp. Mar. Biol. Ecol.* 26:211–224.

Penhale, P. A., and W. O. Smith. 1977. Excretion of dissolved organic carbon and phosphorus by eelgrass (*Zostera marina* L.) and its epiphytes. *Limnol. Oceanogr.* 22:400–407.

Pennock, J. R., and J. H. Sharp. 1986. Phytoplankton production in the Delaware estuary: temporal and spatial variability. *Mar. Ecol. Prog. Ser.* 34:143–155.

Perry, H. M., and J. Y. Christmas. 1973. Estuarine zooplankton, Mississippi. In J. Y. Christmas (ed.), *Cooperative Gulf of Mexico Estuarine Inventory and Study*: Mississippi. Ocean Springs, MS: Gulf Coast Res. Lab., pp. 198–241.

Peterson, B. J., R. W. Howarth, and R. H. Garritt. 1985. Multiple stable isotopes used to mark the flow of organic matter in estuarine food webs. *Science* 277:1361–1363.

Peterson, C. H. 1979. Predation, competitive exclusion and diversity in the soft-sediment benthic communities of estuaries and lagoons. In R. J. Livingston (ed.), *Ecological Processes in Coastal and Marine Systems*. New York: Plenum, Press, pp. 223–264.

Peterson, C. H. 1982. Clam predation by whelks (*Busycon* spp.): experimental tests of the importance of prey size, prey density and seagrass cover. *Mar. Biol. (Berlin)* 66:159–170.

Peterson, C. H., and N. M. Peterson. 1979. The ecology of intertidal flats of North Carolina: a community profile. *U.S., Fish Wildl. Serv., Off. Biol. Serv.* [*Tech. Rep.*] *FWS/OBS* FWS/OBS/79-39:1–73.

Pfeiffer, W. J., and R. G. Wiegert. 1981. Grazers on *Spartina* and their predators. In L. R. Pomery and R. G. Wiegert (eds.), *The Ecology of a Salt Marsh*. New York: Springer, pp. 87–112.

Philips, R. J., W. D. Burke, and E. J. Keener. 1969. Observations on the trophic significance of jelly fishes in Mississippi Sound with quantitative data on the associative behavior of small fishes with medusae. *Trans. Am. Fish. Soc.* 98:703–712.

Phillips, J. 1972. Chemical processes in estuaries. In R. S. K. Barnes and J. Green (eds.), *The Estuarine Environment*. London: Applied Science, pp. 33–50.

Phillips, R. C. 1960. *Observations on the Ecology and Distribution of the Florida Seagrasses*. Prof. Pap. Ser. No. 2. Florida State Board of Conservation Marine Lab.

Phillips, R. C. 1974. Temperate grass flats. In H. T. Odum, B. J. Copeland, and E. A. McMahan (eds.), *Coastal Ecological Systems of the United States: A Source Book for Estuarine Planning*, Vol. 2. Washington, DC: Conservation Foundation, pp. 244–299.

Phillips, R. C., C. McMillan, H. F. Bittaker, and R. Heiser. 1974. *Halodule wrightii* Ascherson in the Gulf of Mexico. *Contrib. Mar. Sci.* 18:257–261.

Pomeroy, L. R. 1960. Residence time of dissolved phosphate in natural waters. *Science* 131:1731–1732.

Pomeroy, L. R., and J. Imberger. 1981. The physical and chemical environment. In L. R. Pomeroy and R. G. Wiegert (eds.), *The Ecology of a Salt Marsh*. New York: Springer, pp. 21–36.

Pomeroy, L. R., and R. G. Wiegert (eds.). 1981. *The Ecology of a Salt Marsh*. New York: Springer.

Pomeroy, L. R., L. R. Shenton, R. D. H. Jones, and R. J. Reimold. 1972. Nutrient flux in estuaries. *Spec. Symp. —Am. Soc. Limnol. Oceanogr.* 1:274–291.

Pomeroy, L. R., W. M. Darley, E. L. Dunn, J. L. Gallagher, E. B. Haines, and D. W. Whitney. 1981. Primary production. In L. R. Pomeroy and R. G. Wiegert (eds.), *The Ecology of a Salt Marsh*. New York: Springer-Verlag, pp. 39–68.

President's Council on Environmental Quality. 1984. *15th Annual Report of the Council on Environmental Quality*. Washington, DC: Executive Office of the President.

Price W. A. 1954. Shorelines and coasts of the Gulf of Mexico. *Fish. Bull.* 55:39–65.

Pritchard, D. W. 1955. Estuarine circulation patterns. *Proc. Am. Soc. Civ. Eng.* 81:1–11.

Pritchard, D. W. 1967. What is an estuary: a physical viewpoint. In G. H. Lauff (ed.), *Estuaries*, Publ. No. 83. Washington, DC: Am. Assoc. Adv. Sci., pp. 3–5.

Provosts, M. W. 1973. Mean high water mark and use of tidelands in Florida. *Fla. Sci.* 36:50–66.

Rader, D. N. 1984. Salt-marsh benthic invertebrates: small scale patterns of distribution and abundance. *Estuaries* 7:413–420.

Ragotzkie, R. A. 1959. Plankton productivity in estuarine waters of Georgia. *Publ. Inst. Mar. Sci., Univ. Tex.* 6:146–158.

Ralph, R. D. 1977. Myxophyceae of the marshes of southern Delaware. *Chesapeake Sci.* 18:208–221.

Randall, J. E. 1964. Contributions to the biology of the queen conch, *Strombus gigas. Bull. Mar. Sci. Gulf Caribb.* 14:246–295.

Ranwell, D. S. 1972. *Ecology of Salt Marshes and Sand Dunes*. London: Chapman & Hall.

Reames, R., and A. B. Williams. 1984. Mud crabs of the *Panopeus herbstii* H. M. Edw. complex in Alabama, U.S.A. *Fish. Bull.* 81:885–890.

Recher, H. F. 1966. Some aspects of the ecology of migrant shorebirds. *Ecology* 47:393–407.

Reeve, M. R. 1970. Seasonal changes in the zooplankton of south Biscayne Bay and some problem assessing the effects on the zooplankton of natural and artificial thermal and other fluctuations. *Bull. Mar. Sci.* 20:894–921.

Reeve, M. R. 1975. The ecological significance of the zooplankton in the shallow subtropical waters of South Florida. In L. E. Cronin (ed.), *Estuarine Research*, Vol. 1. New York: Academic, pp. 352–371.

Reimold, R. J., and F. C. Daiber. 1970. Dissolved phosphorous concentrations in a natural saltmarsh of Delaware. *Hydrobiologia* 36:361–371.

Reimold, R. J., J. L. Gallagher, C. A. Linthurst, and W. J. Pfeiffer. 1975. Detritus production in coastal Georgia salt marshes. In L. E. Cronin (ed.), *Estuarine Research*, Vol. 1. New York: Academic, pp. 217–228.

Rey, J. R., and E. D. McCoy. 1982. Terrestrial arthropods of northwest Florida salt marshes: Hemiptera and Homoptera (Insecta). *Fla. Entomol.* 65:241–248.

Rey, J. R., and E. D. McCoy. 1983. Terrestrial arthropods of northwest Florida salt marshes: Araneae and Pseudoscorpiones (Arachnida). *Fla. Entomol.* 66:497–503.

Rey, J. R., and E. D. McCoy. 1986. Terrestrial arthropods of northwest Florida salt marshes: Diptera (Insecta). *Fla. Entomol.* 69:197–205.

Rhoads, D. C., and Young, D. K. 1970. The influence of deposit-feeding organisms on sediment stability and community trophic structure. *J. Mar. Res.* 28:150–179.

Rice, J. D., R. P. Trocine, and G. N. Wells. 1983. Factors influencing seagrass ecology in the Indian River Lagoon. *Fla. Sci.* 46:276–286.

Richards, G. P. 1985. Shellfish-associated enteric virus illness in the United States. *Estuaries* 8:94a.

Richardson, F. D. 1980. Ecology of *Ruppia maritima* L. in New Hampshire (USA) tidal marshes. *Rhodora* 82:403–439.

Richardson, J. P. 1986. Additions to the marine macroalgal flora of coastal Georgia. *Ga. Acad. Sci.* 44:131–135.

Richardson, J. P. 1987. Floristic and seasonal characteristics of inshore Georgia macroalgae. *Bull. Mar. Sci.* 40:210–219.

Richardson, J. P. 1988. Effects of sedimentation, grazing and predation on subtidal macroalgae of Radio Island jetty, North Carolina. *ASB Bull.* 35:97–110.

Riley, G. A. 1956. Oceanography of Long Island Sound, 1952–1954, 1: introduction. *Bull. Bingham Oceanogr. Collect.* 15:5–14.

Roessler, M. 1965. An analysis of the variability of fish populations taken by otter trawl in Biscayne Bay, Florida. *Trans. Am. Fish. Soc.* 94:311–318.

Rogers, B. D., and W. H. Herke. 1985. *Temporal Patterns and Size Characteristics of Migrating Juvenile Fishes and Crustaceans in a Louisiana Marsh*, Res. Rep. No. 5. Baton Rouge: School of Forestry, Wildlife and Fisheries, Louisiana State University Agricultural Center.

Rowe, G. T., C. H. Clifford, K. L. Smith, and P. L. Hamilton. 1975. Benthic nutrient regeneration and its coupling to primary productivity in coastal waters. *Nature (London)* 255:215–217.

Rublee, P. A. 1982. Seasonal distribution of bacteria in salt marsh sediments in North Carolina. *Estuarine, Coastal Shelf Sci.* 15:67–74.

Rusnak, G. A. 1967. Rates of sediment accumulation in modern estuaries. In G. H. Lauff (ed.), *Estuaries*, Publ. No. 83. Washington, DC: Am. Assoc. Adv. Sci., pp. 180–184.

Ryan, E. P. 1956. Observations on the life histories and the distribution of the Xanthidae (mud crabs) of Chesapeake Bay. *Am. Midl. Nat.* 56:138–162.

Sage, W. W., and M. J. Sullivan. 1978. Distribution of bluegreen algae in a Mississippi gulf coast salt marsh. *J. Phycol.* 14:333–337.

Sanders, H. L. 1958. Benthic studies in Buzzards Bay, 1: animal-sediment relationships. *Limnol. Oceanogr.* 7:63–70.

Sand-Jenson, K. 1977. Effects of epiphytes on eelgrass photosynthesis. *Aquat. Bot.* 3:55–63.

Sand-Jenson, K., and J. Borum. 1983. Regulation of growth of eelgrass (*Zostera marina* L.) in Danish waters. *Mar. Tech. Soc. J.* 17:15–21.

Santos, S. L., and J. L. Simon. 1974. Distribution and abundance of the polychaetous annelids in a south Florida estuary. *Bull. Mar. Sci.* 24:669–689.

Santos, S. L., and J. L. Simon. 1980a. Marine soft-bottom community establishment following annual defaunation: larval or adult recruitment. *Mar. Ecol.: Prog. Ser.* 2:235–241.

Santos, S. L., and J. L. Simon. 1980b. Response of soft-bottom benthos to annual catastrophic disturbance in a south Florida estuary. *Mar. Ecol.: Prog. Ser.* 3:347–355.

Schelske, C. L., and E. P. Odum. 1961. Mechanisms maintaining high productivity in Georgia estuaries. *Proc. Annu. Gulf Caribb. Fish. Inst.* 14:75–80 (Univ. Ga. Coll. Rep. Vol. 3, Misc. No. W).

Schmidley, D. J. 1980. Marine animals of the Southeastern United States coast and Gulf of Mexico. *U.S., Fish Wildl. Serv., Off. Biol. Serv. [Tech. Rep.] FWS/OBS* FWS/OBS/80-41:1–163.

Schreiber, R. W., and R. W. Risebrough. 1972. Studies of the brown pelican, 1: status of brown pelican populations in the United States. *Wilson Bull.* 84:119–135.

Schreiber, R. W., G. E. Woolfenden, and W. E. Curtsinger. 1975. Prey capture by the Brown Pelican. *Auk* 92:649–654.

Schroeder, W. W. and W. J. Wiseman, Jr. 1986. Low-frequency shelf-estuarine exchange processes in Mobile Bay and other estuarine systems on the northern Gulf of Mexico. In V. S. Kennedy (ed.), *Estuarine Variability*. New York: Academic Press, pp. 355–367.

Schubel, J. R., and D. J. Hirschberg. 1978. Estuarine graveyards, climatic change and the importance of the estuarine environment. In M. Wiley (ed.), *Estuarine Interactions*. New York: Academic, pp. 285–303.

Seaman, W. 1985. *Florida Aquatic Habitat and Fishery Resources*. Fla. Chptr., Am. Fish. Soc., Kissimmee, FL.

Sellers, M. A., and J. G. Stanley. 1984. *Species Profiles: Life Histories and Environmental Requirements of Coastal Fishes and Invertebrates (North Atlantic)—American Oyster*, U.S. Fish Wildl. Serv. FWS/OBS-82/11.23, TR EL-82-4. Washington, DC: U.S. Army Corps of Engineers.

Sellner, K. G., and R. G. Zingmark. 1976. Interpretations of the ^{14}C method of measuring the total annual production of phytoplankton in a South Carolina estuary. *Mar. Bot.* 19:119–125.

Shanholtzer, G. F. 1974. Relationship of vertebrates to salt marsh plants. In R. J. Reimold and W. H. Queen (eds.), *Ecology of Halophytes*. New York: Academic Press, pp. 463–474.

Shanholtzer, S. F. 1973. Energy flow, food habits and population dynamics of *Uca pugnax* in a salt marsh system. Thesis, University of Georgia, Athens.

Sharp, H. F., Jr. 1967. Food ecology of the rice rat *Oryzomys palustris* (Harlan) in a Georgia salt marsh. *J. Mammal.* 48:267–278.

Sheridan, P. F., and R. J. Livingston. 1979. Cyclic trophic relationships of fishes in an unpolluted, river-dominated estuary in north Florida. In R. J. Livingston (ed.),

Ecological Processes in Coastal and Marine Systems. New York: Plenum, pp. 143–161.

Sheridan, P. F., and R. J. Livingston. 1983. Abundance and seasonality of infauna and epifauna inhabiting a *Halodule wrightii* meadow in Apalachicola Bay, Florida. *Estuaries* 6:407–419.

Shier, D. E. 1969. Vermetid reefs and coastal development in the Ten Thousand Islands, southeast Florida. *Geol. Soc. Am. Bull.* 80:485–508.

Singletary, R. C., and H. B. Moore. 1974. A redescription of the *Amphioplus coniortodes-Ophionepthys limicola* community of Biscayne Bay, Florida. *Bull. Mar. Sci.* 24:690–699.

Smalley, A. E. 1959. The growth cycle of *Spartina* and its relation to the insect populations in the marsh. In *Proceedings of the Salt Marsh Conference*. Sapelo Island: Marine Institute, University of Georgia, pp. 96–100.

Smalley, A. E. 1960. Energy flow of a salt marsh grasshopper population. *Ecology* 41:785–790.

Smith, L. D., and B. C. Coull. 1987. Juvenile spot (Pisces) and grass shrimp predation on meiobenthos in muddy and sandy substrata. *J. Exp. Mar. Biol. Ecol.* 105:123–136.

Smith, N. P. 1986. The rise and fall of the estuarine intertidal zone. *Estuaries* 9:95–101.

Smith, R. F., A. H. Swartz, and W. H. Massman (eds.). 1966. *A Symposium on Estuarine Fisheries*, Spec. Publ. No. 3. Washington, DC: Am. Fish. Soc.

Smith, S. M., J. G. Hoff, S. P. O'Neil, and M. P. Weinstein. 1984. Community and trophic organization of nekton utilizing shallow marsh habitats, York River, Virginia. *Fish. Bull.* 82:433–467.

Smith, S. V. 1981. Responses of Kaneohe Bay, Hawaii to relaxation of sewage stress. In B. J. Neilson and L. E. Cronin (eds.), *Estuaries and Nutrients*. Clifton, NJ: Humana Press, pp. 391–410.

Sogard, S. M., G. V. N. Powell, and J. G. Homquist. 1987. Epibenthic fish communities on Florida Bay banks: relations with physical parameters and seagrass cover. *Mar. Ecol.: Prog. Ser.* 40:25–39.

Soniat, T. M. 1985. Changes in levels of infection of oysters by *Perkinsus marinus*, with special reference to the interaction of temperature and salinity on parasitism. *Northeast Gulf Sci.* 7:171–174.

South Carolina Water Resources Commission. 1970. South Carolina Tidelands Report.

Sprunt, A., Jr. 1955. *North American Birds of Prey*. New York: Harper Bros.

Spurrier, J. D., and B. Kjerfve. 1988. Estimating the net flux of nutrients between a salt marsh and a tidal creek. *Estuaries* 11:10–14.

Stalter, R. 1968. An ecological study of a South Carolina salt marsh. Ph.D. Dissertation, University of South Carolina, Columbia.

Stalter, R. 1974. The vegetation of the Cooper River estuary. Pp. 41–45. In F. P. Nelson (ed.), *The Cooper River Environmental Study*, Rep. No. 117. Columbia: South Carolina Water Resources Commission, State Water Plan.

Stalter, R., and W. T. Batson. 1969. Transplantation of salt marsh vegetation, Georgetown, South Carolina. *Ecology* 50:1087–1089.

Stancyk, S. E., and R. J. Feller. 1986. Transport of non-decapod invertebrate larvae in estuaries: an overview. *Bull. Mar. Sci.* 39:257–268.

Steidinger, K. A. 1973. Phytoplankton ecology: a conceptual review based on eastern Gulf of Mexico research. *CRC Crit. Rev. Microbiol.* 3:49–68.

Steidinger, K. A. 1985. Phytoplankton of Tampa Bay: a review. In S. F. Trent, J. L. Simon, R. R. Lewis III, and R. L. Whitman, Jr. (eds.), *Proceedings of the Tampa Bay Area Scientific Information Symposium*, Rep. No. 65. Bellwether Press, Gainesville, FL, Florida Sea Grant College, pp. 147–174.

Stevenson, H. L., and R. R. Colwell (eds.). 1973. *Estuarine Microbial Ecology.* Columbia: University of South Carolina.

Stickney, R. R. 1984. *Estuarine Ecology of the Southeastern United States and Gulf of Mexico.* College Station: Texas A & M UP.

Stoner, A. W. 1980a. Abundance, reproductive seasonality and habitat preferences of amphipod crustaceans in seagrass meadows of Apalachee Bay, Florida. *Contrib. Mar. Sci.* 23:63–77.

Stoner, A. W. 1980b. The role of seagrass biomass in the organization of benthic macrofaunal assemblages. *Bull. Mar. Sci.* 30:537–551.

Stoner, A. W. 1983a. Distribution of fishes in seagrass meadows: role of macrophyte biomass and species composition. *Fish. Bull.* 81:837–846.

Stoner, A. W. 1983b. Distributional ecology of amphipods and tanaidaceans associated with three seagrass species. *J. Crustacean Biol.* 3:505–518.

Stout, J. P. 1979. Marshes of the Mobile Bay estuary: status and evaluation. In H. A. Loyacano, Jr. and J. P. Smith (eds.), *Symposium on the Natural Resources of the Mobile Bay Estuary, Alabama.* Mobile, AL: U.S. Army Corps of Engineers, Mobile District, pp. 113–122.

Stout, J. P. 1984. The ecology of irregularly flooded salt marshes of the northeastern Gulf of Mexico: a community profile. *U.S., Fish Wildl. Serv., Biol. Rep.* 85(7.1):1–98.

Street, M. W. 1982. Fisheries resources and trends of the Albemarle Sound area. Pp. 57–60. In *Albemarle Sound Trends and Management*, UNC-SG-82-02. Raleigh: University of North Carolina, Sea Grant College Program..

Stross, R. G., and J. R. Stottlemyer. 1965. Primary production in the Pawtuxent River. *Chesapeake Sci.* 6:125–140.

Stroud, L. M., and A. W. Cooper. 1968. *Color Infrared Aerial Photographic Interpretation and Net Primary Productivity of a Regularly Flooded N.C. Salt Marsh*, Rep. No. 14. Raleigh: Water Resources Research Institute, North Carolina State University.

Subrahmanyam, C. B., and C. L. Coultas. 1980. Studies on the animal communities in two north Florida salt marshes, Part 3: seasonal fluctuations of fish and macro-invertebrates. *Bull. Mar. Sci.* 30:790–818.

Subrahmanyam, C. B., and S. H. Drake. 1975. Studies of the animal communities in two north Florida salt marshes, Part 1: fish communities. *Bull. Mar. Sci.* 25:445–465.

Subrahmanyam, C. B., W. L. Kruczynski, and S. H. Drake. 1976. Studies on the animal communities in two north Florida salt marshes, Part 2: macroinvertebrate communities. *Bull. Mar. Sci.* 26:172–195.

Sullivan, M. J. 1978. Diatom community structure: taxonomic and statistical analysis of a Mississippi salt marsh. *J. Phycol.* 14:468–475.

Summerson, H. C., and C. H. Peterson. 1984. Role of predation in organizing benthic communities of a temperate-zone seagrass bed. *Mar. Ecol.: Prog. Ser.* 15:63–77.

Sutherland, J. P., and R. H. Karlson. 1977. Development and stability of the fouling community at Beaufort, North Carolina. *Ecol. Monogr.* 47:425–446.

Swingle, H. A. 1971. Biology of Alabama estuarine areas-cooperative Gulf of Mexico estuarine inventory. *Ala. Mar. Res. Bull.* 5:1–123.

Taft, J. L., and W. R. Taylor. 1976. Phosphorus dynamics in some coastal plain estuaries. In M. Wiley (ed.), *Estuarine Processes*, Vol. 1. New York: Academic, pp. 79–89.

Tagatz, M. E. 1967. Fishes of the St. Johns River. *Fla. Acad. Sci.* 30:25–50.

Tanner, W. F. 1960. Florida coastal classification. *Trans.—Gulf Coast Assoc. Geol. Soc.* 10:259–266.

Taylor, J. D., and M. S. Lewis. 1970. The flora, fauna and sediments of the marine grass beds of Mahe, Seychelles. *J. Nat. Hist.* 4:199–220.

Taylor, J. L. 1974. The Charlotte Harbor estuarine system. *Fla. Sci.* 37:205–216.

Taylor, S. E. 1965. Phaeophyta of the eastern Gulf of Mexico. Thesis, Duke University, Durham, NC.

Teal, J. M. 1958. Distribution of fiddler crabs in Georgia salt marshes. *Ecology* 39:185–193.

Teal, J. M. 1962. Energy flow in the salt marsh ecosystem of Georgia. *Ecology* 43:614–624.

Teal, J. M., and J. Kanwisher. 1966. Gas transport in the marsh grass, *Spartina alterniflora. J. Exp. Bot.* 17:355–361.

Tenore, K. R. 1972. Macrobenthos of the Pamlico River estuary, North Carolina. *Ecol. Monogr.* 42:51–69.

Tenore, K. R. 1977. Food chain pathways in detrital feeding benthic communities: a review, with new observations on sediment resuspension and detrital recycling. In B. C. Coull (ed.). *Ecology of Marine Benthos.* Columbia: University of South Carolina Press, pp. 37–53.

Tenore, K. R., and B. C. Coull (eds.), 1980. *Marine Benthic Dynamics.* Columbia: University of South Carolina.

Tenore, K. R., L. Cammen, S. E. G. Findley, and N. Phillips. 1982. Perspectives of research on detritus: do factors controlling the availability of detritus to macro-consumers depend on its source? *J. Mar. Res.* 40:473–490.

Terrell, T. T. 1979. Physical regionalization of coastal ecosystems of the United States and its territories. *U.S., Fish Wildl. Serv., Off. Biol. Serv.* [*Tech. Rep.*] *FWS/OBS* FWS/OBS/78-80:1–30.

Thayer, G. W. 1971. Phytoplankton production and distribution of nutrients in a shallow unstratified estuarine system near Beaufort, N.C. *Chesapeake Sci.* 12:240–253.

Thayer, G. W., and H. H. Stuart. 1974. The bay scallop makes its bed of eelgrass. *Mar. Fish. Rev.* 36:27–39.

Thayer, G. W., and J. F. Ustach. 1981. Gulf of Mexico wetlands: Value, state of knowledge and research needs. In D. K. Atwood (convener), *Proceedings of a Symposium on Environmental Research Needs in the Gulf of Mexico (GOMEX)*, Vol. 2B. Miami, FL: U.S. Department of Commerce, Atlantic Oceanographic and Meteorological Laboratories, pp. 1–30.

Thayer, G. W., D. E. Hoss, M. A. Kjelson, W. F. Hettler, and M. W. LaCroix. 1974. Biomass of zooplankton in the Newport River estuary and the influence of postlarval fishes. *Chesapeake Sci.* 15:9–16.

Thayer, G. W., S. M. Adams, and M. W. Lacroix. 1975a. Structural and functional aspects of a recently established *Zostera marina* community. In L. E. Cronin (ed.), *Estuarine Research*, Vol. 1. New York: Academic Press, pp. 517–540.

Thayer, G. W., D. A. Wolfe, and R. B. Williams. 1975b. The impact of man on a seagrass system. *Am. Sci.* 63:288–296.

Thayer, G. W., D. W. Engel, and M. W. Lacroix. 1977. Seasonal distribution and changes in the nutritional quality of living, dead and detrital fractions of *Zostera marina* L. *J. Exp. Mar. Biol. Ecol.* 30:109–127.

Thayer, G. W., H. H. Stuart, W. J. Kenworthy, J. F. Ustach, and A. B. Hall. 1979. Habitat values of salt marshes, mangroves and seagrasses for aquatic organisms. In P. E. Greeson, J. R. Clark, and J. E. Clark (eds.), *Wetland Functions and Values: The State of Our Understanding*. Minneapolis, MN: Am. Water Resour. Assoc., pp. 235–247.

Thayer, G. W., W. J. Kenworthy, and M. S. Fonesca. 1984a. The ecology of eelgrass meadows of the Atlantic coast: a community profile. *U.S., Fish Wildl. Serv., Off. Biol. Serv.* [*Temp. Rep.*] *FWS/OBS* FWS/OBS/84-02:1–147.

Thayer, G. W., K. A. Bjorndal, J. C. Ogden, S. L. Williams, and J. C. Zieman. 1984b. Role of larger herbivores in seagrass communities. *Estuaries* 7:351–376.

Thorhaug, A., and M. A. Roessler. 1977. Seagrass community dynamics in a sub-tropical estuarine lagoon. *Aquaculture* 12:253–277.

Tietjen, J. H. 1980. Microbial-meiofaunal interrelationships: a review. *Microbiology* 1980:335–338.

Tiner, R. W., Jr. 1977. *An Inventory of South Carolina's Coastal Marshes*, Tech. Rep. No. 23. Columbia: South Carolina Marine Resources Center.

Tiner, R. W., Jr. 1984. *Wetlands of the United States: Current Status and Recent Trends*. Washington, DC: U.S. Dept. of the Interior, U.S. Fish Wildl. Serv.

Tomkins, J. R. 1947. The oystercatcher of the Atlantic coast of North America and its relation to oysters. *Wilson Bull.* 59:204–208.

Trocine, R. P., J. D. Rice, and G. N. Wells. 1981. Inhibition of seagrass photosynthesis by ultraviolet-B radiation. *Plant Physiol.* 68:74–81.

Turner, J. T. 1981. Latitudinal patterns of calanoid and cyclopoid copepod diversity in estuarine waters of eastern North America. *J. Biogeogr.* 8:369–382.

Turner, R. E. 1976. Geographic variations in salt marsh macrophyte production: a review. *Contrib. Mar. Sci.* 20:47–68.

Turner, R. E. 1979. Louisiana's coastal fisheries and changing environmental conditions. In J. W. Day, Jr., D. D. Culley, Jr., R. E. Turner, and A. J. Mumphrey, Jr. (eds.), *Proceedings of the Third Coastal Marsh and Estuary Management Symposium*. Baton Rouge: Louisiana State University, pp. 363–373.

Turner, R. E., and J. G. Gosselink. 1975. A note on standing crops of *Spartina alterniflora* in Florida and Texas. *Contrib. Mar. Sci.* 19:113–118.

Uhler, F. M., and H. Hotchkiss. 1968. Vegetation and its succession in marshes and estuaries along South Atlantic and Gulf coasts. *Proc. Marsh Est. Manage. Symp.*, LA State Univ., pp. 26–32.

U.S. Army Corps of Engineers. 1982. Final report *Benthic Macroinfauna Community Characterizations in Mississippi Sound and Adjacent Waters*. Final Rep., Contract No. DACW01-80C427. Mobile, AL: U.S. Army Corps of Engineers, Mobile District.

U.S. Department of Commerce. 1983. *Local Climatological Data, Annual Summaries 1982*, Parts 1 and 2. Ashville, NC: U.S. Dept. of Commerce, National Oceanic and Atmospheric Administration, National Environmental Satellite Data Information Service, National Climatological Data Center.

U.S. Department of Commerce. 1985b. *Marine Environmental Assessment, Gulf of Mexico, Annual Summary 1983*. Washington, DC: U.S. Dept. of Commerce, National Oceanic and Atmospheric Administration, National Environmental Satellite Data Information Service.

U.S. Department of Commerce. 1985a. *National Estuarine Inventory, Data Atlas*, Vol. 1. *Physical and Hydrologic Characteristics*. Washington, DC: U.S. Dept. of Commerce, National Oceanic and Atmospheric Administration, National Ocean Survey.

U.S. Environmental Protection Agency (USEPA). 1975. *Environmental and Recovery Studies of Escambia Bay and the Pensacola Bay System Florida*, 904/9-76-016. Washington, DC: USEPA.

U.S. Environmental Protection Agency (USEPA). 1986. *Near Coastal Waters Strategic Options Paper*. Washington, DC: U.S. Environ. Prot. Agency.

Valiela, I. 1984. Mechanisms linking producers and consumers in salt marsh estuarine ecosystems. In B. J. Copeland, K. Hart, N. Davis, and S. Friday (eds.), *Research for Managing the Nation's Estuaries: Proceedings of a Conference in Raleigh, North Carolina*, UNC-SG-84-08. Raleigh: University of North Carolina, Sea Grant College Program, pp. 265–294.

Valiela, I., and J. M. Teal. 1979a. The nitrogen budget of a salt marsh ecosystem. *Nature (London)* 280:652–656.

Valiela, I., and J. M. Teal. 1979b. Inputs, outputs and interconversions of nitrogen in a salt marsh ecosystem. In R. L. Jeffries and A. J. Davy (eds.), *Ecological Processes in Coastal Environments*. Cambridge, UK: Blackwell, pp. 399–419.

Valiela, I., J. M. Teal, and W. G. Deuser. 1978. The nature of growth forms in the salt marsh grass *Spartina alterniflora*. *Am. Nat.* 112:461–470.

van Montfrans, T., R. L. Wetzel, and R. J. Orth. 1984. Epiphyte-grazer relationships in seagrass meadows: consequences for seagrass growth and production. *Estuaries* 7:289–309.

Vernberg, F. J., and W. B. Vernberg (eds.). 1981. *Functional Adaptations of Marine Organisms* (Physiol. Ecol. Ser.). New York: Academic.

Virnstein, R. W. 1979. Predation on estuarine infauna: response patterns of component species. *Estuaries* 2:69–86.

Virnstein, R. W. 1987. Seagrass-associated invertebrate communities of the southeastern U.S.A.: a review. *Fla. Mar. Res. Publ.* 42:89–116.

Virnstein, R. W., and R. K. Howard. 1987. Motile epifauna of marine macrophytes in the Indian River Lagoon, Florida, 2: comparisons between drift algae and three species of seagrasses. *Bull. Mar. Sci.* 41:13–26.

Virnstein, R. W., P. S. Mikkelson, K. D. Cairns, and M. A. Capone. 1983. Seagrass beds versus sand bottoms: the trophic importance of their associated benthic invertebrates. *Fla. Sci.* 46:363–381.

Virnstein, R. W., W. G. Nelson, F. G. Lewis III, and R. K. Howard. 1984. Latitudinal patterns in seagrass epifauna: do patterns exist, and can they be explained? *Estuaries* 7:310–330.

Voors, A. W., W. D. Johnson, S. H. Steele, and H. Rothschild. 1978. Relationship between respiratory cancer and wetlands residency in Louisiana. *Arch. Environ. Health*, May/June: 124–149.

Voss, G. L. 1976. The invertebrates of Biscayne Bay. In A. Thorhaug and A. Volker (eds.), *Biscayne Bay: Past/Present/Future*. Spec. Rep. No. 5. Coral Gables, FL: University of Miami, Sea Grant Program, pp. 173–179.

Voss, G. L., and N. A. Voss. 1955. An ecological survey of Soldier Key, Biscayne Bay, Florida. *Bull. Mar. Sci. Gulf Caribb.* 5:203–229.

Waits, E. D. 1967. Net primary productivity of an irregularly flooded North Carolina salt marsh. Thesis, North Carolina State University, Raleigh.

Wanless, H. R. 1976. Geologic setting and recent sediments of the Biscayne Bay region, Florida. In A. Thorhang (ed.), *Biscayne Bay, Past/Present/Future*. Coral Gables, FL: University of Miami, Sea Grant Program, pp. 1–32.

Warme, J. E. 1971. Paleoecological aspects of a modern coastal lagoon. *Univ. Calif. Publ. Geol. Sci.* 87:1–131.

Warren, J. H. 1985. Climbing as an avoidance behaviour in the salt marsh periwinkle, *Littorina irrorata* (Say). *J. Exp. Mar. Biol. Ecol.* 89:11–28.

Watzin, M. C. 1985. Interactions among temporary and permanent meiofauna: observations on the feeding and behavior of selected taxa. *Biol. Bull. (Woods Hole, Mass.)* 169:397–416.

Webb, K. L. 1981. Conceptual models and processes of nutrient cycling in estuaries. In B. J. Nielson and L. E. Cronin (eds.), *Estuaries and Nutrients*. Clifton, NJ: Humana Press, pp. 25–46.

Weiland, R. T., T. H. Chrzanowski, and H. L. Stevenson. 1979. Influence of freshwater intrusion on microbial biomass in salt-marsh creeks. *Estuaries* 2:126–129.

Weinstein, M. P. 1979. Shallow marsh habitats as primary nurseries for fishes and shellfish, Cape Fear River, North Carolina. *Fish. Bull.* 77:339–357.

Weinstein, M. P. 1982. Commentary: a need for more experimental work in estuarine fisheries ecology. *Northeast Gulf Sci.* 5:59–64.

Weinstein, M. P., and H. A. Brooks. 1983. Comparative ecology of nekton residing in a tidal creek and adjacent seagrass meadow: community composition and structure. *Mar. Ecol.: Prog. Ser.* 12:15–27.

Weinstein, M. P., S. L. Weiss, R. G. Hodson, and L. R. Gerry. 1980. Retention of three taxa of postlarval fishes in an intensively flushed tidal estuary, Cape Fear River, North Carolina. *Fish. Bull.* 78:419–435.

Wells, B. W. 1928. Plant communities of the coastal plain of North Carolina and their successional relations. *Ecology* 9:230–242.

Wells, H. W. 1961. The fauna of oyster beds, with special reference to the salinity factor. *Ecol. Monogr.* 31:239–266.

Wells, R. S., A. B. Irvine, and M. D. Scott. 1980. The social ecology of inshore odontocetes. In L. H. Herman (ed.), *Cetacean Behavior: Mechanisms and Functions*. New York: Wiley, pp. 263–317.

Wenner, E. L., M. H. Shealy, Jr., and P. A. Sandifer. 1982. A profile of the fish and decapod crustacean community in a South Carolina estuarine system prior to flow alteration. *NOAA Tech. Rep.* NMFS SSRF-757:1–17.

West, D. L., and A. H. Williams. 1986. Predation by *Callinectes sapidus* (Rathbun) within *Spartina alterniflora* (Loisel) marshes. *J. Exp. Mar. Biol. Ecol.* 100:75–95.

West, T. L. 1985. Abundance and diversity of benthic macrofauna in subtributaries of the Pamlico River estuary. *J. Elisha Mitchell Sci. Soc.* 101:142–159.

Whitledge, 1985. *Nationwide Review of Oxygen Depletion in Estuarine and Coastal Waters.* New York: Brookhaven National Lab.

Whitney, D. M., A. G. Chalmers, E. B. Haines, R. B. Hanson, L. R. Pomeroy, and B. Sherr. 1981. The cycles of nitrogen and phosphorus. In L. R. Pomeroy and R. G. Wiegert (eds.), *The Ecology of a Salt Marsh.* New York: Springer, pp. 163–182.

Whitten, H., H. Rosene, and J. Hedgpeth. 1950. The invertebrate fauna of Texas coast jetties: a preliminary survey. *Publ. Inst. Mar. Sci., Univ. Tex.* 1:53–88.

Wiegert, R. G., R. R. Christian, J. L. Gallagher, J. R. Hall, R. D. H. Jones, and R. L. Wetzel. 1975. A preliminary ecosystem model of a coastal Georgia *Spartina* salt marsh. In L. E. Cronin (ed.), *Estuarine Research*, Vol. 2. New York: Academic Press, pp. 583–601.

Wiegert, R. G., A. G. Chalmers, and P. F. Randerson. 1983. Productivity gradients in salt marshes: the response of *Spartina alterniflora* to experimentally manipulated soil water movement. *Oikos* 41:1–6.

Williams, A. B. 1955. A contribution to the life histories of commercial shrimps (Penaeidae) in North Carolina. *Bull. Mar. Sci. Gulf Caribb.* 5:116–146.

Williams, A. B. 1969. A ten-year study of meroplankton in North Carolina estuaries: cycles of occurrence among penaeidean shrimps. *Chesapeake Sci.* 10:36–47.

Williams, A. B. 1971. A ten-year study of meroplankton in North Carolina estuaries: annual occurrence of some brachyuran developmental stages. *Chesapeake Sci.* 12:53–61.

Williams, A. B. 1972. A ten-year study of meroplankton in North Carolina estuaries: juvenile and adult *Ogvides* (Caridea:Ogyrididae). *Chesapeake Sci.* 13:145–148.

Williams, A. B., and K. H. Bynum. 1972. A ten-year study of meroplankton in North Carolina estuaries: Amphipods. *Chesapeake Sci.* 13:175–192.

Williams, A. B., and E. E. Deubler. 1968. A ten-year study of meroplankton in North Carolina estuaries: assessment of environmental factors and sampling success among bothid flounders and penaeid shrimp. *Chesapeake Sci.* 9:27–41.

Williams, A. B., and H. J. Porter. 1971. A ten-year study of meroplankton in North Carolina estuaries: Occurrence of postmetamorphal bivalves. *Chesapeake Sci.* 12:26–32.

Williams, A. B., G. S. Posner, W. J. Woods, and E. E. Deubler, Jr. 1973. *A Hydrographic Atlas of Larger North Carolina Sounds*, UNC-SG-73-02. Chapel Hill: University of North Carolina, Sea Grant Program.

Williams, J. B., B. J. Copeland, and R. J. Monroe. 1986. Population dynamics of an r-selected bivalve, *Mulinia lateralis* (Say) in a North Carolina estuary. *Contrib. Mar. Sci.* 29:73–89.

Williams, R. B. 1962. The ecology of diatom populations in a Georgia salt marsh. Thesis, Harvard University, Cambridge, MA.

Williams, R. B. 1966. Annual phytoplankton production in a system of shallow temperate estuaries. In H. Barnes (ed.), *Some Contemporary Studies in Marine Sciences*. London: Allen & Unwin, pp. 699–716.

Williams, R. B., and M. B. Murdoch. 1966. Phytoplankton production and chlorophyll concentration in Beaufort Channel, North Carolina. *Limnol. Oceanogr.* 11:73–82.

Williams, R. B., and M. B. Murdoch. 1968. Compartmental analysis of production and decay of *Juncus roemerianus*. *Abstr. ASB Bull.* 15:59.

Williams, S. L. 1981. *Caulerpa cupressoides*: the relationship of the uptake of sediment ammonium and of algal decomposition for seagrass bed development. Thesis, University of Maryland, Baltimore.

Wilson, C. A., and L. H. Stevenson. 1980. The dynamics of the bacterial population associated with a salt marsh. *J. Exp. Mar. Biol. Ecol.* 48:123–138.

Wilson, K. A. 1962. *North Carolina Wetlands: Their Distribution and Management*. North Carolina Wildlife Resources Commission.

Wiltse, W. I., K. H. Foreman, J. M. Teal, and I. Valiela. 1984. Effects of predators and food resources on the macrobenthos of salt marsh creeks. *J. Mar. Res.* 42:923–942.

Wiseman, D. R. 1978. Benthic marine algae. In R. G. Zingmark (ed.), *An Annotated Checklist of the Biota of the Coastal Zone of South Carolina*. Columbia: South Carolina UP, pp. 23–26.

Wolaver, T. G., R. F. Dame, J. D. Spurrier, and A. B. Miller. 1988. Sediment exchange between a euhaline salt marsh in South Carolina and the adjacent tidal creek. *J. Coastal Res.* 4:17–26.

Wolf, R. B., L. C. Lee, and R. R. Sharitz. 1986. Wetland creation and restoration in the United States from 1970 to 1985: an annotated bibliography. *Wetlands* 6:1–88.

Wolff, W. J. 1977. A benthic food budget for the Grevelinger Estuary, the Netherlands and a consideration of the mechanisms causing high benthic secondary production in estuaries. In B. C. Coull (ed.), *Ecology of Marine Benthos*. Columbia: South Carolina UP, pp. 267–280.

Wood, E. J. F. 1965. *Marine Microbial Ecology*. New York: Reinhold.

Woodin, S. A. 1976. Adult-larval interactions in dense infaunal assemblages: patterns of abundance. *J. Mar. Res.* 34:25–41.

Woodin, S. A. 1981. Disturbance and community structure in a shallow water sand flat. *Ecology* 62:1052–1066.

Woodin, S. A. 1986. Settlement of infauna: larval choice. *Bull. Mar. Sci.* 39:401–407.

Woodmansee, R. A. 1958. The seasonal distribution of the zooplankton off Chicken Key in Biscayne Bay, Florida. *Ecology* 39:247–262.

Woolfenden, G. E. 1956. Comparative breeding behavior of *Ammospiza caudacuta* and *A. maritima*. *Univ. Kan. Publ. Mus. Nat. Hist.* 10:47–75.

Wright, R. T., R. B. Coffin, C. P. Ersing, and D. Pearson. 1982. Field and laboratory measurements of bivalve filtration of natural marine bacterioplankton. *Limnol. Oceanogr.* 27:91–98.

Young, D. K., and D. C. Rhoads. 1971. Animal-sediment relations in Cape Cod Bay, Massachusetts, I. a transect study. *Mar. Biol. (Berlin)* 11:242–254.

Young, D. K., and M. W. Young. 1977. Community structure of the macrobenthos associated with seagrass of the Indian River, Florida. In B. C. Coull (ed.), *Ecology of Marine Benthos*. Columbia: South Carolina UP, pp. 359–382.

Young, D. L. 1974. Salt marshes and thermal additions at Crystal River, Florida. In *Crystal River Power Plant: Environmental Considerations*, Final report to Interagency Res. Advisory Comm. Florida Power Corp., St. Petersburg, pp. 1–91.

Youngbluth, M. J., and H. S. Gamble. 1976. Studies of the zooplankton of the Indian River region. In K. Young (ed.), *Indian River Coastal Zone Study*. Ft. Pierce, FL: Harbor Branch Consortium, pp. 46–56.

Zeitzschel, G. 1980. Sediment-water interactions in nutrient dynamics. In K. R. Tenore and B. C. Coull (eds.), *Marine Benthic Dynamics*. Columbia: University of South Carolina Press, pp. 195–218.

Zieman, J. C., Jr. 1970. The effects of thermal effluent stress on the seagrass and macroalgae in the vicinity of Turkey Point, Biscayne Bay, Florida. Thesis, University of Miami, Coral Gables, FL.

Zieman, J. C., Jr. 1975. Quantitative and dynamic aspects of the ecology of turtle grass, *Thalassia testudinum*. In L. E. Cronin (ed.), *Estuarine Research*, Vol. 1. New York: Academic Press, pp. 541–562.

Zieman, J. C., Jr. 1982. *The Ecology of the Seagrasses of South Florida: A Community Profile*, FWS/OBS-82/25. Washington, DC: U.S. Fish Wildl. Serv., Off. Biol. Serv.

Zieman, J. C., Jr., G. W. Thayer, M. B. Robblee, and R. T. Zieman. 1979. Production and export of seagrasses from a tropical bay. In R. J. Livingston (ed.), *Ecological Processes in Coastal and Marine Systems*. New York: Plenum, pp. 21–34.

Zimmerman, M. S., and R. J. Livingston. 1976. The effects of Kraft mill effluents on benthic macrophyte assemblages in a shallow bay system (Apalachee Bay, North Florida, USA). *Mar. Biol. (Berlin)* 34:297–312.

Zimmerman, R., R. Gibson, and J. Harrington. 1979. Herbivory and detritivory among gammaridean amphipods from a Florida seagrass community. *Mar. Biol. (Berlin)* 54:41–47.

Zimmerman, R. J., and T. J. Minelo. 1984. Densities of *Penaeus aztecus, Penaeus setiferus* and other natant macrofauna in a Texas salt marsh. *Estuaries* 7:421–433.

Zingmark, R. G. 1978. *An Annotated Checklist of the Biota of the Coastal Zone of South Carolina*. Columbia: University of South Carolina.

Zobell, C. E. 1938. Studies on the bacterial flora of marine bottom sediment. *J. Sediment. Petrol.* 8:10–18.

PART V
Historical Perspective

15 Aquatic Communities of the Southeastern United States: Past, Present, and Future

COURTNEY T. HACKNEY

Department of Biological Sciences, University of North Carolina, Wilmington, NC 28403

S. MARSHALL ADAMS

Environmental Sciences Division, Oak Ridge National Laboratory, Oak Ridge, TN 37831

GEOLOGIC HISTORY OF THE SOUTHEAST

By the beginning of the Holocene (10 000 years B.P.) the environment in the SE United States had begun to approach present conditions. Ice sheets, which had covered large parts of the North American continent 15 000 years B.P., were in full retreat, and even though the Southeast was not covered by ice, there had been indirect impacts. Although there remains some doubt as to the extent and magnitude of climatic changes in the Southeast during glaciation (see Wright (1983) and Porter (1983) on Holocene environments in the United States during the Holocene), the general consensus is that the environment was cooler and drier (Knox 1983). Post-glacial sediment discharge in rivers associated with glacial meltwater was about ten times today's norm in the southeast flow of the Mississippi and Susquehanna rivers (Baker 1983). With ice sheets melting, sea level rose rapidly (approximately 1 cm/year) and by 10 000 years B.P., resulted in an ocean level that was only 30–60 m below its present level (Bloom 1983).

Vegetation in the SE United States was probably altered by the cooler and drier climate prior to the Holocene, but the exact degree of alteration and the location of various vegetative community types remain uncertain. Most studies that describe vegetation in the early Holocene are based on pollen preserved in sediments of fresh wetlands and lakes. However, these sediment records may be incomplete and may not accurately reflect vegetative types and patterns for the entire landscape. Major vegetative types were established

Contribution 033, Center For Marine Science Research, University of North Carolina at Wilmington.

in the Southeast by the beginning of the Holocene, but the gradually warming climate with increasing precipitation allowed expansion and migration of these ecosystems (Watts 1983). By the middle of the Holocene, for example, southern pines had extended their range and were a dominant ecosystem type (Knox 1983, Webb et al. 1987).

Because the majority of southeastern drainage basins were well south of the ice sheet, the tremendous flow of melt water did not create deep river channels and fill river basins with glacial sediments as was the case farther north. Glaciers did not physically modify the landscape, abandon chunks of ice to produce kettle lakes, or directly impact the southeastern corner of North America at the close of the last ice age. With the exception of the Mississippi River, most southeastern rivers probably carried less water when the Holocene began than they do today because the climate was drier and cooler (Knox 1983).

Except for modification of coastal features such as barrier islands and the drowning of river valleys resulting from rising sea level, the ice age had very little influence on the creation or loss of most types of aquatic ecosystems in the SE United States. Lotic communities owed their existence to the same physical forces that produce them today: gravity and erosion. Some lentic systems result from these same forces, with oxbow lakes being a common feature in the river floodplain and coastal plains. Lakes, however, were not a common feature of the southeastern landscape at the beginning of the Holocene, with some exceptions.

Carolina Bay lakes were common on the southeastern landscape primarily in the Carolinas at the beginning of the Holocene. These shallow, elliptical depressions are abundant on the Coastal Plain today and may reach several miles in diameter (Wells and Boyce 1953; also see Chapter 12 for other explanations of their origin). Owing their origin to a meteor shower 40 000–100 000 years ago (Wells and Boyce 1953), many of what were formerly lakes are now filled with peat and fully vegetated, usually with an assortment of evergreen shrubs that typify pocosins or bays (see Chapter 7, *Terrestrial Communities* volume). The aquatic systems of the Florida peninsula are unique because of the near-surface limestone deposits which have produced the karst topography. Florida was apparently much drier as the Holocene began (Knox 1983) and may have contained fewer lakes and less surface water. Lower sea level undoubtedly confined much of the water to the limestone itself, leaving only an occasional deep sinkhole with surface water (See Chapter 12 for more on Florida lakes and Chapter 3 on endemic cave species in Florida).

ENGINEERING OF THE SOUTHEASTERN LANDSCAPE

American Indians

Despite the early description of the American landscape as wilderness by early explorers, the southeastern United States was well settled by indigenous people when Europeans arrived (Hughes 1983). It is clear from the travels of William Bartram in the late 18th century (Harper 1958) that American Indians and their ancestors had dramatically altered the landscape. Fire was used in many different ways: for hunting, for clearing land to plant crops, and to maintain certain favored habitats (Hughes 1983). Such management practices would not have altered the general nature of aquatic communities because swamps and marshes adjacent to most water courses were resistant to fires except during extreme droughts. When fires did burn to the water's edge there would have been increased nutrient inputs but less organic inputs (see chapter 1). In headwater streams in the Piedmont and mountains, such events were probably more frequent. Extreme droughts on the Coastal Plain probably helped maintain many shallow lakes by allowing fires to mineralize peat in these areas, as has been documented for the Okefenokee swamp in southern Georgia (Cypert 1961).

Beavers

American Indians used streams and rivers for commerce and transportation as well as a source of food. Even though many aboriginal communities were located adjacent to water bodies, they did not attempt to manage or manipulate the nature of these water systems. The one power of manipulation they did have came as a result of their interaction with the other great engineer on the planet, the beaver (*Castor canadensis*). Beavers have probably created more aquatic and wetland habitats than human efforts have ever done. Most of their construction takes place on small tributaries; nevertheless, huge dams were known or reported by many early explorers. Millions of such dams would have dramatically slowed flow rates after storms and created an enormous amount of backwater habitat for fish and other animals adapted to shallow aquatic habitats. Beavers are currently reinvading much of the Southeast and their dams frequently flood large areas of upland habitat, killing trees and allowing the invasion of herbaceous plants. Beaver dams eventually fill with debris and are abandoned (Johnston and Naiman 1987) to the forces of erosion and eventual recolonization in a never ending cycle.

American Indians regularly killed beavers in the southeastern United States, but little information is available as to how much impact hunting actually had on beaver populations and their aquatic habitat. When Lewis and Clark made their historic expedition to the northwestern United States, beavers were wherever there was water and few Europeans (Burroughs 1961). The extent of the beaver's abundance and alteration of the landscape is found throughout the expeditions records. "I proceeded on about two miles crossing those different channels all of which were dammed with beaver in such a manner as to render passage impracticable" (Burroughs 1961). Thus, American Indians and beavers seemed to coexist into the 19th century until Europeans arrived to trap or trade for beaver skins. Beavers can be easily eliminated by trapping due to their slow rate of reproduction (Hill 1982), and by the late 1700s beaver were extirpated from much of the SE United States. In fact, the famous American naturalist William Bartram barely mentions them in his travels through the Carolinas, Georgia, and Florida (Harper 1958). Bartram is known for his elaborate and detailed descriptions of plants, animals, and American Indians and it is hard to imagine him not saying more about beavers if they were present in any numbers.

The alteration that the loss of beavers (and the habitat they produced) had on ecological processes cannot be estimated. Shallow aquatic communities dominated by herbaceous vegetation were replaced by infrequently flooded swamps and upland communities. Leaves and twigs that once may have decomposed behind beaver dams began to be carried downstream, and water from sudden storms now rapidly moved into the main tributaries of rivers, resulting in increased frequency of flooding of the adjacent riparian zone. Clearing of the land for farms added large quantities of sediments that quickly moved into lotic systems. It is hard to comprehend the landscape of the southeastern United States that greeted the first European explorers or the quantum alteration caused by the demand for beaver pelts. This was just the first of many large-scale changes European culture would make across the North American landscape, and even the reintroduction of beavers into the southeast in the 20th century has not, and will not, return the landscape to its original condition.

European Settlers

When Europeans arrived along the eastern coast of the United States they not only brought with them a technology vastly superior to that of the American Indians, but they aggressively and indiscriminately utilized this technology to their advantage. Would the American Indian culture have used technology as readily or as indiscriminately as the early settlers? They certainly were quick to use firearms, iron, and other trappings of European culture, but European diseases quickly eliminated their civilization and culture. Colonial attitudes essentially allowed the use of technology whenever it suited their economic interests without regard to the environment or long-term consequences. Any engineering that calmed or modified nature was both encouraged and praised. This "slash and burn" attitude continued to govern engineering practices and governmental policies in the United States and around the world until the 1960s. Impacts of these attitudes and practices were minimal at first, but increased exponentially as technology improved, culminating in the most massive environmental alteration in our history when such attitudes and activities became sanctioned and financed by the U.S. government and resulted in modifications to just about every river and stream in the SE United States.

Early modifications to aquatic communities were governed by economic needs, for example, small dams to run mills, and simple activities for waterborne commerce. As commerce was more easily moved via water born transportation, modifications to rivers and streams were initially limited by technology and available capital. Snagging, the removal of debris from streams to make them more passable, evolved into more massive projects such as canal building and dredging of natural waterways. The Erie Canal and similar projects are examples and prominent pieces of early American history.

While history books discuss the construction of canals mostly in the northeastern part of the United States, the southeastern region of the country also was altered. Our first president, George Washington, for example, mounted an attempt to drain the Great Dismal Swamp in southeastern Virginia and northeastern North Carolina with canals that moved water from the swamp and eventually to Chesapeake Bay. From southeastern North Carolina to southern Georgia large dikes were built along the freshwater, tidal portions of estuaries to impound water for rice culture after the American Revolution (Clifton 1973). Virtually all tidal swamps in this part of the southeastern coast were converted from cypress/gum tidal swamp to rice monoculture. Today we would view the removal of vast tidal areas and their associated detrital input from the estuary as a major ecological disaster. We can only speculate on the concomitant ecological changes that followed such massive alteration of the landscape.

Army Corps of Engineers

It is difficult to find anyone or a single publication that presents an unbiased view of the U.S. Army Corps of Engineers and their activities in U.S. Waters. In the beginning, 16 March 1802, the Corps was devoted exclusively to military operations. As the one organized group of engineers "on call" for the U.S. government, they quickly became associated with the construction and maintenance of waterways and harbors through which the U.S. military could rapidly move ships, troops, and supplies. The lack of a national policy related to transportation and defense became obvious to many American leaders after the war of 1812 with Britain. In 1824 the U.S. Army Corps of Engineers was officially given legislative authority to participate in civil engineering projects (see Holt 1923, for details of early COE history). Of perhaps greatest importance was the fact that the Corps of Engineers not only undertook projects directed by the military, but planned and directed projects that were primarily related to civilian commerce. Clearing rivers of snags, building canals and roads, erecting piers and breakwaters all became part of the role of the Corps of Engineers before the Civil War.

After the Civil War both the limited accepted role of the Corps in civilian projects and the annual appropriations from Congress expanded dramatically. The Rivers and Harbors Act of 1899 further expanded the Corps of Engineers' authority by granting them regulatory authority of all construction activities in navigable waters. This not only gave them authority over individual projects, but also gave them preeminence over all other agencies and boards when it came to potentially navigable waters.

All U.S. Army Corps of Engineers activities are mandated by Congress. Although the Corps may recommend certain activities (usually after a directive from Congress for study), their activities are mostly driven by various individuals and agencies through their elected officials (see Morgan 1971, for some exceptions). Almost from the beginning civilians have had an influence in initiating what later became Corps of Engineers projects. While some projects were suggested by community-spirited individuals, many had the potential to bring large profits to individuals or certain industries. Congress, however, ultimately directs all such projects through annual appropriations. Corps of Engineers projects were often used to bring jobs to an area and became pork barrel projects for elected officials. For example, after the U.S. Congress overruled the U.S. Army Corps of Engineers opinion that reservoirs were ineffective for flood control (through the 1936 Flood Control Act), the Corps of Engineers and the Tennessee Valley Authority directed the construction of 300–400 reservoirs for that purpose during the next 50 years. Brought on by the Great Flood along the Mississippi in 1927 and the Great Depression, economic costs became a secondary factor to the production of jobs and the control of nature. This same act separated responsibilities for controlling watershed problems. The Corps of Engineers was given control of projects that prevented floods, and the Department of Agriculture was given the authority to prevent erosion and soils runoff (Anonymous 1986).

The impact of this shift in the Corps of Engineers directive totally transformed the landscape of the southeastern United States. While previously dedicated to building levees and canals and dredging rivers and harbors, the Corp of Engineers enthusiastically began turning rivers into a series of impoundments. Since their original goal was to hold water, little thought was initially given to the environment. Table 1 in Chapter 11 of this volume lists 144 major reservoirs in the SE United States, most located outside the Coastal Plain (see Figs. 1–5 in Chapter 11). These impoundments were primarily for flood control, although recreation and power generation were important for the economic justification of each. Many were constructed under the auspices of the Tennessee Valley Authority. Add to these major impoundments an additional 1 million farm ponds built primarily for erosion control and subsidized by the U.S. Agriculture Department (see Small Impoundments, Chapter 10) and it is clear that the U.S. Government, through its agencies, has dramatically altered the pre-European landscape. Today, virtually all large rivers and streams that could be dammed profitably have been. In fact, impoundments are the dominant aquatic feature of the SE U.S. landscape (see Fig. 2, Chapter 11). Despite large numbers of southeastern reservoirs, they store only between 15 and 24% of the annual rainfall, of which eventually 98% flows into the Gulf of Mexico and the Atlantic Ocean (Hirsch et al. 1990). The timing of flow, however, has been altered.

The completion of the Tennessee–Tombigbee Waterway marked the end of an era of giant water projects. In one sense the Tennessee–Tombigbee Waterway also began a new era, one that included a more balanced view of the value of an unaltered landscape and one that seriously questioned the long-term economic viability of huge dams and other large engineering projects.

Megascale Modification of the Landscape

Virtually every chapter in this volume refers to human modifications of aquatic systems that have dramatically altered their biota and functions. What follows are two case studies, one in the Coastal Plain and a second which involves connecting two major river systems. While each has some unique features, they are symptomatic of the types of water projects that have been completed in the Southeast.

The Tennessee–Tombigbee Waterway   This waterway is the largest civil works project ever undertaken by the U.S. Army Corps of Engineers. First envisioned more than a century before its completion in 1984, it provides a connection between the Tennessee River and the Black Warrior System which enters the Gulf of Mexico at Mobile Bay. Authorized by the Rivers and Harbors Act of 1946 this engineering project was a tremendous environmental disaster (Palmer 1986), a monetary boondoggle (Miller 1978), or an economic miracle for the states of Mississippi and Alabama, depending on one's point of view. Without doubt, however, it was the greatest modification of the landscape ever accomplished during one project by the Corps of Engineers. The scale of this project has raised many water-related issues that either have been or should have been addressed in other smaller projects. The remainder of this section will concentrate on those landscape modifications. Those interested in the historical and complex legal issues raised by this project are referred to Green (1985) and McClure (1985a) for a summary and for additional references.

The waterway connects to the Tennessee River at Pickwick Lake and enters Demopolis Lake after dropping 341 ft (104 m) over 234 miles (377 km) (Fig. 1). Over the course of this waterway there are 10 locks, five of which are associated with reservoirs. The lower 149 miles (240 km) was constructed by dredging a 300-ft-wide (91.5-m) by 9-ft-deep (2.7-m) channel in the Tombigbee River. Many sections of this part of the waterway altered large shallow riffle areas that once contained a high diversity of fish (115 species) and mussels (51 species) (Palmer 1986). Today, water levels are maintained by a series of locks (Fig. 1), although a few sections of riffles have been preserved and are maintained alongside the main channel.

The Tennessee and Tombigbee river basins are naturally separated by a divide. To cross this divide the COE constructed a chain of lakes east of the headwaters of the Tombigbee by constructing a levee on the west side of the canal. Here the channel is 12 ft (3.7 m) deep and 300 ft (91.5 m) wide, with water levels maintained by 5 locks. The final section, 39 miles (63 km) long, actually cuts through the divide between the two basins (Fig. 1). At the deepest point (the highest part of the divide), 175 ft (53 m) of material was removed along a 1500-ft (457-m) corridor. The entire divide cut required the removal of 150 million cubic yards (114 million m³). In total, the construction of the waterway involved the removal, transportation, and placing of 282 million yards (216 million m³) of material, one-third more than the Panama Canal

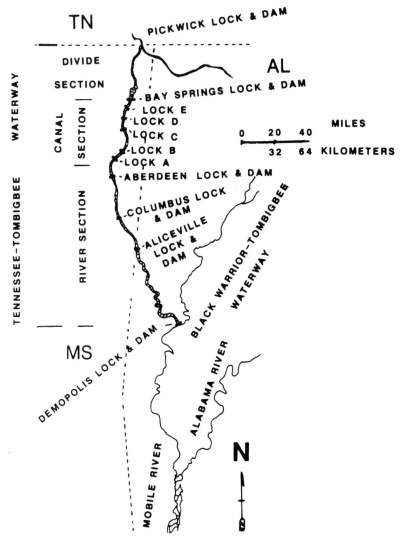

FIGURE 1. Tennessee–Tombigbee Waterway located in the northeastern corner of Mississippi. The waterway connects the Tennessee River, which is part of the Mississippi drainage system, to the Mobile River system, which empties into the northern Gulf of Mexico. River and waterway widths are not drawn to scale.

(McLindon 1985), altered 85 579 acres (34 634 ha) of land, and produced 10 lakes and 44 000 acres (17 806 ha) of surface water. Besides the alteration of the Tombigbee River and the flooding of large areas of uplands, over 14 000 acres (5666 ha) were covered with excavated materials (McLindon 1985). Almost 6000 acres (2428 ha) of bottomland hardwood forests were filled, 343 acres (139 ha) of cypress/gum swamp, 116 acres (47 ha) of marsh, and 123 acres (48 ha) of lakes and ponds. The remaining acreage filled consisted mostly of pine forest and croplands (McLindon 1985). Both advocates and critics of the Tennessee–Tombigbee project agree that this was a major alteration of the landscape. The arguments develop when discussions of economic benefits versus environmental costs are raised.

Besides the more obvious changes to the surface landscape, a number of other important issues were raised during the litigation that accompanied the project. Aquifer drawdown, waterlogging of soil caused by elevated lakes, cross contamination of water and biota between drainage basins, endangered and threatened species, water quality in the waterway, sedimentation and erosion along the waterway, and destruction of adjacent wildlife habitat were additional issues raised during the project (McClure 1985b). While these issues were not addressed to the satisfaction of all who followed construction of the Tennessee–Tombigbee Waterway, there is little doubt that the process of attempting to minimize environmental damage and the documentation of the megascale changes that occurred during the project have probably doomed any future projects of this scale. Unfortunately, despite these potential environmental problems, dozens of smaller-scale projects are being planned that could impact some of the last free-flowing, high-diversity streams in the southeastern United States (Benke 1990).

Cooper River Rediversion Project   What happens when the engineering plan and the economic benefits do not meet expectations or the environmental impact is larger than anticipated? As is often the case, no action is taken and unfulfilled economic promises are quickly forgotten. In the case of the Cooper River project, additional costs demanded that the project be redesigned to reverse part of the original intent of the project. Unlike many large water projects which were constructed mostly on the Piedmont and higher gradient landscapes, modifications of the Santee and Cooper rivers occurred on the Coastal Plain (Fig. 2). Prior to 1941 the Santee River had the fourth highest freshwater discharge into the Atlantic Ocean south of the St. Lawrence Seaway. Modifications to the river began with a 22-mile-long (35-km) canal begun in 1793 and completed in 1800. Canals with locks for waterborne transportation were common during this period of U.S. history. Few additional major modifications occurred to either river until 1915 when the Santee–Cooper project began. When completed in 1941, the Santee River was dammed at river mile 87 creating Lake Marion (155 square miles or 401 km²). A nearby dam on the Cooper River created Lake Moultrie (95 square miles or 246 km²). Both lakes are relatively shallow compared to Piedmont reservoirs. Because the primary justification for the project was power generation, the higher gradient along the more southern Cooper River was more applicable for this purpose. Thus, water was diverted from Lake Marion to Lake Moultrie via a 7.5-mile diversion canal, with most of the water from the Santee River reaching the coast through the Cooper River and Charleston Harbor.

Beginning in 1942 a phenomenal increase in shoaling occurred in Charleston Harbor that was directly related to the Santee–Cooper Diversion project.

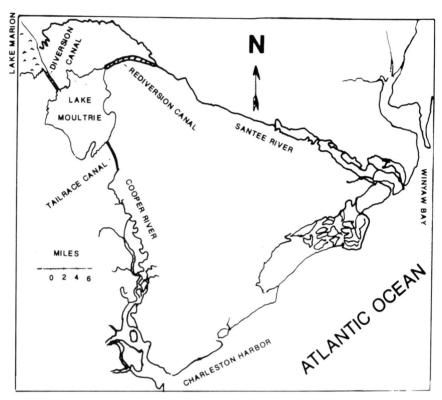

FIGURE 2. Cooper and Santee rivers located in southeastern South Carolina. The map shows the structures that were a part of the original Santee–Cooper Diversion project and the canal dug as part of the Santee–Cooper Rediversion project. Canal width is not to scale.

Freshwater discharge from the Cooper River had been increased from 72 cfs (cubic feet per second) to 15 600 cfs. Even after traveling through two reservoirs, water reaching Charleston Harbor carried a high sediment load primarily of Piedmont origin. The cost of dredging Charleston Harbor escalated annually and reached a critical point when environmental regulations prohibited open water disposal of dredged material or disposal on adjacent wetlands. In 1975 a final Environmental Impact Statement (U.S. Army Corps of Engineers 1975) proposed to redivert the majority of freshwater back to the Santee River, saving Charleston Harbor from the rapid shoaling, but reducing hydroelectric generation. With the diversion completed the lower Santee River is again experiencing fresh and oligohaline conditions, while the lower Cooper River is undergoing salinization of both aquatic and adjacent tidal wetlands. Presumably, this is the reverse of the original modification of the two rivers and represents a "more natural" condition.

THE FUTURE OF SOUTHEASTERN AQUATIC COMMUNITIES

The Tennessee–Tombigbee Waterway and the Santee–Cooper Diversion and Rediversion projects are but two large-scale examples where humans have dramatically altered southeastern aquatic communities. Smaller water projects also impact aquatic systems to varying degrees. In fact, all chapters in this volume address the type and magnitude of changes that have taken place in the various aquatic systems due to human activities. All of these alterations to southeastern aquatic systems have dramatically affected the nature and patterns of biodiversity in the southeastern United States. Benke (1990) calculated that 98% of the free-flowing freshwater communities in the United States have been dramatically altered and that only 20% have high enough quality to be worthy of federal protection. To date, 16 000 km of streams have been afforded a conservation status. Only 10% of these are east of the Mississippi. Southeastern streams have been afforded little preservation (Benke 1990), despite having the highest diversity of fish species in the United States (McAllister et al. 1986). (See Chapter 2 for the underlying reasons for this diversity.) Benke (1990) noted that many of the remaining free-flowing streams, especially in the Piedmont, are threatened with small-scale hydroelectric projects. Although significant, relatively pristine segments remain (Table 1), most larger rivers have already been dammed (see Chapter 11).

Virtually all of the lotic system chapters in this volume emphasize the modified nature of these waterbodies. Many smaller streams on the Coastal Plain have escaped megascale reservoir projects, but they have been modified by dredging, levees, and locks, resulting in the loss of many commercial, anadromous fish stocks. Humans have also succeeded in mixing faunas between drainage basins by passive means such as release of bait fish and bait clams and by large scale water projects (e.g., the Tennessee–Tombigbee and the Santee–Cooper projects). Virtually all SE U.S. rivers (through their

TABLE 1 Total Lengths of Relatively Pristine Rivers Identified by the Nationwide Rivers Inventory in the Different Physiographic Provinces in the Southeastern United States

Province	Total Length (km)
Ridge and Valley	6 586
Blue Ridge	1 908
Piedmont	6 267
Coastal Plain	19 688

Note. Some designations also include areas adjacent to the southeastern states; thus, these are overestimates for the Southeast.

Source. Frome Benke (1990).

estuarine portions) are now connected by the Atlantic Intracoastal Waterway (Parkman 1983) and the Gulf Intracoastal Waterway (Alperin 1983).

Previous chapters have emphasized the fact that biodiversity has been reduced in most if not all of the major aquatic communities in the Southeast. Much of this loss of diversity is caused by the inability of highly specialized stream and flowing water species to live in the lentic habitats produced by reservoirs and/or dredged channels. Lotic species, especially those from high gradient streams, show some extremely high species diversity. Lentic aquatic species are mostly generalized species supplemented by introduced, exotic species which predominate in reservoirs and ponds. At the other end of the spectrum from fast-moving streams are estuaries (Chapters 13 and 14). Despite the major modifications to most estuaries there has been little loss of species diversity. Most estuarine species are excellent colonizers, are extremely fecund, and disperse rapidly from estuary to estuary. This may be augmented by the intracoastal waterways, which also have the potential to aid the movement of introduced species and disease-causing organisms. While these connections may not cause the loss of large numbers of species, there is the possibility that genetic variations between populations in estuaries may be reduced (see Parker et al. 1981, for an example of genetic divergence in a single estuarine species).

Human populations in almost all southeastern states are growing rapidly, and aquatic ecosystems will be subjected to increasing pressure. The fate of wetlands adjacent to rivers, streams, and lakes is of great importance as these communities process and purify water moving through them (Whigham et al. 1988). Unfortunately, headwater ecosystems, which are best able to process many pollutants (Brinson 1988), are the very ones that have been most modified (Cashin 1990).

It is difficult to avoid pessimism when evaluating the future of southeastern aquatic ecosystems. Large impoundments are more abundant then ever and very small headwater beaver ponds are becoming more numerous. These two habitats, however, do not harbor or characterize the unique and diverse communities of the southeastern United States. These unique systems are found in the free-flowing streams and rivers, few of which have not been modified. It is not enough to simply place such streams in some conservation status; it is also necessary to control upstream activities and streamside landscape alterations. The prohibition of wetland filling by the Clean Water Act has now been extended to palustrine wetlands, the Swampbuster Act eliminated government subsidies for farmers that destroy wetlands, and the reduced use of tilling and fertilizers by farmers offers some hope for improvement. The U.S. Agriculture Department has also initiated a variety of programs that aid the quality of runoff waters. The ultimate answer, of course, is landscape planning on a scale previously reserved for landscape alteration (dams). For purposes of the preservation of free-flowing streams, management needs to be by watershed. The objective of such plans should be the preservation not only of the habitat itself, but of the adjacent uplands and wetlands as well. It will require extensive cooperation between all levels of government, as well as conservation groups and landowners. To be successful, such a plan has to provide economic incentives to governments and landowners alike and society must be willing to bear the short-term economic costs. Preservation of a functioning landscape usually affects society and individuals in a positive fashion when viewed on a long-term basis. Short-term losses can be large because of a decline in values of land zoned for conservation and because of a diminution of the tax base of local governments. These two factors are and will be the major economic hurdle to the establishment of landscape management in the Southeast and the United States as a whole, but must be solved if we are to leave a fully functioning landscape as a legacy to future generations.

REFERENCES

Alperin, L. M. 1983. *History of the Gulf Intracoastal Waterway*, NWS-83-9. Vicksburg, MS: U.S. Army Corps of Engineers.

Anonymous. 1986. *The History of the U.S. Army Corps of Engineers*, EP 360-1-21. Vicksburg, MS: U.S. Army Corps of Engineers, Waterways Experiment Station.

Baker, V. R. 1983. Late-Pleistocene fluvial systems. In S. C. Porter (ed.), *Late-Quaternary Environments of the United States*, Vol. 1. Minneapolis: Minnesota UP, pp. 115–129.

Benke, A. C. 1990. A perspective on America's vanishing streams. *J. North Am. Benthol. Soc.* 9:77–88.

Bloom, A. L. 1983. Sea level and coastal changes. In H. E. Wright, Jr. (ed.), *Late-Quaternary Environments of the United States*, Vol. 2. Minneapolis: Minnesota UP, pp. 42–50.

Brinson, M. M. 1988. Strategies for assessing the cumulative effects of wetland alteration on water quality. *Environ. Manage.* 12:655–662.

Burroughs, R. D. (ed.). 1961. *The Natural History of the Lewis and Clark Expedition.* Ann Arbor: Michigan State UP.

Cashin, G. E. 1990. Wetland development in the North Carolina coastal plain. Thesis, Duke University, Durham, NC.

Clifton, J. M. 1973. Golden grains of white: Rice planting on the lower Cape Fear River North Carolina. *Hist. Rev.* 50:356–393.

Cypert, E. 1961. The effects of fire on the Okefenokee swamp in 1954 and 1955. *Am. Midl. Nat.* 66:485–503.

Green, S. R. 1985. An overview of the Tennessee-Tombigbee Waterway. *Environ. Geol. Water Sci.* 7:9–13.

Harper, F. (ed.). 1958. *The Travels of William Bartram.* New Haven, CT: Yale UP.

Hill, E. P. 1982. Beaver. In J. A. Chapman and G. A. Feldhamer (eds.), *Mammals of North America*. Baltimore, MD: Johns Hopkins Press, pp. 256–281.

Hirsch, R. M., J. F. Walker, J. C. Day, and R. Kallio. 1990. The influence of man on hydrologic systems. In M. G. Wolman and H. C. Riggs (eds.), *Surface Water Hydrology*. Boulder, CO. Geol. Soc. Am., pp. 329–359.

Holt, W. S. 1923. *The Office of the Chief of Engineers of the Army, Its Non-Military History, Activities and Organization*. Baltimore, MD: Johns Hopkins Press.

Hughes, J. D. 1983. *American Indian Ecology*. El Paso: Texas Western Press, University of Texas.

Johnston, C. A., and R. J. Naiman. 1987. Boundary dynamics at the aquatic-terrestrial interface: the influence of beaver and geomorphology. *Landscape Ecol.* 1:47–57.

Knox, J. C. 1983. Response of river systems to Holocene climates. In H. E. Wright, Jr. (ed.), *Late Quaternary Environments of the U.S.*, Vol. 2. Minneapolis: Minnesota UP, pp. 26–41.

McAllister, D. E., S. P. Platania, F. W. Schueler, M. E. Baldwin, and D. S. Lee. 1986. Ichthyofaunal patterns on a geographic grid. In C. H. Hocutt and E. O. Wiley (eds.), *The Zoogeography of North American Freshwater Fishes*. New York: Wiley, pp. 17–51.

McClure, N. D., IV. 1985a. A major project in the age of the environment: out of controversy, complexity and challenge. *Environ. Geol. Water Sci.* 7:15–24.

McClure, N. D., IV. 1985b. A summary of environmental issues and findings: Tennessee-Tombigbee Waterway. *Environ. Geol. Water Sci.* 7:109–124.

McLindon, G. J. 1985. Creative spoil: design concepts, construction techniques and disposal of excavated materials. *Environ. Geol. Water Sci.* 7:91–108.

Miller, J. N. 1978. Trickery on the Tenn–Tom. *Readers Dig.* (Sept).

Morgan, A. E. 1971. *Dams and Other Disasters: A Century of the Army Corps of Engineers in Civil Works*. Boston, MA: Porter Sargent.

Palmer, T. 1986. *Endangered Rivers and the Conservation Movement*. Los Angeles: California UP.

Parker, E. D., Jr., W. D. Burbanck, M. P. Burbanck, and W. W. Anderson. 1981. Genetic differentiation and speciation in the estuarine isopods *Cyathura polita* and *Cyathura burbancki*. *Estuaries* 4:213–219.

Parkman, A. 1983. *History of the Waterways of the Atlantic Coast of the United States*, NWS-83-10. Vicksburg, MS: U.S. Army Corps of Engineers.

Porter, S. C., (ed.). 1983. *Late-Quaternary Environments of the United States*, Vol. 1. Minneapolis: Minnesota UP.

U.S. Army Corps of Engineers. 1975. *Final Environmental Impact Statement*, Cooper River Rediversion Project, Charleston Harbor. Charleston, SC: U.S. Army Corps of Engineers, Charleston District.

Watts, W. A. 1983. Vegetational history of the eastern United States. In S. C. Porter (ed.), *Late-Quaternary Environments of the United States*, Vol. 1. Minneapolis: Minnesota UP, pp. 294–310.

Webb, T., III, P. J. Bartlein, and J. E. Kutzbach. 1987. Climatic changes in eastern North America during the past 18,000 years; comparisons of pollen data with model results. In W. F. Ruddiman and H. E. Wright (eds.), *North America and Adjacent Oceans During the Last Deglaciation*. Boulder, CO: Geol. Soc. Am., pp. 447–462. V- K-3.

Wells, B. W., and S. G. Boyce. 1953. Carolina Bays: additional data on their origin, age and history. *J. Elisha Mitchell Sci. Soc.* 69:117–141.

Whigman, D. F., C. Chitterling, and B. Palmer. 1988. Impacts of freshwater wetlands on water quality: a landscape perspective. *Environ. Manage.* 12:663–651.

Wright, H. E. (ed.). 1983. *Late-Quaternary Environments of the U.S.*, Vol. 2. Minneapolis: Minnesota UP.

APPENDIX

Reviewers

Dr. Arthur C. Benke
Department of Biology
University of Alabama
University, AL 35486

Dr. Mark Brinson
Department of Biology
East Carolina University
Greenville, NC 27858

Dr. Arthur Brown
Department of Zoology
University of Arkansas
Fayetteville, AR 72701

Dr. W. D. Burbanck
Department of Biology
Emory University
Atlanta, GA 30307

Dr. Robert Cashner
Department of Biology
University of New Orleans
New Orleans, LA 70122

Dr. Mark Clark
Department of Geological
   Sciences
University of Tennessee
Knoxville, TN 37996-1410

Dr. Carroll Cordes
U.S. Fish and Wildlife Service
Route 2, Box 760
LaCombe, LA 70445

Dr. David C. Culver
Department of Ecology and
   Evolutionary Biology
Northwestern University
Evanston, IL 60201

Dr. Mary G. Curry
3404 Tolmas Drive
Metarie, LA 70002

Dr. Richard Dame
4811 Belvedere Lane
North Litchfield Beach
Pawleys Island, SC 29585

Dr. Neil H. Douglas
Department of Biology
NE Louisiana University
Monroe, LA 71209

Dr. J. W. Elwood
Environmental Sciences Division
Oak Ridge National Laboratory
Oak Ridge, TN 37831

Dr. Thomas D. Forsythe
Land Between The Lakes
Golden Pond, KY 42231

Dr. John Hains
80 Palmetto Avenue
Newry, SC 29665

Dr. John Holsinger
Department of Biology
Old Dominion University
Norfolk, VA 23508

Dr. Jack Jones
School of Forestry, Fish and
   Wildlife
112 Stephans Hall
University of Missouri
Columbia, MO 65201

Dr. Rudolf H. Kiefer
Department of Earth Sciences
University of North Carolina at
  Wilmington
601 South College Road
Wilmington, NC 28403

Dr. B. L. Kimmel
Environmental Sciences Division
Oak Ridge National Laboratory
Oak Ridge, TN 37831

Dr. Mark LaSalle
CE WES ER-C
P.O. Box 631
Vicksburg, MS 39180

Dr. Richard Lowrance
USDA-ARS
SE Watershed Research
  Laboratory
P.O. Box 946
Tifton, GA 31793

Dr. Richard Marzolf
Hancock Biological Station
Murray State University
Murray, KY 42701

Dr. Judy L. Meyer
Institute of Ecology
University of Georgia
Athens, GA 30602

Dr. John Moring
Marine Coop. Fisheries Research
  Unit
Department of Zoology
University of Maine
Orono, ME 04469

Dr. Samuel C. Mozely
Department of Zoology
North Carolina State
Raleigh, NC 27650

Dr. P. J. Mulholland
Environmental Sciences Division
Oak Ridge National Laboratory
Oak Ridge, TN 37831

Dr. Seth Reice
Department of Biology
202 Wilson Hall
University of North Carolina
Chapel Hill, NC 27514

Dr. Leonard A. Smock
Department of Biology
Virginia Commonwealth
  University
Richmond, VA 23284

Dr. D. M. Soballe
South Florida Water Management
  District
P.O. Box 24680
West Palm Beach, FL 33416

Dr. Royal Suttkus
Tulane University
Museum of Natural History
Bell Chasse, LA 70037

Dr. M. Van Den Avyle

Dr. J. Bruce Wallace
Department of Entomology
University of Georgia
Athens, GA 30602

Dr. Varley Wiedeman
Biology Department
University of Louisville
Louisville, KY 40292

Dr. William Woolcott
Department of Biology
University of Richmond
Richmond, VA 23173

Dr. Victor A. Zullo
Department of Earth Sciences
University of North Carolina
Wilmington, NC 28403

INDEX